ENVIRONMENTAL ENGINEERING

BALKEMA – Proceedings and Monographs
in Engineering, Water and Earth Sciences

PROCEEDINGS OF THE SECOND NATIONAL CONGRESS OF ENVIRONMENTAL ENGINEERING, LUBLIN, POLAND, 4–8 SEPTEMBER 2005

Environmental Engineering

Edited by

Lucjan Pawłowski, Marzenna Dudzińska & Artur Pawłowski
Institute of Environmental Protection Engineering,
Lublin University of Technology,
Lublin, Poland

Taylor & Francis
Taylor & Francis Group
LONDON / LEIDEN / NEW YORK / PHILADELPHIA / SINGAPORE

Taylor & Francis is an imprint of the Taylor & Francis Group, an informa business

Typeset by Charon Tec Ltd (A Macmillan Company), Chennai, India

Published by: Taylor & Francis/Balkema
P.O. Box 447, 2300 AK Leiden, The Netherlands
e-mail: Pub.NL@tandf.co.uk
www.balkema.nl, www.taylorandfrancis.co.uk, www.crcpress.com

British Library Cataloguing in Publication Data
A catalogue record for this book is available from the British Library

Library of Congress Cataloging in Publication Data
A catalog record for this book has been requested

ISBN13: 978-0-415-40818-9 (Hbk)

Environmental Engineering – Pawłowski, Dudzińska & Pawłowski (eds)
© 2007 Taylor & Francis Group, London, ISBN13 978-0-415-40818-9

Table of Contents

Solid waste and sludge treatment

Air pollution control

Miscellaneous

Environmental Engineering – Pawłowski, Dudzińska & Pawłowski (eds)
© 2007 Taylor & Francis Group, London, ISBN13 978-0-415-40818-9

Preface

The central goals of the 2nd National Congress on Environmental Engineering are to summarize research carried out in Poland, and to improve technology transfer and scientific dialogue in this time of economical transformation from a planned to a free market economy, thereby leading to a better comprehension of solutions to a broad spectrum of environmentally related problems.

The 1st congress was held in 2001, and major discussion was focused on environmental engineering aspects related to Poland accession to EU (Findings were published in the book entitled: *Environmental Engineering Studies. Polish Research on the Way to EU*, by Kluwer Academic/Plenum Publisher, 2002).

Poland, like other post-communist countries, is undergoing transformation into a capitalist system. This transformation brings many problems – economical, social, psychological and also ecological. Ecological problems are strongly connected with the political, economic and psychological inheritance of the past as well as with changes in the post-communist society.

To understand these problems it is necessary to consider the following issues:

- The geographic situation of Poland,
- The political transformations that occurred after World War II – forced development of heavy industry combined with neglect of its effects on the environment, and
- The economic problems

Its geographical position in the European lowland, with mountains in the south and the Baltic Sea to the north, gives Poland some advantages such as trading and transportation opportunities. On the other hand, Poland's geography creates excellent conditions for pollution migration. Since 85% of the winds are from the west or south-west, about 50% of the sulphur dioxide in Poland comes from former East Germany and Czechoslovakia. Therefore, the western parts of Poland are much more heavily polluted by sulphur than are the eastern ones. The largest Polish rivers – the 1047-km Vistula (Wisła) and the 845-km Oder (Odra) – originate in the mountains of the highly industrialized southern part of Poland, and flow to the northern lowlands where rural areas and beautiful lakes prevail. Most of the Polish water supplies in the north are highly affected by contamination of the upper rivers.

Deposits of coal, cooper, zinc and other metals are found in the southern and southwestern parts of Poland. As a result of these raw materials, heavy industry developed in the region and caused significant degradation of air, water and soil in that area.

After the Second World War, Poland came under communist rule and heavy industrial development was forced for political reasons – to form a labour class. Most decisions on the localization of new enterprises were based on purely political reasoning, irrespective of economics or environmental health. The most typical examples are the steel works near the historical capital of Poland – Cracow, a city where the intelligentsia had previously had a strong position. Along with industrialization, a policy of neglecting ecological and psycho-social factors was developed. The ecological or human costs of living in the degraded environment were never taken into account.

Ongoing political and social changes in Poland have caused some environmental improvements, but also some new problems, both expected and unpredicted. We have observed "ecological fashion". This "fashion" for environmental protection and "ecology" has resulted in a plethora of information in the media. This situation causes social pressure on pro-ecological behaviour. However, there are also new conflicts, often associated with job losses that accompany the closing of polluting industries.

Money at the local level is now distributed by local, democratically elected councils. Because of the "ecological fashion" it is easier to make the decision of spending local funds on building a new sewage treatment plant. Such decisions are popular among the local populace and this is a positive result of democracy. Democratic mechanisms are less satisfactory when considering the possibility of convincing people about the necessity of locating a landfill in their neighborhood or building a waste incinerator.

Increased use of motor vehicles is one of the most serious problems in Poland today. No incentives or economic stimulation for buying pro-ecological cars have yet been introduced.

A similar situation is observed in the case of detergents. A huge selection of washing powders imported or produced in Poland is available on the market. The main chemical factories producing washing powders have been bought up by big international companies. But Polish law does not prohibit polyphosphates with the result that producers or importers of the detergents are not interested in more ecologically sensitive products. Foreign companies, which produce phosphate-free detergents for use in their own countries, sell phosphate-based detergents in Poland.

Nevertheless, due to EU pro-ecological programs, during first two years of membership in the EU, a lot of very important environmentally oriented projects have been realized in Poland in which also international companies are participating.

The number of multinational consortia with participation of Polish partners is steadily growing.

Therefore, a presentation of the scientific findings and technical solutions created by the Polish research community ought to be of interest not only for Polish institutions, but also for international specialists, seeking solutions for environmental problems in emerging new democracies, especially those who plan to participate in numerous projects sponsored by the European Union.

For this reason we have decided to publish in this book seventy six carefully selected papers from the total of two hundred and fifteen presented during the 2nd National Congress on Environmental Engineering.

Finally, I would like to express my appreciation to all who have helped to prepare this book. Dr Gordon Filby of innovEdit performed a herculean task working with great patience, aiding many authors to improve the linguistic side of their papers. Anonymous reviewers who not only evaluated papers, but very often made valuable suggestion helping authors and editors to improve the scientific standard of this book. And finally, last but definitely not least Mrs Katarzyna Wójcik – Oliveira for her invaluable help in preparing a lay out of all papers.

Lublin, September 2006

Lucjan Pawłowski

President of the Environmental Engineering Committee
of the Polish Academy of Science

Environmental Engineering – Pawłowski, Dudzińska & Pawłowski (eds)
© 2007 Taylor & Francis Group, London, ISBN13 978-0-415-40818-9

About the Editors

LUCJAN PAWŁOWSKI

Lucjan Pawłowski, was born in Poland, 1946. Dean of Faculty of Environmental Engineering and Director of the Institute of Environmental Protection Engineering of the Lublin University of Technology, Member of the European Academy of Science and Arts, honorary professor of China Academy of Science. He got his Ph.D. in 1976, and D.Sc. (habilitation in 1980 both at the Wrocław University of Technology). He started research on the application of ion exchange for water and wastewater treatment. As a result he together with B. Bolto from CSIRO Australia, has published a book "Wastewater Treatment by Ion Exchange" in which they summarized their own results and experience of the ion exchange area. In 1980 L. Pawłowski was elected President of International Committee "Chemistry for Protection of the Environment". He was Chairman of the Environmental Chemistry Division of the Polish Chemical Society from 1980 to 1984. In 1994 he was elected the Deputy President of the Polish Chemical Society and in the same year, the Deputy President of the Presidium Polish Academy of Science Committee "Men and Biosphere". In 1999 he was elected a President of the Committee "Environmental Engineering" of the Polish Academy of Science. In 1991 he was elected the Deputy Reactor of the Lublin University of Technology, and this post he held for two terms (1991–1996). He has published 19 books, over 128 papers, and authored 68 patents, and is a member of the editorial board of numerous international and national scientific and technical journals.

MARZENNA R. DUDZIŃSKA

Marzenna R. Dudzińska, D.Sc., Ph.D., M.Sc., received her master degree in chemistry in 1983 from Marie Curie-Sklodowska University (UMCS) in Lublin, Poland. Her master thesis was awarded by Polish Chemical Society. In 1987 she worked in Institute of Spectroscopy and Applied Chemistry in Dortmund, FRG. She got Fulbright scholarship in 1989 and spent 1989/1900 at University of Houston, Texas, in Civil and Environmental Engineering Department working on experimental part of her doctoral research. She received Ph.D. in environmental chemistry from Marie Curie-Sklodowska University in 1992. Since then, she is a staff member in the Faculty of Civil and Sanitary Engineering of Lublin University of Technology. In 1995, her interest moved to solid waste utilisation and incineration in cement kilns, focusing on the emissions from waste incineration (PCDD/Fs). Now she works on PCDD/Fs transportation and degradation in soil/water environment. In 2003 she fulfilled the habilitation procedure and got associate professor position in the Institute of Environmental Protection Engineering at Lublin University of Technology. She is an author or co-author of 65 papers on different aspects of physical and environmental chemistry and waste utilisation, two books on PCDD/Fs in the environment, 2 patents and co-editor of 5 volumes, published Plenum Publishing/Kluwer. She is a member of Polish Chemical Society (in 1984–1988 as President of Student Section of Polish Chemical Society), member of international committee "Chemistry for Protection of the Environment" and Secretary of Polish Academy of Science Committee "Environmental Protection Engineering".

ARTUR PAWŁOWSKI

Artur Pawłowski, Ph.D., was born in 1969 in Poland. In 1993 he received M.Sc. of the philosophy of nature and protection of the environment at the Catholic University of Lublin. Since that time he has been working in the Lublin University of Technology in the institute of Environmental Protection Engineering. In 1999 he defended Ph.D. thesis "Human's responsibility for the nature" in The University of Card. Stefan Wyszyński in Warsaw. Now he works on landscape ecology and problems connected with the idea of sustainable development. He has published 30 articles (in Polish, English and Chinese), 5 books, and has been an editor of further 11 books.

General problems

Environmental Engineering – Pawłowski, Dudzińska & Pawłowski (eds)
© 2007 Taylor & Francis Group, London, ISBN13 978-0-415-40818-9

Environmental engineering in the protection and management of the human environment

Lucjan Pawłowski
Faculty of Environmental Engineering, Lublin University of Technology, Lublin, Poland

ABSTRACT: Considering the current situation in Poland; comprehensive research is needed in the following areas: on the development of a strategy for the management of waste and sewage sludge, on a strategy for the short- and long-term utilisation of different elements of the environment (energy supply, the role of alternative energy sources, water management, land management, management of resources); on a better description of anthropogenic and natural sources of pollutants, as well as their transformations, pathways and dispersion through geoecosystems; on the search for means to shape and manage socioeconomic relationships through appropriate legal regulation which will lead to rational utilisation of the environment (i.e. through rationalised consumption of resources and land use and the minimization of anthropopressure). It is stressed that environmental engineering may be one of the most important tools in the implementation of a concept for sustainable development in the country.

Keywords: Environmental engineering, sustainable development, waste, water and wastewater management.

1 INTRODUCTION

Once again, after three years, the Polish environmental engineering community gather to present their most important achievements in this discipline of such importance to the existence of modern civilisation. That civilisation is irrevocably linked with a negative influence on the environment. Mostly because of the side effects of mining raw materials and their processing for useable products. Such activity exerts a particularly negative impact upon air and water, which have great effects for the whole biosphere. If we proceed on the assumption that the negative impacts of human activity can never be entirely eliminated, it is then on the minimising of effects that we must focus all our many and varied human undertakings.

Environmental engineering has a particular role to play in this respect, taking the lead when it comes to the elimination of environmental threats, and hence also greatly influencing human health. This is an interdisciplinary venture, very varied in its subject matter since, in seeking to neutralize pollution in all compartments of the environment (i.e. the hydrosphere, atmosphere and lithosphere), it builds upon knowledge of processes and phenomena from a number of basic sciences such as biology, chemistry, meteorology and physics.

Polish environmental engineering originated in the Faculties of Sanitary and Water Engineering of the country's technical universities, which have brought together scientific research and teaching at all levels. Where the Polish Academy of Sciences' structure is concerned, the subject emerged in 1961, when the Zabrze-based Department of Scientific Research of the Upper Silesian Industrial District was set up. The discipline's development began to accelerate greatly in the 1970s, following conversion of the above mentioned Faculties at the Technical Universities into Faculties of Environmental Engineering, and the designation of significant amounts of funding for appropriate research projects.

A first definition of environmental engineering and its scope came with a formulation arrived at in the Second (1973) Polish Scientific Congress. It was accepted there that: "environmental engineering is the science of the methods and technological means by which the environment particular compartments thereof are protected and transformed, with a view to ensuring optimisation of social and existential conditions and human wellbeing, as well as stimulating rational utilisation of natural resources".

It was stressed at the 3rd Polish Scientific Congress (1986) that environmental engineering is an interdisciplinary venture embracing, in a comprehensive manner, technical, technological, biological, economic, socio-economic, scientific and political aspects impinging upon the state, protection and management of the human environment. In line with a resolution of the 3rd Congress, the most urgent tasks of environmental engineering include "development of methods for reduction of the burden pollution imposes upon the environment, as well as methods of limiting emissions of particulate and gaseous air pollutants on the basis of low-waste technologies, improved methods for the removal of dusts, desulphurisation and removal of nitrogen oxides and hydrocarbons from flue-gases, the devising and introduction of wastewater treatment technologies for industry and agriculture, and the development of nationwide environmental monitoring, an early warning system for pollution and a system by which to assess the value of natural resources and the quality of the environment on the basis of economic accounting".

In April 1991, the Environmental Engineering Committee of the Polish Academy of Sciences in cooperation with the Governing Council of Deans of Faculties of Environmental Engineering of the Polish Technical Universities defined the most important areas of scientific research in environmental engineering, which were taken to include:

1. the protection and management of the indoor environment (ventilation, air-conditioning and heating installations).
2. the protection, management and utilisation of the external natural environment: water supply, the removal and treatment of wastewaters and wastes, waste management, water engineering and management, air protection, land and soil protection, pathways of pollutants in the environment, environmental monitoring and environmental impact assessment.

Environmental engineering engages in pro-environmental activity within the above scope, as well as developing or putting in place the appropriate technical conditions and technological methods allowing for the maintenance of the natural environment in a state of balance, the neutralization of the effects of natural or manmade disasters (flood, drought and the contamination of water, the air, land and soil) and the removal or limitation of the negative consequences of economic activity.

As of the 1970s, there were only just over 100 people engaged in the field of environmental engineering in Poland. However, a further phase of institution-building within the discipline took place in the late 1980s and early 1990s, the ultimate result being more than 2000 scientific employees now dealing with different aspects of environmental engineering and protection.

As far as theoretical study is concerned, Polish work in the sector competes at the highest international level. In contrast, the domestic situation does depart somewhat from that internationally as far as experimental studies and fieldwork are concerned. This situation primarily reflects financial outlays on science that are less than one-tenth that of some Western countries. Where financial resources are in short supply, work of a theoretical nature may still achieve a reasonable quality, but experimentation requiring specialist equipment is very much obstructed.

The development of science was always affected by the twin factors of a desire on the part of the learned to better understand nature, and a need for solutions to be found for the problems considered important to progress. Bearing in mind the fact that the pursuit of contemporary science requires ever greater resources, attempts are being made to prioritize research objectives that reflect a need to forecast directions to civilisational development. One such priority concerns sustainable development, of which environmental engineering is an integral part.

A turning point as regards approaches to environmental matters was the famous U. Thant report of 1969, which spelled out the threats attendant upon environmental degradation. There was a long period of time through which environmental protection was mainly understood in terms of nature conservation, and this approach remains the prevalent one in Poland.

Meanwhile, despite the success in decelerating further degradation of the natural environment over major areas of the globe, no such success is visible where socioeconomic relations are concerned.

For this reason, the concept hitherto understood as the protection of the natural environment would need to be replaced by the concept of the protection of the environment for human existence. Environmental protection understood in this way takes in, not only the well-known issues geared primarily towards nature conservation, but also the whole matter of managing the earth's resources. It is also imperative that reference be made to the social context, with account being taken of the socioeconomic relations that exert such a major influence on the quality of life. Unemployment has just as destructive an effect upon the human being as does life in a degraded environment.

Environmental protection understood in this way encourages unavoidable changes on our planet – unavoidable since there is no alternative. This allows for a somewhat more optimistic look into the future, since the ongoing changes are becoming irrevocably linked with the need to ensure environmental conditions sufficient to allow people to live with at least minimal human dignity.

The overriding aim in protecting the environment for human life is to ensure that present and future generations enjoy healthy living conditions. Similar objectives are set for the development of techniques and technologies. Furthermore, the development of the latter supplies new tools which, if used in the right way, may exert a significant influence in improving the state of the environment. Simplifying somewhat, we may say that techniques and technologies are tools allowing for the transformation of raw materials into utilisable products as human civilisation operates. Their abrupt and accelerated development (particularly in the 20th century), through a geometric increase in humankind's capacity to produce goods, led to a marked increase in living standards across large parts of the world. However, an open question in this context concerns whether the encroachment upon this of a marketing system stirring up constant demand for new goods through advertising is, genuinely raising the quality of life.

This question is made sensible enough by the fact that the growth in output is accompanied by the accelerated utilisation of resources, itself linked on the one hand with the possibility of these being used up sooner or later, and on the other with an undesirable ongoing increase in the level of pollution in the environment. It is also certain that humankind's future will be very much dependent upon the way in which flows of the Earth's resources within the environment for human life are managed.

The regulation of the flows in question is determined by the adopted concepts of the socioeconomic functioning of civilisation. These concepts should be shaped by a knowledge of the Earth's resources and their availability, as well as of the influence on the environment for human existence that methods used to convert resources into products exert. The functioning of the entire biosphere – and its human component in particular – is mainly decided by the means of utilizing the Earth as a whole.

It was a growing awareness of these issues that led to the formulation of the sustainable development concept set out in the 1987 "Bruntland Report", officially entitled Our Common Future. According to this, sustainable development is that kind of development which guarantees the meeting of current generations' needs, without limiting the possibilities for future generations to satisfy needs of their own. The proper management of the Earth assumes key importance in this context, since it is upon the rational utilisation of resources that the guaranteed meeting of future generations' own needs will depend.

In going beyond the questions of a purely nature-related character, attention will also need to be paid to the many problems of a general civilisational kind, e.g. the growing disparities between the rich and poor nations, or the increase in numbers of people going hungry and lacking access to clean water.

Data from the UN and WHO show that, on average, somebody dies every eight seconds as a result of having drunk polluted water. More than a billion people have no access to any source of water whatever in their immediate vicinity. What is worse, the situation has actually deteriorated over recent years. While international programmes over a number of years gave support for the construction of public water-supply systems in the developing countries, recent World Bank pressure has led to a number of privatisations, bringing with them significant increases in the costs of water supplied. It is thus more and more common to find that, while sources of clean water may be readily at hand, they are rendered inaccessible to a larger and larger group of people who cannot afford to purchase water for their own needs. These are major problems of today's world.

It needs to be emphasized that the above state of affairs has political and economic underpinning, since environmental engineering now has all the knowledge it needs to treat water effectively. Alas, polluted sources of water are most often found in areas of poverty and overpopulation. Those who live there cannot afford to have the treatment installations installed. Even in Poland, the inhabitants of rural areas often have nothing but low-quality water at their disposal.

In the course of the upcoming Congress there will be detailed discussion of many problems of importance to the development of Poland today. In my presentation, I would like to signal just some of these.

In my opinion, there is a need to carry out comprehensive research in the following areas:

1. On the development of a strategy for the management of wastes and sewage sludges;

   Waste management exerts a significant influence on the degradation of the environment on the one hand (through land degradation and the generation of secondary pollutants to the soil – water environment and the air) and on the functioning of the economy on the other. Too liberal a policy will lead to excessive environmental degradation, while too restrictive a policy may obstruct economic development.

   There is a need to better understand the consequences of waste dumping, also in the case of abandoned mines, pathways and quantities of pollutants generated connected with dumping, as well as to gain greater insight into the mechanisms by which these pollutants are transferred from landfills to the different components of the environment, and to determine the effects that migrating pollutants impose upon different elements of geoecosystems.
2. On the devising of a strategy for the short- and long-term utilisation of different elements of the environment (energy supply, the role of alternative energy sources, water management, land management, management of resources);

   The aim of this research should be to gain a better understanding of the environmental conditioning underpinning Poland's development, with account taken of its own natural resources. It is commonly believed that import of primary energy resources, such as gas, is an ecological undertaking. However, no attention is paid to the fact that gas can only be purchased if paid for by exporting other products whose manufacture may increase environmental degradation considerably.

Principles for the protection of natural resources need to be set out, with particular attention to underground and surface waters.

3. On a better description of anthropogenic and natural sources of pollutants, as well as their transformations in pathways and movements through geoecosystems;

   The negative impact of each pollutant is manifested when it passes from the place at which it is generated into the living organism, wherein it is able to affect life processes.

   Civilisational development is associated with the mining of raw materials from geoecosystems and their processing into utilisable products. Following use, these return to geoecosystems in the form of pollution. The process is inseparable from the exerting of anthropopressure via pollutants introduced into the environment. Some of these are chemical compounds already present in nature. However, as chemistry has developed, a whole array of new chemical compounds unknown to nature have appeared, these sometimes displaying an exceptionally high level of biological activity. Chemical Abstracts listed in excess of 5 million chemical compounds, with around 50,000 new ones being registered annually.

   In this situation, it becomes impossible to understand precisely the behaviour of all known chemical compounds in the environment. Hence there is a need to better understand pathways and transformations of different groups of chemical compounds through the environment, as well the influence these exert – especially on the biosphere and by way of the different food chains. Such information is indispensable if remedial action is to be taken to limit the negative impacts of chemicals introduced into the environment.

   To simplify analyses it would be helpful to define the most important pathways of chemicals in geoecosystems.

4. On the search for means to shape and manage socioeconomic relationships through appropriate legal regulation leading to rationalised utilisation of the environment (i.e. through rationalised consumption of resources and land use and the minimization of anthropopressure). It would seem of importance to obtain a better understanding of the attitudes representative of Polish society, and to look for means by which to shape such attitudes in order to favour implementation of the sustainable development concept.

Table 1. Pathways of pollutants through geoecosystems.

Stages of pathway of chemical compounds in the environment	Mechanisms important to the flow of chemicals
Sources	• Mechanisms by which chemical compounds enter different compartments of the environment. • Mechanisms slowing down the appearance of chemical compounds in different elements of the environment
Movement through geoecosystems	• Flow in streams: – of air – of water • Spread across the Earth's surface via movement of materials • Spread across the Earth's surface via the utilization of products
Entry into the human food chain	• From the air stream air → man air → plant → man air → plant → animal → man • From the water stream water → man water → plant → man water → plant → animal → man

The aim of this work should be to attain a better understanding of the socioeconomic and legal mechanisms shaping relationships between humankind and the environment.

It results from all this that rational action to ensure Poland's harmonious development will depend first and foremost on knowledge of the functioning of geoecosystems, as well as the skill to limit the negative impacts on their functioning that human civilisation exerts. Assuming particular significance in this context is the work being conducted in environmental engineering, on methods indispensable in the protection and appropriate shaping of the environment for human life.

Environmental Engineering – Pawłowski, Dudzińska & Pawłowski (eds)
© 2007 Taylor & Francis Group, London, ISBN13 978-0-415-40818-9

The scientific, educational, economic and social meanings of sustainable development

Jerzy Błażejowski
University of Gdańsk, Faculty of Chemistry, Gdańsk, Poland

ABSTRACT: Origin and meaning of sustainable development in nowadays civilization is discussed. Importance of ethic, which justify preservation of natural resources for future generation is stressed. Thermodynamic lows may be used to support man obligation for environment and recourses protection.

Keywords: Environmental engineering, sustainable development, environmental ethic.

1 INTRODUCTION

The term sustainable development was coined in the report "Our Common Future", compiled in 1987 as a result of the work of the UN World Commission for the Environment and Development chaired by Dr Gro Harlem Brundtland, then Prime Minister of Norway [1]. In that document, sustainable development was defined as follows: "The world economic system should meet the needs of the present without compromising the ability of the future generations to meet their own needs" [2,3]. The formulation of the term sustainable development had been preceded by a series of earlier reports highlighting the threats to the global environment from fast economic development [2–4]. Two World Summits for Sustainable Development – in Rio de Janeiro (1992) and Johannesburg (2002) – substantially extended the meaning of the term [5]. Since then, sustainable development has taken on a number of meanings.

2 THE WORLDWIDE DIMENSION OF SUSTAINABLE DEVELOPMENT

Sustainable development is the way towards improving the quality of human life all over the world without the wasteful exploitation of the Earth's natural resources [5]. If this aim is to be reached, then activities in three crucial areas need to be integrated:

- Economic growth through cooperation between nations and between societies, links between global systems, and equal sharing of profits;
- Protection of natural resources and the environment for present and future generations;
- Social development guaranteeing jobs, health care, access to education, food, energy and water, and appropriate living conditions (including the protection of cultural heritage, social diversity, and equipping people with the knowledge and tools enabling them to shape their future).

But these goals are achievable only if the mechanisms of international cooperation are improved: decisions taken in one region influence the existence of human beings elsewhere in the world.

Sustainable development is a subject of concern of numerous important institutions and persons. The late Pope John Paul II, for example, raised many aspects of this problem in his address to the Pontifical Academy of Sciences [6]. This now ubiquitous term is used in legal and official documents at national and international levels. It is invoked in Article 5 of the Polish constitution [7]; it appears in programs for elementary, junior high and high schools, as well as universities [8] and other forms of education [9] in Poland. Sustainable development has been the topic of international conferences [10–12], monographs [4,13–15] and research [3,16], and is a major research priority in the European Union (EU) [17].

Sustainable development is a fundamental issue that is being addressed by national governments and European or international institutions. Meeting at the spring summit in Brussels in March 2005, the leaders of EU countries underlined the importance of environmental conservation in Europe and indicated that the Lisbon strategy for development should go hand in hand with the ideas of sustainable development. It was recalled at this summit that global temperature rise must be slowed down and the emission of greenhouse gases reduced. Much greater support must be given to pro-environmental innovations and environmentally friendly investments in the transportation and energy sectors. The decline in Europe's biodiversity must be halted. Despite general agreement among EU countries as to the main strategies towards sustainable development, there are differences in individual approaches between the old and new EU members. Sustainable development tops the list of priority development strategies advocated by the administrations of European countries and regions [18,19].

An important element of sustainable development is the promotion of environmentally friendly economic activities. Well-known companies use the term in advertisements of their activities and products. Networks of research institutions [20] and companies [21] have been created, whose activities are focused on sustainable development.

It emerges from the above that sustainable development has many meanings. There have been in the past attempts to describe sustainable development by using various models [22]. This article is intended to demonstrate how the simple formalism of thermodynamics can be used to describe the functioning of Nature and of the human presence in it [23,24], and how sustainable development can be understood in terms of thermodynamics.

3 BASIC RELATIONS OF THERMODYNAMICS

Thermodynamics is based on phenomenological relationships reflecting the fundamental mechanisms by which Nature functions [25]. Observing the surrounding world, humans perceived two opposing trends – the tendency for matter and energy on the one hand to be organized, and on the other to be dissipated. The measures of these tendencies are thermodynamic state functions that depend on the condition of the system. The basic state parameters are pressure (p), volume (v), temperature (T), and the composition of the system (or the concentration of components in the system). The state functions reflecting the tendency for matter and energy to be organized are energy (e – a function of v and T) and enthalpy (h – a function of p and T), and the state function reflecting the tendency for matter and energy to be dissipated is entropy (s^v – a function of v and T, or s^p – a function of p and T). The tendency towards the organization of matter and energy is most pronounced at the temperature of absolute zero (the lowest physically attainable temperature), since dissipation cannot occur under such conditions (the entropy is then equal to zero). At any temperature above absolute zero, both tendencies determine the properties and behavior of the system. This is reflected in state functions known as free energy (f) ($f = e - Ts^v$, a function of v and T) and Gibbs free energy (g) ($g = h - Ts^p$, where $h = e + pv$, a function of p and T). The system becomes more ordered if the energy factor predominates, but less ordered if the entropy factor prevails.

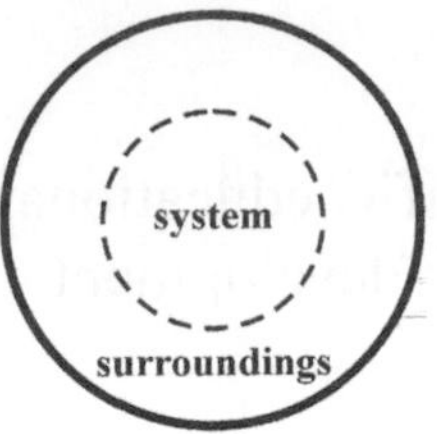

Figure 1. System and surroundings.

Let us assume that natural processes take place in a system that in our mind we have separated from the environment surrounding it (Figure 1).

A system and its surroundings together form a new system, which can be regarded as isolated if it retains matter and energy, closed if it retains matter, or open if it exchanges matter and energy. The world surrounding us is changing continuously. These natural changes occur in finite times and the processes involved are irreversible. In such processes, the free energy (Gibbs free energy) changes of a system do not keep up with those of the surroundings and are thus not thermodynamically equilibrated. On the other hand, we can conceive other, reversible processes, in which the free energy (Gibbs free energy) of a system and surroundings follow one another. Although these latter processes are thermodynamically equilibrated, their realization requires an infinite length of time. So to describe the behavior of Nature we need to consider the sum of the energy and entropy changes in the system and its surroundings (extended so far as to include all changes). The exchange of energy and mass in such an isolated system takes place in accordance with the law of conservation, regardless of whether the processes concerned are reversible or irreversible. In actual fact, however, the conservation law applies to the sum of mass (m) and energy, and allows for their equivalence in the sense of the equation $e = mc^2$ (where c is the velocity of light in a vacuum). Entropy exchange in this isolated system would take place in accordance with the conservation law only if the processes were reversible – and such do not occur in reality. Natural, irreversible processes are always accompanied by an increase in global entropy. In other words, irreversible processes are the source of entropy – that creates the foundation of the thermodynamics of irreversible processes [26,27]. This branch of thermodynamics distinguishes two categories of quantities: thermodynamic forces (X), reflecting the forces driving natural processes or the causes of natural phenomena, and thermodynamic fluxes (J), which describe the changes taking place in the system at a given time. The basic quantity of irreversible thermodynamics is the entropy source (σ), which is defined as the sum of the products of thermodynamic forces and the thermodynamic fluxes directly caused by them:

$$\sigma = \sum_i X_i J_i$$

A general description of processes based on irreversible thermodynamics rests on the assumption that each thermodynamic force acting on the system can influence any of the thermodynamic fluxes. In the simplest approach – known as linear irreversible thermodynamics – each thermodynamic flux is proportional to each thermodynamic force influencing the system: this is expressed by the equation

$$J_i = \sum_j \sum_i L_{j,k} X_k$$

where L is the phenomenological coefficient and i (above), j and k indicate the quantities.

Thermodynamic fluxes do not occur in the absence of thermodynamic forces and the entropy source has zero value. Such a situation applies to a time-invariable equilibrium state. If a force acts temporarily, it gives rise to a flux that eventually disappears. After an infinite time, the system attains, under the conditions determined by T, v or p, a state of equilibrium. It can, however, happen that a force is constant in time. Then, with the passage of time, the system will tend towards the steady state. The order of the steady state depends on the number of constant thermodynamic forces acting on the system: a steady state is zero-order when it is in equilibrium, first-order when one force is constant, second-order when two forces are constant, and so on. The system is in a non-stationary state if all the acting thermodynamic forces are variable in time.

From this analysis it emerges that a completely stable state – an equilibrium state – does not often occur in Nature. States exhibiting certain hallmarks of stability can be identified with stationary states. Non-stable states are non-stationary states.

4 THERMODYNAMICS IN THE DESCRIPTION OF THE WORLD AROUND US

The phenomenological thermodynamics presented so far (Greek *phainómenon* = phenomenon) describes the behavior of a macroscopic systems. It can thus be used to describe the behavior of large systems such as the Earth or the solar system [23,24,28]. In the case of the Earth, it is reasonable to assume that it is a thermodynamically closed system exchanging energy with the Sun (this exchange is in one direction – from the Sun to the Earth). The solar system can, on the other hand, be treated as an isolated system retaining mass and energy. In the closed system of the Earth, irreversible processes proceed continuously and are the

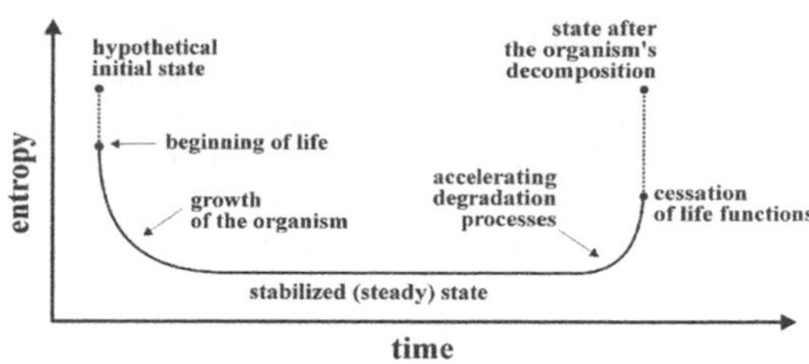

Figure 2. How living matter comes into being, functions, and disappears in the stationary state of the Earth.

ineluctable future of Nature. Even though such processes take place within finite periods of time, there is no way of completely foreseeing their effects or of returning the system to its initial state. A complete prediction of the future is therefore impossible, and the unintentional and unwanted effects of irresponsible activities cannot be undone. Irreversible processes are invariably accompanied by increasing disorder, and consequently, entropy. If there were no counteracting mechanisms, the Earth would be in a non-stationary state, in which conditions would be unpropitious to the existence of living matter. Since, however, the Earth is under the constant influence of the Sun, its state, formed during periods far exceeding the lifetime of organisms, can be considered stationary. This state is attained because the increase in entropy caused by the spontaneous processes in living matter is balanced by the decrease in entropy that takes place during the Sun-driven process of photosynthesis. In other words, the Sun restores order after the disorder produced by the irreversible processes taking place on the Earth.

How living matter comes into being, functions, and disappears in the stationary state of the Earth is illustrated in Figure 2.

This shows the entropy changes with time in a closed system containing the requisite amount of matter (simple molecules or atoms) for an organism to be created and to function. Before life is conceived, matter is in chaos, a state characterized by a high entropy. Initially, simple molecules are converted to complex ones: these processes precede the conception of life. But once life has been conceived, these complex molecules combine to form ordered structures like the macromolecules of living matter: this process reduces the organism's entropy. As the organism evolves and its mass increases, its entropy decreases yet further. On reaching a certain age – in the case of the humans, between 16 and 20 years – the organism stops growing, but continues to live at a more or less stable level of entropy. Throughout its entire existence, the organism exchanges matter and energy with its surroundings. But when it reaches a certain greater age, degradation processes set in. Gradually accelerating, they increase the organism's entropy. The end result is the cessation of life functions and death. Together with this, the organism ceases to exchange matter and energy with its surroundings and finally breaks down into the same small molecules from which it was created.

Figure 2 illustrates an intrinsic aspect of the world in which man exists – the finiteness of everything that comes into existence, either naturally or through human agencies. Each organism can be born and exist once and once only, although after its death and decomposition, small fragments of it may be incorporated into other organisms. Every organism is unique, since its existence, growth and demise are realized under different conditions and in different surroundings. So the re-creation of once existing organisms is impossible owing to the irreversible character of events in the world around us.

What we have just said about irreversible processes also applies to the physical and social sides of human activity. Emotions, ever present in the human condition, have their beginning, period of full bloom, and end. An emotion (like love) may well outlive an individual human being, but it can also last a shorter time. All human communities (e.g. political parties, states, empires) come into existence, enjoy a period of prosperity, and then disappear from the world's stage. Creation is usually accompanied by euphoria, but extinction ensues from uncontrolled events like revolutions or wars. Of the vast empires that once existed few memories have survived. This transience will also apply to existing or inchoate communities. It is, however, of unpredictable duration, and intense efforts directed towards the organization and functioning of a given community can delay the natural tendency towards dissipation. Through their rational activities, humans are able to influence the course of physical and social processes. There is also reason to believe that people's mental spheres can affect their behavior.

The processes occurring in the life of a human being or other organism cannot be considered in isolation, but only against the background of changes taking place on our planet and beyond it in a much longer time perspective. Even if during the lifetime of a given person or during several generations the state of the Earth can be treated as stationary, in the much longer term, going back to the beginnings of our planet, this state will have changed. Since the Earth's temperature is continuously falling, we may presume that its entropy is also gradually decreasing (Figure 3). The decrease in temperature has been accompanied by the appearance of ordered forms, first of non-living, and later of living matter. Thermodynamics is unable to explain what caused life to come into existence. That life happens to exist on the Earth does not mean that a level of entropy identical to that existing on the Earth now or in the past guarantees the presence of living matter elsewhere in the universe. Our observations show that the appearance of a species is followed by its development, and ultimately, by its extinction. Of the great

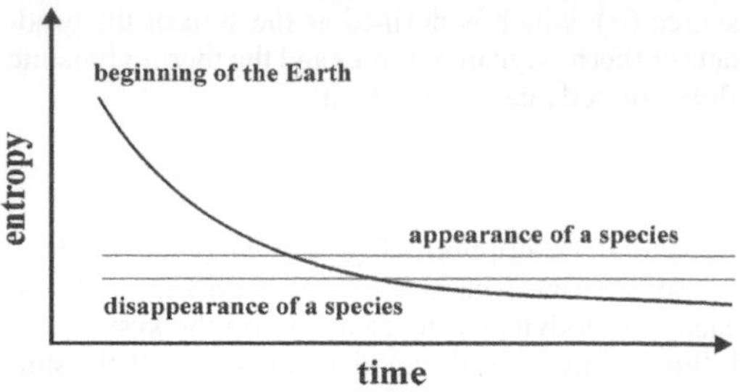

Figure 3. Entropy changes since the beginning of the Earth.

reptiles that once existed on Earth there remain only vestiges. Even animals living not so long ago, like the mammoths, have vanished. Every day new species or forms of living matter appear while others disappear. With the passage of time, or to put it another way, with the gradual lowering of the Earth's entropy, the organisms newly appearing are increasingly on a higher level of organization (like inorganic matter). This process is called evolution and is more a consequence of the entropy changes on the Earth than of the progressive development of living matter.

It is an intrinsic property of irreversible processes, indicated in Figure 3, that species have a beginning and an end. We can assume that each species has a characteristic entropy "window" in which it can exist. If the entropy moves beyond the window relevant to a given species, this will no longer be able to exist. The existence of living matter is possible given the stationary state of the Earth, which is governed by the Sun's activity and Nature's regulatory mechanisms. Natural changes in the Earth's stationary state are slow and present no danger to the functioning of Nature or man. Importantly, however, man's intentional and unintentional activities introduce forces that are altering the steady state of the Earth. These anthropogenic forces generally arise from activities that are totally unlike natural processes. One anthropogenic force is thus the combustion of coal and crude oil resources, which accumulated the carbon dioxide once present in the atmosphere. The resulting emission of carbon dioxide into the atmosphere is causing the greenhouse effect. A second anthropogenic force is the excessive production of nuclear energy. A third anthropogenic force is the production, application, and release into the environment of chlorine-containing compounds and other synthetic substances. Armed conflicts wreak havoc in the environment, and given the present advanced state of military technology, they constitute a completely unpredictable anthropogenic force. Yet another such force is the exploitation of the environment beyond its regenerative capacities. These and further anthropogenic forces are bringing about global changes in entropy, and therefore in the conditions for the existence of human beings and the whole of Nature.

Are humans a threat to the environment? Yes, to a certain extent they are. Nature, like the organisms that are part of it, has developed protective mechanisms. Their main task is to maintain the stationary (sustainable) state of the Earth or an organism. Up to a point, therefore, Nature is able to neutralize people's unintentional or irresponsible activities. But is this enough to guarantee the uninterrupted existence of present and future generations? There is no simple answer to this question.

We know that one of Nature's defense mechanisms is the elimination of individuals or species that are upsetting the natural equilibrium. Such a scenario is not very likely, however, in view of the defense mechanisms that man now possesses. It is thus probable that man's activities will bring about changes in the Earth's entropy beyond the window in which he can exist.

5 CONCLUSIONS

Sustainable development (a better term would be sustainable existence) is not an idea that we can or cannot accept. Sustainable development needs to be invoked at once; it is indispensable to man's continued existence on this Earth. The simplest recipe for sustainable development would be the non-action of anthropogenic forces on the environment, that is, the use of renewable energy resources, the use of readily degradable materials, and the economical exploitation of natural resources. In order to incorporate all the rules of sustainable development (sustainable existence) into our lives – so that we can live in complete harmony with Nature – we must understand Nature's laws, disseminate knowledge of them to others, and adapt our economic and social activities to comply with them. Unfortunately, it is not always possible to bring all these factors to bear together.

Nevertheless, we must strive towards this with all our might. All people, especially the intellectually gifted and enlightened, are duty-bound to care for the state of Nature, so that our uninterrupted existence on the Earth can be guaranteed now and in the future.

Man's great passion is learning about Nature and her laws. Having discovered these laws, he sets about applying them: this much seems obvious. The problem is that of this knowledge man chooses those parts which are profitable for him, disregarding others, which are inconvenient for him. Now is the time to accept these laws unconditionally, such as they are, and to adjust our activities to a harmonious coexistence with the environment – our highest Earthly good.

REFERENCES

[1] Rio+10, World Summit for Sustainable Development. http://www.mos.gov.pl/rio10/1_2.html.

[2] A.J. Scott. *Journal of Environmental Management*, 54 (1998) 291–303.

[3] J.C. Bridger, A.E. Luloff. *Journal of Rural Studies*, 15 (1999) 377–387.

[4] N.A. Robinson (ed.). Agenda 21: Earth's Action Plan. Oceana Publications (USA), 1993.

[5] Sustainable development. http://www.unic.un.org.pl/johannesburg/zrow_rozwoj.php.

[6] John Paul II. Sustainable development. http://www.opoka.org.pl/biblioteka/W/WP/jan_pawel_ii/przemowienia/pan_12031999.html.

[7] Constitution of Poland from 2nd April of 1997. Dz.U. Nr 79 poz. 483 from 16 July of 1997.

[8] J. Ciechowicz-McLean. Program of the subject "Legal aspects of sustainable development". http://www.prawo.univ.gda.pl/gospodarcze/2.htm.

[9] http://www.unic.un.org.pl/edukacja_rozwoj/index.php

[10] *Atmospheric Environment*, 29 (1995) 1879.

[11] *Renewable Energy*, 23 (2001) 697–699.

[12] *Energy*, 28 (2003) 1041.

[13] *Biological Conservation*, 93 (2000) 278–279.

[14] *Resources, Conservation and Recycling*, 32 (2001) 83–84.

[15] J. Bohdanowicz. Towards Civilization of Ecodevelopment. University of Gdańsk, Gdańsk, 1998.

[16] M.J. Terpstra. *American Journal of Infection Control*, 29 (2001) 211–217.

[17] http://www.pwcglobal.com/pl/pol/about/svcs/abas/ue/ramowy_rozwoj.html.

[18] K. Bovin, S. Magnusson. Local initiatives for sustainable development. The Swedish Society for Nature Conservation, Stockholm, 1997.

[19] M. Kistowski. Regional model of sustainable development and environmental protection in Poland, and strategy of development of provinces. University of Gdańsk, Gdańsk, 2003.

[20] South-south networking and cooperation on renewable energy and sustainable development. *Renewable Energy*, 29 (2004) 2273–2275.

[21] First EU network on sustainable product-service development launched. *Journal of Cleaner Production*, 11 (2003) 703–704.

[22] C.V. Matrossova. *Nonlinear Analysis*, 47 (2001) 5403–5413.

[23] J. Błażejowski. Thermodynamic view on functioning of Nature – ethic behavior in environmental context, Monografie Komitetu Inżynierii Środowiska PAN, 26 (2004) 149–160.

[24] J. Błażejowski. Sustainable development in scientific, educational, economic, and social context. Monografie Komitetu Inżynierii Środowiska PAN, 32 (2005) 41–51.

[25] P.W. Atkins. Physical Chemistry, Oxford (U. K.), 1998.

[26] K. Gumiński. Irreversible Thermodynamics, PWN, Warsaw, 1992.

[27] Chan Eu Byong. Kinetic Theory and Irreversible Thermodynamics, New York, 1992.

[28] I. Prigogine, G. Nicolis and A. Babloyante. Thermodynamics of evolution, Postępy fizyki, 26(1975) 253–281.

Environmental Engineering – Pawłowski, Dudzińska & Pawłowski (eds)
© 2007 Taylor & Francis Group, London, ISBN13 978-0-415-40818-9

Water management in Poland on the verge of the XXI Century

Zdzisław Kaczmarek
Institute of Geophysics, Polish Academy of Sciences, Księcia Janusza, Warsaw

ABSTRACT: Poland is relatively scarce in water in some regions of the country. Due to variability of hydrological processes the reliable water resources available at 95% of time are equal to about 22 km^3, twice of the average annual water withdrawals. These resources are sufficient for satisfying objectively motivated needs of people, industry, energy and agriculture, if managed in efficient way. Local and seasonal shortages of water in Poland result from its contamination and underdeveloped infrastructure. To eliminate these problems and ensure protection against floods and droughts, large amounts of financial means must be spent in coming decades in order to implement the EU Water Framework Directive by the year 2030. The efficient water management requires the ability to quantify its objectives, proper forecasting of changing physical and economic conditions and application of reasonable decision-making procedures. The multidisciplinary research potential of Poland is adequate to meet educational and technological needs required for sustainable water management.

Keywords: Water management, sustainable development.

1 INTRODUCTION

Water management integrates various geophysical, biological and socio-economic processes within spatial systems whose boundaries are determined by river catchments. Rational management of resources demands the definition and quantification of targets, good recognition of present and future tasks and needs, efficient management and application of decision-making procedures to ensure social participation of water users.

Contrary to common opinions, the surface water and groundwater resources of Poland are sufficient for satisfying objectively motivated needs of people, industry, energy production and agriculture, along with the preservation of the natural environment in our country. Local, seasonal shortages of water result from its contamination or underdeveloped infrastructure. In order to eliminate these items and ensure sufficient protection against the catastrophic effects of extreme hydrological events, large amounts of public and private financial means must be spent. It will be also necessary to radically rearrange and improve the systems for administering and managing this branch of the national economy.

2 VISION ON THE WATER MANAGEMENT IN EUROPE IN THE YEARS 2001–2003

The Water Framework Directive 2000/60/WE, established in 2000, imposed on the participating countries a series of obligations in order to coordinate their actions towards preservation, protection and improvement of the quality of water environment. Its realization in Poland will demand large organizational and financial effort, including the adaptation of national legislation to the common legal rules of the European Union. Vision on the water management in Europe in 2030 was outlined in the program document of the Water Supply and Sanitation Technology Platform, WSSTP, established in 2004 in the framework of European Commission planning (Vision: Water Safe, Strong, and Sustainable, 2005). The priorities stipulated in this document embrace several key problems, including:

(a) Full implementation of the Water Framework Directive until the year 2030 by the EU member countries, including the concept of integrated water management whose aim is to ensure balance between the users' needs and preservation of a good ecological status of water objects.
(b) Rationalization of water consumption in households, industry and agriculture. The use of water-saving technologies and closed systems in economically sound cases. Limitation of water loss in water-supply systems. Implementation of instruments for irrigation systems ensuring optimal dosage of water in agriculture.
(c) Intensification of research and engineering studies, as well as investments aiming at considerable limitation of discharge of harmful substances to surface water and groundwater from point and disperse sources by 2030. Widespread monitoring

Table 1. Groundwater and surface water resources (without inflow from abroad) and water demands (P – population, I – industry, E – energy production, A – agriculture) (Source: Statistical Yearbook of Poland for 2003, Central Statistical Office).

	Resources (R)		Demand (D)		Water use by sectors in %			
Country	km^3	m^3/cap	km^3	D/R %	P	I	E	A
Belgium	12.4	1220	7.44	60.0	19.2	18.8	59.0	3.0
France	180.2	3070	32.32	17.9	18.2	11.5	60.3	10.0
Greece	60.0	6870	9.44	15.7	9.1	10.4	0.02	80.5
Spain	111.1	2800	40.90	36.8	9.3	3.1	35.4	52.2
Germany	111.0	1350	37.91	34.1	14.6	15.4	68.9	1.1
Poland	54.8	1420	11.75	21.4	23.6	7.0	59.6	9.8
Sweden	170.0	19180	2.58	0.02	35.8	54.4	5.8	4.0
Great Britain	157.8	2690	15.89	10.0	48.5	15.0	21.0	15.5
Italy	167.0	2570	56.17	32.1		No data available		

and alert systems against ecological catastrophes (accidents, terrorist threat) associated with pollution transport in water.

(d) Extensive social participation in decision-making processes concerning planning and exploitation of the water-management systems through ecological education, social consultations, dissemination of data and information, and participation of users in water management.

While discussing the vision of water management in Europe, attention has been drawn to thediversity of situations in various EU countries, as demonstrated in Table 1 presenting water situation in some European countries, characterised by different population, economic and climatic conditions.

The above data show that, when making international comparisons, the water consumption structure may be of greater importance than the factors determining the need-resource relation. It has been stressed in the WSSTP document that the rules for using water in Europe should be adjusted to regional conditions. A comparative analysis is of cognitive value only if it is related to water systems of similar structure of tasks. In Spain, for instance, the water resources per capita are double those of Poland, while water consumption is – owing to the necessity of agricultural irrigation – four times greater than in our country.

Still, the WSSTP's analyses give grounds for formulating general recommendations for most of the European countries, including Poland. The implementation of 2000/60/WE Directive means a change in the traditional paradigm of water management, according to which the planning and technological actions had been exclusively, or predominantly, determined by financial profits. This was the basis for formulating previous Polish national and regional plans of water resources infrastructure, the main objective of which was to meet the users' expectations, usually formulated without proper consideration of environmental conditions. The Directive targets at promoting actions toward sustainable use of water based on long-term protection of the quantity and quality of the available water resources.

One such action is a rational approach to water consumption by gradually replacing the "supply water management" by "supply/demand water management". This should lead to reduction of water intake in regional scales, including, in particular, the lessening of the pressure on the ecosystems in water-deficit areas. It should be stressed that water consumption in Poland has dropped by about 30% in the 1990s. However, this was not a consequence of purposeful action, but rather of industrial transformation and increasing costs of water intake and sewage disposal.

One of the universal recommendations stipulated in both the WSSTP platform documents and the EU Directive, applicable by all Union members, is the previously mentioned postulate of increasing social participation in decision-making water management procedures.

3 WATER MANAGEMENT IN THE 2007–2013 NATIONAL DEVELOPMENT PLAN (NDP)

In accordance with the assumptions of the Water Management Strategy[1], the main objectives related to the water use are the following:

- To fulfill reasonable water needs of people and economy, accounting for the principle of sustainable water usage.
- To achieve and maintain good quality of water and water-dependent ecosystems.
- To protect people and economy against the threat of floods and droughts.

[1] Working document prepared by the Water Resources Department of the Ministry of Environment (June 2005).

The instruments for programming and financing these objectives in the next 10 years will be the sectoral and regional operational programs of the 2007–2013 National Development Plan. In the NDP proposal it is not foreseen to establish a separate Sectoral Operational Plan (SOP) to deal with the water resources of the country. This enhances the importance of cooperation between the water administration units and the institutions responsible for preparation and realization of those SOPs and ROPs, whose priorities are more or less closely related to the water supply, protection of water ecosystems and ecological safety of the country. Of particular importance will be the cooperation with local administrative units in preparing regional plans for their regions. The efficient realization of the 2007–2013 NDP in water management domain and full use of European Union sectoral financial means makes it necessary to establish an efficient coordination system on the central state administration level.

Organizational actions and investments to be realized in the framework of 2007–2013 National Development Plan should be a fragment of a many-year program of re-arranging and improving the efficiency of water management of Poland, which now suffers from the following drawbacks:

- Local difficulties with water supply, resulting from insufficient quality of water; excessive exploitation of underground water resources, non-rational use of water and unsatisfactory technical infrastructure;
- Limited abilities of water storing and displacing, if need occurs, of surface water resources which are unevenly distributed in time and space;
- Insufficient protection of large areas of the country against losses due to catastrophic floods and droughts;
- Little use of the energetic and navigational potential of rivers.

In spite of the above drawbacks, Poland is not threatened by a water supply crisis in the near future, provided that the resources are properly used.

In the past, Polish planners have erroneously estimated tendencies of economic growth and demographic changes in our country, which has led to considerable overestimation of the expected water consumption. In 2000, the water intake in Poland for household and economic purposes (except for cooling systems in thermal power stations) amounted to about 5.3 km^3, i.e., three times less than the estimate for this year foreseen in the forecasts elaborated in 1970s and 1980s. Table 2 shows the author's estimate for the water consumption in Poland for the next decades. The forecast is based on the following assumptions: (a) 180 liter of water per day and capita, for household needs of urban and rural inhabitants, (b) low growth of water intake by the industry, under the assumption of its rational use, (c) stabilization of water needs for open cooling systems of generators in thermal power stations, (d) participation of irrigated areas in the amount of 4.0% of the total arable land. The demographic forecast for Poland for the years 2001–2050 was based on the estimate of the European statistical bureau Eurostat. After Czyżewski et al. (2000), the annual gross national product growth of 4.0% was adopted.

Table 2. Expected water demands in Poland in year characterised by average hydrological conditions (Source: own estimate by Z. K.).

General data/Sector of economy	2000	2025	2050
Population, in thousands	38644	36836	33665
GNP per capita in thousand PLN (prices for the year 2000)	18.3	48.8	130.0
Water consumption for household use [km^3]	2.35	2.42	2.21
Water consumption by industrial production [km^3]	0.85	1.39	2.29
Water intake for power stations cooling [km^3]	6.79	6.79	6.79
Water consumption by agriculture (irrigation and fishery) [km^3]	1.06	1.76	2.18
Other (uncontrollable) water intakes [km^3]	1.00	1.00	1.00
Total water consumption in Poland [km^3]	12.05	13.36	14.47

The estimate of water demand in Poland in the first half of the present century, as proposed here, will need periodic refinements and updating, notably after the tasks of 2007–2013 National Development Program have been completed. It is to be noted that the expectations as to the future water needs of agriculture are highly uncertain because of the lack of a long-term development strategy for this sector of the economy, and the possibly considerable impact of the expected climate change on the water balance of soil and the agricultural production.

On the basis of the 2007–2013 NDP assumptions and the water need forecast presented here it can be concluded that the major future problems of water management do not follow from the increasing water consumption. Instead, they are connected with the necessity of improving the living standard and security level of people and the fulfillment of the demands of the EU Framework Water Directive in terms of attaining and preserving good conditions of surface water and groundwater.

4 SUSTAINABLE DEVELOPMENT OF WATER MANAGEMENT IN A CHANGING ENVIRONMENT

The Constitution of the Republic of Poland stipulates that natural resources management should be carried out in a sustainable manner. The concept of sustainable development, first formulated in the UN Report of the World Commission of Environment and Development (Brundtland et al., 1987), has not yet been transformed in a convincing way into the language of practical actions, and the adjective "sustainable" is sometimes a general slogan added to water resources management carried out in a traditional manner.

The sustainable policy may cover actions in various branches of the edge economy and should refer to the future, e.g., the future technological and organizational changes. This also concerns the water management, which should be carried out in accordance with the following principles (Loucks, 2000): "Sustainable development is the management and conservation of the natural resource base and the orientation of technological and institutional change in such manner as to ensure the attainment and continued satisfaction of human needs for present and future generations. Such sustainable development conserves land, water, plant and animal genetic resources, is environmentally sound, technically appropriate, economically viable and socially acceptable".

Water management is a function of many time-varying factors and limitations: social, economic, technological, and natural. Decision-making processes should take into account possible conflicting situations associated with such an approach. It is necessary to envision not only the dynamics of future changes in water resources and demands, but also the preferences of future beneficiaries (or victims) of the present decisions. As experience teaches, this is possible only if the time-scale of planning is not very long.

The process of efficient water management involves: the ability to define and quantify the objectives; proper forecasting of future needs and tasks; and reasonable choice of decision-making procedures. Sustainable development is associated with the use of various measures of profits, losses and risk in the future situations, under conditions which are not known, yet may be affected by the present-day decisions. It is a requirement of sustainable development to work out a strategy for making the water management adjustable to changes.

In the publication by Young et al. (1994), the principles for sustainable development in relation to water management have been formulated as follows:

- Criterions of sustainable development should be applied to all categories of water systems: local and multifunctional.
- The development of technical infrastructure and operational control of water resources should be flexibly adjustable to changing needs, the prediction of which is associated with great uncertainty.
- Changes introduced by water resources investments should not lead to a degradation of environment.
- It should be ascertained that the water systems be immune to barely predictable changes in resources due to anthropogenic factors, including climate changes and investments in river catchments.
- Users, including local communities, should participate in decision-making processes as regards the water management.

The above postulates may constitute a consensus platform for the communities of engineers and natural scientists, who often use different criteria in their discussions on the water system planning. The diversity of views on water management undertakings may be seen in disputes about the construction of water reservoirs and controversies as to flood-effects prevention measures.

In the last thirty years, there were implemented numerous methodological innovations, applying to the water management decision-making processes the formalized mathematical methods taken from the general system theory. Such rational water management principles can be used when the aims, means and limitations are quantified. Evaluation and optimization of water management systems by means of measurable criteria is a standard method of aiding the decision-making process. However, in the 1980s critical views began to appear responding to the overformalism in the planning of water systems which are to operate properly in future conditions that are uncertain in terms of both natural and socio-economic conditions.

The concept of sustainable development was created by politicians and publicized by the media as a slogan responding to social needs. In order to make it useful as a decision-making tool, it will be necessary to combine the paradigm of care for meeting expectations of present and future generations with the methodological tools of natural and technical sciences. Without quantifiable criteria of equilibrium as a function of time, this concept will remain in the sphere of general slogans. If the water system to be designed is to operate in a sustainable manner, the reliability, risk, regeneration, resistance, and vulnerability indices determined in the phase of its planning should not deteriorate in time. Our abilities to foresee the natural (e.g., climate changes), technological, and socio-economic processes are limited, so the forecast of these indices is only possible for short time horizons. This leads to the conclusion that the water-management systems should be designed in iterative mode, and the conditions of their functioning should be updated every few years.

Analyzing the actions in the water management domain undertaken in Poland in the last 50 years we can state that the constitutional principles of sustainable development have not been substantially violated. The need for re-establishing the water management strategy is a consequence of the following factors: the technological and structural changes in industry and the resulting restrictions on water demands as compared to earlier expectations, the previously mentioned updating of demographic forecasts, a possibility of applying new research results, and the necessity of adjusting the Polish water legislation to the European Union directives. Work on preparation of such a strategy should be carried out by interdisciplinary teams, including, along with hydro-technologists, ecologists, regional planners, economists and lawyers.

Much attention should be paid to potential threats resulting from the impact of economic activity on climate and water resources. Although the forecasts of climatic changes and their consequences are associated with large uncertainty, the economic effects of such changes may be substantial. At the initiative of international scientific communities and many non-governmental institutions, international climatic study projects were established, the UN Convention on Climate Protection was signed, and intergovernmental (EU, FAO, UNEP, UNESCO, WHO, WMO, World Bank) programs were implemented to counteract adverse consequences of climate changes. The Intergovernmental Panel on Climate Change makes periodic evaluations of the indicators of such changes and their possible consequences (Arnell et al., 2001; Kaczmarek et al., 1996; Kundzewicz et al., 2001; Liszewska (ed.), 2004). In view of the uncertain prediction of climatic processes, expensive economic decisions, including those in the domain of water management, are often against the immediate interest of politicians.

The conclusions following from the studies on potential impact of climate changes on water management carried out in Poland and in other countries, can be summarized as follows:

- There is a large degree of uncertainty following from inadequacy of climatic models, and partly from difficulties in assessing the efficiency of adaptive measures.
- Non-climatic factors (demographic, technological, economic development and urbanization) are often of greater importance than the non-stationarity of hydrological processes.
- In well-organized countries it is easier to adjust water management to changing conditions than in countries with a lower level of knowledge, technological culture and possibly an inadequately developed water infrastructure.
- Non-investment actions (changes in legislation and economic principles, efficient organization and competent control of water systems in situation of extreme threat) are decisive to the efficiency of adaptation to changes in hydrological processes.

5 RESEARCH PRIORITIES AND CONDITIONS ENABLING IMPLEMENTATION OF RESEARCH RESULTS

Water management as a branch of science integrates hydrological, meteorological and hydrogeological investigations, work connected with water supply and sewage cleaning, hydrotechnics, and problems related to planning, designing and management of water resources. It belongs to the scientific disciplines which, owing to the natural resources needs of the country, deserve particular attention by the state authorities. In the past, this was acknowledged, at least formally, by enacting, in the 1970s, the National Research-Development Program devoted to shaping and exploitation of the Polish water resources. It was continued through the 1980s by means of central governmental programs for basic and applied research. Although the practical realization of these programs has not been fully satisfying, they brought into being water management as a "respectable" scientific discipline. The programs constituted an efficient platform for scientific contacts, exchange of thoughts, new research initiatives, popularization of science, and integration of specialists dealing with water management.

The basic and applied research in the field of water management is carried out, with various degrees of engagement, by five institutes of the Polish Academy of Sciences, eight universities, nine technical universities, four agricultural academies and several research-development units. These institutions employ over 150 professors (full professors and Doctor Habilitatus) declaring their interest in the above-mentioned research fields. Their organization and activity should be adapted to the changing conditions of sustainable water resources management, taking into account the interdisciplinary character of such studies. The research potential of Poland in studies related to water management can be considered adequate for meeting the educational needs and the present, financially constrained expectations of economic practice.

Drawbacks of water management research include its modest and dispersed financing, difficulties with performing experimental studies and lack of adequate coordination of research. Of concern also is the age structure of researchers, resulting, among other things, from the lack of inflow of young scientists to the institutes and universities. As an attempt to stop this tendency, scientific schools in the field of hydraulics, hydrology and water management have been systematically organized by the Committee of Water Management of the Polish Academy of Sciences over the last dozen years or so.

The present state of water management research basis makes it difficult to undertake studies that would tackle technological and organizational problems in a complex manner. What distinguishes the situation in Poland from that in other European countries is the lack of inspiration and financing from the water management administration authorities. The participation of the National Fund for Environmental Protection and Water Management, as well as other extra-budgetary funds, in the funding of this research is also disappointingly small.

The results obtained thus far can be evaluated in terms of their contribution to the general knowledge on the studied phenomena, as well as in terms of their usefulness for solving practical problems. The achievements of Polish scientific institutions in basic research have been documented in recent years by the large number of citations of Polish authors by the Institute for Scientific Information (ISI) in Philadelphia. Let us enumerate some examples to illustrate this observation.

Internationally acknowledged results were obtained in Poland concerning the expected climate change on quantitative and qualitative characteristics of water resources. For this reason Polish scientists have been assigned to play an important role in the preparation of future IPCC reports. Experience from the droughts in Poland at the turn of 1980s and 1990s, and the floods in 1997–2002, give grounds for promoting studies aiming at preparation of protective measures against consequences of climate changes on the threat related to weather extremes.

Of international impact were also Polish studies in the field of eco-hydrology, aiming at the modelling and control of hydrological and biological processes in river catchments in order to enhance the resistance of water ecosystems to perturbations and pollution (Zalewski, 2004). In the framework of the UNESCO International Hydrological Program, eco-hydrology will be one of the key research areas. Achievements in this field have been acknowledged by establishing in Poland the European Regional Centre for Eco-hydrology under the auspices of UNESCO.

Worth mentioning are also the world recognized achievements of Polish scientists in the field of probabilistic assessment of the threats related to extreme hydrological events, notably the droughts and floods that have been plaguing our country (see e.g., Kaczmarek, 2005; Singh and Strupczewski, 2002).

Although these examples concern mainly basic research, many of the obtained results can be useful in practice. In the case of basic research, the topics of study are being chosen by the research teams, and the role of governmental institutions is mostly restricted to sponsoring. The choice of the topics is in this case inspired by the logic of the advancement of knowledge, creating seasonal priorities for some topics; this can be easily seen in the scientific literature, in the lectures delivered at annual meetings of the European Geophysical Society, or at meetings of the American Geophysical Union. In many cases, the division into basic and applied research is arbitrary (Opportunities, 1991).

It is more difficult, however, to point out important applied studies, carried out in the recent years as a result of targeted activity inspired and financed by governmental funding institutions or water administration. The decisive factors for choosing the research problems of applicative nature should be the actual needs of economy, formulated by practitioners or scientists who understand these practical needs. These can be, for instance, the following tasks, vital for our country:

- Updating the assessment of water resources in Poland, with the account for new measurement results, changes in quality factors and the state of catchment management; recognition of interrelations between the surface waters and groundwater; redefinition of usable resources and their assessment for individual river basins.
- Updating the assessment of risk associated with the occurrence of floods and droughts, development of principles for water management in extreme situations; evaluation of the role of retention in shaping of hydrological processes.
- Adjustment of water management plans to changes in river basins management and climatic scenarios; assessment of how the changes in land use affect the water balance and floods (e.g., by construction of highways),
- Development of studies on ecological aspects of water management, taking into account environmental demands associated with the use of water resources (program NATURA 2000), and identification of areas in which the resources are to be used to preserve their recreational status.
- Updating of water demands in accordance with demographic and macroeconomic indices of the National Development Plan, designing of plans for sensible water usage norms for household and production demands, including the impact of climate changes on the water needs in agriculture.

In spite of the lack of proposals from the water administration for research projects to support water management in Poland, such projects have been partly undertaken at the initiative of research teams, sometimes supported by the governmental institutions.

To give an example of how such initiatives are implemented, let us quote the project of designing a methodological basis for integrated management on the water catchment scale, which has been established in year 2003 by the Ministry of Science and coordinated by the Cracow Technical University. The studies were carried out by scientists from a number of institutes and technical universities, in cooperation with water administration in pilot catchments.

Another project financed by PHARE, whose aim is to provide technical aid to teams preparing the implementation of Directive 2000/60/WE, is now under way at the Institute of Environmental Engineering of the Warsaw Technical University. Studies on conditions of sustainable development of rural regions, supported by the Ministry of Science, have been completed by the Institute of Land Reclamation and Grassland Farming.

Polish scientific institutions take also part in various international programs in the field of hydrology, hydraulics and water management, coordinated by international organizations (UNESCO, United Nations Environment Programme, World Meteorological Organization, and the non-governmental organizations: International Union of Geodesy and Geophysics, International Association of Hydrological Sciences, International Association of Hydraulic Research, and others). However, as regards the European Union framework projects, the participation of Polish scientists is limited. With the financial support of the European Commission, the Ministry of Science has created the following excellence centres in the field of water sciences:

- Centre for Environmental Engineering and Mechanics at the Institute of Hydroengineering of the Polish Academy of Sciences,
- Centre of Wetland Hydrology at the Warsaw Agricultural University,
- Centre on Geophysical Methods and Observations for Sustainable Development at the Institute of Geophysics, Polish Academy of Sciences.

The organizational activity of these centres is evidence of the growing participation of Polish science within the European Research Area. It involves broader cooperation of Polish scientific institutions with partners abroad, organization of scientific workshops in our country, and financing of personal exchange between the scientific institutions in Poland and other European countries.

In the "traditional" European Union countries, cooperation in the field of environmental research is a result of many-year contacts. It now becomes a timely question: what are the aspirations of Poland as a new member of the Union; are we going to play a marginal role in the research initiated by the European Commission or will we engage our full intellectual potential in it? It is of great importance for the progress of science in Poland to facilitate our access – on a partnership basis – to the research projects of the European Union.

6 CONCLUSIONS

Social, technological and organizational problems of water resources management in Poland can be successfully solved in terms of tasks to be fulfilled in the XXI century. The groundwater and surface water resources and the present and future water demands of people and economy have been fairly well recognized. In this century, these resources and demands can be subject to substantial changes resulting from transformation of both geophysical (climate change) and social-economic processes. The research that is presently carried out on rational control of integrated water management systems should account for dynamics of water management and the requirements of European Union directives. What is needed is the determination to act and the provision by political authorities, governmental and semi-governmental administration of the means that are necessary for sustainable water use. Polish science is able to provide the methodological basis for such actions.

REFERENCES

Arnell N.W., Chunzhen Liu, Compagnucci R., Da Cunha L., Howe C., Hanaki K, Mailu G., Shiklomanov I., Stakhiv E.Z., 2001, *Hydrology and Water Resources*, in: Climate Change 2001: Impacts, Adaptation and Vulnerability, Cambridge University Press, 191–233.

Brundtland G.H., 1987, *Our Common Future*, Report of the World Commission on Environment and Development, Oxford University Press, 400 pp.

Czyżewski A.B., Orłowski W.M., Zienkowski L., 2000, *Macroproportions in the years 2000–2020*, in: Strategia rozwoju Polski do roku 2000, Polska Akademia Nauk, Komitet Prognoz "Polska 2000 PLUS" (in Polish).

Water Framework Directive 2000/60/WE of the European Parliament of 23 October 2000 determining the framework of water management (OJ L 327, 22.12.2000).

EUROSTAT, European Statistical Office, Demographic Indicators Publication Year.

Kaczmarek Z., Arnell N.W., Stakhiv E.Z., 1996, *Water resources management*, in: IPCC 1995 – Impacts, Adaptation and Mitigation of Climate Change, Cambridge University Press, 471–486.

Kaczmarek Z., Jurak D., 2003, *Assessment and prediction of hydrological droughts*, Papers on Global Change – IGBP, Polish Academy of Sciences, **10**, 79–96.

Kaczmarek Z., 2005, *Risk and uncertainty in water management*, Acta Geophysica Polonica, **53**, 4, 343–355.

Kundzewicz Z.W., Parry L.M., Cramer W., Holten J.I., Kaczmarek Z., Martens P., Nicholls R.J., Öquist M., Rounsevell M.D.A., Szolgay J., 2001, *Europe*, in: Climate Change 2001 – Impacts, Adaptation and Vulnerability, Cambridge University Press, 641–692.

Liszewska M. (ed.), 2004, *Potential Climate Changes and Sustainable Water Management*, Publs. of the Institute Geophysics, E-4 (377), 92 pp.

Loucks D.P., 2000, *Sustainable water resources management*, Water International, **25**, 1, 3–10.
Opportunities in the Hydrological Sciences, 1991, National Academy Press, Washington D.C., 348 pp.
Singh V.P., Strupczewski W.G., 2002, *On the status of flood frequency analysis,* Hydrological Processes, **16**, 3737–3740.
VISION: Water safe, strong and sustainable – European Water Supply and Sanitation in 2030, 2005, European Commission, Water Supply and Sanitation Technology Platform, Brussels.
Young G.J., Dooge J.C.I., Rodda J.C., 1994, *Global Water Resources Issues*, Cambridge University Press, 194 pp.
Zalewski M., 2004, *Ecohydrology as a system approach for sustainable water, biodiversity and ecosystem services*, Ecohydrology and Hydrobiology, **4**, 3, 229–235.

Environmental Engineering – Pawłowski, Dudzińska & Pawłowski (eds)
© 2007 Taylor & Francis Group, London, ISBN13 978-0-415-40818-9

The historical aspect to the shaping of the sustainable development concept

Artur Pawłowski
Politechnika Lubelska, Wydział Inżynierii Środowiska, Nadbystrzycka, Lublin

ABSTRACT: While the concept of sustainable development was propounded in 1987, as the report Our Common Future was released, the discussion on sustainability had begun much earlier. And while environmental protection only really emerged in the 19th century, diverse motives for the environment to be protected appear in documents stretching far back into history. The present article discusses some of the most important initiatives and legal instruments (both international and the less well-known solutions adopted domestically in Poland), which first led towards environmental protection, and then also took in sustainable development. The analysis extends, not only to contemporary documentation, but also to items from history which can be taken from certain points of view as having led to the preservation of natural resources.

Keywords: Sustainable development, environmental legislation, Agenda 21.

1 INTRODUCTION

The sustainable development concept is now the subject of wide discussion, both in general terms and as regards the detailed subject-matter (Daly 1989). In truth, the concept – which aims to foster such a utilisation of the environment as will leave it in undiminished condition for future generations to use[1] – represents a development of previously-existing thought as regards conservation and preservation. With a view to this being better appreciated, it is worth going back to the subject's historical aspect, which was after all what led up to the formulation of the sustainable development concept. Actions taken in this line were in fact dispersed, arising in different regions at different times. Furthermore, they were characterised by marked diversity in terms of both the legal provisions involved and the motivations underpinning their introduction.

The oldest ordinances relating to environmental protection are most probably those from China, introduced more than a thousand years BC in order that the more valuable types of tree and forest might be preserved. Indeed, the classical civilisations were often associated with tree cults (e.g. revolving around the oak in Europe or the baobab in Africa), even in association with an understanding of the economic value of forests. Furthermore, the understanding of how significant forests were, did, at times, represent a premise upon which protective measures could be introduced.

Having marched into some conquered town, the victorious armies of old would not only plunder and pillage, but also cut down trees as a way of subjugating those they had conquered. While burnt-down homes could be rebuilt with relative ease, a new generation of trees would demand a much longer wait.

The original motives behind nature conservation trace back to old beliefs. Back then, honour needed to be done – and tribute paid – to the inexplicable forces of nature. The protected centres of such religious practices were often mounds or elevations, wetlands, canyons and gullies, or individual trees. As legends grew up around these places, their status as untouchable was merely enhanced further. Appropriate names, sometimes of a purely religious nature, were also assigned to the special places.

Historical and/or patriotic motives for conservation appeared much later, as protection began to be extended to individual sites or even larger areas associated with important events that had played out at or within them.

Another motive – the aesthetic – played a particular role in the Romantic period. The beauty of nature unsullied by human hand was then frequently compared with the greatest works of art. One of the first "aesthetic" instruments of law was one adopted in Switzerland in 1353, which extended protection to songbirds on account of the beauty of their voices.

Today, it is the need for environmental protection for its own sake that is in the ascendancy as

[1] This was the way in which sustainable development was defined in the famous 1987 report entitled "Our Common Future".

a justification, most especially as part of so-called non-anthropocentric environmental ethics.

From the historical point of view, however, the oldest motivation was of a different kind. Leaving aside the special care lavished upon cult sites, nature was also protected as part of the assets of a given ruler – which were to be left undisturbed. This was the situation in Poland at least.

2 THE POLISH PERSPECTIVE

The beginnings of the legal protection of the environment in Poland were associated with the adoption of different forms of law (either through official enactment or in line with respect generally afforded to widespread custom). From whatever point of view these laws were passed, their effect was to prevent destruction of the diverse resources of nature.

The earliest initiatives of this kind include the so-called *Regalia* prerogatives enjoyed exclusively by Poland's rulers, but nevertheless at times made subject to delegation, the powers to do certain things being conferred upon others. The restrictions arising from the *Regalia* related, *inter alia*, to hunting, fisheries and mining rights. While the primary justification and motivation behind the use of these instruments was to safeguard the assets of the ruler, the practical consequences extended more broadly. Employing modern terminology, they can be said to have represented measures serving the conservation of nature – in all its varied dimensions – against uncontrolled exploitation. For it was appreciated even then that the resources in question could be destroyed or exhausted. What a pity, then, that this attitude came to be abandoned over time, most of all as the scientific revolution dawned.

Table 1. Polish initiatives launched in the name of environmental protection – a historical perspective. Source: author's own compilation.

Year	Initiative or measure
11th c.	*Regalia* of rulers (restricted use of the environment based around the premise that assets needed protecting).
1346	Statute from King Kazimierz the Great (protecting forests, but still solely from the asset-protection point of view).
1420, 1423	The Warka Statutes of King Władysław Jagiełło (protecting forests and individual yew trees).
1529	The Statute of Lithuania (protecting beavers).
1778	The Universal Decree on Forests issued by King Stanisław August Poniatowski (protecting forests).
1869	In Galicia (the Polish lands partitioned by Austria): the Act on the capture, extermination and sale of Alpine animals unique to the Tatra Mountains.
1874	In Galicia: the Act on the protection of certain animals useful in the cultivation of land.
1934	Poland's first comprehensive act on nature conservation.
1944	The new Nature Conservation Act (also relating to environmental protection).
1980	The Environmental Protection and Management Act.

While the *Regalia* can not be said to have represented nature conservation for its own sake, the real consequences of their being exercised nevertheless led in the right direction.

The system of *Regalia* came in as the structure of authority at state level was being put in place. It was already quite well-developed by the 11th century. The legal protection of nature in Poland can thus be said to have a history of nearly 1000 years behind it. The environmental dimensions of the *Regalia* were varied. In the cases of fisheries or mining rights, the levels of technology at the time were anyway inadequate to the task of exhausting the resources concerned. The situation was different in the case of hunting, however. As early as in the Middle Ages, a distinction began to be drawn between large game (*venatio magna*) and small (*venatio parva*) – with the hunting of the former being left as the sole preserve of rulers. Furthermore, it is worth emphasizing that – in the cases of the species considered valuable (and rare even at that time), like the European bison and aurochs, hunting could only take place in the presence of the King himself. Since any violation of this principle could result in the most severe punishments, the practical result was a constant and largely adhered-to protection of the species in question. However, this was still not nature (or species) protection for its own sake – at least not at the level of the stated reasons for the regulations being introduced.

It was in 1346 that a Statute of King Kazimierz the Great was announced, including orders mainly relating to the protection of forests. However, the motivation remained the preservation of a ruler's assets.

The Warka Statutes of King Władysław Jagiełło were adopted in 1420 and 1423. The first was another instrument relating to forest protection and was treated – as before – as a confirmation of the rights of owners. The second statute in turn postulated the essential need for yew trees to be protected, these being subject to excessive cutting even then. For yew wood was an outstanding raw material in the manufacture of crossbows and bows, the leading weapons of their day. It was to this end that Polish yews were cut to meet domestic needs as well as for export.

The second Warka Statute was thus the first document to pay such unambiguous attention to the need to protect, not only larger units (like forests in their entirety), but also the individual species comprising them. It should be recalled that protection with a view to addressing the threat of the final disappearance of one or more species remains a key pillar of contemporary nature conservation. It is clear from the Warka

Statute that this kind of thinking can trace its roots back to the 15th century!

More fully-developed protective provisions are to be found in a 1529 Statute of Lithuania, which was adopted by the Diet of Vilnius, and then made subject to signature by Grand Duke Zygmunt. The role of the Statute was to reaffirm legal principles relating to forests.

Very interesting and innovative legal provisions were developed in respect of the protection of beavers. In or around the feeding areas for the species, even landowners were not permitted to engage in works, local fields could not be ploughed and meadows had to be left unmown. So this was not just a simple ban on hunting, but rather the comprehensive protection of the beaver's biotope! Account was even taken of situations in which beavers left their original feeding grounds to set up elsewhere. Here the motive may no longer be said to be the maintenance of a ruler's assets (which are actually made subject to statutory restrictions), but a perception that there is value in animals of a species remaining at large in its natural habitat! This kind of thinking made this a unique legal instrument for its day . . . even on the world scale!

A further legal instrument of significance, adopted in 1778, was the Universal Decree on Forests issued by last King of Poland Stanisław August Poniatowski. Making reference to previous pieces of law, this introduced a clear prohibition on unmanaged and excessive felling of trees, highlighting at the same time the clear danger of near-complete deforestation on the national scale (a concept that is still the subject of the law in force!). There is thus justification for claiming that the motif of forest being protected as a good of the whole Polish nation appeared in the country's law as early as in the 18th century. What is also interesting here is that the standard proclamation of this law, and its publication as a document, was augmented by a recommendation that provisions be announced "from the pulpit". This was thus an educational premise that would, in today's parlance, be considered to fall within a programme of environmental education.

The Universal Decree on Forests was the last instrument of law of this rank – and relating to environmental protection – before Poland was partitioned out of existence by the Austrian, Prussian and Russian Empires. What is interesting in that regard is that the loss of Polish statehood did not stand in the way of further interesting legal instruments being adopted. This is seen to be particularly true where the legislation adopted in Galicia (the Austrian part) was concerned. The route by which these successive legal instruments became binding was nevertheless a rather complicated one. The National Parliament in Lvov first passed the law, the text then being passed on to Vienna, where the final stage in the procedure involved receipt of the Emperor's confirmatory signature.

Such a signature was received in 1869, by a statute adopted by Parliament a year earlier "on the capture, extermination and sale of Alpine animals unique to the Tatra Mountains, i.e. the Alpine Marmot and Wild Goat". Fines for violation of the bans on capture and hunting were set, a person finding him/herself unable to pay was made subject to arrest. This was therefore species protection of a very direct kind!

It was also in 1868 that the National Parliament in Lvov passed an Act to protect certain animals described as of use in the cultivation of land. This Act had to wait quite a long time (until 1874) to receive the Emperor's signature. It ushered in a ban on the "collection and damaging of eggs and nests of all wild-living birds that are harmless", as well as on the capture and killing of birds. The only exception applied where scientific purposes could be invoked. Furthermore, the Act detailed those birds that were to enjoy protection, as well as the level of punishment that could be anticipated for non-compliance with its provisions, and an indication as to those services that would be responsible for enforcement of the law. In all these ways, the Act did not depart greatly in structure from contemporary statues. However, in some important ways it went further than today, setting out solutions that would be considered innovative even now. We have seen that the Universal Decree on Forests of 1778 refers to the need for and importance of acquainting the public with legal provisions. The Galicia Act made this matter more precise and expanded upon it. Among its provisions is one to the effect that: "teachers at peasant schools are obliged to explain to both day-school and Sunday-school pupils the harmfulness of collecting nests and of the capture or killing of useful birds. They should present the provisions of this Act each year, specifically in advance of the onset of the breeding season". If we were to employ modern terminology, we would say this was an obligation in respect of environmental education! And education detailed not only in terms of the scope of its subject matter, but also as regards when it was to be presented and by what means.

These ideas did not always meet with success in practice, however. For example, in late 1903 and early 1904, the Austrian Ministry of Education in Galicia ordered that something be done in the matter of monuments of nature. The issue had arisen with Hugon Conwentz's decision to establish a movement in the name of these objects[2]. It was planned for monuments to be looked for and inventoried across the Empire, and then for care to be extended to them. However, this goal and task went unimplemented on account of shortages of both manpower and money. It is hard not to notice the similar problems that afflict many

[2] The actual concept of the "monument of nature" was introduced by Alexander von Humboldt (1769–1859).

modern initiatives launched in regard to environmental protection and sustainable development.

In Poland, the concept of the monument of nature was first presented more fully by Marian Raciborski in 1903. The failed attempt to inventory monuments of nature across the Austrian Empire coincided with attempts to establish the first Nature Reserves. A proposal on this from November 1910 did not gain the support of the sitting of the National Sejm in Lvov that considered it, however, and the idea was therefore abandoned.

Thinking as regards conservatorial protection had come a long way by the time Poland regained its independence in 1919. Having begun with the conservation of nature as the property of the ruler, it had passed through the realisation that protection was an activity worth pursuing for its own sake and that individual species were of significance, including in terms of their being public goods belonging to the whole nation.

When the first Polish government was established, matters of environmental protection came within the remit of the Ministry of Religious Denominations and Public Education. As early as September 16th 1919, the responsible Minister issued an Ordinance on the protection of Monuments of Nature, as well as a still-modest one on species protection (this taking in just 5 kinds of plant and 10 of animal).

It was in that very same year that the Interim State Nature Conservation Commission was called into being. It was proposed that many areas under legal protection should be established in Poland, including in the form of Nature Reserves and the first National Parks.

1920 saw the emergence of a draft Regulation calling into being a State Nature Conservation Council, though this was not actually established until 1925, under the chairmanship of Władysław Szafer.

The first comprehensive Nature Conservation Act took even longer to bring about – until May 13th 1934. This was conservatorial in its approach, providing for the legal protection of natural forms (caves, waters, waterfalls, animals, plants, minerals and fossils) where preservation could be thought of as in the public interest in scientific, aesthetic, historical or commemorative terms. It was this Act that brought in the institution of the nature conservator, as well as clearing the way in legal terms for the establishment of the country's first National Parks.

It is worth asking why Poland's first Nature Conservation Act did not come into force until 1934. In fact, such a long period of discussion preparing for it was associated with difficulties in finding a compromise between the permissible level of burdening of the environment through Poland's inter-War development and industrialisation on the one hand, and the requirements of nature conservation on the other. It is interesting to note that a proper – and steadily updateable – solution to this problem remains to this day one of the most serious challenges put before environmental protection and sustainable development strategies.

As it prepared the Act, the State Nature Conservation Council was thinking in terms of introducing very restrictive legislation. The priority was to be the establishment of very large natural protected areas, within which all kinds of development would be prohibited. Industrial development in general was frowned upon. A compromise with such extreme conservatorial radicalism could not be reached, the Act was never to generate any delegated Ministerial regulations and the entire Council resigned a year after it had been adopted. The official reason given was opposition to the construction of a cable-car on to Kasprowy Wierch (Tatra Mountains). Nevertheless, the work of the Council was not wholly forgotten, it being used in later documents – including some in force at the present time.

Also worth noting from the inter-War period is the activity of Adam Wodziczko, who was responsible for the recognition of nature conservation as a new branch of knowledge and learning. He was also the author of the now forgotten physiotactic concept. In the author's opinion, this was to have been a science of the correct relationship between humankind and nature. The idea later became a foundation for the Polish scientific concept of "sozology", as well as assuming the form of an ethical premise. Moreover, Wodziczko drew attention to the weakness of any strategy that confined itself solely to the kind of conservatorial nature conservation that the State Nature Conservation Council had been in favour of. He was thus to speak of the need for rational management of natural resources – this being more or less a key premise in the later establishment of the sustainable development concept!

1945 had not come to an end before the State Nature Conservation Council convened once more, and began to put in place some new laws on environmental protection. 1949 brought the enactment of a new Nature Conservation Act, while environmental issues in general came within the remit of the Ministry of Forestry and the Timber Industry.

Of necessity, the rapid post-War industrialization of the country entailed a departure from the purely conservatorial concept of environmental protection – all the more so since the rapid development of heavy industry was to become communist Poland's single greatest priority. The inevitable and unfortunate result was a steady worsening of the state of the country's environment.

1972 saw the establishment of a Ministry of the Field Economy and Environmental Protection, which was later transformed into the Ministry of Environmental Protection, Natural Resources and Forestry.

Table 2. Selected contemporary acts of law relating to environmental protection. Source: compilation by author.

Year	Name of Act
	Basic Acts
1991	Nature Conservation Act
2001	Environmental Protection Law Act
	Selected supplementary Acts
1991	Forests Act
2000	Atomic Law Act
2001	Water Law Act
2003	Planning and Spatial Management Act

Table 3. Selected international legal commitments assumed by Poland. Source: author's own compilation.

Year	Commitment
1971	The Ramsar Convention on Wetlands of International Importance Especially as Waterfowl Habitat
1972	The Convention on the Prevention of Marine Pollution by Dumping of Wastes and Other Matter
1973	The Washington Convention on International Trade in Endangered Species of Wild Flora and Fauna
1974	The Helsinki Convention on the Protection of the Marine Environment of the Baltic Sea Area
1979	The Geneva Convention on Long-range Transboundary Air Pollution
1985	The Vienna Convention for the Protection of the Ozone Layer
1987	The Montreal Protocol (to the aforesaid Vienna Convention) on Substances that Deplete the Ozone Layer
1989	The Basel Convention on the Control of Transboundary Movements of Hazardous Wastes and Their Disposal

The abbreviated name in use today is "simply" Ministry of the Environment.

In the post-War period, general issues connected with environmental protection were linked up with those of nature conservation, with the result that a distinct Environmental Protection and Management Act only appeared in 1980.

The consequences of the transformations began in 1989 included the adoption of many new legal instruments, as well as numerous amendments to the Acts already in existence. Today, the most important pieces of legislation relating to environmental matters are the Environmental Protection Law Act of 2001 (Polish Official Journal of Laws of 2001. 62.627), as well as the earlier Nature Conservation Act 1991 (consolidated text: *Dz.U.* of 2001: 99.1079).

However, the above cannot be seen in isolation from the international perspective, with reference to the documents adopted at the 1992 "Earth Summit" (UN Conference on the Environment and Development) in Rio – most especially "Agenda 21"; or else to the legislation of the European Community – most especially in regard to the Environmental Action Programmes (Kiss, Shelton 1993).

Moreover, Poland is also party to many conventions and other international agreements dealing with matters of environmental protection.

Contemporary Polish legislation as regards matters of the environment has a broad historical context, but is also deeply integrated with agreements of an international nature. Since the latter plays a particular role where the formulation of the sustainable development concept is concerned, it is treated in more detail below.

3 THE INTERNATIONAL PERSPECTIVE

One of the themes dealt with during every environment-related study course is smog (Sorlin 1997). Yet this is in no way a contemporary problem, having been first noted in 13th-century London, in association with the increasing popularity of the burning of coal to heat homes and cook food. Yet the first scientific studies of the composition of London's air had to wait until the 18th century, while concrete action to deal with the problem was not taken until the 1950s. The time interval between the smog issue becoming a reality and the first attempts at a solution thus lasted 700 years! It is worth emphasizing that the motivation behind the introduction of concrete measures was not so much the information that became available about health as the work of scientists, plus the taking up of the problem by the media.

The beginnings of contemporary environmental protection can be traced back to the early 19th century. Of a symbolic nature in this regard was the movement for the care of monuments of nature set up by Hugon Conwentz (1855–1922).

The real breakthrough came as late as 1969, with the famous address given by UN Secretary-General U'Thant. Support from the popular press also proved important in the case of the aforementioned speech. This took place at the 23rd session of the UN General Assembly and has passed into history under the heading of the U'Thant Report. It included discussion about an arising global crisis (touching both developed and developing world), related to humankind's responsibility for the nature. Its symptoms had been visible for a long time: population explosion, inadequate integration of technology with environmental requirements, deterioration of arable lands, expansion of cities, decreasing amount of untouched land, and extinction of many animal and plant species.

Also of significance was the conclusion that we all live in a biosphere, in which space and resources

Table 4. Selected contemporary initiatives of an international nature serving environmental protection. Source: author's own compilation.

Year	Initiative
1969	Speech by UN Secretary-General U'Thant
1970	Establishment of the UNESCO "Man and Biosphere" Programme
1972	UN Stockholm Conference (on the Human Environment) and Stockholm Declaration. First report from the Club of Rome. Establishment of UNEP (the UN Environment Programme)
1980	Elaboration and publication of the "World Conservation Strategy" commissioned by the IUCN
1987	Report "Our Common Future" from the World Commission on the Environment and Development, which presents the sustainable development concept
1992	Rio "Earth Summit" (UN Conference on Environment and development) developing the sustainable development concept and including "Agenda 21" among its adopted documents

are huge, however limited[3]. This thesis was to prove of fundamental significance for environmental protection.

U'Thant also pointed to the need for people to assume responsibility for nature. That said, he concentrated mainly on responsibility of a political nature, emphasising the action in the fora of international organisations, and at regional and local level. The formulated conviction as to the existence of limits in nature is also an important pillar on the road to the devising of the sustainable development concept.

An important consequence of U'Thant's address was the attempt to usher in international cooperation on environmental protection at the UN forum. The UNESCO Man and Biosphere (MaB) Programme had already appeared in 1970 (www.portal.unesco. org, www.unesco.org), while the Stockholm Conference was held in 1972, attracting participants from 112 countries (Helsinki Commision 1994). In this way, humankind's responsibility for nature gained a socio-political context. The 26-principle "Stockholm Declaration" was signed. Of particular significance was its Principle 21, whereby "States have, in accordance with the Charter of the United Nations and the principles of international law, the sovereign right to exploit their own resources pursuant to their own environmental policies, and the responsibility to ensure that activities within their jurisdiction or control do not cause damage to the environment of other States or of areas beyond the limits of national jurisdiction (Bergstrom 1992)".

It is interesting and important to note that – along with other communist-bloc countries – Poland chose not to appear at the Conference, on account of a Western refusal to recognise the existence of the German Democratic Republic. It was thus in protest over this that Eastern delegates were not present. This did not stand in the way of the inclusion, within the Polish Constitution drafted four years later, of two pro-environmental provisions declaring themselves in line with the spirit of the Stockholm Declaration.

A further important moment in 1972 came with the establishment of the United Nations Environment Programme, UNEP. The *Club of Rome* was also founded, commissioning and receiving its first report before the year's end[4]. This was concerned with global environmental problems, and effectively broadened the ongoing discussion to encompass economic aspects. It thus represented a further step towards the formulation of the sustainable development concept (www.clubofrome.org).

Entitled *Limits to Growth* (Meadows et al. 1972), the 1972 report to the Club of Rome echoed U'Thant in its references to the existence of limits in nature whose exceedance (be it through overexploitation of natural resources or excessive pollutant emissions) might lead to a breakdown in the biosphere's biological balance.

Further reports emerged every few years and represented a response to continuing changes in the world situation.

Two works from the 1980s seem worthy of particular attention.

The *World Conservation Strategy* commissioned by IUCN was published in 1980 (www.iucn.org, IUCN 1980). Its recommendations could be summarised in three points:

1. maintaining essential ecological processes and life support systems (place could thus be found here for the protection of soils, vegetation and forests);
2. preserving genetic diversity;
3. ensuring the sustainable development of species and ecosystems.

The introduction to the strategy was designed to lead to such an integration of conservation and development-related assumptions as would allow for the existence and wellbeing of all human beings. This general aim is thus close to the principle of sustainable development. What is more, the third point even contained the word "sustainable".

[3] It is worth adding that such well-known contemporary problems as global warming or ozone depletion were not signalled at all by U'Thant, even though the first of these at least was undoubtedly in operation by that time.

[4] The Club brought together an international group of entrepreneurs, leading statesmen and women and scientists, among whom a leading role was played by those based at the Massachusetts Institute of Technology (MIT).

However, a proper formulation of the concept only came along in 1987, as the report of the committee under Gro Harlem Bruntland was being compiled under the title *Our Common Future*. This study, prepared by the World Commission on the Environment and Development, was an attempt to sum up humankind's successes and failures in the 20th century. It is noteworthy that, leaving aside the ecological factors, much attention was paid to social issues of the kind usually glossed over or ignored in studies devoted to environmental protection.

Among the successes mentioned in the report were:

- a decline in rates of infant mortality,
- an extension of average life expectancy,
- an increase in the share of literate young people,
- an increase in the share of the world's children attending school,
- growth in global food production continuing to outstrip population increases.

In turn, the failures were taken to include:

- an increase in the numbers hungry or starving,
- an increase in the numbers of illiterate people,
- an increase in the numbers of people lacking access to clean water,
- an increase in the numbers of people not having somewhere healthy and safe in which to live,
- an increase in the number of people lacking fuelwood,
- a deepening of the divide between the rich and poor nations.

The route by which the problems indicated above were supposed to be influenced and ultimately resolved was sustainable development, the definition of which was set out in the now classic report. In the view of the Bruntland Commission, "*sustainable development* [...] meets the needs of the present without compromising the ability of future generations to meet their own needs".

This concept met with considerable interest around the world and was a centrepiece at the UN Conference on Environment and Development convened in Rio de Janeiro in 1992. Known popularly as the "Earth Summit", this brought together representatives of 180 states (www.ecouncil.ac.cr, Holmberg et al. 1993). Admittedly, not all of them signed all of the documents, there being resistance to some projects on the part of the United States, for example. Likewise, only some of what was adopted at Rio has proved capable of implementation, though this in no way diminishes the significance of the Conference, which was massive[5].

In the first place, it brought the subject of sustainable development to the attention of the media, ensuring its popularisation, and bringing it down from the ivory towers of academic speculation. In the second place, by way of its *Agenda 21* document (United Nations 1992), the Earth Summit supplied both a model and a methodology by which the world, regions, countries and localities could set about preparing strategies for sustainable development.

Altogether, Rio ultimately led to the adoption of five important documents/instruments (www.ecouncil.ac.cr):

1. the Rio Declaration, as a set of principles and code of conduct on how humankind should proceed in relation to the natural environment;
2. the aforesaid *Agenda 21*, referring to action that would extend into the 21st century only recently started: this is a 900-page action programme (usually accessible as shorter studies) concerning environmental protection and sustainable development, as well as a set of recommendations for governments and international organisations, having as its aim the integration of global policy and decisions taken at the state or international levels;
3. the UN Framework Convention on Climate Change, setting out tasks by which to combat climatic warming and its effects, most especially through limitation of emissions of so-called greenhouse gases to the atmosphere;
4. the Convention on Biological Diversity, relating to the preservation and protection of existing species of flora and fauna, and paying particular attention to endangered species;
5. the Forest Principles – forming a Declaration on international cooperation in the protection of forests.

Also worthy of particular attention from the 1990s are two reports compiled for the aforementioned *Club of Rome*. First published in 1994, the report entitled *Beyond the Limits* (Meadows et al. 1993), sought to summarise all that had happened since 1972 (when the original *Limits to Growth* report to the Club had been published). In it, the same authors note that the trends outlined back in 1972 had in no way faded away, but had rather assumed even greater intensity. They even point to the existence of "overshoot", the ad hoc, unintentional exceeding of environmental limits. The report also states that: "Human use of many essential resources and generation of many kinds of pollutants have already surpassed rates that are physically sustainable." However, still we have technical means which can make our survival possible.

Then *Factor Four:* Doubling Wealth, Halving Resource *Use* appeared in 1997 (Weizsacker et al. 1997), dealing *inter alia* with matters of the growing opposition between rich and poor countries.

Discussions at international level are also augmented by solutions approved by the European Union

[5] See also information about Earth Summit in Johannesbourg in 2002, www.earthsummit2002.org

or its predecessor communities (Kiss et al. 1993, Jachtenfuchs et al. 1992). These began with remarks regarding the environment set out in the Single European Act of 1987, and then developed through Directives and Framework Programmes up to ISO and EMAS environmental management systems. The activity of the European Environment and Sustainable Development Advisory Council is important, this being responsible for the 2005 publication *Sustaining Sustainability* (www.eeac-network.org) amongst other things. Nevertheless, there is no concealing the fact that the EU was called into being for its anticipated economic benefits, rather than to save the natural environment.

4 CONCLUSIONS

The presented selection of documents and legal instruments relating to environmental protection and sustainable development make clear the strong grounding these have in solutions known from earlier history. Examples here might be the statutes enacted in Galicia which were a breakthrough in their time still continue to make a big impression. What is also important, their provisions were widely respected in practice – which is not something that can be said of many of the contemporary initiatives, even the better-prepared among them.

A further problem is the emphasis on environmental conditioning, with frequent neglecting or even ignoring of the other aspects to sustainable development. The reasons for this need to be looked for in history: it is environmental protection that has been most widely spoken about since the mid 20th century, while sustainable development only began to be promoted at the end of the 1980s.

Meanwhile, the sustainable development concept is much more multidimensionsal than environmental protection. There is not just nature here, or technology, but also social matters and the problem of growing unemployment that has become more and more tangible in recent years. It is not only the natural environment that may become subject to environmental degradation, for the same can be true of the social environment. In turn, a lack of respect for the environment is in fact a consequence of a lack of respect for fellow human beings.

There is no doubt that an action programme encompassing all aspects of sustainable development is difficult to prepare, and still more difficult to bring into effect. Detailed programmes relating to just a selected group of issues seem to prevail, the disadvantage being the major risk of losing sight of the "big picture".

The practical aspects of the subject in question remain a major problem. It is not just a matter of whether or not to create ever more legal instruments (be these better or worse), but one of lifestyle and the shaping of ecological or environmental "awareness". Quite different situations apply where people engage in a given activity because the law requires them to, and where they act in the given way out of their own convictions.

This is a task for environmental education, both formal and informal. The document of importance here from the Polish point of view is the National Environmental Education Strategy (of 1997, as amended in 2001, Kozlowski 2005).

Churches can also play an important role. Faith is a matter of personal conscience for each human being, and is often dear to the heart. Churches could therefore exert a greater impact on matters of individual choice than can the standard educational system.

Indeed, documents adopted within the Catholic Church stand out here. Reflections on the future of civilisation first appeared at the Vatican II Ecumenical Council. The Papal Encylicals (especially those of the late John Paul II) are also of importance, especially *Redemptor Hominis* (John Paul II 1979), *Sollicitudo Rei Socialis* (John Paul II 1987), *Centesimus Annus* (John Paul II 1991), *Evangelium Vitae* (John Paul II 1995) and *Fides et Ratio* (John Paul II 1998). Even the first of these read as follows: "Man often seems to see no other meaning in his natural environment than what serves for immediate use and consumption. Yet it was the Creator's will that man should communicate with nature as an intelligent and noble master' and 'guardian', and not as a heedless 'exploiter' and 'destroyer' ".

However, irrespective of the motivation underpinning the frameworks to environmental protection or sustainable development strategies, the most important issue must surely be the implementation of their theses and recommendations in practice, at global and international levels, via the strategies adopted at the levels of individual countries and ultimately at local level. In the latter case, a number of decisions boiling down to the making of choices will be in the hands of each and every one of us.

REFERENCES

Bergstrom G.W. 1992. Environmental Policy & Cooperation in the Baltic Region. Uppsala University. Uppsala.

Club of Rome. www.clubofrome.org.

Daly H. E. 1989. Sustainable development: from concept and theory towards operational Principles in: Population and Development Review. Hoover Institution Conference.

Earth Summit in Johannesburg in 2002. www.earthsummit 2002.org.

European Environmental Advisory Councils: www.eeac-network.org.

Helsinki Commision 1994. 20 Years of International Cooperation for the Baltic Marine Environment 1974–1994. HELCOM. Helsinki.

Holmberg J. Thomson K. Timberlake L. 1993. Facing the Future. Beyond the Earth Summit. Earthscan. London.

IUCN (International Union for Conservation of Nature and Natural Resources). www.iucn.org/en/about.
IUCN 1980. The World Conservation Strategy. Gland.
Jachtenfuchs M. Struber M. 1992. Environmental Policy in Europe: Assessment, Challenges and Perspectives. Nomos Verlagsgesellschaft. Baden Baden.
John Paul II 1979. Redemptor Hominis. Libreria Editrice Vaticana. Vatican.
John Paul II 1987. Sollicitudo Rei Socialis. Libreria Editrice Vaticana. Vatican.
John Paul II 1991. Centesimus Annus. Libreria Editrice Vaticana. Vatican.
John Paul II 1995. Evangelium Vitae. Libreria Editrice Vaticana. Vatican.
John Paul II 1998. Fides et ratio. Libreria Editrice Vaticana. Vatican.
Kiss A. Shelton D. 1993. Manual of European Environmental Law, Cambridge University Press. Cambridge.
Kozłowski S. 2005. Przyszłość ekorozwoju (The Future of Eco-development). KUL. Lublin.
MaB (Man and Biosphere Programme). www.unesco.org./MAB.
Meadows D. H. et al. 1972. The Limits to Growth. Universe Books. New York.
Meadows D. H. Meadows D.L. Randers J. 1993. Beyond the Limits, Global Collapse or a Sustainable Future. Chelsea Green Publishing Company. Reprint edition. London.
Narodowa Strategia Edukacji Ekologicznej (National Environmetal Education Strategy) 2005. Liber. Warsaw.
Our Common Future. The Report of the World Commission on Environment and Development 1987. Oxford University Press. Oxford.
Rio Summit Documents: www.ecouncil.ac.cr/about/ftp/riodoc.htm.
Sorlin S. 1997. The Road Towards Sustainability – a Historical Perspective. Uppsala University. Uppsala.
Sustainable Development: From Concept and Theory to Operational Principles in: Resources, Environment and Population, New York.
UNESCO. www.portal.unesco.org.
United Nations Division for Sustainable Development, Agenda21. www.un.org/esa/sustdev/documents/agenda21/index.htm.
Weizsäcker E. U. von et al. 1997. Factor four, Doubling Wealth – Halving Resource Use. Earthscan. London.

Environmental Engineering – Pawłowski, Dudzińska & Pawłowski (eds)
© 2007 Taylor & Francis Group, London, ISBN13 978-0-415-40818-9

Alleviation of greenhouse effect by reduction of methane emission from anthropogenic sources

Witold Stępniewski
University of Technology, Chair of Landsurface Protection Engineering, Lublin, Nadbystrzycka, Poland

ABSTRACT: The paper presents an analysis of sources and sinks of atmospheric methane and possibilities to reduce its emission form anthropogenic sources such as rice fields, ruminants, biomass burning, municipal landfills and exploitation of natural gas and coal. The analysis showed a possibility of substantial limitation of the emission, if the influence of environmental factors on particular processes is better recognized. A special attention has been paid to factors influencing methane oxidation in soil and the possibilities of limitation of methane emission from municipal landfills by its biochemical oxidation in the remediation soil cover.

Keywords: Greenhouse effect, methane emission, anthropogenic sources, mitigation strategy.

1 INTRODUCTION

Methane is the third, after water vapor and carbon dioxide, most important greenhouse gas in the atmosphere and simultaneously the most abundant organic greenhouse gas. The atmospheric methane level has varied in time and the last thousand years were characterized by a concentration of ca. 700 ppbv with fluctuations of the order of 10–15 ppbv and a gradient between the polar zones of the order of 30–60 ppbv before the industrial era (Wuebbles and Hayhoe, 2002). After 1800 a monotonic increase of atmospheric methane to the present level of about 1750 ppbv has been observed. The increase of methane concentration rate before 1990 was about 1% per year, and in the nineties it decreased to about 0.3% annually (Wayne, 2002, Wuebbles and Hayhoe, 2002). Due to increased emission in the northern hemisphere the gradient between polar zones increased in the last century to 150 ppmv (Wuebbles and Hayhoe, 2002).

Annual emission of methane to the atmosphere is currently assessed as 410–660 Tg (Table 1). The radiative activity of methane is about 20–30 times higher compared to that of carbon dioxide (Leliveld et al. 1993, Samarkin et al. 1995), and the residence time in the atmosphere is estimated to be 10–17 years (Crutzen, 1994; Mosier, 1998). The present day contribution of methane to the greenhouse effect is assessed as 0.8 K, with the total greenhouse effect of the earth's atmosphere being 32 K (Kożuchowski and Przybylak, 1995). This contribution can double during the XXI century, and due to this there is a great interest in the possibilities to slowly stabilize it. It is suggested that stabilization of the methane concentration at the present day level would require reduction of its emission rate by about 10% i.e. by 40–70 Tg a year (Leliveld et al. 1993).

2 SOURCES AND SINKS OF ATMOSPHERIC METHANE

The contribution of particular sources of atmospheric methane is presented in Table 1 while the contribution of particular sinks is shown in Fig. 2. As can be concluded from Table 1, the emission sources are mainly of biological and anthropogenic character.

Soil is the most important source of atmospheric methane, as this gas is produced as a result of anoxic decomposition of soil organic matter. Such conditions exist in natural wetlands and in rice paddy soils. A significant contribution comes also from municipal landfills.

Table 2 shows that methane sinks are located mainly in the troposphere, and to a smaller extent (one order of magnitude less) in the stratosphere. In addition to the atmosphere soil also can be a sink for methane and its contribution is assessed to be similar to that of with the stratosphere. Thus the soil, being the largest methane source, may also oxidize it depending on oxygen conditions. The distribution of methane in the atmosphere, against the background of other greenhouse gases is presented in Fig. 1.

Table 1. Sources of atmospheric methane according to different authors (Tg y^{-1}).

Literature source*	A	B		C	D	E	F
CH_4 source	Value	Value	Interval	Value	Value	Value	Value
Natural sources together					345	145	
Natural wetlands	125 ± 70	115	55–150	100		100	92–232
Other natural sources		50	25–140				
– termites	30 ± 30					20	2–22
– oceans	10 ± 5					4	0.2–2
– marine sediments						5	0.4–12.2
– geological	5 ± 5					14	12–36
– fresh waters	5 ± 5						
Agricultural sources together						141	
Rice fields	70 ± 50	60	20–100	50		60	25–90
Ruminants and animal wastes	105 ± 30	105	85–130	100		81	65–100
Anthropogenic non-agricultural sources together						217	
Energy production and use		100	70–120	100		30	25–50
– gas	80 ± 45					46	15–64
– carbon	35 ± 10					30	6–60
Landfills	40 ± 25	30	20–70	22		61	40–100
Biomas burning	30 ± 15	40	20–80			50	27–80
Municipal waste waters	25 ± 10	25					
CH_4 sources together	560 ± 90	525		540		503	410–660

* Explanations: A. Lelieveld and Crutzen, 1993; B: IPCC, 1995; C: Mosier, 1998; D: Etiope and Klusman, 2002; E: Khalil, 2000; F: Different sources, after Wuebbles and Hayhoe, 2002.

Table 2. Sinks of atmospheric methane according to different authors (Tg $year^{-1}$).

Reference*	A	B		D	E	F
CH_4 sink	Value	Value	Interval	Value	Value	Value
Decomposition in atmosphere		470	421–520			
– reactions with OH radicals in troposphere	455 ± 50			490	445	360–530
– reactions with OH, Cl, O• in stratosphere	45 ± 10			40	40	32–48
Oxidation of atmospheric methane in soil	30 ± 25	30	15–45	38	30	15–45
CH_4 sinks together	530 ± 85	500		568	515	430–600
Annual increase in the atmosphere	+30	+25			−12	+20–60

* Explanations: A. Lelieveld and Crutzen, 1993; B: IPCC, 1995; C: Mosier, 1998; D: Etiope and Klusman, 2002; E: Khalil, 2000; F: Different sources, after Wuebbles and Hayhoe, 2002.

3 POSSIBILITIES TO REDUCE METHANE EMISSION FROM ANTHROPOGENIC SOURCES

Stabilization of atmospheric methane level is possible both by lowering its emission as well as by an increase of the activity of its sinks. The latter is possible only with respect to methane oxidation in soil as the influence on methane oxidation in the troposphere and stratosphere (90% of total methane oxidation) through hydroxyl radicals seems to be at present beyond human possibilities. Thus, for the strategy of methane stabilization in the atmosphere limiting its anthropogenic emission seems to be more promising and feasible.

In practice the measures under consideration can be related to rice fields, ruminants, natural gas and coal exploitation, as well as municipal landfills and waste water treatment plants.

As far as the rice fields are concerned a significant increase in their cultivation area took place during last several decades. Similarly an increase of the population of ruminants took place due to increased

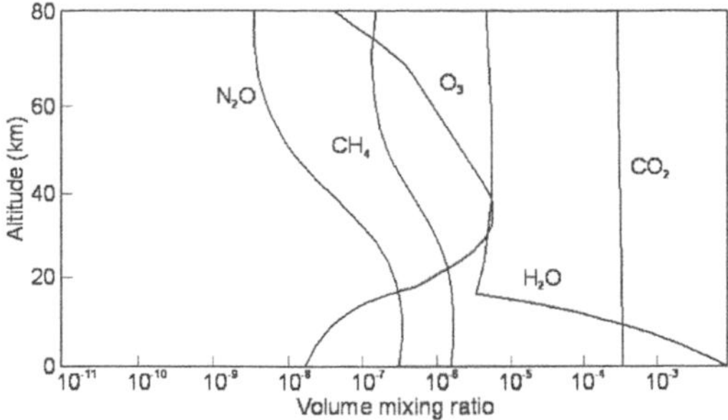

Figure 1. Distribution of concentration of methane and of other greenhouse gases in the atmosphere (after Salby, 1996 – changed).

food demand. Emission of methane from rice fields depends on such factors as temperature (optimum in the range from 37 to 45°C, Boone, 2000), soil properties, cultivation method, water management, rice variety and others. It was found that periodical dewatering of rice fields, introduction of oxidants and other mineral fertilizers, as well as a selection of varieties of low methane emission allows methane emission to be reduced by 20–70%, while application of organic fertilization elevates methane emission to 50% (Wuebbles and Heyhoe, 2002).

Thus a better recognition of factors affecting methane emission from rice fields gives a chance of substantial lowering of the global emission of this gas without lowering the area of lowland rice cultivation. Strategies aiming at lowering the emission of methane from rice fields can be oriented towards: 1° – reduction of its production, 2° – intensification of its oxidation, and 3° – reduction of its transport rate through the tissues of the plant itself (Le Mer and Roger, 2001). It should be emphasized that the effect of temperature on methane emission from rice fields and natural wetlands is a cause of additional positive feedback between methane emission and climate changes, as global warming increases production of methane, what in turns intensifies the global warming.

Methane emission by ruminants depends on the diet and is a byproduct of incomplete digestion. Better feeding of animals improves their protein uptake and reduces methane emission by even 40% (Wuebbles and Heyhoe, 2002). An additional factor of methane production is animal manure management. Emission of methane increases substantially under the conditions of long term storage of large amounts of manure.

Exploitation of fossil fuels creates another possibility to reduce methane emission. It is assessed that 1–2% of the natural gas produced escapes to the atmosphere in the developed countries, while in the developing countries this contribution can be several time higher, depending on the quality of the technology applied (Wuebbles and Heyhoe, 2002). In the case of hard coal mining there is the possibility of the capturing methane in order to produce energy.

The next source of methane emitted to the atmosphere is biomass burning. During incomplete, combustion formation of methane occurs; the amount of it being dependent on the kind of the biomass, its moisture content and on the burning conditions especially on oxygen availability and temperature (Levine et al. 2000).

Another significant (and tending to increase) source of atmospheric methane are municipal landfills. Their contribution is assessed as 8–20% of the total methane emission, which provides 20–100 Tg of the total annual methane emission. Methane production in landfills depends, among other things, on the composition of the wastes, on temperature and water content. Emission of methane produced in municipal landfills can be reduced by its energetic utilization, by flaring and by its biochemical oxidation in the landfill soil cover or in biofilters. According to Bogner et al. (1995) and Borjesson and Svenson (1997) in this way the methane emission can be reduced by 90%. Since the soil plays an important role as methane sink both in the open field and in the landfills this issue will be illustrated in more detail. Special emphasis will be placed on municipal landfills, because methane emission in this case is concentrated in a fairly limited area, where the possibility to regulate the conditions is much more feasible compared to the vast areas of cultivated fields and of natural ecosystems.

4 METHANE OXIDATION IN SOIL

An important role in limiting methane concentration in the atmosphere is played by its microbial oxidation in soil. Methane occurrence in soil as well as the contact of soil with atmospheric methane is connected with the occurrence of methanotrophic microorganisms (mainly bacteria) characterized by the possibility to oxidize methane and to utilize it as their only source of carbon and energy. Most of these organisms are obligate methylotrophs, capable of utilising methane and the products of its partial oxidation such as methanol, formaldehyde and methyl formate i.e. compounds without C-C bonds (Conrad, 1996; Rożej et al. 1999; Stępniewski and Rożej, 2000). They are obligate aerobes, often with their optimum at subatmospheric concentrations of oxygen, and their activity depends on the oxygenation status (Jones & Nedwell, 1993; King, 1994; Kightley et al. 1995).

Methanotrophic activity in soil depends on numerous physical, physicochemical and biological factors.

According to Le Mer and Roger (2001) two basic types of methanotrophic activity can be distinguished in soil. The first one is "high affinity oxidation" (i.e. of very low Michaelis constant value) which occurs at low methane concentration <12 ppm. This type of

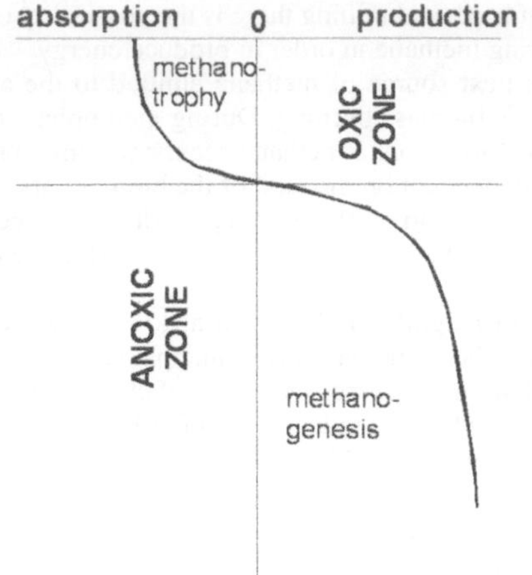

Figure 2. Idealized distribution of methane sinks and sources in soil profile containing oxic and anoxic zone. According to Stępniewski and Stępniewska (2000).

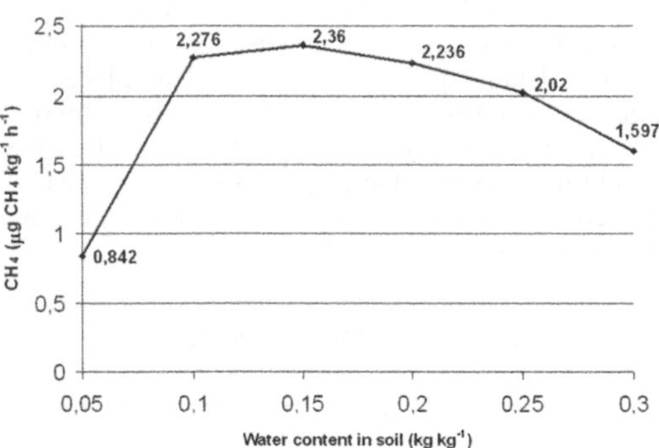

Figure 3. The influence of water content on methane uptake by a landfill cover soil containing 1.79% organic matter, 9.3% clay, 14.5% silt, 76.2% sand. The soil was classified according to USDA system as a sandy loam (modified from Boeckx and Van Cleemput, 1996).

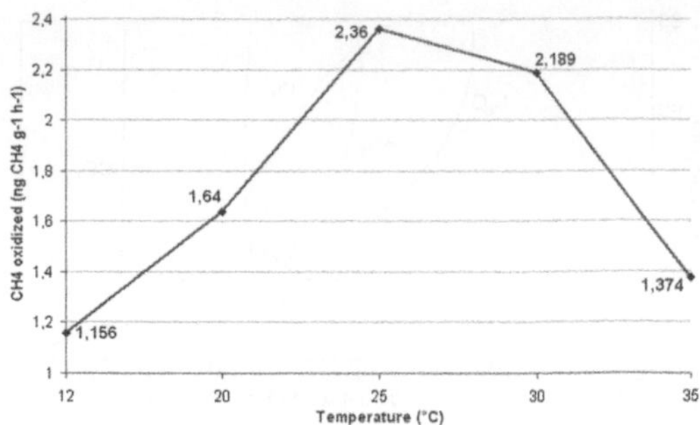

Figure 4. The effect of temperature on methane uptake rate by the same landfill cover soil as in Figure 3 (modified from Boeckx and Van Cleemput, 1996).

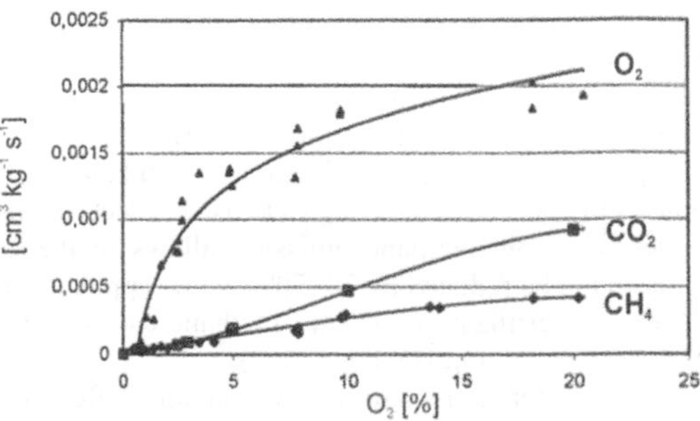

Figure 5. The effect of oxygen concentration on methane oxidation rate as measured by oxygen and methane uptake, as well as by carbon dioxide production in a sandy material incubated at a room temperature in the atmosphere containing 8% of methane and different concentrations of oxygen (modified from Pawłowska and Stępniewski, 2004).

methanotrophy takes place in soils, which were not exposed to high methane concentration. The bacteria performing this process are recognized only to a low extent. The second type of methanotrophy called "low affinity oxidation" (i.e. with a high value of the Michaelis constant) occurs at methane concentrations >40 ppm. This type of oxidation is considered the proper methanotrophic activity and is performed by methanotrophic bacteria. It occurs in methanogenic media such as peat lands, rice fields, wetlands and waste deposits. Usually in soil possessing a deeply situated anoxic zone and the oxic zone at the soil surface, there is a production of methane by methanogens living in the anoxic zone and consumption of methane by methanotrophs living in the oxic zone close to the soil surface. An idealized distribution of methane sources and sinks within the soil is presented in Fig. 2. It should be emphasized that over 90% of methane produced in the rice fields can be oxidized by the methanotrophs of the oxic zone. (Patrick and Jugsujinda, 1992; Le Mer and Roger, 2001). This creates an opportunity to reduce the global methane emission from the rice fields.

The optimum pH for methanotrophs is close to neutral (Dunfield et al. 1993), but they are capable of adapting to a wide pH range from 3.5 to 8.0 (Borne et al. 1990). An important effect on methanotrophic activity of soil is exerted by its water content and temperature (Boeckx and Van Cleemput, 1996; Le Mer and Roger, 2001; Wuebbles and Heyhoe, 2002). Examples of such relationships are presented in Figs. 3 and 4. The dependence of methanotrophic activity on water content shows a maximum. Lowering the methanotrophic activity at low water contents is a result of limited availability of water for microorganisms,

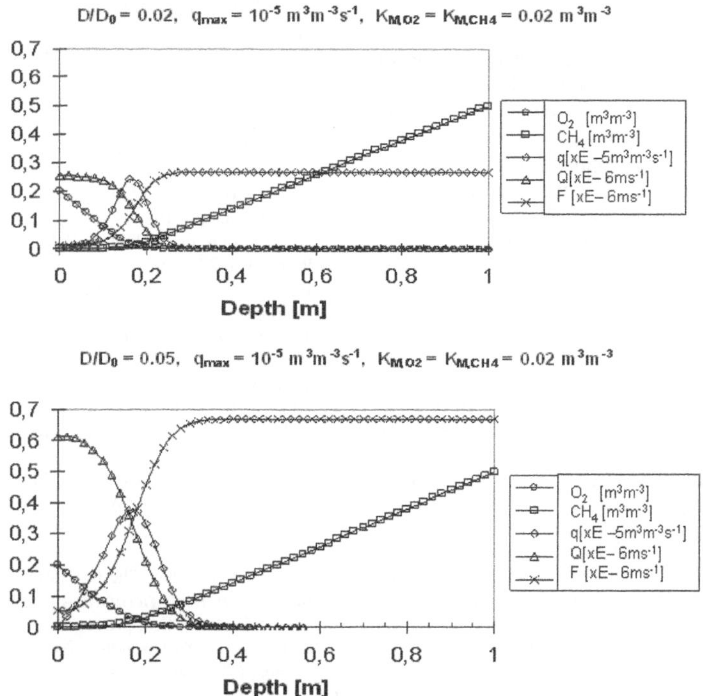

Figure 6. Calculated (by finite element method) distribution of the concentration of methane (CH_4) and oxygen (O_2), as well as of methanotrophic activity (q), cumulative methane consumption (Q) and methane flux (F) in a homogenous layer of landfill cover soil at different values of relative gas diffusion coefficient D/D_0 (D – diffusion coefficient of the given gas in soil, D_0 – diffusion coefficient of the same gas in the atmospheric air at the same pressure and temperature conditions. For methane at 20°C $D_{0,CH_4} = 2.23 \times 10^{-5}\ m^2\ s^{-1}$ was assumed). Upper plot $D/D_0 = 0.02$, lower plot $D/D_0 = 0.05$. Boundary conditions: $q_{max} = 10^{-5}\ m^3\ m^{-3}\ s^{-1}$, $K_{M1} = K_{M2} = 0.02\ m^3\ m^{-3}$, $C_{1,0} = 0$, $C_{1,L} = 0.50\ m^3\ m^{-3}$, $C_{2,0} = 0.20\ m^3\ m^{-3}$, $C_{2,L} = 0$, L (soil profile depth) = 1 m, q_{max} and K_M–: maximum methanotrophic activity and Michaelis constant, respectively (after Stępniewski and Zygmunt, 2000a – modified).

while its decrease at water contents above the maximum is caused by the limited availability of substrates i.e. of oxygen and methane due to lowering of their transport rate (Omal'cenko et al. 1992; Mancinelli, 1995).

Boeckx and Van Cleemput (1996) found maximum methanotrophic activity at 25°C (Fig. 4). This value is somewhat lower than the optimum (31°C) found by Whalen et al (1990) for a landfill cover soil. Both these values are within the interval 19–38°C found by Adamson and King (1993) for pure cultures of the methanotrophic bacteria Methylomonas rubra.

Dependence of methanotrophic activity on oxygen concentration is presented in Fig. 5. This relationship resembles that of Michaelis – Menten, especially when the activity is measured by the oxygen uptake. The influence of methane concentration on methanotrophic activity was studied by Bender and Conrad (1995). They observed an increase in the population of methanotrophs above 0.7% of methane; while at 20% of methane a 100-fold increase in methanotrophic activity and 10–100 fold of methanotroph counts took place. Pawłowska (1999) observed a stabilization of methanotrophic activity in sandy materials at methane concentrations above 6%. Investigation of methanotrophic activity of different size fractions of sandy materials (Pawłowska, 1999; Pawłowska et al. 2003) showed that the highest activity was measured for the 0.5–1.0 mm fraction.

Stępniewski and Zygmunt (2000a, 2000b) modelled the effect of gas diffusion coefficient on methantrophic activity, on oxygen and methane concentration distribution as well as on the methane flux within the landfill cover soil. They found that physical conditions have a decisive influence on the efficiency of utilization of methanotrophic potential of methanotrophs (see also Stępniewski et al. 2002). An example of the influence of gas diffusion coefficient is presented in Fig. 6.

5 SUMMARY

The presented analysis of the ways of mitigation of methane emissions to the atmosphere showed a possibility of substantial limitation of the emission, if the influence of environmental factors on the process is better recognized. Special emphasis has been placed on the possibilities of limitation of methane emission from municipal landfills, being a methane source of limited area, much smaller compared, for instance, to rice fields.

REFERENCES

Adamsen, A.P.S. King, G.M. 1993. Methane consumption in temperate and subarctic forest soils: rates, vertical zonation, and responses to water and nitrogen. *Appl. Environ. Microbiol.* 59, 485–490.

Bender, M. Conrad, R. 1995. Effect of methane concentrations and soil conditions on the induction of methane oxidation activity. *Soil Biol. Biochem.* 27, 1517–1527.

Boeckx, P. Van Cleemput, O. 1996. Methane oxidation in a neutral landfill cover soil: influence of moisture content, temperature, and nitrogen-turnover. *J. Environ. Qual.* 25, 178–183.

Bogner, J. Spokas, K. Burton, E. Sweeney, R. Corona, V. 1995. Landfills as atmospheric methane sources and sinks. *Chemosphere* 31, 4119–4130.

Borjesson, G. Svensson, B. 1997. Effects of a gas extraction interruption on emissions of methane and carbon soil. *J. Environ. Qual.* 26, 1182–1190.

Boone, D. 2000. Biological formation and consumption of methane. In Khalil, M (Ed). Atmospheric Methane: Its Role in the Global Environment. *Springer Verlag*, New York, NY, pp 42–62.

Borne, M. Dorr, H. Levin, I. 1990. Methane consumption in aerated soils of the temperate zone. *Tellus* 42, 2–8.

Conrad, R. 1996. Soil microorganisms as controllers of atmospheric trace gases (H_2, CO, CH_4, OCS, N_2O, and NO). *Microbiol. Rew.* 60, 609–640.

Crutzen, P.J. 1994. Global budgets for non-CO_2 greenhouse gases. *Environmental monitoring and assessment*, 31, 1–15.

Dunfield, P. Knowles, R. Dumont, R. Moore, T.R. 1993. Methane production and consumption in temperate and subarctic peat soils: response to temperature and pH. *Soil Biol. Biochem.* 25, 321–326.

Etiope, G. Klusman, R. 2002. Geological emissions of methane to the atmosphere. *Chemosphere*, 49, 777–789.

Hanson, R.S. Hanson, T.E. 1996. Methanotrophic Bacteria. *Microbiol. Rev.* 60, 439–471.

Intergovernmental Panel for Climate Change, 1995. W: Houghton, J.T. Meira Filho, L. Bruce, J. Lee, H. Callander, B. Haites, E. Harris, H. Maskell, K. (Eds). *Climate Change*, 1994. Cambridge University Press, Cambridge, UK, 339 ss.

Jones, H.A. Nedwell, D.B. 1993. Methane emission and methane oxidation in land-fill cover soil. *FEMS Microbiol. Ecol.* 102, 185–195.

Khalil, M.A.K. Atmospheric methane: an introduction. W: Khalil, M. (Ed). Atmospheric Methane, Its Role in the Global Environment, *Springer Verlag*, New York, pp. 1–8.

Kightley, D. Nedwell, D.B. Cooper, M. 1995. Capacity for methane oxidation in landfill cover soils, measured in, laboratory-scale soil microcosms. *Appl. Environ. Microbiol.* 61, 592–601.

King, G.M. 1994. Methanotrophic associations with the roots and rhizomes of aquatic vegetation. *Appl. Environ. Microbl.* 60, 3220–3227.

Kożuchowski, K. Przybylak, R. 1985. Efekt cieplarniany (Greenhouse Effect-in Polish), *Wiedza Powszechna*, Warszawa, 220 pp.

Leliveld, J. Crutzen, P.J. 1993. Methane emission into the atmosphere. An overview, *Proceedings of IPCC Workshop Methane and Nitrous Oxide*, Amersfoort, 17–25.

Leliveld, J., Crutzen, P.J. Bruehl, C. 1993. Climate effects of atmospheric methane, *Chemosphere* 26, 739–768.

Le Mer, J. Roger, P. 2001. Production, oxidation, emission and consumption of methane in soil. *Eur. Journal Soil Biol.* 37, 25–50.

Levine, J.S. Cofe, III W.P. Pinto, J.P. Biomass Burning. W: Khalil, M (Ed). 2000. Atmospheric Methane: Its Role in the Global Environment. *Springer Verlag*, New York, NY, pp 190–201.

Mancinelli, R.L. 1995. The regulation of methane oxidation in soil. *Ann. Rev. Microbiol.* 49, 581–605.

Mosier, A.R. 1998. Soil processes and global change, *World Congress of Soil Science* 20–26 August, Montpellier, France, CD proceedings, nr 1940.

Omal'chenko, L. Savel'eva, N.D. Vasil'eva, V. Zavarzin, G.A. 1992. A psychrophilic methanotrophic community from tundra soil. *Mikrobiologiya* 61, 1072–1076.

Patrick, W.H. Jugsujinda, A. 1992. Sequential reduction and oxidation of inorganic nitrogen, manganese, and iron in flooded soil. *Soil Sci. Soc. Am. J.* 56, 1071–73.

Pawłowska, M. 1999. Możliwość zmniejszenia emisji metanu z wysypisk na drodze jego biochemicznego utleniania w rekultywacyjnym nadkładzie glebowym – badania modelowe. Politechnika Lubelska (A possibility to reduce the methane emission from landfills by its biochemical oxidation in recultivation soil cover – model investigations- in Polish), ISBN 83-881110-36-5.

Pawłowska, M. Stępniewski, W. 2004. The effect of oxygen concentration on the activity of methanotrophs in sand material, *Envir. Prot. Engineering*. 30, 81–91.

Pawłowska, M. Stępniewski, W. Czerwiński, J. 2003. The effect of texture on methane oxidation capacity in a sand layer – a model laboratory study. W: Environmental Engineering Studies (Pawłowski et al. – Eds), *Kluver Academic/Plenum Publishers*, New York, 339–354.

Rożej, A. Stępniewski, W. Małek, W. 1999. Bakterie metanotroficzne w ekosystemach, Wydawnictwo *Postępy Mikrobiologii*, 38, 4, 295–313.

Salby, M.L. 1996. *Fundamentals of Atmospheric Physics*. Academic Press.

Samarkin, V.A. Vecherskaya, M.S. Rivkina, E.M. 1995. Methane in permafrost soils of Cryolithozone of North –Eastern Siberia. *Ecological Chemistry*, 1, 24–30.

Stępniewski, W. Horn, R. Martyniuk, S. 2002. Managing soil biophysical properties for environmental protection. *Agriculture Ecosystems and Environment*, 88, 175–181.

Stępniewski, W. Rożej, A. 2000. Methanotrophic bacteria and the impact of soil physical conditions on their activity. *Int. Agrophysics*, 14, 135–139.

Stępniewski, W. Stępniewska, Z. 2000. Oxygenology of treatment wetlands and its environmental effects. 7th International Conference on Wetland Systems for Water Pollution Control, November 11–16, Lake Buena Vista, Florida, vol. II, 671–678.

Stępniewski, W. Zygmunt, M. 2000a. Methane oxidation in homogenous soil covers of landfills. A finite element analysis of the influence of gas diffusion coefficient. *Int. Agrophysics*. 14, 449–456.

Stępniewski, W. Zygmunt, M. 2000b. Mitigation of methane emission from landfills. W: Sustainable Development – a European View. *Zeszyty Naukowe Komitetu Człowiek I Środowisko*, 27, 79–93.

Wayne, R.P. 2002. *Chemistry of Atmospheres*, Oxford University Press. 775 ss.

Whalen, S.C. Reeburgh, W.S. Sandbeck, K.A. 1990. Rapid methane oxidations in a landfill cover soil. *Appl. Environ. Microbiol.* 56, 3405–3411.

Wuebbles, D.J. Hayhoe, K. 2002. Atmospheric methane and global change. *Earth-Science Reviews*, 57, 177–210.

Environmental Engineering – Pawłowski, Dudzińska & Pawłowski (eds)
© 2007 Taylor & Francis Group, London, ISBN13 978-0-415-40818-9

Heat and water transfer in building materials

Henryk Sobczuk
Faculty of Environmental Engineering, Lublin University of Technology, Nadbystrzycka Lublin, Poland

ABSTRACT: In the current paper some basic problems of measuring heat and water transfer in building materials and structures are described. The number of buildings which need renewal in Europe justifies introduction of sophisticated methods of measurement and modelling for assessment of performance of new materials in connection with old ones in retrofitted buildings. Some of these methods are briefly described, especially those connected with moisture and heat transfer processes in building materials and structures.

Keywords: Heat transfer, moisture transfer, measurement technology.

1 INTRODUCTION

Rising anthropopression on the environment comes, to a major part, from less than optimal building construction and heating systems. According to the scale of the problem, any progress in technology that curbs excess energy demand, will give rise to large scale economical savings while limiting greenhouse gas emission and the use of energy carriers and building materials. A high percentage of buildings existing in Poland, but also in other European countries, were built after World War II and do not fulfil today's heat isolation standards. The low quality of thermal insulation of a large number of these buildings means, in our climate, that high costs of heating in Winter are inevitable.

This causes a tough pression on the environment due to use of energy resources, carbon dioxide emission and as a result, climate change caused by the greenhouse effect.

2 MATERIALS AND METHODS

There exist about 56 mln. flats in the EU, even without new accessing countries, that were build after year 1950. They are not up to today's heat standard nor do they provide good indoor climate quality due to unsatisfactory thermal insulation of the applied materials. These flats need urgent renovation, their exploitation in this state is economically infeasible, while they not provide satisfactory standard of internal climate parameters. In Poland the situation is even worse due to slow changes of regulations on flat rental prices. The market for building renovation, and especially thermo-renovation has developed very quickly in recent years. It has been forced by the rapid increase in energy prices.

The total area of office building is around 1200 million square meters in Europe [Caccavelli i Gugerli, 2002]. Only about 25% of this area consists of buildings constructed more than 25 years ago, this means that most of them are relatively new. However, the exploitation time span for office buildings is shorter than residential buildings. Older buildings do not fulfil expected standards and need renovation. The costs of renovation of such buildings is much lower than the cost of destroying and building them from scratch (a rough assessment of renovation costs is about half to one third of a complete new building.

Renovation of buildings, comprises a significant percentage of the building market in all European countries. As the illustration we can use in the data presented at Figure 1. [Balaras et al., 2000].

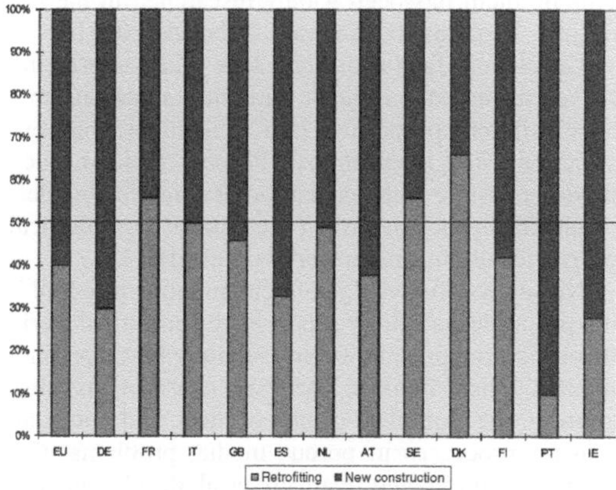

Figure 1. Distribution of costs in European countries for renovation and new constructions for the year 1996.

Most existing hotels were built in the seventies and eighties of the last century. Today's expectations are much higher concerning comfort, technical equipment and the effectiveness of installations. Most of those buildings, after 15–25 years of exploitation need complete renovation. The yearly growth in the number of hotels was around 8% in 1980 but now has dropped to around 1% per year. The most important factor causing this is the integration off the European hotel market. As exploitation costs of ineffective buildings are cumulating in time, it is clear that the necessity of renovating buildings is inevitable and will have growing character.

Renovation of buildings and especially thermorenovation is playing a major role in the decrease of energy use in the EU; the building sector consumes around 30%–40% of the energy, in which is responsible for the emission off around 40%–45% of the carbon dioxide. The total yearly energy demand in an "old" EU country can be estimated to be 150–230 kWh/m^2 while in central and eastern Europe it reaches 250–400 kWh/m^2. For comparison, it is interesting to point out that in Scandinavian countries, where the climatic conditions are rather more demanding, well isolated buildings need 120–150 kWh/m^2 and the most effective can reach 60–80 kWh/m^2. Comparison of these numbers and understanding the scale of the problem shows the reality of possible energy saving.

Historical buildings comprise a separate problem due to the need for preservation of historical facades in unchanged form. This is a real barrier against the technical possibilities of renovation and it generates additional costs. Numerous historical buildings are utilized for residential, office and hotel purposes. Their exploitation is usually more expensive than modern buildings. Due to their historical facades only internal thermal insulation is feasible for application. This method has been used for a long time but it often fails due to the exploitation problems caused by improper thermal and moisture properties of applied materials. This problem has been widely researched in the EU project "Insumat" [Grunewald, Sobczuk. 2005].

Developers face a difficult task when interfacing old elements of buildings with new materials that have different properties, into well interacting construction with a long life span. Proper solution of such tasks is possible with application of numerical models of physical processes when the database on physical properties of building materials is available.

Nowadays, knowledge of heat and moisture building properties is solidly supported by empirical work; theoretical support, however, is mainly based on phenomenological models. The main cause of this situation is the complex nature of heat and moisture transfer processes in porous media, problems with their description, and relatively weak development of testing methods that allow measurement of complex processes and characteristics. Lack of a consistent system of knowledge about heat and moisture transfer in building porous materials is the main cause of the situation when new materials are introduced into the market on the basis of incomplete testing in various exploitation conditions. Because of this, many constructions are not built optimally. This raises the costs of construction, renovation and exploitation, diminishes durability and the quality of the resulting construction.

Thermal renovation of buildings is connected with considerable costs and as such should be based on an optimal project analysis. The proper analysis will allow project leaders to find weak points in the construction, discover places in which condensation may occur and calculate and predict moisture content and time of condensation in given climatic conditions. It is especially important in residential buildings built with materials incompatible to the presence of long term moisture. Prediction of places at which water condensation may occur in the project phase makes prevention possible by optimal utilisation of thermal insulation for specific climatic conditions.

The possibility of using a physical model for the prediction of heat properties of building elements depends on the knowledge of moisture parameters and characteristics of applied building materials. This is the important reason why advanced measurement methods are being so intensively developed.

Thermal and moisture properties of building materials decide their applicability, durability in normal exploitation and preservation of a comfortable interior climate. Moisture content is an important factor that modifies heat exchange parameters in building materials. Temperature distribution in building structures change according to internal and external climate conditions and according to moisture distribution within the structure. The water flow in porous media can be described, in a simplified manner, by the Richards equation, which in one dimensional version takes the form:

$$\frac{\partial \theta}{\partial t} = \frac{\partial}{\partial z}\left[K(\theta)\left(\frac{\partial \psi}{\partial z} + 1\right)\right] \tag{1}$$

Where:
ψ – water potential [cm H_2O], θ – moisture content [cm^3/cm^3], $K(\theta)$ – moisture conductivity function [cm/s], t – time [s], z – coordinate (height) [cm].

Even such a simplified equation is a nonlinear partial differential equation which can be generally solved with the application of numerical methods, but only with the knowledge of retention characteristics $\psi(\theta)$, and water conductivity function $K(\theta)$. The measurement of these characteristics is however complicated and time consuming. Methods for prediction of their shape are rather inaccurate; one should concentrate on their measurement in well equipped laboratories in order to fill a database containing data for readily available materials.

More complicated models of heat and moisture transport have to take into account the phase change of water to vapour and heat transfer via different mechanisms. Some of them are heat transport with vapour, with liquid phase and by solid phase of the material. The description of these kinds of processes is much more complex than the flow of water. For this, other properties and parameters must also be known. The necessary parameters have to be measured, calculated from other properties or estimated using theoretical analysis.

The basic properties for the description of heat and moisture transport in a porous medium are temperature, moisture content, and the energy of interaction of water with the material together with transport parameters. Measurement of these basic properties allows us to characterize building materials with respect to heat and moisture transport on an engineering level.

Moisture content measurement can be realized with relatively simple equipment using a gravimetric method. The gravimetric method relys on measuring the amount of water by comparing the wet and dry sample weight. The basic disadvantage is the need to take a sample for each measurement from the measured object. Repetition of such measurements will destroy the object being measured.

Electrical methods of measuring moisture content developed over the years, confront the basic problem that moisture in building structures is never pure water but rather a solution of salts which are present in the material. Such a solution usually has a relatively large electrical conductivity, which is moisture content and material structure dependent. This is the reason why most electrical methods fail in measuring the moisture of building materials.

Fortunately, methods exist which are largely independent of the electrical conductivity of solute within the porous body. These methods rely on the measurement of the speed of an electrical pulse that propagates within the wet porous material. This method, called TDR (time domain reflectometry), has become one of the essential methods of moisture content measurement. The working principle on which the TDR method relys is that water has very high relative dielectric permittivity (81), in the liquid state, while most of solid materials have permittivities of the order 2 to 5. The speed of propagation of an electromagnetic wave in the medium depends on its relative dielectric permittivity according to the formula:

$$v = \frac{c}{\sqrt{\varepsilon}} \tag{2}$$

Where: v – speed of electromagnetic pulse within the material, cm/s, c – speed of light in vacuo, cm/s, ε – effective relative dielectric permittivity of the material.

Knowledge of the speed of electromagnetic pulse propagation in the investigated medium allows the calculation of dielectric permittivity and, from this, the respective moisture content after conducting a calibration. The technical problem is the accurate measurement of the short time span -in practical application one has to measure time of an order of nanoseconds (10^{-9}s) with the accuracy of few picoseconds (10^{-12}s). Devices that realize this measurement are produced and used in many laboratories (Fig. 2). Probes applied to this kind of measurement are of different form according to the need. Especially interesting are probes that apply two rods. Figures 3–5 show three types of probes applicable to measure moisture content in various building materials.

Temperature measurement methods are highly developed in many branches of engineering. There are many types of temperature sensors employing different physical phenomena, starting from the simplest, like thermocouples, through thermistors with positive and negative thermal coefficients of resistivity and ending with complicated semiconductor circuits that

Figure 2. TDR device in laboratory.

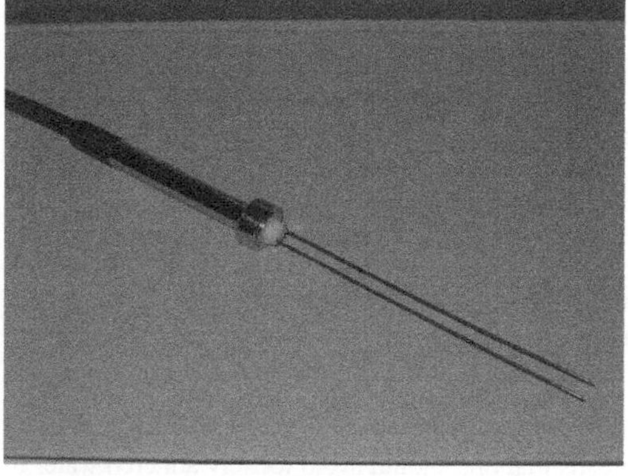

Figure 3. TDR probe with thin rods.

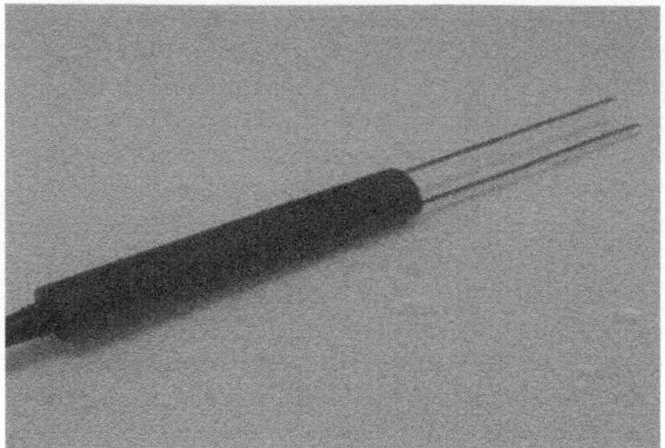

Figure 4. TDR probe with intermediate rods.

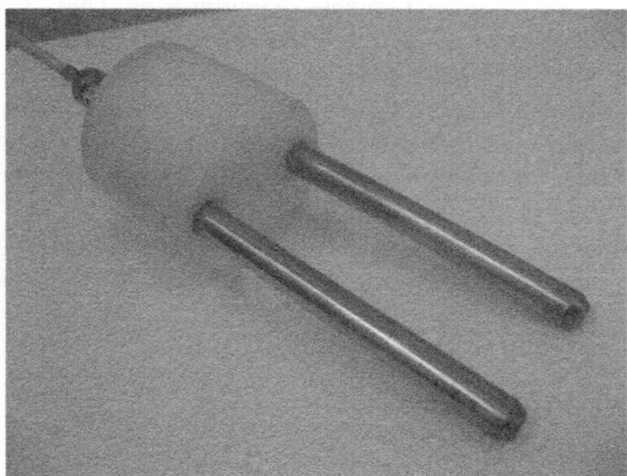

Figure 5. TDR probe with thick rods to apply to hard materials.

measure temperature and at the same time translate it to a digital form. Measurement of the temperature in the building material should not disturb the temperature field in it and this is why the sensor should be properly constructed. A temperature sensor applied in the laboratory is presented in Fig. 6.

Sensors of this type make it possible to accurately measure temperature while preventing disturbance in the temperature field. The plastic shield pipe with low heat transfer coefficient which is applied does not significantly disturb the thermal field. Thin signal wires prevent heat transfer through metal.

Water potential is an indication of how strongly the water is attracted by the solid phase of the building material. If the water is tightly adhered, as in the case of air dry building materials, it does not pose any problem in exploitation. If it is loosely adhered to the material it can move within it, easily evaporating and freezing. It can also support microbial development. This means that one should avoid loosely adhered water in the material. Prolonged presence of such water is dangerous for his durability of the construction. Especially

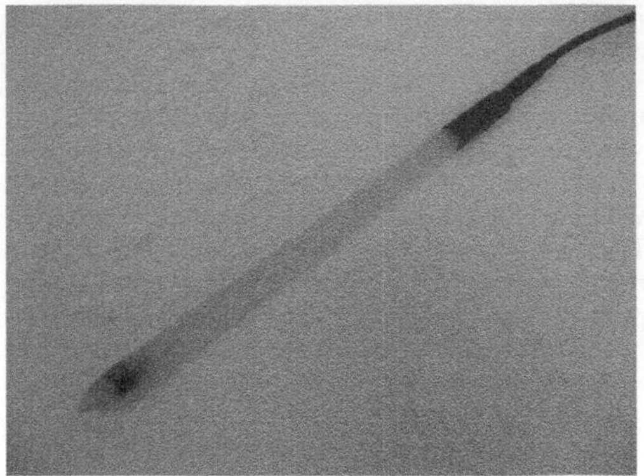

Figure 6. Laboratory thermometer with semiconductor sensor.

inconvenient is water vapour condensation that takes place during colder periods.

The relative humidity of the air inside the material can be measured using a thermocouple psychrometer.

The thermocouple psychrometer allows measurement of the relative humidity of air that is within the building material and thus in equilibrium with water bound to the building material itself. As a result one can draw conclusions about the water potential.

The thermocouple is constructed from two thin wires with the diameter 0.0025 cm made of different materials. The electrical current flowing in the wires connection causes excess heat generation when flowing in one direction and absorption when flowing in the opposite direction. This effect allows one to manipulate the temperature of the wires (Peltier effect). Note that the inverse phenomenon is also known. In this the thermocouple generates an electrical current which depends on its temperature (Seebeck effect). Both of these effects are employed for the measurement.

Application of the mentioned phenomena allows the measurement of the relative humidity of air and eventual calculation of water potential with equation (3):

$$\psi = \frac{RT}{V} \ln \frac{e}{e0} \tag{3}$$

In this equation R is the universal gas constant ($8.3143\,\mathrm{J \cdot mol\,K^{-1}}$), e/e0 relative air humidity, V molar water volume (1.8′10−5 m3′mol-1) and T temperature (K).

The probe shown in Fig. 7 is made of constantan and chromel alloys. The thermocouple is cooled at the beginning of the measurement by electrical current till water vapour reaches its point of condensation. Some water condenses on the couple. After that the current is stopped and the condensed water is allowed to vaporise, thus diminishing the couple temperature. When the temperature drops, one can measure a voltage that is proportional to the temperature drop.

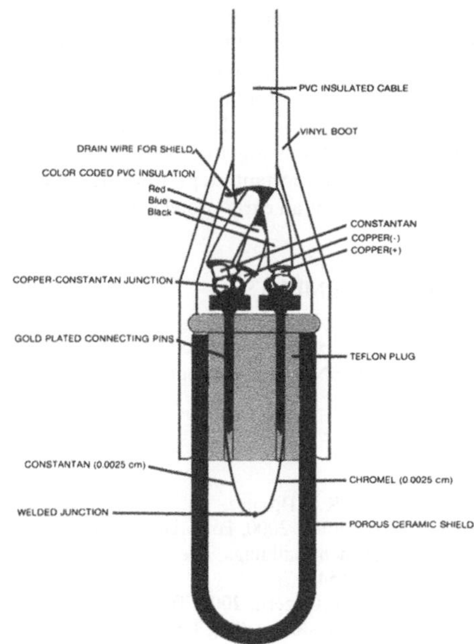

Figure 7. Psychrometer probe [WESCOR, 1998].

The intensity of vaporization and temperature drop depend on the relative humidity of ambient air. Measurement of thermocouple response to this process allows the measurement of relative air humidity.

An example of the thermocouple signal during measurement is shown on Fig. 8.

It presents the thermocouple signal change after switching off the cooling current. Water condensed at the thermocouple evaporates causing a reduction in temperature and measurable electrical output from the thermocouple. After some time the condensed water evaporates completely and the temperature difference will tend to zero and then vanish completely. The useful signal is of some tens of microvolts. The measuring device has to be stable and accurate to reliably measure such small signals. A microprocessor readout device capable of doing that has been built and applied for this purpose [Skierucha et al., 2005]. The noise level is usually only of the order of a few nanovolts, so that measurement of the water potential in a range equivalent to humidity between 0.95 to 1.0, is possible. A calibration curve, established with an NaCl solution of known concentration, is presented in Fig. 9. Using this calibration one can measure the water potential in building materials in the range close to condensation. This range is especially important in building physics.

2.1 *Application of thermal boundary conditions*

The possibility of applying different boundary condition for heat and moisture measurement is inevitable

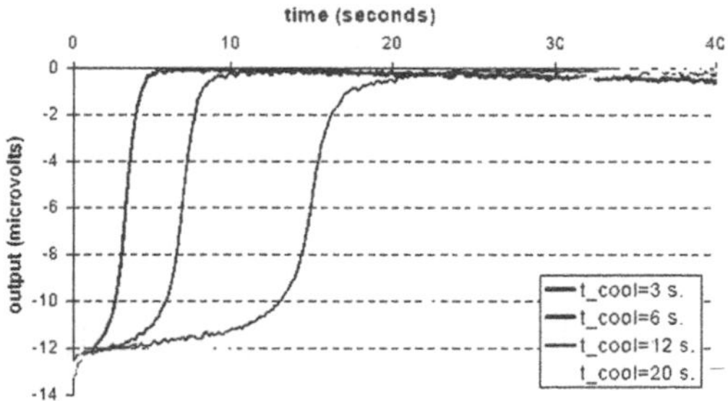

Figure 8. Course of thermocouple psychrometer signal during measurement (microvolts).

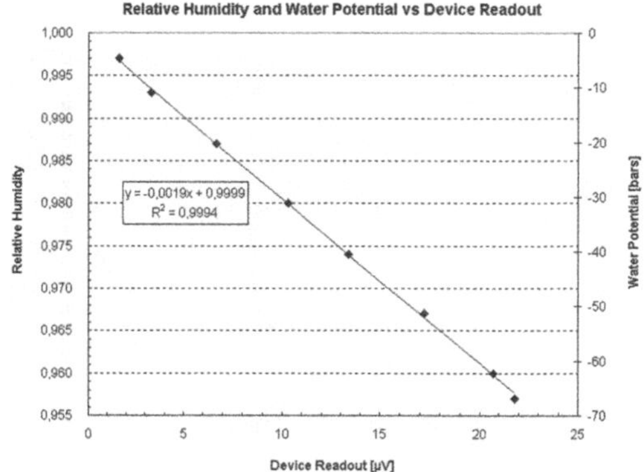

Figure 9. Calibration of the thermocouple device.

for laboratory investigations of building materials and their structures. One of the basic devices used in such cases is the climatic chamber for simulation of natural conditions on a sample of a building structure. This allows the simulation of climatic conditions that are normally present on the inside and outside layers of materials. The internal climate can be simulated by proper application of heat and moisture conditions comparable to those existing in the building. Outside climate condition should be in agreement with the climatic region being simulated. The experimental setup (Fig. 10) allows the simulation of natural conditions acting on the building sample, both external and internal.

[Carmeliet J., et al., 2005].

Application of multiprobe devices for moisture content and temperature measurement allows the investigation of the time variability of moisture and temperature. Critical points of the building structure that may undergo moisture condensation can be investigated. The results of the measurement can be used for verification of numerical models of heat and water flow within porous building materials or structures.

Application of the verified models for modelling structures at the stage of project development allows

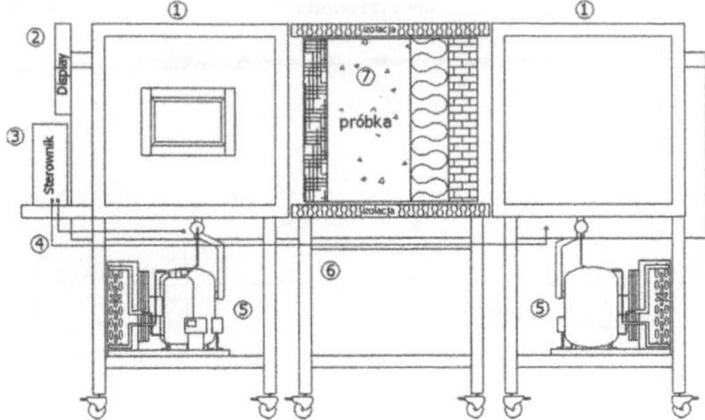

Figure 10. Climatic chamber for boundary condition application.

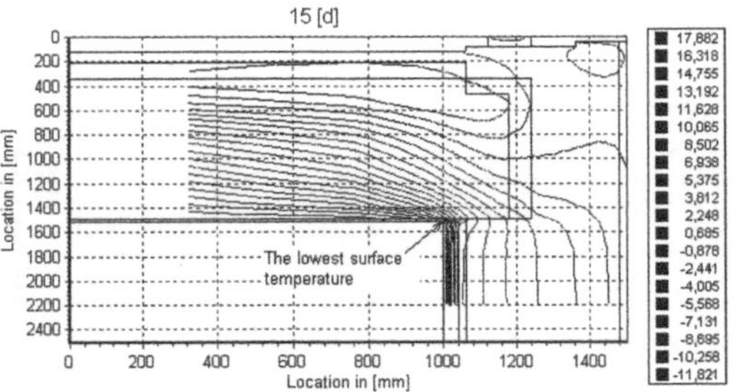

Figure 11. Example of numerical modelling results.

verification of the most demanding details by numerical simulation.

An example of the results obtained by the numerical model is shown in Fig. 11. Calculations were done using the program DELPHIN4. The results are taken from the INSUMAT project report, [Grunewald J. (editor), 2003].

3 CONCLUSIONS

Thermorenovation is an important task from the point of view of environmental protection. In the climate of Poland and eastern Europe it can cause a substantially lower energy demand.

Thermal insulation, properly carried out, demands interconnection of old, traditional materials with newly introduced thermal insulation. Knowledge of water and heat transfer in building materials is needed for proper analysis of projected solutions. Further development of methods for the measurement of moisture and heat transfer in building materials will need specific laboratory devices. Development of numerical simulation methods together with characteristics measurement are important steps towards assessing construction plans at an early project stage.

ACKNOWLEDGEMENTS

The measurement setup presented in the paper was developed in a KBN grant; number 4 T09D 052 22.

REFERENCES

Balaras C. A. K. Droutsa, A. A. Argiriou and D.N. Asimakopoulos, 2000, Potential for energy conservation in apartment buildings. *Energy & Buildings*, Vol. 31(2): pp. 134–154.

Caccavelli D. and H. Gugerli. 2002. TOBUS – A European diagnosis and decision making tool for office building upgrading,. *Energy & Buildings*, Vol. 34, pp. 113–119.

Carmeliet J, H. Hens, S. Roels, O.Adan, H. Brocken, R. Cerny, Z. Pavlik, C. Hall, K. Kumaran and L. Pel, Determination of the Liquid Water Diffusivity from Transient Moisture Transfer Experiments, *Journal of THERMAL ENV. & BLDG. SCI.*, Vol. 27, No. 4—April 2004

Grunewald J., 1999. Numerical Simulation of Heat and Mass Transport – Theoretical Background and Simulation Tool. *Proceedings of Seminar on Deterioration of stone monuments, historical tombs, cave sites, and brick buildings and their protective measures, pp. 78–105*, Tokyo National Research Institute of Cultural Properties, Tokyo

Grunewald J. and H. Sobczuk, "Development of insulation materials with specially designed properties for building renovation", *Monografie Komitetu Inżynierii Środowiska PAN*, vol. 34, 2005, ISBN 83-89293-21-8

Skierucha W, H. Sobczuk and M. Malicki, Applicability of psychrometric probes to soil water potential measurement 2002, *Acta Agrophysica, vol.* 53, 2001

Suchorab Z, D. Barnat-Hunek and H. Sobczuk, Adaptation of reflectometric techniques for moisture measurement of rock walls on the example of Janowiec Castle, *City of Tomorrow and Cultural Heritage, Pomerania Outlook, International Workshop, December 08–09, 2005, Gdańsk, Poland, Volume I, 71–74, ISBN 83-60261-02-4*

Water Potential Systems. *WESCOR Scientific Products Catalog*. 1998

Waste water treatment and disposal

Environmental Engineering – Pawłowski, Dudzińska & Pawłowski (eds)
© 2007 Taylor & Francis Group, London, ISBN13 978-0-415-40818-9

Estimating the efficiency of wastewater treatment in activated sludge systems by biomonitoring

Agnieszka Montusiewicz, Jacek Malicki, Grzegorz Łagód & Mariola Chomczyńska
Lublin University of Technology, Department of Environmental Engineering, Nadbystrzycka, Lublin, Poland

ABSTRACT: A method using saprobionts for the biomonitoring of wastewater treatment efficiency, ignoring the precise taxonomic identification of the counted individuals is discussed. A number of structural indices (Is, S, H, H_{max}, V, E and P indices) were calculated. Their values were determined on the basis of species abundances for activated sludge communities present in wastewater with different pollution levels. These values were also calculated using abundances of morphological-functional groups representing easily identified taxonomic levels higher than species level. The results show that biomonitoring based on indicator species and on measurements of community structure with identification of species enables wastewater pollution levels to be distinguished, but it requires careful classification of individuals into many different species. Bioindication based on abundances of morphological-functional groups is generally easier and also allows wastewater pollution levels to be distinguished.

Keywords: Biomonitoring, activated sludge system, saprobes system, biodiversity.

LIST OF ABBREVIATIONS USED IN TEXT, TABLES AND FIGURES

Πi – relative abundance of taxon
Πc.s. – relative abundance of swimming ciliates
Πc.a. – relative abundance of attached ciliates
Πc.c. – relative abundance of crawling ciliates
Πro. – relative abundance of rotifers
Πfl. – relative abundance of flagellates
Πam. – relative abundance of amoebas
Πmet. – relative abundance of metazoans
Πc.all – relative abundance of all ciliates
n_T – total number of all individuals
Sc.s. – number of swimming ciliate species
Sc.a. – number of attached ciliate species
Sc.c. – number of crawling ciliate species
Smet. – number of metazoan species
Sfl. – number of flagellate species
Sro. – number of rotifer species
Sam. – number of amoeba species
Sc.all – number of all ciliate species
S – number of all species
Sk – number of indicator species
H – the Shannon-Wiener index
H_{max} – the maximum value of Shannon-Wiener index
V – the evenness index
E – the MacArthur index
P – the proportionality index
Is – the saprobic index
x – xenosaproby
o – oligosaproby
βm – β-mesosaproby
αm – α-mesosaproby
p – polysaproby

1 INTRODUCTION

The vital activity of many saprobionts (saprophile and saprotrophe microorganisms) enables the treatment of municipal (and some industrial) wastewater in activated sludge systems (Curds et al. 1968, Curds & Cockburn 1970a, Klimowicz 1972, Genoveva et al. 1991, Martin-Cereceda 1996). Activated sludge is generally formed by bacteria, protozoans, rotifers, fungi and sometimes by representatives of *Metazoa*, for example nematodes (Klimowicz 1970, Genoveva et al. 1991, Martin-Cereceda et al. 1996). Protozoans not only perform wastewater treatment from the live and dead fraction of disintegrated suspended solids, but can also be useful as indicators of activated sludge condition and wastewater treatment efficiency (Curds & Cockburn 1970b, Madoni 1994, Lee et al. 2004, Łagód et al. 2004, Chomczyńska et al. 2005, Puigagut et al. 2005).

Bioindication of wastewater quality, just as biomonitoring the quality of other environments, can be based on indicator species, on combinations of indicator species or on characteristics of community structure (Buchs 2003, Iliopoulou-Georgudaki et al. 2003, Gorzel & Kornijów 2004, Jiang & Shen 2005). For bioindication needs representatives of stenotopic species with narrow tolerance ranges can be used. If stenotopic species were to be universally useful, they should be ubiquitous. Otherwise they can only have local importance. Unfortunately, most species are rather eurytopic and local endemic species verify themselves best in bioindication. For the above reasons bioindication methods based not on species but on species communities are searched. Such methods can use biocenotic indices, for example: biodiversity indices, species richness or species variation indices (Gorzel & Kornijów 2004). Application of activated sludge microorganisms for the bioindication of wastewater quality based on biocenotic indices requires establishing their abundances, as well as taxonomic identification, and can therefore lead to a number of difficulties (Pantle & Buck 1955, Curds & Cockburn 1970b, Washington 1984, Madoni 1994, Martin-Cereceda et al. 1996). This being the case, a possible solution might be the creation of an automatic counting and identification system for microorganisms in activated sludge (Amaral et al. 1999, Da Motta et al. 2001). However, it is conceivable to bypass the necessity of identifying the counted individuals into species, and confine oneself to the identification and counting of individuals in overspecies or overgenus groups (Łagód et al. 2004, Chomczyńska et al. 2005).

The purpose of this paper is to discuss a method using saprobionts for the biomonitoring of wastewater treatment efficiency, ignoring the precise taxonomic identification of the counted individuals. The paper also aims at the presentation a method for graphic comparison of the essential indices calculated for activated sludge communities.

2 MATERIALS AND METHODS

The material used for the purposes of this paper comes from Klimowicz's study of microfauna in activated sludge systems – one of the most complete description of active sludge biocenosis (Klimowicz 1970, 1972, 1973, 1983). The author presented 190 taxa of activated sludge microfauna, including: 89 ciliate species (*Ciliata*) with 34 species from the Kolkwitz-Marrson system, 52 rotifer species (*Rotatoria*) among which 19 are indicator species, 26 flagellate species (*Flagellata*) with 14 belonging to the saprobic system, 14 amoeba species (*Rhizopoda and Actinopoda*) with 7 indicator species, and 9 groups of organism of the following categories: phylum, subphylum, class etc., among which representatives of the saprobic system are absent. Among these 190 identified taxa of organisms occupying activated sludge reactors with different wastewater pollution levels, Klimowicz qualifies 74 species to the saprobic system, from which only 11 can be recognized as stenotopic ones. Information presented by Klimowicz dates back 25 years. No doubt over this time wastewater characteristics, wastewater treatment technologies and the construction of wastewater treatment devices have changed. It is obvious that the species composition of activated sludge has also changed. However, nowadays significant changes in species composition in a bioreactor of any wastewater treatment plant, in parallel with changes in wastewater quality, are observed on a weekly scale. Thus, the phenomenon was a reason for searching the biomonitoring method presented in this article.

For biomonitoring purposes the structural indices specified below were used. Their values were calculated for communities of activated sludge microfauna described by Klimowicz (1970, 1972, 1973, 1983) on the basis of species abundances and on the abundances of the following morphological-functional groups, representing easily identified taxonomic levels higher than the species level:

1. swimming ciliates (*Holotricha*)
2. attached ciliates (*Peritricha*)
3. crawling ciliates (*Hypotricha*)
4. rotifers
5. flagellates
6. amoebas
7. nematodes
8. oligochaetes
9. tardigrades
10. gastrotriches
11. arachnids
12. copepods
13. cladocers
14. turbellarians

Representatives of all groups can be found in activated sludge reactors if wastewater is treated to a significant degree and effluent BOD_5 values do not exceed $20\, gO_2\, m^{-3}$ (Curds et al. 1968, Curds & Cockburn 1970a).

As mentioned above, for biomonitoring of wastewater treatment efficiency, structural indices such as: saprobic index Is, taxon richness S, Shannon-Wiener biodiversity index H, evenness index V, MacArthur's index E and the authors' own proportionality index P were calculated. The values of the indices were determined for saprobiont communities present in wastewater with different extents (levels) of treatment. The latter were defined, according to Curds and Cockburn (1970a, b) as well as Klimowicz (1970, 1972, 1973, 1983) based on the BOD_5 range: 0–10, 11–20, 21–30 and $>30\, gO_2\, m^{-3}$.

The saprobic index Is was calculated on the basis of the following equation (Pantle & Buck 1955):

$$Is = \frac{\sum_{i=1}^{Sk} k_i s_i}{\sum_{i=1}^{Sk} k_i} \qquad (1)$$

where: S_k = number of indicator species from a saprobic system; k_i = weighted mean of individuals' number of the "i-th" indicator species; and s_i = saprobic value of the "i-th" species (from the range of 0–4).

The taxon richness S was determined by simply summing all taxa belonging to a community (Gove et al. 1994).

The Shannon-Wiener index, H was calculated according to the equation (Gove et al. 1994):

$$H = -\sum_{i=1}^{S} \Pi_i \log_2 \Pi_i \qquad (2)$$

where: S = species (or morphological-functional group) richness, number of species (or number of morphological-functional groups); and Π_i = relative abundance of the "i-th" species (or the "i-th" morphological-functional group).

The evenness index, V was calculated according to the following equation (Hurlbert 1971, Magurran 1988):

$$V = \frac{H}{H_{max}} \qquad (3)$$

where: H = observed value of the Shannon-Wiener index for the studied saprobiont community; and H_{max} = value of the Shannon-Wiener index when all taxa are equally abundant in the community.

The value of index V expressed in per cent after multiplication by 100, can be indirectly treated as an index of the "evenness deficiency" in relative abundances of taxa (Chomczyńska et al. 2005). If the value of V equals 100%, the distribution of individuals into species is the most even.

The value of H_{max} was calculated as follows (Hurlbert 1971, Magurran 1988):

$$H_{max} = \log_2 S \qquad (4)$$

where: S = number of species (or morphological-functional groups).

MacArthur's index, E was calculated according to the following equation (MacArthur 1965):

$$E = 2^H \qquad (5)$$

where: 2 = the base of the logarithm.

The value of MacArthur's index, E is the taxon richness of a community for which the observed value of H equals H_{max}.

The proportionality index, P was calculated using equation (Chomczyñska et al. 2005):

$$P = \tfrac{E}{S} \cdot 100 \qquad (6)$$

where: E = value of MacArthur's index; and S = species richness or morphological-functional group richness for studied community.

The index P can express "shortage in the taxa number" in the investigated community.

Relative abundances, necessary for the calculation of the Shannon-Wiener index and derived indices, were determined on the basis of the following equation (Gove et al. 1994):

$$\Pi_i = \frac{n_i}{n_T} \qquad (7)$$

where: n_i = number of individuals in the "i-th" species or in "i-th" morphological-functional group; and n_T = total number of individuals in a sample.

Relative abundances take values in the range 0–1; after multiplication by 100 they are expressed as percentages.

The structure of the activated sludge communities under study was presented using "radar" plots, also called "AMOEBAs" since the publication of Ten Brink's paper (Ten Brink et al. 1991, Lane & Peters 1993). "AMOEBA" is an acronym for "a general method of ecosystem description and assessment" (Ten Brink et al. 1991). This method can be used for presenting both the relative abundances of taxa and the quantitative parameters characterizing community structure. Thus, "AMOEBAs" are suitable tools for the graphic comparison of the indices determined for activated sludge communities and community structures. During preparation of "AMOEBA" plots, for better presentation of study results, values of Is, H and H_{max} indices were marked without changes and values of the remained indices and relative abundances were multiplied by 0.1.

3 RESULTS AND DISCUSSION

The study results are presented in Tables 1–6 and in Figures 1–4. Table 1 contains species richness S, numbers of indicator species from the Kolkwitz-Marrson system and abundances of morphological-functional groups present in specified extents of wastewater pollution. It can be seen that species richness S for swimming ciliates, attached ciliates, all ciliates, rotifers and all species decreases along with an increase in wastewater pollution level. Such distinct tendency is not observed for the richness of indicator species. The data presented in Table 1 shows that abundances of activated sludge microorganisms are in the range of

Table 1. Species numbers and species abundances in specified extents of wastewater quality.

Organisms	Indices	$BOD_5(gO_2\,m^{-3})$ 0–10	11–20	21–30	>30
Swimming ciliates	S	27	25	18	12
	S_k	14	14	11	8
	n_T*	1969	3712	1477	1392
Attached ciliates	S	35	29	26	18
	S_k	13	13	10	7
	n_T	5862	5284	5198	1606
Crawling ciliates	S	14	14	12	7
	S_k	4	5	5	3
	n_T	1449	1535	1263	1142
All ciliates	S	76	68	56	37
	S_k	31	32	26	18
	n_T	9280	10531	7938	4140
Rotifers	S	45	40	35	24
	S_k	16	16	16	11
	n_T	6354	2318	503	962
Flagellates	S	13	15	19	26
	S_k	5	9	12	14
	n_T	4785	4199	4463	6399
Amoebas	S	14	14	14	7
	S_k	7	7	7	4
	n_T	3914	3507	4385	4850
Metazoans**	S	8	9	1	1
	S_k	0	0	0	0
	n_T	4214	1328	1100	50
All	S	152	146	125	95
	S_k	59	64	61	47
	n_T	28547	21883	18389	16401

* n_T – total number of individuals in cm^3 of wastewater (after multiplication by 10^3 in dm^3).

** Metazoans – the rest of organisms representing *Metazoa*: nematodes, oligochaetes, tardigrades, gastrotriches, arachnids, copepods, cladocers, turbellarians.

$1.64 \cdot 10^7$–$2.8 \cdot 10^7$ individuals per dm^3 of wastewater. Abundances of all organism groups cover abundances of metazoans (in the range of $5 \cdot 10^4$–$4.2 \cdot 10^6$ individuals per dm^3) and abundances of protozoans, for which the dominance structure changes along with the wastewater pollution level. Changes of dominance structure for all organism groups are presented in Figure 1. This figure shows the dominance of ciliates in a wide range of changes of wastewater quality level from 0 to $30\,gO_2\,m^{-3}$. This pattern disappears when $BOD_5 > 30\,gO_2\,m^{-3}$ whereafter flagellates are dominants and amoebas are subdominants.

Many methods of rapid microscopic assessment of wastewater treatment efficiency are based on community dominance observations. The application of dominance structure changes of particular taxonomic

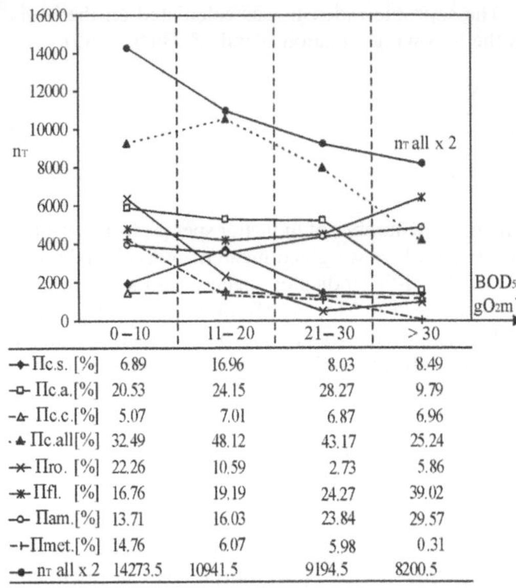

Figure 1. Dominance structure of activated sludge microfauna in specified extents of wastewater quality.

groups for wastewater quality evaluation is mentioned in many publications (Curds et al. 1968, Curds & Cockburn 1970a, b, Klimowicz 1970, 1972, 1973, Genoveva et al. 1991, Madoni 1994, Martin-Cereceda et al. 1996, Da Motta et al. 2001). Unfortunately, observations of this nature do not allow treatment process efficiency to be univocally determined. For this reason, values of the saprobic index Is for ciliates, rotifers, flagellates, amoebas and all species in specified extents of wastewater quality were calculated (Tab. 2). For ciliates, flagellates and amoebas no clear changes of the saprobic index values are observed with the changing levels of effluent pollution, indicating the change of saprobic zone. An increase in saprobic index values in parallel with an increase in wastewater pollution level is observed only for a community of all species of activated sludge. Thus, it could be said that application of the saprobic index is a good method for biomonitoring wastewater treatment efficiency just as it has been for the biomonitoring of the self-purification process of water bodies for over 100 years. However, two crucial problems remain: (i) the know-how required to classify individuals into species from a saprobic system and (ii) the knowledge of indicator species for particular saprobic zones which is required. Usually several thousands of species from many taxonomic groups are present in saprobic systems. Thus, for reliable calculations of the saprobic index the involvement of many taxonomic specialists will be inevitable.

Tables 3–5 contain biocenotic indices calculated on abundances of species representing ciliates (Tab. 3),

Table 2. Values of saprobic index Is in specified extents of wastewater quality.

	$BOD_5 (gO_2\ m^{-3})$			
Organisms	0–10	11–20	21–30	>30
Ciliates	2.3375 βm	2.3779 βm	2.5302 βm	2.531 βm
Rotifers	2.191 βm	2.7539 αm	2.635 αm	2.8151 αm
Flagellates	3.6259 p	3.6779 p	3.7543 p	3.714 p
Amoebas	3.2362 p	3.2278 p	3.3613 p	3.2909 p
All	2.5522 βm	2.6622 αm	2.9958 αm	3.2981 p

Is ε 0.0–1.0 – xenosaproby
Is ε 1.1–2.0 – oligosaproby
Is ε 2.1–2.5 – β-mesosaproby
Is ε 2.6–3.0 – α-mesosaproby
Is ε 3.1–4.0 – polysaproby

Table 3. Indices calculated on abundances of ciliate species in specified extents of wastewater quality.

		$BOD_5 (gO_2\ m^{-3})$			
Organisms	Indices	0–10	11–20	21–30	>30
Swimming ciliates	S	27	25	18	12
	H_{max}	4.7549	4.6438	4.1699	3.9849
	H	3.1695	2.543	1.963	1.4755
	V	66.7%	54.8%	47.1%	41.2%
	E	8.9973	5.8279	3.8987	2.7808
	P	33.3%	23.3%	21.7%	23.2%
Attached ciliates	S	35	29	26	18
	H_{max}	5.1293	4.8579	4.7004	4.1699
	H	3.5538	3.3403	2.9636	2.05
	V	69.3%	68.8%	63.1%	49.2%
	E	11.7436	10.1282	7.8007	4.141
	P	33.6%	34.9%	30%	23%
Crawling ciliates	S	14	14	12	7
	H_{max}	3.8073	3.8073	3.5849	2.8073
	H	1.1387	1.3217	0.5689	0.1636
	V	29.9%	34.7%	15.9%	5.8%
	E	2.2018	2.4996	1.4834	1.1201
	P	15.7%	17.9%	12.4%	16.0%
All ciliates	S	76	68	56	37
	H_{max}	6.248	6.087	5.807	5.21
	H	4.425	4.197	3.714	2.351
	V	70.8%	69.0%	64.0%	45.1%
	E	21.48	8.34	13.12	5.1
	P	28.3%	27.0%	23.4%	13.8%

rotifers, flagellates, amoebas, metazoans (Tab. 4) and all taxa (Tab. 5) encountered in activated sludge. Analysis of the data presented in the tables indicates that the values of species richness S, H index, H_{max} index and E index change the most frequently and consecutively in parallel with changes in wastewater pollution level. It is also seen that for all ciliates the values of calculated indices decrease together with an increase in wastewater pollution level (Tab. 3). Thus, ciliates as a whole are good taxa for biomonitoring purposes, and the exclusion of crawling ciliates would probably give the best results for bioindication. It is interesting that for attached and swimming ciliates more biocenotic indices decrease their values in parallel with an increase in wastewater pollution level than

Table 4. Indices calculated on abundances of rotifer, flagellate, amoeba and metazoan species in specified extents of wastewater quality.

		$BOD_5 (gO_2\ m^{-3})$			
Organisms	Indices	0–10	11–20	21–30	>30
Rotifers	S	41	40	35	24
	H_{max}	5.357	5.322	5.129	4.585
	H	3.684	3.591	4.475	2.87
	V	68.8%	67.5%	87.3%	62.6%
	E	12.85	12.05	22.24	7.31
	P	31.4%	30.1%	63.5%	30.5%
Flagellates	S	13	15	19	26
	H_{max}	3.7	3.907	4.248	4.7
	H	2.815	2.883	2.833	3.424
	V	76.1%	73.8%	66.7%	72.9%
	E	7.037	7.377	7.125	10.733
	P	53.1%	49.2%	37.5%	41.3%
Amoebas	S	14	14	14	7
	H_{max}	3.807	3.807	3.807	2.807
	H	2.926	2.845	2.725	2.538
	V	76.9%	74.7%	71.6%	90.4%
	E	7.6	7.185	6.612	5.808
	P	54.3%	51.3%	47.2%	83.0%
Metazoans	S	8	9	1	1
	H_{max}	3	3.17	0	0
	H	2.913	1.097	0	0
	V	97.1%	34.6%	–	–
	E	7.532	2.193	1	1
	P	94.2%	23.8%	100%	100%

Table 5. Indices calculated on abundances of all activated sludge species in specified extents of wastewater quality.

		$BOD_5 (gO_2\ m^{-3})$			
Organisms	Indices	0–10	11–20	21–30	>30
All species	S	152	146	125	95
	H_{max}	7.248	7.19	6.966	6.57
	H	5.743	5.453	4.945	4.806
	V	79.2%	75.8%	71.0%	73.2%
	E	53.564	43.804	30.803	27.974
	P	35.2%	30.0%	24.6%	29.5%

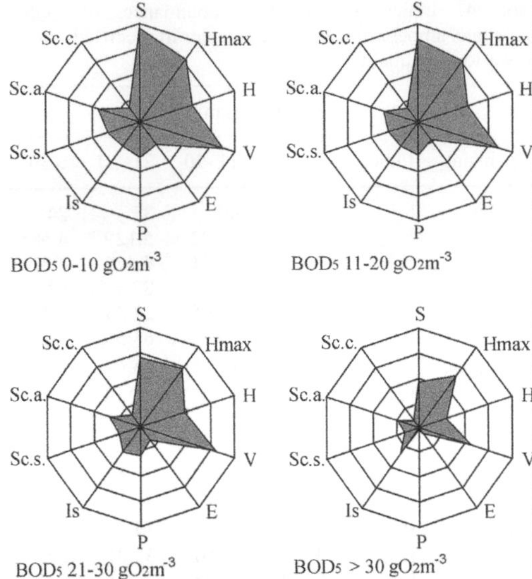

Figure 2. AMOEBA presentation of index values for the community of all ciliates in specified extents of wastewater quality.

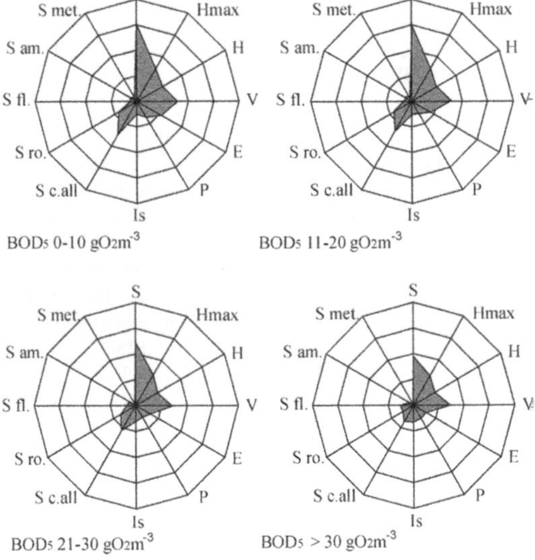

Figure 3. AMOEBA presentation of index values for the community of all species in specified extents of wastewater quality.

in the case of all saprobic species (Tabs 3, 5). Thus, biomonitoring based on identification and counting only of attached and swimming ciliates gives even better results than the application of all saprobic species of activated sludge. Results of calculations based on species abundances are also presented in AMOEBA

Table 6. Indices calculated on abundances of morphological-functional groups in specified extents of wastewater quality.

		$BOD_5(gO_2\ m^{-3})$			
Organisms	Indices	0–10	11–20	21–30	>30
Morphological-functional groups	S	14	14	7	7
	H_{max}	3.807	3.807	2.807	2.807
	H	2.853	2.655	2.447	2.209
	V	74.9%	69.7%	80.2%	78.7%
	E	7.225	6.298	5.453	4.623
	P	50.6%	45.0%	77.9%	66.1%

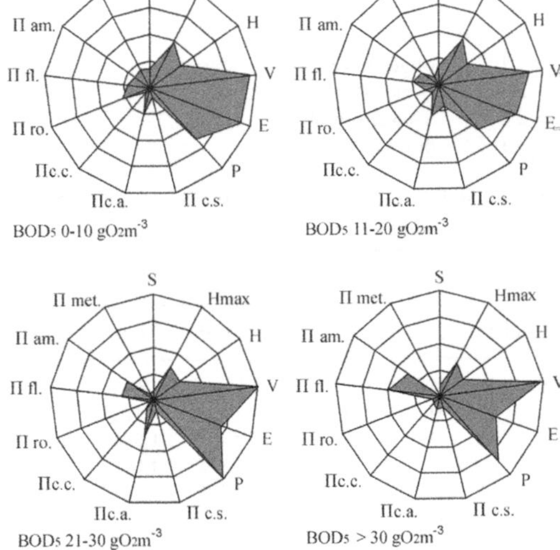

Figure 4. AMOEBA presentation of index values and relative abundances for the morphological-functional groups in specified extents of wastewater quality.

figures (Figs 2, 3). They show clear changes of index values calculated for all ciliate species and for all saprobic species listed in Klimowicz's publications (1970, 1972, 1973, 1983).

The application of the above-mentioned indices in bioindication just as in the application of the saprobic index requires the, generally difficult, classification of individuals into species. Thus, the values of biocenotic indices were calculated on the basis of easily identified morphological-functional groups. Table 6 and Figure 4 present values of these indices.

It is seen that only the values of the Shannon-Wiener index and MacArthur's index change together with changes in the extent of wastewater quality. Figure 4 shows that shapes of amoeba plots for morphological-functional groups are more variable in extents of wastewater pollution than those for ciliate species

(Fig. 2). Thus, biomonitoring based on a small number of morphological-functional groups (14) allows wastewater quality levels to be distinguished much more easily than other methods based on abundances of ciliate species (Fig. 2). For the latter, the necessary number of species is at least thirty and may often exceed sixty.

4 CONCLUSIONS

Based on the results presented herein the following conclusions are offered:

Biomonitoring based on indicator species requires considerable skill in classifying individuals into species belonging to many different higher taxa.

Biomonitoring based on measures of structure and dominance of communities with identification of species can be a problem for persons without knowledge of the taxonomic features of taxa.

Biomonitoring based on measures of structure and dominance of communities with identification of morphological-functional groups is generally easy.

Among the many saprobiont taxa present in activated sludge, species representing swimming and attached ciliates together are best for biomonitoring based on species abundances.

REFERENCES

Amaral, A.L., Babtiste, C., Pons, M.N., Nicolan, A., Lima, N., Ferreira, E.C., Mota, M. & Vivier, H. 1999. Semi automated recognition of protozoa by image analysis. *Biotechnology Techniques* 13: 111–118.

Buchs, W. 2003. Biodiversity and agri-environmental indicators-general scopes and skills with special reference to the habitat level. *Agriculture Ecosystems & Environment* 98: 35–78.

Chomczyńska, M., Malicki, J., Łagód, G. & Montusiewicz, A. 2005. Interpretation of the results of wastewater quality biomonitoring using microfaunal saprobionts. (Interpretacja wyników biomonitoringu jakości ścieków z wykorzystaniem mikrofauny saprobiontów). *Monografie Komitetu Inżynierii Środowiska PAN* 33: 347–355.

Curds, C.R., Cockburn, A. & Vandike, J.M. 1968. An experimental study of the role of ciliated protozoa in the activated-sludge process. *Water Pollution Control* 67: 312–329.

Curds, C.R. & Cockburn, A. 1970a. Protozoa in biological sewage-treatment processes – I. A survey of the protozoan fauna of British percolating filters and activated – sludge process. *Water Research* 4(3): 225–236.

Curds, C.R. & Cockburn, A. 1970b. Protozoa in biological sewage-treatment processes – II, Protozoa as indicators in the activated-sludge process. *Water Research* 4(3): 237–249.

Da Motta, M., Pons, M.N., Vivier, H., Amaral, A.L., Ferreira, E.C., Roche, N. & Mota, M. 2001. Study of protozoa population in wastewater treatment plants by image analysis. *Brazilian Journal of Chemical Engineering* 18: 103–111.

Genoveva, E., Tellez, C. & Bantista, L.M. 1991. Dynamics of Ciliated Protozoa Communities in Activated-Sludge Process. *Water Research* 25: 967–972.

Gorzel, M. & Kornijów, R. 2004. Biologiczne metody oceny jakości wód rzecznych. *Kosmos* 53(2): 183–191.

Gove, I.H., Patil, G.P., Swindel, B.F. & Taille, C. 1994. Ecological diversity and forest management. In G.P. Patil & C.R. Rao (eds), *Handbook of Statistic* 12, 409–462. Amsterdam, London, New York, Tokyo: Elsevier Science B. V., North-Holland.

Hulbert, S.H. 1971. The nonconcept of species diversity: a critique and alternative parameters. *Ecology* 48: 577–586.

Iliopoulou-Georgudaki, J., Kantzaris, V., Katharios, P., Kaspiris P., Georgiadis, Th. & Montesantou, B. 2003. An application of different bioindicators for assessing water quality: a case study in the rivers Alfeios and Pineios. *Ecological Indicators* 2: 345–360.

Jiang, J.G. & Shen, Y.F. 2005. Use of the aquatic protozoa to formulate a community biotic index for an urban water system. *Science of the Total Environment* 346: 99–111.

Klimowicz, H. 1970. Microfauna of activated sludge. I Assemblage of microfauna in laboratory models of activated sludge. *Acta Hydrobiolgica* 12: 357–376.

Klimowicz, H. 1972. Microfauna of activated sludge. II. Assemblage of microfauna in block aeration tanks. *Acta Hydrobiologica* 14: 19–36.

Klimowicz, H. 1973. Microfauna of activated sludge. III. The effect of physico-chemical factors on the occurrence of microfauna in the annual cycle. *Acta Hydrobiologica* 15: 167–188.

Klimowicz, H. 1983. *The facility of microfauna in activated sludge systems of wastewater treatment. (Znaczenie mikrofauny przy oczyszczaniu ścieków osadem czynnym)*. Warszawa: Zakład Wydawnictw Instytutu Kształtowania Środowiska.

Lane, W.E.M. & Peters, J.S. 1993. Ecological objectives for management purposes: applying the amoeba approach. *Journal of Aquatic Ecosystem Health* 2: 277–286.

Lee, S., Basu, S., Tyler, C., & Wei, I. 2004. Ciliate populations as bio-indicator at Deer Island Treatment Plant. *Advances in Environmental Research* 8: 371–378.

Łagód, G., Malicki, J., Montusiewicz, A. & Chomczyńska, M. 2004. Application of saprobionts for bioindication of wastewater quality in sewage systems. (Wykorzystanie mikrofauny saprobiontów do bioindykacji jakości ścieków w systemach kanalizacyjnych). *Archives of Environmental Protection* 30(3): 3–12.

MacArthur, R.H. 1965. Patterns of diversity. *Biological Review* 40: 510–533.

Madoni, P. 1994. A Sludge Biotic Index (SBI) for the evaluation of the biological performance of activated sludge plants based on the microfauna analysis. *Water Research* 28: 67–75.

Magurran, A. 1988. *Ecological diversity and its measurements*. London, Sydney: Croom Helm.

Martin-Cereceda, M., Serrano, S. & Guinea, A. 1996. A comparatives study of ciliated protozoa communities in activated – sludge plants. *FEMS Microbiology Ecology* 21: 267–276.

Pantle, R. & Buck, H. 1955. Biological monitoring of water bodies and the presentation of results. (Die biologische Überwachung der Seewasser und die Darstellung der Ergebnisse). *Gas und Wasserfach* 96: 604.

Puigagut, J., Salvado, H. & Garcia, J. 2005. Short-term harmful of ammonia nitrogen on activated sludge microfauna. *Water Research* 39: 4397–4404.

Ten Brink, B.J.E., Hosper, S.H. & Colijn, F. 1991. A quantitative method for description and assessment of ecosystems: the AMOEBA – approach. *Marine Pollution Bulletin* 23: 265–270.

Washington, H.G. 1984. Diversity, biotic and similarity indices: a review with special relevance to aquatic ecosystems. *Water Research* 18: 653–694.

Environmental Engineering – Pawłowski, Dudzińska & Pawłowski (eds)
© 2007 Taylor & Francis Group, London, ISBN13 978-0-415-40818-9

Different procedures for Mg and Ca role assessment during enhanced biological P removal

Ireneusz Zdybek
Institute of Environmental Protection Engineering, Wroclaw University of Technology, Wroclaw, Poland

ABSTRACT: In this paper, results of enhanced biological phosphorus removal (EBPR) analysis, from the standpoint of Mg and Ca concentration changes, are presented. During the experiment significant analytical problems in magnesium and calcium measurement were noticed. Magnesium and calcium measurements, obtained by commonly recommended and used methods showed significant inconsistencies in the results obtained. These problems are generally linked with the unfavourable influence of phosphorus. As a consequence, an alternative gravimetric method was applied. This allowed much more consistent results to be obtained for the whole process cycle. Therefore, the method for Mg and Ca measurement may also be a promising tool for assessment of the true role of both Mg and Ca during anaerobic-aerobic phosphorus removal as well as in other areas of research where P, Mg and Ca interactions are examined.

Keywords: Wastewater treatment, phosphorus removal, magnesium, calcium.

1 INTRODUCTION

The wide application in recent years of Enhanced Biological Phosphorus Removal (EBPR) processes into the practice of modern wastewater treatment systems proved an important role for the cations of certain metals in phosphate elimination mechanisms (Bashan & Bashan 2004). The transportation of phosphate anions to/from cells of Phosphate Accumulating Organisms (PAO) is connected with the transportation of cations for maintaining the electroneutral potential of the cell (Smolders et al. 1994, Mulkerrins et al. 2004). Currently, the requirements of different cations and their influence on EBPR processes seem not to be clearly understood, as can be seen by the diversity of literature on this subject. Most sources suggest that potassium K and magnesium Mg cations play the major role in EBPR mechanisms because of their observed simultaneous (with phosphate) anaerobic release and aerobic uptake (Wentzel et al. 1991, Jonsson et al. 1996, Pattarkine & Randall 1999, Liu et al. 2005, Aguado et al. 2006). Based on the measured release of P, K and Mg, the stoichiometric formula of polyphosphates accumulated by PAO's composition is proposed as $Mg_{1/3}K_{1/3}PO_3$ (Smolders et al. 1994). Thus, the shortage of these metals in wastewater can influence the phosphorus elimination efficiency which could explain the temporary deterioration of EBPR sometimes observed in full-scale Wastewater Treatment Plants (WTP), for example after long rainfalls (Brdjanovic et al. 1996, Liu et al. 2006).

Co-transport of other cations, including sodium Na and calcium Ca , is observed much less frequently. In numerous studies co-removal of calcium and phosphorus, as a result of their accumulation in sludge, has been observed (Groenestijn et al. 1988, Carlsson et al. 1997, Sosnowska 2003). This phenomenon is identified as a Biologically-Induced Chemical Phosphate Precipitation (BICPP) which can occur under some specific conditions. In particular it can take place in sequencing anaerobic-aerobic conditions as a consequence of increased P and Ca concentration and pH value in the volume of treated wastewater or in local areas of activated sludge floccules structure (Arvin 1983, Carlson et al. 1997, Maurer & Boller 1999).

Magnesium can also take part in phosphate chemical binding. The possibility of magnesium ammonium phosphate (struvite) crystallization at high pH values was proved for industrial wastewater (Suzuki et al. 2002). Ndegwa (2004), showed that depletion of Ca ions (or/and other similarly acting metal ions) is most probably one of main factors limiting removal of soluble ortho-P during aeration treatments of piggery slurry. Formation of struvite may also occur in municipal WTP operation practice, though it applies rather to sludge management units within anaerobic digestion and postdigestion processes in wastewater treatment (Battistoni et al. 2000, Quintana et al. 2005). Thus, because of the magnesium issue, the explanation of BICPP mechanisms explicitly with the influence of calcium ions is rather unconvincing.

The unclear role of different metals in the mechanisms of EBPR and BICPP processes as well as the undefined scope of interactions between these two processes and their role in a total observed phosphate removal efficiency in anaerobic-aerobic wastewater treatment conditions aroused interest in this subject. This interest was intensified by analytical problems with calcium and magnesium concentration measurement during extensive research on the influence of chloride and sulphate ions on the EBPR process. During this research, the results of comparative analyses of phosphate, calcium and magnesium concentrations were seen to strongly depend on the measurement procedure used. These differences influenced conclusions about stoichiometry, kinetics and even the role of these metals in the observed phosphate elimination mechanisms. This indicates that the analytical procedures could be an important factor in the assessment of real phosphate binding mechanisms in anaerobic-aerobic wastewater treatment process conditions.

In this paper, results of comparative analyses of the changeability of phosphate, calcium and magnesium concentrations while using three different analytical methods for calcium and magnesium are presented. The results presented herein, indicate that interpretation of the phosphorus removal mechanisms may differ, depending on the analytical method used. In opinion of the author, more attention should be given to the real reliability of concentration measurements.

2 MATERIALS AND METHODS

As mentioned above, analyses of the changeability of calcium and magnesium concentration were carried out during research on the influence of chloride and sulphate ions on the EBPR process. The data describe experiments that were spread in time and taken from different series, thus the phosphorus removal efficiency can not be directly compared. Results are mainly discussed from the viewpoint of influence of the calcium and magnesium measurement method on assessing their part in anaerobic-aerobic conditions of the lab-scale wastewater treatment process. Measurement methods were changed after realising the impossibility of obtaining a coherent mass balance for calcium and magnesium. The experimental setup presented below describes only the part of experiment utilized in this paper.

2.1 *Experimental setup*

The experiment was carried out in a laboratory scale sequencing batch reactor (SBR). The total volume of reactor was 4 dm^3. The treated wastewater volume was 1.5 dm^3, thus the sedimentation volume was 2.5 dm^3. The control algorithm was adjusted to the EBPR process requirements. After a short feed period (0.1 h) there was 2.5 h anaerobic mixing phase followed by a 2.5 h aeration phase and 0.75 h of sedimentation phase. After sedimentation, the treated wastewater was discharged from the reactor and there was a short stoppage period to the full 6 hours of the total wastewater treatment cycle. Then the next cycle was started.

To keep the designed solids retention time (SRT) at the level of 4 days, at the end of the aerobic phase, 0.25 dm^3 of mixed wastewater was discharged. The real SRT was computed by taking into account the escape of solids with treated wastewater. In the research no pH adjustment during phases was applied. All data presented herein describe bioreactors with a fully developed steady state.

2.2 *Wastewater composition*

The composition of the synthetic wastewater used is presented in Table 1. The organic carbon fraction consisted mainly of bullion and peptone in concentrations of 200 and 300 g/m^3, respectively. Additionally, for supporting the EBPR process, sodium acetate in a concentration of 212.5 g/m^3 was used which increased the COD value to 100 gO_2/m^3. The mineral solution was composed of 30 g/m^3 NH_4Cl, 2 g/m^3 $MgSO_4$, 65 g/m^3 KH_2PO_4, 160 g/m^3 NaH_2PO_4, 70 g/m^3 NaCl and 30 g/m^3 $NaHCO_3$. The presence of most microelements was ensured by the tap water composition that was the base for the preparation of the wastewater.

The use of a very high phosphorus concentration (about 50.0 gP/m^3) was due to its total removal (100% efficiency) observed in the lower concentrations in the experimental conditions. Thiourea CH_4N_2S in a concentration of 5 g/m^3 was used to achieve a total inhibition of nitrification processes, thereby eliminating possible negative effects of nitrate and nitrite on the EBPR process.

2.3 *Analytical procedures*

The range of analytical procedures used in the work included the monitoring of the main phosphorus forms (phosphates and total phosphorus), nitrogen forms (ammonia, Kjeldahl nitrogen, nitrate and nitrite), COD, calcium and magnesium concentrations. The

Table 1. Average composition of used synthetic wastewater.

Parameter	Unit	Value
pH	–	6.4
Alkalinity	$mgCaCO_3/dm^3$	175
COD	gO_2/m^3	570
BOD_5	gO_2/m^3	450
Ammonia	gN/m^3	14.0
Kjeldahl nitrogen	gN/m^3	84.0
Phosphate	gP/m^3	50.0
Total phosphorus	gP/m^3	51.0

analyses were carried out in raw and treated wastewater, and in the sludge. The changeability of this parameter's concentration was also determined during particular phases of a treatment cycle (anaerobic an aeration phase). Moreover, the extended sludge concentrations and suspended solids in wastewater after sedimentation were measured. The use of the multiparameter instrument WTW Multi 340i enabled the online control of the pH, dissolved oxygen concentration, temperature, conductivity and redox potential. The physicochemical analyses were carried out according to the standard procedures set out by EPA (Eaton et al. 1995).

As mentioned earlier, the calcium and magnesium measurements procedures were changed twice during progress of research. In the beginning, the calcium was measured by Phlapho 4 Carl Zeiss Jena flame photometer and the magnesium by AAS-30 Carl Zeiss Jena atomic absorption spectrometer. Both methods were used without sample conditioning. For these methods acronyms FP1 and AAS1, respectively for Ca and Mg will be used in following sections of this paper. The next methods used for calcium and magnesium measurements differed only in the addition of lanthanum chloride solution, which is required for the unfavourable phosphorus influence compensation on the results. These methods will be called FP2 and AAS2 further in this paper. In the final part of the research a completely different – gravimetric method (GrM) for both calcium and magnesium were used. This method is fully described by Gomółka & Gomółka (1998)/in polish/. Eaton et al. (1995) described this method only for magnesium.

2.4 *Procedures of results quality control*

The results of the phosphate, calcium and magnesium concentration measurements were checked to determine their conformity with the use of two numerical analyses.

The first one was based on comparison of P, Ca and Mg concentration phase changeability (release and uptake) observed parallel in wastewater and sludge. The magnitude of changeability for particular process phase was calculated as a difference between the final and initial concentration values. Based on the mass balance statement, magnitudes of P, Ca and Mg concentrations in wastewater and sludge should be exactly the same. Differences of computed values for wastewater and sludge (counted as the sum of changeability in wastewater and sludge) indicates possible inaccuracy of measurement methodology.

The second analysis is based on comparison of the inflow and outflow loads to/from SBR of P, Ca and Mg in the whole treatment cycle. The inflow load was produced only by raw wastewater while outflow load was discharged with treated wastewater and excess sludge. Such an analysis among others, performed for phosphorus, was an indicator of achieving fully developed steady state conditions for each stage. Under SBR steady state conditions the lack of conformity in this analysis for waste constituents that can be neither stripped nor oxidized (such as P, Ca and Mg) may also indicate possible inaccuracies of the measurement methodology.

Discrepancies and interpretation problems of results of these two analyses for calcium and magnesium were the main reason for changing the methodology. It has to be stressed, that in each stage, data were taken from only one SBR work cycle. Despite this, we feel that using the fully developed steady state for sample analyses makes the results obtained, reliable.

3 RESULTS AND DISCUSSION

3.1 *Mg and Ca determination by AAS1 and FP1 methods, respectively*

The examination of calcium and magnesium concentration changeability in EBPR was initiated using the commonly recommended and used analytical methods atomic absorption spectrometry (Mg) and flame photometry (Ca) without sample conditioning. Concentrations of P, Mg and Ca were measured at three different points of the treatment cycle: at the beginning of anaerobic phase (0 h), the end of the anaerobic phase (2.5 h) and at the end of the aerobic phase (5 h). These are presented in Figure 1 for both wastewater and sludge.

Data collected on Figure 1 were the next subject of assessment using the procedures presented earlier. Results of changeability analysis are presented in Table 2, while results of loads analysis are gathered in Table 3.

The phosphorus concentration variability (Fig. 1a) shows a pattern typical for EBPR. In the anaerobic phase, P release from sludge is equal to its concentration increase in wastewater (Tab. 2). Also, aerobic P uptake has a similar value in sludge and wastewater. The consistency of these results is also confirmed by the mass balance (Tab. 3). The phosphorus inflow load is practically the same as the outflow load.

Magnesium shows the same pattern as phosphorus (Fig. 1b). It reveals the importance of Mg ions in the P removal mechanism as it is commonly considered. The comparative molar ratios of Mg/P concentration changeability (molMg/molP) computed for wastewater were 0.29 for anaerobic release and 0.26 for aerobic uptake which clearly correspond with the average formula of polyphosphates accumulating by PAO's – $Mg_{1/3}K_{1/3}PO_3$ (Smolders et al. 1994). The same coefficients estimated for sludge analysis were 0.42 and 0.29, respectively for anaerobic and aerobic

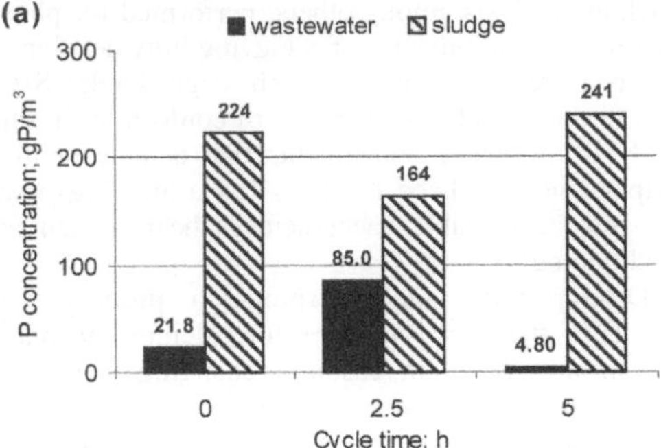

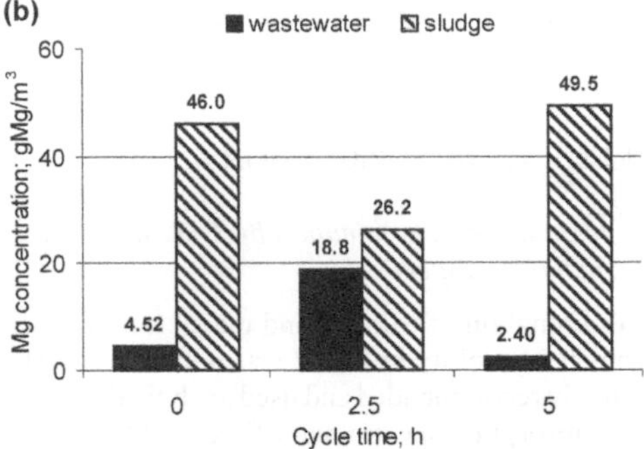

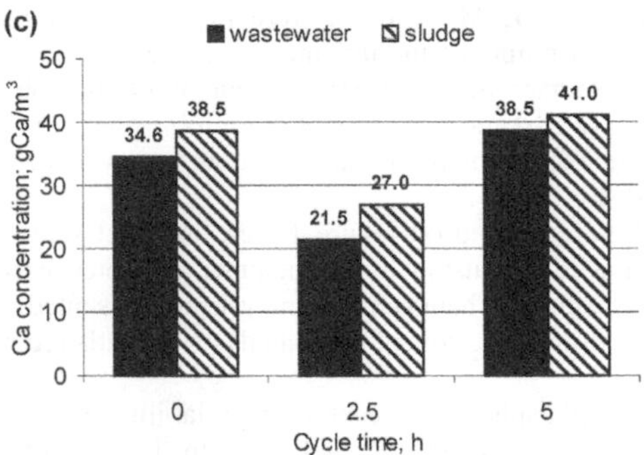

Figure 1. Constituent concentrations in wastewater and sludge. (a) phosphorus; (b) magnesium by AAS1; (c) calcium by FP1.

phases. The inconsistencies in these coefficients raises questions about the correctness of results. Moreover, a detailed mass balance (Tab. 3) of Mg shows a discrepancy of 5.4 g Mg/cycle which was more than 40%.

Results obtained for calcium were difficult to interpret. First of all, calcium concentration values in raw wastewater (about 32 g Ca/m^3) were far lower than expected. In tap water, the basis for wastewater preparation, the Ca content, examined by different methods, was about 80.0 g Ca/m^3. Next, the changeability of Ca concentration observed for wastewater was completely incoherent with regard to sludge

Table 2. Comparison of constituents phase changeability during research stage with the use of FP1 and AAS1 methods.

Constituent	Changeability	Anaerobic phase	Aerobic phase
P	Wastewater	63.2	−80.2
g/m^3	Sludge	−60.1	77.0
	Σ	3.1	−3.2
Mg	Wastewater	14.3	−16.4
g/m^3	Sludge	−19.8	23.3
	Σ	−5.5	6.9
Ca	Wastewater	−13.1	17.0
g/m^3	Sludge	−11.5	14.0
	Σ	−24.6	31.0

Table 3. Constituents mass balance during research stage with the use of FP1 and AAS1 methods.

Constituent	Inflow	Outflow mg/cycle	Difference
P	75.0	74.9	−0.1
Mg	12.1	17.5	5.4
Ca	42.0	69.3	27.3

changeability (Fig. 1c). Decrease of calcium content in sludge, in the anaerobic phase, should bring about the increase in its concentration in wastewater, while it was different (Tab. 2). The same, during the aerobic phase, increase in calcium content in sludge is not reflected in wastewater analysis. The observed pattern in wastewater analysis could be explained by chemical binding as a consequence of phosphate concentration increase due to anaerobic release (Maurer et al. 1999, Pattarkine & Randall 1999). Next, the Ca concentration increase, in the aerobic phase, to the initial value (characteristic for raw wastewater), suggested the rather unstable nature of this phenomenon, as there was also no confirmed calcium removal from wastewater. All these results obtained for wastewater are contrary to the fact that there was noticeable calcium content in a sludge (41.0 g Ca/m^3) which was more than 2% of TSS.

A significant discrepancy between the calcium and magnesium mass balances for each reactor as well as the inconsistent results of wastewater and sludge changeability analysis revealed limitations of the FP1 and AAS1 methods for the examination of this metal concentration in wastewater. According to the literature (Eaton et al. 1995) phosphate presence in analyzed probes can disrupt the quantitative examination of calcium and magnesium in these analytical methods. To compensate for this influence, the addition of a lanthanum chloride solution is recommended. Such a modification was conducted in the next experimental series and its results are presented below.

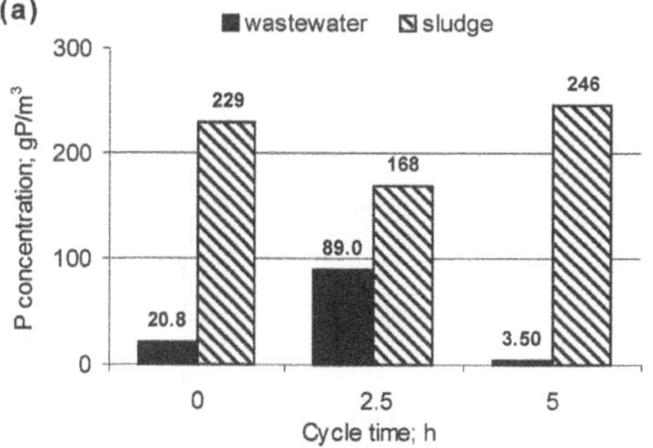

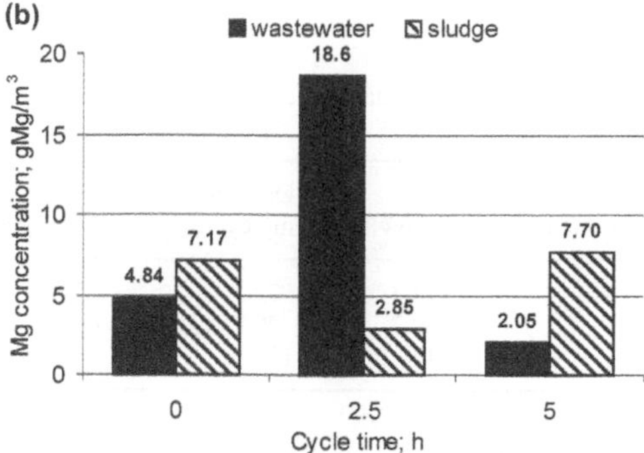

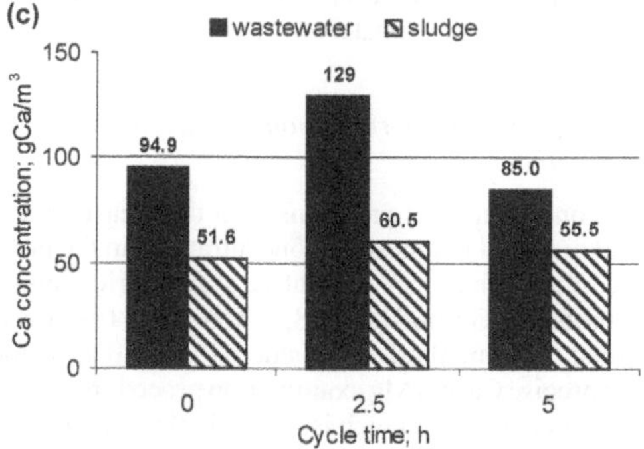

Figure 2. Constituents concentrations in wastewater and sludge. (a) phosphorous; (b) magnesium by AAS2; (c) calcium by FP2.

3.2 *Mg and Ca determination by AAS2 and FP2 methods, respectively*

The results of the phosphate, magnesium and calcium analyses, for the second stage of this experiment, for wastewater and sludge are presented on Figure 2.

As above, data collected in Figure 2 were the subject of assessment using the same control procedures. Results of changeability analysis are presented in Table 4, while results of loads analysis are gathered in Table 5.

Table 4. Comparison of constituent phase changeability during research stage with the use of FP2 and AAS2 methods.

Constituent	Changeability	Anaerobic phase	Aerobic phase
P	Wastewater	68.2	−85.5
g/m^3	Sludge	−60.9	78.0
	Σ	7.3	−7.5
Mg	Wastewater	13.8	−16.6
g/m^3	Sludge	−4.3	4.9
	Σ	9.4	−11.7
Ca	Wastewater	34.1	−44.0
g/m^3	Sludge	8.9	−5.0
	Σ	43.0	−49.0

Table 5. Constituents mass balance during research stage with the use of FP2 and AAS2 methods.

Constituent	Inflow	Outflow mg/cycle	Difference
P	74.3	73.5	−0.8
Mg	14.3	5.2	−9.1
Ca	167	143	−24

Phosphorus concentration changes (Fig. 2a) are similar to those obtained during the first stage of the experiment which shows a similar efficiency of the EBPR.

The magnesium concentration changes (Fig. 2b) generally correspond with results of phosphate analysis. There is also anaerobic release and aerobic uptake of Mg. However, there were quite significant differences in the extents of Mg release and uptake noticed during sludge and wastewater analysis (Tab. 4). Also the mass balance analysis for Mg reveal very high inconsistency (Tab. 5). Survey of analyses for magnesium indicate that the addition of lanthanum chloride solution does not improve the coherence of results obtained by the AAS method.

Sample conditioning with lanthanum chloride solution caused an increase in the measured Ca content in raw wastewater. Obtained values were on average 115 g Ca/m^3 which seems to be rather high when compared to the tap water used for wastewater composition (80.0 g Ca/m^3). As shown in Figure 2c, modification of the analytical procedure revealed completely different calcium concentration changes in wastewater during the whole EBPR cycle. Ca concentration increases in the anaerobic phase and decreases in the aeration period. This pattern is similar to observed phosphate concentration changes; this indicates the possible importance of calcium in the observed phosphorus removal mechanisms. Nevertheless, there was still some discrepancy in Ca examination in wastewater and sludge. As shown in Table 4, in both phases changes in

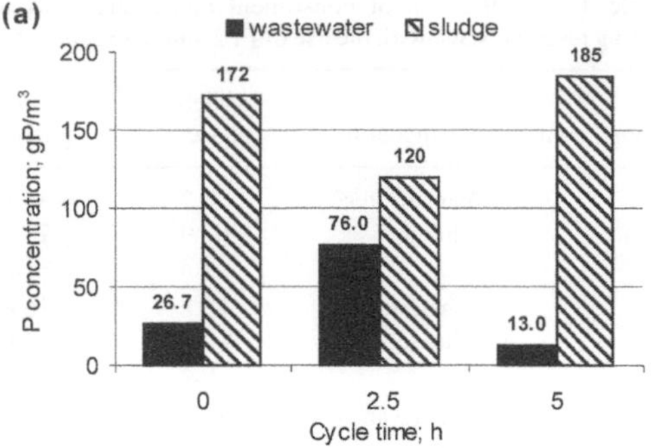

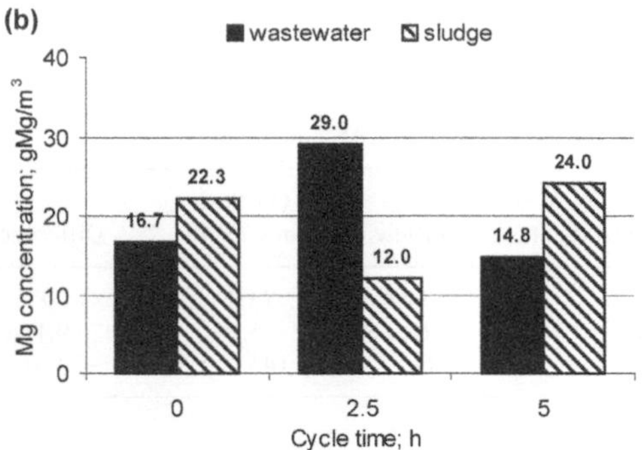

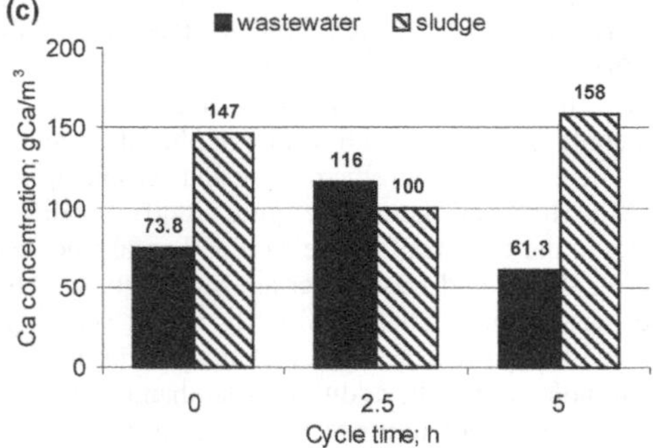

Figure 3. Constituents concentrations in wastewater and sludge. (a) phosphorous; (b) magnesium by GrM; (c) calcium by GrM.

the Ca wastewater concentrations and in sludge have the same direction while they should be opposite. The reliability of these measurements is also questioned by the low mass balance consistency (Tab. 5).

Further problems of interpretation of the calcium and magnesium results obtained led to experiments with different lanthanum chloride doses (results not presented in this paper). They generally showed increasing values of concentration of both metals with higher lanthanum doses. Thus, the procedure leads to the serious possibility of erroneous results. This situation induced us to search for an alternative method for Ca and Mg examination, which could eliminate the interference by phosphate.

Table 6. Comparison of constituents phase changeability during research stage with the use of GrM.

Constituent	Changeability	Anaerobic phase	Aerobic phase
P	Wastewater	49.3	−63.0
g/m^3	Sludge	−51.8	65.0
	Σ	−2.5	2.0
Mg	Wastewater	12.3	−14.2
g/m^3	Sludge	−10.3	12.0
	Σ	2.0	−2.2
Ca	Wastewater	42.2	−54.7
g/m^3	Sludge	−46.7	58.0
	Σ	−4.5	3.3

Table 7. Constituents mass balance during research stage with the use of GrM.

Constituent	Inflow	Outflow mg/cycle	Difference
P	74.3	72.3	−2.0
Mg	29.7	29.1	−0.6
Ca	142	137	−5

3.3 *Mg and Ca determination by gravimetric method*

As a consequence of problems with the measurement of calcium and magnesium concentrations in the presence of phosphate, an alternative gravimetric method (Gomółka & Gomółka. 1998, Eaton et al. 1995) was applied. This method is considered to be one of the most precise Ca and Mg examination procedures, but it is time-consuming, which is probably the main reason for its rare application. The advantage of this method is the low pH value of calcium oxalate precipitation allowing for the elimination of the negative influence of phosphate presence. Magnesium examination is performed by precipitation of magnesium ammonium phosphate at high pH after preliminary calcium removal from the sample to be analyzed.

Comparative results of phosphate magnesium and calcium concentrations in wastewater and sludge, obtained by these gravimetric methods, are illustrated in Figure 3.

Data collected on Figure 3 were the subject of assessment with the use of the same control procedures. Results are presented in Tables 6, 7.

As illustrated in Figure 3, the gravimetric method shows the identical character of P, Mg and Ca concentration changes during anaerobic and aerobic phases.

In wastewater examination as well as in sludge analysis simultaneous anaerobic release from sludge to solution and aerobic uptake from solution to sludge, for phosphate, magnesium and calcium ions, were observed. It definitely proves an important role of both metals in phosphate removal mechanisms. Also the extents of Mg and Ca release and uptake (Tab. 6), observed in parallel sludge and wastewater analysis, are very similar and demonstrate the essential correctness of this analytical procedure. Clearly, differences between sludge and wastewater metal contents changes are much smaller (Tab. 6) when compared to those obtained with the two other measurement methods (Tab. 2, 4). The accuracy of the gravimetric method is additionally confirmed by the very good mass balance (Tab. 7) for which the error obtained is the smallest of all analytical methods used.

4 SUMMARY AND CONCLUSIONS

All the experimental series presented are similar from the perspective of phosphorus removal. The use of different measurement methods for magnesium and calcium causes anomalous patterns of concentrations for these elements. Using atomic absorption spectrometry for magnesium and flame photometry for calcium, both without and with sample conditioning, causes inconsistencies in the obtained concentration values. In both cases, higher inconsistencies were observed for calcium. Problems with Mg and Ca measurements in FP and AAS are connected with high phosphorus concentrations. As shown, sample conditioning by the addition of lanthanum chloride solution as recommended does not improve the quality of the results. High instantaneous values of phosphorus concentration are typical for the anaerobic stage of the EBPR process, so Mg and Ca measurements problems become significant.

Especially for calcium, inconsistencies in the results obtained with both FP1 and FP2 methods are very high. Using sample conditioning may even change observed direction of the phasic changes. Incorrectness of results (identified by their inconsistencies) makes it impossible to interpret the real role of calcium in phosphorus removal during anaerobic-aerobic treatment. It may cause problems with the assessment of range, kinetics, stoichiometry and even the occurrence and nature of the real phosphorus elimination mechanisms: Biologically Induced Chemical Phosphate Precipitation (BICPP) and Enhanced Biological Phosphorus Removal (EBPR).

It seems to be reasonable to presume that the proposed, gravimetric method gives reliable results of magnesium and calcium analyses even in samples with high phosphorus concentration. Both analyses for this method are highly consistent. Thus, using gravimetric method for Mg and Ca measurement may be a good choice for assessing the real role of both Mg and Ca during the anaerobic-aerobic phosphorus removal process.

The problems arising with the reliable assessment of Mg and Ca role in EBPR process also raise questions about the possibility of application of these methods in other research areas such as rock, soils, sludge or plants and animals tissues examination – where one can expect significant amounts of phosphorus in samples.

The present work may thus give some important clues, not only for the verification of the real role of Ca and Mg in mechanisms of phosphorus removal from sludge but also for further research concerning Ca, Mg and P interactions.

REFERENCES

Aguado D., Montoya T., Ferrer J. & Seco A. 2006. Relating ion concentration variation to conductivity variation in sequencing batch reactor operated for enhanced biological phosphorus removal. *Environmental Modelling & Software* 21: 845–851

Arvin E. 1983. Observations supporting phosphate removal by biologically mediated chemical precipitation – a review. *Wat. Sci. Tech.* 15: 43–63

Bashan L.E. & Bashan Y. 2004. Recent advances in removing phosphorus from wastewater and its future use as fertilizer (1997–2003). *Water Research* 38: 4222–4246

Battistoni P., Pavan P., Prisiciandaro M. & Cecchi F. 2000. Struvite crystallization: a feasible and reliable way to fix phosphorus in anaerobic supernatants. *Wat. Res.* 34(11): 3033–3041

Brdjanovic D., Hooijmans C.M., van Loosdrecht M.C.M., Alaerts G.J. & Heijnen J.J. 1996. The dynamic effects of potassium limitation on biological phosphorus removal. *Wat. Res.* 30(10): 2323–2328

Carlson H., Aspegren H., Lee N. & Hilmer A. 1997. Calcium phosphate precipitation in biological phosphorus removal systems. *Wat. Res.* 31(5): 1047–1055

Eaton A.D., Clesceri L.S. & Greenberg A.E. 1995. *Standard methods for the Examination of Water and Wastewater – 19th Edition.* American Public Health Association. Washington

Gomółka B. & Gomółka E. 1998. *Ćwiczenia laboratoryjne z chemii wody*. Oficyna Wydawnicza Politechniki Wrocławskiej. Wrocław 1998 /in Polish/

Groenestijn van J.W., Vlekke G.J.F.M., Anink D.M.E., Deinema M.H. & Zehnder A.J.B. 1988. Role of cation in accumulation and release of phosphate by Acinetobacter strain 210A. *Appl. Environ. Microbiol.* 54: 2894–2901

Jonsson K., Johansson P., Christensson M., Lee N., Lie E. & Welander T. 1996. Operational factors affecting enhanced biological phosphorus removal at the waste water treatment plant in Helsingborg, Sweden. *Wat. Sci. Tech.* 34(1–2): 67–74

Liu Y., Lin Y.M. & Tay J.H. 2005. The elemental composition of P-accumulating microbial granules developed in sequencing batch reactors. *Process Biochemistry* 40: 3258–3262

Maurer M., Abramovich D., Siegrist H. & Gujer W. 1999. Kinetics of biologically induced phosphorus precipitation in wastewater treatment. *Wat. Res.* 33(1): 147–163

Maurer M. & Boller M. 1999. Modelling of phosphorus precipitation in wastewater treatment plants with enhanced biological phosphorus removal. *Wat. Sci. Tech.* 39(2): 159–165

Mulkerrins D., Dobson A.D.W. & Colleran E. 2004. Parameters affecting biological phosphate removal from wastewaters – review. *Environmental International* 30: 249–259

Ndegwa P.M. 2004. Limitation of orthophosphate removal during aerobic batch treatment of piggery slurry. *Biosystems Engineering* 87(2): 201–208

Pattarkine V.M. & Randall C.W. 1999. The requirement of metal cations for enhanced biological phosphorus removal by activated sludge. *Wat. Sci. Tech.* 40(2). 159–165

Quintana M., Sanchez E., Colmenarejo M.F., Barrera J., Garcia G. & Borja R. 2005. Kinetics of phosphorus removal and struvite formation by the utilization of by-product of magnesium oxide production. *Chemical Engineering Journal* 111: 45–52

Smolders G.J.F., van der Meij J. & van Loosdrecht M.C.M. 1994. Model of anaerobic metabolism of the biological phosphorus removal process: stoichiometry and pH influence. *Biotechnology and Bioengineering* 43: 461–470

Sosnowska B. 2003. *Kinetyka i efektywność biologicznego oczyszczania ścieków w beztlenowo-anoksycznym reaktorze SBR, z wydzieloną nitryfikacją na złożu biologicznym.* Ph.D. thesis. Institute of Environmental Protection Engineering. Wrocław University of Technology. Poland/ in Polish/

Suzuki K., Tanaka Y., Osada T. & Waki M. 2002. Removal of phosphate, magnesium and calcium from swine wastewater through crystallization enhanced by aeration. *Wat. Res.*(36): 2991–2998

Wentzel M.C., Lotter L.H., Ecama G.A., Loewenthal R.E. & Marais G. 1991. Evaluation of biochemical models for biological excess phosphorus removal. *Wat. Sci. Tech.* 23(4–6): 567–576

Environmental Engineering – Pawłowski, Dudzińska & Pawłowski (eds)
 ISBN13 978-0-415-40818-9

Present state and future of wetland technology in environmental protection in Poland

Hanna Obarska-Pempkowiak, Magdalena Gajewska & Ewa Wojciechowska
Faculty of Civil and Environmental Engineering, Gdańsk University of Technology

ABSTRACT: Polish experience with implementation and operation of constructed wetland systems are presented. Wetland technology can be applied for the removal of contaminants from aerial sources, point sources and for utilization of sewage sludge.

From 1994–2002 monitoring of three different facilities for water protection was carried out. Substantial improvement of water quality after implementation of the facilities was observed.

From 1995–2004 measurements of contaminant removal in twenty two constructed wetlands were carried out. The average efficiency of BOD_5 removal in individual plants was 76.4% and 86.7% for hybrid wetland systems. The efficiencies of N_{tot} removal, however, changed from 31.0% for individual plants to 56.5–75.6% for local plants and up to 90.0% for hybrid systems.

In Northern Poland three macrophyte facilities were constructed: for sewage sludge dewatering. The investigations proved that utilisation in reed beds results in considerable dewatering (up to 40% dry mass) and partial stabilisation of the sludge.

Keywords: Constructed wetlands, water protection, sewage treatment, sludge dewatering.

1 INTRODUCTION

Unfortunately, Poland is a region rather poor in water resources thus water protection is the most important issue and sewage treatment requires the most effective technologies. According to the postulate of Agenda 21 technologies should be used that are both inexpensive and not harmful to the environment, as well as flexible, so that they can be easily adapted to local conditions.

In Poland in the regions with dispersed households – villages and recreation sites, municipal sewage has been often treated insufficiently carefully. These areas usually are not equipped with sewer systems. In recent years pollution of water with sewage from village areas has become a serious problem. In 2002 only 14.0% of the rural population was using wastewater treatment plants, while in the towns it was 83.2%. The share of the rural population using sewer systems almost doubled in the years between 1995 and 2002 (it increased from 5.9 to 12.3%). These numbers indicate that the rest of the rural community use household and local wastewater treatment plants, often septic tanks and percolating drainage systems, that do not provide effective sewage treatment. These solutions cannot be accepted in the long-term, since mechanically treated sewage should not be discharged directly to the receivers. Village areas in Poland often have small surface water resources with low capacity for receiving contaminant loads.

Application of constructed wetlands is being considered as a possible solution to these problems. These systems simulate the hydraulic and habitat conditions of natural marsh ecosystems. Organic substances, nutrients as well as heavy metals and organic micropollutants are removed as a result of natural processes, supported by heterotrophic microorganisms and aquatic and hydrophite plants grown in specially designed soil filters or ponds. Constructed wetlands are used as the II or the III stage of sewage treatment in the village areas (either for individual households or for whole villages with up to 2000 pe – person equivalent) and for dewatering and stabilization of sewage sludge from primary stage of treatment. They can also be designed as surface flow systems for removal of contamination from non-point sources or used for landfill leachate or mining drainage treatment. Constructed wetlands are considered to be an effective and economical solution to the problem of sewage treatment in the rural areas, where construction of sewer systems and transport of sewage to large conventional WWTPs is too expensive. Easy maintenance and operation, low energy consumption, tolerance to fluctuating sewage inflow and high potential of fitting these facilities in the landscape make constructed wetlands an attractive treatment method for rural areas. In addition to this, constructed wetlands, in contrast to conventional treatment plants, do not produce sewage sludge.

In Poland the first hydrophite plants were constructed in the 1980s and there are now >150 in operation. Unfortunately, most of them do not provide sufficient nitrogen removal.

The entry of our country to the EU necessitated alteration of the Regulation of Environmental Protection. Following an EU trend, it was calculated that the load of pollutants discharged from small WWTPs is low anyway. Thus, the permissible pollutant concentrations were increased. The tenets of the former regulation, which did not differentiate the WWTP size, were difficult to fulfil in the small plants. The new guidelines, implemented in 2004, create an opportunity for further intensive development of constructed wetlands in the rural areas.

2 METHODS

In the period from 1995 to 2004 measurements of contaminant removal in eleven individual pilot households and four local community plants: four equipped with one-stage horizontal flow bed and three with hybrid reed wetland systems (composed of HF-CW and VF-CW filters), were carried out.

In the period from 1994 to 2002 the measurements of contamination removal and influence on water quality of three facilities built for surface water protection were carried out.

In the case of water and sewage, averaged samples of influent (sewage after the mechanical stage and before the constructed wetland), effluent (sewage after the constructed wetland) and after a subsequent stage of biological treatment in all systems were analysed. Samples were collected once or twice a month. In order to evaluate the efficiency of organic matter and nitrogen removal in subsequent stages of treatment the following parameters were measured: organic matter (COD and BOD_5), total nitrogen, ammonium nitrogen ($N-NH_4^+$), nitrate, nitrite according to Polish standards and guidelines given by Hermanowicz et al. (2000).

Measurements of the sludge were recorded in Darżlubie for six years. The sludge bed was divided into four sections along the symmetry axes. The samples of sludge were collected from four sampling points located in the centre of each section. An average sample was obtained by mixing equal volumes of collected material. The samples were collected at six week intervals, following the frequency of sludge loading. In Swarzewo the thickness of the sludge layer was measured after a three year period. The average samples of non-stratified sludge were collected once a month during the period of investigation. In Zambrów the layer of sludge discharged to the bed and remaining in the bed was measured only once every three month in the three year period (1995–1998).

The following parameters were determined in the sludge samples: moisture, organic matter, total nitrogen and total phosphorus content.

All analyses were carried out using standard methods. The removal efficiency was calculated as a quotient of difference in concentration of contaminants in influent (C_o) and effluent (C) after subsequent stages of constructed wetland and concentration in influent (C_o), $\eta = (C_o - C)/C_o$.

3 RESULT AND DISCUSSIONS

3.1 *Removal of contamination from surface sources*

At the beginning of the 1990s, in order to improve the water quality of the Bay of Gdansk, treatment of the most polluted streams (Rynaszewski Stream in Oliwa – the main tributary of the Jelitkowski Stream and Swelina Strem in Sopot) inflowing to the Bay was undertaken. At the zoo in Oliwa, along the bed of the Rynaszewski Stream ($Q_z \cong 70$ l/s), the main inflow of the Jelitkowski Stream ($Q_z \cong 250$ l/s), a set of hydrophite facilities was constructed for decreasing the load of organic nitrogen and elimination of faecal coli bacteria. Two sand filters, a vegetation filter with an area of $3100\,m^2$ and five willow buffer zones of total area of $6650\,m^2$ were built (Obarska-Pempkowiak et al. 2001).

The willow buffer zones were located at the edges of the ponds and along the stream bed. The zones were built in the natural stream bed, since it was the most easily accessible area and the costs were minimized. In order to provide better retention of rainwater, the zones are cross-cut by furrows and antislopes (Fig. 1), parallel to the flow.

The buffer zones captured and removed the pollutants outflowing from the animal cages and sheds.

The mean load of contaminants removed by the wetland system at the zoo was as follows: 45,8 kg/d Ntot (36,8 kg/d: Norg and 9 kg/d $N-NH_4^+$), 2,7 kg/d PO_4^{-3} and 31,5 kg/d COD_{Mn}(assuming that the average stream flow was 70 l/s) (Obarska-Pempkowiak et al., 2001).

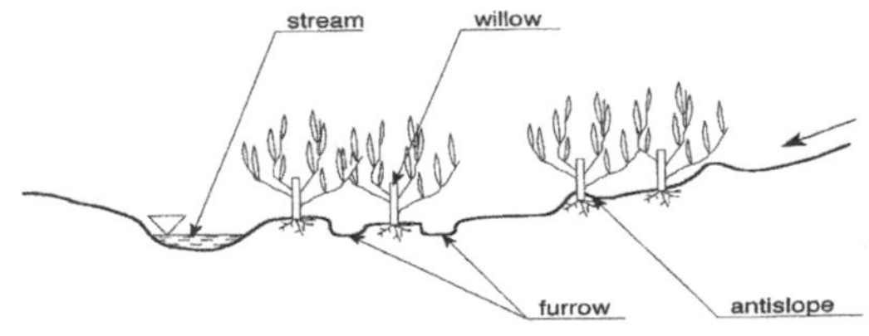

Figure 1. A schematic of the willow buffer zone.

In 1994, a sedimentation – retention reservoir ($500\,m^3$ volume; active layer depth 1 m) and a vegetation sand filter ($870\,m^3$ volume; the active layer depth 1 m; retention time 2 h at the flow $Q_z = 30\,l/s$) were constructed at Swelina Stream. The facilities were financed by the local authorities of Sopot in order to protect the water of the Bay of Gdańsk against nutrient compounds and microbiological contaminations of faecal origin carried by the Swelina Stream.

In order to protect the surface water intake at the Goszyn Lake (Straszyn Reservoir), a wetland system was constructed at the Stream receiving the waters from Bielkowo village, which directly inflows to the Lake. The wetland system was designed as a reservoir surrounded with ground slopes, consolidated with fascine and turf. The total area of the reservoir is $6200\,m^2$; the volume is ca. $5000\,m^3$. Inside the reservoir a set of filtration dykes was constructed. The system consists of two sub-units:

- wet unit (pond) filled with water all the time (retention time 24 h and water flow 32 l/s)
- dry unit designed for storm water (maximal flow 640 l/s and retention time 0.5 h)

The results proved that the hydrophite system ensured reduction in the concentrations of total suspended solids, total nitrogen and total phosphorus (Table 1). The BOD_5 and COD removal effectiveness was lower (Obarska-Pempkowiak et al., 2002, Obarska-Pempkowiak et. al., 2004).

It was found that mineralization of organic substances took place before the inflow to the hydrophite system. The lower concentration of contaminants in the vegetation season confirms the potential of hydrophite plants for contaminant removal (Obarska-Pempkowiak et al., 2002). Surface flow CWs inhabited with several macrophytes species are capable of heavy metals removal from storm water runoff. The highest contents of heavy metals were found in ground parts of plants. In case of *Glyceria maxima* the roots and the rhizomes seem to contain the maximum amounts of heavy metals (Wojciechowska et.al., 2004).

Table 1. Efficiency of contamination removal in years 2000–2001 in Bielkowo [%].

	η_d		η_w		η_T	
Parameter	2000	2001	2000	2001	2000	2001
s.s	39.3	66.6	81.4	95.6	20.4	62.2
BOD_5	35.3	34.5	75.3	82.3	10.0	16.1
COD	33.0	27.2	76.0	78.3	8.4	5.9
N_{Tot}	34.4	47.5	75.6	86.9	10.0	31.3
P_{Tot}	38.8	38.7	77.4	73.8	16.2	12.5
O_2	26.4	33.2	73.8	78.3	24.3	11.5

η_d – efficiency removal in dray units, η_w – efficiency removal in wet units, η_T – efficiency removal of the entire system.

3.2 *The removal of pollutants from point sources*

Most of CWs in Poland were built in the 1990s as one-stage HF-CW beds for II stage treatment of domestic sewage. The results of monitoring of individual household plants (near Ciechanów, Lublin and Ostrołęka) and local plants (in the Gorzów Province) indicated that HF-CW beds working as the II stage treatment provided effective removal of organic substances (BOD_5 and COD_{Cr}) as well as suspended solids. The BOD_5 removal effectiveness varied from 45.6 to 99.1% (76.4% on average) in the range of loading from 15 to 115 kg/ha·d (Fig. 2).

Based on removed loads of organic matter (BOD_5) and total nitrogen (about 70 measurements) and adequate cross-section the rates of contamination removal were calculated from the analysed facilities. Table 2 presents the removal of organic matter (BOD_5) and total nitrogen for $1\,m^2$ of the bed for the analysed HF-CW plants. The removal of organic matter from $1\,m^2$ in varies very much from only $0.1\,g\,m^{-2}day^{-1}$ to as much as $10.2\,g\,m^{-2}day^{-1}$. Such big differences in the rate of contamination removal in one stage CWs proved that their effectiveness depended mainly on proper design and operation.

The effectiveness of the total nitrogen removal in these plants was lower and varied from 22.4 to 84.2% (44.5% on average) in the range of loading from 2.5 to $37.0\,kg\,ha^{-1}\cdot day^{-1}$ (Fig. 2). The removal of total nitrogen from $1\,m^2$ of analysed HF-CW plants

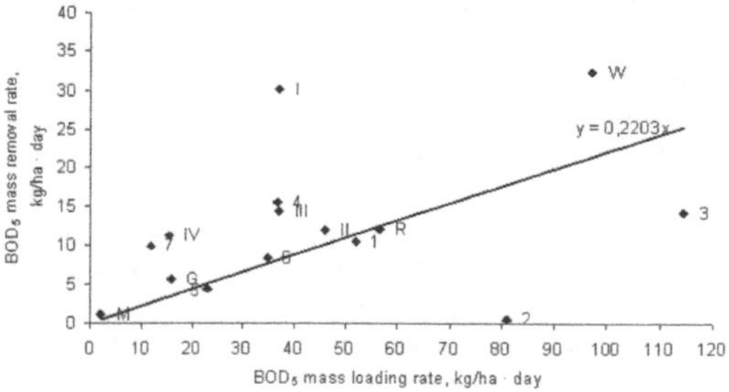

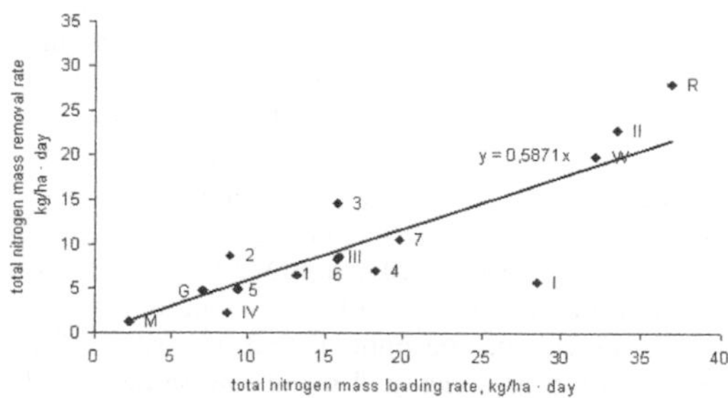

Figure 2. The load of BOD_5 and total nitrogen at the inflow and the outflow of HF-CW systems. The individual plants: I – IV plants located near Ciechanów, 1 – 4 plant located near Lublin; 5–7 plants located near Ostrołęka. The local plants: G, W, M, R located near Gorzów Wielkopolski.

Table 2. The average BOD_5 and total nitrogen removal with standard deviation from 1 m^2 of analysed HF-CWs.

Local plants	Individual plants		
Near Gorzów Wielkopolski	Near Ciechanów	Near Lublin	Near Ostrołęka
BOD_5 g/m^2d			
G – 6,5 ± 1.9	I – 0,7 ± 0.2	1 – 4,2 ± 1.0	5 – 1,9 ± 0.4
W – 1,0 ± 0.25	II – 3,4 ± 0.8	2 – 8,2 ± 2.1	6 – 2,7 ± 0.6
M – 0,1 ± 0.05	III – 2,3 ± 0.9	3 – 10,8 ± 2.6	7 – 0,2 ± 0.04
R – 4,5 ± 1.1	IV – 0,4 ± 0.1	4 – 2,1 ± 0.6	
N_{tot} g/m^2d			
G – 0.2 ± 0.07	I – 2.2 ± 0.4	1 – 0.6 ± 0.2	5 – 0.8 ± 0.2
W – 1.2 ± 0.2	II – 1.1 ± 0.3	2 – 0.2 ± 0.05	6 – 0.4 ± 0.1
M – 0.1 ± 0.05	III – 0.7 ± 0.2	3 – 0.1 ± 0.06	7 – 0.9 ± 0.3
R – 0.9 ± 0.08	IV – 0.6 ± 0.1	4 – 1.1 ± 0.3	

varied from 0.1 to 2.2 g $Nm^{-2}day^{-1}$; thus the scope is much narrower than the removal of organic matter. Such facilities are not hence predestined to nitrogen removal even in very good conditions (Table 2).

Many of the analysed plants did not work properly. The main reason for this was the inadequate operation of the septic tanks, resulting in the inflow of grease and suspended solids to the hydrophite beds, which caused the decrease of hydraulic conductivity of the beds. In such cases, due to clogging, the subsurface flow facility started to work as a surface one. The unit area of these plants was: 2.7 m^2 pe^{-1} in Wawrów, 3.0 m^2 pe^{-1} in Gralewo, 4.0 m^2 pe^{-1} in Małyszyno and 6.1 m^2 pe^{-1} in Rokitno.

Up to the present vertical flow beds (VF-CW) have not been used in the II stage treatment in Poland. However, pilot studies were performed by Soroko (2001) and Kowalik et al. (2004). The average efficiencies of BOD_5 and N_{tot} removal in Soroko's experiments were 97.4% and 41.6%, respectively. According to Kowalik et al. (2004), the efficiencies of BOD_5 and N_{tot} removal in the II stage of treatment were equal to 89.1% and 76.1%, respectively, and in the III stage of treatment – 93.8% and 79.1%. Comparing these reports to the results obtained by Brix et al. (2002) we see that VF-CW facility can provide 99.9% efficiency of the total nitrogen removal. In France two-stage facilities composed only of VF-CW beds treating raw sewage have been working for over 20 years (Molle et al., 2004). Based on the maintenance of the systems it was possible to determine the optimal unit area of VF-CW beds working in series and using raw sewage feed. The minimal area of the first VF-CW bed was determined as 1.5 m^2/pe, while for the second VF-CW bed was equal to 0.8 m^2/pe. The hydraulic loading should not exceed 0.6 m^3 m^{-2} h^{-1}. Following these guidelines at the designing stage ensures the 90.0% removal of organic substances, 95.0% removal of suspended solids and 85.0% nitrification of ammonia nitrogen (Boutin et al. 1997, Molle et al. 2004). In Germany the VF-CW beds were applied for treatment of overflow rainwater from combined sewerage systems. The beds proved to be tolerant to irregular sewage inflow. They provided high efficiency of suspended solids removal and mineralization of organic substances (Uhl and Dittmer, 2004).

In the second part of the 1990s several so-called hybrid systems, composed of the vertical-flow as well as horizontal flow beds, were constructed in Poland. In Europe two configurations of the hybrid wetland systems (HWS) are used, with either vertical-flow bed or horizontal-flow bed at the beginning of biological treatment process. In Poland only the configurations proposed by Johansen and Brix (1996), with a horizontal-flow bed at the beginning, were used.

Four facilities, located in Darżlubie, Sarbsk and Wiklino (in two measurement periods Wiklino I and Wiklino II) in the Pommerania Province and in Sobiechy in the Podlasie Province, were analysed. Mechanically treated sewage was directed to biological treatment at reed beds. In Sarbsk and Sobiechy part of the effluent from the vertical-flow bed was recirculated to the beginning of the treatment system (to the horizontal-flow bed). In Darżlubie and Wiklino another horizontal-flow bed followed the vertical-flow bed.

The pollutant removal efficiencies of hybrid wetland systems (HWS) were higher if compared to one-stage systems – it varied from 59.2% to 92.3% for BOD_5 (mean 86.7%) in the range of loadings from 21 kg ha^{-1}·day^{-1} to 123 kg ha^{-1}·day^{-1} (Fig. 3).

Table 3 presents the average removal of BOD_5 and total nitrogen from 1 m^2 of bed in the analysed hybrid wetlands systems (HWSs). The removal rates were

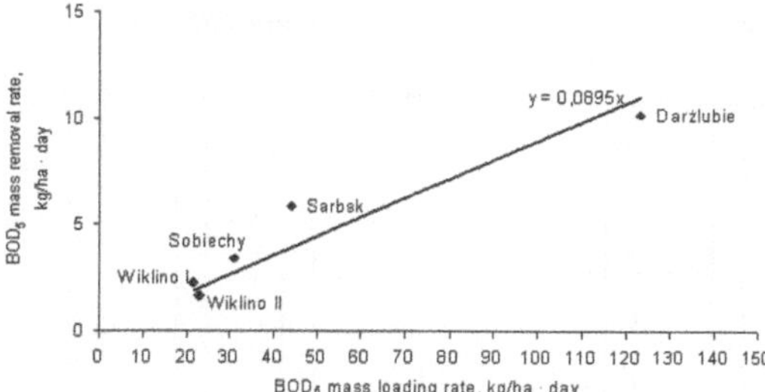

Figure 3. The load of BOD_5 at the inflow and at the outflow of the hybrid systems.

Table 3. Removal of BOD_5 and total nitrogen from 1 m^2 of bed in the analysed HWSs.

HWS configuration	
HF/VF/HF	HF/VF + rec.
BOD_5 g/m^{2}d	
Wiklino – 2.7 ± 0.7	Sarbsk – 3.9 ± 0.9
Darżlubie – 3.3 ± 0.8	Sobiechy – 5.6 ± 1.1
N_{tot} g/m^{2}d	
Wiklino – 0.8 ± 0.2	Sarbsk – 0.6 ± 0.2
Darżlubie – 0.4 ± 0.1	Sobiechy – 0.6 ± 0.3

calculated for about 70 measurements of average removed loads of organic matter (BOD_5) and total nitrogen and adequate cross-section. The system in Wiklino was the most effective in the total nitrogen removal but organic matter rate was the smallest.

The capability of nitrogen removal in the hybrid systems (from 4.2 to 14.6 kg/ha·d) was higher and more stable if compared to one-stage systems. The average load of removed nitrogen was 7.8 kg/ha·d. Similar results of nitrogen removal were obtained in Danish constructed wetland systems (3.0–7.0 kg/ha·d) (Brix, 1996). In spite of the high efficiency of pollutants removal, the nitrogen removal in the Darżlubie plant was unsatisfactory. This resulted from the inflow of liquid manure from several farms in the vegetation season (Obarska-Pempkowiak and Gajewska 2003, Gajewska et al. 2004).

3.3 *Sewage sludge dewatering*

During the last twenty five years the natural method of sewage sludge dewatering using hydrophite plants has been developed. Lienard et al. (1990) report that the first experiences with using reed for dewatering of sludge were performed by Seidel and Kickuth in Germany at the Nuclear Research Center in Karlsruhe in the 1960s. In the 1980s increasing interest in this new method of sludge drying reed beds (SDRB) was observed. The growing popularity of the method was obviously correlated with numerous successfully working plants constructed in Denmark (Nielsen 1990, 1993, 2002), in the eastern USA (Kim, 1994) and in France (Lienard et al. 1990). In the years 1988–1996 27 SDRBs were built in Denmark and a further 56 – in the years 1997–2000. At present around 105 reed beds for sewage sludge dewatering are working in Denmark. In the USA reed beds are mostly applied for dewatering of sludge from the rural areas. Local SDRBs are also applied in the rural areas in France, Belgium, Germany and Austria. The loading of most of these beds is usually 200 t d.m./year, while the beds constructed in Denmark in the last decade were designed to work under the loading of 300–1000 t d.m./year.

The SDRB method provides effective sludge dewatering (final dry matter content can reach 40%) as well as stabilization. The final product is a full-value fertilizer with a high content of humic compounds.

In the municipality Wienhousen near Hanover, Lower Saxony (Germany) an interesting solution to water-sewage management has been developed. The sewage from individual farms is treated at the site of its generation – at individual household wastewater treatment plants, while the sludge produced in the treatment process is transported to the local Central Utilization System (CUS). About 100 individual farms have joined this program.

Before it inflows to the reed beds the sewage is treated in a septic tank. The sludge produced there is periodically removed and transported to the CUS. The sludge processing technology at the CUS incorporates intensified natural processes since the SDRBs were applied there. The effluent separated in the natural drying process is clarified at specially designed VF-CW systems and then discharged to the environment. Once processing is over, the sludge becomes a valuable humus substance, which is applied as a fertilizer at local farms. In this way the cycle of contaminants generated by the municipality is closed.

In Poland the qualitative and quantitative transformations of sewage sludge in the reed beds have been analysed by Zwara and Obarska-Pempkowiak (2000).

In Northern Poland three macrophite facilities were constructed: in Darzlubie near Gdansk (loaded with primary sludge), in Swarzewo near Gdansk and in Zambrow near Suwalki (loaded with secondary sludge). In the village Darzlubie on the coast of the Bay of Puck, in the Pommerania Province, suspended solids are removed from sewage in household sedimentation tanks. Then sewage (ca. 140 m^3/day) is directed to an Imhoff tank. Further treatment takes place in a hybrid constructed wetland. Digested sludge containing 90.0–96.0% of water, in a volume of 36 m^3, is removed from the Imhoff tank eight times a year and discharged into the reed bed. There are two beds

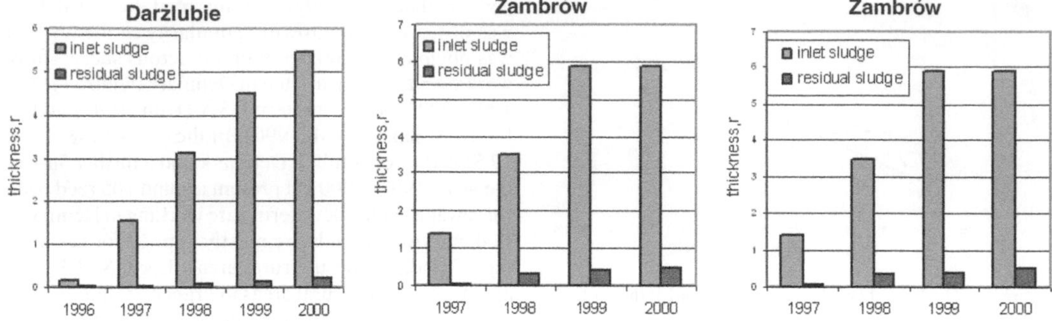

Figure 4. Dewatering of sewage sludge in reed beds in Darżlubie and in reed lagoons in Swarzewo and Zambrów.

with the total area of 480 m^2 (12.0 × 20.0 m each) in the facility. In January 1998, a small control plant (0.8 × 1.2 m), separated from contact with discharged sludge was established in the reed bed.

In the WWTP in Zambrow (Podlasie Province), the average amount of treated domestic sewage equals 3500.0 m^3/day. Domestic sewage and rainwater are collected separately. Domestic sewage undergoes treatment on screens, a sand trap and in biological reactors with activated sludge. The excess secondary sludge (ca. 150.0 m^3/day), containing 99% water, is collected in two traditional lagoons in the non-vegetation period. In the vegetation season it is discharged directly to a reed lagoon (total area of 5500 m^2). The amount of sludge utilized in the reed lagoon is 87.0% of the total volume of produced sludge. The remaining part of sludge (13.0%) is directed to the vermiculture beds during the summer season, and then, in autumn, it is used in agriculture.

The yearly average amounts of sludge loaded to the reed beds and remaining in the analyzed beds are presented in the Fig. 4.

The total thickness of the sludge layer discharged to the bed in Darżlubie was 3.15 m and the thickness of remaining layer of sludge was only 0.18 m. The 15 cm thick layers of primary, aerobically stabilized sludge, were discharged to the bed once every six weeks . Thus, the annual amount of sludge was equal to 1.2 m^3/m^2·year (Fig. 4). The US EPA suggests the following hydraulic loadings of the beds: 0.78 m^3/m^2·year for anaerobically stabilized sludge with the dry matter content of 5% and 2.35 m^3/m^2·year for aerobically stabilized sludge. According to De Maeseneer (1996), the wetland systems operating in Western Europe were fed with sludge 8, 16 or 24 times a year. The hydraulic loadings varied from 0.4 to 1.6 m^3/m^2·year for anaerobically stabilized sludge and from 0.7 to 5.7 m^3/m^2·year for aerobically stabilized sludge. Thus, the hydraulic loadings of the beds in Darzlubie were similar to those applied in other countries.

The volume of the sludge decreased by 94.3% due to the transformations taking place in the beds. Similar results were obtained by Nielsen (1993) during investigations in Allerslev and Regstrup (90.3%). It was shown that the main reason for the decrease of the sludge volume was dewatering and, to a smaller extent, biochemical decomposition.

The total volume of drainage water was only 4 m^3, which is 10% of the volume of loaded sludge (36 m^3).

The reed lagoon in Swarzewo was in operation between May 1995 and September 1998. In this period it was loaded with 10.5 m layer of secondary sludge (Fig. 4). When the utilization process was completed, the thickness of the sludge layer was 1.1 m. This rather high thickness of remaining sludge layer resulted from loading huge volumes of sludge, leading to devastation of the reeds in several places.

Dewatering of sewage sludge in the plantation in Zambrów was significantly slower than dewatering in reed beds in Darżlubie or Swarzewo. The average loading was equal to 34 kg d.m./m^2 year. In the operation period, the volume of drained off effluent was approximately 100–120 m^3/day. The moisture content decreased by 28% within two years. The low effectiveness of dewatering was due to lack of a draining system to remove effluent and poor aeration of the rhizosphere resulting from low porosity.

The average volume of the dewatered sludge decreased significantly during treatment in reed beds. The highest reduction of the sludge volume was observed in Darżlubie about 94.3%, very similar in Swarzewo 93.4% and slightly lower for the reed lagoons in Zambrów 80.1%.

4 CONCLUSIONS

1. The constructed wetland systems can be applied for removal of pollutants from point and surface sources and for sewage sludge dewatering and

stabilization. They provide effective removal of suspended solids and organic matter.
2. Effective and stable nitrogen removal in the local plants is only possible in the so-called hybrid constructed wetlands, consisting of at least two beds, with vertical and horizontal flow of sewage.
3. Constructed wetlands consisting of VF-CW beds only, popular in the rest of Europe, seem to be an interesting solution also for Poland
4. Constructed wetlands for sewage sludge dewatering provide effective water removal (final d.m. content reaches 40%) and stabilization. The final product is a full-value fertilizer containing large amount of humus substances.

ACKNOWLEDGEMENTS

Funding support from the Committee of Scientific Research in Poland for constructed wetland study (3 T09D 017 27) is acknowledged.

REFERENCES

Boutin C., Lienard A., Esser D., 1997. Development of a new generation of reed-beds filters in France: First results: Wat. Sci. Tech., 35(5): pp 315–322.

Brix H. 1996. Role of macrophytes in constructed wetlands, Proceedings of 5th International Conference on Wetland System for Water Pollution Control, Universitaet fuer Bodenkultur Wien and International Association on Water for water pollution control, Universitaet fuer Bandenkutur Wien and International Association on Water Quality, Vienna 1996 I/4–8.

Brix H., Arias A., 2002. BOD and nitrogen removal from municipal wastewater in an experimental two-stage vertical flow constructed wetland system with recycling. Proceedings of 8th International conference on wetland system for water pollution control. Univerity of Dar es Salaam, Arusha, Tanzania, Vol II: 400–423.

De Maeseneer J.L. 1996. Sludge dewatering by means of constructed wetlands. In: Proc. of 5th Int. Conf. on Wetland System for Water Pollution Control. Universität fuer Bodenkultur Wien and International Association on Water Quality, Vienna 1996, chapter XIII/2.

Gajewska M., Tuszyńska A., Obarska-Pempkowiak H. (2004). Influence of Configuration of the Beds on Contaminations Removal in Hybrid Constructed Wetlands, Polish Journal of Environmental Studies. Vol 13, ss 149–153.

Hermanowicz W., Dożańska W., Dojlido W., Koziorowski B., 2000. Fizyczne i chemiczne badania wody i ścieków. Arkady , Warszawa, 2000.

Johansen N.H., Brix H., 1996 Design criteria for a two-stage constructed wetland. Proceedings of 5th International conference on wetland system for water pollution control, Universitaet fuer Bondenkutur Wien and International Association on Water Quality, Vienna 1996 IX/3–7.

Kim B. 1994. Field evaluations of reed bed sludge dewatering technology: summary of benefits and limitations based on 4 years experience at Fort Campbell, Kentucky. Paper presented at the 67th annual Conference of the water and Environment Federation, Chicago, October, 1994.

Kowalik P., Mierzejewski M., Randerson P. F., Williams H. G. 2004. Performance of subsurface vertical flow constructed wetlands receiving municipal wastewater. Skuteczność pracy oczyszczalni hydrofitowych z pionowym przepływem ścieków komunalnych. Archives of Hydro-Engineering and Environmental Mechanics, p. 349–370,

Lienard A., Esser D., Deguin A., Virloget F., 1990. Sludge dewatering and drying in reed beds: an interesting solution? General investigation and first trials in France. Proceedings of the conference, "Use of Constructed Wetlands in Water Pollution Control". Cambridge, 1990: 257–267.

Molle P., Lienard C., Boutin G., Iwema A., 2004. How to treat raw sewage with constructed wetland: an overview of the French systems. Proc. 9th International Conference on Wetland Systems for Water Pollution Control. Avignon (France), 26–30th of September 2004:11–21.

Nielsen S.M. 1990. Sludge dewatering and mineralisation in reed bed systems. Proceedings of the conference, "Use of Constructed Wetlands in Water Pollution Control". Cambridge, 1990: 245–256.

Nielsen S. 2002. Sludge drying reed beds. 8th International Conference on Wetland systems for Water Pollution Control. Arusha International Conference Centre (AICC), 2002.University of dear Salaam. Vol I, p. 24–39.

Nielsen S.M. 1993. Biological sludge drying in constructed wetlands. Constructed Wetlands for Water Quality Improvement (ed. Moshihiri G.A.) Boca Raton, Florida. Lewis Publishers 1993, pp. 549–558.

Obarska-Pempkowiak H., Ozimek T., Chmiel W. 2001. Protection of surface water against contamination by wetland systems in Poland. Water Science & Technology 44(11–12): 325–331.

Obarska-Pempkowiak H., Ozimek T., Haustein E. 2002. The removal of biogenic compounds and suspended solids in a constructed wetland system. Polish Journal of Environmental Studies 11(3): 261–266.

Obarska-Pempkowiak H., Gajewska M., 2003. The removal of nitrogen compounds in hybrid wetland systems. Polish Journal of Environmental Studies 12(6),ss. 739–746.

Obarska-Pempkowiak H., Haustein E., Wojciechowska E. 2004 Distribution of heavy metals in vegetation of constructed wetlands in agricultural catchments Ed. J. Vymazal Natural and Constructed Wetlands, Nutrients, Metals and Management, Backhuys Publishers, Leiden, The Netherlands, pp. 125–134.

Polish standards according limits for discharged sewage and environmental protection from 8 July 2004 (Dz.U. Nr 168, poz.1763.

Soroko M. 2001. Effectiveness of organic matter and nutrients removal at several constructed wetland systems. Water – Environment – Rural areas 1, IMUZ 2001, vol.1: 173–186.

Uhl M., Dittmer U., 2004. Constructed wetlands for CSO treatment-an overview of practice and research in Germany, Proc. 9th International Conference on Wetland Systems for Water Pollution Control. Avignon (France), 26–30th of September 2004: 21–31.

Wojciechowska E., Haustein E., Obarska-Pempkowiak H. 2004. Distribution of heavy metals in ecosystem of free water surface system in Bielkowo. In: [Materials] VI-th International Conference, XVIII-th National Conference Water Supply and Water Quality. Poznań 2004 r. Ed. M. Elektorowicz, Marek M. Sozański. Poznań: Pol. Zrzesz. Inż. i Tech. Sanit. Oddz. Wlkp. 2004 vol. 2, p. 413–430.

Zwara W., Obarska-Pempkowiak H. (2000). Polish experience with sewage sludge utilisation in reed beds. Wat. Sci. Tech. 41(1): 65–68.

Environmental Engineering – Pawłowski, Dudzińska & Pawłowski (eds)
 ISBN13 978-0-415-40818-9

Nitrogen pathways during sewage treatment in constructed wetlands: seasonal effects

Magdalena Gajewska & Hanna Obarska-Pempkowiak
Gdansk University of Technology, Faculty of Civil & Environmental Engineering, Narutowicza, Gdansk, Poland

ABSTRACT: Wetland systems remove many contaminants, including organic matter (BOD, COD), suspended solids, nitrogen, trace metals and pathogens and therefore this method has recently become very popular for treatment of domestic wastewater. Although the main mechanisms of organic matter and nitrogen removal are quite well known, their kinetics and rate processes are not well documented. The objectives of this paper is to estimate the role and quantity of the unit process responsible for removal or/and retention of nitrogen in different types of constructed wetlands within and without the vegetation season. The decrease of effectiveness of nitrogen transformation process outside the vegetation season was confirmed by the values of the reaction rate constant which, in the case of the facilities studied were equal to 60% of the values reached in the vegetation season.

Keywords: Domestic sewage, treatment, reed, willow, nitrogen.

1 INTRODUCTION

Constructed wetland systems can secure removal of organic matter (BOD, COD), suspended solids, nitrogen, trace metals and pathogens. For these reasons this method became popular for treatment of domestic wastewater. Removal of pollutants is accomplished by diverse biochemical unit processes. Although the main mechanisms of organic matter and nitrogen removal are quite well known, their kinetic and unit rates processes and balance are still not well recognized (Reddy and D'Angelo, 1996; Vymazal, 1999; Cooper, 2001). It was proved that in nitrogen removal, emissions of N_2O and N_2 (from both sequential microbiological nitrification and denitrification) play the most important role. Up till now it was considered that the most common unit processes responsible for nitrogen transformation are sequence of ammonification, authotrophic nitrification and heterotrophic denitrification (Mander et al, 2003, Kadlec and Knight,1996). Other mechanisms such as plant uptake, ammonia volatilisation and sorption in the soil matrix are generally of less importance. Also the role of plant uptake depended on the species and measurements on sorption in the soil matrix are still being carried out (Reddy and D'Angelo,1996; Langergraber, 2004).

Sewage discharged to constructed wetlands are rich in nitrogen mainly in dissolved forms. Under the conditions of Polish rural regions nitrogen flows into constructed wetlands with domestic sewage mainly as ammonium nitrogen (2/3) and organic nitrogen (1/3), and is retained or removed in several physical and biochemical processes (Gajewska et al, 2004). Depending on the type of constructed wetlands and the configuration of the beds the role of unit treatment processes varied significantly. In constructed wetlands with subsurface flow of sewage (Horizontal Flow and Vertical Flow) sorption and filtration are the dominating processes. **H**orizontal **F**low **C**onstructed **W**etlands (HF-CW) ensure efficient removal of suspended solids, organic matter and, due to mainly anoxic condition create appropriate conditions for the denitrification process. While **V**ertical **F**low **C**onstructed **W**etlands (VF-CW) discharged with sewage periodically create better conditions for the nitrification process. Also, the different species of hydrophytes plants may play an important role in the removal of contamination due to the particular conditions they create (Reddy and D'Angelo, 1996; Brix, 1996; Ozimek and Obarska-Pempkowiak, 2000; Vymazal et al, 1999).

The objective of this paper is to estimate the role of plants (reed and willow), kinetics and rate of the unit process responsible for removal or/and retention of nitrogen in different types of constructed wetlands as a function of season (i.e. in the vegetation season and outside)

2 STUDY AREAS AND METHODS

The studies were carried out at two constructed wetlands situated in the villages of Wiklino near Słupsk

Table 1. The characteristics of facilities.

WWTPs	Type of plant	Flow [m^3/day], (pe)	Configuration	Area [m^2]	Depth [m]	Hydraulic load, [mm day^{-1}]	Unit area [$m^2 pe^{-1}$]
Wiklino	Reed	18.6, (200)	HF*	1050	0.6	17.7	7.0
			VF**	624	0.4	46.9	2.0
			HF	540	0.6	25.7	3.4
				Total 2214			Total 12.4
Lubliejewo	Willow	0.036, (6)	HF	Total 60	1.0	0.6	Total 6.0

* Horizontal Flow, ** Vertical Flow.

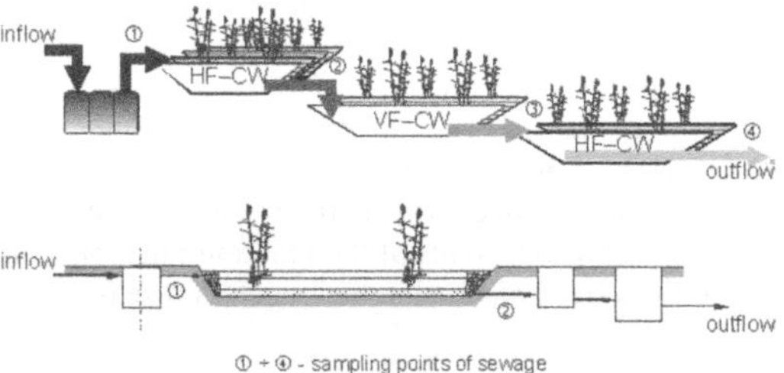

Figure 1. The schemes of analysed facilities in Wiklino (a) and Lubliejewo (b).

and Lubiejewo near Ostrołeka in Poland. The characteristics of the applied solution are given in Table 1. Schemes of the systems are given in Figure 1. Plants in Wiklino were supplied with sewage from about 200 pe (local system) while at Lubiejewo sewage was supplied from individual constructed households from 6 pe (household system). In Wiklino, the sewage after mechanical treatment was pumped into the wetland responsible for biological treatment. This system consisted of two beds: HF-CW I and HF-CW II and one VF-CW bed.

Such systems with mixed flow (horizontal and vertical) are called **H**ybrid **C**onstructed **W**etlands. The HCW in Wiklino was inhabited with common reed (*Phragmites australis*) in 1995. In Lubiejewo sewage after treatment in a septic tank was discharged to HF - CW inhabited with willow (*Salix viminalis*) in 1994. To minimize the impact of the facilities on underground water all bottoms of beds were equipped with an impermeable layer.

Averaged samples of Wiklino sewage at the beginning of biological treatment (after mechanical stage) and effluent (sewage after the constructed wetland) as well as after subsequent stages of biological treatment were analysed. Samples were collected once per month during a five year period (April 1998–March 2001 and July 2003–May 2005); and thus the number of collected samples was 60 in Wiklino. In case of Lublijewo during three years period (April 1999–September 2003) the number of collected samples was 40. The quality of the sewage was based on estimation

Table 2. Number of samples and temperature of sewage and air in vegetation and outside vegetation seasons.

Parameter	Vegetation season		Outside vegetation season	
	Lubliejwo	Wikliono	Lubliejwo	Wikliono
Temperature of air (T_a)	12.6°C	11.7°C	1.7°C	2.2°C
Temperature of sewage (T_s)	15.3°C	14.5°C	3.3°C	3.8°C
Number of samples (n)	23	37	17	23

of following parameters: temperature of sewage and air, total suspended solids, BOD_5, COD_{Cr}, ammonium nitrogen ($N\text{-}NH_4^+$), nitrate, nitrite and organic nitrogen. All analyses were performed at Gdańsk University of Technology, according to the Polish Standard Methods (1999). The measurement period was divided into vegetation (from April to October) and non-vegetation seasons (from November to March) for both facilities.

In the case of the constructed wetlands under analysis, the removal efficiency was calculated as the quotient of contaminant concentration difference in influent (C_o) and effluent (C) after subsequent stages of treatment and concentration in influent (C_o), $\eta = (C_o - C)/C_o$.

In order to estimate removal and transformation rate, calculations of the decay rates of organic matter (k_{pBOD}) and both total nitrogen removal (kp_{Ntot}) and organic nitrogen mineralization (kp_{Norg}) were done on the sewage. It was assumed that the decay can be described by a first-order reaction constant (Cooper 1998, Reed et al, 1995). Based on the retention time and concentrations of organic matter and nitrogen, the constant decay rates for the sewage treated in HF beds were found and corresponding modified decay rates for sewage treated in VF-CW bed (Brix and Johansen, 1999). The values of the above mentioned decay rates ($k_{p(T)}$) for 20°C, taking into account average monthly temperatures, were calculated using the relationship $k_{p(T)} = k_{p(20)}(1.1)^{T-20}$.

Table 3. Average values of characteristic parameters in inflowing and outflowing sewage in HCW in Wiklino.

Parameter	Unit	Vegetation season $n = 37$, $T_s = 14.5°C$		Outside vegetation season $n = 23$, $T_s = 3.8°C$	
		Inflow	Outflow	Inflow	Outflow
Flow	m^3/day	19.0 ± 2.1*	12.9 ± 0.8	18.4 ± 1.6	13.9 ± 1.1
SS	mg/l	183.1 ± 32.4	27.6 ± 3.4	207.5 ± 42.5	29.0 ± 5.2
BOD_5	mg O_2/l	246.8 ± 56.8	23.8 ± 4.7	284.2 ± 64.3	32.1 ± 6.7
COD	mg O_2/l	370.9 ± 81.3	59.4 ± 7.5	444.4 ± 88.6	78.7 ± 15.5
N_{Tot}	mg/l	90.2 ± 14.5	15.8 ± 3.1	103.9 ± 20.4	25.9 ± 4.5
Norg	mg/l	16.8 ± 1.4	4.1 ± 0.6	26.8 ± 4.7	7.5 ± 1.3
$N\text{-}NH_4^+$	mg/l	73.1 ± 12.3	10.6 ± 2.4	76.9 ± 17.1	17.1 ± 2.9

* – standard deviation, n – number of samples, T_s – average temperature of sewage.

In order to determine the dry mass of the plants inhabiting these systems the procedure given by Ozimek and Obarska-Pempkowiak, (2003) was applied. A $0.25\,m^2$ quadrant frame was used to determine the density of above-ground shoots. The dry mass of stems and leaves was determined by weighing after drying for 24 hours at 105°C. Determinations of nitrogen contents in above-ground biomass were made three times during the growing season: at the beginning (in June), at the highest intensity of growth (in August) and in the end (in October) of the growing season.

In order to determine the role of the soil matrix in the removal of contaminants in filling material of beds the following parameters were performed: moisture, dry matter, nitrogen contents and loss on ignition as a organic matter content. The filling media of the beds were sampled at several sites along the Profile. All sampling and measurements were done according to the procedure given by Obarska-Pempkowiak (1999).

3 RESULT AND DISCUSSIONS

3.1 *Local constructed wetland – Wiklino*

During the vegetation season the average amount of sewage discharged into the Wiklino facilities was slightly higher (about 3.1%) than the average amount discharged in the non-vegetation season. While the average amounts of sewage outflowing from the facility in the vegetation season were about 7.7% smaller in comparison to the outside vegetation one, probably due to the high evapotranspiration, which could reach 5 mm/day (Brix,1996). The average concentration of organic matter (expressed by BOD_5, COD_{Cr} and SS) discharged to the facilities in Wiklino was higher in the outside vegetation season than in the vegetation one. The average values with standard deviations were calculated from about 60 samples and are given in Table 3. The average concentrations of

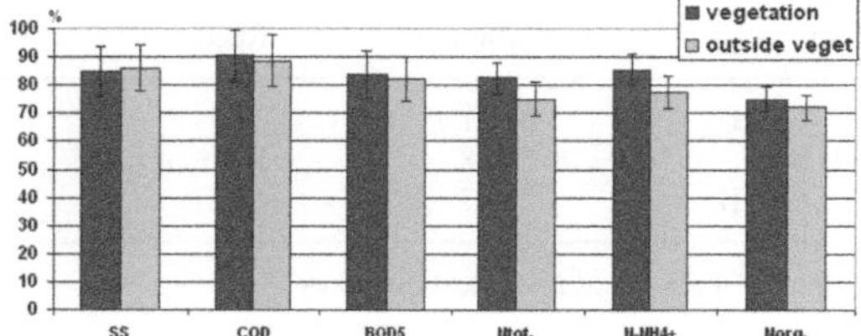

Figure 2. Comparison of average efficiency of contaminations removal in vegetation and outside vegetation seasons in CWs in Wiklino.

nitrogen compounds in sewage flowing in to the facilities during vegetation were lower (about 10.0%) than outside the vegetation season. Also in discharged sewage from the facilities the concentration of all contamination was slightly higher but still below the permissible value (SS = 50 mg/l, $BOB_5 = 40$ mgO_2/l, COD = 150 mgO_2/l and $N_{Tot} = 30$ mg/l). Due to the higher concentration of contaminations and lower amount of sewage in the outside vegetation season the load of contaminations discharged to the facilities in this season was higher than in loads during the vegetation one. The efficiency of different contamination removal did not vary very much in both seasons (Fig. 2).

The biggest differences were observed for the removal of nitrogen compounds. In the case of ammonium nitrogen the efficiency in the vegetation season was equal to 85.1% and in the outside vegetation season decreased to 78.3%.

Based on changes in the different forms of nitrogen concentration, it was calculated that the rate of the nitrification process, in conditions of the constructed wetland at Wiklino was equal to $0.56\,gm^{-2}\,day^{-1}$ and denitrification was equal to $0.59\,gm^{-2}\,day^{-1}$. The ammonification process was the slowest and the rate was $0.15\,gm^{-2}\,day^{-1}$. It was also proved that aminification of organic nitrogen took place mainly in the mechanical treatment under the conditions of the

Table 4. The average values of reaction rate constants for analysed unit processes at 20°C.

	HF-CW I [day^{-1}]	HF-CW II [day^{-1}]	VF-CW [$m\,day^{-1}$]	
Vegetation season				
$k_{pBOD(20)}$	0.122	0.27	$k_{BOD(20)}$	0.031
$k_{pNtot(20)}$	0.094	0.124	$k_{Ntot(20)}$	0.025
$k_{pNorg(20)}$	0.061	0.061	$k_{Norg(20)}$	0.024
Outside vegetation season				
$k_{pBOD(20)}$	0.071	0.111	$k_{BOD(20)}$	0.019
$k_{pNtot(20)}$	0.045	0.062	$k_{Ntot(20)}$	0.019
$k_{pNorg(20)}$	0.048	0.052	$k_{Norg(20)}$	0.018

Description of reaction rate constant is given in study areas and methods.

Table 5. Distribution of nitrogen in elements of HCW in Wiklino (kg/year).

Bed	Inflow	Reed	Soil matrix	Outflow	Difference*
HF-CW I	646.1	30.7	10.4	346.8	257.6
VF-CW	346.8	Without reed	2.5	211.7	132.6
HF-CW II	211.7	14.6	1.6	94.9	100.6

*Denitrification and analytical mistakes.

Wiklino constructed wetland (Gajewska et al, 2004). The achieved rates for nitrification and ammonification were similar to data given in the literature by Reddy and D'Angelo (1996) (ammonification $0.004 - 0.357\,gm^{-2}day^{-1}$ and the rate of denitrification $0,003 - 1,02\,gm^{-2}\,day^{-1}$) In case of nitrification rate the data were much lower: $0,01\,d - 0,161\,gm^{-2}\,day^{-1}$.

The differences could be caused by different operation conditions of the analysed facilities. According to Reddy and D'Angelo (1996) the measurements and calculation were carried out in USA, mainly in natural wetlands or in one stage constructed wetlands with horizontal flow of sewage. Those type of wetland do not create favourable condition for nitrification process. While the facility in Wiklino is hybrid constructed wetland. The second bed of treatment in biological stage took place in VF-CW, which could create appropriate environment for nitrification process due to periodical discharged with sewage.

The obtained average values of the reaction rate constants for sewage treated in HF-CW I and HF-CW II beds and corresponding modified constant rates for sewage treated in VF-CW bed are given in Table 4.

The obtained values of reaction rate constants indicate that the decomposition of organic substances in sewage was the fastest process, while total nitrogen removal was slightly slower in both seasons. Mineralization of organic nitrogen in sewage was the slowest process, both during the vegetation season and outside it. According to Gajewska & Obarska-Pempkowiak (2001) it was also shown that a substantial part of the organic nitrogen in sewage was ammonified in the septic tank. The temperature influence on the ammonification process effectiveness was negligible. The average values of the constant rates $k_{pBOD(20)}$ and $k_{pNtot(20)}$ were higher by 42.0 and 52.0% in the vegetation season than outside in case of sewage treated in the HF-CW I bed. For sewage treated in the HF-CW II the average values of the constant reaction rates were over 50% higher for the vegetation season than outside it.

The average values of the modified constants $k_{BOD(20)}$and $k_{Ntot(20)}$ in the vegetation season were equal to 0.031 and $0.025\,m\,day^{-1}$, respectively and were higher than in the outside vegetation season. Similar results were obtained by Brix & Johansen (1999) for 37 VF-CW beds in Denmark. The average value of the modified $k_{Ntot(20)}$ constant rate was $0.0247\,m\,day^{-1}$. The values of $k_{BOD(20)}$ obtained during investigation carried out in Austria and Great Britain varied from 0.067 to $0.1\,m\,day^{-1}$ and were higher than the values calculated for the VF-CW beds in Wiklino. This could probably result from the higher average air temperature (Harbel et al, 1998, Cooper & de Maeseneer, 1996).

The measurements carried out enables the qualitative assessment of nitrogen speciation (distribution) between the different elements of the whole constructed wetland system.

The average annual load of nitrogen inflowing and outflowing from the constructed wetland and the accumulation of nitrogen in above-ground parts of reed and in soil in the Wiklino hybrid system are given in Table 5.

Table 6. Average values of characteristic parameters in inflowing and outflowing of sewage in HF-CWs in Lublijewo.

		Vegetation season $n=23$, $T_{sc}=15.3°C$		Outside vegetation season $n=17$, $T_{sc}=3.3°C$	
Parameter	Unit	Inflow	Outflow	Inflow	Outflow
Flow	m^3/day	0.38 ± 0.09	0.23 ± 0.08*	0.36 ± 0.08	0.38 ± 0.1
SS	mg/l	125.0 ± 22.4	17.0 ± 1.8	153.1 ± 32.5	43.0 ± 7.2
BOD5	mgO_2/l	298.3 ± 67.8	20.1 ± 6.7	301.2 ± 68.3	35.1 ± 6.9
COD	mgO_2/l	508.2 ± 93.3	80.4 ± 11.5	549.2 ± 102.6	135.1 ± 35.5
N_{Tot}	mg/l	165.9 ± 44.5	84.1 ± 21.1	171.5 ± 50.4	88.4 ± 24.5
N_{org}	mg/l	48.9 ± 11.4	15.3 ± 4.6	49.3 ± 11.7	22.3 ± 7.3
$N\text{-}NH_4^+$	mg/l	114.6 ± 32.3	53.7 ± 12.6	118.9 ± 37.1	71.8 ± 22.9

*from May to September there was no outflow thus the average was calculated only for April and October in measurements periods, n – number of samples, T_s – the average temperature of sewage.

Based on the nitrogen balance already carried out it was observed that the amount of nitrogen accumulated in the above-ground part of reed was equal to 7.0% of the inflowing load, while in the soil matrix about 5.9% was accumulated probably due to sorption processes of inflowing nitrogen. The investigation carried out and described by Obarska-Pempkowiak and Ozimek (2000), proved that bioaccumulation of nitrogen in above-ground reed can amount to up to 10% of the inflowing load of nitrogen.

The efficiency of the processes responsible for emission of nitrogen in both forms (N_2O and N_2) was equal to 72.4% of the supplied nitrogen load. According to Mander et al, 2003 the fluxes of the N_2O and N_2 are very high and vary significantly depending on the kind of flow in the CWs. For HF-CW beds the emission rate of N_2 was up to 17fold higher in comparison to emission of N_2O from VF-CV and varied 1710-4680 kg N-N_2/ha year and 2.4- 32 kg N-N_2O/ha year, respectively. This proved that condition of HF-CW beds favorite the denitrification process. VF-CW beds showed lower values of N_2 emission (40–44 kg N-N_2/ha year) and relatively high values for N_2O fluxes (850–1600 kg N-N_2O/ha year). It has been shown that in the well-aerated but still moist conditions of VF-CW beds emission of N_2O could substantial result from nitrification of ammonium nitrogen. Thus emission of nitrogen from HCWs could be the result of sequential authotrophic nitrification and heterotrophic denitrification or other unit processes which lead to so-called "shorten ways" of nitrogen removal. Hybrid constructed wetlands, due to different oxygen conditions, could create suitable conditions for simultaneous nitrification and denitrification, which occurs in conventional WWTPs with activated sludge (Collivignarelli iand Bertranza, 1999). Up till now there is too little date to judge which unit process contributes significantly to emission of nitrogen.

On the basis of the nitrogen balance carried out, the average rate of nitrogen removal ranged up to 0.7 g N_{tot} m^{-2} day^{-1}. This value exceeds the highest values given by Brix and Johansen in 1999 (0.3–0.7 g N_{tot} m^{-2} day^{-1}).

3.2 *Household constructed wetland – Lubliejewo*

During the vegetation season the average quantity of sewage flowing in to the system was equal to 0.38 m^3 day^{-1} i.e. it was slightly higher in comparison with the quantities discharged during outside the vegetation season (Table 6). The amount of sewage outflowing from the facility depended on the season. In the period from May to September there was no outflow from the system due to the intensive transpiration of willow. In the other seasons (especially in autumn and winter months) the outflow was by 14 to 50% higher than in the inflow.

Due to the higher concentration of contaminations and the slightly lower amount of sewage in the outside vegetation season, the load of contaminations discharged to the facilities in this season was higher than in corresponding loads in the vegetation season. Outside the vegetation season the concentrations of outflowing contamination were from 17.1 to 48.5% higher in comparison with corresponding concentrations in sewage discharged in the vegetation season.

The efficiency of different contaminations removal did not vary very much in both seasons (Fig. 3). The biggest difference were observed for nitrogen compounds removal. In the case of ammonium nitrogen the efficiency in the vegetation season was equal to 53.1% and in the outside vegetation season decreased to 39.6% (Fig. 3).

Based on changes in the different forms of nitrogen concentrations, it was calculated that the rate of denitrification process, in conditions of the facility was equal to 0.84 gm^{-2} day^{-1} and nitrification was equal to 0.58 gm^{-2} day^{-1}. The ammonification process was the slowest one with a rate of 0.19 gm^{-2} d^{-1}.

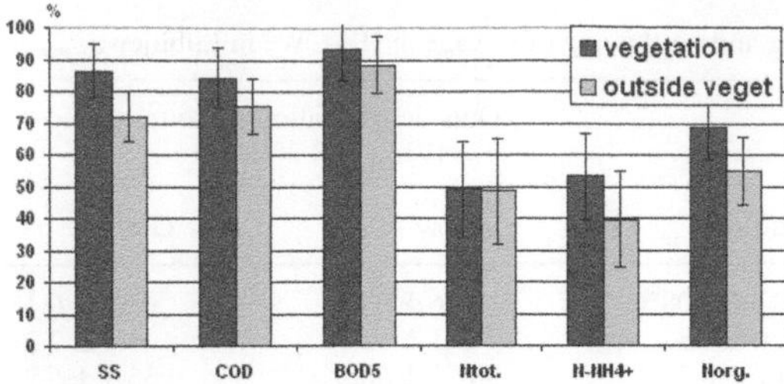

Figure 3. Comparison of average efficiency of contaminations removal in vegetation and the outside vegetation seasons in HF-CWs in Lublijewo.

Table 7. The average values of reaction rate constants for analysed unit processes in the temperature 20°C, day^{-1}.

Constants	Vegetation season	Outside vegetation season
$k_{pBOD(20)}$	0.179	0.097
$k_{pNtot(20)}$	0.051	0.023
$k_{pNorg(20)}$	0.051	0.041

Table 8. Distribution of nitrogen in household HF-CWs in Lubljiewo, $kg \cdot year^{-1}$.

Inflow	Willow	Soil matrix	Outflow	Differences*
21.23	0.786	1.894	6.02	12.65

*Denitrification and analytical mistakes.

The measured values of unit processes responsible for the removal of nitrogen in case of the facility in Lublijewo were very similar to the values in Wiklino, which suggests that significant differences in the type of wetlands (local-household, one stage-multi stage) inhabited with different species of hydrophytes did not have an influence on the rate of unit processes. The average values of the analysed reaction rates constant for 20°C are given in Table 7.

Reaction rate constants in sewage indicate that the decomposition of organic substances was the fastest process, while total nitrogen removal was much slower in both seasons. Temperature did not significantly affect the mineralization process. The average values of the rate constants$k_{pBOD(20)}$ and $k_{pNtot(20)}$ were 54.1 and 45.1% higher in the vegetation season than outside it. The determined values of reaction rate constant for organic matter decomposition (in sewage) for the facility inhabited with willow were about 30% higher in comparison with corresponding value determined for the facility inhabited with reed. In the case of $k_{pNtot(20)}$ and $k_{pNorgt(20)}$ rate they are from 20.6% to 49.3% higher for the facility inhabited with reed.

The annual load of nitrogen inflowing and outflowing from the constructed wetland and accumulation of nitrogen in above-ground parts of willow and in soil in Lubliejwo are given in Table 8.

Basing on the nitrogen balance carried out it was evident that the amount of nitrogen accumulated in the above-ground part of willow was equal to 3.7% of the inflowing load, while in sorption processes in soil matrix was accumulated about 8.9% of inflowing nitrogen. The results indicate that in the above-ground part of willow much less nitrogen was bioaccumulated in comparison to above-ground part of reed, while the soil matrix with roots of willow created probably better conditions for sorption of ammonium nitrogen. The most intensive removal of nitrogen was by emission of N_2O or N_2. The efficiency of processes responsible for emission of nitrogen was 59.3% of the supplied nitrogen load.

4 CONCLUSIONS

1. The concentrations of organic matter and nitrogen compounds discharged to the both analysed systems outside the vegetation season were higher (from 10% to 20%) in comparison with the vegetation season.
2. The average removal efficiency of nitrogen in both seasons did not vary significantly. The biggest difference was observed for ammonium nitrogen and was, in both facilities, 10% higher in the vegetation season than outside it.
3. The main process responsible for removal of nitrogen in both constructed wetlands was release of nitrogen to the atmosphere. In the case of local CWs this was 72.4% and for household CWs 59.3% of the discharged load of nitrogen.
4. Indications are that bioaccumulation of nitrogen in reed is nearly twofold higher in comparison to that in willow.
5. Based on values of the reaction rate constant it was established that organic matter decomposition was the fastest process, while removal of total nitrogen was slightly slower in both vegetation and outside vegetation seasons.
6. The reduced effectiveness of nitrogen transformation outside the vegetation season is confirmed by the values of reaction constant rates which, in the case of our facilities were 60% of the values reached in vegetation season.

REFERENCES

Brix H. 1996. Role of macrophytes in constructed wetlands, *Proceedings of 5th International Conference on Wetland System for Water Pollution Control*, Universitaet fuer Bodenkultur and International Association on Water for water pollution control, Universitaet fuer Bodenkutur

Wien and International Association on Water Quality, Vienna 1996 I/4–8

Brix H., Johansen N.H., 1999, Treatment of domestic sewage in a two-stage constructed wetland – design principles, *Nutrient cycling and retention in natural and constructed wetland* (Ed.) J. Vymazal, Backhuys Publishers, Leiden, The Netherlands:155–165

Collivignarelli C., Bertranza G., 1999, Simultaneous nitrification-denitryfication processes in activated sludge plants: Performance and applicability, Wat. Sci. Tech., 40: 187–194

Cooper P., De Maeseneer J., 1996, Hybrid systems – what is the best way to arrange the vertical and horizontal-flow stage? IAWQ Specialist Group on Use of Macrophytes in Water Pollution Control, December: 8–13

Cooper P., 1998, A review of the design and performance of vertical flow and hybrid reed bed treatment systems, *Proceedings of 6th International Conference on Wetland System for Water Pollution Control*, Brazil 1998, chapter 4-*Design of Wetland Systems*: 229–242

Cooper P., 2001, Nitrification and Denitrification in Hybrid Constructed Wetlands Systems, Transformation of Nutrients in Natural and Constructed Wetlands (Ed) J. Vymazal, Bachuys Publishers, Leiden, The Netherlands:257–271

Gajewska M., Tuszyńska A., Obarska-Pempkowiak H., 2004, Influence of configurations of the beds on contaminations removal in hybrid constructed wetlands. *Pol. J. Environm. Stud.* vol. 13 Suppl. 3s. 149–152

Haberl R., Perfler R., Laber J., Grabher D., 1998 Austria. Vymazal J., Brix H., Cooper P.F., Green M.B., & Haberl R., (Ed.) In: *Constructed wetlands for wastewater treatment in Europe.* Backhuys Publishers, Leiden:67–76

Langergraber G., 2004, The role of plant uptake on removal of organic matter and nutrient in subsurface flow constructed wetlands – a simulation study. *Proceedings of 9th International Conference on Wetland System for Water Pollution Control* (IWA), Avignion, France, September 2004, II 491–501

Mander U., Teiter S., Lohmus K., Mauring T., Nurk K., Augustin J., 2003 Chapter, Emission rates of N_2O, N_2 and CH_4 in riparian alder forests and subsurface flow constructed wetlands In. *Wetlands-nutrients, metals and mass cycling* (Ed.) J. Vymazal Backhuys Publishers, Leiden, The Netherlands 2003: 259–279

Obarska-Pempkowiak H., Chapter 4 (1999). *Nutrient Cycling and Retention in Natural and Constructed Wetlands.* Backhuys Publishers, Leiden, The Netherlands 1999: 31–49.

Obarska-Pempkowiak H., Ozimek T., (2000) Efficiency of wastewater treatment in an improved constructed wetland, Polish Arch. Hydrobiology 47, 247–256

Obarska- Pempkowiak H., Ozimek T., 2003. Chapter: Comparison of Usefulness of Three Emergent Macrophytes for Surface Water Protection against Pollution and Eutrophication: Case study, Bielkowo, Poland. *Wetlands-nutrients, metals and mass cycling* (Ed.) J. Vymazal Backhuys Publishers, Leiden, The Netherlands 2003: 215–226

Polish standards (1999) according chemical and physical analyses

Regulation Of The Environment Department from 8.VII.2004 r regarding the conditions of sewage discharge to water and soil and substances toxic for aquatic environment. Dz. U. Nr 168, poz. 1763

Reddy K.R., D'angelo E.M., 1996, Biochemical indicator to evaluate pollutant removal efficiency in constructed wetlands. *Proceedings of 5th International Conference on Wetland System for Water Pollution Control*, Universitaet fuer Bondenkutur Wien and International Association on Water Quality, Vienna 1996 I 1–21

Reed S.C., Crites R.W., Middlebrooks E.J., 1995, Natural systems for waste management and treatment. Second Ed. New York: McGraw Hill Chapter 6, *Wetland Systems*: 173–281

Vymazal J., Brix H., Cooper P.F., Haberal R., Perfler R., Laber J.,1998, Constructed Wetlands for Wastewater Treatment in Europe, Removal mechanisms and types of constructed wetlands., Backhuys Publishers, Lieden, The Netherlands:17–66

Vymazal J., 1999, Nitrogen removal in constructed wetland with horizontal sub-surface flow- can we determine the key process? *Nutrient cycling and retention in natural and constructed wetlands.* (Ed.) J. Vymazal. Backhuys Publishers, Leiden, The Netherlands.: 1–17

Vymazal J., Dusek J., Kvet J., 1999. Nutrient uptake and storage by plants in constructed wetlands with horizontal sub-surface flow: a comparative study. *Nutrient cycling and retention in natural and constructed wetlands.* (Ed.) J Vymazal. Backhuys Publishers, Leiden, The Netherlands: 87–100.

Environmental Engineering – Pawłowski, Dudzińska & Pawłowski (eds)
© 2007 Taylor & Francis Group, London, ISBN13 978-0-415-40818-9

Separation of detergents from industrial effluents using membrane processes

Izabela Kowalska, Malgorzata Kabsch-Korbutowicz & Katarzyna Majewska-Nowak
Wroclaw University of Technology, Institute of Environment Protection Engineering, Wybrzeze Wyspianskiego, Wroclaw, Poland

Maciej Pietraszek
PP-EKO sp. z o.o., ul. Agatowa, Warsaw, Poland

ABSTRACT: The suitability of ultrafiltration for purification and concentration of effluents from detergent production was evaluated. The results obtained indicate that ultrafiltration membranes of high cut-off values enable 63–76% decrease in COD-Cr of the treated wastewater. The reduction of COD-Cr by more than 80% can be reached for membranes with a cut-off lower than 10 kDa. Application of a membrane with a cut-off of 0.5 kDa yields the best separation efficiency: the decrease of COD-Cr value is over 85%, corresponding to a COD-Cr of permeate of 8800 g O_2/m^3. It was found that UF modules are suitable for concentration of highly-polluted effluents containing detergents; they are characterised by stable transport and separation properties. In the course of a long-term concentration of effluents, no noticeable decrease in permeability was observed and the permeate quality remained almost constant, although a systematic increase in pollution load of the concentrate occurred.

Keywords: Ultrafiltration, membrane, surfactant, wastewater, treatment.

1 INTRODUCTION

In recent years considerable attention has been focused on the problems caused by laundry and cleaning chemicals. The environmental risks associated with manufacture, use and disposal of these substances is of great interest. Detergent products are used in large quantities in industrial application such as textiles, food, paints, polymers, cosmetics, pharmaceuticals, mining and paper production. The typical components of detergents are anionic and nonionic surfactants, builders and bleachers. Other components are polymers, enzymes, bleaching agent activators (TAED) and perfumes (Hutzinger et al. 1992). The consumption of surfactants for both industrial and domestic purposes has resulted in a worldwide production of approximately 17 million tonnes in 2000 (including soap), with expected future growth rates of 3–4% per year globally and of 1.5–2.0% in the EU. Anionic surfactants are commercially the most important product, and the market share is about 50% of total production (Patel 2004).

One of the main consequences of the high production level is the increase in the pollution of wastewaters produced during the washing processes in plants manufacturing toiletries and detergents (Papadopoulos et al. 1997). The high and varied pollution loads of these effluents are, in the main, due to the residual products in the reactor, which have to be removed in order to use the same facility for manufacturing other products. The majority of detergent products reach the environment with domestic and industrial wastewaters. Detergent effluents can cause significant environmental problems because detergent products and their ingredients can be relatively toxic to aquatic life. A direct result of this production is the necessity of the manufacturer to assess the effluent characteristics of their wastewaters and consequently define pollution control methodologies. In order to meet more stringent legislative requirements in discharging the wastewaters to the environment or to the sewage system, an effective treatment process must be applied. Environmental, as well as economic reasons (e.g. increasing water prices), necessitate the search for new methods in wastewater treatment.

Membrane-based separation processes (Goers et al. 2000, Vatai et al. 2000) may be attractive alternatives in wastewater treatment with relation to various physicochemical processes, i.e., coagulation (Mahvi et al. 2004), foaming (Boonyasuwat et al. 2003), advanced oxidation processes (Adams et al. 2000, Schröder et al. 2005) and adsorption on different types of activated carbons (González-García et al. 2004) and polyelectrolytes (Lapitsky et al. 2004). The processes – because of the membrane selectivity – have the potential to recover resources and process water as well as to reduce the high organic load of the wastewaters.

The objectives of the present study were two-fold i) to evaluate the suitability of different types of ultrafiltration membranes for the treatment of industrial wastewaters containing detergents, and ii) the selection of membrane modules for concentration cross-flow tests.

2 MATERIALS AND METHODS

2.1 *Characteristics of the wastewater*

The experiments were conducted using wastewater derived from detergent production. The characteristics of the effluent treated are given in Table 1.

Due to the low pH of the wastewater, the feed stream was neutralized using NaOH solution. Prefiltration using a bag filter (pore diameter: 150 μm) was followed by effluent concentration experiments.

2.2 *Membranes*

The preliminary tests aimed at determining the suitability of ultrafiltration for purification and concentration of detergent effluents were performed using the flat-sheet Amicon (cut-off 100, 50, 1 and 0.5 kDa) and Intersep (cut-off 30, 10 and 4 kDa) membranes. The characteristics of the membranes investigated are given in Table 2.

2.3 *Membrane modules*

The cross-flow experiments were performed using Koch/Romicon capillary ultrafiltration modules with polysulfone membranes of a cut-off equal to 2 and 5 kDa. Technical parameters of the experimental modules are listed in Table 3.

2.4 *Experimental systems*

2.4.1 *Experimental system with Amicon cell*

The experiments with the flat-sheet membranes were carried out in the laboratory set-up presented in Figure 1. The main part of the system consisted of an Amicon 8400 UF cell with a total volume of 350×10^{-6} m^3 and a diameter of 0.076 m. The effective surface area of the membrane was 4.54×10^{-3} m^2. The applied transmembrane pressure amounted to 0.30 MPa.

2.4.2 *Semi-pilot experimental system*

The effluent concentration experiments, as well as the determination of transport properties of ultrafiltration modules, were carried out using the system presented in Figure 2.

The set-up of the cross-flow system incorporated the following major parts: ultrafiltration module (1), hydraulic membrane pump Milroyal C – Dosapro Milton Roy (3), 25 μm prefilter (5), feed tank (2) and permeate tank (10). Additionally, due to the high concentration of suspended solids, effluents were

Table 1. Characteristics of the effluent from detergent production.

Parameter	Value
pH	1.5–3.3
COD-Cr, g O_2/m^3	40,132–59,072
Anionic surfactants (AS), g/m^3	1552–1650
TOC, g C/m^3	11,650

Table 2. Characteristics of the membranes investigated.

Membrane	Producer	Cut-off kDa	Mean pore radius* $\times 10^{-9}$ m
Regenerated cellulose (YC)	Amicon	0.5	0.3
Regenerated cellulose (YM)	Amicon	1	0.5
		50	
		100	15.7
Polyethersulfone (PES)	Intersep	4	0.6
		10	2.04
		30	8.38

* (Majewska-Nowak et al. 1996).

Table 3. Characteristics of Koch/Romicon® ultrafiltration modules.

	PM2	PM5
Cut-off, kDa	2	5
Membrane area, m^2	0.09	0.09
Number of capillaries	66	66
Internal capillary diameter, m	1.1×10^{-3}	1.1×10^{-3}

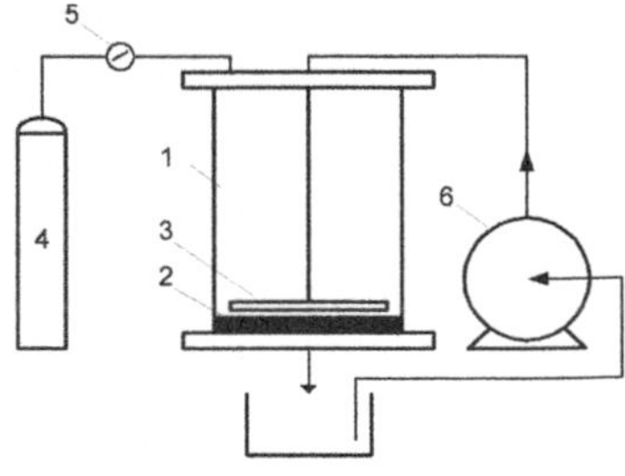

Figure 1. Schematic set-up of ultrafiltration unit (1 – Amicon 8400 ultrafiltration cell, 2 – membrane, 3 – stirrer, 4 – gas cylinder, 5 – reducer, 6 – recirculation pump).

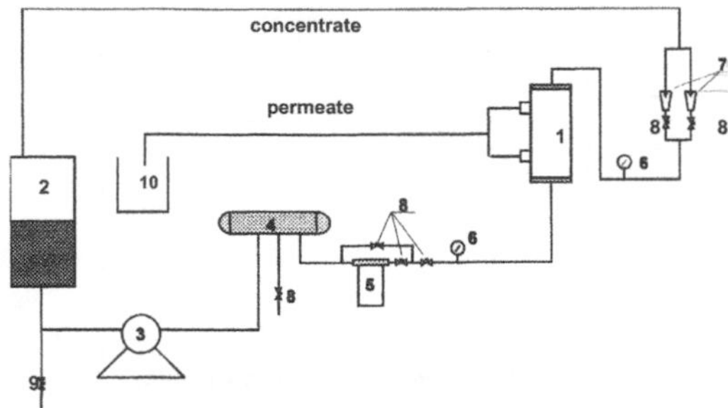

Figure 2. Semi-pilot UF set-up (1 – membrane module, 2 – feed tank, 3 – pump, 4 – pressure accumulator, 5 – preliminary filter 25 μm, 6 – manometers, 7 – rotameters, 8 – control valves, 9 – drain valve, 10 – permeate tank).

passed through a 150 μm filter bag prior to membrane filtration. The total volume of the set-up was 0.08 m^3.

2.5 *Analytical methods*

In order to evaluate the efficiency of the membrane separation, anionic detergents and COD-Cr were determined in the feed and permeate. The concentration of anionic surfactant (AS) was determined using a colour reaction with the indicator Rhodamine G6. The absorbance at 565 nm was measured using a spectrophotometer (Shimadzu UV-MINI-1240) and 0.5×10^{-2} m glass cuvettes.

2.6 *Determination of transport and separation properties of the membranes*

2.6.1 *Preliminary tests*

Ultrafiltration of distilled water and pretreated wastewater was carried out at the transmembrane pressure of 0.30 MPa. In order to maintain a constant concentration of the substances in the feed solution, the permeate was recirculated to the UF cell. At this stage of the work only the separation efficiency of membranes was determined (the value of COD-Cr in the permeate was measured).

2.6.2 *Experiments in the semi-pilot system*

Prior to each cycle, the membrane module was treated with distilled water at 0.10 MPa, until a constant permeate volume flux was established. The transport properties with respect to water and industrial effluents were investigated at transmembrane pressures of 0.05, 0.10 and 0.15 MPa, respectively, and a cross-flow velocity corresponding to 2 m/s.

The permeate volume flux (J) was calculated as follows:

$$J = \frac{V}{t \cdot A}, \quad m^3/m^2 \cdot day \tag{1}$$

where V is permeate volume (m^3), t stands for time (day), and A denotes the effective membrane surface area (m^2).

The wastewater volume of 0.055 m^3 was concentrated at a transmembrane pressure of 0.10 MPa. After each hour of the test the permeate flux was measured. Periodically, the concentration of the anionic surfactants and the COD-Cr values were analysed in the permeate and concentrate.

The degree of effluent concentration was determined according to the following formula:

$$CF = \frac{V_0}{V_t} \tag{2}$$

where CF is the concentration factor, V_0 and V_t denote initial volume of feed and volume of concentrate after time t, respectively.

The efficiency of the ultrafiltration process was determined using the following expression:

$$R = \frac{c_0 - c_p}{c_0}, \quad \% \tag{3}$$

where R is the retention coefficient, and c_0 and c_p are the parameter values of the feed and permeate, respectively.

3 RESULTS AND DISCUSSION

3.1 *Preliminary evaluation of the UF process in the wastewater treatment*

The results of the wastewater treatment experiments using membranes varying in the molecular weight cut-off are shown in Figure 3. The aim of the preliminary tests was to assess the usability of ultrafiltration process in the treatment of industrial wastewater arising from detergent production. With this in mind, only the separation efficiencies of several flat membranes were verified during the first stage of investigation. The tested membranes were characterised by the different cut-off values (from 0.5 kDa to 100 kDa).

The results obtained (Fig. 3) indicate that lower membrane cut-off values improve the effluent treatment efficiency. Ultrafiltration membranes of the high cut-off values enable a 63–76% decrease in the COD-Cr level of the treated wastewaters. The reduction of COD-Cr by more than 80% can be achieved in the UF process for the membranes with a cut-off lower than 10 kDa. The application of the membrane with a cut-off of 0.5 kDa yields the best separation efficiency: the decrease of COD-Cr value is >85%, which corresponds to COD-Cr permeate equal to 8800 g O_2/m^3.

As it was shown by the preliminary tests, the concentration process of the detergent wastewaters can be performed by applying membranes of relatively

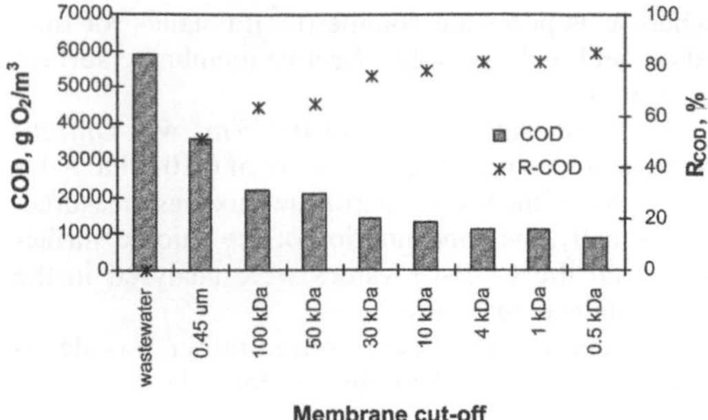

Figure 3. Efficiency of detergent wastewater treatment by various UF membranes ($\Delta P = 0.30$ MPa).

low cut-off values (0.5–4 kDa). These membranes are characterised by a high COD-Cr reduction.

3.2 *Transport properties of membrane modules*

On the basis of the preliminary tests, the capillary modules made by Koch/Romicon were chosen. The modules of this type were characterised by the internal capillary diameter of 1.1×10^{-3} m. It was anticipated that the capillary size was large enough to avoid intensive fouling and membrane clogging. The modules with the cut-off values equal to 2 and 5 kDa were applied.

The transport properties of the PM2 and PM5 modules were determined at the following transmembrane pressures: 0.05, 0.10 and 0.15 MPa. Distilled water and detergent effluents were passed through the modules. The obtained results are shown in Figure 4.

Taking into account the results obtained, it can be concluded that the treated wastewaters showed a strong affinity toward the membrane material. The five-fold and two-fold decrease in the membrane permeability (in comparison to the water flux) found for the PM5 and PM2 modules, respectively, was probably caused by the sorption of contaminants in the membrane pores. The shape of the curves describing a relationship between the effluent volume flux and the applied pressure, indicate that further increase of the transmembrane pressure is not economically justified. A slight increase in the membrane permeability will not compensate the increment of energy demand.

3.3 *Wastewater concentration process*

The suitability of the PM2 and PM5 modules for the concentration of detergent effluents was verified in the next part of investigation. The results obtained are shown in Figures 5 and 6.

The applied modules were characterised by stable properties during a long-term concentration process. The PM5 module provided a greater than three-fold concentration of the initial volume of the detergent wastewaters, whereas for the PM2 module the

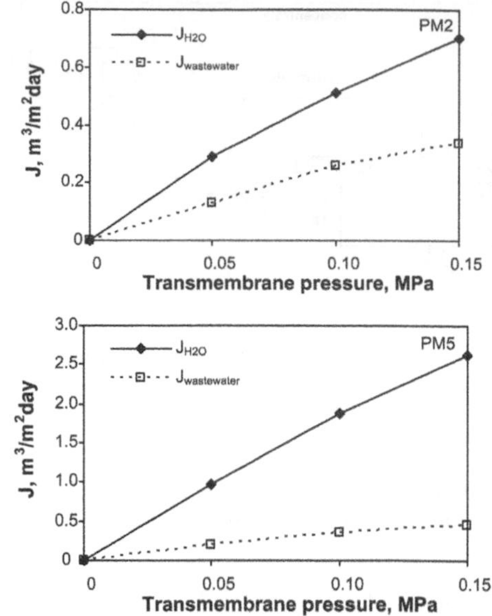

Figure 4. Transport properties of the PM2 and PM5 modules in relation to water and detergent effluents.

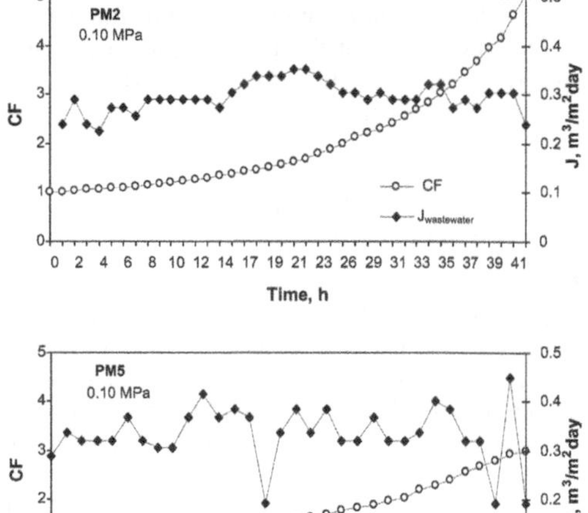

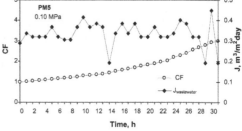

Figure 5. Concentration factor and volume flux versus time of wastewater concentration for PM2 and PM5 modules.

concentration factor amounted to 5. The membrane permeability remained almost on a constant level approaching 0.20–0.30 m^3/m^2day (PM2 module) and 0.25–0.40 m^3/m^2day (PM5 module). The permeate quality was also satisfactory in the course of the wastewater concentration. The permeate COD-Cr value remained in the range of 10,000–20,000 g O_2/m^3

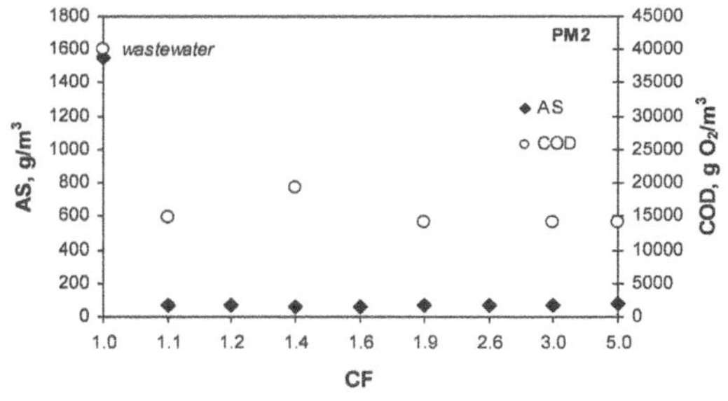

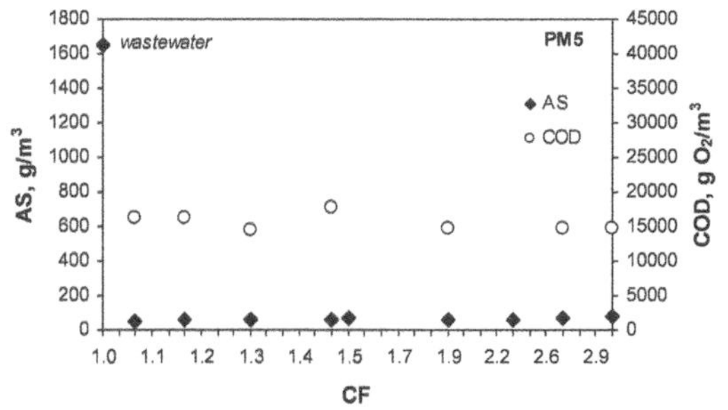

Figure 6. Permeate quality in the course of wastewater concentration for PM2 and PM5 modules.

for both modules. It can be anticipated that concentration factors higher than 5 can be reached in technical installations.

It is worth noting that the effluent permeate is characterised by a low detergent concentration. High retention coefficient of detergents is probably caused by creation of high-molecular-weight pre-micellar aggregates and micelles, which are well separated by UF membranes. At the same time, the interactions between contaminant particles and membrane material can be regarded as a factor supporting the efficiency of the detergent wastewater treatment by ultrafiltration.

4 CONCLUSIONS

Taking into account the presented results, it can be stated that ultrafiltration – low pressure membrane process, enables efficient treatment of industrial wastewaters arising from detergent production.

The results obtained indicate that the UF capillary modules (PM2 and PM5) made by Koch/Romicon are suitable for concentration of highly-polluted effluents containing detergents. The applied modules are characterised by stable transport and separation properties. In the course of a long-term effluent concentration, only a negligible drop in permeability was observed and the permeate quality remained almost constant, despite a systematic increase in pollution load of the concentrate. The separation efficiency for both modules tested was similar: the modules yielded 65–85% reduction in the COD-Cr value and >95% retention of the anionic detergents. The PM5 module seems to be more economically viable when applied to concentration of detergent effluents by UF. This module showed substantially better (by 20%) hydraulic permeability than the PM2 module.

ACKNOWLEDGEMENTS

The work was partly supported by Polish Ministry of Education and Science, Grant #3 T09D 025 26.

REFERENCES

Adams, C.D., Kuzhikannil, J.J. 2000. Effect of UV/H_2O_2 preoxidation on the aerobic biodegradation of quaternary amine surfactants. *Water Research* 34(2): 668–672.

Boonyasuwat, S., Chavadej, S., Malakul, P., Scamehorn, J.F. 2003. Anionic and cationic surfactant recovery from water using a multistage foam fractionator. *Chemical Engineering Journal* 93(3): 241–252.

Goers, B., Mey, J., Wozny, G.W. 2000. Optimised product and water recovery from batch-production rinsing waters. *Waste Management* 20(8): 651–658.

González-García, C.M., González-Martín, M.L., Denoyel, R., Gallardo-Moreno, A.M., Labajos-Broncano, L., Bruque, J.M. 2004. Ionic surfactant adsorption onto activated carbons. *Journal of Colloid and Interface Science* 278(2): 257–264.

Hutzinger, O. 1992. Volume editor: N.T. de Oude. The handbook of environmental chemistry. Detergents.

Lapitsky, Y., Kaler, E.W. 2004. Formation of surfactant and polyelectrolyte gel particles in aqueous solutions. *Colloids and Surfaces A: Physicochemical and Engineering Aspects* 250(1–3): 179–187.

Mahvi, A.H., Maleki, B. 2004. Removal of anionic surfactants in detergent wastewater by chemical coagulation. *Pakistan Journal of Biological Science* 7(12): 2222–2226.

Majewska-Nowak, K., Kabsch-Korbutowicz, M., Bryjak, M., Winnicki, T. 1996. Separation of organic dyes by hydrophilic ultrafiltration membranes, in Proceedings of the 7th World Filtration Congress, Budapest, Hungary, 885–889.

Papadopoulos, A., Savvides, C., Loizidis, M., Haralambous, K.J., Loizidou, M. 1997. An assessment of the quality and treatment of detergent wastewater. *Water Science Technology* 36(2–3): 377–381.

Patel, M. 2004. Surfactant based on renewable raw materials. Carbon dioxide reduction potential and policies and measures for the European Union. *Journal of Industrial Ecology* 7: 46–62.

Schröder, H.Fr., Meesters, R.J.W. 2005. Stability of fluorinated surfactants in advanced oxidation processes – A follow up of degradation products using flow injection–mass spectrometry, liquid chromatography–mass spectrometry and liquid chromatography – multiple stage mass spectrometry. *Journal of Chromatography A* 1082(1): 110–119.

Vatai, Gy., Békássy-Molnár, E., Goers, B., Magda, S., Wozny, G. 2000. CIP water treatment and reuse with membrane separation. *Hungarian Journal of Industrial Chemistry* 28(4): 305–310.

Environmental Engineering – Pawłowski, Dudzińska & Pawłowski (eds)
 ISBN13 978-0-415-40818-9

Application of membrane bioreactors to the treatment of the meat industry wastewater

Jolanta Bohdziewicz & Ewa Sroka
Institute of Water and Wastewater Engineering, Silesian University of Technology, ul. Konarskiego Gliwice, Poland

ABSTRACT: The work aimed at determining the effectiveness of the treatment of the wastewater coming from the meat industry in four hybrid systems:

- combining the biological methods of activated sludge (SBR reactor) and reverse osmosis.
- combining the biological methods of activated sludge (aerobic bioreactor) and ultrafiltration.
- combining the biological methods of activated sludge (bioreactor with separated chambers of denitrification and nitrification), ultrafiltration and reverse osmosis.
- combining coagulation, ultrafiltration and reverse osmosis.

The tests, carried out on the wastewater from the Meat Processing Plant Uni-Lang in Wrzosowa, showed that the biological treatment resulted in satisfactory removal of contaminants from the wastewater, allowing them to be discharged into receiving water. In order to make it possible for the wastewater to be reused in the production cycle, additional treatment with reverse osmosis was necessary.

Keywords: Membrane bioreactor, activated sludge, wastewater produced by the meat industry.

1 INTRODUCTION

One of the branches of the food industry which has the greatest impact on the degradation of the natural environment is the meat industry. Over 90% of the water taken for production needs is discharged as wastewater characterized by considerable content of organic matter, high COD and BOD_5, high concentration of etheric extract, suspension, biogenic and dissolved substances. For this reason, the wastewater should be thoroughly treated prior to its discharge into receiving water (Bickers & Oostrom 2000, Boursier & Beline 2005). This process, however, requires the application of several complementary technologies which enable the wastewater to be purified to such an extent that will allow it to be reused in industry or discharged into receiving water.

Pressure-driven membrane operations have many advantages which make them highly recommended for wastewater treatment. Membranes, notably microfiltration and ultrafiltration ones, can be used in membrane bioreactors to enhance the transport of substances in biochemical reactions (Stephenson, Brindle, Judd, Jefferson 2000, Cicek 2002).

There are two ways to combine pressure-driven membrane operations with biological treatment of wastewater using activated sludge. The first one uses a bioreactor from which a membrane module is physically separated, while in the other the membrane constitutes an integral part of the bioreactor.

In both techniques, the biological treatment of wastewater is based on the rules used in traditional biological treatment plants in which the membrane functions as a secondary settling tank. The filtration module enables a complete separation of biomass from treated wastewater and this considerably increases the concentration of activated sludge in the aeration tank. The concentration of the retained sludge may be even tenfold, compared to the amount of biomass used in the traditional technologies which employ secondary settling tanks. The membrane also enables the retention of swollen sludge. Owing to the retention of suspension particles in the system and long retention time of scarcely biodegradable substances, a decrease in contamination indicators of the treated wastewater is found. The decrease in COD and BOD_5 reaches 96% with complete removal of suspension (Visvanathan, R. Ben Aim, Parameshwaran, 2000, Rosenberger, Kruger, Witzig, Manz, Szewzyk, Kraume 2002).

2 MATERIALS AND METHODS

The wastewater was sampled from the Meat-Processing Plant "UNILANG" in Wrzosowa whose activity covers the slaughter and processing of pigs.

Table 1. Characteristics of raw wastewater and permissible loadings.

Pollution indices	Concentration of pollution in raw wastewater, g/m^3 (Range)	Load, kg/d (Mean value)	Permissible standards, g/m^{3*}
pH	6.0–7.2	–	6.5–9.5
COD	2780–6720	309.2	125
BOD_5	1200–3000	126.8	25
Total nitrogen	49–287	13	30
Total phosphorus	15–70	2.1	3
Total suspension	112–1743	26.1	30

* Regulation of the Minister of the Environment of 8 July 2004 concerning the requirements of wastewater discharge into receiving water and ground.

The values of the basic and eutrophic pollution indicators were high and ranged widely during the whole production cycle. The wastewater was of red and brown colour, smelled unpleasant and tended to foam and putrefy. The characteristics of the raw wastewater are presented in Table 1.

The wastewater pre-treated on screens, sieves, grease trap and finally a flotation unit was subsequently treated in four hybrid systems which combined:

1 activated sludge system (SBR) and reverse osmosis (Bohdziewicz, Sroka, Lobos 2002, Sroka, Kaminski, Bohdziewicz 2004, Bohdziewicz & Sroka 2005),
2 coagulation with ultrafiltration and reverse osmosis (Bohdziewicz, Sroka, Korus 2003, Bohdziewicz, Sroka 2005),
3 activated sludge system (aerobic bioreactor) and ultrafiltration,
4 activated sludge (bioreactor equipped with separate denitrification and nitrification tanks) with ultrafiltration and reverse osmosis. The investigations were aimed at comparing the efficiency of wastewater treatment in the above systems.

The effectiveness of membrane operations was assessed on the basis of the observed volume permeate fluxes (J_v [$m^3/m^2 \times s$]) and removal of particulate contaminants (R [%]).

The tests were carried out on flat ultrafiltration membranes (DS-CQ – cellulose, DS-GH 2K and DS-GH 8K – composite) and osmotic membrane (SS-10 – cellulose acetate) manufactured by Osmonics (GE Osmonics Labstore, www.osmolabstore.com). Polysulfone ultrafiltration membranes, prepared in the laboratory (Institute of Water and Wastewater Engineering, Silesian University of Technology) by means of SPE i.e. PSf-12 and PSf-15, were also tested. Membrane filtration was carried out at a pressure of 0.1–0.3 MPa for UF and 2.0 MPa for RO. Both processes were conducted at the linear flow velocity of wastewater over the membrane of 2.0 m/s.

The biological processes enabled the determination of the most favourable substrate loading of activated sludge. The measurements tested the range from $0.002\ gCOD/g_{TS} \times d$ to $0.75\ gCOD/g_{TS} \times d$. The tests were conducted applying the most favourable process parameters determined earlier i.e. concentration of dry weight in the tank of 5 g/dm^3 and aeration rate of 800^3 air/h. The ratio of stirring time to the total of stirring and aeration times was 0.3, while the residence time for wastewater reached 12 h. The efficiency of wastewater treatment was assessed on the basis of the decrease in BOD_5 and COD, and the concentration of biogenic substances i.e. nitrogen and phosphorus. The determinations of COD, phosphorus and total nitrogen concentrations were carried out using test kits for a MERCK SQ 118 photometer, while BOD_5 was assayed with OxiTOP measuring cylinders produced by WTW (Wissenschaftlich-Technische Werkstätten, e-mail: info@wtw.com). The weight of the sludge was determined gravimetrically.

3 RESULTS AND DISCUSSION

3.1 *Wastewater treatment in an SBR reactor combined with reverse osmosis*

The treatment of the wastewater was carried out biologically applying the activated sludge method in a 40 dm^3 chamber. The chamber was equipped with two aeration pumps MAXIMA R manufactured by Elite whose average capacity was 420 dm^3air/h and an RZR 2020 stirrer manufactured by "Heidolph" with adjustable rotation velocity ranging from 40 to 2000 rpm.

Reverse osmosis was conducted in a GH 100–400 high-pressure apparatus, capacity – 400 cm^3, produced by the same company. The system operated in the dead-end mode on flat membranes whose active area was 36.3 cm^2.

Figs 1 and 2 show the correlation between COD, BOD_5, phosphorus and total nitrogen removal and the activated sludge loading applied.

The increasing activated sludge loading caused a gradual decrease in contaminant removal. The highest COD removal was observed for the loading of $0.05\ gCOD/g_{TS} \times d$ which reached 98.9% (raw wastewater: 5200 gO_2/m^3, treated wastewater 57.2 gO_2/m^3), while the lowest one of 90.2% was found for the loading of $0.75\ gCOD/g_{TS} \times d$, COD of treated wastewater being 509.6 gO_2/m^3. No clear correlation between BOD_5 removal and the activated sludge loading applied was observed. The removal was very high, reaching 99% (30 mgO_2/dm^3).

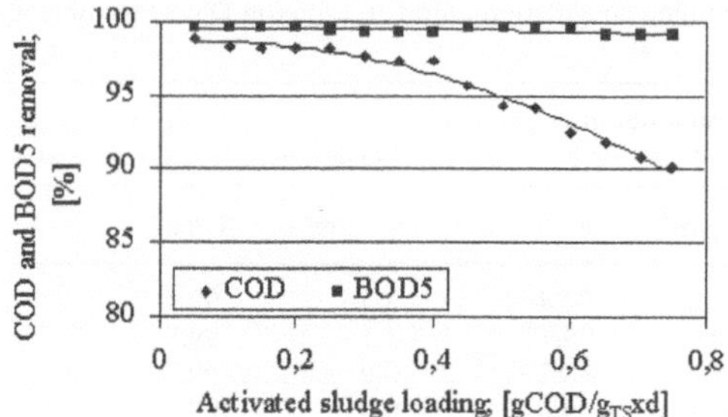

Figure 1. Correlation between COD and BOD_5 removal and activated sludge loading.

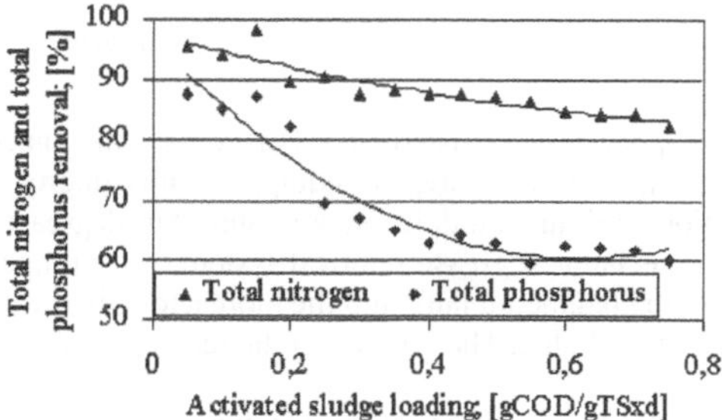

Figure 2. Correlation between removal of biogenic substances and activated sludge loading.

The removal of total nitrogen also depended on the substrate loading of activated sludge applied. It was the highest for the loading of 0.15 $gCOD/g_{TS} \times d$, reaching 98.2% (raw wastewater – 530 gN_{tot}/m^3, treated wastewater – 9.5 gN_{tot}/m^3). The lowest removal of that biogene was 82.1% (raw wastewater 236 gN_{tot}/m^3, treated wastewater – 42.2 gN_{tot}/m^3) for the loading of 0.75 $gCOD/g_{TS} \times d$.

The highest phosphorus removal of 92.6% (treated wastewater – 2.8 gP/m^3) was found for the loading of 0.15 $gCOD/g_{TS} \times d$, while the lowest one of 59.4% (treated wastewater – 20.9 gP/m^3) was observed for 0.55 $gCOD/g_{TS} \times d$. The results obtained enabled the determination of the most favourable activated sludge loading of 0.15 $gCOD/g_{TS} \times d$ and are compiled in Table 2.

It has been found that meat industry wastewater can be treated biologically using the activated sludge system, however, only to the extent which enables its discharge into receiving water. Since the meat industry uses large quantities of water, most of which is usually discharged as highly loaded wastewater, an attempt to treat it additionally so that it could be reused in the production cycle, was made. Thus, the wastewater was additionally treated by reverse osmosis following its biological treatment.

Table 2. Effectiveness of wastewater treatment using activated sludge system for the activated sludge loading of 0.15 $gCOD/g_{TS} \times d$.

Pollution indices	Raw wastewater, g/m^3	Retention %	Wastewater after activated sludge bioreactor, g/m^3
COD	5300	98.1	102
BOD_5	2900	99.6	10
Total nitrogen	557	98.2	9.5
Total phosphorus	37.8	87.3	4.8

The tests determined the correlation between the volume flux of treated wastewater and its recovery. It has been found that the membrane efficiency at the 20% permeate recovery decreased negligibly and was lower by only 10.6% compared to the flux of deionized water. The average permeate flux was $1.6 \times 10^{-6}\ m^3/m^2 \times s$.

The concentrations of contaminants in treated wastewater after the membrane process were as follows: phosphorus – 0.09 gP/m^3, total nitrogen – 1.3 gN_{tot}/m^3, COD and BOD_5 were also low – 10.8 gO_2/m^3 and 5 gO_2/m^3 respectively. They enabled the wastewater to be reused in the production cycle.

3.2 *Wastewater treatment in the hybrid process of coagulation, ultrafiltration and reverse osmosis*

Ultrafiltration and reverse osmosis were carried out applying a SEPA CF-HP pressure apparatus equipped with a plate-frame module produced by Osmonics, membrane active area – 155 cm^2. The system operated in the cross-flow mode.

Raw wastewater, after preliminary coagulation with an aqueous solution of $Fe_2(SO_4)_3$ (PIX-113), conc. – 40–42%, (stirring time – 45 s, slow stirring and sedimentation times – 30 min.), was treated on DSCQ, DSGH-2K, DSGH-8K, PSf-12 and PSf-15 ultrafiltration membranes followed by an additional treatment with reverse osmosis on SS-10 membrane. The effectiveness of wastewater treatment in this system is shown in Table 3.

It is clearly seen that the pollution indexes were still high after coagulation, which prevented wastewater discharge into receiving water. Therefore, it was additionally treated with UF. The highest removal of contaminants was observed for DSCQ and PSf-15 membranes (Fig. 3), which was respectively: COD – 70.6% (159 gO_2/m^3) and 65.5% (165 gO_2/m^3), BOD_5 – 72.7% (140 gO_2/m^3) and 62.5% (130 gO_2/m^3), total nitrogen – 64% (52 g/m^3) and 67% (48 g/m^3), phosphorus – 98.2% (0.1 g/m^3) and 96% (0.1 g/m^3).

Table 3. Effectiveness of wastewater treatment in the system combining coagulation, ultrafiltration (on DSCQ membrane) and reverse osmosis.

Pollution indices	Raw wastewater g/m^3	Wastewater after coagulation process		Wastewater after ultrafiltration process		Wastewater after RO process	
		C, g/m^3	R, %	C, g/m^3	R, %	C, g/m^3	R, %
COD	2839	542.0	80.1	159.0	94.4	3.5	99.9
BOD$_5$	1890	490.0	74.1	130.0	93.1	3.1	99.8
Total nitrogen	447.5	144.9	67.6	52.7	88.2	0.9	99.7
Total phosphorus	27.6	3.4	87.7	0.1	99.6	0.0	100

C – Concentration, R – Retention.

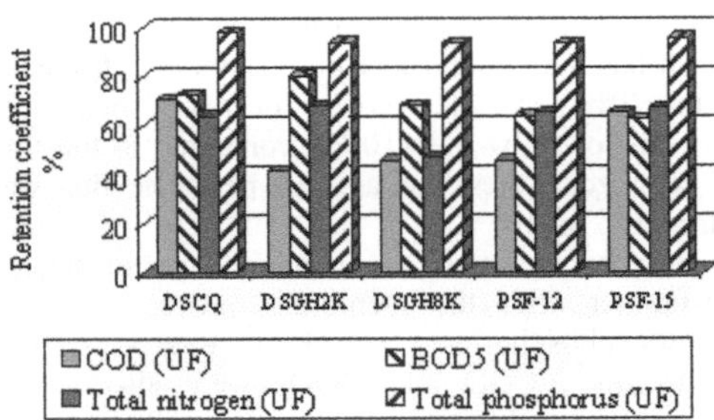

Figure 3. Effect of membrane type on contaminant removal.

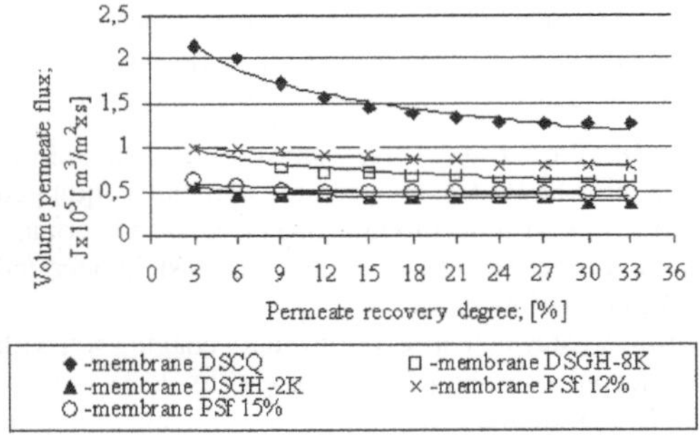

Figure 4. Correlation between volume permeate flux and recovery.

DSCQ was considered to be a better membrane since it was characterized by a higher permeate flux (Fig. 4). The removal of contaminants for the other membranes was lower by several per cent.

All the pollution indices were characterized by very high removal during reverse osmosis (over 99%). The highest volume permeate flux of 0.67×10^{-6} [m^3/m^2 $\times$ s] was observed when the wastewater was filtered following coagulation and ultrafiltration on DSCQ membrane. This enabled a reuse of the wastewater in the production cycle.

3.3 *Wastewater treatment in aerobic membrane bioreactor*

The apparatus consisted of a balancing tank joined with a membrane bioreactor equipped with a microfiltration capillary module. Raw wastewater, degreased in a mechanical grease removal system, was pumped from the balancing tank into the bioreactor filled with activated sludge. The volume of the reaction tank was 25 dm^3.

The bottom part of the bioreactor was filled with air whose bubbles travelled upward along the capillaries caused them to vibrate, making it impossible for the sludge to deposit on the surface. This is important when the module operates at high activated sludge concentration. This method of tank aeration also enabled its thorough mixing. The reaction tanks were equipped with sensors that detected the level of wastewater. It was also necessary to install a balancing tank to ensure a constant, assumed loading of the sludge.

The polymer capillary membranes used in the tests were of Canadian manufacture and were purchased from Zenon System Ltd., Tychy (Potrzebka, Wawrzyńczyk 2002). They were of high mechanical strength and chemical resistance. So far, these types of capillary modules have been used in the Zee Weed® to treat surface waters of different contamination level.

During the tests, the capillary module operated at the underpressure of 2.5×10^{-4} MPa – 11×10^{-4} MPa, created by the filtrate pump. 800 dm^3/h air was pumped into the bottom part of the aerobic tank. The wastewater treated biologically was flowing inside through the walls of the capillaries.

Fig. 5 shows the correlation between the decrease in COD of treated wastewater and activated sludge loading. The highest COD removal of 98.1%, which corresponded to COD concentration of 73 gO_2/m^3 (raw wastewater 3880 gO_2/m^3), was observed for the loading of 0.063 gCOD/g_{TS} $\times$ d. The lowest removal of 93.2% was found for the loading of 0.004 gCOD/g_{TS} $\times$ d. The reduction in organic pollutants (BOD$_5$) in treated wastewater, COD

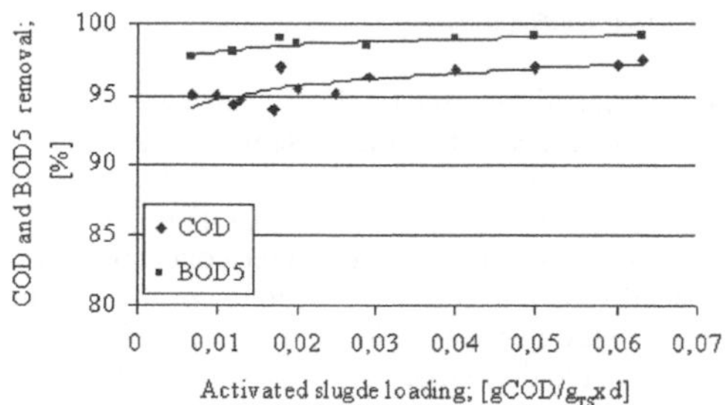

Figure 5. Correlation between COD and BOD_5 removal and activated sludge loading.

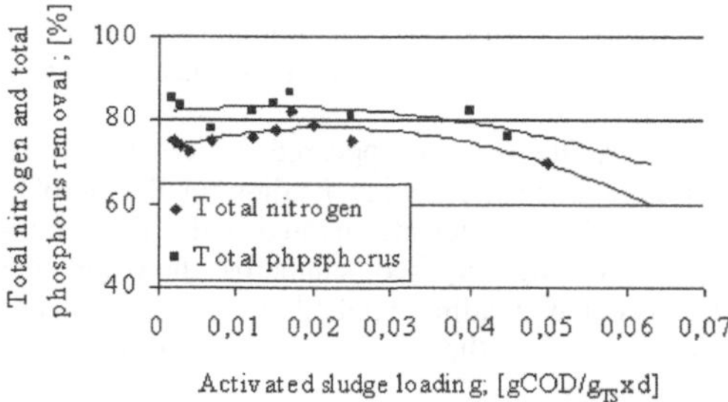

Figure 6. Correlation between total nitrogen and phosphorus removal and activated sludge loading.

likewise, depended on activated sludge loadings (Fig. 5). Its maximum removal was 99.4% for the loading of 0.063 $gCOD/g_{TS} \times d$ (from 1600 gO_2/mn^3 to 10 gO_2/m^3). The lowest reduction in organic substances of 97.6% (BOD_5 of treated wastewater – 20 gO_2/m^3) was revealed at the sludge loading of 0.004 $gCOD/g_{TS} \times d$.

The strong influence of activated sludge loading on the concentration in treated wastewater was also observed for total nitrogen (Fig. 6). The highest reduction in total nitrogen, by as much as 82.1% (treated wastewater – 29 gN_{tot}/m^3), was found for the loading of 0.017 $gCOD/g_{TS} \times d$. The lowest removal of total nitrogen of 65.7% was observed for the loading of 0.05 $gCOD/g_{TS} \times d$ (treated wastewater – 55.6 gN_{tot}/m^3.

Phosphorus was the next element assayed (Fig. 7). The most favourable results were obtained for the sludge loading of 0.003 $gCOD/g_{TS} \times d$ (89.6%, 3.0 gP/m^3 in treated wastewater). The lowest phosphorus removal of 67.9% was found for the loading of 0.063 $gCOD/g_{TS} \times d$ (treated wastewater – 9.3 gP/m^3).

It has been found that not all the pollution indices were the lowest for the same sludge loading. Because total nitrogen concentration in treated wastewater at the loading of 0.017 $gCOD/g_{TS} \times d$ was lower than the permissible one, it was selected as the most favourable

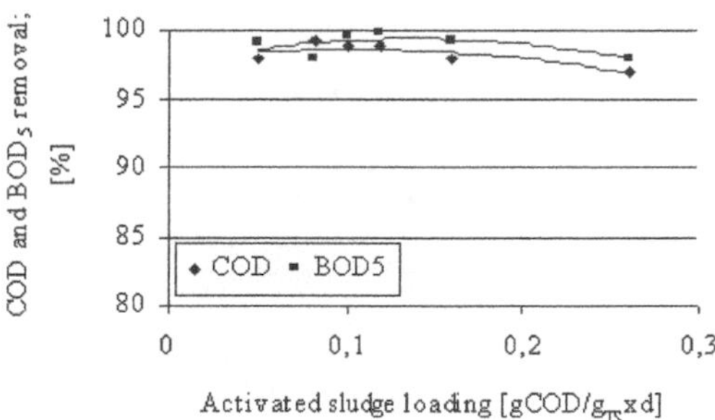

Figure 7. Correlation between COD and BOD_5 removal and activated sludge loading.

Table 4. Reduction in pollution indexes for treated wastewater at the activated sludge loading of 0.017 $gCOD/g_{TS} \times d$.

Pollution indices	Concentration of pollution in raw, wastewater, g/m^3	Retention, %	Concentration of pollution in purified wastewater, g/m^3
COD	3880	96.9	115
BOD_5	1600	99.1	14
Total nitrogen	162	82.1	29
Ammonia nitrogen	48	98.8	0.6
Total phosphorus	29	86.5	3.9

to carry out the process in question. Table 4 shows the pollution indexes which characterize treated wastewater for the optimum activated sludge loading of 0.017 $gCOD/g_{TS} \times d$. Those conditions enabled their treatment to the extent which allowed the wastewater to be discharged into receiving water.

3.4 *Wastewater treatment in membrane bioreactor with separated denitrification and nitrification tanks*

The apparatus used consisted of a balancing tank, anaerobic tank and finally an aerobic one in which an ultrafiltration capillary module was installed. Raw wastewater was pumped from the balancing tank into the 25 dm^3 anaerobic tank and subsequently a 25 dm^3 aerobic tank. The recirculation of the sludge from the aerobic tank into the anaerobic one was 300%. Denitrification took place in the anaerobic tank while nitrification occurred in the aerobic one. The tests employed the same membrane capillary module used in the aerobic membrane bioreactor.

The tests revealed that COD and BOD_5 of treated wastewater depended negligibly on the activated sludge loadings applied (Fig. 7). Their removal was

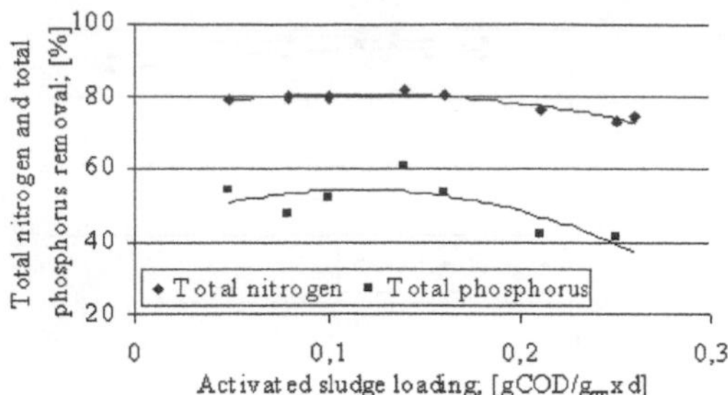

Figure 8. Correlation between biogenic compounds removal and activated sludge loading.

Table 5. Effectiveness of wastewater treatment in membrane bioreactor using activated sludge loading of 0.14 $gCOD/g_{TS} \times d$.

Pollution indices	Concentration of pollution in raw, wastewater, g/m^3	Retention, %	Concentration of pollution in purified wastewater, g/m^3
COD	1480	96.9	115
BOD_5	980	99.1	14
Total nitrogen	130	98.2	24.2
Total phosphorus	7.4	87.3	2.9

stable and ranged from 97% to 99%. It did not exceed the permissible standards with respect to the discharge of wastewater into receiving water (15–28 $gCOD/m^3$, 8–10 $gBOD_5/m^3$).

The next step of the research enabled the determination of the correlation between biogenic compounds removal in treated wastewater and activated sludge loading. Total nitrogen removal was the highest for the sludge loading of 0.14 $gCOD/g_{TS} \times d$ and reached 81.4% (Fig. 8) (raw wastewater – 130 gN_{tot}/m^3, treated wastewater – 24.2 gN_{tot}/m^3).

The lowest removal was 72.7% (raw wastewater 154 gN_{tot}/m^3, treated wastewater – 42.1 gN_{tot}/m^3) for the loading of 0.25 $gCOD/g_{TS} \times d$.

Phosphorus removal was the most dependent on activated sludge loading. Its lowest value was observed at the loading of 0.14 $gCOD/g_{TS} \times d$ when the the phosphorus concentration in treated wastewater reached 2.9 gP/m^3 (removal 60.7%). The lowest removal was found for the loading of 0.26 $gCOD/g_{TS} \times d$ (41.2%) which corresponded to the concentration of phosphorus from raw wastewater of 4.4 gP/m^3 (Fig. 8). The results indicated that the activated sludge loading of 0.14 $gCOD/g_{TS} \times d$ was the most favourable. (Table 5).

4 CONCLUSIONS

It is concluded that meat industry wastewater can be treated using any of the four systems in question.

Recommended parameters for the SBR reactor are: activated sludge loading – 0.15 $gCOD/g_{TS} \times d$, aeration rate – 800 $dm^3 air/h$, constant concentration of activated sludge in the tank – 4 g/dm^3, residence time of the wastewater in the bioreactor – 12 h. These conditions enable discharge of wastewater into receiving water.

The wastewater described may also be treated in the hybrid system of coagulation, UF and RO. The combination of coagulation and ultrafiltration is not sufficient to allow the wastewater to be discharged into receiving water. However, an additional treatment by reverse osmosis enabled its reuse in the production cycle of a plant.

0.017 $gCOD/g_{TS} \times d$ appeared to be the most favourable activated sludge loading in the aerobic bioreactor. The wastewater treated under those conditions met the required standards, however, the extremely low loading applied was highly uneconomical and thus the use of an anaerobic tank was recommended.

The investigations revealed that the membrane bioreactor with separate denitrification and nitrification tanks produced a similar removal of contaminants from treated wastewater, applying, however, a higher activated sludge loading (both membrane bioreactor and SBR reactor operated under the same conditions). Nevertheless, the membrane bioreactor has one decisive advantage in comparison with the conventional wastewater treatment; it does not serve as a secondary settling tank only (which retains suspension, colloids and swollen sludge that partially pass to the receiving water in conventional treatment plants), but also removes macromolecular contaminants (COD and BOD_5). It is possible to use higher concentrations of suspensions in the activated sludge tank (up to 25 kg/m^3, while it is only 1.5 to 3.0 kg/m^3 in conventional treatment plants), which enables the sludge to operate at lower substrate loadings.

When the wastewater needs to be recirculated to the production cycle, an additional treatment using reverse osmosis is recommended.

REFERENCES

Bohdziewicz J, Sroka E, Lobos E, 2002, Application of the system which combines coagulation, activated sludge and reverse osmosis to the treatment of the wastewater produced by the meat industry, *Desalination* 144 (1–3), 393–398.

Bohdziewicz J, Sroka E, Korus I, 2003, Application of ultrafiltration and reverse osmosis to the treatment of the wastewater produced by the meat industry, *Polish Journal of Environmental Studies* 12 (3), 269–274.

Bohdziewicz J, Sroka E, 2005, Integrated system of activated sludge-reverse osmosis in the treatment of the wastewater from the meat industry, *Process Biochemistry* 40 (5), 1517–1523.

Bohdziewicz J, Sroka E, 2005, Treatment of wastewater from the meat industry applying integrated membrane systems, *Process Biochemistry* 40 (3–4), 1339–1346.

Boursier H, Beline F, Paul E, 2005, Piggery wastewater characterisation for biological nitrogen removal process design, *Bioresource Technology* 9, 351–358.

Cicek Nazim, 2002, Membrane bioreactors in the treatment of wastewater generated from agricultural industries and activities, Written for presentation at the AIC 2002 Meeting, Saskatoon, Saskatchewan, July 14–17, Paper No. 02-404.

Paul O. Bickers, Albert J. van Oostrom, 2000, Availability for denitrification of organic carbon in meat-processing wastestreams, *Bioresource Technology* 73, 53–58.

Potrzebka A, Wawrzyńczyk J, 2002, The application of ZeeWeed® microfiltration for the contaminated surface water treatment for potable water production, *Scientific materials of The Silesian University of Technology*, z. 46, 137–146.

Rosenberger S, Kruger U, Witzig R, Manz W, Szewzyk U and Kraume M, 2002, Performance of a bioreactor with submerged membranes for aerobic treatment of municipal waste water. *Water Research* 36 (2): 413–420.

Regulation of the Minister of the Environment of 8 July 2004 concerning the requirements of wastewater discharge into receiving water and ground (Bills and Acts Bulletin 168. 1763).

Sroka E, Kaminski W, Bohdziewicz J, 2004, Biological treatment of meat industry, Desalination 162 (1–3), 85–91.

Stephenson T, Brindle K, Judd S and Jefferson B, 2000, Membrane Bioreactors for Wastewater Treatment. London, UK: IWA Publishing.

Visvanathan C, Ben Aim R and Parameshwaran K, 2000, Membrane separation bioreactors for wastewater treatment, *Critical Reviews in Environmental Science and Technology* 30 (1): 1–48.

Environmental Engineering – Pawłowski, Dudzińska & Pawłowski (eds)
© 2007 Taylor & Francis Group, London, ISBN13 978-0-415-40818-9

Use of dissolved air flotation in the pretreatment of wastewaters arising from vegetable fat processing

Sławomir Żak
Department of Technology and Chemical Engineering, University of Technology and Agriculture, Bydgoszcz, Poland

ABSTRACT: The paper presents the results on the use of dissolved air flotation (DAF) with preliminary oxidation and flocculation in pretreating wastewaters from the vegetable fat processing industry. A specially constructed set was used for this purpose. The relation of the basic indicating values reduction has been determined as a function of hydrogen peroxide concentration at constant levels of: doses of the applied coagulant and flocculent, the saturation pressure and the recirculation rate of pretreated effluents. The use of H_2O_2 in doses 300.0–500.0 g/m^3 effluent in flotation reduced TSS (total suspended solids) and O & G (oil and greases) by more than 95%, whereas COD (chemical oxygen demand) and BOD_5 (biological oxygen demand) were reduced only by ca. 60%. It also significantly changed the relation COD(BOD_5):N:P (N – nitrogen concentration, P – total phosphorus concentration) compared to the same relation in raw wastewaters.

Keywords: Industrial wastewater pretreatment; vegetable oils and greases; DAF.

1 INTRODUCTION

Dissolved air flotation (DAF) is a well known method for pretreating the processing effluents from manufacturing vegetable oils and greases (Vlyssides et al. 2004, Rubio et al. 2002, Schneider et al. 1995a, b, Rüffer & Rosenwinkel 1991, Ho & Tan 1989, Ng et al. 1988, Lovett & Travers 1986, Hemming 1977). Carried out using pressure saturation, the method permits a high reduction of TSS, O & G, COD and BOD_5 (Żak 2005a, Vlyssides et al. 2004, Rubio et al. 2002, Steiner & Gec 1992, Ho & Tan 1989, Ng et al. 1988). Literature data indicate that the reduction levels in installations using pressure flotation enable efficient pretreatment of effluents from fat processing factories. The reduction level of TSS, O & G, BOD_5 and COD exceeds 50% (Żak 2005a, Steiner & Gec 1992). Flotates are characterized by high concentrations of suspended solids, oils, fatty acids, glycerides, phosphatides and lecithins (Beccari et al. 1999, Ho & Tan 1989, Ng et al. 1988). The effectiveness of this technique may be improved by aiding the process with coagulation (and electro-coagulation), flocculation and oxidation (Chen 2004, Inan et al. 2004, Schneider et al. 1995b, Lovett & Travers 1986). Additional flotation aided by hydrogen peroxide oxidation is a relatively little known method of increasing the effectiveness of dissolved air flotation (DAF) (Żak 2005a, Inan et al. 2004, Chen 2004, Maennig & Scherer 1989, Lovett & Travers 1986, http://www.h2o2.com). The use of this reagent increases neither the volume of sludge nor the secondary salinity of the pretreated wastewaters (Maennig & Scherer 1989, Lovett & Travers 1986, http://www.h2o2.com). The reagent provides full oxidation and mineralization as well as selective oxidation of some organic compounds found in the wastewater contamination load, which is reflected in the reduction of COD and BOD_5 (Maennig & Scherer 1989, Lovett & Travers 1986, http://www.h2o2.com). Furthermore, the release of oxygen – a product of hydrogen peroxide auto-decomposition ($H_2O_2 \rightarrow O_2 + 2H^+ + 2e^-$) – considerably aids the contaminant flotation due to formation of gas micro-bubbles, thus intensifying the process (Żak 2005a, Maennig & Scherer 1989, http://www.h2o2.com). The phenomenon of forming micro-bubbles, advantageous for flotation, is also a result of organic substrate decomposition which leads to the formation of carbon dioxide (http://www.h2o2.com). Hydrogen peroxide has bacteriostatic properties, which induce higher biological stability of the after-flotation effluents (Casani et al. 2005, http://www.h2o2.com). Transformations of the pollutant load components induced by hydrogen peroxide may also result in reactions between the organic substrate (R) and H_2O_2, such as: addition (e.g.: $H_2O_2 + R \rightarrow R \cdot H_2O_2$), substitution (e.g.: $H_2O_2 + 2RH \rightarrow ROOH + HR + 2H^+$ or $H_2O_2 + 2RX \rightarrow ROOR + 2HX$) and oxidation (e.g.: $H_2O_2 + R \rightarrow RO + H_2O$) (Talinli & Anderson 1992, http://www.h2o2.com, http://www.hydroxyl.com). Chemical changes proceeding according to the above reactions lead mainly to more degradable substances at the final stage of wastewater biological treatment. The main goal of the work was to assess the effect of H_2O_2 dose in enhancing pressurized flotation in the processing effluents from vegetable fat production. The tests were carried out in an attempt to reduce the load of the pollutants and to change the relations COD/BOD_5, $COD(BOD_5)/N$, $COD(BOD_5)/P$ and N/P by physico-chemical pretreatment to prepare the effluents for non-collision treatment by using biological methods.

2 MATERIALS AND METHODS

The method was tested and the investigations were carried out in the system of our own construction, depicted in the diagram (Fig. 1). Raw processing wastewaters (SS1) were a mixture of after-processing effluents which included wastewaters from raw oil, refinery plants and soap acidification installations. Washing waters from cleaning installations and production areas were also included. The basic parameters of the raw wastewaters used in the flotation process are presented in Table 1. Wastewaters with no mechanical treatment were used for the tests.

2.1 *Testing unit and procedure*

Wastewaters, after stabilizing with sodium benzoate, were stored in an averaging-storage tank (1) and mixed with the use of a wastewater pump (1.1) in order to average and maintain the temperature at $15.0 \pm 1.0°C$ by heat exchange in a coil pipe (1.2). The averaged raw wastewater (SS1), after pipe introduction of a coagulant (10% aqueous solution of lime milk) dosed from the station (3) or/and 30% aqueous solution of hydrogen peroxide dosed from the station (4), was directed through a pipe reactor (6) into a flotator chamber (7) of volume of the decompression chamber $0.15\,m^3$, equipped with a superficial system of flotate outflow (7.1). A pressurized flotation unit was equipped with a saturation tank (8), with a pump (8.1) and an aeration valve (8.2), by which the air from a high-pressure gas cylinder (9) was introduced. The flocculent from the station (5) was directed into a pipeline of the pipe reactor (6) before it was introduced into the flotator decompressing chamber (7).

The averaged raw wastewaters were flotated by using the pretreated effluents saturated with air under pressure of 400 kPa for 7.0–10.0 minutes and treated with a constant concentration of hydrogen peroxide. For the option with preliminary coagulation using lime milk, the dosage of the reagent was adjusted so that its concentration (as CaO) amounted to $1.0\,kg/m^3$ for

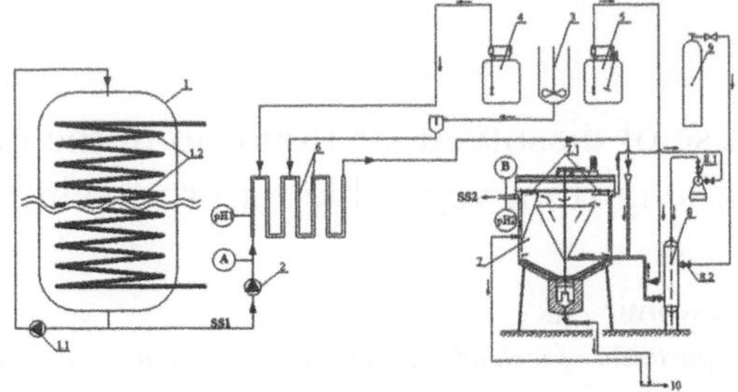

Figure 1. Flow diagram of the testing system. Where: SS1 and SS2) raw wastewater and pretreated effluent, (1) averaging-storage tank, (1.1) circulation pump, (1.2) coil pipe, (2) wastewater pumping station, (3) lime milk preparation and dosing station, (4) hydrogen peroxide dosing station, (5) flocculent preparation station, (6) pipe reactor, (7) flotator, (7.1) flotate skimmer, (8) saturation tank, (8.1) saturation pump, (8.2) aerating valve, (9) compressed air cylinder, (10) to a tank for storing flotate and sediments, pH1 and pH2) pH-meters. A and B – sampling points.

each type of effluent. The process was additionally intensified by using the flocculent Praestol 2340 BC (Stockhausen; 20.0 g per m^3 effluent). Wastewaters remained in the flotator chamber (7) for 30 minutes. 20% of the pretreated wastewater volume was air saturated.

2.2 *Analytical part*

In averaged samples of raw wastewaters (SS1) taken at sampling point A (Fig. 1) and after flotation (SS2) (point B in Fig. 1), the following indicator parameters were determined: reaction (PN-90/C-04540/01), total suspended solids (TSS) (PN-72/C04559/02), chemical oxygen demand (COD) by dichromate method (PN-74/C-04578/03), biochemical oxygen demand (BOD) (PN-EN 1899-1, 2002), ether extract (EE) (PN-74/C04573/02, 03 and 10), total phosphorus (PN-91/C-04537/09), and total nitrogen (PN-73/C-04576/12). The sodium benzoate blank was determined and removed from obtained COD and BOD_5 values. Additionally, an actual chemical oxygen demand was presented after correcting the value by subtracting the fraction induced by hydrogen peroxide, based on the equation $COD_r = COD_p - f \cdot d$ (COD_r – actual, COD_p – determined in the after-reaction sample, c – H_2O_2 concentration in a sample – determined iodometrically, f = 0.25 – a correction coefficient assumed on the basis of the relation defined by Talinli and Anderson (Żak 2005b, Talinli & Anderson 1992).

3 RESULTS AND DISCUSSION

Processing techniques used in vegetable oil plants are inseparably connected with water management and

Table 1. The physico-chemical characteristic of the raw wastewaters*.

No.	Parameter	Unit	Processing wastewaters[(a)]		
			SS1(RO)	SS1(SO)	SS1(PO)
1	Reaction	pH	4.1	4.3	4.1
2	Total nitrogen	gN/m^3	80.5	130.1	157.3
3	Total phosphorus	gP/m^3	98.4	104.7	120.6
4	Ether extract (EE)	g/m^3	453.4	393.6	624.7
5	Total suspended solids (TSS)	g/m^3	466.7	420.7	587.3
6	Chemical oxygen demand (COD)	gO_2/m^3	7773.0	5801.4	8294.7
7	Biochemical oxygen demand (BOD_5)	gO_2/m^3	2980.6	2342.8	3420.2

* The processing wastewaters were determined from the individual productions, respectively: SS1 (RO) – production of rape oil, SS1(SO) – production of soya oil, SS1(PO) production of palm oil.

producing processing effluents. The methods applied to pretreating such effluents basically use pressure flotation sometimes supplemented by (i) chemical enhancement by coagulation (using mainly lime milk) and (ii) flocculation (Vlyssides et al. 2004, Rüffer & Rosenwinkel 1991).

A recent trend in the pretreatment of effluents from the vegetable fat manufacturing industry is the use of flotation supplemented with oxidizers additionally evolving gaseous products which intensify the process of flotation by forming micro-bubbles of the floating gas, e.g. oxygen, in the reaction environment (Steiner & Gec 1992, Maennig & Scherer 1989, http://www.h2o2.com). Since hydrogen peroxide belongs to such a group aiding flotation (Żak 2005a, Inan et al. 2004, Chen 2004, Beccari et al. 1999, Steiner & Gec 1992, Maennig & Scherer 1989, http://www.h2o2.com), the two methods (DAF and oxidation) can be combined, introducing other reagents with no secondary adverse effects after their application. In order to increase the level of reduction in indicating parameters, tests to intensify conventional pressure flotation with hydrogen peroxide were carried out. It was assumed that the flotation would be induced by dissolved air and molecular oxygen evolved from the oxidizer. The practical aim of the research was to find a solution which permitted an increase in the output of the wastewater pretreatment plant where the flotation was applied at the first stage of effluent treatment. To define the quantitative effect of the oxidizer dose on the reduction level of indicating parameters, the following constant values were assumed: raw wastewater temperatures (15 ± 1.0°C), saturation pressures (400 kPa), rate of the pretreated wastewater recirculation (20%), concentrations of flocculent (20.0 g/m^3) and coagulant (1.0 $kgCaO/m^3$), volume inflow into a flotation chamber (0.30 m^3/h) and the time of storing wastewaters in the flotation chamber (30.0 min.). Tested processing effluents were characterized by relatively high total

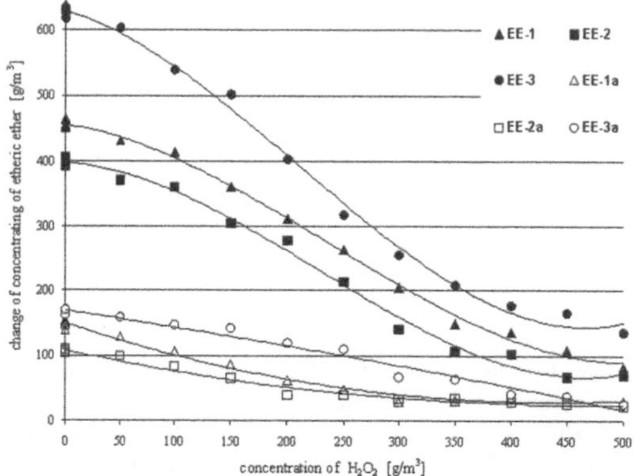

Figure 2. Dependence of ether extract (EE) measurements from hydrogen peroxide concentration. Where: EE-1, EE-2 and EE-3 define processing wastewaters from manufacturing: rape oil, soya oil and palm oil respectively, as well as EE-1a, EE-2a and EE-3a – with additional coagulation and flocculation.

protein concentrations and a significant level of emulsification. In order to define the impact of reagents and their quantity, wastewaters without oxidizer and flocculent were introduced into the pipe reactor (6) of the experimental installation (points on Y-coordinative in Figures 2–4 at doses: 0.0 g of hydrogen peroxide and the flocculent). For flotation with no chemical enhancement, no reduction of total suspended solids occurred, but rather an increase in the stability of the emulsive character of the tested wastewaters was observed. Additionally, owing to a significant contribution of mono- and diglycerides which possess strong emulsifying properties, the use of the simplest version of dissolved air flotation resulted in a stable emulsion which did not respond to further treatment. After determining optimum concentrations, flocculent introduced separately as 0.5% aqueous solution in the stated optimal dose reduced the waste load, e.g. of COD and wastewaters from manufacturing rape, soya and palm oils, respectively by: 5–17, 10–20 and 8–19%. The application of the oxidizer (H_2O_2) and an increase in its concentration enlarged the reduction of the tested indicating parameters, e.g. for a dose of 300.0 g/m^3 it exceeded 50% for TSS, COD_r, BOD_5 and EE, for all the tested wastewaters. The use of the flocculent- and oxidizer- aided pressure flotation additionally reduced the amount of hydrogen peroxide (Fig. 2–4), e.g. for a dose of 100.0 g/m^3, TSS for wastewaters from manufacturing: rape oil was reduced from the initial level of 466.7 to 140.2–147.8 g/m^3; soya oil – from 520.7 to

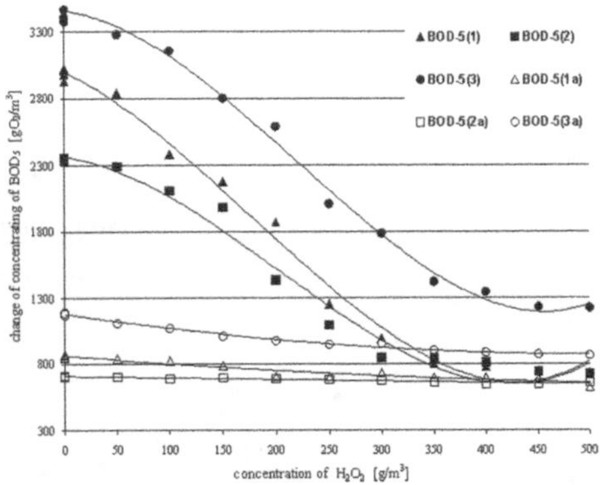

Figure 3. Dependence of BOD_5 measurements from hydrogen peroxide concentration. Where: BOD_5-(1), BOD_5-(2) and BOD_5-(3) indicate flotated and oxidized processing wastewaters from manufacturing vegetable fats, respectively: rape oil, soya oil and palm oil, as well as BOD_5-(1a), BOD_5-(2a) and BOD_5-(3a) with additional coagulation and flocculation.

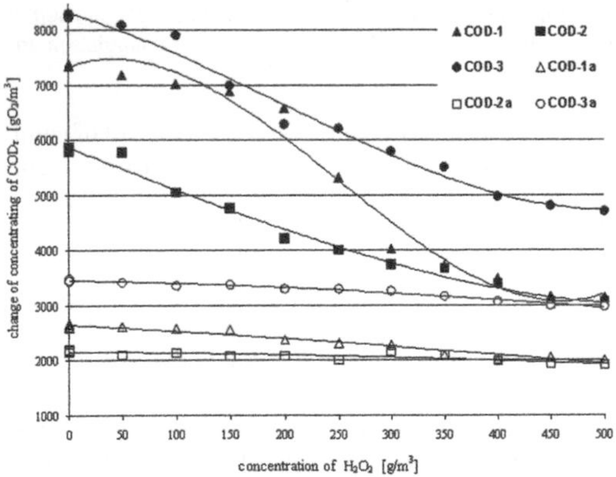

Figure 4. Dependence of COD measurements from used hydrogen peroxide concentration. Where: COD-1, COD-2 and COD-3 indicate flotated and oxidized processing wastewaters from manufacturing vegetable fats, rape oil, soya oil and palm oil respectively, as well as COD-1a, COD-2a and COD-3a with additional coagulation and flocculation.

209.4–223.3 g/m^3; and palm oil – from 487.3 to 174.0–187.5 g/m^3. Using the oxidizer dose of 250.0 g/m^3, the reduction of TSS exceeded 95% for all types of tested wastewaters. The other indicating parameters showed lower levels of reductions, e.g. for COD_r, at 100.0 g/m^3 they were, respectively: 4–9, 12–16 and 4–11% for wastewaters from manufacturing rape oil, soya oil and palm oil. Laboratory tests of the coagulation with lime milk showed that the doses 1.0 kg/m^3 were optimal to reduce pollutants for individual types of wastewater and made it possible to lower the amounts of the after-coagulation sediments to a minimum. The optimal dose of flocculent was introduced at the coagulation step when coagulant was homogenous. To exemplify, for an ether extract (EE) (with no additional oxidant) the reduction was as follows: for wastewaters from manufacturing: rape oil – up to 140.6–154.3 g/m^3, soya oil up to 105.5–109.6 g/m^3, and palm oil up to 163.0–171.4 g/m^3 (Fig. 2).

Similar significant changes were noted for COD and BOD_5. The increase in the oxidant concentration resulted in a successive increase in the reduction of these parameters up to the limit level of 350.0–400.0 g/m^3, above which no further significant increase in pollutant abatement at the outlet from the flotator was found.

Coagulation with lime milk appeared to be the most effective method of eliminating pollutants in our experiments. The introduction of this reagent significantly changed the zeta potential (ζ) and induced the processes of unloading colloid systems, aggregation of the precipitated floccules and co-precipitation of the pollutants forming soluble COD and BOD (COD_S and $BOD_{5(S)}$) and dispersed COD and BOD (COD_D and $BOD_{5(D)}$). Coagulation reduced additional amounts of $COD_{r(CaO)}$, $BOD_{5(CaO)}$, total nitrogen and total phosphorus ($N_{(CaO)}$ and $P_{(CaO)}$), which resulted in a clear increase in the reduction of total COD_T, $BOD_{5(T)}$, N and P in comparison with other operations, e.g. flocculation. This effect was induced mainly by Ca^{2+} ions and, to a lesser degree, by its complex forms: $[CaOH]^+$ and $[CaHCO_3]^+$. Sorption on particles (suspensions) of CaO from lime milk and, to a smaller extent, of $CaSO_{4(s)}$ or $CaCO_{3(s)}$ supported this process (Dojlido 1987).

Moreover, the co-precipitation resulted in closing the dispersed and partly dissolved load of COD and in absorbing the compounds forming sums of COD_S and $BOD_{5(S)}$ of the produced floccules. A considerable effect on pollutant elimination was probably a result of forming calcium soaps of type $(RCOO)_2Ca$ (where R – are mainly: $C_{16}H_{33}$, $C_{16}H_{31}$, $C_{17}H_{33}$, $C_{17}H_{31}$, $C_{17}H_{29}$, $C_{18}H_{33}$, $C_{18}H_{31}$), which intensified flocculation of the pollutants (Żak 2005a). The use of hydrogen peroxide contributed to the formation of oxygen enhancing the process of the pollutant flotation, which caused a bigger reduction and significantly shortened the time of this operation. After mixing raw wastewaters with saturated and pretreated ones, decompression in the flotation chamber was initiated, causing an intense release of gases. This phenomenon was conducive to hydrogen peroxide decay towards molecular oxygen formation according to a simplified reaction (http://www.h2o2.com, http://www.hydroxyl.com): $2H_2O_2 \rightarrow O_{2(g)} + 2H_2O$ (air decompressing slip causes dissolution of H_2O_2). Its decomposition was additionally accelerated by increasing the pH via introduction of lime milk. During and after coagulation, the pH of the reaction mixture increased, leading to more rapid H_2O_2 decomposition according to the reaction: $H_2O_2 + 2OH^- \rightarrow O_{2(g)} + 2H_2O$ (http://www.hydroxyl.com, http://www.h2o2.com). The tests showed that at pH = 8.0–9.0, produced by introducing lime milk (with saturation pressure of 400 kPa), hydrogen peroxide decomposition (300.0–500.0 gH_2O_2/m^3) and the release of gaseous oxygen took place during 20–30 minutes with the efficiency of 64.2–81.4%. The tested wastewaters treated with hydrogen peroxide alone showed COD_r elimination for the effluents from production of rape oil 2.2–3.8%, production of soya oil 1.9–3.2% and production of palm oil 2.1–4.5%. Reductions were at the level close to the limits of analytical errors for the method determining this parameter. Thus, the participation of this compound in the doses applied to eliminate COD by oxidation and mineralizing substances forming the load of pollutants in the treated wastewaters should be considered as a method of secondary importance.

Reduction in COD and BOD_5 was observed at the lower level with reference to the other parameters, e.g. EE. However, the results are comparable with available literature data for those parameters (Steiner & Gec 1992, Maennig & Scherer 1989). Dissoluble COD and BOD_5 , e.g. products of hydrolysis of proteins, lecithins, mucoid and odorous substances which do not co-precipitate during coagulation and do not form stable substances after introducing coagulant, e.g. precipitable salts or complex compounds, thus they are not susceptible to flocculation and flotation. These compounds remain as pollutants in the pretreated effluents leaving the flotation chamber and comprise the load which must be removed by other methods, e.g. by using activated sludge at the stage of final biological wastewater treatment.

Reductions in COD, BOD_5 and EE as a function of hydrogen peroxide concentration in the oxidant-assisted flotation are observed with different rates depending on the methods of chemical pretreatment. Practically, the selection of a suitable oxidizer dose in order to achieve a useful reduction in basic parameters requires exhaustive testing. The established interrelations are helpful in that they permit the selection of the optimal dosage of hydrogen peroxide for the required reduction in indicating values. Analytically determined COD_r, BOD_5 and EE reduction values as a function of oxidant dosage for the individual types of wastewaters were characterized by close distribution, according to a general equation COD_r (BOD_5 and EE) = $Ax^3 - Bx^2 - Cx + D$ (where $x = [H_2O_2]$) for flocculation enhanced by hydrogen peroxide and a flocculent (Praestol 2340 BC Stockhausen) (Table 2). The additional application of lime coagulation changed the character of analytically determined values of the COD_r, BOD_5 and EE reduction for the individual types of effluents. In these

Table 2. Regression equation coefficients for the reduction of COD_r (BOD_5, EE) = f($[H_2O_2]$) by using the oxidation and flocculation enhanced pressure flotation*.

			Values of slops for the equation [(b)] COD_r (BOD_5, EE) = $Ax^3 - Bx^2 - Cx + D$			
No.	Equation [(a)]	Correlation coefficient R^2	A	B	C	D
1	EE-1	0.9976	5E-06	0.0035	0.2169	455.48
2	EE-2	0.9936	6E-06	0.0042	0.0790	396.48
3	EE-3	0.9962	7E-06	0.0047	0.4640	629.74
4	BOD-5(1)	0.9931	4E-05	0.0208	3.6398	2995.5
5	BOD-5(2)	0.9821	4E-05	0.0239	0.9868	2359.5
6	BOD-5(3)	0.9938	4E-05	0.0279	1.0390	3458.3
7	CODr-1	0.9828	0,0001	0.0904	6.8636	7336.6
8	CODr-2	0.9876	9E-06	0.0002	7.7000	5864.3
9	CODr-3	0.9882	3E-05	0.0200	5.7054	8315.2

* experimental values were obtained at 15.0 ± 1.0°C, saturation pressures (400 kPa), rate of the pretreated wastewater recirculation (20%), concentrations of flocculent (20.0 g/m^3) and coagulant (1.0 $kgCaO/m^3$), volume inflow a flotation chamber (0.30 m^3/h) and the time of storing wastewaters in the flotation chamber (30.0 mim.); the equations were marked; [(a)] for effluents from manufacturing rape oil (EE-1, BOD-5(1), COD_r-1), soya oil (EE-2, BOD-5(2), COD_r-2) and palm oil (EE-3, BOD-5(3), COD_r-3) respectively; [(b)] where: $x = [H_2O_2]$ in g/m^3.

Table 3. Regression equation coefficients for the reduction of COD_r (BOD_5, EE) = f($[H_2O_2]$) by using pressure flotation aided with lime milk coagulation, oxidation and flocculation*.

No.	Equation [(a)]	Correlation coefficient R^2	Values of slops for the equation [(b)] COD_r (BOD_5, EE) = $-Ax^2 - Bx + C$		
			A	B	C
1	EE-1a	0.9889	0.0006	0.5598	150.63
2	EE-2a	0.9787	0.0004	0.3666	109.04
3	EE-3a	0.9763	0.0001	0.2520	170.28
4	BOD-5(1a)	0.9323	0.0005	0.6380	862.01
5	BOD-5(2a)	0.9379	2E-05	0.1326	709.28
6	BOD-5(3a)	0.9900	0.0011	1.1832	1177.7
7	COD_r-1a	0.9527	0.0004	1.2020	2656.5
8	COD_r-2a	0.9439	0.0006	0.1863	2167.2
9	COD_r-3a	0.9770	0.0011	0.4324	3457.8

* experimental values were obtained at $15.0 \pm 1.0°C$ and conditions as given in Table 2; [(a)] the equations were marked for effluents from manufacturing, respectively: rape oil (EE-1a, BOD-5(1a), COD_r-1a), soya oil (EE-2a, BOD-5(2a), COD_r-2a) and palm oil (EE-3a, BOD-5(3a), COD_r-3a); [(b)] where: $x = [H_2O_2]$ in g/m^3.

cases, the COD_r, BOD_5 and EE distributions followed an equation of a general form of COD_r (BOD_5 and EE) $= -Ax^2 - Bx + C$ (where $x = [H_2O_2]$) (Table 3). For wastewaters derived from rape oil, soya oil and palm oil manufacture, the total nitrogen concentrations after coagulation with lime milk (1.0 kg CaO/m^3 effluent), flocculation and flotation were respectively: 40.5, 61.7 and 84.2 gN/m^3, which resulted in reductions of: 49.7, 52.6 and 46.5%.

The process of flotation aided by hydrogen peroxide insignificantly decreased this parameter and, for a dose of e.g. 500.0 g/m^3, reductions were respectively: 54.9, 59.6 and 57.3%. For wastewaters from manufacturing rape oil, soya oil and palm oil, the total phosphorus concentrations in wastewaters after lime milk coagulation (1.0 kg CaO/m^3 effluent) flocculation and flotation were respectively: 11.3, 14.4 and 34.6 gP/m^3, which resulted in reductions of: 88.5, 86.2 and 71.3%. The process of flotation aided with hydrogen peroxide insignificantly reduced this parameter and, for a dose of e.g. 500.0 g/m^3, it amounted to, respectively: 91.4, 89.6 and 84.8%. Carrying out the wastewater treatment enhanced by oxidation, an insignificant increase in the reduction of total nitrogen and phosphorus with reference to the results obtained by the method without oxidizer was observed. It was stated that supporting the process with hydrogen peroxide did not significantly change the relation of COD_r/BOD_5 in the pretreated wastewaters with reference to raw wastewaters as a function of the oxidizer increase in the tested range of its concentrations. However, the ratios COD_r/N, BOD_5/N, COD_r/P, BOD_5/P and N/P changed significantly, which was a result of a relatively higher elimination of COD, BOD_5 and total suspended solids with reference to total nitrogen elimination and the lower reduction of these parameters in relation to total phosphorus removal (Table 4).

Analyses of total nitrogen (N) in the pretreated wastewaters showed that these parameters decreased below 50%. The reduction in $N\text{-}N_{NH4}$ can be explained by the sorption of the pollutant floccules formed during coagulation. The elimination by binding into such forms as: $NH_4H_2PO_4$, $(NH_4)_3PO_4$ or $CaNH_4PO_4$ and their sorption on floccules of the coagulated pollutants is also probable. Differences in reduction of N and P are a result of substantially different mechanisms of physico-chemical elimination. The reduction in total nitrogen (mainly in a form of the unhydrolysed organic nitrogen) takes place owing to coagulation of soluble and dispersed protein substances. This process is ineffective as shown by literature data (Suzuki & Maruyama 2002, Bremmell et al. 1994, Halliday & Beszedits 1984). Total phosphorus in raw wastewaters is generally found as HPO_4^{2-} and $H_2PO_4^-$ (in aqueous environment in the range of pH 6.0–7.0, the fraction of HPO_4^{2-} is 16.3–66.1% and $H_2PO_4^-$ 33.9–83.7% of the total sum of phosphates (Dojlido 1987)).

Therefore, the elimination of total phosphorus was probably caused by precipitation of scarcely soluble salts of $Ca_3(PO_4)_2$ and $CaHPO_4$ after dosing the coagulant – lime milk (Dojlido 1987).

The oxidation and aggregation of the released particles, mainly protein-fat fractions, into larger clusters is caused by hydrogen peroxide. The addition of hydrogen peroxide to raw wastewaters separated protective colloids and reduced the dispersion level of particularly protein-fatty and fatty emulsions, which improved the flotation parameters and increased its effectiveness. This compound, even in small concentrations, causes changes in the surface density of

Table 4. Balance of COD, BOD_5, N and P relations as a function of hydrogen peroxide dosage.

No. *	Concentration of H_2O_2 [g/m^3] [(a)]	COD/BOD_5	COD/N	COD/P	BOD_5/N	BOD_5/P	N/P
I	Raw wastewater	2.61	96.56	78.99	37.03	30.29	0.82
II [(b)]	After coagulation	3.03	64.33	230.57	21.25	76.15	3.58
III [(c)]	300.0	3.11	58.28	219.10	18.72	70.38	3.76
IV [(c)]	400.0	2.89	52.83	215.86	18.28	74.71	4.09
V [(c)]	500.0	3.18	55.15	238.31	17.35	74.96	4.29
VI	Raw wastewater	2.48	44.59	55.41	18.01	22.38	1.24
VII [(b)]	After coagulation	3.07	35.31	151.28	11.48	49.20	4.28
VIII [(c)]	300.0	3.22	37.28	170.85	11.58	53.05	4.58
IX [(c)]	400.0	3.09	36.08	170.77	11.84	56.05	4.73
X [(c)]	500.0	2.92	36.20	175.36	12.40	60.06	4.84
XI	Raw wastewater	2.43	52.73	68.78	21.74	28.36	1.30
XII [(b)]	After coagulation	2.94	41.20	100.26	14.03	34.14	2.43
XIII [(c)]	300.0	3.41	43.52	124.42	12.76	36.48	2.86
XIV [(c)]	400.0	3.46	43.28	140.77	12.52	40.72	3.25
XV [(c)]	500.0	3.46	44.36	162.57	12.81	46.96	3.66

* for wastewaters from manufacturing vegetable oils: I–V) rape oil, VI–X) soya oil and XI–XV) palm oil; [(a)] concentration of hydrogen peroxide introduced into the pipe reactor (6) in Fig. 1; [(b)] after coagulation with lime milk, the nitrogen and phosphorus concentrations were respectively, for: No. II) 40.5 gN/m^3 and 11.3 gP/m^3, No. VII) 61.7 gN/m^3 and 14.4 gP/m^3, as well as for No. XII) 84.2 gN/m^3 and 34.6 gP/m^3; [(c)] for No. III–V, VIII–X and XIII–XV, value of COD was corrected $COD_r = COD_p - f \cdot d$ [15] (obtained after coagulation with lime milk in dose of 1.0 $kgCaO/m^3$).

electric charge of colloid systems and in ion contents in the solution, resulting in a variation of ζ potential (Al-Shamrani et al. 2002, Ho & Ahmad 1999). At a fixed volume of aeration and amount of flocculent, it was shown that the reduction in total ether extract correlated with the content of the introduced hydrogen peroxide, equally for BOD_5 and COD. A high percent of COD and BOD_5 decrease was probably the effect of mineralization and oxidation of some groups of compounds susceptible to these processes. Moreover, the release of gaseous reaction products resulted in a more intensive flotation process by forming gas microbubbles (Steiner & Gec 1992, Maennig & Scherer 1989, http://www.h2o2.com). This phenomenon, significant due to the concentration of the oxidizer and its participation in forming gas flotating impurities, leads to an increase in the effectiveness of flotation, enabling micro-bubbles of sizes accepted in literature as typical for pressure saturation: 10–100 μm (Hoigne 1998, Dupre et al. 1996, Schneider et al. 1995b, Steiner & Gec 1992, Rüffer & Rosenwinkel 1991, Kitchener & Gochin 1981, Takahashi et al. 1979). Hydrogen peroxide used in the process of enhanced flotation visibly improved parameters of the flotate, decreasing its volume by ca. 20–30% (for a dose of 400.0–500.0 g/m^3) compared to the unenhanced process, and significantly increased dewatering ability.

4 CONCLUSION

Application of the pressure flotation process accompanied by flocculation and H_2O_2 oxidation brings about a stronger reduction of the parameters such as BOD_5, COD, EE, nitrogen and phosphorus than the process carried out by pressure flotation alone. The larger reduction in TSS, EE, BOD_5, COD and N after applying hydrogen peroxide and a flocculent is due to the elimination of colloid systems: fatty micro-emulsions and protein-oil emulsions and to the intensification of the pollutant flotation by micro-bubbles of the gas arising from oxidant decomposition. Simultaneous flocculation and oxidation enhancement increases the pollutant reduction after exceeding dose of 300.0–400.0 g/m^3 H_2O_2. The pollutant reduction was increased additionally by lime milk coagulation, and optimal doses of flocculent and oxidizer enhance this effect. The method mentioned above may be used at the physico-chemical stage of pretreating wastewaters from manufacturing oils and greases of vegetable origin. The pretreated processing effluents with their advantageous change in the relations COD/BOD_5, COD(BOD_5)/N, COD(BOD_5)/P and N/P may be treated biologically with, e.g. aerobic methods of activated sludge.

REFERENCES

Al-Shamrani, A.A., James, A. & Xiao, H. 2002. Destabilisation of oil–water emulsions and separation by dissolved air flotation. *Water Research* 36: 1503–1512.

Beccari, M., Majone, M., Riccardi, C., Savarese, F. & Torrisi, L. 1999. Integrated treatment of olive oil mill efflunets: effect of chemical and physical pretreatment on anaerobic treatability. *Water Science and Technology* 40: 347–355.

Bremmell, K.E., Jameson, G.J. & Farrugia, T.R. 1994. Agglomeration of fats and proteins. *Chemical Engineering in Australia* 19: 23–26.

Casani, S., Rouhany, M. & Knøchel, S. 2005. A discussion paper on challenges and limitations to water reuse and hygiene in the food industry. *Water Research* 39: 1134–1146.

Chen, G. 2004. Electrochemical technologies in wastewater treatment. *Separation and Purification Technology* 38: 11–41.

Dojlido, J. 1987. *Water chemistry*: 95–96 and 136–145. Warsaw: Scientific Publishers Arkady.

Dupre, V., Ponasse, M., Aurelle, Y. & Secq, A. 1998. Bubble formation by water release in nozzles-I. Mechanisms. *Water Research* 32: 2491–2497.

Halliday, P.J. & Beszedits, S. 1984. Proteins from food processing wastewaters. *Engineering Digest* 30: 24–29.

Hemming, M.L. 1977. The treatment of effluents from the production of palm oil. In: D.A. Earp & W. Newall (eds), *International Development in Palm Oil*: 79–101. Kuala Lumpur.

Ho, C.C. & Tan, Y.K. 1989. Comparison of chemical flocculation and dissolved air flotation of anaerobically treated palm oil mill effluent. *Water Research* 23: 395–400.

Ho, C.C. & Ahmad, K. 1999. Electrokinetic behavior of palm oil emulsions in dilute electrolyte solutions. *Journal of Colloid and Interface Science* 216: 25–33.

Hoigne, J. 1998. *The handbook of environmental chemistry*: 83–141, vol. 5 part C. Berlin: Springer-Verlag.
http://www.h2o2.com/applications/industrialwastewater
http://www.hydroxyl.com

Inan, H., Dimoglo, A., Şimşek, H. & Karpuzcu, M. 2004. Olive oil mill wastewater treatment by means of electrocoagulation. *Separation and Purification Technology* 36: 23–31.

Kitchener, J.A. & Gochin, R.J. 1981. The mechanism of dissolved air flotation for potable water: basic analysis and proposal. *Water Research* 15: 585–590.

Lovett, D.A. & Travers, S.M. 1986. Dissolved air flotation for abattoir wastewater. *Water Research* 20: 421–426.

Maennig, D. & Scherer, G. 1989. Fat, oil and grease flotation treatment of poultry and food industry waste water utilizing hydrogen peroxide. European Patent EP 0324167.

Ng, W.J., Goh, A.C.C. & Tay, J.H. 1988. Palm oil mill effluent treatment liquid-solid separation with dissolved air flotation. *Water Research* 5: 257–268.

Rubio, J., Souza, M.L. & Smith, R.W. 2002. Overview of flotation as a wastewater treatment technique. *Minerals Engineering* 15: 139–155.

Rüffer, H. & Rosenwinkel, K.H. 1991. *Taschenbuch der Industrieabwasserreinigung*: 115–134. München Wien: Oldenbourg Verlag.

Schneider, I.A.H., Manera Neto, V., Soares, A., Rech, R.L. & Rubio, J. 1995a. Primary treatment of a soybean protein bearing effluent by dissolved air flotation and by sedimentation. *Water Research* 29: 69–75.

Schneider, I.A.H., Smith, R.W. & Rubio, J. 1995b. The separation of soybean protein suspensions by dissolved air flotation. *Journal of Resource and Environmental Biotechnology* 1: 47–64.

Steiner, N. & Gec, R. 1992. Plant experience using hydrogen peroxide for enhanced fat flotation and BOD removal. *Environmental Progress* 11: 261–264.

Suzuki, Y. & Maruyama, T. 2002. Removal of suspended solids by coagulation and foam separation using surface-active protein. *Water Research* 36: 295–2204.

Talinli, I. & Anderson, G.K. 1992. Interference of hydrogen peroxide on the standard COD test. *Water Research* 26: 107–110.

Takahashi, T., Miyahara, T. & Mochizuki, H. 1979. Fundamental study of bubble formation in dissolved air pressure flotation. *Journal of Chemical Engineering of Japan* 4: 275–280.

Vlyssides, A.G., Loizides, M. & Karlis, P.K. 2004. Integrated strategic approach for reusing olive oil extraction by-products. *Journal of Cleaner Production* 12: 603–611.

Żak, S. 2005a. Processing of organic waste from vegetable and animal fat production. *Polish Journal of Chemical Technology* 7: 74–79.

Żak, S. 2005b. The use of Fenton's system in the yeast industry wastewater treatment. *Environmental Technology* 26: 11–19.

Environmental Engineering – Pawłowski, Dudzińska & Pawłowski (eds)
© 2007 Taylor & Francis Group, London, ISBN13 978-0-415-40818-9

The assessment of effluent toxicity by ecotoxicological tests

Monika Załęska-Radziwiłł & Dominik Wojewódka
Institute of Environmental Engineering Systems, Warsaw University of Technology Warsaw, Poland

ABSTRACT: Ecotoxicity tests of petrochemical effluents and effluents from the production of erythromycin have been performed. The battery of tests used encompassed Toxkit (Daphtoxkit F, Thamnotoxkit F), growth test on the alga Scenedesmus quadricauda, survival test on the fish Lebistes reticulatus, enzymatic inhibition test on the bacteria Vibrio fischeri (Lumistox), genotoxicity test on the bacteria Escherichia coli (SOS Chromotest). Moreover, for the tested effluent samples, the values of the toxicity index PEEP have been determined. Among the bioindicators used, crustaceans, algae and the bacteria Vibrio fischeri proved to be the most sensitive. In no case was genotoxic activity of the tested effluents observed. The values of the PEEP index for petrochemical effluents exceeded the level of 5 (toxic effluents), while for effluents from the production of erythromycin they exceeded the level of 2 (medium toxic effluents). Considerable negative correlations between the results of toxicological tests and COD values for petrochemical effluents were found.

Keywords: Industrial effluents, toxicity of wastewater, ecotoxicological tests, battery of tests, PEEP index.

1 INTRODUCTION

Polish legislative regulations concerning effluents refer to their physical and chemical properties. It is known, however, that effluents consist of mixtures of various non-organic and organic compounds not mentioned in the Decree by the Ministry of the Environment. In production processes, transient compounds are formed in chemical reactions. Among the components of effluents are toxic compounds which cause adverse effects in aquatic biocoenoses, at concentrations lower than 1 mg/l. However, synergistic effects of poisons on organisms and evidence of bioaccumulation in trophic chains are known. A variety of chemical compounds in effluents and their biological impact make it necessary to apply ecotoxicological tests to complement physical and chemical control of effluents discharged into surface waters or to the soil. Farré and Barcelló (2003) indicated that, in order to assess the total impact and the identity of relevant toxic compounds present in industrial effluents or sewage sludge, bioassays and biosensors, need to be developed and applied. Ecotoxicological endpoints, determining the effects of effluents on organisms, including effects of physical-chemical parameters, should constitute the basis for setting safety levels of effluent discharge into waters.

2 MATERIALS AND METHODS

2.1 *Effluent samples*

Samples of industrial effluents coming from industrial plants were selected for the tests. The effluents originated from the production of erythromycin and from the petrochemical industry. Effluent from erythromycin production is the most polluted of all waste waters arising in the pharmaceutical industry. Erythromycin is produced in fermentors from a strain of the actinomyces Saccaropolyspora erythraea. Because the antibiotic is secreted outside the cells into the culture medium, the isolation of erythromycin is carried out in two stages: the filtration of post-fermentation wort and the extraction of erythromycin.

The petrochemical effluents comprised sewage containing chemical substances dissolved in water, apart from suspensions and oily products. The following installations are connected to the system: catalytic cracking, alkylation, the production of acetobenzene, reforming, asphalt oxidation, the separation of pyrolytic gases, the production of ethylene oxide and ethylene glycol and the transport of ethylene derivatives.

The tested effluent samples were analyzed for the following parameters: oxygen demand – CODCr, BOD_5 [acc. to the Polish Standard PS-74C-04578/03,

Table 1. Classification of effluents according to Persoone (Kahru et al., 2000).

Toxicity units [TUa]	Assessment of effluents toxicity
<1	I – Non-toxic
1–10	II – Toxic
10–100	III – Very toxic
>100	IV – Extremely toxic

Table 2. Assessment of effluents toxicity according to the values of PEEP index.

Values of PEEP index	Assessment of effluents toxicity
0–2	Non-toxic
>2–5	Medium toxic
>5–10	Toxic
>10	Highly toxic

WTW Report 997 230, 997 231], content of ammonium nitrogen N_{NH_4} [acc. to the Polish Standard PS-73C-04576/02], nitrite nitrogen N_{NO_2} [acc. to the Polish Standard PS-73C-04576/06], nitrate nitrogen N_{NO_3} [acc. to the Polish Standard PS-73C-04576/08], phosphorus P_{PO_4} [acc. to the Polish Standard PS-73C-04537/09], content of dissolved substances [acc. to the Polish Standard PS-78C-04541] and pH.

2.2 *Toxicity tests*

Ecotoxocological tests comprised a battery of the following tests: enzymatic Lumistox on the bacteria Vibrio fischeri (Dr Lange, 1994), genotoxity test of SOS Chromotest on the bacteria Escherichia coli (EBPI version 6.0), chronic growth test on the green alga Scenedesmus quadricauda (ISO 8692), acute lethal test on the crustaceans Thamnocephalus platyurus (Thamnotoxkit F) and Daphnia magna (Daphtoxkit F) and acute lethal test on the fish Lebistes reticulatus (PS-90/C-04610/04).

Biodegradation tests to determine the PEEP (Potential Ecotoxic Effects Probe) index, were carried out in beakers under aerobic conditions at constant aeration for five days. No addition of bacterial strains to the assays was made nor any addition of nutrients intensifying biodegradation applied.

The assessment of effluents' toxicity was based on the classification proposed by Persoone shown in table 1 and the PEEP index was calculated according to the following formula (Costan et al., 1993):

$$\text{PEEP} = \log_{10}\left[1 + n\left(\frac{\sum_{i=1}^{N} T_i}{N}\right)Q\right]$$

where:

N – total number of toxicological tests carried out before and after the process of biodegradation,

n – number of tests, which exhibited changes,

Ti – toxicity units (TU) obtained from toxicological tests of effluents before and after the process of biodegradation,

Q – effluent flow rate [m^3/h].

The classification of effluents toxicity according to the values of the PEEP index are presented in table 2.

3 RESULTS

3.1 *Chemical analyses*

The results of chemical analyses for different types of effluents are presented in tables 3 and 4.

The effluents contained considerable loads of organic compounds – confirmed by the high values of COD which varied between 18200 and 33800 mgO_2/l for effluents from production of erythromycin and between 883 and 1179 mgO_2/l for petrochemical effluents. Much lower values of BOD_5 than of COD were found for the effluents. It was noted that the 5-day biodegradation substantially decreased the values of contamination indicators. For effluents from the production of erythromycin, the reduction of COD amounted to 10–40%, whereas of BOD_5, it ranged from 30 to 62%. For petrochemical effluents the reduction amounted 25–59% and 30–80% respectively. In effluents from erythromycin production, an increase in the content of ammonium nitrogen as well as a drop in concentrations of nitrate nitrogen and phosphates were noticed during the biodegradation. In contrast, nitrification accompanied the biochemical process of decay of petrochemical effluents.

3.2 *Toxicological tests*

The results of toxicological tests as well as classification of effluents according to Persoone are given in tables 5 and 6.

Effluents from the production of erythromycin turned out to be toxic to the bioindicators [tab. 5]. The least sensitive proved to be the fish Lebistes reticulatus. The values of TUa ranged between 26.9 and 50.0, which allowed the classification of the effluents as highly toxic. After the process of biodegradation, the toxicity dropped (TUa = 13.7–31.1), but not sufficiently to classify the effluents to another toxicity class. In the case of the growth assay with the algae Scenedesmus quadricauda, the effluents should be classified as extremely toxic (TUa = 189). As a

Table 3. Characteristics of the effluents from the production of erythromycin before and after a 5-day biodegradation.

Number of test	Type of effluent	Parameters [mg/l]						
		COD	BOD_5	N_{NH_4}	N_{NO_3}	P_{PO_4}	Substance dissolved	pH
1	B	24595	18000	297	12.7	874	22161	9.3
	A	20105	10500	377	7.2	15.9	16887	8.9
2	B	25740	21000	230	15.8	20.3	24953	8.8
	A	20405	11115	346	8.0	13.9	22309	8.5
3	B	31720	26000	322	11.1	27.2	24310	8.2
	A	19136	10000	468	7.5	15	19404	8.9
4	B	33800	25000	292	4.8	12.9	25607	9.0
	A	22672	16000	326	3.8	8.03	21146	8.7
5	B	18200	13500	233	33.2	14.2	24083	8.8
	A	16400	9500	326	9.0	3.8	18779	8,8

B – before the biodegradation.
A – after the biodegradation.

Table 4. Characteristics of the petrochemical effluents before and after a 5-day biodegradation.

Number of test	Type of effluent	Parameters [mg/l]						
		COD	BOD_5	N_{NH_4}	N_{NO_2}	P_{NO_3}	P_{PO_4}	pH
1	B	1179	200	4.7	0.0328	0.18	0.404	7.22
	A	484	140	1.0	0.0134	0.20	0.082	8.07
2	B	833	20	10.9	0.051	0.22	0.038	7.55
	A	445	5	5.13	0.024	0.34	0.088	8.17
3	B	1042	10	26	0.213	0.31	0	6.43
	A	584	5	17	0.098	0.44	0	7.37
4	B	1016	5	16.9	0.0026	0.27	0.054	7.3
	A	757	<1	13.7	0.0052	0.36	0.082	7.3

B – before the biodegradation.
A – after the biodegradation.

result of the biodegradation, a rise in toxicity was noticed in three cases.

In relation to the bacteria Vibrio fischeri (Lumistox) all effluents but one exhibited extreme toxicity (TUa = 1429 – 357), while after the biodegradation the values of TUa ranged between 6.71 and 8.06, which classifies them as toxic effluents. The species most sensitive to components of effluent from the production of erythromycin proved to be the crustaceans Thamnocephalus platyurus and Daphnia magna, for which effluents in all cases showed extreme toxicity. In the case of the Thamnotoxkit F assay, the values of TUa ranged between 806 and 2500, and for Daphnia magna between 1923 and 3704. Toxicity decreased after the biodegradation, and consequently, the effluent could be classified as highly toxic. The SOS Chromotest showed that the effluents did not contain any genotoxic, pro- and mutagenic compounds for Escherichia coli and that they did not influence the activity of alkaline phosphatase.

The bioindicators most sensitive to the petrochemical effluents were the bacteria Vibrio fischeri. The values of TUa ranged from 14.4 to 49.3, which classified these effluents as highly toxic according to the Persoone classification. The TUa values for crustaceans Daphnia magna and Thamnocephalus platyurus ranged respectively between 6.41 and 22.3 and between 3.76 and 35.7 [tab. 6]. For the fish Lebistes reticulatus, the values of toxicity units varied from 6.67 to 19.8. In relation to these bioindicators the tested effluents can be classified as toxic/-highly toxic. The most resistant in relation to effluents before the biodegradation turned out to be the algae Scenedesmus quadricauda; the values of toxicity units ranged from 2.40 to 3.04 TUa, which according to the Persoone classification matches toxic effluents. The 5-day long

Table 5. The values of acute toxicity units TUa for effluents from the production of erythromycin and their classification according to Persoone.

Kind of sample	Kind of assay	Acute toxicity units [TU_a] Number of sample					Classes of toxicity
		1	2	3	4	5	
Before biodegradation	Lebistes reticulatus	26.9	31.2	33.3	50.0	33.3	III
	Growth assay	74.6	51.8	20.5	60.6	189	III–IV
	Lumistox	417	357	1429	385	2.64	II–IV
	Thamnotoxkit F	806	2128	1429	2500	2083	IV
	Daphtoxkit F	3704	3448	1923	3571	2564	IV
	SOS Chromotest−S9	0	0	0	0	0	I
	SOS Chromotest+S9	0	0	0	0	0	I
After biodegradation	Lebistes reticulatus	31.1	17.5	13.7	17.2	14.3	III
	Growth assay	22.0	52.6	35.2	130	141	III–IV
	Lumistox	<1.2	8.06	7.75	13.5	6.71	II–III
	Thamnotoxkit F	31.5	62.6	34.8	17.4	14.3	III
	Daphtoxkit F	24.0	12.8	11.5	12.0	15.4	III
	SOS Chromotest−S9	0	0	0	0	0	I
	SOS Chromotest+S9	0	0	0	0	0	I

Table 6. The values of acute toxicity units TUa for petrochemical effluents and classification of effluents according to Persoone.

Kind of sample	Kind of assay	Acute toxicity units [TU_a] Number of sample				Classes of toxicity
		1	2	3	4	
Before biodegradation	Lumistox	14.4	49.3	46.5	37.7	III
	Lebistes reticulatus	19.8	6.67	16.1	10.4	II–III
	Growth assay	3.04	2.40	2.61	2.66	II
	Thamnotoxkit F	3.76	9.34	35.7	31.2	II–III
	Daphtoxkit F	22.3	6.41	10.4	8.80	II–III
	SOS Chromotest−S9	0	0	0	0	I
	SOS Chromotest+S9	0	0	0	0	I
After biodegradation	Lumistox	0	0	0	10.9	I–III
	Lebistes reticulatus	11.1	6.30	13.7	19.8	II–III
	Growth assay	1.16	2.40	6.0	3.57	II
	Thamnotoxkit F	2.72	2.44	4.88	3.85	II
	Daphtoxkit F	27.1	7.30	18.2	13.2	II–III
	SOS Chromotest−S9	0	0	0	0	I
	SOS Chromotest+S9	0	0	0	0	I

biological decay resulted in most cases in a minor decrease of effluents toxicity, particularly in relation to V. fischeri luminescence.

A decrease in toxic effects was noted also for the crustacean Thamnocephalus platyurus; the values of toxicity units after biodegradation amounted to 2.44–4.88 TUa (toxic effluents). For the fish Lebistes reticulatus a single increase in toxicity after biodegradation was noted. The values of TUa were 6.30–19.8 (toxic/-highly toxic effluents). Tests on the algae Scenedesmus quadricauda showed in three cases out of four a small increase in adverse effects of the effluents after biodegradation; TUa amounted to 1.16–6.0, which classifies them as toxic effluents, as before biodegradation. The harm by the tested effluents after the biological decay for Daphnia magna increased. The toxicity values ranged from 13.2 to 27.1 TUa, which classified them as highly toxic in three cases, and in one case – with TUa = 7.30, as toxic. The SOS Chromotest proved that the effluents did not contain genotoxic and pro-mutagenic compounds and that they did not influence the survival rate of microorganisms based on the alkaline activity of phosphatase.

The statistical assessment of chemical and ecotoxicological results for effluents from the production of

Table 7. Pearson correlations coefficient values between the values of COD and results of toxicological tests.

	COD [mgO_2/l]	
Toxicological tests [L(E)C 50-t]	From production of erythromycin	Petrochemical
Fish	−0.57	−0.95
Daphtoxkit	0.09	−0.99
Thamnotoxkit	−0.14	0.5
Algae	0.6	−0.97
Lumistox	−0.78	0.79

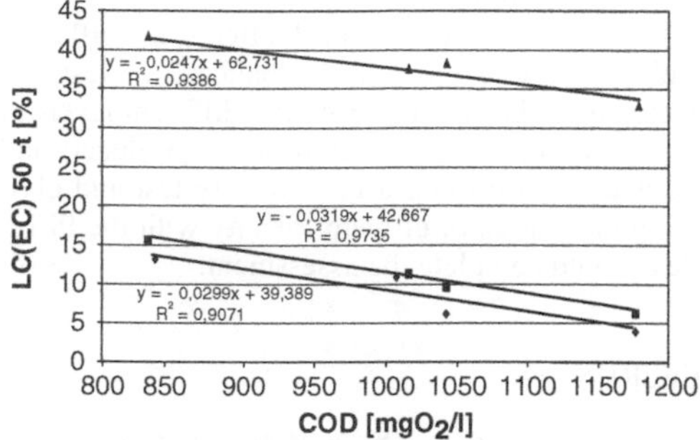

Figure 1. Relationships between the COD and toxicity (LC (EC) 50-t) in the petrochemical effluents (linear regressions).

erythromycin showed generally low values of Pearson correlation coefficients, which indicates the necessity of searching for the specific chemical compounds responsible for toxicity in relation to the bioindicators [tab. 7]. In the case of raw petrochemical effluents the Pearson correlation coefficients between the values of COD and results of toxicological tests on daphnia, fish and algae were very high (−0.99; −0.95; −0.97 respectively). Also, linear regressions between the values of COD and results of toxicity tests showed significant relationships (fish R2 = 0.907; algae R2 = 0.939; daphnia R2 = 0.973). Along with the increase in the value of COD, the adverse effects of effluent on test organisms increased [Fig. 1].

The calculated values of the PEEP index [tab. 8] for effluents from the production of erythromycin showed that they are medium toxic (PEEP = 4.8–4.9), and petrochemical effluents belong to the toxic category (PEEP = 5.5–5.8).

The classifications according to the PEEP index and according to Persoone differ because the PEEP index takes into account the process of effluent biodegradation and its magnitude. Hence the criteria for assessing the toxicity of effluents according to the PEEP index are different from those according to Persoone.

Table 8. The PEEP index values for effluents from the production of erythromycin and petrochemical effluents.

	Type of effluent			
	Erythromycin production		Petrochemical	
Number of sample	Flow rate [m^3/h]	PEEP	Flow rate [m^3/h]	PEEP
1	18	4.8	5610	5.6
2	17	4.9	6030	5.5
3	17	4.8	6060	5.8
4	14	4.8	6360	5.8
5	16	4.8	–	–

4 DISCUSSION

Industrial effluents influence the quality of surface waters and their biocoenoses. In the last ten years a number of papers have been published describing relationships between results of toxicological tests and physical-chemical composition of effluents discharged into surface waters. Some authors also compared the sensitivity of bioindicators used in tests to the effects of specific components of effluents.

Wang et al. (2003) introduced the bioluminescence test with Vibrio fischeri to assess the potential ecotoxicological effects of the selected effluents from a municipal treatment plant. They stated that the bioassays could be developed as a promising technique for discharge control of the treated wastewater. In China, a comparison of municipal effluent and effluent wastewater toxicities to Daphnia magna was undertaken to determine the most representative aqueous fraction for future toxicity identification evaluation (TIE) studies (Hongxia et al., 2004).

In this research, the toxicity assessment of chosen industrial effluents was conducted using a battery of tests, which took into account luminescent bacteria, algae, crustaceans and fish. It should be noted that, in the Polish Decree by the Ministry of the Environment dated 29 November 2002 concerning the conditions of discharging effluents into waters and into the soil, only a test on fish was introduced as a compulsory toxicity assay for effluents. As it was proved in our study, this test is not adequate to estimate the toxicity of effluents. The fish Lebistes reticulatus appeared to be generally the least sensitive to the action of effluents among the bioindicators used.

In the next Decree by the Ministry of the Environment dated 8 June 2004, the fish test as the toxicological indicator was removed and was not replaced by any other ecotoxicity test. These actions are not compliant with the code of practice of EU countries, the USA and Canada, where industrial (and also municipal) effluents are examined for toxicity and ecotoxicity. This also applies to purified effluents discharged into waters, whereas in Poland about 50% of raw effluent accesses water reservoirs.

It should also be noted that, as in the case of effluent from the production of erythromycin, the results of toxicity tests do not depend on changes in the content of organic carbon, even in AO processes in which various metabolites are formed, which was confirmed by Fernández-Alba et al. (2002).

Already in the 1980s the US EPA proposed the use of at least three test organisms for the assessment of toxicity of effluents – apart from the fish Brachydanio rerio also the algae Selenastrum capricornutum and the crustacean Ceriodaphnia dubia. Development of sensitive, simple and cheap test methods such as Microtox – Lumistox with the luminescent bacteria Vibrio fischeri and Toxkit with crustaceans and rotiferas increased the chances of more objective assessment of the effluent toxicity in relation to aquatic biocoenosis. The test results obtained could be compared with the use of new bioindicators and a general conclusion was a statement that a "battery of tests" is more useful for toxicity assessment than single tests. Neumegen et al. (2005), Latif et al. (2004), Blaise et al. (2004), and Howard et al. (2004), also proved the usefullness of a battery of microbiotests for effluent toxicity testing.

Although generally the greatest sensitivity is displayed by crustaceans, algae, and luminescent bacteria Vibrio fischeri, the results presented in this paper confirm the necessity of using a battery of tests because of the wide range of TUa values obtained for different bioindicators. The increase of adverse effects caused by one effluent after the process of biological decay confirms the sensitivity of these bioindicators to the action of many sources of contamination, including transient metabolites of biodegradation. The results also indicate that the absolute values of COD do not reflect the toxicity of effluent.

The genotoxicity tests on effluents without increasing concentration, using the SOS-Chromotest according to the ISO guidelines did not prove the presence of any genotoxic components in the effluents tested or any pro-mutagenic compounds. However, the use of genotoxicity tests for the assessment of effluent toxicity with the use a battery of tests seems useful and legitimate (Lah et al., 2004).

5 CONCLUSIONS

Among the tested samples of industrial effluents, the most toxic turned out to be the effluent from the production of erythromycin (maximum TUa = 3704), whereas the PEEP index, taking into account the process of biodegradation of contaminants, showed the highest hazard from petrochemical effluents. No genotoxic or pro-mutagenic components were detected in the effluents using the SOS-Chromotest, without increasing concentration of samples.

The most sensitive to the effluents toxicity were crustaceans, algae, and the bacteria Vibrio fischeri. A 5-day biodegradation of the effluents generally decreased their toxicity; however, in a few cases, an increase of adverse effects after this process was observed.

Significant Pearson correlation coefficients between the COD values and toxicity test results on the petrochemical effluent were found. These correlations were not observed, however, in waste from erythromycin production.

The assessment of harmful effects by effluents should be carried out with a "battery" of tests on selected bioindicators representing different levels of the aquatic trophic chain – destruents, producers and consumers. Performing a genotoxicity test and classifying the degree of effluent toxicity with the PEEP index should complete the assessment.

REFERENCES

Blaise C., Gagne F., Chevre N., 2004. Toxicity Assessment of Oil – Contaminated Freshwater Sediments, Environmental Toxicology, vol. 19 (4), 267–273.

Costan G., Bermingham N., Blaise C., Ferard J.F., 1993. Potential Ecotoxic Effects Probe (PEEP): A Novel Index to Assess and Compare the Toxic Potential of Industrial Effluents, Environmental Toxicology and Water Quality 8, 115–140.

Chunxia W., Wang Y., Kiefer F., Yediler A., Wang Z., Kettrup A., 2002. Ecotoxicological and chemical characterization of selected treatment process effluents of municipal sewage treatment plant, Ecotoxicology and Environmental Safety, 56, 211–217.

Decree by Ministry of the Environment dated 29 November 2002 concerning the requirements to be met when dumping wastes into the waters or soil, and concerning substances extremely harmful to the aquatic environment (Dziennik Ustaw No 212, p. 1799).

Decree by Ministry of the Environment dated 08 July 2004 concerning the requirements to be met when dumping wastes into the waters or soil, and concerning substances extremely harmful to the aquatic environment (Dziennik Ustaw No 168, p. 1763).

Farré M., Barceló D., 2003. Toxicity testing of wastewater and sewage sludge by biosensors, bioassays and chemical analysis, Trends in Analytical Chemistry, vol. 22 (5), 299–310.

Fernández-Alba A.R., Hernando D., Agüera A., Cáceres J., Malato S., 2002. Toxicity assays: a way for evaluating AOPs efficiency, Water Research 36, 4255–4262.

Hongxia Y., Cheng J., Yuxia C., Huihua S., Zhonghai D., Hongjun J., 2004. Application of toxicity identification evaluation procedures on wastewaters and sludge from a municipal sewage treatment works with industrial inputs, Ecotoxicology and Environmental Safety, 57, 426–430.

Howard I., Espigares E., 2004. Evaluation of Microbiological and Physicochemical Indicators for Wastewater Treatment, Environmental Toxicology, vol. 19 (3), 241–249.

ISO 8692 "Water quality-Algal inhibition test", 1994.

Kahru A., Põlluma L., Reiman R., Rätsep A., 2000. Microbiotests for evaluation of the pollution from the oil shale industry, New Microbiotests for Routine Toxicity Screening and Biomonitoring, 357–365.

Lah B., Malovrh S., Narat M., Cepeljinik T., Marisek R., 2004. Detection and Quantification of Genotoxicity in Wastewater – Treated Tetrahymena thermophila using the Comet Assays, Environmental Toxicology, vol. 19 (6), 545–553.

Latif M., Licek E., 2004. Toxicity Assessment of Wastewater, River Waters, and Sediments in Austria using Cost – Effective Microbiotests, Environmental Toxicology, vol. 19 (4), 302–309.

Lumistox – Bedienungsanleitung manual, Dr Lange, 1994.

Neumegen R., Fernández-Alba A.R., Yusuf C., 2005. Toxicities of Triclosan, Phenol, and Copper Sulfate in Activated Sludge, Environmental Toxicology, vol. 20 (2), 160–164.

Polish Standard PS-90C-04610/04. Water and waste water. Tests for toxicity of pollutants to aquatic organisms. Determination of acute toxicity to Lebistes reticulatus Peters.

Polish Standard PS-73C-04576/02. Water and waste water. Tests for content of nitrogen compounds. Determination of ammonium nitrogen by the titration method.

Polish Standard PS-73C-04576/06. Water and waste water. Tests for content of nitrogen compounds. Determination of nitrite nitrogen by colorimetric method with sulphanilic acid and 1-nafthyloamine.

Polish Standard PS-73C-04576/08. Water and waste water. Tests for content of nitrogen compounds. Determination of nitrate nitrogen by colorimetric method with sodium salicilate.

Polish Standard PS-74C-04578/03. Water and waste water. Tests for oxygen demand and organic carbon content. Determination of chemical oxygen demand (COD) by dichromate method.

Polish Standard PS-73C-04537/09. Water and waste water. Tests for content of phosphorus compounds. Determination of dissolved orthophosphates(V) by colorimetric or extract-colorimetric method.

Polish Standard PS-78C-04541. Water and waste water. Determination of solid residue, residue after ignition, ignition loss, dissolved matter, dissolved mineral matter and dissolved volatile matter.

Standard Operational Procedure "Thamnotoxkit F", Freshwater Toxicity Screening Test, Microbiotest Inc., Belgium.

Standard Operational Procedure "Daphtoxkit F", Freshwater Toxicity Screening Test, Microbiotest Inc., Belgium.

The SOS – Chromotest Kit, version 6.0, Environmental BioDetection Inc. (EBPI), Canada.

WTW Application Report 997 230. BOD measurement in household wastewater.

WTW Application Report 997 231. BOD measurement in organically heavily contaminated wastewater.

Environmental Engineering – Pawłowski, Dudzińska & Pawłowski (eds)
© 2007 Taylor & Francis Group, London, ISBN13 978-0-415-40818-9

Electrochemical treatment of reactive dyes – Helaktyn Red F5B

Elzbieta Kusmierek & Ewa Chrzescijanska
Institute of General and Ecological Chemistry, Technical University of Lodz, ul. Zeromskiego, Poland

ABSTRACT: Industrial wastewater arising from a dyeing process using reactive dyes is hazardous to the environment. Conventional wastewater treatment methods do not bring about the complete destruction of dyes. Electrochemical methods can be an interesting solution of this problem. The electrochemical oxidation of Helaktyn Red F5B at a platinum electrode starts at the potential of 0.7 V vs. SCE and proceeds in at least one step before potentials reach the value at which oxygen evolution starts. Kinetic parameters (e.g., half-wave potential of the first step, anodic transfer coefficient, heterogeneous rate constant) were determined. Under the optimal parameters of the process, the dye conversion calculated as a change in COD was 100%. The decrease in TOC was lower. The solution was completely decolourised but not mineralized. The effectiveness of the process was improved by application of a $Ti/TiO_2/RuO_2$ electrode and a combination of the electrochemical oxidation with a photochemical process.

Keywords: Reactive dyes; Helaktyn Red F5B; Electrochemical treatment.

1 INTRODUCTION

Reactive dyes, which occur in industrial wastewaters, are hazardous to the environment. The textile industry produces around 700,000 t of about 10,000 different types of dyes and pigments each year (Carneiro et al. 2004). Reactive dyes are an important group among the commercial synthetic pigments and they account for approximately 12% of the worldwide production of dyes (Pellegrini et al. 1999).

Reactive dyes are used in dyeing and printing processes of cellulose fibres as well as polyamide and protein fibres. The dyeing process involving reactive dyes consists of two steps: in the first step the fabric is immersed in a dyeing bath (causing adsorption and diffusion of a dye inside the fibre), in the second step the excess of dye is removed (Zanoni et al. 2003). In the second step, the fabric is washed several times in order to remove the excess of dye or its hydrolysis products. Wastewater from the dyeing process is produced mainly in the second step. This wastewater is either discharged to a sewer and treated by municipal sewage treatment plants or released directly to a watercourse (Carneiro et al. 2003). A release to the sewer system can disrupt treatment processes in the municipal sewage plant. A release to the watercourse can have undesirable effects on biocenosis. Only 60–90% of reactive dyes are effectively used in the dyeing process (Camp et al. 1990). Consequently, a significant amount of unfixed dyes is released in wastewater. Approximately, 70 to 200 l of water are used in the dyeing process for every kilogram of finished fabric (Allegre et al. 2004, Carneiro et al. 2004). The dyeing wastewater is characterized by a low biological oxygen demand (BOD), high chemical oxygen demand (COD), high color, fluctuation of pH and high content of salts (Alinsafi et al. 2005, Kim et al. 2002). Moreover, low $BZT_5/ChZT$ ratio values (usually lower than 0.1) indicate that this wastewater is resistant to conventional methods of biological treatment (Liakou et al. 1997). Reactive azo dyes can be reduced in aqueous media which results in formation of toxic and carcinogenic amines (Swaminathan et al. 2003). Concentrations of reactive dyes in the industrial wastewater vary from 5 to 1500 mg/l (Pierce 1994).

Many treatment methods including biological, chemical and physico-chemical methods have been tested in the treatment of industrial wastewater containing reactive azo dyes. However, conventional methods do not cause destruction of dyes but only their transformation to another form which also needs treatment (Daneshvar et al. 2003). Recently, an application of advanced oxidation processes (AOP's) including chemical processes (H_2O_2/UV, ozone, Fenton, UV, wet air oxidation and natural sunlight) in the treatment of textile dyes, has attracted a lot of interest (Al-Momani et al. 2002). However, electrochemical treatment methods, especially electrochemical oxidation, are more friendly to the environment than other methods because they often do not require an application of additional chemicals and use "environmentally clean" electrons. Moreover, electrochemical oxidation of organic pollutants often results in their transformation to simple inorganic compounds, e.g., CO_2, H_2O, N_2 and HCl, or simple organic compounds which can be easily biodegraded. These methods are efficient, but not always the cheapest. Electrochemical oxidation of azo dyes results in formation of such products as CO_2, N_2 and Na_2SO_4. The formation of aromatic esters, phenols, aliphatic carboxylic acids, cyclic and aliphatic hydrocarbons, aromatic amines etc., is also possible (Scott et al. 1995). In the presence of chloride ions, the oxidation of azo dyes proceeds directly or indirectly through the formation of "active" chlorine.

This paper presents results of electrochemical treatment of the reactive dye – Helaktyn Red F5B with triazyne group. The effectiveness of the process was evaluated through removal of total organic carbon (TOC) and chemical oxygen demand (COD), and decrease in absorbances determined at various wavelengths in the UV-VIS region.

2 MATERIALS AND METHODS

The azo reactive dye Helaktyn Red F5B (Reactive Red 2, C.I. 18,200) was obtained from Dyestuff Industry Works Boruta S.A. in Zgierz (Poland). Its main characteristics are presented in Figure 1.

An aqueous solution of Helaktyn Red F5B (500 mg/l) was prepared by dissolution of the dye in 0.1 mol/l $NaClO_4$ (Fluka). Dye concentrations of 500 mg/l or higher commonly occur in textile effluents (Tang et al. 2004).

Chemicals necessary to determine the chemical oxygen demand (COD): H_2SO_4, Ag_2SO_4, $K_2Cr_2O_7$, $HgSO_4$ and $Fe(NH_4)_2(SO_4)_2$ were reagent grade for analysis and were all obtained from POCh Gliwice, Poland.

An electrochemical investigation of kinetic parameters of Helaktyn Red F5B electrooxidation was carried out using cyclic and differential pulse voltammetry. Voltammetric curves were recorded on an Autolab electrochemical workstation (EcoChemie,

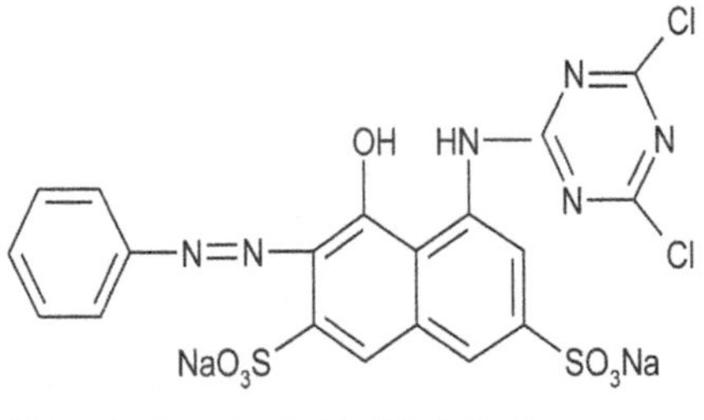

Molecular formula: $C_{19}H_{10}N_6O_7S_2Na_2Cl_2$,
Molecular weight: 615.33 g/mol,
Water solubility: 70 g/l.

Figure 1. Characteristics of Helaktyn Red F5B (Reactive Red 2, C.I. 18,200).

Holland). The measurements were carried out in a three-electrode cell. A platinum electrode was used as an anode. The cathode was also platinum or hanging mercury drop (HMDE). The potential of the working electrode was measured vs. saturated calomel electrode (SCE). Prior to the measurements the solutions were de-oxygenated by purging with argon. During the measurements, a blanket of argon was kept over the solutions.

The scope of the investigation included determination of the reaction steps of the electrochemical oxidation and reduction of Helaktyn Red F5B, the influence of concentration, scan rate and pH on the oxidation reaction, and determination of the basic kinetic parameters of the first step of the electrode reaction.

An investigation of the electrochemical treatment of Helaktyn Red F5B solutions was carried out in an electrochemical cell with undivided electrode compartments. A platinum electrode was used as an anode. A platinum cathode in the form of a net surrounded the anode. The potential of the working electrode was measured vs. saturated calomel electrode (NEK). The volume of the electrolysis solution was 100 ml.

The effectiveness of the process was determined as conversion of the substrate calculated as a change in the chemical oxygen demand (COD), total organic carbon (TOC) and absorbances in UV-VIS spectra. TOC was analysed with TOC 5050A Shimadzu Total Organic Carbon Analyser. UV-VIS spectra were recorded in the wavelength range from 190 to 800 nm using a UV-VIS spectrophotometer (Shimadzu UV-24001 PC). The solution samples were diluted 5 times before spectrophotometric measurements.

The effect of the current intensity, the surface area of the anode, the electrolysis time and the temperature on the substrate conversion was investigated. The results of the electrochemical oxidation of the dye were analysed by determining the COD, TOC and the absorbances from UV-VIS spectra.

An investigation of the electrochemical treatment combined with photoelectrochemical oxidation of the substrate was carried out in an electrochemical cell made of quartz glass. The electrochemical cell was placed in the photochemical reactor Rayonett RPR-200 (Southern New England Ultraviolet Co.). The reactor was equipped with eight lamps emitting radiation at the wavelength of 254 nm. A titanium electrode covered with TiO_2 (70%) and RuO_2 (30%) was used as the anode in the photoelectrochemical experiments.

3 RESULTS AND DISCUSSION

3.1 *Determination of kinetic parameters of Helaktyn Red F5B electrooxidation*

The dependence of the reaction current on the electrode potential provides preliminary information about the character of an electrode reaction. A sample voltammogram of the substrate oxidation is presented in Figure 2.

The oxidation of Helaktyn Red F5B proceeds at a platinum electrode in at least one step before potentials reach the value at which the oxygen evolution starts and is irreversible. The process starts from the potential of 0.7 V vs. SCE. In the reverse scan, a current peak is observed at the potential of 0.25 V. This peak is due to the reduction of products formed during the oxidation process (Fig. 2).

The substrate is reduced at the platinum electrode also in at least one step before potentials reach the value at which the hydrogen evolution starts. The reduction starts from the potential of −0.4 V vs. SCE and is irreversible (Fig. 3, curve 1). The substrate is also reduced at HMDE in at least two steps before the hydrogen evolution starts. The reduction starts

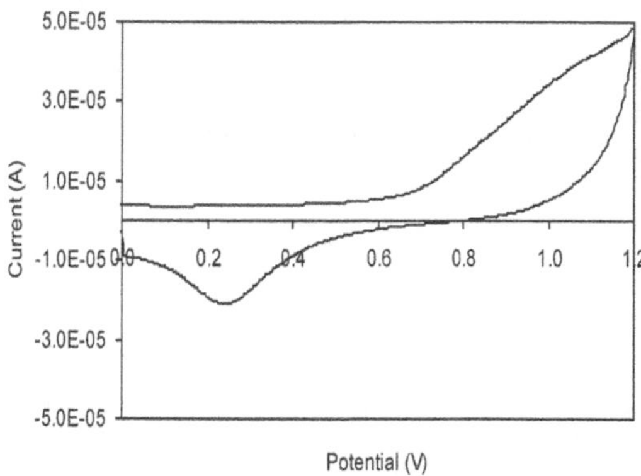

Figure 2. Cyclic voltammogram of Helaktyn Red F5B oxidation recorded at the platinum electrode; c = 500 mg/l in 0.1 mol/l $NaClO_4$, v = 0.01 V/s.

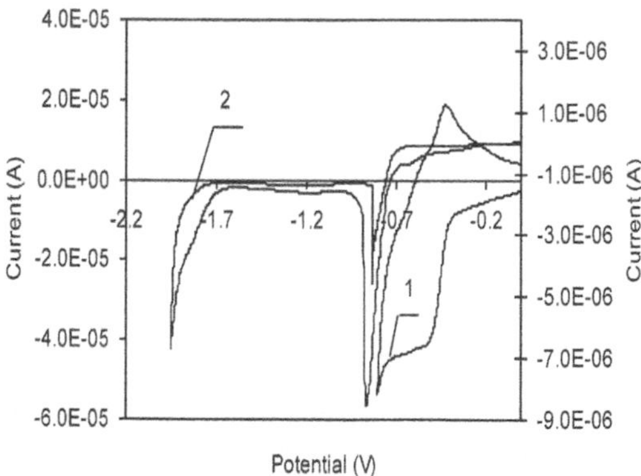

Figure 3. Cyclic voltammogram of Helaktyn Red F5B reduction at the platinum electrode – curve 1 (left axis) and hanging mercury drop electrode (HMDE) – curve 2 (right axis); c = 500 mg/l in 0.1 mol/l $NaClO_4$, v = 0.01 V/s.

from the potential of 0.75 V and is irreversible (Fig. 3, curve 2).

In the potential range where a peak related to the first step of the dye oxidation is formed, the dependence of the peak current on the square root of the scan rate is linear and crosses the origin of the coordinates (Fig. 4).

This fact indicates lack of adsorption and diffusion control of the process in this potential range. The lack of Helaktyn Red F5B adsorption at the platinum electrode, in the potential range where a faradaic reaction is observed, indicates that the peak current should be analyzed the as current resulting from the substrate diffusion to the electrode.

In order to determine the electrode kinetics of the first step of Helaktyn Red F5B oxidation, cyclic voltammograms were recorded at various scan rates and in solutions with various substrate concentrations.

The peak current (Table 1) and the half-wave potential (Table 2) were determined from cyclic voltammograms recorded at various scan rates and in solutions with various dye concentrations. The peak current

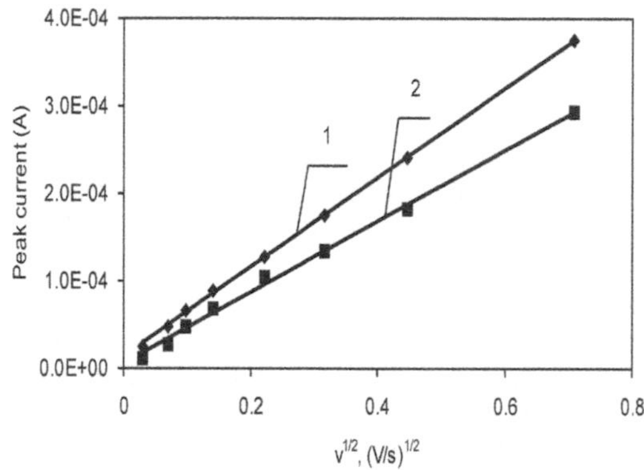

Figure 4. Change of the peak current of Helaktyn Red F5B with the change of the square root of the scan rate; curve 1 – c = 500 mg/l, 2 – c = 1000 mg/l.

Table 1. Peak current values of the first step oxidation of Helaktyn Red F5B at the platinum electrode in the solutions with various substrate concentrations.

C (mg/l)	$I_p \cdot 10^5$ (A) v = 5 mV/s	$I_p \cdot 10^5$ (A) v = 10 mV/s
10	0.50	0.90
50	0.60	1.02
100	0.81	1.26
250	1.20	1.86
500	2.20	3.03
750	2.75	4.12
1000	5.55	5.05
2000	5.31	6.66

Table 2. Values of $E_{1/2}$, $\beta_{n\beta}$ and k_{bh} for the electrode reaction of the first step in Helaktyn Red F5B oxidation at the platinum electrode.

C (mg/l)	E_p(V)	$E_{1/2}$(V)	$\beta_{n\beta}$	$k_{bh} \cdot 10^4$ (cm/s)
250	1.023	0.895	0.28	4.52
500	1.073	0.904	0.27	4.62
750	1.090	0.908	0.29	4.52
1000	1.102	0.928	0.28	4.53
2000	1.102	0.929	0.29	4.52

Table 3. Values of E_p, $E_{1/2}$, I_p for the first step of Helaktyn Red F5B oxidation in solutions with various pH values; c = 500 mg/l, v = 10 mV/s, platinum electrode.

pH	E_p (V)	$E_{1/2}$ (V)	$I_p \cdot 10^4$ (A)
0.91	0.88	0.79	0.368
2.43	0.87	0.77	0.396
4.75	0.82	0.73	0.362
6.09	0.65	0.59	0.165
7.29	0.59	0.52	0.126
9.82	0.57	0.45	0.140
11.85	0.54	0.44	0.142
13.70	0.48	0.41	0.247

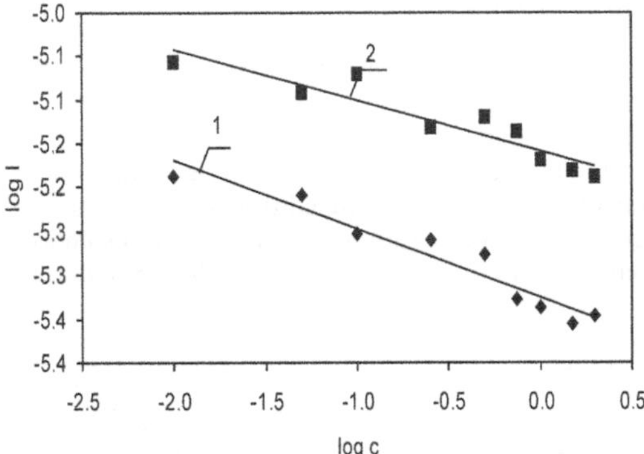

Figure 5. Dependence of the logarithm of oxidation current at a constant electrode potential on the logarithm of the dye concentration; curve 1 – E = 0.70 V, 2 – 0.75 V; v = 0.01 V/s.

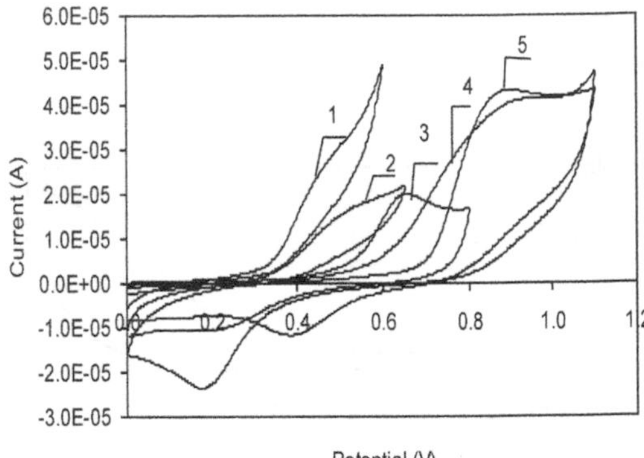

Figure 6. Cyclic voltammograms of Helaktyn Red F5B oxidation recorded at the platinum electrode; curve 1 – pH = 13.7, 2 – pH = 11.8, 3 – pH = 6.1, 4 – pH = 4.8, 5 – pH = 2.4; c = 500 mg/l.

increases linearly with the increase in the substrate concentration to 1.0 g/l. Further increase in the concentration causes a slight increase in the peak current. The anodic transfer coefficient, which characterises symmetry of the activation barrier of an electrode reaction, totals 0.3 for the process. This fact indicates an exchange of two electrons. The heterogeneous rate constant (Galus et al. 1994) calculated at the half-wave potential totals $4.5 \cdot 10^{-4}$ cm/s (Table 2).

Concentration of the dye in wastewater can vary. Thus, the effect of the concentration on the oxidation rate is determined. Under the conditions similar to conditions of the linear diffusion, the order of the reaction was determined at the constant potential (Fig. 5). Its value is 0.065. This proves that the substrate undergoes a direct electrode reaction (the reaction is not preceded by another reaction).

The wastewater formed during the dyeing process is characterised by pH in the range from 6 to 9. Therefore, the effect of pH on the oxidation of Helaktyn Red F5B was investigated. The pH of the solutions used in the experiments varied from 0.3 to 13.7. Peak potentials, half-wave potentials and peak currents (Table 3) for the first step of the dye oxidation, were determined from cyclic voltammograms (Fig. 6).

The first step of the dye oxidation proceeds easier at higher pH values. The rate of the reaction is the highest in pH range from 0 to 5.0. Further increase in pH causes a decrease in the reaction rate.

3.2 *Electrochemical oxidation of Helaktyn Red F5B*

The effectiveness of the electrochemical treatment of Helaktyn Red F5B solutions was evaluated, among other factors, by determination of absorbances from UV-VIS spectra. Example UV-VIS spectra recorded in the dye solutions before and after electrolyses are presented in Figure 7.

Within the visible region (380–780 nm), two adsorption bands are observed at the wavelengths of 512 and 538 nm. They correspond to the π-π^* transition of electrons in the azo-group connecting phenyl and naphthyl. A decrease in the absorbances at these wavelengths is connected with the decolourisation of the dye solution. Within the near ultraviolet region (280–380 nm), three adsorption bands are observed at

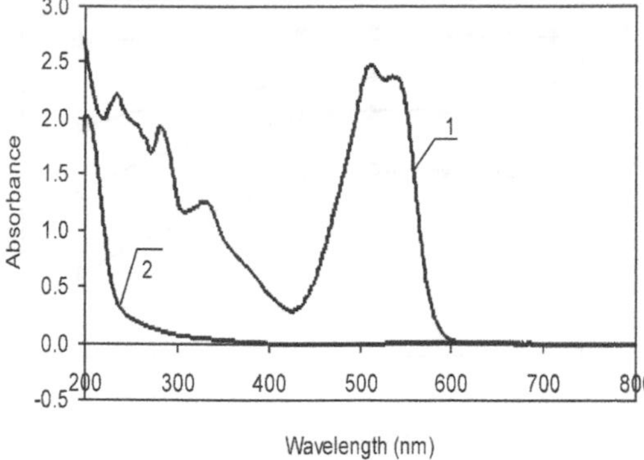

Figure 7. UV-VIS spectra recorded in the solution of Helaktyn Red F5B before (curve 1) and after electrolysis (curve 2).

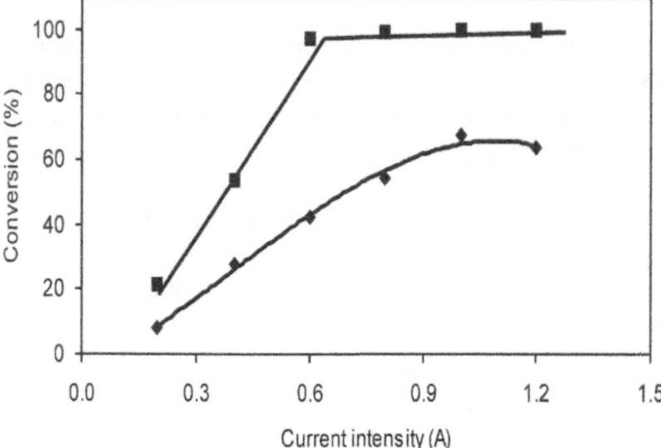

Figure 8. The effect of the current intensity on the conversion of Helaktyn Red F5B. The conversion was calculated as the change in COD (■) and TOC (♦).

the wavelengths of 234, 282 and 328 nm. They probably result from the unsaturated system of benzene and naphthalene rings (Fan et al. 2004). In the solution after electrolysis, the decrease in the absorbances at wavelengths <380 nm was not total, which indicates incomplete mineralization of the dye.

The effect of the current intensity on the effectiveness of the process was investigated using a platinum anode with surface area of 20 cm^2. The electrolysis time was 2 hours. The results of the experiments are presented in Figures 8–9.

Electrolyses at the current intensity of 0.6 V or higher caused a substrate conversion of 100% calculated as a change in COD value (Fig. 8). However, the highest conversion calculated as the change in TOC was achieved at the current intensity of 1.0 V.

UV-VIS spectra show that in the entire current intensity range, a total decolourisation of the dye solutions was achieved (Fig. 9 – absorbances at 512 and

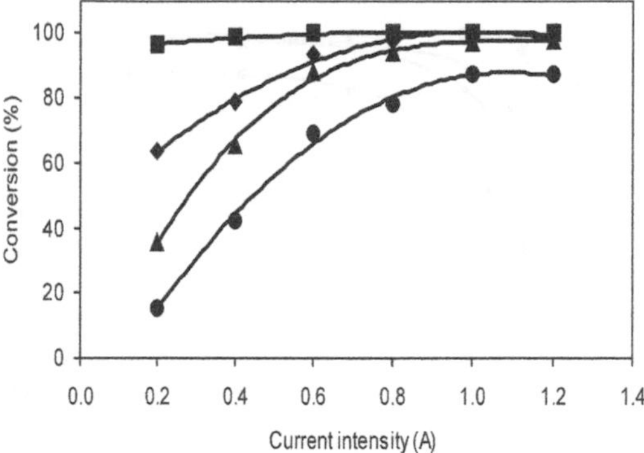

Figure 9. The effect of the current intensity on the dye conversion. The conversion was calculated as the change in the absorbance at 234 nm (●), 282 nm (▲), 328 nm (♦), 512 and 535 nm (■).

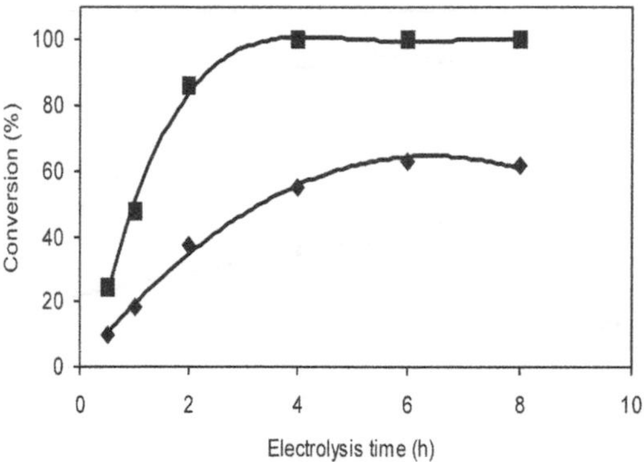

Figure 10. The effect of electrolysis time on conversion of the substrate calculated as a change in COD (■) and TOC (♦).

535 nm). The values of the absorbances at the wavelengths <330 nm prove that the solutions were not completely mineralized. The decolourisation of the solution is faster than the removal of COD and TOC.

The effect of the electrolysis time on the conversion of the substrate was determined during electrolyses carried out at the current intensity of 0.5 A and the same platinum electrode. The results of the experiments are presented in Figures 10–11.

Results of the electrolyses show that the complete removal of COD value could be achieved during a 4 h electrolysis (Fig. 10). The highest conversion calculated as a change in TOC was observed during a 6 h electrolysis. Higher electrolysis time did not cause significant changes in COD and TOC values.

A total decolourisation of the dye solution was observed during the first two hours of the process (Fig. 11). During a longer electrolysis (6 and 7 h), the absorbance at the wavelength of 234 nm decreased

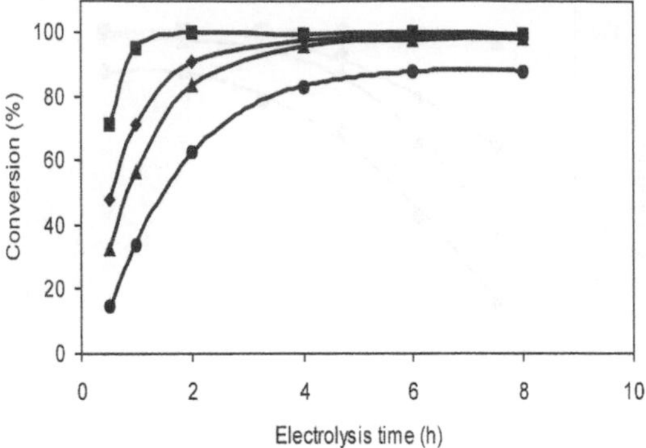

Figure 11. The effect of the electrolysis time on the dye conversion calculated as a change in the absorbance at various wavelengths; • – 234 nm, ▲ – 282 nm, ♦ – 328 nm, ■ – 512 and 535 nm.

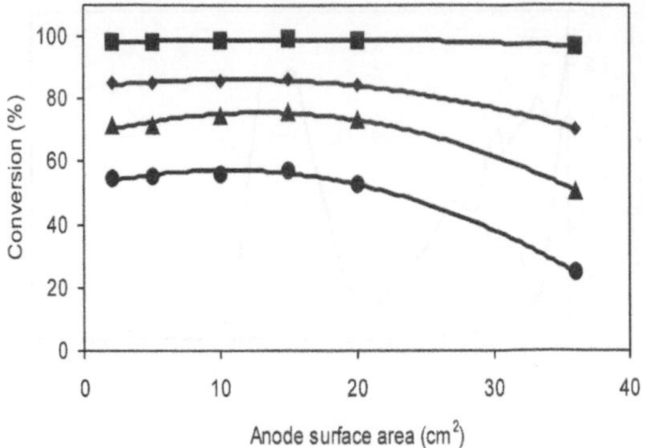

Figure 13. The effect anode surface area on the dye conversion calculated as a change in the absorbance at various wavelengths; • – 234 nm, ▲ – 282 nm, ♦ – 328 nm, ■ – 512 and 535 nm.

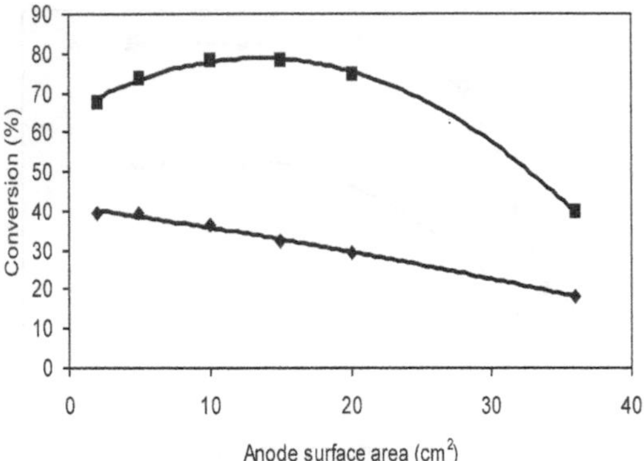

Figure 12. The effect of the anode surface area on the conversion of the substrate calculated as a change in COD (■) and TOC (♦).

by about 90%. Electrolysis times >4 hours did not increase the effectiveness of the process.

The effect of the anode surface area on the results of the electrochemical treatment of the substrate was investigated at the current intensity of 0.5 A and electrolysis time of 1.5 h. The results of such electrolyses are presented in Figures 12–13.

The highest conversion of Helaktyn Red F5B calculated as a change in COD was achieved during the electrolysis carried out at the platinum anode with surface area of 10 and 15 cm^2 (Fig. 12). The conversion calculated as a change in TOC decreases with a decrease in the anode surface area. Thus, an application of a platinum anode with lower surface area seems to be more desirable due to the surface effects.

Anode surface areas of 20 cm^2 or higher had disadvantageous effects on the decrease in the absorbances in the near ultraviolet region (Fig. 13). A decrease in the dye conversion calculated as a change in the absorbance at 234, 282 and 328 nm was observed. In all tested ranges of the anode surface area, the solutions after electrolysis were completely decolourised but not mineralized.

The results of the investigation show that the best effects of the electrochemical treatment of Helaktyn Red F5B were achieved under the following conditions:

- current intensity of 1.0 A,
- electrolysis time of 6 h,
- anode surface area of 2 cm^2.

These conditions correspond to a current density of 0.5 A/cm^2.

Under these conditions the results of the process were as follows:

- conversion calculated as a change in COD: 100%,
- conversion calculated as a change in TOC: 71.6%.

The solution was completely decolourised and the absorbance at 234, 282 and 328 nm decreased by more than 90%.

These results prove a significant but incomplete degradation of Helaktyn Red F5B. The effectiveness of the process can be improved by combining the electrochemical process with the photochemical degradation, with an application of titanium electrodes covered with titanium and ruthenium oxides. Because the temperature inside the photochemical reactor increases up to 70°C, it was important to determine the effect of the temperature on the effectiveness of the electrochemical process. The experiments were carried out at the current intensity of 0.5 A, electrolysis time of 1.5 h and with an application of a platinum anode with surface area of 2 cm^2. The results of electrolyses are presented in Figures 14–15.

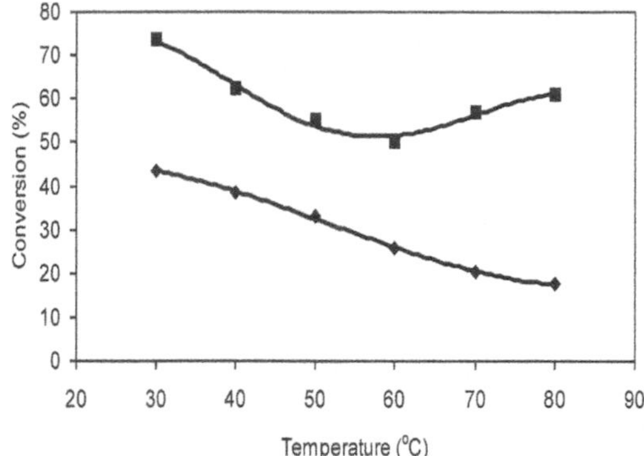

Figure 14. The effect of the temperature on the dye conversion. The conversion was calculated as a change in COD (■) and TOC (♦).

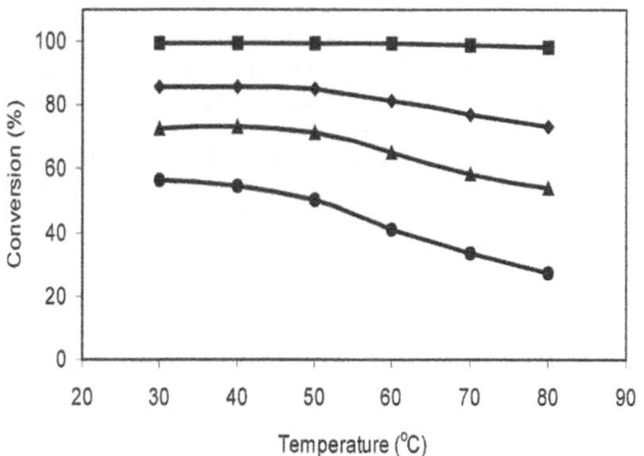

Figure 15. The effect of the temperature on the substrate conversion. The conversion was calculated as a change in the absorbance at various wavelengths; • – 234 nm, ▲ – 282 nm, ♦ – 328 nm, – 512 and 535 nm.

Increasing the temperature to 70°C caused a twofold decrease in the dye conversion calculated as a change in TOC (Fig. 14). Simultaneously, the dye conversion calculated as a change in COD decreased with an increase in the temperature to 60°C. Further increase in the temperature resulted in a slight increase in the dye conversion.

In the entire temperature range tested, the solutions were completely decolourised (Fig. 15). However, the dye conversion calculated as a change in the absorbances at the wavelengths lower than 330 nm decreased with an increase in the temperature.

At 70°C, the results of the electrolyses carried out at the platinum anode with surface area of 2 and 20 cm^2 were almost the same.

A significant improvement of the results can be achieved by application of a $Ti/TiO_2/RuO_2$ electrode in the electrochemical oxidation and by a combination of this process with photochemical oxidation. The results of various experiments presented in Table 4 were obtained during the process carried out for 2 hours at anodes with surface area of 20 cm^2 and a current intensity of 0.5 A.

In the process of photochemical degradation of Helaktyn Red F5B, the substrate did not undergo a photochemical reaction. An application of titanium covered with TiO_2 (70%) and RuO_2 (30%) in the photochemical process caused only a slight change in the conversion of the dye. A significant improvement of the results obtained at the $Ti/TiO_2/RuO_2$ electrode was observed during the electrochemical oxidation and its combination with a photochemical process. The improvement was due to the electrocatalytic properties of RuO_2 and the photocatalytic properties of TiO_2. The dye conversion (in the photoelectrochemical process at $Ti/TiO_2/RuO_2$) calculated as a change in COD and TOC increased almost twice in comparison with the electrochemical process at the platinum electrode with a comparable surface area (Table 4). A decrease in the absorbance at the wavelengths lower than 330 nm shows higher mineralization of the solution but not its complete degradation. In the case of the electrochemical oxidation at $Ti/TiO_2/RuO_2$ and its combination with photochemical oxidation, the conversion calculated as a change in the absorbance at 512 and 535 nm was 100% which corresponds to a total decolourisation of the solution. An increase in the process time may result in a further improvement of the process effectiveness. The process carried out for four hours yielded a dye conversion calculated as a change in the absorbance at 234, 282 and 328 nm higher than 90% (Table 4). Also, the conversion calculated as the change in COD and TOC increased.

4 CONCLUSIONS

Electrochemical oxidation of Helaktyn Red F5B at a platinum electrode starts at the potential of 0.7 V vs. SCE and proceeds in at least one step before potentials reach the value at which the oxygen evolution starts. In the reverse anodic scan, a new peak is observed at 0.25 V. This fact indicates the reduction of products formed during the oxidation.

Helaktyn Red F5B is reduced at the platinum electrode in at least one step before potentials reach the value at which the hydrogen evolution starts. The reduction begins from the potential of −0.4 V vs. SCE and is irreversible. The substrate is also reduced at the hanging mercury drop electrode in at least two steps. The reduction process starts at the potential of −0.75 V and is also irreversible.

The half-wave potential of the first step of the dye oxidation does not change with a change in the dye concentration. Its value is 0.92. The anodic transfer coefficients total 0.3. The heterogeneous rate constant

Table 4. Dependence of Helaktyn Red F5B conversion on the process type.

Process type	η_{TOC} (%)	η_{TOC} (%)	$\eta_{Abs.}$ (234 nm) (%)	$\eta_{Abs.}$ (282 nm) (%)	$\eta_{Abs.}$ (328 nm) (%)	$\eta_{Abs.}$ (512 & 535 nm) (%)
Photochemical oxidation	0.9	5.0	0.4	1.6	2.3	2.6
Photochemical oxidation $+Ti/TiO_2$	0	2.9	3.0	4.5	4.8	4.5
Electrochemical oxidation at Pt	20.9	47.8	34.3	55.1	71.5	91.5
Electrochemical oxidation at Ti/TiO_2	31.2	63.6	25.0	75.4	86.3	99.8
Photochemical + Electrochemical oxidation at Ti/TiO_2	54.5	87.7	76.9	92.1	92.2	99.8
Photochemical + electrochemical oxidation at Ti/TiO_2*	69.1	98.6	90.8	96.1	97.1	99.9

* Twofold higher time of the process.

is about $4.5 \cdot 10^{-4}$ cm/s. The reaction order (0.065) does not vary with a change in the potential. The dye undergoes a direct electrode reaction.

The oxidation of Helaktyn Red F5B proceeds more easily at higher pH values. The rate of the first step of the dye oxidation is at its highest in the pH range from 0 to 5.

The best results for substrate electrooxidation at the platinum electrode were achieved under the following experimental conditions: ambient temperature, current intensity – 1.0 V, electrolysis time – 6 h and anode surface area – 2 cm^2 which correspond to the current density of 0.5 A/cm^2. Under such conditions the dye solution was completely decolourised but not mineralized. The dye conversion calculated as a change in COD and TOC reached the value of 100 and 71.6%, respectively. The decolourisation of the solution was faster than the decrease in COD and TOC.

Two concurrent reactions, an electrochemical and a photochemical step using a $Ti/TiO_2/RuO_2$ electrode, result in an increase in the effectiveness of the substrate destruction. Electrochemical oxidation of the dye during a 2 h process in the quartz cell resulted in a decrease in COD and TOC by 63.8 and 31.2%, respectively. A combination of the electrochemical oxidation with the photochemical process causes a decrease in COD by 87.7% and TOC by 54.5%. These results, as well as a comparison of the results obtained during the electrochemical oxidation at Pt and $Ti/TiO_2/RuO_2$ electrodes with comparable surface areas, indicate the combined effects, electrocatalytic (RuO_2) and photocatalytic (TiO_2), of the electrode materials.

ACKNOWLEDGEMENTS

This study was supported by public funds for science in 2005-2008, grant No. 3 T09B 017 28.

REFERENCES

Alinsafi, A., Khemis, M., Pons, M.N., Leclerc, J.P., Yaacoubi, A., Benhammou, A. & Nejmeddine, A. 2005. Electrocoagulation of reactive textile dyes and textile wastewater. *Chem. Eng. Proc.* 44: 461–470.

Allegre, C., Moulin, P., Maisseu, M. & Charbit, F. 2004. Savings and re-use of salts and water present in dye house effluents. *Desalination* 162: 13–22.

Al-Momani, F., Touraud, E., Degorce-Dumas, J.R., Roussy, J. & Thomas, O. 2002. Biodegradability enhancement of textile dyes and textile wastewater by VUV photolysis. *J. Photochem. Photobiol. A: Chemistry* 153: 191–197.

Camp, R. & Sturrock, P.E. 1990. The identification of the derivatives of C.I. Reactive Blue 19 in textile wastewater. *Water Res.* 24: 1275–1278.

Carneiro, P.A., Fugivara, C.S., Nogueira, R.F.P., Boralle, N. & Zanoni, M.V.B. 2003. A comparative study on chemical and electrochemical degradation of reactive blue 4 dye. *Portugaliae Electrochim. Acta* 21: 49–67.

Carneiro, P.A., Osugi, E.M., Sene, J.J., Anderson, M.A. & Zanoni, M.V.B. 2004. Evaluation of color removal and degradation of a reactive azo dye on nanoporous TiO_2 thin-film electrodes. *Electrochim. Acta* 49: 3807–3820.

Daneshvar, N., Salari, D. & Khataee, A.R. 2003. Photocatalytic degradation of azo dye acid red 14 in water: investigation of the effect of operational parameters. *J. Photochem. Photobiol. A: Chemistry* 157: 111–116.

Fan, L., Yang, F. & Yang, W. 2004. Performance of the decolorization of an azo dye with bipolar packed bed cell. *Separation and Purification Technology* 34: 89–96.

Galus, Z. 1994. Fundamentals of electrochemical analysis. 2nd ed. New York: Ellis Horwood & PWN.

Kim, T., Park, C., Shin, E. & Kim, S. 2002. Decolorization of disperse and reactive dyes by continuous electrocoagulation process. *Desalination* 150: 165–175.

Liakou, S., Pavlou, S. & Lyberatos, G. 1997. Ozonation of azo dyes, *Wat. Sci. Tech.* 35(4): 279–286.

Pellegrini, R., Peralta-Zamora, P., de Andrade, A.R., Reyes, J. & Duran, N. 1999. Electrochemically assisted photocatalytic degradation of reactive dyes. *Appl. Cat. B: Environ.* 22: 83–90.

Pierce, J. 1994. Colour in textile effluents – the origins of the problem. *J. Soc. Dyers Colourists* 110: 131–133.

Scott, K. 1995. Electrochemical processes for clean technology, Published by the Royal Society of Chemistry 1995, Thomas Graham House, Cambridge, UK.

Swaminathan, K., Sandhya, S., Carmalin Sophia, A., Pachlade, K. & Subrahmanyam, Y.V. 2003. Decolorization and degradation of H-acid and other dyes using ferrous-hydrogen peroxide system. *Chemosphere* 50: 619–625.

Tang, C. & Chen, V. 2004. The photocatalytic degradation of reactive black 5 using TiO_2/UV in an annular photoreactor. *Wat. Res.* 38: 2775–2781.

Zanoni, V.B., Jeosadaque, J.S. & Anderson, M.A. 2003. Photoelectrolytic degradation of Remazol Brilliant Orange 3R on titanium dioxide thin – film electrodes. *J. Photochem. Photobiol. A: Chemistry* 157: 55–63.

Environmental Engineering – Pawłowski, Dudzińska & Pawłowski (eds)
© 2007 Taylor & Francis Group, London, ISBN13 978-0-415-40818-9

Removal of As(V) from solutions by precipitation of mimetite $Pb_5(AsO_4)_3Cl$

Tadeusz Bajda, Ewelina Szmit & Maciej Manecki
AGH-University of Science and Technology; Faculty of Geology, Geophysics and Environment Protection; Department of Mineralogy, Petrography, and Geochemistry; al. Mickiewicza, Kraków Poland

ABSTRACT: The proposed method of As(V) removal from solutions is based on two sequential reactions: 1) reaction with Pb(II) and Cl(I) ions resulting in precipitation of mimetite $Pb_5(AsO_4)_3Cl$, followed by 2) precipitation of pyromorphite $Pb_5(PO_4)_3Cl$ by reaction with phosphates (PO_4^{3-}) to remove excess Pb(II). This work concentrated on the efficiency of As(V) removal and the stability of synthetic mimetite. The only solid product of reaction 1 is rapidly precipitating crystalline mimetite. When a 1.4 excess of Pb was used in the pH range from 4 to 10 mimetite formation was complete within the first two hours of reaction and more than 99.99% of As(V) was removed from the solution: [As] was lowered from an initial concentration of $4000\,mg/dm^3$ to $0.4\,mg/dm^3$. The lowest solubility of mimetite was determined at the pH range from 6 to 7. The calculated $\log K_{sp}$ and free energy of formation, $\Delta G^{o}_{f.298}$, for the mimetite are -21.69 ± 1.05 and -2640 ± 6 kJ/mol, respectively.

1 INTRODUCTION

Increasing presence of mobile arsenic compounds in soils and waste waters has become an environmental concern in recent years. Arsenic pollution comes from ore production and treatment, metal processing, high-temperature combustion, electronics and chemical industries and many other sources. Arsenic in the environment occurs usually in the As(III) or As(V) oxidation state; both forms are toxic (Adriano 2001).

Sorption and precipitation techniques are commonly utilized for As(V) removal from waste waters. More effective are the sorption techniques: ferrihydrite, aluminum hydroxides, and activated carbon are often used as sorbents (Robins et al. 1987, Twidwell et al. 1994). Precipitation is achieved with the use of calcium hydroxide, Fe(III), and hydroxyapatites (Narasaraju et al. 1977, Piver 1983, Nishimura 1988, Harper & Kingham 1992, Bothe & Brown 1999, Dixit & Hering 2003). Existing precipitation methods have several disadvantages. Slowly agglomerating precipitate is often difficult to separate by settling, and the product is not very stable when stored over extended period of time (Comba 1987).

The method proposed here for removal of As(V) from solutions by precipitation of mimetite $Pb_5(AsO_4)_3Cl$ is based on a sequence of two reactions: 1) controlled addition of Pb^{2+} and Cl^- ions to the polluted solutions resulting in immediate precipitation of crystalline mimetite, followed by 2) removal of excess Pb(II) by addition of PO_4^{3-} and Cl^-, resulting in immediate precipitation of crystalline lead chlorophosphate ($Pb_5(PO_4)_3Cl$; pyromorphite). Removal of As(V) with Pb(II) was proposed earlier (Comba 1987, Twidwell et al. 1999 and literature therein). At that time, however, there was concern that precise control of the stoichiometric lead concentration would be difficult practically leading to either incomplete As(V) removal (insufficient Pb) or lead pollution (excess Pb).

The objective of this work is to establish the optimal conditions for effective As(V) removal by precipitation of mimetite with respect to element concentrations, pH, and time. Additionally, the solubility of mimetite at various pH values is evaluated and the apparent solubility constant is calculated. The details regarding the removal of potential excess of Pb(II) (reaction 2) are the topic of ongoing research and will be discussed in a separate paper.

2 MATERIALS AND METHODS

Experiment I. Three series of experiments involving different ratios of Pb to As were designed to test the effectiveness of As removal. Each experiment was run for 24 hours and repeated at six different pH values: 2, 4, 6, 8, 10, and 12. The initial arsenic (0.053 M) and chlorine (0.018 M) concentrations were intentionally selected in the 3:1 molar proportion, identical to that in mimetite $Pb_5(AsO_4)_3Cl$. Three different initial lead concentrations were used: 0.089 M, 0.107 M, and 0.125 M. These provided the stoichiometric ratios (Pb:As:Cl equal to 5:3:1), 1.2x excess of Pb over As (Pb:As:Cl equal to 6:3:1), and 1.4x excess of Pb over As with respect to the stoichiometry of mimetite (Pb:As:Cl equal to 7:3:1). Equal volumes ($20\,cm^3$) of solutions were mixed gradually using a peristaltic pump. The initial pH was adjusted using 0.5 M HNO_3 and 2 M NaOH. After 24 hours of aging, suspensions were filtered, washed, and dried at 105°C.

Experiment II. In the experiments testing the effects of aging time, 0.125 M Pb solution was used, providing a 1.4x excess of Pb over As; all other parameters were the same as above. Experiments were conducted for 2, 6, 15, and 48 hours and repeated at initial pH values of 2, 4, 6, 8, 10, and 12.

Experiment III. The stability of precipitated mimetite was assayed by dissolution in water at initial pH 2, 4, 6, 8, 10, and 12, adjusted with 0.1 M HNO_3 and 0.1 M NaOH. The pH was monitored but not adjusted during the experiment. Mimetite (0.5 g) was placed in solution ($0.5\,dm^3$) for up to two months. Small portions of solution ($<10\,cm^3$) were periodically filtered for As, Pb and pH analysis. The concentration of Cl was determined only at the end of experiment.

In all experiments $Na_2HAsO_4 \cdot 7H_2O$, $Pb(NO_3)_2$, and NaCl were used. Experiments were run in triplicate (Exp. I) and duplicate (Exp. II and III). The results are expressed as arithmetical averages with standard deviation as experimental error.

The precipitates were identified using powder diffraction patterns (XRD) obtained using a Philips PW 3020 X'Pert-APD diffractometer system (with Cu anode and graphite monochromator) at a step scan mode with step size of 0.02° 2Θ and a rate of 1 s per step. Infra-red (FTIR) spectra were recorded on a Bio-Rad FTS-60 spectrometer from 4000 to $100\,cm^{-1}$ with $2\,cm^{-1}$ resolution. Positions and intensities of bands were determined after decomposition of the spectra into component bands. Spectra decomposition was carried out following the method proposed by Handke et al. (1994). Scanning electron microscopy (SEM) was performed with a JEOL 5200 microscope on gold-coated samples. For wet chemical analysis, mimetite was dissolved in EDTA. EDTA was used due to the low solubility of mimetite in mineral acids. Standards were also prepared with EDTA – no interference was observed (Wojnar et al. 2005). In solutions, Pb was determined by atomic absorption spectrometry (AAS spectrometer Philips PU-9100x), and Cl by titration (Mohr's method). As was analyzed using colorimetry (molybdene blue method, Lenoble et al. 2003) to avoid large dilutions.

3 RESULTS AND DISCUSSION

In all experiments the reaction of Pb(II) with arsenate in the presence of chloride resulted in immediate formation of a white, crystalline precipitate. The initial arsenic concentration $[As] = 4000\,mg/dm^3$ was reduced to below $50\,mg/dm^3$ (Fig. 1).

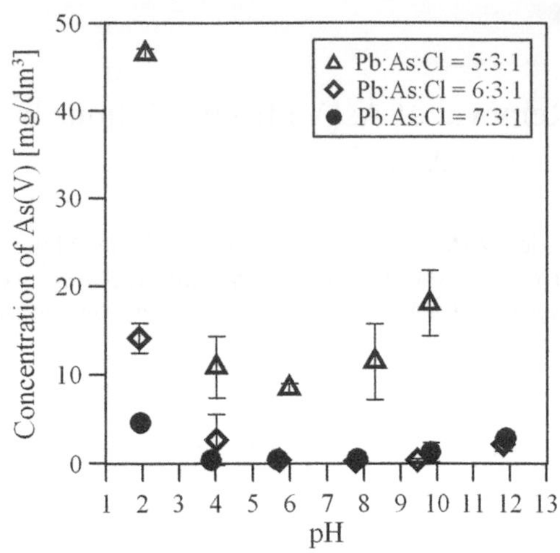

Figure 1. Concentration of As(V) after 24 hours of reaction at various pH and Pb:As ratios.

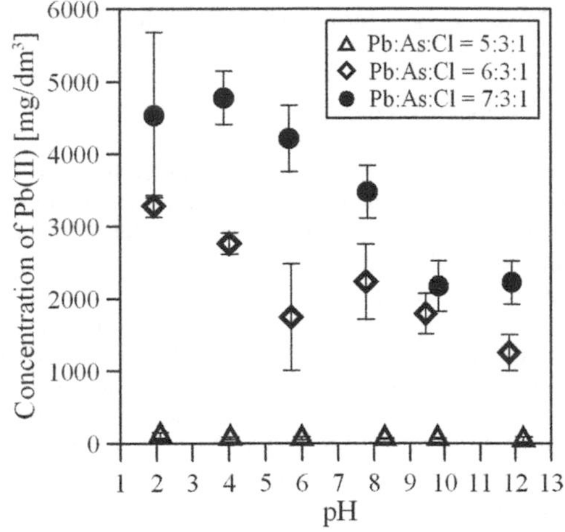

Figure 2. Concentration of Pb(II) after 24 hours of reaction at various pH and Pb:As ratios.

The highest reduction was measured in the experiments with the highest excess of lead (1.4x; Pb:As:Cl = 7:3:1) and in the pH range from 3.9 to 7.8. The lowest final [As] was equal to $0.37 \pm 0.04\,mg/dm^3$ and was measured at pH 3.9 for 1.4x excess of Pb (99.99% reduction of initial [As]). For 1.2x excess of Pb, the lowest $[As] = 0.41 \pm 0.15\,mg/dm^3$ was measured at pH 5.7 (99.99% reduction). For stoichiometric Pb the lowest $[As] = 8.47 \pm 5.1\,mg/dm^3$ was measured at pH 6.0 (99.79% reduction). The equilibrium [Pb] is $4780 \pm 450\,mg/dm^3$ for a 1.4x initial excess of Pb and pH 6.0, $1750 \pm 740\,mg/dm^3$ for a 1.2x excess and pH 5.7, and $70 \pm 16\,mg/dm^3$ for the stoichiometric initial Pb:As ratio and pH 3.9 (Fig. 2).

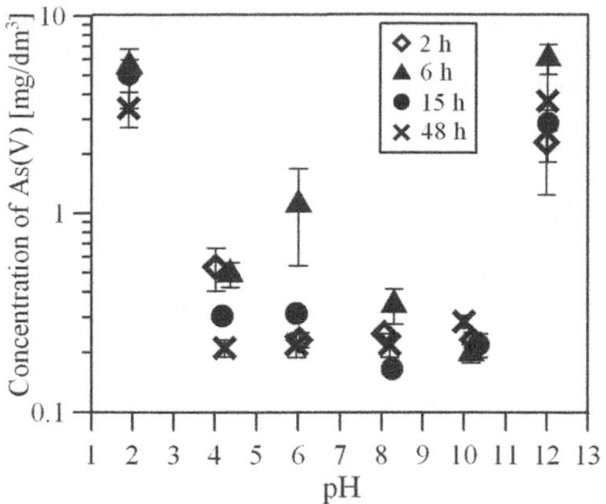

Figure 3. Final concentration of As(V) at various pH measured at different time of reaction with 1.4x excess Pb (Pb:As:Cl = 7:3:1).

In the experiments assessing the dynamics of As removal, an excess of Pb was used (initial lead concentration 1.4x higher than stoichiometric) because this proportion results in the most effective As removal in this experimental setup. A crystalline precipitate forms immediately after solutions are mixed. At pH 6.0 after two hours, the As concentration is reduced from initial 4000 to 0.23 ± 0.02 mg/dm^3 (99.99% reduction, Fig. 3). Further progress of the reaction is insignificant and often within the experimental error. For example, after 48 hours, the As concentration at pH 6.0 was 0.22 ± 0.03 mg/dm^3.

In all experiments, the white crystalline precipitate was identified with XRD as mimetite $Pb_5(AsO_4)_3Cl$. Calculated unit cell parameters are a = 10.247(2) Å and c = 7.448(2) Å, close to synthetic mimetite ICDD standard number 19–683 (a = 10,251 Å and c = 7.442 Å). Wet chemical analysis reflects the composition of mimetite with molar proportions Pb:As:Cl of 5.000 ± 0.080:2.927 ± 0.001:0.976 ± 0.001, close to the theoretical 5:3:1. Mid-infrared and far-infrared spectra of mimetite show characteristic bands originating from the vibrations of As-O bonds in the $[AsO_4]$ tetrahedron (Bartholomai 1978, Bajda et al. 2004): $\nu_1 = 730\,cm^{-1}$; $\nu_2 = 318\,cm^{-1}$; $\nu_3 = 783$, 804, 816 cm^{-1}; $\nu_4 = 370$, 411 cm^{-1} (Fig. 4, far infrared results not shown). Mimetite forms aggregates of euhedral crystals 1–2 μm in size (Fig. 5). The morphology does not vary with pH, time, or initial Pb:As proportions.

Dissolution of mimetite at various pH values was conducted for 62 days until the apparent equilibrium was attained. The solubility of mimetite as determined via the concentration of As(V) released to the solution is lowest at the pH values of 6.1 and 7.2 (Fig. 6); equilibrium [As] equal to 0.83 ± 0.03 and 0.93 ± 0.04 mg/dm^3, respectively. Similarly, the lowest [Pb] (<0.1 mg/dm^3) was measured for pH 6.1 and 7.2 (Fig. 7). These concentrations are lower than expected from the stoichiometry of mimetite and indicate incongruent dissolution for these pH conditions. Equlibrium modeling using the Phreeqc computer code (Parkhurst 1995) indicates that Pb released during dissolution may hydrolyse at near neutral and alkaline pH. Adsorption of Pb on fine-grained mimetite may also be partly responsible for the lower than stoichiometric concentrations of Pb at equilibrium.

The dynamics of mimetite dissolution as measured by the evolution of [As] and [Pb] with time is presented in Figures 8, 9. Dissolution is relatively slow and concentrations close to equilibrium are reached only after seven days.

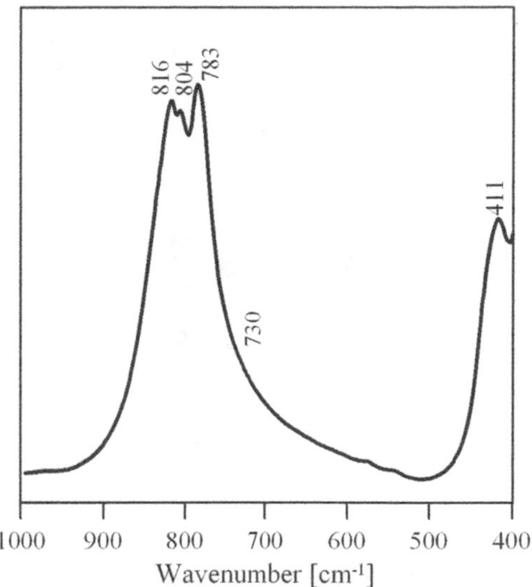

Figure 4. FTIR spectrum of mimetite precipitate.

Figure 5. SEM image of mimetite precipitate.

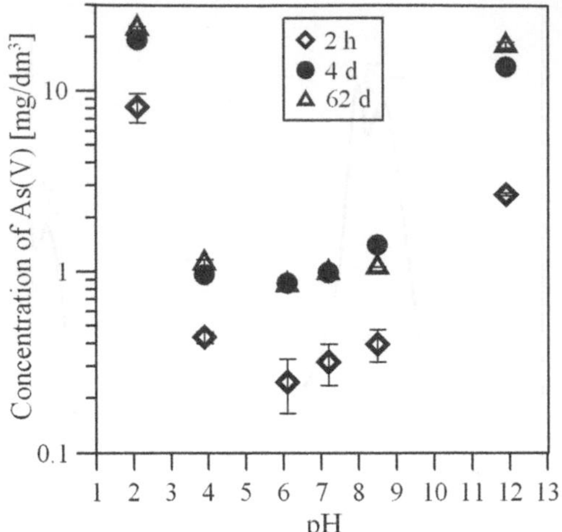

Figure 6. Concentration of As(V) with pH at various times of dissolution of mimetite.

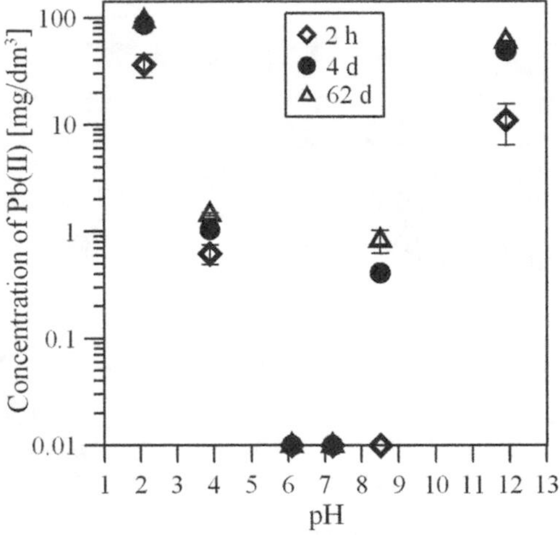

Figure 7. Concentration of Pb(II) with pH at various times of dissolution of mimetite.

In the pH range 3.9–8.5, the concentration of As(V) was $0.79 \pm 0.06 - 1.28 \pm 0.02$ mg/dm^3 at 48 hours and stabilized at $0.84 \pm 0.03 - 1.05 \pm 0.01$ mg/dm^3 within 62 days. More than ten-fold higher concentrations were noted at pH 2.1 and 11.9. Similarly, lead is released to the solution in significant amounts only in strongly acidic and alkaline solutions. In neutral solutions practically all lead is bound in the solid phase while at pH values in the range 3.9 to 8.5, Pb(II) concentrations of 1.41 ± 0.04 and 0.88 ± 0.20 mg/dm^3, were measured, respectively.

Based on the measured concentrations of Pb, As, and Cl at equilibrium at pH 2.1 and 3.9 the solubility constant K_{sp} and Gibbs free energy of formation $\Delta G^o_{f.298}$ were calculated. At equilibrium, K_{sp} is equal to the ion activity product (IAP) which, for the reaction of mimetite dissolution in acidic solutions:

$$Pb_5(AsO_4)_3Cl + 6H^+ \leftrightarrow 5Pb^{2+} + 3H_2AsO_4^- + Cl^- \quad (1)$$

can be written:

$$\log IAP = 5\log\{Pb^{2+}\} + 3\log\{H_2AsO_4^-\} + \log\{Cl^-\} + 6pH \quad (2)$$

where brackets denote the activities.

Based on concentrations measured at equilibrium, the activities of Pb(II), As(V), and Cl(I) were calculated using the speciation model Phreeqc (Parkhurst 1995). The empirical solubility product for mimetite determined from experiments at pH 2.1 and 3.9 is $\log K_{sp} = -21.69 \pm 1.05$. The Gibbs free energy of the reaction (1)

$$\Delta G^o_{reaction} = 5\Delta G^o_{f.298}(Pb^{2+}) + 3\Delta G^o_{f.298}(H_2AsO_4^-) + \Delta G^o_{f.298}(Cl^-) - \Delta G^o_{f.298\ mimetite} \quad (3)$$

is, at equilibrium, related to a solubility constant

$$\Delta G^o_{reaction} = -RT\ln K_{sp} \quad (4)$$

where R is the gas constant and T is temperature. This allows the Gibbs free energy of formation of mimetite to be calculated

$$\Delta G^o_{f.298\ mimetite} = 5\Delta G^o_{f.298}(Pb^{2+}) + 3\Delta G^o_{f.298}(H_2AsO_4^-) + \Delta G^o_{f.298}(Cl^-) + RT\ln K_{sp} \quad (5)$$

Based on thermodynamic data for Pb^{2+}, $H_2AsO_4^-$, and Cl^- according to Magalhães & Silva (2003) and the empirical K_{sp} calculated above, the value of $\Delta G^o_{f.298\ mimetite} = -2640 \pm 6$ kJ/mol was calculated. This value compares well with that of Comba (1987) $\Delta G^o_{f.298\ mimetite} = -2643$ kJ/mol and Robie et al. (1978) $\Delta G^o_{f.298\ mimetite} = -2674 \pm 3$ kJ/mol.

4 CONCLUSIONS

Removal of As from polluted wastes by addition of Pb(II) leading to precipitation of lead chloroarsenate – mimetite – can be used as a remediation method. The results of experiments presented herein indicate that when an excess of Pb is used, even high concentrations of As(V) can be reduced by 99.99% to levels comparable with the regulatory limits. The reaction is very fast. Mimetite precipitates immediately and the majority of As(V) is removed from the solution within minutes or hours, depending on pH. Precipitating mimetite forms aggregates of 1–2 μm crystals. Their well-defined crystalline character, high density,

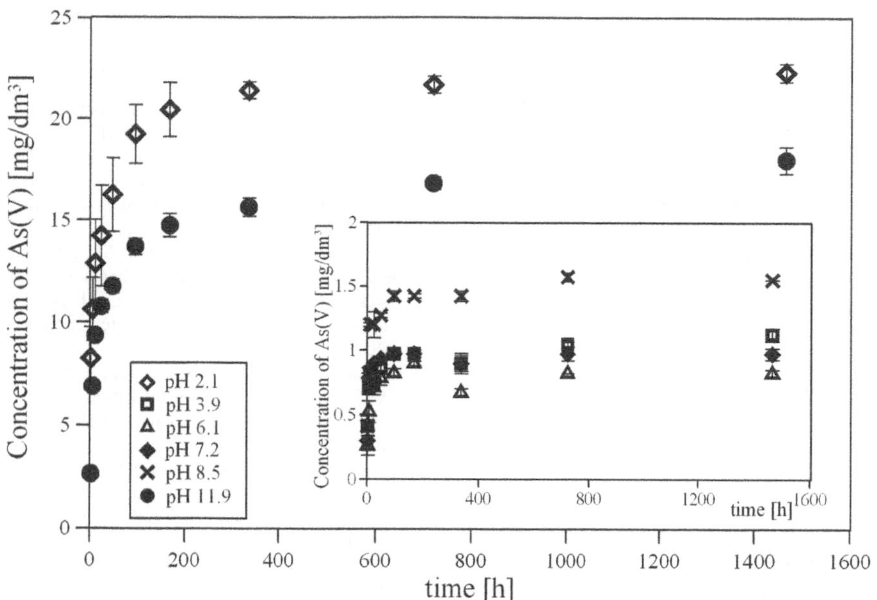

Figure 8. Evolution of As(V) concentration during mimetite dissolution at various pH.

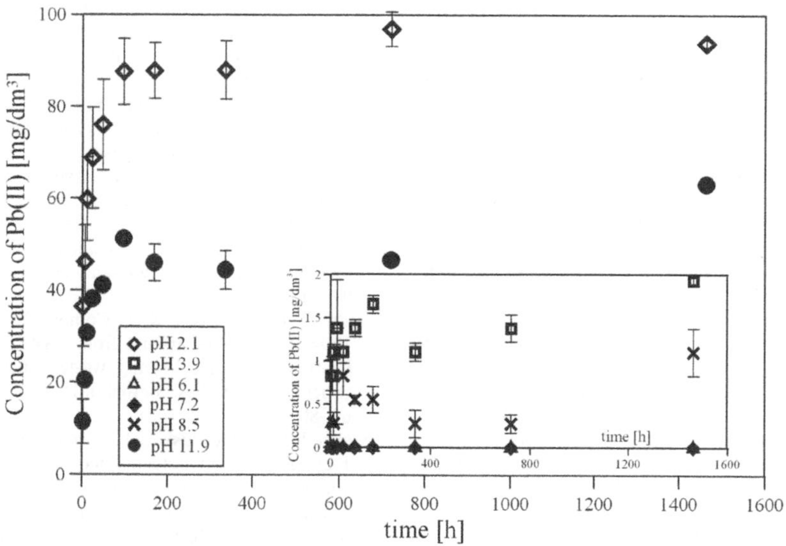

Figure 9. Evolution of Pb(II) concentration during mimetite dissolution at various pH.

and a tendency to form aggregates assures ease of separation from the suspension by settling or filtration, which is important for technological implementations. Also, mimetite is much safer for storage in typical natural pH ranges (4–9) than products of other As removal methods. This is a result of its relatively high thermodynamic stability.

The application of the method presented requires introduction of an excess of Pb(II) which is itself a toxic heavy metal. However, a very effective and relatively cheap method of lead immobilization was introduced and patented in the last decade. It is based on the reaction with phosphate (PO_4^{3-}) in the presence of chloride ions Cl^- resulting in precipitation of crystalline lead chlorophosphate – pyromorphite $Pb_5(PO_4)_3Cl$ (Cotter-Howells & Capron 1996, Manecki et al. 2000). Pyromorphite is isostructural with mimetite and also exhibits all its advantageous properties: crystallinity, high density, a tendency to form aggregates, and extremely low solubility resulting in high stability in the environment. Therefore, applications of stoichiometrically controlled amounts of phosphates (for example in the form of apatites) in the second stage of sequential treatment guarantees (at pH range 5–9) removal of excess lead to below the drinking water limit (Ma et al. 1993, Manecki et al. 2000). More research is needed on the levels of phosphate present in the wastewater as a result of this treatment.

ACKNOWLEDGEMENTS

The authors wish to thank M. Sikora (AGH) for his help in AAS analyses. This research was partially financed by MEN grant 18.25.140.365 and partially as the AGH-University of Science and Technology grant 11.11.140.158.

REFERENCES

Adriano, D.C. 2001. Trace Elements in Terrestrial Environments: Biogeochemistry, Bioavailability and Risks of Metals. New York: Springer.

Bajda, T., Mozgawa, W., Manecki, M., Szmit, E., Sikora, M. 2004. Vibrational spectra of pyromorphite-mimetite solid solutions. Proc. XXVII European Congress on Molecular Spectroscopy. Kraków, Poland, 5–10 September 2004.

Bartholomai, G. & Klee, W.E. 1978. The vibrational spectra of pyromorphite, vanadinite and mimetite. Spectrochimica Acta 34A: 831–843.

Bothe, J.V. & Brown, P.W. 1999. Arsenic immobilization by calcium arsenate formation. Environmental Science & Technology 33: 3806–3811.

Comba, P.G. 1987. Removal of arsenic from process and wastewater solutions. M.S. Thesis, Montana College of Mineral Science and Technology. Butte, Montana.

Cotter-Howells, J.D. & Capron, S. 1996. Remediation of contaminated land by formation of heavy metal phosphates. Applied Geochemistry 11: 335–342.

Dixit, S. & Hering, J.G. 2003. Comparison of Arsenic(V) and Arsenic(III) Sorption onto Iron Oxide Minerals: Implications for Arsenic Mobility. Environmental Science & Technology 37: 4182–4189.

Handke, M., Mozgawa, W., Nocuń, M. 1994. Specific features of IR spectra of silicate glasses. Journal of Molecular Structures 325: 129–136.

Harper, T.R. & Kingham, N.W. 1992. Removal of arsenic from wastewater using chemical precipitation methods. Water Environment Research 64: 200–203.

Lenoble, V., Deluchat, V., Serpaud, B., Bollinger, J.C. 2003. Arsenite oxidation and arsenate determination by the molybdene blue method. Talanta 61: 267–276.

Ma, Q.Y., Logan, T.J., Traina, S.J. 1993. Lead immobilization from aqueous solution and contaminated soils using phosphate rocks. Environmental Science & Technology 29: 1118–1126.

Magalhães, M.C.F. & Silva, M.C.M. 2003. Stability of lead(II) arsenates. Monatshefte für Chemie 134: 735–743.

Manecki, M., Maurice, P.A., Traina, S.J. 2000. Kinetics of aqueous Pb reaction with apatites. Soil Science 165: 920–933.

Narasaraju, T.S.B., Rao, K.K., Rai, U.S., Kapoor, B.K. 1977. Preparation, characterization and solubility of arsenic hydroxyapatite. Indian Journal of Chemistry 15A: 1014–1015.

Nishimura, T. 1988. Stabilities and solubilities of metal arsenites and arsenates in water and effect of sulfate and carbonate ions on their solubilities. In: R.G. Reddy, J.L. Hendricks & P.B. Queaneau (eds), Arsenic metallurgy fundamentals and applications: 77–98.Warrendale, PA.

Parkhurst, D.L. 1995. User's guide to PHREEQC – A computer program for speciation, reaction-path, advective-transport, and inverse geochemical calculations. U.S. Geological Survey Water-Resources Investigations Report: 95–4227.

Piver, W.T. 1983. Mobilization of arsenic by natural and industrial processes. In B.A. Fowler (ed.), Biological and environmental effects of arsenic: 1–50. Amsterdam: Elsevier Verlag.

Robie, R.A., Hemingway, B.S., Fisher, J.R. 1978. Thermodynamic properties of minerals and related substances at 298.15 K and 1 bar (10^5 pascal) pressure and at higher temperatures. US Geological Survey Bulletin 1452.

Robins, R.G., Huang, J.C.Y., Nishimura, T., Kothe, G.H. 1987. The adsorption of arsenate ion by ferric hydroxide. In R.G. Reddy, J.L. Hendricks & P.B. Queaneau (eds), Arsenic metallurgy fundamentals and applications: 99–112.Warrendale, PA.

Twidwell, L.G., Plessas, K.O., Comba, P.G., Dahnke, D.R. 1994. Removal of arsenic from wastewaters and stabilization of arsenic bearing waste solids: Summary of experimental studies. Journal of Hazardous Materials 36: 69–80.

Twidwell, L.G., McCloskey, J., Miranda, P., Gale, M. 1999. Technologies and potential technologies for removing arsenic from process and mine wastewater. In I. Gaballah, J. Hager & R. Ssolozabal (eds), Proceedings, Global Symposium on Recycling, Waste Treatment and Clean Technology, TMS: 1715–1726. Warrendale, PA.

Wojnar, A., Manecki, M., Bajda, T. 2005. Bioaccessibility of As(V) and Pb(II) from mimetite. Proc. 15th Annual Goldschmidt Conference. Moscow, Idaho, USA, 20–25 May 2005. Geochimica et Cosmochimica Acta: 69(10): A628–A628.

Environmental Engineering – Pawłowski, Dudzińska & Pawłowski (eds)

Effect of nitrification-filter packing material on the time to reach its operation capacity

Dorota Papciak
Department of Water Protection and Purification, Technical University of Rzeszów, Powstanców Warszawy Ave, Rzeszów, Poland

ABSTRACT: The paper contains an assessment of the effect of biofilter packing properties on the time to bring the nitrifying beds to operational capability. Tests were carried out with biological reactors, filled with diatomite, clinoptylolite, active carbon and a pumice filtration substance named 'Hydro-Filt'. The effect of the specific surface, sorptive, ion-exchange and buffering capabilities, as well as those of stimulating the development of a biological film, of the applied packing, were analysed. It was found that the properties of the material constituting the bio-filter packing determine the time to bring the nitrifying beds to operational capability and the effectiveness of the nitrification process. The presence of a biological film was noted earliest on the clinoptylolite bed, but its effectiveness in removing ammonium ion was the lowest of all applied filtration materials. The nitrification process proceeded with the highest efficiency on the diatomite packing, but the best water quality was obtained using an active carbon bed.

Keywords: Biofilters, packing material, drinking water, nitrification.

1 INTRODUCTION

Removal of nitrogen compounds from water to be used for drinking constitutes a serious problem in water-treatment technology. Because of the costs of removing nitrogen-containing contaminants, biological methods of achieving this goal are being proposed more and more often. A number of processes and systems utilizing nitrification have been developed and implemented wherever there is a necessity to remove ammonium ions without removing all nitrogeneous material. Bio-filtration is numbered among several highly-effective processes, but it is influenced by many factors (Fdz.-Polanco et al. 1995): column diameter, hydraulic loading rates, and the concentration of total ammonia nitrogen (Sandu et al. 2002), concentration of oxygen (Graaf et al. 1995, Bock et al. 1998, Chadran & Smets 2005), concentration of free ammonia (Villaverde et al. 2000), concentration of nitrite (Nijhof & Klapwijk 1995) and temperature (Fdz.-Polanco et al. 1994, Anderson et al. 2001). The extended time period needed for bio-film formation and that for a suitable concentration of biomass to be reached, are problems that still remain to be solved. The effectiveness of the bio-process and the time for bio-filter to reach full capacity are determined by the properties of the filter packing material (Pujol et al. 1994, Green et al. 1996, Menoud et al. 1999, Anderson et al. 2001) and the specific activity of nitrifying biofilm (Liu & Capdeville 1996). The important parameters are not only its granulation or specific surface, but also its capability to generate durable combinations between microorganisms and the carrier (Emtiazi et al. 2004, Yun et al. 2005). Filter packing material may play a role not just as the substrate for bio-film formation but also of adsorbent, ion-exchanger, nutrient and a buffer medium for biochemical reactions (Green et al. 1996, Anderson et al. 2001, Lipponen et al. 2004). The bio-filter packing material is selected not only because of its effectiveness but also because of its price. The most sought-after packing materials are therefore those which are both cheap and effective, as well as being able to stimulate the biomass development.

This paper presents the results of a study on the possible application of four different natural materials, with different physical and chemical properties, as packing material for nitrification filters. They were evaluated with respect to their respective effect on the nitrification process and on bio-film formation time.

2 MATERIALS AND METHODS

2.1 *Experimental set up*

The research was carried out using four biological reactors operated as nitrification filters with

Table 1. Selected parameters of filter packing materials.

Parameter, unit	Diatomite	Natural pumice	Clinoptylolite	Active carbon
Specific surface, m^2/g	20	–	21	980
Granulation, mm	1.2–2.0	1.2–2.0	1.2–2.0	1.2–2.0
Specific weight, g/cm^3	2.25	2.44	1.18	0.43
Ignition losses, %	7–10	1.0	1.16	–
SiO_3, %	68–73	56	71.8	–
Al_2O_3, %	9–12	23	12.8	–
Fe_2O_3, %	4–6	3	1.16	–
CaO, %	2–3	1	7.1	–
P_2O_5, %	0.2	–	0.05	–

Table 2. Components of the synthetic influent.

Parameter	Unit	Min	Max
Ammonia nitrogen	gN/m^3	1.74	3.24
Nitrite nitrogen	gN/m^3	0.00	0.006
Nitrate nitrogen	gN/m^3	0.58	2.24
Dissolved oxygen	gO_2/m^3	7.8	8.4
pH	–	6.62	8.12
Color	gPt/m^3	2	6
Turbidity	NTU	0	2
Temperature	°C	16	18

upward flow and filled with: active carbon (WG-15), clinoptylolite, diatomite, natural pumice (HYDRO-FILT filtration material) (Table 1). All packing materials were flushed with de-chlorinated tap water, in order to remove the dust fraction which would otherwise raise the filtrate turbidity and color. The parameters of filters studied were: bed height – 1.1 m, diameter – 35 mm, the filtration rate reached approx. 3 m/h.

The experiments were conducted with model solutions prepared from de-chlorinated tap water enriched with ammonia nitrogen (2–3 mgN/dm^3) and a nitrification-bacteria containing bio-preparation (Table 2). Bio-preparation was added for ten days to water to be directed to biofilters. Bed performance was estimated on the basis of familiar physical and chemical indices, i.e. ammonia nitrogen, nitrite and nitrate nitrogen, pH, color, turbidity, dissolved oxygen.

2.2 *Analysis*

The analytical methods used to measure the different forms of nitrogen concentration were taken from Standard Methods for the Examination of Water and Wastewater 17-th ed. (1989).

3 RESULTS

Our results indicate that the time of bio-film formation depends on the properties of bio-filter packing material. If we assume a reference-point situation in which the ammonia nitrogen removal is accompanied by the rise in nitrite content, the nitrification starting time is comparable for all studied materials. During the first 48 hrs of the experiment, ammonia nitrogen was removed by sorption and ion exchange resulting from the properties of the respective minerals. The first phase of the nitrification process could be observed from the third day. The rise in nitrite content remained in correlation with the ammonia-ion removal process. Diatomite proved to be most effective. A 100% effectiveness of the process, with the ammonia-ion content in the filtrate reduced to zero, was achieved as early as only 28 days (Fig. 1A). In the case of active carbon, the lowest ammonia-ion content achievable was 0.64 mgN/dm^3 (on the 36th day of bed operation, Fig. 1B). Clinoptylolite and HYDRO-FILT enabled water with ammonia-ion content not lower than 1.6 mgN/dm^3 and 1.1 mgN/dm^3, on the 9th and 24th day, respectively to be obtained (Figs 1C, D).

The most rapid nitrite content growth took place in water after bio-filtration on a diatomite-packed filter (Fig. 2A). The maximum content of this form of nitrite (3 mg $N\text{-}NO_2^-/dm^3$) was recorded on the 26th day of bed operation. With the remaining filter packing materials the nitrites grew much slower. Their growth became noticeable after only twenty days of bed operation. In the case of active carbon (the 30th day) and clinoptylolite (the 36th day) their highest concentration reached 1.3 mg $N\text{-}NO_2^-/dm^3$ (Figs 2B, C). In the case of HYDRO-FILT, the nitrite content did not exceed 0.5 mg $N\text{-}NO_2^-/dm^3$ (Fig. 2D).

The nitrate content variations in water subjected to bio-filtration were irregular. In most cases their content in the filtrate was higher than that in water directed onto beds (Fig. 3). Also, in some recorded cases, their concentration in the filtrate was lower. Lowering the content of that nitrogen form in treated water could be slowed by the denitrification process.

Variations in the remaining, controlled parameters confirm the mechanism of ammonia nitrogen removal by nitrification. The treated-water reaction decreased steadily from the 20th day of biofilter operation onwards (Fig. 4A). The lowest pH value was recorded in water after bio-filtration on active carbon and diatomite. The variation in oxygen content confirmed the biological processes taking place. The most noticeable lowering of its concentration was observed with active carbon (Fig. 4B). After 20 days of bed operation, water treated on the diatomite bed had a higher oxygen content than the respective filtrate from the

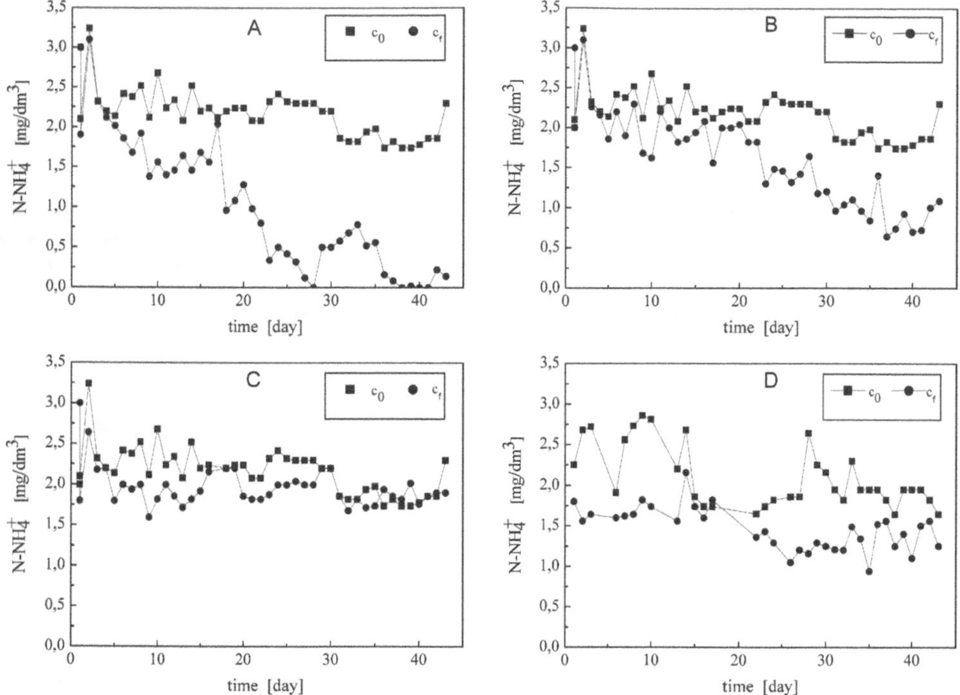

Figure 1. Changes in ammonia nitrogen contents in water treated on biofilters: A – diatomite, B – active carbon WG-15, C – clinoptylolite, D – natural pumice HYDRO-FILT, c_0 – initial concentration, c_f – concentration in efflux.

active carbon bed, despite its higher effectiveness in ammonia ion removal (Fig. 4B).

Color and turbidity variations of treated water may provide an answer to the question of which packing material generates the most durable combinations with microorganisms and exhibits characteristics which stimulate bio-film formation. The growth in turbidity of treated water may indicate a growing number of microorganisms (if the turbidity is not caused by the low mechanical strength of the packing material). The highest values of treated-water turbidity occurred in case of the clinoptylolite bed (Fig. 4C). It may be assumed that too weak bonds between clinoptylolite and the bio-film lead to microorganisms being washed out of the bed and caused the decreasingly effective removal of the ammonia nitrogen. Despite good ion-exchange and buffering properties it was not possible to obtain a bio-film of sufficient enzymatic activity. Higher color value of water after bio-filtration confirmed leaching of the biological film from the biofilter and, similar to turbidity, it was the highest for the clinoptylolite bed (Fig. 4D). A slight rise in turbidity and a lowering of water color value, was obtained in the case of active carbon (Figs 4C, D).

4 DISCUSSION

By carrying out a nitrogen balance, we may obtain information on the character of processes taking place on nitrifying beds as well as being able to assess their percentage share in the complex bio-sorption process.

In addition to biological processes, i.e. assimilation, nitrification and denitrification, sorption processes may also take place, depending on the type and properties of biofilter filling material(s). Thus, physico-chemical processes may have a considerable share in the first period of biofilter operation in case of fillings, which are sorbents and ion exchangers. Therefore, when discussing processes taking place in nitrifying beds, the possible occurrence of the following cases need to be taken into account:

1. Sorption occurs when:

$$C_{0\ N\text{-}NH4} < C_{k\ N\text{-}NH4},\ C_{0\ N\text{-}NO2} = C_{k\ N\text{-}NO2} \quad (1)$$

and

$$C_{0\ N\text{-}NO3} = C_{k\ N\text{-}NO3} \quad (1a)$$

where indices '0' and 'k' denote concentration of nitrogen species in influent and effluent respectively.

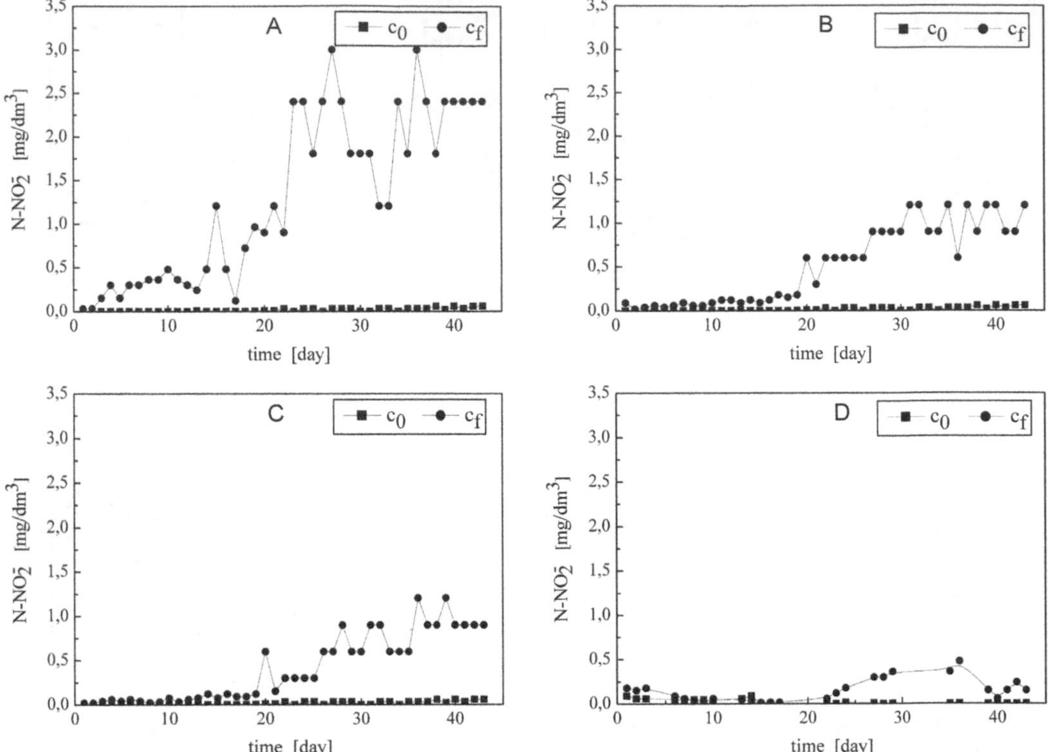

Figure 2. Changes in nitrite nitrogen contents in water treated on biofilters: A – diatomite, B – active carbon WG-15, C – clinoptylolite, D – natural pumice HYDRO-FILT, c_0 – initial concentration, c_f – concentration in efflux.

2. Sorption and N-NH_4^+ oxidation (first stage of nitrification) occur when:

$$C_{0\ N\text{-}NH4} + C_{0\ N\text{-}NO2} + C_{0\ N\text{-}NO3} > C_{k\ N\text{-}NH4} + C_{k\ N\text{-}NO2} + C_{k\ N\text{-}NO3} \quad (2)$$

and

$$C_{0\ N\text{-}NO3} = C_{k\ N\text{-}NO3} \quad (2a)$$

and quantity of nitrogen removed by ion exchange, physical and chemical adsorption, and assimilation ($C_{S\ N}$) can be calculated by following equation:

$$C_{S\ N} = (C_{0\ N\text{-}NH4} + C_{0\ N\text{-}NO2}) - (C_{k\ N\text{-}NH4} + C_{k\ N\text{-}NO2}) \quad (3)$$

3. Sorption and N-NH_4^+ and N-NO_2^- oxidation (first and second stage of nitrification) occur when:

$$C_{0\ N\text{-}NH4} + C_{0\ N\text{-}NO2} + C_{0\ N\text{-}NO3} > C_{k\ N\text{-}NH4+} + C_{k\ N\text{-}NO2} + C_{k\ N\text{-}NO3} \quad (4)$$

and

$$C_{S\ N} = (C_{0\ N\text{-}NH4} + C_{0\ N\text{-}NO2} + C_{0\ N\text{-}NO3}) - (C_{k\ N\text{-}NH4} + C_{k\ N\text{-}NO2} + C_{k\ N\text{-}NO3}) \quad (4a)$$

4. First stage of N-NH_4^+ oxidation occurs when:

$$C_{0\ N\text{-}NH4} + C_{0\ N\text{-}NO2} + C_{0\ N\text{-}NO3} = C_{k\ N\text{-}NH4} + C_{k\ N\text{-}NO2} + C_{k\ N\text{-}NO3} \quad (5)$$

and

$$C_{0\ N\text{-}NO3} = C_{k\ N\text{-}NO3} \quad (5a)$$

5. Processes of N-NH_4^+ and N-NO_2^- oxidation occur when:

$$C_{0\ N\text{-}NH4} + C_{0\ N\text{-}NO2} + C_{0\ N\text{-}NO3} = C_{k\ N\text{-}NH4} + C_{k\ N\text{-}NO2} + C_{k\ N\text{-}NO3} \quad (6)$$

6. Nitrification and N-NO_3 to N-NO_2 reduction (denitrification) occur when:

$$C_{0\ N\text{-}NH4} + C_{0\ N\text{-}NO2} + C_{0\ N\text{-}NO3} = C_{k\ N\text{-}NH4} + Ck_{\ N\text{-}NO2} + C_{k\ N\text{-}NO3} \quad (7)$$

and

$$C_{k\ N\text{-}NO3} < C_{0\ N\text{-}NO3} \quad (7a)$$

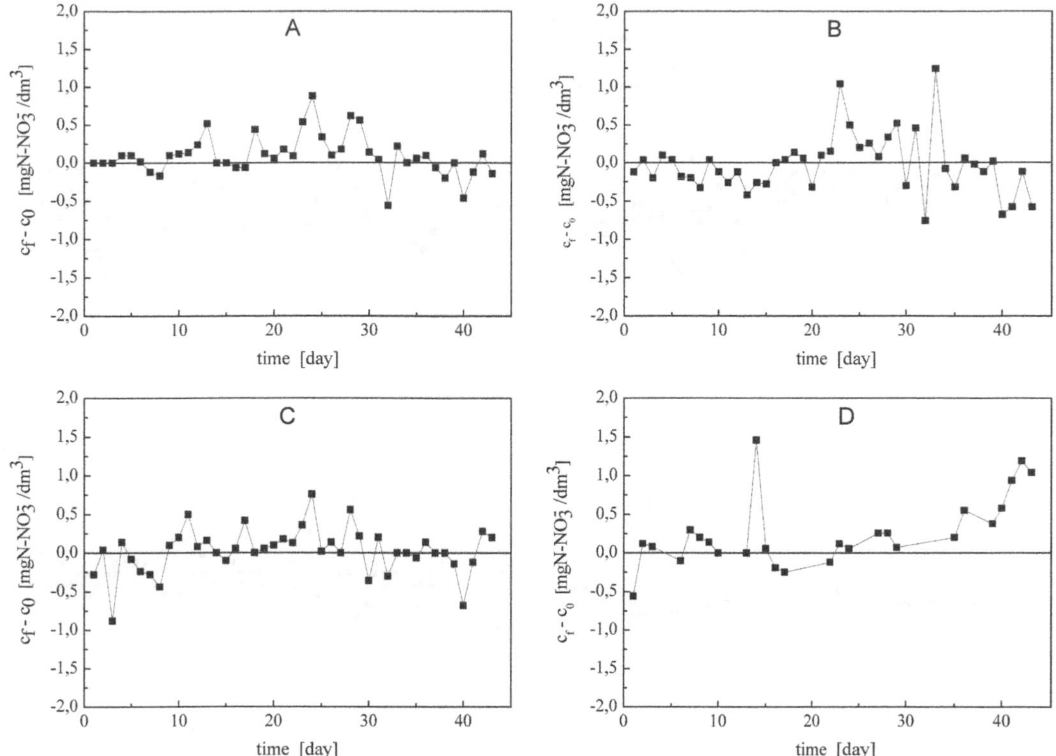

Figure 3. Changes in nitrate nitrogen contents in water treated on biofilters: A – diatomite, B – active carbon WG-15, C – clinoptylolite, D – natural pumice HYDRO-FLT, c_0 – initial concentration, c_f – concentration in efflux.

and

$$C_{k\ N\text{-}NO2} = C_{0\ N\text{-}NH4} - C_{k\ N\text{-}NH4} + C_{0\ N\text{-}NO3} - C_{k\ N\text{-}NO3} + C_{0\ N\text{-}NO2} \quad (7b)$$

The balance of nitrogen species carried out according to equations (1–7b) makes possible to estimate the contributions of sorption and nitrification processes on the packing materials investigated (Table 3).

In the first stage of experiment the highest share of sorption process was observed in the case of clinoptylolite. In the first stage of the experiment the role of sorption was highest in the case of clinoptylolite. The effect probably results from ion exchange properties of the clinoptylolite and its high selectivity towards NH_4^+. The other packing materials showed a lower contribution of sorption – 15, 9, and 7% for active carbon, natural pumice and diatomite respectively. After the 17th day of the experiment decreases in the role of sorption were observed for the active carbon (7%) and the clinoptylolite and pumice (1%) beds, while, in the diatomite bed, the proportion of sorption was increased (18%). These changes could be caused by exhaustion of the sorption capacity of the packing materials as well as the structure of the biofilm formed. The penetration of substrate is better when there are many pores, channels and biopolymers in the biofilm and consequently sorption becomes more important. The opposite effect is observed for a biofilm with high density. It has been reported that the tendency to an increase in porosity increases with biofilm thickness (Lewandowski 2000). The short time for biofilm formation and the rapid increase of biomass deposited on filter filler are not favorable; the decrease in nitrification and accumulation of nitrite ions may result.

Nitrite-ion accumulation is an unfavorable phenomenon because it suppresses the growth of nitrification bacteria. The nitrous acid generated from NO_2^- ions may be toxic for both *Nitrosomonas* and *Nitrobacter* (Klimiuk and Łebkowska 2003). Moreover, the high concentration of nitrite ions disqualifies this water for drinking.

The accumulation of nitrite might be caused by differences in growth of *Nitrosomonas* and *Nitrobacter* and an insufficient quantity of dissolved oxygen, which is necessary for the proper functioning of the nitrification process as well as to the structure and thickness of the biofilm. Biofilm thickness and density

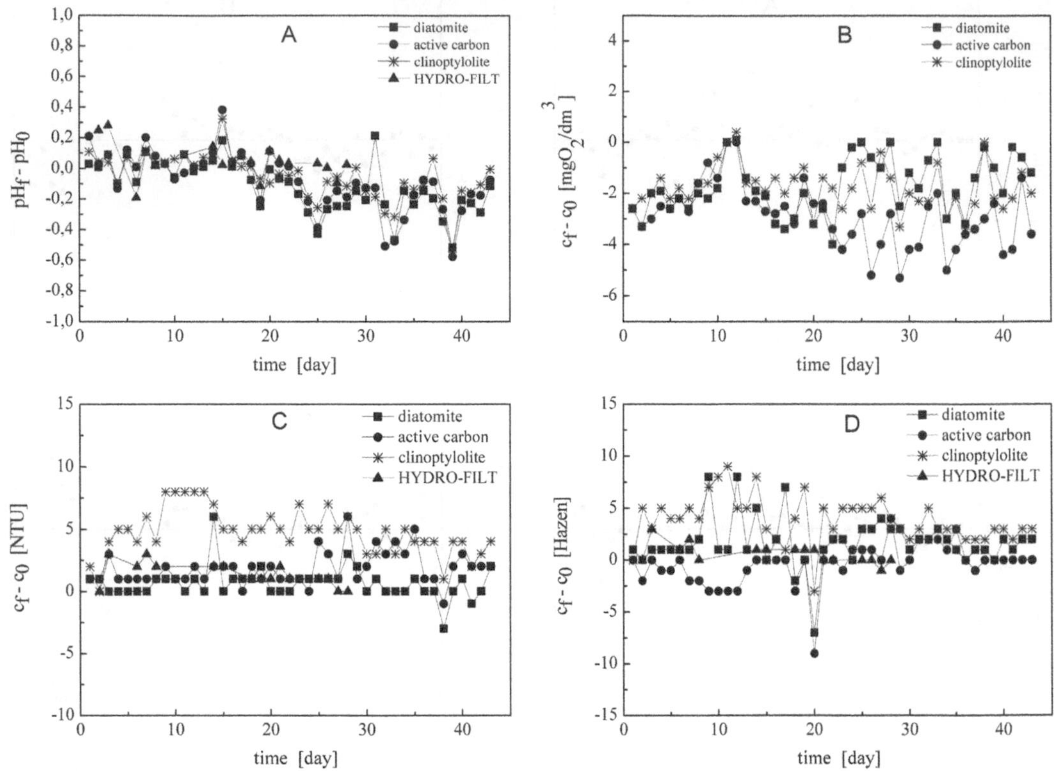

Figure 4. Changes of selected parameters of water treated on biofilters: A – pH value, B – oxygen content, C – turbidity, D – color; index '0' and 'f' denotes initial and final value of parameters respectively.

Table 3. Participation of sorption in ammonia nitrogen removal on biosorption filter beds.

Packing material	Working time of bed [day]	Sorption [%]
Diatomite	Until 17	0–7
	After 17	0–18
Active carbon	Until 17	0–15
	After 17	0–7
Clinoptylolite	Until 17	0–26
	After 17	0–1
Natural pumice	Until 36	0–9
	After 36	0–1

are two major parameters that determine the performance of biofilm reactors. The biofilm thickness affects the mass transfer of substrates from the bulk fluid to the biofilm. Hibiya et al. (2004) studied oxygen penetration in biofilms for wastewater treatment and put forward an empirical equation relating the effective distribution coefficient of oxygen. Oxygen diffused completely into the bottom layer of biofilms, with thicknesses <300 μm, and the oxygen penetration depth increased with increasing biofilm thickness. As a consequence, microbial distribution changes with the biofilm thickness. In an aerobic biofilm heterotrophs and nitrifiers dominate in the top layer, while denitrifying bacteria dominate in the lower layer. Therefore, thin biofilms with thickness of 300–400 μm are appropriate for simultaneous oxidation and nitrification while a denitrification zone was found on a deep biofilm.

As much as 3.43 g of oxygen is needed for the transformation of 1 g of ammonia nitrogen into nitrites, and 1.14 g of oxygen – to transform 1 g of nitrites into nitrate. In all the stoichiometric oxygen consumption in nitrification reactions is therefore 4.57 g of O_2 per g of ammonia nitrogen. In experimental conditions, i.e. in the process of single-stage bio-filtration, the oxygen content was theoretically sufficient for removing max. 2 g of ammonia nitrogen. Thus, it would seem that the process should proceed without disturbance and the nitrites should be completely oxidized to nitrates. The reason for nitrite accumulation may be sought in the structure of the bio-film, where two zones – oxygen and anoxic ones – may be distinguished. Lack of oxygen causes reactions involving heterotrophic bacteria to take place in pores and on grain surfaces, while the biochemical activity of nitrifying bacteria becomes inhibited. The relationship between substrate diffusion and microorganism growth also decides the final effect of the nitrification.

The analysis of oxygen content in filtrate did not reveal any substantial reduction of dissolved oxygen. Its concentration was sufficient for full nitrification to take place.

The reduction in oxygen content was not stoichiometric with the quantity of removed ammonia nitrogen. It was caused by the fact that oxygen is liberated during the bio-synthesis. Moreover, it was shown that in the first stage of ammonia nitrogen oxidation to hydroxylamine, oxygen may stem from water molecules and not just from molecular oxygen (Graaf et al. 1995). If we additionally take into account the fact that part of the nitrogen is built into microorganism cells during the biomass synthesis, the oxygen demand ought be lower than that predicted from chemical equations.

The presence of denitrification bacteria in the investigated beds has already been pointed out. On the other hand, the balance of nitrogen species (eq. 7b) indicates that the denitrification process occurs in our systems.

Moreover, the studies with pure *Nitrobacter* cultures have shown the presence of nitrite oxidoreductase in some species. This enzyme catalyzes the reaction of nitrite oxidation to nitrates in aerobic conditions, whereas in anoxic conditions nitrates become reduced. The enzyme was found to exhibit more affinity to nitrates than to nitrites (Bock et al. 1988). Oxidoreductase presence may therefore be a cause of the excessive accumulation of nitrites and hinder the oxygen uptake.

For the results of water conditioning process both the numbers and the activity of microorganisms inhabiting the bed are of great importance. According to Wang, the metabolic activity of the biomass rather than the numbers determines the effectiveness of biodegradation. Many methods have been developed for testing biosorptive beds. Most take into account the manner of sampling, biomass extraction and assessment of its activity or intensity of its development. Unfortunately, they are time-consuming and frequently quite complicated. As the parameters, i.e. biomass concentration and metabolic activity of bacteria correlate with technological effect (0.5 mg $N\text{-}NH_4/dm^3$, 0.1 mg $N\text{-}NO_2/dm^3$) the assessment of bed performance was based on that effect. The assessment of activity of nitrifying bacteria was based on the analysis of inorganic forms of nitrogen, whereas the durability of combinations/bonds of microorganisms with biofilter packing – was monitored through observation of typical water parameters such as turbidity and color.

Despite its 33-times lower specific surface than that of active carbon (Table 1), the diatomite filter packing was more effective in ammonia nitrogen removal (Figs 2A, B). Slight variations in turbidity and color testify to its efficiency in stimulation bio-film formation (Figs 4C, D). The HYDRO-FILT packing permitted preparation/production of water characterized by a slight growth in color and turbidity compared to water directed to the bed, but still it exhibited low effectiveness in nitrification (Figs 4C, D, Fig. 2D). The high effectiveness of the nitrification process on diatomite bed may be explained by properties stimulating multiplication of nitrification bacteria. Presumably, the carbonate and phosphate containing diatomite becomes a nutrient substance for them, leading to much accelerated synthesis of biomass. Clinoptylolite has similar chemical composition but, compared to diatomite, it contains less phosphorus and, though having a comparable specific surface, its internal structure is not fully utilized.

Research on biological films forming on various filling materials showed that surface features result from bed characteristics and the structure of pores, are of significant importance. The character of the external surface of grains affects the colonization by microorganisms. Microorganisms grow in numbers both in the spaces between grains and in macro-pores, which are of the size that enables penetration of microorganism cells into them to create niches protected against disadvantageous effects of water flow forces (Ventresque et al. 1994).

The cause of inferior colonization of some filling materials may well be the excessive smoothness of the surfaces of capillaries and the poorer availability of substrates inside, not through-flow macro-pores. Moreover, one suggestion holds that the growth of biomass on the bed is stimulated by the number of protected niches (Ventresque et al. 1994), another claims that it is the character of the external surface of grains that is important for the colonization (Mołczan 1999). In sand filters the bacterial growth was found to take place in spaces between bed grains, whereas in active carbon beds bacteria developed inside carbon grain pores, and greater numbers of *Nitrosomonas* bacteria than *Nitrobacter* bacteria were found in the filtrate.

The fraction of meso- and macropores in the WG-15 active carbon reaches 75%. Given the much higher specific surface of that packing material compared to the studied minerals, satisfactory results in the nitrification process may be achieved despite a lack of substances stimulating bio-film formation. However, that requires a longer time to bring the filtration bed to operational capacity.

5 CONCLUSIONS

- The properties of biofilter packing material determine the time needed to bring the nitrification beds to operational capacity and to reach effective nitrification.

- The chemical composition of biofilter packing that stimulates the growth of microorganisms has a decisive effect on the time of biofilm formation.
- A large specific surface, and especially the secondary porosity combined with sorptive properties, permit satisfactory results in the nitrification process but the time required for bio-film formation is longer.
- The nitrification process proceeded most effectively on diatomite and carbon beds but only the active carbon bed (WG-15), was capable of providing the best quality of water in terms of controlled parameters.
- An unfavorable phenomenon is the accumulation of nitrites, which disqualify water for human consumption at concentrations exceeding 0.5 mg NO_2/dm^3. Therefore, attempts will be carried out at clarifying and eliminating this phenomenon.
- Attention to increased turbidity and value color of treated water is necessary. Therefore, additional filtration methods will be devised to remove suspensions washed out from biofilters.

REFERENCES

American Public Health Association, 1989, Standard Methods for the Examination of Water and Wastewater, 17th ed. American Public Health Association, Washington, DC, USA.

Anderson A. & Laurent P. & Kihn A. & Prevost M. & Servais P., 2001, Impact of temperature on nitrification in biological Activated carbon (BAC) filters used for drinking water treatment, Wat. Res. 35: 2923–2934.

Bock E. & Wilderer P.A. & Freitag A., 1988, Growth of Nitrobacter in the absence of dissolved oxygen, Wat. Res. 22: 245–250.

Chadran K. & Smets B.F., 2005, Optimizing experimental design to estimate ammonia and nitrite oxidation biokinetic parameters from batch respirograms, Wat. Res. 39: 4969–4978.

Emtiazi F. & Schwartz T. & Marten S.M. & Krolla-Sidenstain P. & Obst U., 2004, Investigation of natural biofilms formed during the production of drinking water from surface water embankment filtration, Wat. Res. 38: 1197–1206.

Fdz-Polanco F. & Garcia P. & Villaverde S., 1994, Temperature effect over nitrifying bacteria activity in biofilters: activation and free ammonia inhibition, Wat. Sci. Tech. Vol. 30, No 11: 121–130.

Fdz-Polanco F. & Mendez E. & Villaverde S., 1995, Study of nitrifying biofilms in submerged biofilters by experimental design methods, Wat. Sci. Tech. Vol. 32, No 8: 227–233.

Graaf van de A.A. & Mulder A. & Bruijn de P. & Jetten M.S.M. & Kuenen J.G., 1995, Anaerobic oxidation of ammonium is a biologically mediated process, Appl. Environ. Microbiol. 53: 754–760.

Green M. & Mels A. & Lahav O. & Tarre S., 1996, Biological-ion exchange process for ammonium removal from secondary effluent, Wat. Sci. Tech. Vol. 34, No 1–2: 449–458.

Hibiya K. & Nagai J. & Tsuneda S. & Hirata A., 2004, Simple prediction of oxygen penetration depth in biofilms for wastewater treatment, Biochem Eng. J. 19: 1–68.

Klimiuk E. & Łebkowska M., 2003, Biotechnology in environment protection (in polish), PWN Warszawa, ISBN 83-01-14067-4.

Lewandowski Z., 2000, Notes on biofilm porosity, Wat. Res. 34: 2620–2624.

Lipponen M.T.T. & Martikainen P.J. & Vasara R.E. & Servomaa K. & Zacheus O. & Kontro M.H., 2004, Occurrence of nitrifiers and diversity of ammonia-oxidizing bacteria in developing drinking water biofilms, Wat. Res. 38: 4424–4434.

Liu Y. & Capdeville B., 1996, Specific activity of nitrifying biofilm in water nitrification process, Wat. Res. 30: 1645–1650.

Łomotowski J. & Haliniak J., 1997, Ammonia Nitrogen Removal from Groundwater via Biofilters (in polish), Ochrona Środowiska 3(66): 15–17.

Menoud P. & Wong C.H. & Robinson H.A. & Farquhar A. & Barford J.P. & Barton G.W., 1999, Simultaneous nitrification and denitrification using Siporax packing, Wat. Sci. Tech. Vol. 40, No 4–5: 153–160.

Mołczan M., 1999, Organic Matter Removal by Adsorption and Biodegradation on GAC Filter Beds (in polish). Ochrona Środowiska 3(74): 19–25.

Nijhof M. & Klapwijk A., 1995, Diffusional transport mechanism and biofilm nitrification characteristics influencing nitrite levels in nitrifying trickling filter effluents, Wat. Res. 29: 2287–2292.

Pujol R. & Hamon M. & Kandel X. & Lemmel H., 1994, Biofilters: flexible, reliable biological reaktors, Wat. Sci. Tech. 29(10–11): 33–38.

Sandu S.I. & Boardman G.D. & Watten B.J. & Brazil B.L., 2002, Factors influencing the nitrification efficiency of fluidized bed filter with a plastic bead medium, Aquacultural Engineering 26: 41–59.

Ventresque C. & Dagois D.G. & Pillard M. & Delkominette A. & Bablon G., 1994, Comparing two GACs for adsorption and biostabilization, Journal AWWA, Vol. 86, No 3: 91–102.

Villaverde S. & Fdz.-Polanco F. & Garcia P.A., 2000, Nitrifying biofilm acclimation to free ammonia in submerged biofilters. Start-up influence, Wat. Res. 34: 602–610.

Wang J.Z. & Summers R.S. & Miltner R.J., 1995, Biofiltration Performance; Part 1, Relationship to biomass. Jour. AWWA, 87(12): 55–63.

Yun M. & Yeon K. & Park J. & Lee C. & Chun J. & Lim D.J., 2006, Characterization of biofilm structure and its effect on membrane permeability in MBR for dye wastewater treatment, Wat. Res. 40: 45–52.

Environmental Engineering – Pawłowski, Dudzińska & Pawłowski (eds)

Pretreatment of sewage by heavy metal sorption onto natural zeolite

Henryk Wasag
Lublin University of Technology, Faculty of Environmental Engineering, Nadbystrzycka, Lublin, Poland

ABSTRACT: The paper assesses the potential of natural zeolites for heavy metals removal from sewage. In equilibrium studies the influence of pH on the removal of heavy metals and the selectivity of zeolites have been evaluated. Work on the sorption of heavy metals from aqueous solutions of high Ca-Mg content showed zeolites to be appropriate sorbents for sewage pretreatment processes. The effectiveness of heavy metal removal from sewage by natural zeolites was examined using a batch-type method. Granular zeolites of particle size 1–2 mm added to sewage at doses $<0.1\ g/dm^3$ take up heavy metals efficiently in contact times of ca. 10–15 minutes. In this way, 5–40% of the initial content of heavy metals can be removed before separating the primary sludge. Such a reduction in heavy metal content would diminish the concentration of the metals in sewage sludge which, in turn, should simplify their final disposal.

Keywords: Heavy metals, natural zeolites, municipal wastewater, sewage sediments.

1 INTRODUCTION

Heavy metals and their compounds, both organic and inorganic, are released to the environment as a result of anthropogenic activities (Karvelas et al. 2003). The increasing levels of heavy metals in the environment represent a serious threat to human health, living resources and ecological systems (Álvarez-Ayuso et al. 2003). The pollution and toxicity associated with them are commonly known (Ahmed et al. 1998). A very important characteristic of these metals is that they are non-degradable and therefore persistent (Donat et al. 2005). From the standpoint of potential hazards the following elements are considered as priority heavy metals: Cr, Ni, Cu, Zn, Cd and Pb (Álvarez-Ayuso et al. 2003). These elements are commonly used in daily life and therefore they are always present in municipal wastewater. Concentrations of heavy metals in sewage can vary over a wide range and depend on many factors, mainly on the kind of industry in the region and living standards of the society (Sörme & Lagerkvist 2002). The known toxic influence of heavy metals on the environment also has a bearing on processes in wastewater treatment plants (Madoni et al. 1996). Most sewage treatment plants treat wastewater by primary sedimentation followed by a biological process and secondary sedimentation (Kangala & Chipasa 2003). Heavy metals present in wastewater during the treatment processes are distributed between the two final products of wastewater treatment plant (WTP): sewage sludge and treated effluent. The amounts of heavy metals in sewage sludge and in treated effluent depend on many factors (Santarsiero et al. 1998). The effectiveness of purifying processes depends on heavy metal concentration in sewage and consequently affects the amount of heavy metals discharged into the receivers (Chua 1998).

It is necessary to develop simple methods allowing for control of heavy metals flow in a sewage treatment plant. An application of ion exchange seems to be a very promising solution (Bolto & Pawlowski 1987). Because of the very large volume of municipal wastewater only non-conventional ion exchange techniques can be applied with the use of low cost ion exchangers such as natural zeolites. Zeolites have many applications, mainly based on their structural characteristics, their sorbent properties and their relatively high specific surface area. Zeolites, for example, have been used as: chemical sieves, water softeners, removers of ammonia from wastewater, removers of toxic gases from gaseous emissions, filters for odor control, fertilizers and pesticide carriers (Álvarez-Ayuso et al. 2003). Zeolites appear also as suitable sorbents for heavy metals from aqueous solution. There are many reports in the literature concerning application of natural zeolites to the adsorption of d-electron metals (Erdem et al. 2004, Majdan et al. 2003). Some sorption characteristics of heavy metals could be improved by applying modified zeolites or zeolites in Na form (Ahmed et al. 1998, Ćurković et al. 1997). Due to their attractive selectivity for certain heavy metal ions such as Pb^{2+}, Zn^{2+}, Cd^{2+}, Ni^{2+},

Mn^{2+} zeolites are especially applicable for purification of metal electroplating wastewater (Ouki & Kavannagh 1999). The removal of soluble heavy metals from motorway storm water to substantially lower levels was investigated by means of natural and synthetic zeolites, but their uptake appeared to be partially inhibited by other dissolved components in motorway storm water (Pitcher et al. 2004). Even in case of more difficult media, such as sewage sludge, zeolites may be useful for heavy metals uptake (Zorpas et al. 2000). However, in spite of the promising results, the real applicability of these minerals to purify metal wastewater is still fairly unknown (Ouki & Kavannagh 1997) and industrial scale effluents treatment plants are rather seldom (Chojnacki et al. 2004).

The main objectives of the presented research were to study the sorption behavior of heavy metals onto natural zeolites with respect to municipal wastewater pretreatment methods. In an equilibrium study, the influence of pH, experimental sorption isotherms and uptake of heavy metals from solution of high concentration of Ca and Mg were examined. Batch mode experiments showed the influence of zeolite dose and contact (mixing) time on heavy metals sorption from municipal wastewater. On the basis of a series of laboratory tests, average removal degrees of heavy metals from sewage were estimated. The possibility of heavy metals flow control during sewage treatment processes is very important for the minimization of environmental and health risk. The accumulation and partial removal of heavy metals (Cd, Cr, Cu, Ni, Pb, Zn) from municipal wastewater before separating sewage sludges diminishes the content of the metal in the sludges which, in turn, should simplify their final disposal.

2 MATERIALS AND METHODS

Zeolite used for heavy metals sorption was obtained from the Sokornit mine in Ukraine and was classified as clinoptylolite. For equilibrium studies powdered zeolite of particle size below 0.1 mm was used, while subsequent laboratory experiments on heavy metals removal from sewage were done with zeolites of particle size 1–2 mm. Samples of sewage were collected from Lublin Wastewater Treatment Plant (Hajdów).

The value of pH was measured using a combined glass electrode HI 1131B connected to a pH meter HANNA HI 932. Heavy metal concentrations were determined by atomic absorption spectrophotometer (HITACHI Z-2000). Sewage samples prior to analysis were prepared using a microwave digestion system, a CEM MDS-2000. Standard solutions of heavy metals were supplied by Merck. All other chemicals were of analytical grade. Ultrapure water was obtained using the Millipore Milli-Q system.

2.1 *Influence of pH on heavy metals sorption*

To 50 ml of solution containing 2.5 ml of 2 M HCl and 0.5 ml of 0.01 M solution of the heavy metal to be studied, 1 g of zeolite was added. After mixing for 20 minutes the pH of the solution was measured, 0.25 ml sample taken for heavy metal analysis and to the remaining solution a portion of 2 M KOH was added. Such experiments were done for all heavy metals studied (Cu, Cd, Ni, Pb, Cr, Zn).

2.2 *Adsorption isotherm*

Adsorption isotherms were obtained in static conditions from acetic buffer solution of pH 7. To 100 ml samples containing the studied heavy metal (0.00001 to 0.0005 M) 1 g of the zeolites was added. After 2 hours of mixing, samples of the equilibrated solutions were analyzed for heavy metal content. Experimental isotherms were done for adsorption of Cu, Cd, Ni, Pb, Cr, and Zn onto powdered zeolites.

2.3 *Sorption of heavy metals from solution of high Ca-Mg content*

Sorption of heavy metals from solutions of high Ca-Mg content was investigated to assess the ability of natural zeolites for removal of the metals from municipal wastewater. As a model solution tap water were used. A series of 100 ml samples containing the heavy metal (0.00001 to 0.0001 M) under study were adjusted to pH 7 by means of 0.1 M solution of HNO_3or 0.1 M of NaOH. Afterwards to each sample 1 g of the zeolites was added. After 2 hours of mixing, samples of the equilibrated solutions were analyzed for heavy metal content. Similar experiments were performed to study sorption the following heavy metals: Cu, Cd, Ni, Pb, Cr, Zn.

2.4 *Influence of the zeolites dose on sorption of heavy metals*

The degree of sorption of heavy metals from municipal wastewater by zeolites was investigated to determine the optimum dose of the zeolites. To 1 dm^3 of the sewage zeolites of grain size 1–2 mm was added. After 60 minutes of mixing 5 ml samples of the solution were collected for heavy metal content analysis. Such experiments were done for doses of the zeolites from 0.02 to 0.5 g/dm^3.

2.5 *Influence of contact (mixing) time on sorption of heavy metal*

The dependence of contact (mixing) time of the zeolites with sewage was studied to determine the minimum time required for efficient sorption of heavy metals from municipal wastewater. To 1 dm^3 of the sewage 0.1 g of zeolites was added. After 1, 3, 5, 10, 15, 20, 25, 30, 45 and 60 minutes of mixing 5 ml samples of the solution were collected for heavy metal content analysis.

2.6 *Removal of heavy metals from sewage*

Batch experiments were used as a simple method to test whether zeolites would be suitable for pretreatment of municipal wastewater with respect to minimizing the heavy metals load. In laboratory research to 1 dm^3 of the sewage 0.1 g of the zeolites was added. After 10 minutes of mixing a sample of 50 ml from the equilibrated solution was taken for heavy metal analysis. Removal of heavy metals from sewage was examined using zeolites of particle size 1–2 mm. This kind of experiment was carried out every two weeks for raw sewage collected from Lublin WTP in the period of 01-06.2005.

3 DISCUSSION

It is well known, that pH has a significant impact on metals removal by zeolites since it can influence both the character of the exchanging ions and the character of the zeolites themselves (Ouki & Kavannagh 1999). Figures 1–2 exhibits the sorption degree of heavy metals versus pH of the equilibrated solution. The effect of the pH on the metal adsorption by natural zeolites was studied in a wide pH region between 2.0 and 11.0. For all studied heavy metals, except Ni, a rapid decrease in sorption degree was observed for higher values of pH. This is probably caused by heavy metal precipitation; this is in accordance with the observation that precipitation of heavy metals occurred at higher pH and therefore metal precipitation during the ion exchange mechanism is negligible (pH 3–8) (Ouki & Kavannagh 1999). The observed behavior of Ni sorption onto zeolites is quite understandable as its sorption degree increases with an increase in the equilibrium pH. The next exception among the studied heavy metals is Pb for which the highest sorption degree was observed in the pH region between 3.0 and 6.0 and for which at higher pH, sorption decreases. This fits to the observed changes in adsorption of Pb on bentonite (Donat et al. 2005). For other studied heavy metals (Cd, Cu, Cr and Zn) the influence of pH on sorption degree seems to be very similar. The maximum sorption of the metals onto natural zeolites is observed when the solution pH is close to 6.0. The sorption of Cd, Cu, Cr and Zn from more acidic solutions becomes less effective. From the presented results (see figs 1 and 2) one can observe that the optimum condition for heavy metals sorption onto natural zeolites is in the pH region between 5.0 and 7.0.

Experimental adsorption isotherms of heavy metals on the zeolites are presented in Figures 3–4. The results clearly show differences in the selectivity of

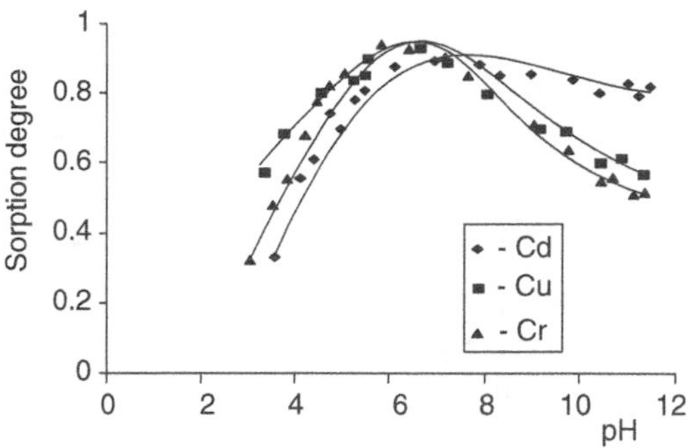

Figure 1. Influence of pH on sorption degree of Cd, Cr, Cu onto natural zeolites.

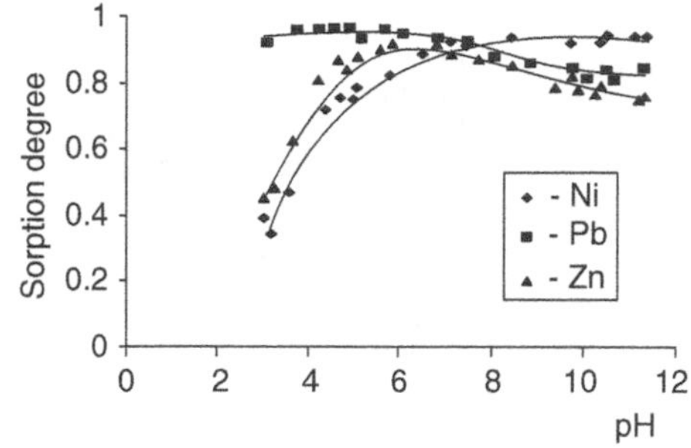

Figure 2. Influence of pH on sorption degree of Ni, Pb, Zn onto natural zeolites.

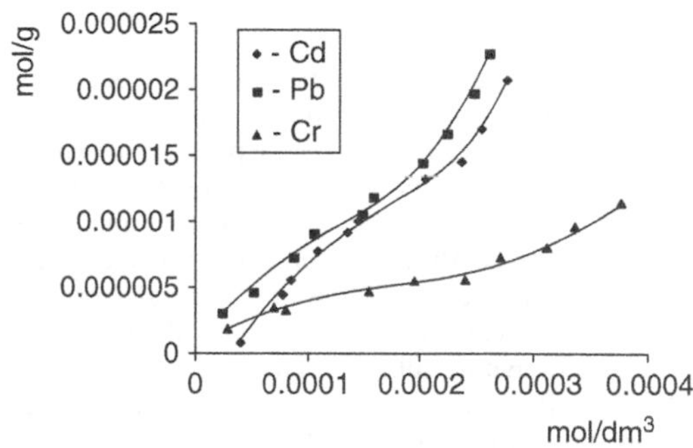

Figure 3. Experimental adsorption isotherm of Cd, Cr, Pb onto zeolites from buffer solution of pH = 7.0.

zeolites towards studied heavy metals. Furthermore, changes in concentration of heavy metals change the selectivity to such an extent that selectivity series for small and high concentrations are often quite different. When the initial concentration of any studied heavy metal is not >0.0001 M the following selectivity series is observed Pb > Cr > Cd > Cu > Ni > Zn. For higher concentrations of heavy metals (>0.003 M) zeolites become highly selective towards Cd and

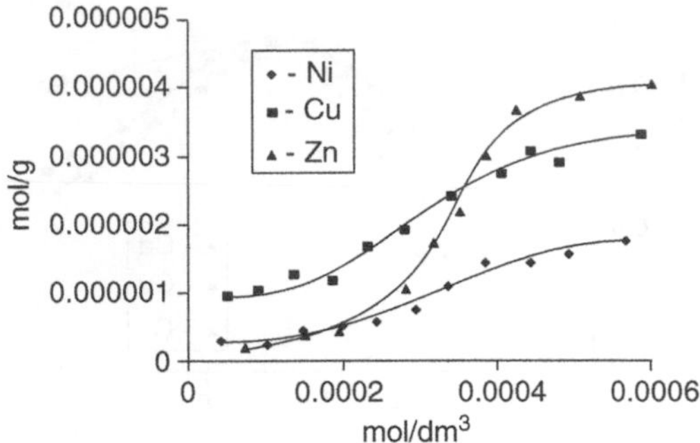

Figure 4. Experimental adsorption isotherm of Cu, Ni, Zn onto zeolites from buffer solution of pH = 7.0.

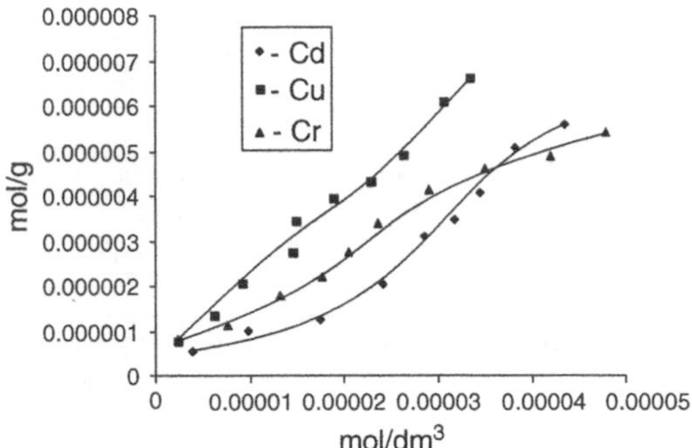

Figure 5. Sorption of Cd, Cu, Cr onto zeolites from solution of high Ca-Mg matrix (pH = 7.0).

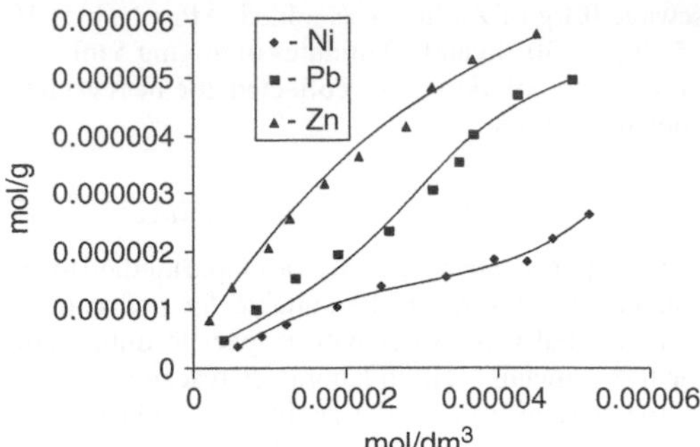

Figure 6. Sorption of Ni, Pb, Zn onto zeolites from solution of high Ca-Mg matrix (pH = 7.0).

Zn and then the selectivity sequence becomes: Pb > Cd > Cr > Zn > Cu > Ni. On the basis of the obtained laboratory results it is not possible to explain such phenomena. One of the reasons for it could be unknown effects of the buffer solution applied in the experiments. But, from practical point of view, much more interesting is the heavy metals sorption onto natural zeolites in the presence of high quantities of competing alkali metals, especially Ca and Mg. The effect of Ca and Mg on heavy metals sorption is shown in Figures 5–6. The selectivity sequence in this case nearly does not depend on the initial heavy metal concentration and could be expressed as follows: Zn ≈ Cu > Cr > Cd > Pb > Ni.

A more quantitative study of heavy metals sorption onto zeolites yields the distribution constant K_d calculated as the ratio of the amount of metal in adsorbent to the amount of metal in solution (Donat et al. 2005). The average values of the distribution constant for competitive sorption of heavy metals onto natural zeolites are listed in Table 1. From the presented results it could be stated that high concentrations of Ca and Mg in solution did not remarkably affect sorption properties of the zeolites towards studied heavy metals. Even in the case of Pb the distribution constant is still high enough to assure efficient its uptake from solutions in which the ratio Pb:Ca is $1{:}10^3$. This observation is of a great value especially when natural zeolites are to be used for removal of heavy metals from sewage.

Table 1. Values of distribution constant (K_d) for competitive adsorption of heavy metal onto natural zeolites from solution of Ca and Mg (124 and 14.4 mg/dm^3 respectively).

	Heavy metal					
Constant	Cd	Cr	Cu	Ni	Pb	Zn
K_d	128	145	219	51	102	225

From the obtained results it is evident that sorption of heavy metals from municipal wastewater depends on the dose of zeolites and the contact time. As an example, the sorption of selected heavy metals versus dose of zeolites is plotted in Figure 7. When the dose of zeolites is increased, the sorption degree increases for each heavy metal studied. For low doses of zeolites, <0.1 g/dm^3, it is possible to obtain a g/g_{max} ratio higher than 0.6. Sorption on the level of 50% of the maximum sorption is obtained for Cd, Ni and Cr with zeolite doses of 0.02 g/dm^3, 0.04 g/dm^3 and 0.06 g/dm^3 respectively. Figure 8 exhibits the influence of contact time of zeolites with sewage on degree of sorption of heavy metals. From the presented results, it can be seen that the sorption degree of heavy metals from sewage increases as the contact (mixing) time is increased. For the same values of contact time, different sorption degrees are observed for different heavy metals. In the cases of Zn and Cu 50% of the maximum sorption were obtained within a contact time of less than 10 minutes, while for Pb the same effect was observed for contact time about 15 minutes. On the base of my experiments I can state that 0.1 g/dm^3 is the most effective dose of zeolites for heavy metals uptake from the tested sewage and that a contact time

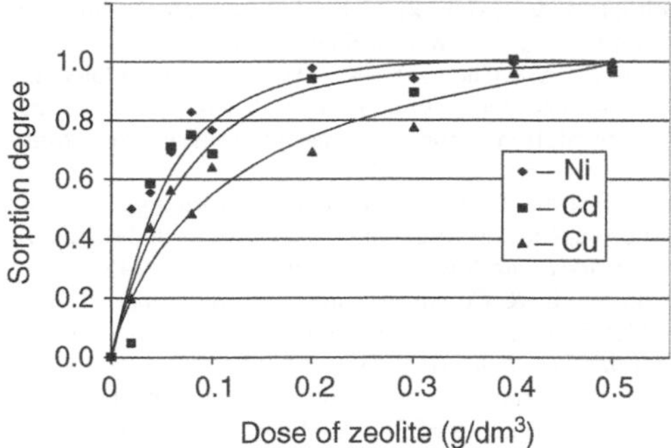

Figure 7. Dependence of heavy metals sorption on dose of zeolites (particle size 1–2 mm).

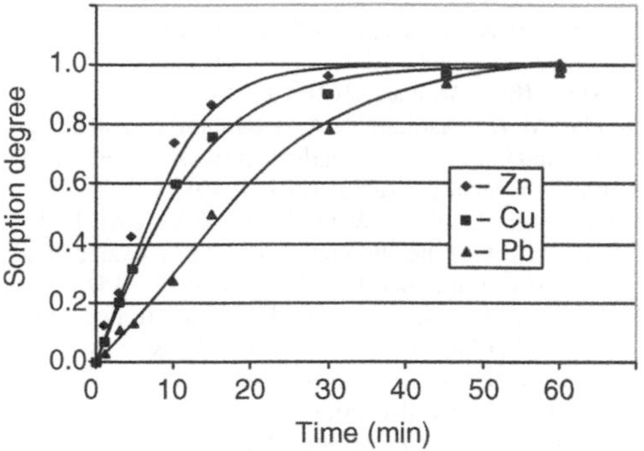

Figure 8. Influence of contact time of zeolites with sewage on sorption of heavy metals (dose of zeolites 0.1 g/dm^3, particle size 1–2 mm).

equal to 15 minutes should assure the metal sorption to a satisfactory extent.

The average results obtained during a half a year of studies on removal of heavy metals from sewage are presented in Figure 9 while, as an example, Table 2 shows the results for a single laboratory test. An application of zeolites of particle size 1–2 mm allows reduction of the content of heavy metals in sewage but the obtained removal degrees are different for particular metals. The highest removal efficiency is observed in the case of Zn where a zeolite dose equal to 0.1 g/dm^3 reduces the metal content by about 30–40%. At the other extreme, the lowest removal degrees were noted for Pb.

The applied zeolites in a batch mode process remove only about 5% of the Pb from municipal wastewater. In the case of other metals, it can be stated that granular zeolites, of a particle size 1–2 mm, applied in batch process removed about 8% of Cd and Cr, 11% of Cu, 15% of Ni from sewage. From the obtained laboratory results one can conclude that natural zeolites, due to their sorption properties, could be applied for control of heavy metals flow in sewage treatment plants. The

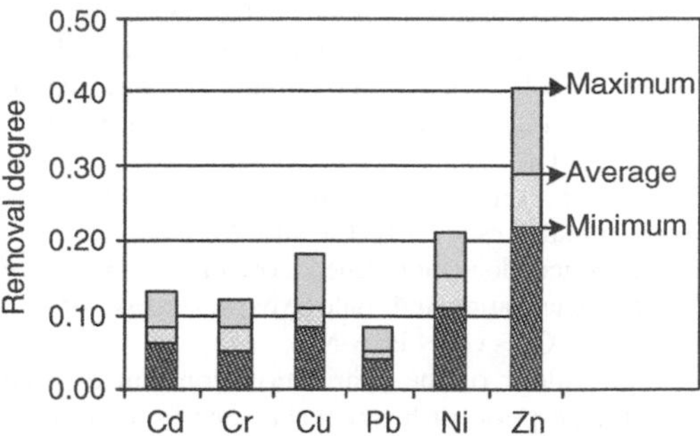

Figure 9. Average removal degrees of heavy metal from sewage achieved by means of natural zeolites (dose of zeolites 0.1 g/dm^3, mixing time 15 minutes).

Table 2. Concentrations of heavy metals in raw sewage and in wastewater treated by means of natural zeolites (mixing time 15 minutes, dose of zeolites 0.1 g/dm^3, sampling date 2005-05-11).

	Concentration (ppb)	
Heavy metal	Raw sewage	Wasterwater treated by means of zeolite
Cd	29.90	27.70
Cr	37.40	34.10
Cu	68.20	57.90
Ni	62.30	51.20
Pb	3.21	2.98
Zn	496.40	327.30

zeolites, added to sewage at their inlet to the WTP, should reduce the content of heavy metals and after that must be removed from sewage, preferably in a grit chamber, before the primary treatment processes. Such a partial removal of heavy metals before separating sewage sludges should diminish the content of the metals in sediments and simplify their final disposal.

4 CONCLUSIONS

With the results of laboratory studies on the sorption of heavy metals on natural zeolites and removal of the metals from sewage, the following conclusions can be drawn:

Zeolites from the Sokornit mine in Ukraine classified as clinoptylolite show a high selectivity towards heavy metal ions.

The metal sorption on natural zeolites is pH dependent, being higher in the pH region between 6.0 and 7.0.

The selectivity of the zeolites used depends slightly on the concentration of heavy metals. For concentrations lower than 0.0001 M the following selectivity sequence was observed: Pb > Cr > Cd > Cu > Ni > Zn while for higher concentrations of heavy metals (above 0.003 M) zeolites become highly selective towards Cd and Zn and then the selectivity sequence is as follows: Pb > Cd > Cr > Zn > Cu > Ni.

In the presence of high quantities of competing alkali metals, especially Ca and Mg, the selectivity sequence does not depend on the initial heavy metal concentration and could be expressed as follows: Zn ≈ Cu > Cr > Cd > Pb > Ni.

High values of the distribution constant assured effective sorption of heavy metal ions from solution for which the ratio of HM:(Ca + Mg) is ca. $1{:}10^3$ or even lower.

The efficiency of heavy metal removal from municipal wastewater by means of zeolites depends on contact time and zeolite dose. It was noticed that 0.1 g/dm^3 is the most effective dose of zeolites for heavy metals uptake from tested sewage. Contact times of ca. 15 minutes should assure satisfactory metal sorption.

Granular zeolites, of particle size 1–2 mm, applied in batch process removed from sewage about 8% of Cd and Cr, 11% of Cu, 15% of Ni, only 5% of Pb and nearly 30% of Zn.

Natural zeolites, due to their sorption properties, can be applied for control of heavy metals flow in sewage treatment plant. A partial removal of heavy metals before separating sewage sludges diminishes the content of the metals in sediments, which should simplify their final disposal.

ACKNOWLEDGMENTS

The author acknowledges the financial support received from Polish Ministry of Education and Science. The presented data are part of research carried out in project No 4 T09D 019 25.

REFERENCES

Ahmed, S. & Chughtai, S. & Keane, M.A. 1998. The removal of cadmium and lead from aqueous solution by ion exchange with Na-Y zeolite. *Separation and Purification Technology* 13: 57–64.

Álvarez-Ayuso, E. & Garcia-Sánchez, A. & Querol, X. 2003. Purification of metal electroplating wastewaters using zeolites. *Water Research* 37: 4855–4862.

Bolto, B.A. & Pawlowski, L. 1987. Wastewater Treatment by Ion Exchange, E. & F.N. Spon, London.

Chojnacki, A. & Chojnacka, K. & Hoffmann, J. & Górecki, H. 2004. The application of natural zeolites for mercury removal: from laboratory tests to industrial scale. *Mineral Engineering* 17: 933–937.

Chua, H. 1998. Effects of trace chromium on organic adsorption capacity and organic removal in activated sludge. *The Science of the Total Environment* 214: 239–245.

Ćurković, L. & Cerjan-Stefanović, Š. & Filipan, T. 1997. Metal ion exchange by natural and modified zeolites. *Water Research* 31(6): 1379–1382.

Donat, R. & Akdogan, A. & Erdem, E. & Cetisli, H. 2005. Thermodynamics of Pb^{2+} and Ni^{2+} adsorption onto natural bentonite from aqueous solutions. *Journal of Colloid and Interface Science* 286: 43–52.

Erdem, E. & Karapinar, N. & Donat, R. 2004. The removal of heavy metal cations by natural zeolites. *Journal of Colloid and Interface Science* 280: 309–314.

Kangala, K.B. & Chipasa, B. 2003. Accumulation and fate of selected heavy metals in a biological wastewater treatment system. *Waste Manag.* 23(2): 135–143.

Karvelas, M. & Katsojiannis, A. & Samara, C. 2003. Occurrence and fate of heavy metals in the wastewater treatment process. *Chemosphere* 53: 1201–1210.

Madoni, P. & Davoli, D. & Gorbi, G. & Vescovi, L. 1996. Toxic effect of heavy metals on the activated sludge protozoan community. *Wat. Res.* 30(1): 135–141.

Majdan, M. & Pikus, S. & Kowalska-Ternes, M. & Gładysz-Płaska, A. & Staszczuk, P. & Fuks, L. & Skrzypek, H. 2003. Equilibrium study of selected divalent d-electron metals adsorption on A-type zeolite. *Journal of Colloid and Interface Science* 262: 321–330.

Ouki, S.K. & Kavannagh, M. 1997. Performance of natural zeolites for the treatment of mixed metal-contaminated effluents. *Water Management and Research* 15: 383–394.

Ouki, S.K. & Kavannagh, M. 1999. Treatment of metals-contaminated wastewaters by use of natural zeolites. *Water Science and Technology* 39(10–11): 115–122.

Pitcher, S.K. & Slade, R.C.T. & Ward, N.I. 2004. Heavy metal removal from motorway stormwater using zeolites. *Science of the Total Environment* 334–335: 161–166.

Santarsiero, A. & Veschetti, E. & Donati, G. & Ottaviani, M. 1998. Heavy Metal Distribution in Wastewater from a Treatment Plant. *Microchemical Journal* 59: 219–227.

Sörme, L. & Lagerkvist, R. 2002. Sources of heavy metals in urban wastewater in Stockholm. *The Science of the Total Environment* 298: 131–145.

Zorpas, A.A. & Constantinides, T. & Vlyssides, A.G. & Haralambous, I. & Loizidou, M. 2000. Heavy metal uptake by natural zeolite and metals partitioning in sewage sludge compost. *Bioresource Technology* 72: 113–119.

Environmental Engineering – Pawłowski, Dudzińska & Pawłowski (eds)

Purification of petrochemical wastewater of heavy metals using zeolites

Barbara Gworek
Institute of Environmental Protection, Warsaw, Poland,
Department of Soil Environment Sciences, Warsaw Agricultural University, Warsaw, Poland

Edward Polubiec, Paulina Chaber & Igor Kondzielski
Institute of Environmental Protection, Warsaw, Poland

ABSTRACT: Continuing industrial development results in the increased production of sewage of differing chemical composition that enters the environment through different routes. Therefore, much work has focussed on examining the possibility of elimination of some of these pollutants, in particular, the heavy metals present in sewage. One possibility is the use of zeolites. The aim of the present study was to check the effectiveness of zeolites in the elimination of heavy metals from petrochemical sewage. For this purpose the sewage from Płock Petrochemical Plant was used. The tested zeolites were synthetic K-Na zeolite (3A) and natural zeolite – a rock containing about 90% of clinoptilolite. The results show that both natural and synthetic zeolites may be successfully used to purify even such specific sewage as petrochemical sewage, which contains a high content of oily substances. The percent reduction in heavy metals from examined matrices is follows mainly the order: Ni (up to 98%) > Pb (up to 89%) > Cd (up to 80%) > Cu (up to 75%).

Keywords: Petrochemical sewage, heavy metals, zeolite.

1 INTRODUCTION

Continuing industrial development results in increasing amounts of sewage of differing chemical composition, entering the environment. This sewage, often rich in biogenic material and thus potentially useful in agriculture, contains considerable amounts of toxic substances, such as the heavy metals: Pb, Cu, Ni, Cd. This, in turn, limits their potential use in agriculture. For this reason much ongoing research is focused on counteracting this negative aspect. Until now research projects on the use of sewage in agriculture were limited to the determination of the doses and methods of application of these products that would cause the least environmental damage. However, the danger of exceeding the doses still exists. This is a reason why the aim of many ongoing projects is the total or at least the partial elimination of heavy metals from sewage. One possibility is the use of selective sorbents, such as natural an synthetic zeolites (Gworek et al., 1992, Zorpas A.A. et al., 2000, Alvarez-Ayuso et al., 2003).

Zeolites are three-dimensional aluminosilicates, which, owing to their specific crystal structure, have a very strong and selective sorption capability (so-called molecular sieve). These specific physicochemical properties together with their potential use were decisive factors in initiating the industrial synthesis of zeolites. Nowadays, around the world about one hundred types of zeolites are commonly synthesised. Some of them do not occur naturally. However, natural zeolites are cheap materials which are readily available in large quantities in many parts of the world. All zeolites, due to their specific properties, are widely used in such activities as drying and purification of gases and liquids, separation of mixtures, in catalysis, gas chromatography, purification of wastewaters, filtration of drinking water and inactivation of heavy metals in polluted soils (Mondale K.D. et al. 1995, Ghobarkar H. et al., 1999, Gworek B., Kozera-Sucharda B. 1999, Lebedynets M. et al., 2004).

In many previous research projects zeolites were used to eliminate the heavy metals from soils, wastewaters and sewage sludge or to immobilise these elements in polluted soils (Gworek B., Borowiak M. 1991, Gworek B., Borowiak M., Brogowski Z. 1991, Gworek B., Borowiak M., Brogowski Z. 1992, Gworek B. 1992, Mondale K.D. et al., 1995, Shanableh A., Kharabsheh A., 1996, Sabeha Kesraoui Ouki, Mark Kavannagh 1997, Zorpas A.A. et al., 2000, Alvarez-Ayuso E. et al., 2003). The types of zeolites used in these studies were chosen especially for their particular sorption capacity towards heavy metals. Also, being natural compounds present in small amounts in soils they are safe for the environment.

The main aim of the present study was to assess the usefulness of natural and synthetic zeolites in eliminating heavy metals from a very specific sewage i.e. that arising from the petrochemical industry.

2 MATERIALS AND METHODS

The petrochemical sewage used in this study was collected from the Petrochemical Plant, which is part of a refinery in Płock.

Samples were taken from:

- system I – sewage from the refinery part of the plant
- system II – sewage from the petrochemical part of the plant
- pond – water from the sedimentation pond, where sewage from systems I and II are mixed in the ratio 3:1, flow out.

Sampled sewage from each sampling point was analysed for the main physicochemical properties: pH, density, dry weight and organic matter content. Also, the content of some alkaline (Na, K, Ca) and of some heavy metals (Pb, Cu, Ni, Cd) samples prior to the purification process was examined; henceforth known as "control".

The following zeolites were used in the purification process:

- 3AP-1 – synthetic Na – K zeolite activated at 200°C
- 3AP-2 – synthetic Na – K zeolite activated at 600°C
- ZN-M – natural zeolite – ground natural zeolite – bearing rock containing about 90% of clinoptilolite, grain size < 0,02 mm
- ZN-K – natural zeolite – ground natural zeolite-bearing rock containing about 90% of clinoptilolite, grain size about 3 mm.

The experiment was performed in a static system, with ratios zeolite:sewage 1:10 and 1:50. The contact times of zeolites with sewage were 1 h and 18 h. After that time the sewage was filtered and analysed for alkaline and heavy metals. The heavy metals were quantified using AAS.

3 RESULTS

The basic physicochemical properties of the examined sewage are summarised in Table 1. In system I and II, the pH of the sewage was similar and ranged from 7.65 to 7.81. In the pond it was lower –5.8. Some differences were also observed in other measured parameters. The liquid density was highest in the pond and slightly higher in system I compared to system II. On the other hand, the comparison of results for system I and II revealed that in the system I the organic matter contents and dry residue were higher.

Table 1. Basic physicochemical properties of the examined sewage.

Sampling point	pH	Density [g/cm^3]	Dry weight [%]	Organic matter [%]
System I	7.65	0.9764	0.0828	34.42
System II	7.71	0.9698	0.1425	37.89
Pond	5.80	1.0021	n.d.*	n.d.*

n.d.* – not determined.

The sewage from all sampling points was examined for content of alkaline (Na, K, Cu) and heavy (Pb, Cu, Ni, Cd) metals prior to and after the experiment. The results are summarized in Tables 2 and 3 as well as in graphic form in Figures 1–3. The sorption times as well as the different sorbent:sewage ratios are not considered as they were found to have no influence on the obtained results.

The concentrations of the alkaline metals determined in sewage before the experiments were highest in systems I and II. The highest concentration of calcium in comparison to concentration of sodium and potassium was determined in system II and in the sedimentation pond. However, the concentration of calcium in sewage from system II was higher than in the sedimentation pond, but lower than the sodium concentration in this sewage. The content of sodium was highest in sewage from system II. In the sedimentation pond, the concentration of sodium in comparison to sewage from system II was lower; however, in comparison to sewage from system I it was higher. The lowest content of sodium was in system I. In comparison to other alkaline metals, the concentration of potassium was the lowest. The highest concentration of this metal was in system II. The content of potassium was higher in the sedimentation pond than in sewage from system I.

The concentration of heavy metals in examined samples followed the order (Table 3):

system I: Ni > Pb = Cd = Cu
system II: Ni > Pb > Cd > Cu
pond: Cd > Ni = Cu > Pb

The highest concentration of heavy metals was in sewage from system II. The concentration of nickel in comparison to other heavy metals was the highest in systems I and II, but in system I the concentration of nickel was higher in comparison to system II. The lowest concentration of nickel was in the sedimentation pond, where the highest concentration in comparison to other metals was cadmium. The content of copper was similar in sewage from system I and II and higher in comparison to water from the sedimentation pond. The concentration of lead was the highest in sewage from system II and the lowest in water from the sedimentation pond. The content of cadmium was

Table 2. Mean concentration (n = 12) of alkaline metals [$mg \cdot l^{-1}$] in sewage before (control) and after the experiment as a function of zeolite type.

Sampling point	Type of zeolite	Na	K	Ca
System I	Control	55.4 ± 12.0	6.1 ± 1.7	123.6 ± 24.7
	3AP-1	178.2 ± 54.2	205.8 ± 49.1	6.8 ± 1.4
	3AP-2	182.3 ± 61.7	148.4 ± 29.3	5.7 ± 1.2
	ZN-M	19.8 ± 4.7	17.5 ± 3.8	143.4 ± 26.4
	ZN-K	20.8 ± 5.1	15.8 ± 3.2	134.0 ± 28.3
System II	Control	235.4 ± 49.2	19.6 ± 4.9	123.0 ± 26.1
	3AP-1	338.8 ± 71.0	365.6 ± 74.1	5.3 ± 1.6
	3AP-2	345.5 ± 68.2	333.4 ± 68.4	5.0 ± 1.2
	ZN-M	93.5 ± 21.0	32.8 ± 7.0	6.6 ± 1.4
	ZN-K	91.0 ± 18.6	29.1 ± 6.8	6.0 ± 1.6
Pond	Control	87.4 ± 19.8	7.9 ± 1.6	88.6 ± 20.4
	3AP-1	202.0 ± 43.7	224.4 ± 48.1	6.5 ± 1.7
	3AP-2	194.4 ± 39.7	206.8 ± 46.4	4.6 ± 1.0
	ZN-M	25.0 ± 6.1	19.0 ± 3.2	150.2 ± 34.2
	ZN-K	26.8 ± 5.9	18.1 ± 2.9	156.0 ± 36.7

Table 3. Mean concentration (n = 12) of heavy metals [$mg \cdot l^{-1}$] in sewage before (control) and after the experiment as a function of zeolite type.

Sampling point	Type of zeolite	Cu	Pb	Ni	Cd
System I	Control	0.04 ± 0.008	0.04 ± 0.008	0.12 ± 0.03	0.04 ± 0.007
	3AP-1	0.01	0.01	0.01	0.01
	3AP-2	0.01	0.01	0.01	0.01
	ZN-M	0.01	0.01	0.01	0.01
	ZN-K	0.01	0.01	0.01	0.01
System II	Control	0.04 ± 0.007	0.09 ± 0.001	0.11 ± 0.02	0.05 ± 0.001
	3AP-1	0.01	0.01	0.01	0.01
	3AP-2	0.01	0.01	0.01	0.01
	ZN-M	0.01	0.01	0.01	0.01
	ZN-K	0.01	0.01	0.01	0.01
Pond	Control	0.02 ± 0.004	0.01 ± 0.0002	0.02 ± 0.004	0.03 ± 0.007
	3AP-1	0.01	0.01	0.01	0.01
	3AP-2	0.01	0.01	0.01	0.01
	ZN-M	0.01	0.01	0.01	0.01
	ZN-K	0.01	0.01	0.01	0.01

the highest in system II and lowest in water from the sedimentation pond.

After application of zeolites 3AP-1 and 3AP-2 there was general increase, in all systems, in the concentration of sodium and potassium. On the other hand, the application of these zeolites resulted in a general decrease of concentration of calcium in sewage from each system. The opposite tendency could be observed when natural zeolites – ZN-M and ZN-K were applied. The concentration of sodium decreased while that of calcium increased. Only in the case of potassium was an increase in concentration observed after application of both types of sorbent.

The percent reduction of heavy metals in sewage from different sampling points, depending on the type of applied zeolite were as follows:

- for zeolites 3AP-1 and 3AP-2:
- system I – Ni – 98%, Pb – 75%, Cd – 75%, Cu – 75%
- system II – Ni – 91%, Pb – 80%, Cd – 80%, Cu – 75%
- pond – Cd – 67%, Ni and Cu – 50%, Pb – concentration has not changed
- for zeolites ZN-K and ZN-M
- system I – Ni – 92%, Pb – 75%, Cd – 75%, Cu – 75%

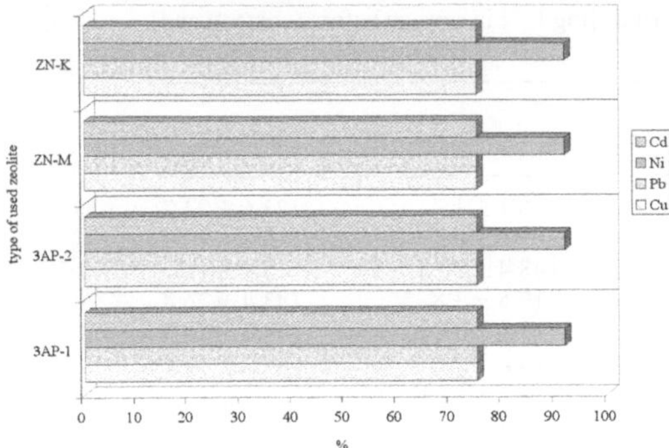

Figure 1. Percentage reduction of heavy metals in sewage from system I after application of zeolites.

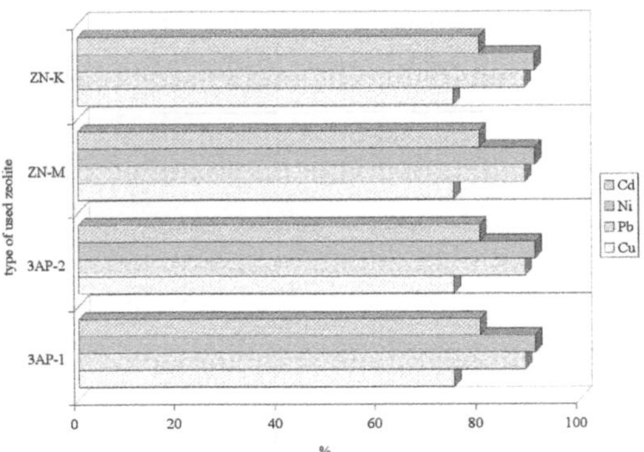

Figure 2. Percentage reduction of heavy metals in sewage from system II after application of zeolites.

- system II – Ni – 91%, Pb – 80%, Cd – 80%, Cu 75%
- pond – Cd – 66%, Ni and Cu – 50%, Pb – concentration has not changed.

4 DISCUSSION

The model study presented above confirms the high efficacy of zeolites in eliminating heavy metals from petrochemical sewage. This statement is true for both the natural (ZN) and the synthetic (3A) zeolites that were tested here.

The contact time (1 h and 18 h) had no influence on the sorption of metal ions by the zeolites. This confirms the observation that the sorption equilibrium is established very rapidly; according to Grabas, Steinger and Zieliński (Grabas et al.) in <30 minutes. However, others report (Alvarez-Ayuso et al., 2003) that for metal cations >90% of equilibrium was attained with 1 h of interaction.

In these experiments the method of activation of the synthetic zeolite (3A) had also no influence on

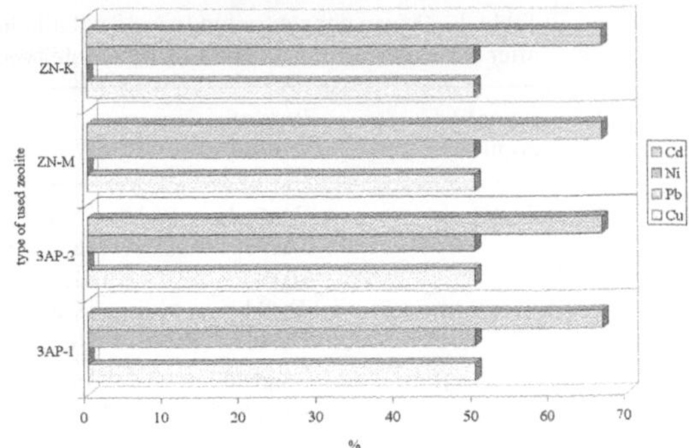

Figure 3. Percentage reduction of heavy metals in water from sedimentation pond after application of zeolites.

Table 4. Radii of hydrated ion and hydration energies for metals (from Ouki S.K. & Kavanagh M. (1997) after Semmens (1981).

Metal	Hydrated radius [Å]	Free energy of hydration [kcal g-ion^{-1}]
Ni	4.04	−494.2
Cu	4.19	−498.7
Cd	4.26	−430.5
Pb	4.01	−357.8

the amount of any metal sorbed. The unique_sorptive properties of zeolites are determined by their structure of cavities (channels and cages) and their inner and outer surface (Ghobarkar H. et al., 1999, Gworek B., Kozera-Sucharda B., 1999). Probably the activation method is only of minor importance. Also, there was no significant difference in the sorption of heavy metals by natural zeolites differing in grain size. The results of the examination of physicochemical properties of zeolites (Gworek B., Kozera-Sucharda, 1999, Zorpas et al., 2002) suggest that the zeolites with higher inner and outer surface area will have higher sorption and cation exchange capacities despite irrespective of their natural or synthetic origin; interestingly it was observed that the inner and outer surface of natural zeolite is independent of the grain size.

In sewage from systems I and II, the examined metals, independent of the zeolites used, were sorbed in following order: Ni > Pb > Cd > Cu (Figure 1 and 2). The order of sorption in the sedimentation pond, regardless of the type of zeolite, was the following: Cd > Ni ≥ Cu (Figure 3). The amount of lead remained unchanged. It is difficult to explain this order but it may be an effect of the metal hydration free energy and/or hydrated ion sizes (Ouki S.K. and Kavanagh M., 1997, Erdem E. et al., 2004). According to the hydration energies listed in Table 4 the selectivity series for the metal considered should be: Pb > Cd > Ni > Cu.

The metal having the highest hydration free energy should remain mostly in a soluble form and thus be most easily sorbed. This theory might explain the predominance of the sorption of Cd and Ni from waters from the sedimentation pond, but it does not explain the reasons for lack of lead sorption (see Figure 3). By comparison of the hydrated ion size of the metals examined (Table 4) the following selectivity sequence is generated: Pb > Ni > Cu > Cd. When the radius of a hydrated metal ion is greater than the zeolite pores the sorption should be automatically lower. This theory, however, in the scope of this experiment, cannot be applied.

Natural zeolites demonstrated their highest sorptive capacity towards lead; while the results for cadmium and nickel were much lower (Shanableh A., Kharabsheh A., 1996). In our experiments however, in system I and II the highest sorption was for nickel, then lead and cadmium while in water from the sedimentation pond lead was not sorbed at all. This may be caused by the specific character of petrochemical sewage – rich in organic compounds that may complex lead and thus prevent its sorption by zeolites. The lack of observed sorption of lead in the sedimentation pond may result from its very low initial concentration, which is close to the detection limit of the analytical method applied.

The results of the examination of sorption of heavy metals by natural and synthetic zeolites (Alvarez-Ayuso E. et al., 2003) suggest that copper is mainly sorbed by means of precipitation of hydroxides while the cation exchange process may be the main mechanism responsible for cadmium and nickel sorption. This mechanism is the most probable explanation for the predominance of the sorption of nickel and cadmium over copper. The elimination of heavy metals from sewage by zeolites occurs via sorption and desorption. This statement can be supported by the increasing content of alkaline metals (Na, K, Ca) after the application of zeolites in all examined sewage. The synthetic zeolite (3A) contains large amounts of exchangeable Na and K – this may result in an increase in the content of these two metals in all sewage samples purified using this type of zeolite. The natural zeolite (ZN) contains exchangeable K and Ca and this is the reason for the increase of these metals in sewage after application of this zeolite (Table 2) (Gworek B et al., 1991, Chojnacki A et al., 2004). However, the zeolites do not only sorb heavy metals. There is a possibility that the cations of alkaline metals, if not exchangeable, can also be sorbed by zeolites as in the case of heavy metals. This is why, after the application synthetic zeolite (3A) the Ca^{2+}, cations were sorbed by this sorbent (Table 2). The same behaviour can be observed in the case of sodium after using natural zeolites – its content decreases.

In other work on zeolites (Ouki S.K., Kavannagh M., 1997) it is suggested that, especially in the case of natural zeolite with a high content of clinoptilolite, the sorption of heavy metals depends mainly on their concentration in the examined matrix. According to these authors the higher the concentration of given metal in the solution the smaller is its sorption. However, in our work the opposite is seen. This may be due to the very low concentration (never > $0.12\ mg{\cdot}l^{-1}$) of all heavy metals in the examined sewage.

Therefore, it is possible that the greater sorption occurs for the higher concentrations of heavy metals (effect of dilution). This may also explain why, in the case of water from the sedimentation pond, in contrast to sewage from system I and II, cadmium was sorbed in a greater amount than nickel. This is due to the fact that in systems I and II the concentration of nickel was higher than cadmium.

Generally, sorption of metals from system I and II sewage was higher than that of the sedimentation pond. One possible explanation may be the difference in pH of the various sampling locations. It is well known that pH has a strong influence on the sorption of metals by zeolites (Alvarez-Ayuso E et al., 2003, Gworek B., Kozera-Sucharda B., 1999, Ouki S.K., Kavannagh M., 1997, Shanableh A., Kharabsheh A., 1996). High pH values induce higher sorption of heavy metals. In the sewage from system I and II the value pH was limited to ca. 7 while in the sedimentation pond it was only 5.8 (Table 1).

5 CONCLUSIONS

The experiments clearly demonstrate that both natural and synthetic zeolites can be successfully used to remove heavy metals from petrochemical sewage containing large amounts of oily residues. The decrease in content of heavy metals in sewage after the application of zeolites mainly followed the order: Ni (up to 98%) > Pb (up to 89%) > Cd (up to 80%) > Cu (up to 75%). The effectiveness of the sorption did not depend on either the grain size in the case of natural zeolite (ZN), or the method of activation of the synthetic zeolite (3A). The higher sorption observed for systems I and II, compared to the sedimentation pond, was probably caused by the higher pH of the sewage.

REFERENCES

Alvarez-Ayuso E., Garcia-Sanchez A., Querol X. (2003) Purification of metal electroplating waste waters using zeolites, Water Research 37, 4855–4862.

Chojnacki A., Chojnacka K., Hoffmann J., Górecki H. (2004), The application of natural zeolites for mercury removal from laboratory tests to industrial scale, Minerals Engineering 17, 933–937.

Erdem E., Karapinar N., Donat R. (2004), The removal of heavy metal cations by natural zeolites, Journal of Colloid and Interface Science 280, 306–314.

Ghobarkar H., Schäf O., Guth U. (1999) Zeolites – from kitchen to space, Prog. Solid St. Chem., Vol. 27, 29–73.

Grabas K., Steininger M., Zieliński S. (1998) Heavy metal removal on mineral – carbon adsorbent, Och. Środ. 1(68), 17–20.

Gworek B. (1992) Use of synthetic zeolites of 3A and 5A type for lead immobilization in anthropogenic soils, Polish Journal of Soil Science, Vol. XXV/1, 35–39.

Gworek B., Borowiak M. (1991) Model studies on immobilization of certain heavy metals by synthetic zeolite, Rocz. Gleb., Vol. XLII, 1–2, 27–35.

Gworek B., Borowiak M., Brogowski Z. (1991) The possibility of purification of industrial sewage sludge of heavy metals with use of synthetic zeolites, Polish Journal of Soil Science, Vol. XXIV/2,147–152.

Gworek B., Borowiak M., Brogowski Z. (1992) Use of zeolites for removal of heavy metals from sludge, Arch. Ochr. Środ., Vol. 1, 193–199.

Gworek B., Kozera-Sucharda B. (1999) Zeolite – origin, structure and basic physical properties, Ochr. Środ. i Zas. Nat., Vol. 17, 157–169.

Lebedynets M., Sprynskyy M., Sakhnyuk I., Zbytniewski R., Golembiewski R., Buszewski B. (2004) Adsorption of ammonium ions onto a natural zeolite: Transcarpathian Clinoptilolite, Adsorption Science & Technology, Vol. 22, No. 9, 731–741.

Mondale K.D., Carland R.M., Aplan F.F. (1995) The comparative ion exchange capacities of natural sedimentary and synthetic zeolite, Minerals Engineering, Vol. 8, No. 4/5, 535–548.

Ouki S.K., Kavannagh M. (1997) Performance of natural zeolite for the treatment of mixed metal-contaminated effluents, Waste Management & Research 15, 383–394

Shanableh A., Kharabsheh A. (1996) Stabilization of Cd, Ni, Pb in soil using natural zeolite, Journal of Hazardous Materials, Vol. 45, 207–217.

Zorpas A.A., Constantinides T., Vlyssides A.G., Haralambous I., Loizidou M. (2000) Heavy metals uptake by natural zeolite and metal partitioning in sewage sludge compost, Bioresource Technology 72, 113–119.

Zorpas A.A., Vassilis I., Loizidou M., Grigoropoulou H., (2002), Particle Size Effects on Uptake of Heavy Metals from Sewage Sludge Compost Using Natural Zeolite Clinoptilolite, Journal of Colloid and Interface Science 250, 1–4.

Environmental Engineering – Pawłowski, Dudzińska & Pawłowski (eds)
© 2007 Taylor & Francis Group, London, ISBN13 978-0-415-40818-9

Distributions of pump work times in an operating vacuum sewerage system

Marian Kwietniewski
Department of Water Supply and Sewerage, Faculty of Environmental Engineering, Lublin University of Technology, Lublin

Katarzyna Miszta-Kruk
Department of Water Supply and Wastewater Treatment, Faculty of Environmental Engineering, Warsaw University of Technology, Warsaw

ABSTRACT: As an alternative to gravitational sewerage systems, the vacuum sewerage system (VSS) is mainly used in dispersed development areas (detached housing, rural development, etc.), areas of small height diversification, areas such as camping-sites and holiday centres, for instance, where the sewage is only generated in some time periods, areas characterised by unfavourable ground and water conditions, protection zones, etc.

The present article discusses results of an investigation into the randomness of work states of the vacuum station pumps as the dominating element in VSS. The basic goal of the investigation was recognition of the random nature of VSS operation.

Keywords: Vacuum sewerage system, vacuum station, vacuum pump, work time.

1 INTRODUCTION

The vacuum sewerage system belongs to a group of so-called unconventional sewerage systems, in which the flow is forced by negative pressure generated in the pipe network, unlike traditional sewerage systems with the sewage flowing under gravity forces in a partially, as a rule, filled channels. At present, continuously growing interest is observed in vacuum sewerage systems, both in Poland and abroad (Germany, England, Japan, USA).

The research results published so far on vacuum sewerage systems are very limited. They are represented by individual articles, books, and manuals prepared by device producers. Rare cases of vacuum sewerage system examination mainly refer to model solutions and concern hydraulic conditions of system operation (Kalenik 2004), simulation of operation of pipe network and system components (Latoszek & Kraśkiewicz 2002; Otterpohl 2002; Fabry & Fabry 2004, Gibbs 2005, Tjandraatmadja 2005) and, in part, operating tests of the examined systems (Naret 1988, Gray 1988; Rodhead 1999, Sullivan et al 2003; Kapcia & Lubowiecka 2004, Fabry 2004; Little 2004; Kalenik & Ćwiek 2005, Kwietniewski et al 2006).

From the point of view of improvement in design principles and operation of the vacuum sewerage systems, of high significance are the results of examination performed in natural operational conditions of those systems. Attempts made in this field, were oriented on recognising failure patterns (Kapcia & Lubowiecka 2004).

2 THE ROLE OF THE VACUUM STATION IN THE VACUUM SEWERAGE SYSTEM

The principle of operation of the vacuum sewerage system consists in forcing sewage flow in the pipes by generating proper negative pressure inside them. The sewage is sucked from the collecting chambers to the vacuum sewers, through which it flows down to the vacuum sewage tank located in the vacuum station building, or to the open sewage sump (siphon-type collecting well) located in the vacuum-siphon station building. Then, the sewage is transported by a gravitational or pressure system of pipes to the treatment plant[1].

[1] Here, two types of design solutions can be named, the vacuum sewerage system and the vacuum-siphon sewerage system. In the vacuum system the sewage is collected in a closed sewage tank with negative pressure inside, while in the vacuum-siphon system the sewage is collected in the siphon-type collecting well (tank), into which it flows through the vacuum sewer linked with the siphoned pipe.

The investigations reported in the present article refer to the vacuum sewerage system, in which the following basic components can be named:

- collecting chambers with interface valves, that collect sewage from apartment buildings,
- vacuum sewerage network,
- vacuum station.

The vacuum sewerage system reveals:

- periodicity of work, resulting from non-uniformity, both in time and space, of sewage flow to the system,
- short work times and long idle times.

Due to the above characteristics, working conditions of VSS are kept steady by the vacuum station, which may be considered the key component of the system. Moreover, the station is a decisive factor for operational reliability of VSS, as it:

- is connected in series with the remaining system components,
- plays a double role, namely generating the negative pressure necessary for collecting and transporting sewage in the sewerage network, and then using pressure to transport the collected sewage to the treatment plant.

The basic components of the pumping station are the following:

- vacuum pumps, creating negative pressure in the sewerage network,
- sewage pumps, transporting sewage collected in the vacuum tank to the treatment plant.

3 THE NATURE AND RANGE OF VACUUM STATION EXAMINATION

The research activities reported in the article fall into the category of operational reliability tests, i.e. the collection of reliable data under normal operational conditions for further processing and analyses. This is the reason why great attention is paid to monitoring objects of the vacuum sewerage system. The scheme of the system is shown in Figure 1.

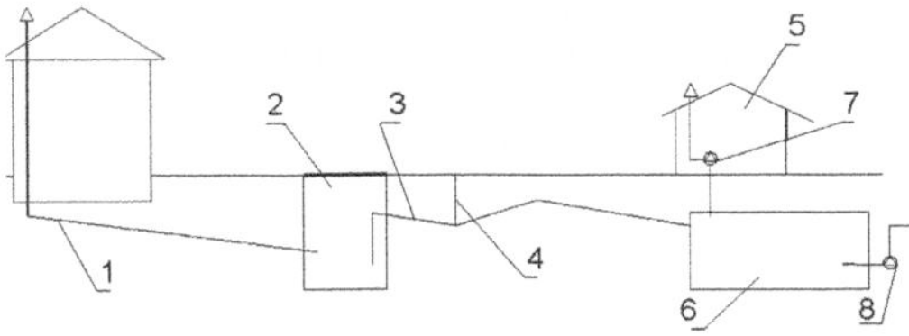

Figure 1. The scheme of the vacuum sewerage system. 1-gravitational sewer, 2-collecting chamber with interface valve, 3-vacuum sewers network, 4-access eye, 5-vacuum station building, 6-vacuum tank for collecting sewage, 7-vacuum pump, 8-sewage pump.

The examined vacuum sewerage system, constructed in Iseki technology, served about 600 buildings occupied by 2420 people. In this system, the sewage from the buildings is transported using gravitational sewers to the collecting chambers. These chambers are connected with the network of vacuum sewers, which transport the sewage to the vacuum tank in the vacuum station. The role of the station, equipped with three vacuum pumps and two sewage pumps, is to create and keep negative pressure within the range between 50 and 65 kPa in the sewerage network in order to take the sewage and transport it to the treatment plant.

The vacuum sewerage system was equipped with a computer system monitoring pumps' work times, alarm signalling, and software that allows failure data transfer. The values of the work states, treated as events during which a pump executes its tasks (the vacuum pump generates the negative pressure, while the forwarding pump transports the sewage) were recorded automatically by the monitoring system. During the examination, work times describing work states of the pumps were recorded, and transmitted to the central control point located in the sewage treatment plant. The investigations were carried out during five months, from September 29, 2002 to March 01, 2003.

4 METHODS AND RESULTS OF WORK STATES ANALYSIS OF VACUUM PUMPS

Taking into account all recorded work times of the vacuum pumps, a cumulative statistical analysis of those states was carried out as the first step. Its results are given in Table 1.

Table 1 reveals that work time characteristics of all three vacuum pumps are very similar to each other. Average work times for particular pumps differ by as little as 0.09 min, i.e. by 5 s. (the extreme values are respectively equal to 1 min 37 s. and 1 min 42 s.). Noteworthy is the central value – median, a symbolic value

Table 1. Results of cumulative statistical analysis of vacuum pump work times.

	Vacuum pump		
Statistics	I	II	III
Number of recorded values	2838	2826	2797
Maximum value [min]	21.0	25.0	25.0
Minimum value [min]	0.0	0.0	0.0
Average value [min]	1.61	1.65	1.7
Central value [min]	1.0	1.0	1.0
Standard deviation [min]	2.06	2.20	2.16

for the ordered series, with equal number of observations below and above it. Here, the median is equal to 1 minute for all pumps.

To make the presentation of work time distribution (density) more readable, empirical histograms and box diagrams were prepared for each pump. Their analysis confirmed the earlier observation that, statistically, the pumps work in an almost identical manner (histograms and box diagrams obtained for particular pumps reveal very slight differences). As an example, the work time histogram (Fig. 2) and box diagram (Fig. 3) are shown for vacuum pump No. I.

The diagrams shown in Figs. 2 and 3 reveal that the overwhelming majority of work times range from 0.5 to 1.5 min. For the box diagram (Fig. 3), 47 external points and 65 distant external points were obtained. Taking into consideration that the total number of recorded work times equalled 2838, the data corresponding to external points and distant external points make up only a small percentage of the entire set and can be neglected in a limited-accuracy analysis.

In order to select an appropriate probability distribution model for vacuum pump's work times, an exponential and Weibull's distribution were examined. As a result, better consistency was obtained for Weibull's distribution at the significance level of 0.05. Then, it was assumed that the work time probability of the vacuum pump is governed by the two-parameter Weibull's distribution in the form:

$$f(t)=\frac{\alpha}{\beta^{\alpha}}\cdot t^{(\alpha-1)}\cdot e^{-\left(\frac{t}{\beta}\right)^{\alpha}} \tag{1}$$

(Fig. 4) with parameters of shape $\alpha = 1.85$ and scale $\beta = 1.3$.

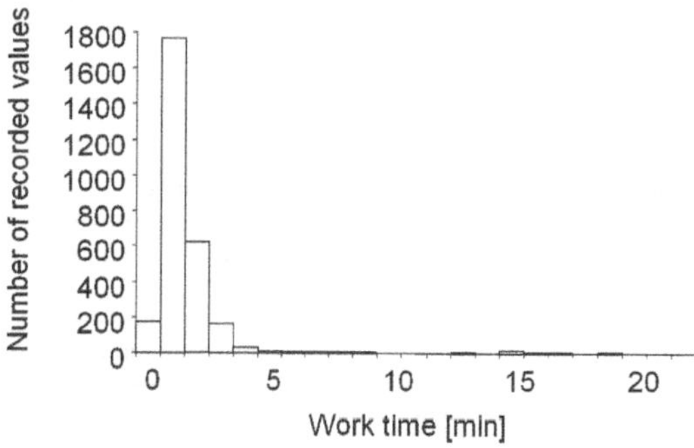

Figure 2. Empirical work time distribution for vacuum pump No. I.

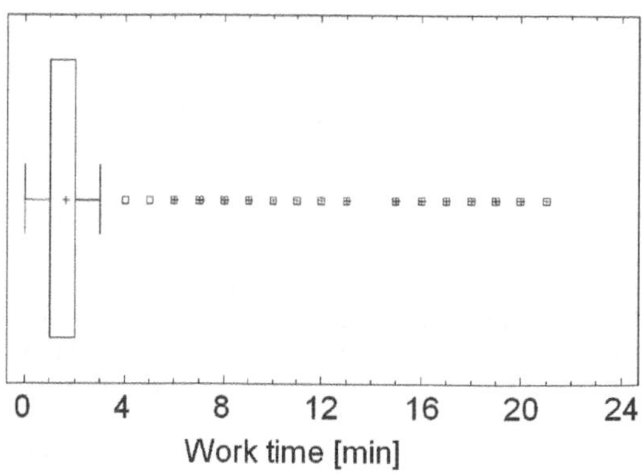

Figure 3. Work time box diagram for vacuum pump No. I.

5 METHODS AND RESULTS OF WORK STATES ANALYSIS OF SEWAGE PUMPS

Taking into account all data obtained from the monitoring, a cumulative statistical analysis of sewage pump work times was carried out. Its results are given in Table 2.

Comparing work time statistics of the two examined pumps leads to the conclusion that the installed

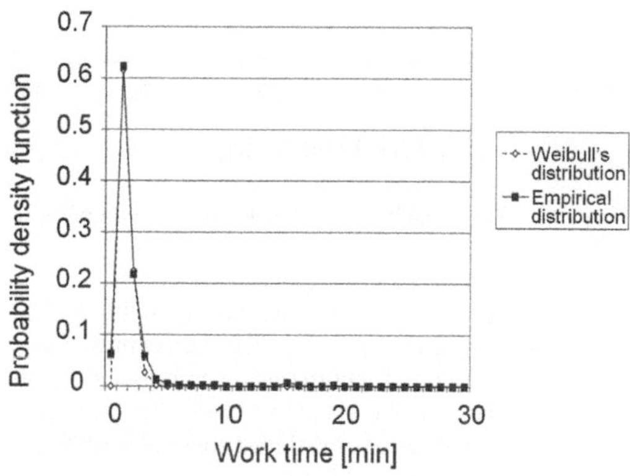

Figure 4. Adapting the work time distribution model of vacuum pump No. I to measured results at significance level of 0.05, $f(t) = 1.1386 \cdot t^{0.85} \cdot e^{-(t/1.3)^{1.85}}$.

Table 2. Results of cumulative statistical analysis of sewage pump work times.

	Forwarding pump	
Statistics	I	II
Number of recorded values	311	307
Maximum value [min]	9.0	10.0
Minimum value [min]	0.0	0.0
Average value [min]	3.10	2.96
Central value [min]	3.0	3.0
Standard deviation [min]	0.83	0.69

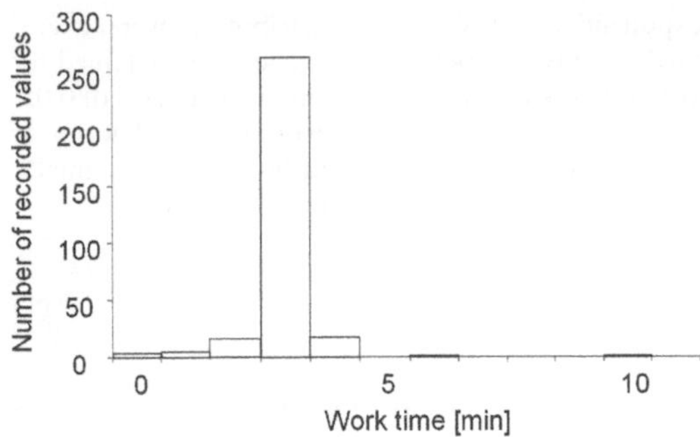

Figure 5. Empirical work time histograms for sewage pump No. II.

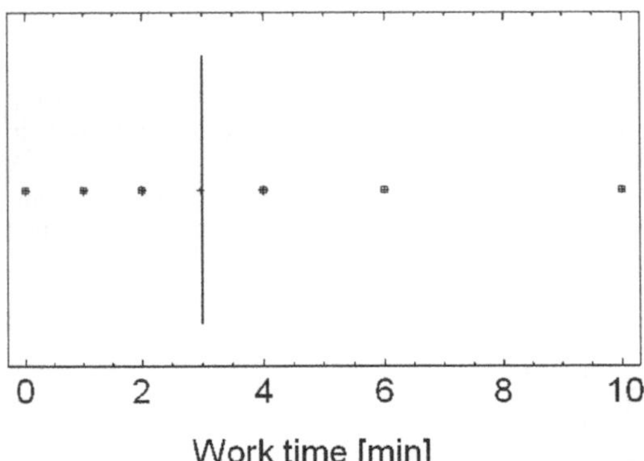

Figure 6. Work time box diagrams for sewage pump No. II.

pumps reveal very similar changeability of the random variables. One can even see that the central value (median) is the same for both sewage pumps and equals 3 min. Similarly to the vacuum pumps, here the work time average values also differed insignificantly, i.e. by as little as 0.14 min. (about 8 s.).

The sewage pump work time histograms and box diagrams reveal very similar distribution densities of random variables for the both examined pumps. To illustrate the nature of those distributions, the next figures present, as examples, the work time histograms (Fig. 5) and box diagrams (Fig. 6) of the sewage pump No. II.

In case of work times of the examined sewage pump, it is easily seen that their values are mainly located within the range between about 2.5 and 3.5 min, (Fig. 5). More precisely, half of the data in the sample fall within a very narrow interval −50% of work time data are equal to 3 minutes (Fig. 6).

As a result of statistical verification at the significance level of 0.05 it was assumed that the work time probability for the sewage pump is governed by the two-parameter Weibull's distribution

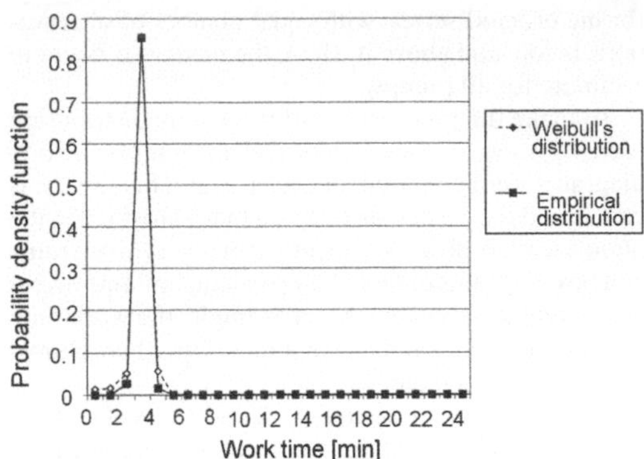

Figure 7. Adapting the work time distribution model of sewage pump No. II to measured results at significance level of 0.05; $f(t) = 3.54 \cdot 10^5 \cdot t^{9.5} \cdot e^{-(\frac{t}{3.32})^{10.5}}$.

Table 3. Vacuum pump utilisation per 24 hours.

Average work time of vacuum pump per 24 hours [h]	Average idle time of vacuum pump per 24 hours [h]	Pump utilisation factor [Average 24 – hour work time/24 h]
0.61 (36 min)	23.39	0.025

Table 4. Sewage pump utilisation per 24 hours.

Average work time of sewage pump per 24 hours [h]	Average idle time of sewage pump per 24 hours [h]	Pump utilisation factor [Average 24 – hour work time/24 h]
0.12 (7 min)	23.88	0.005

with parameters of shape $\alpha = 10.5$ and scale $\beta = 3.32$ (Fig. 7).

6 METHODS AND RESULTS OF TOTAL WORK TIMES ANALYSIS AND TOTAL IDLE TIMES OF THE PUMPS

The reported investigations also included an analysis of proportions between the total work times and the total idle times of the pumps in the vacuum station. For this purpose, average values of the total 24-hour values of work times and idle times, recorded during the entire time of the investigations, were calculated for the vacuum pumps and sewage pumps. Then the pump utilisation factor, corresponding to the average work time of the pump in 24 hours, was determined for each type of pump (Tables 3 and 4).

The utilisation factor for the vacuum pump is approximately 0.025. For the sewage pumps this factor is about 0.005. This allows a conclusion that:

- the average total work times for the both types of pumps are very short, compared to their average idle times,
- the average total work time of the sewage pumps is only 1/5 of the corresponding average corresponding average total work time of the vacuum pumps.

7 CONCLUSIONS

The reported investigation has allowed a deeper insight into the nature and range of changes of work states of vacuum and sewage pumps in the vacuum station, based on the vacuum sewerage system selected for study. The analysis of the obtained results allows the following final conclusions to be formulated:

- work characteristics of particular pumps are very similar to each other, this refers to both the set of vacuum pumps and to the group of sewage pumps, and is testified by very similar distributions of work and idle times for those pumps,
- noteworthy is the strong difference between the work times and the idle times, recorded both for the vacuum and sewage pumps; the average idle times are approximately forty times as long as the average work times of the vacuum pumps, while the ratio between the average idle time and the average work time of the sewage pumps is approximately 170:1,
- the majority of the work times of the vacuum pumps is within the interval between 0.5 and 1.5 min (62% of all values),
- the work times of the sewage pumps are mainly within the interval between about 2.5 and 3.5 min (88% of all values),
- the vacuum pump utilisation factor is approximately equal to 0.025 which means that the pump works for a total of 36 min, during the period of 24 hours, while for the sewage pump its utilisation factor is much lower and equals 0.005 i.e. about 7 min of work during 24 hours,
- the work time probability distributions for the both vacuum and sewage pumps can be modelled by the two-parameter Weibull distribution.

ACKNOWLEDGEMENTS

The authors would like to express their sincere gratitude to the personnel of the system operation service for their assistance in collection of the data used in the reliability tests. The authors are also grateful to reviewers for their constructive comments.

REFERENCES

Fabry, G. & Fabry, G. 2004. Vacuum technology gains momentum *Water and Wastewater International, v 19, n 2, pp 43*

Fabry, G. 2004. Preparing the operators of vacuum sewerage systems *Water and Wastewater International, v 19, n 9. pp 29–31*

Foremna, B.E. 1985. Wastewater collection by vacuum. *Proceedings of International Symposium on Urban Hydrology, Hydraulic Infrastructures and Water Quality Control.*, Lexington, Kentucky, USA, pp 37–42

Gibbs, S. 2005. Vacuums save time, money. *Public Works, v 136, n 11. pp 53–56*

Gray, D.D.M. 1988. Vacuum Sewers: Fundamentals and design methods. *Proceedings of the 15th Annual Water Resource Conference, Critical Water Issues and Computer Applications,* Norfolk, ASCE. pp 348–351

Kalenik M. & Ćwiek A. 2005. Badania eksploatacyjne sieci kanalizacji podciśnieniowej w systemie AIRVAC. (Examination of the vacuum sewerage network in operation in the AIRVAC system). *Gas, Water and Sanitary Technique. No. 3. pp. 26–30 (in Polish).*

Kalenik, M. 2004. Hydraulic conditions of vacuum sewerage system's operation, *Gas, Water and Sanitary Technique, No. 4. pp.125–130 (in Polish).*

Kapcia, J. & Lubowiecka, T. 2004. Analiza funkcjonowania systemu kanalizacji niekonwencjonalnej na przykładzie systemu w Stanisławowicach (The analysis of operation of the unconventional sewerage system on an example of Stanislawice),*Scientific Journal of the Rzeszow University of Technology*, Faculty of Civil and Environmental Engineering, Rzeszow. pp.187–192 (in Polish).

Kwietniewski, M., Miszta-Kruk, K. & Pienkowska, J. 2006. Results of the vacuum station monitoring in a selected vacuum sewerage system. *Proceedings of SPIE – The International Society for Optical Engineering, v 6159 II, Photonics Applications in Astronomy, Communications, Industry, and High-Energy Physics Experiments IV, pp 61593B*

Latoszek, B. & Kraśkiewicz, A. 2002. Przydomowe węzły opróżniające ROEDIGER stosowane w kanalizacji podciśnieniowej (Near-house evacuating centres ROEDIGER used in vacuum sewerage systems), *Gas, Water and Sanitary Technique. No. 7. pp. 240–244 (in Polish)*

Little, C.J. 2004. A comparison of sewer reticulation system design standards gravity, vacuum and small bore sewers. *Water SA, v 30. n 5. pp 685–692*

Naret, R.P.E. 1988. Vacuum Sewers: Construction and operating experience. *Proceedings of the 15th Annual Water Resource Conference, Critical Water Issues and Computer Applications,* Norfolk, Virginia. ASCE. pp 352–355

Otterpohl, R. 2002. Options for alternative types of sewerage and treatment systems directed to improvement of the overall performance. *Water Science and Technology, v 45, n 3. pp 149–158*

Rodhead, P.A. 1999. The ultimate vacuum. *Vacuum 53, pp 137–149*

Sullivan, J.F. & Harrington, B. & Johnson, S. & Bergman, K.A. 2003. Advent of vacuum sewers in New England. *Journal of New England Water Environment Association, v 37, n 2. pp 145–162*
Thomas, J.H., Hoffman, D. & Singh B. 1998. *Handbook of vacuum science and technology.* Academic Press
Tjandraatmadja, G. Burn, S.; McLaughlin, M. & Biswas, T. 2005. Rethinking urban water systems – Revisiting concepts in urban wastewater collection and treatment to ensure infrastructure sustainability. *Water Science and Technology: Water Supply, v 5, n 2. pp 145–154*

Environmental Engineering – Pawłowski, Dudzińska & Pawłowski (eds)
© 2007 Taylor & Francis Group, London, ISBN13 978-0-415-40818-9

Fault tree reliability evaluation method for a vacuum sewerage system

Katarzyna Miszta-Kruk
Department of Water Supply and Wastewater Treatment, Faculty of Environmental Engineering, Warsaw University of Technology, Warsaw

Marian Kwietniewski
Department of Water Supply and Sewerage, Faculty of Environmental Engineering, Lublin University of Technology, Lublin

ABSTRACT: This paper presents a reliability evaluation method for Vacuum Sewerage Systems (VSS), using a fault tree. The method is based on the analysis of a combination of causes (states of elements) described with Boolean algebra and leading to the outcome event representing the state of the sewerage system. The following main issues have been addressed: the analysis of the state of a sewerage system, which allows the elements important for VSS reliability evaluation to be separated, an overview of logical operators and rules governing their use in VSS reliability evaluation. The evaluation method has been illustrated with an example calculation.

Keywords: Fault tree, reliability, vacuum sewerage system.

1 INTRODUCTION

Faults and other malfunctions of sewerage systems, including vacuum sewerage systems, are a nuisance for their users, disrupt the ecological balance in the natural environment, cause material losses and increase the operating cost of these systems. As sewerage systems are expected to reliably collect and convey wastewater, a need arises for methods to evaluate the reliability of vacuum sewerage systems in order to facilitate their design and increase efficiency.

Reliability evaluation of a vacuum sewerage system is a new issue, not yet fully explained. Apart from a few papers addressing the issue of sewerage failure (Naret 1988, Kalenik & Ćwiek 2005) the issue of reliability evaluation for this type of sewerage system has not yet been subjected to scientific study, be it on-site testing or theoretical considerations. Papers addressing reliability evaluation published so far focus mainly on traditional gravitational sewerage systems, and specifically on the analysis of types, causes and consequences of failures as well as the reliability evaluation of a gravitational sewerage system using Markov processes theory (Dippold & Jedlitschka 1976, Ermolin 2001, 2002, Yan-Gang 2001).

2 VACUUM SEWERAGE SYSTEM – OVERVIEW

The vacuum sewerage system falls under the category of the so-called unconventional systems, along with pressure and low-diameter systems, and is an alternative to the traditional gravitational means of sewage collection (ATV-A116P 1992, Thomas et al. 1998, PN_EN 1091 2002).

The vacuum system forces wastewater flow by creating vacuum in the pipes. Wastewater from an installation situated in a building is collected under gravity in a common vacuum pot with a discharge valve. From the common vacuum pot the wastewater is sucked into a vacuum pipeline system and transported to a sewage tank in a vacuum-pressure pumping station or to an open sewage tank (siphon collector well) in a vacuum-siphon pumping station. From there, wastewater is transported in a gravitational or pressure pipeline system to a wastewater treatment plant.

Thus, the following two types of solutions are currently in use: vacuum and vacuum-siphon sewerage systems. In a *vacuum* system the wastewater is collected in a closed vacuum sewage tank. In a vacuum-siphon system this tank is replaced with an open siphon collector well, containing a vacuum pipe linked with the tank by another siphoned pipe (Olszewski 2002).

This paper discusses a widely used vacuum sewerage system. This system (Fig. 1) comprises the following typical elements:

- common vacuum pots with discharge valves collecting wastewater from individual buildings,
- a system of vacuum pipelines,
- vacuum-pressure pump station with a sewage tank,
- a system of pressure pipelines.

3 THE DEFINITION AND A RELIABILITY MODEL OF A VACUUM SEWERAGE SYSTEM

The definition of a reliability model of a vacuum sewerage system has to take into consideration both its purpose and specific properties of this system. Bearing in mind the general properties of sewerage systems' reliability as in (Mays 1996, 1989, Kwietniewski et al. 1993, Wieczysty 1990) and taking into account the factor.forcing wastewater flow in a vacuum sewerage system, the following definition can be adopted:

The reliability of a vacuum sewerage system is its capability to collect a predicted amount of wastewater from a given area into a given collector, under the necessary vacuum level in the pipeline and under specific conditions of use and within the predicted lifecycle of the system.

The definition above allow us to infer that the following events, occurring independently of each other, symptomatize the unreliability of a vacuum sewerage system:

- wastewater collection failure due to e.g. insufficient vacuum level in the pipelines, mechanical damages to discharge valves etc.,
- incomplete/partial wastewater collection due to e.g. full/partial pipeline leakage (only part of the system is operating), wastewater is collected only from some of the common vacuum pots.

The model assumes that the operation process of the vacuum sewerage system consists of consecutive *operational* states, when the system performs its function, and states of *failure*, when the system does not perform its function according to the requirements. Thus, the two-state vacuum sewerage system reliability model distinguishes between:

1. operational state,
2. failure (malfunction) state.

An operational state is a situation in which the system meets all the technical criteria imposed on it, e.g. all wastewater which enters the system is conveyed under the required vacuum level.

Failure (malfunction) state is a situation in which the system does not meet the required technical criteria.

Elements of the system are also assumed to take either of the two states: operational or failure state.

For the purpose of the reliability evaluation it is assumed that malfunctions in different elements of the vacuum sewerage system occur independently.

4 RELIABILITY EVALUATION METHOD FOR A VACUUM SEWERAGE SYSTEM

System reliability depends on the functioning of its elements. In a vacuum sewerage system the following

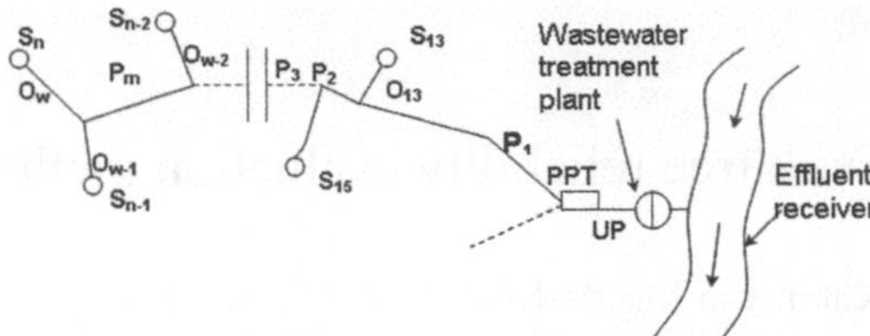

Figure 1. Schematic representation of a sample vacuum sewerage system with selected elements for the purpose of reliability evaluation. S1,...– a common vacuum pot with discharge valve (m – the number of vacuum pots), O1,...– a section of vacuum pipeline connecting the common vacuum pot to the main pipeline (m – the number of pipeline sections), P1,...– a section of main vacuum pipeline (n – the number of main vacuum pipeline sections), PPT – vacuum-pressure pumping station with a vacuum sewage tank, UP – a pressure pipeline system.

elements have been selected as playing a role in its reliability (the symbols in brackets will be used in calculations):

- common vacuum pot with a discharge valve – collection chamber (S),
- a section of a vacuum pipeline connecting the common vacuum pot to the main pipeline system (O),
- a section of the main vacuum pipeline between adjacent junctions (P),
- in a vacuum-pressure pumping station (PPT),
  - vacuum sewage tank (ZPS),
  - vacuum pump (PP),
  - sewage pump (PS),
  - automatic controls (A),
  - manual controls (R),
- in pressure pipelines (PT),
  - pressure pipe (PT).

Figure 1 shows a schematic representation of a vacuum sewerage system prepared for the purposes of reliability evaluation including the elements listed above.

From the technical point of view the vacuum sewerage system takes the form of a branching "tree" structure composed of common vacuum pots, (S), sections of vacuum pipelines (O and P), a vacuum-pressure pumping station (PPT) and a pressure pipeline system (UP). The analysis of the vacuum sewerage system allows three subsystems affecting the overall reliability, to be distinguished:

I. Vacuum pipeline system with common vacuum pots, for the evaluation purposes referred to as the Network (SP),
II. Vacuum-Pressure Pumping Station (PPT),
III. Pressure Pipeline System (UP).

A fault tree is a structured graphical representation of conditions and factors affecting the occurrence of a specified undesirable event referred to as "top event", i.e. failure of the sewerage system. This makes it possible to determine and analyse:

- factors affecting reliability and characteristics describing the functioning of the system, e.g. the kind of an element's failure, external conditions;
- conflicting requirements potentially affecting reliable functioning;
- related events, affecting more than one element of the system or subsystem.

The fault tree is a useful tool when analysing different, especially complex, systems which include several functionally related or independent subsystems serving different purposes (Liudong 2004, Jang-Soo & Sung-Deok 2005).

The analysis of a vacuum sewerage system with the fault tree is basically a deductive, logical (top-down) method, aimed at pointing out the cause or set of causes leading to a particular "top event". In such cases the analysis concentrates on undesirable events (failure states) associated with each element in order to characterize the top event, which is a failure of sewerage system, i.e. its inability to transport wastewater from a given area to a wastewater treatment plant or another receiver.

The description of a vacuum sewerage system analysed with a fault tree should contain the information about the way it operates, the specification of its elements and their connections. The fault tree may cover events resulting from all failure causes, which can include the effects of technical factors, external conditions and others, as necessary. The methodology proposed here takes into account causes related to the technical operation of sewerage systems, which have so far been discovered in the on-site study of reliability.

The tree construction begins with the specification of the top event. Generally, this is the state of the sewerage system in which it is capable of discharging all collected wastewater from a given settlement unit within the area covered by the system. The top event is the output of the top logical gate, while particular input events determine the possible causes and conditions for the occurence of the top event. Each input event may, at the same time, be the output for a lower level logic gate. If the output event of a gate determines the inability to perform the intended function, the corresponding input event may describe the failure of a given element or its limitations.

The tree construction stops when one or more of the following events occur:

- basic event; i.e. an independent event, the important characteristics of which can be obtained by other means than the fault tree;

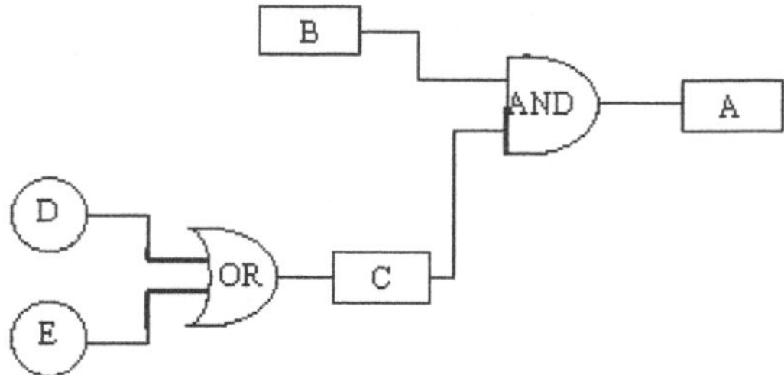

Figure 2. A sample fault tree with events B, C, D, E and top event A. AND – event conjunction operator (gate), OR – event disjunction operator (gate).

- event which does not require to be expanded further;
- event which has been or will be expanded further in another fault tree; if the event is expanded further, it has to bear the same identification features as the corresponding event in another tree – in this way the tree constructed later is a correct continuation of the earlier tree.

The analysis of a fault tree is conducted in steps. A sample tree is shown in Figure 2.

Direct visual analysis is possible only for small trees, not exceeding 70 events (PN-IEC-1025, Amari & Akers 2004). All events connected to the top event by an uninterrupted chain of logic disjunction (OR) operators (gates) cause the occurence of the top event. The complexity of the fault tree analysis grows quickly with the size of the tree, but an in-depth analysis allows independent branches to be distinguished, which can be analysed separately. In calculating numerical values describing consecutive states in the tree one may make use of non-failure and readiness forecasting methods or operating data.

The reliability evaluation method for vacuum sewerage systems using a fault tree structure employs Boolean algebra, which allows us to calculate the influence of different events and their combinations on the final event. It is assumed that the occurence of the top event is not a function of the time or sequence of occurence of other events. In this case a logical tree of events is constructed, using special operator symbols or so-called logic gates, where operability states of sewerage system elements constitute elementary events (cf. Table 1). Thus, the final result describes the operability state of the vacuum sewerage system.

Under the assumption that $w_1, w_2, \ldots, w_n$, are events such that a given element is either in the operability or failure state and denoting them with a zero-one vector as input events (Tchórzewska-Cieślak & Rak 2006), i.e

Table 1. Basic logical operators (gates) and their descriptions.

Gate	Graphical symbol	Logical operation in Boolean algebra	Operation generalized with a non-failure coefficient (readiness coefficient K)
AND	w_1, w_2, w_M → W	$W = w_1 \cap w_2 \cap \ldots \cap w_M$	$K_g = K_1 \cdot K_2 \ldots K_M$, (1) M_A – the number of events (states of elements) in the structure of AND gate
OR	w_1, w_2, w_M → W	$W = w_1 \cup w_2 \cup \ldots \cup w_M$	$K_g = 1 - \prod_{j=1}^{M}(1 - K_j)$, (2) M_O – the number of events (states of elements) in the structure of OR gate
NOT	w1 → $\overline{w1}$	$w_1 = \overline{w1}$	$K_g = 1 - K_1$, (3)

All these texts fit in a frame which should not be changed (Width: Exactly 187 mm (7.36"); Height: Exactly 73 mm (2.87") from top margin; Lock anchor).

$$w_1, w_2, \ldots, w_M - input \quad event = \begin{cases} 1 - if & element \quad i \quad is \quad operational \\ 0 - if & element \quad i \quad has \quad failed \end{cases}$$

then input event W as a result of the application of Boolean algebra will be denoted as:

$$W - input \quad event = \begin{cases} 1 - if & whole \quad system/subsystem \quad is \quad operational \\ 0 - if & whole \quad system/subsystem \quad has \quad failed \end{cases}$$

The application of probability and reliability theories to the logical event tree allows the characteristics of the system reliability to be determined. Basic dependencies shown in equations (1–3) in Table 1 are used to this end. It should be noted that a given tree may have only one top event.

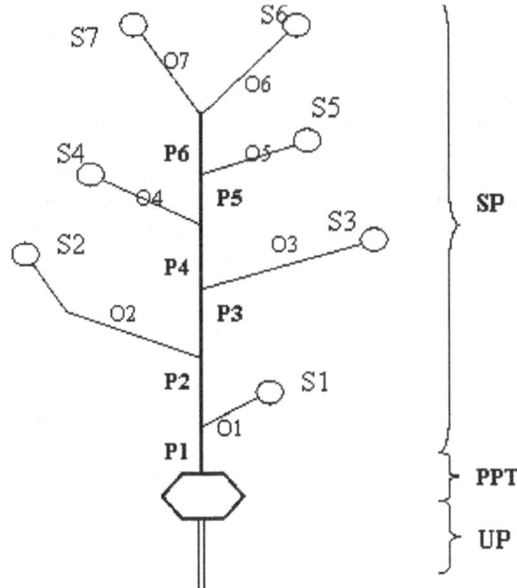

Figure 3. Schematic representation of a vacuum sewerage system with 3 subsystems (calculation example). (Symbols as in Fig 1).

5 CALCULATION EXAMPLE

The reliability evaluation method is illustrated by a simplified example of a vacuum sewerage system (Fig. 3).

This vacuum sewerage system comprises seven common vacuum pots with discharge valves (S1,…,S7) connected by pipeline sections (O1,…,O7) to sections of main vacuum pipeline (P1,…,P6), which convey wastewater to vacuum sewage tank ZPS1 or ZPS2 (tanks operate interchangeably) by vacuum generated in vacuum pumps PP1 or PP2 (operating interchangeably). From there wastewater is conveyed by sewage pumps PS 1 or PS 2 (operating interchangeably) to pressure pipelines PT 1 or PT 2 and further to a wastewater treatment plant. To simplify the description of the fault tree each common vacuum pot with a discharge valve is considered as one element (two serially connected elements), denoted as SO.

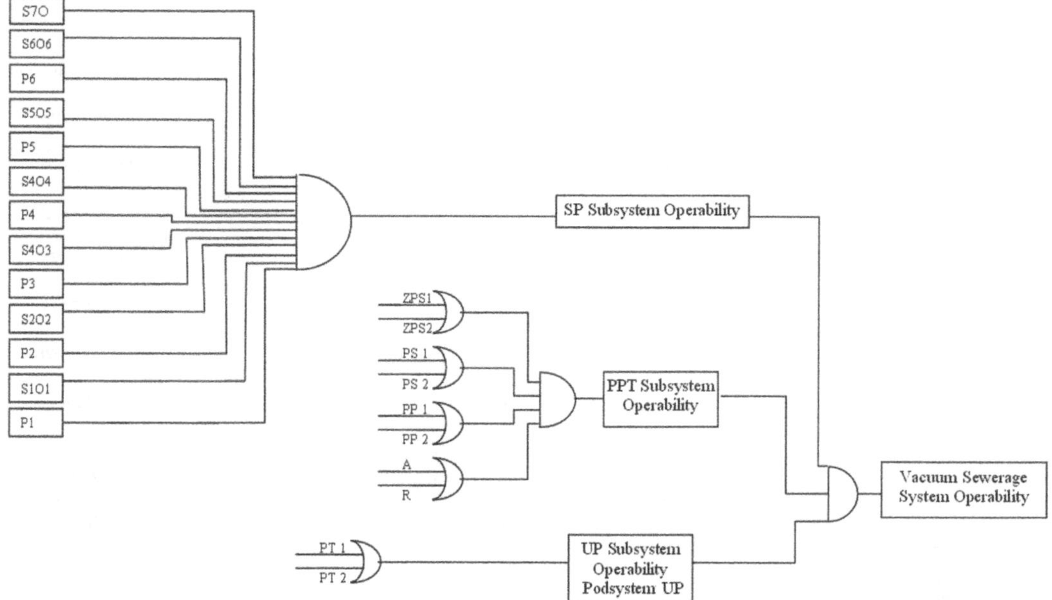

Figure 4. Representation of a logical structure of the event tree for the purposes of vacuum sewerage system reliability evaluation.

Figure 4 shows a schematic representation of the logic event tree for the vacuum sewerage system constructed in accordance with the principles mentioned above.

The reliability of the vacuum sewerage system and its elements has been denoted using readiness coefficient, for simplicity denoted K (Han, X-T. et al. 2004, Mays 1989, Kwietniewski et al. 1993, Surendran et al. 2005, Tanyimboh et al. 1999, Wieczysty 1990), which expresses the probability of the object (element, system) being in an operational state at any time. Stationary form of the coefficient has been used, denoted by:

$$\lim_{t\to\infty} K(t) = K = \frac{T_p}{T_p + T_o} \tag{3}$$

where: K – readiness coefficient, T_p – mean operability time, T_o – mean failure time.

Whereas the probability of an adverse event, i.e. such that the sewerage system is transformed to a failure state can be denoted as the complement-to-one of the readiness coefficient:

$$\bar{K} = 1 - K \tag{4}$$

where $\overline{K}$ denotes the unreliability of the system/element.

The numerical values of readiness coefficients for sewerage system elements have been obtained in onsite study of several systems operating in Poland.

Table 2. Operability and failure states denoted as event conjunction or disjunction.

Operator (gate) AND			Operator (gate) OR		
w_1	w_2	W	w_1	w_2	W
1	1	1	1	1	1
1	0	1	1	0	1
0	1	1	0	1	1
0	0	0	0	0	0

Table 3 shows readiness coefficient values for elements of sewerage system from Figure 4.

Using the logic structure of the event tree and associated equations describing individual gates (Table 2), readiness coefficients have been calculated for consecutive levels of the analysis specificity and eventually the overall reliability of the vacuum sewerage system.

A. Calculations for the first level of specificity – the analysis of the elements' states

1. Network Subsystem (SP)
One applies equation (1) from table 2, which gives the readiness coefficients for common vacuum pots (S) with pipeline sections connecting them to the main pipeline (O) and for main pipeline sections (P), M_A being the sum of these elements, i.e. (m + m + n) respectively. Thus:

$$K_{SP} = K_{S7O7} \cdot K_{S6O6} \cdot K_{P6} \cdot K_{S5O5} \cdot K_{P5} \cdot K_{S4O4} \cdot K_{P4} \cdot K_{S3O3} \cdot K_{P3} \cdot K_{S2O2} \cdot K_{P2} \cdot K_{S1O1} \cdot K_{P1} = 0.99906779$$

Table 3. Readiness coefficient values for the elements of the vacuum sewerage system in the calculation example.

Element name	Symbol	Readiness coefficient symbol	Numerical value
Common vacuum pot S7 and pipeline segment O7	S7O7	K_{S7O7}	*0.99988995*
Common vacuum pot S6 and pipeline segment O6	S6O6	K_{S6O6}	*0.99981420*
Common vacuum pot S5 and pipeline segment O5	S5O5	K_{S5O5}	*0.99987512*
Common vacuum pot S4 and pipeline segment O4	S4O4	K_{S4O4}	*0.99987475*
Common vacuum pot S3 and pipeline segment O3	S3O3	K_{S3O3}	*0.99987453*
Common vacuum pot S2 and pipeline segment O2	S2O2	K_{S2O2}	*0.99989250*
Common vacuum pot S1 and pipeline segment O1	S1O1	K_{S1O1}	*0.99984638*
Main pipeline segment	P1,...,P6	$K_{P1} = K_{P2} = K_{P3} = K_{P4} = K_{P5} = K_{P6}$	1.0
Vacuum-sewage tank	ZPS1, ZPS2	$K_{ZPS1} = K_{ZPS2}$	*0.99939904*
Vacuum pump	PP1	K_{PP1}	*0.99817493*
Vacuum pump	PP2	K_{PP2}	*0.99506579*
Sewage pump	PS1	K_{PS1}	*0.99920195*
Sewage pump	PS2	K_{PS2}	*0.99914429*
Automatic controls	A	K_A	*0.99902850*
Manual controls	R	K_R	1.0
Pressure pipeline	PT	K_{PT}	1.0

2. PPT Subsystem
To calculate readiness coefficients for individual units, i.e. vacuum-pressure tanks (ZPS), vacuum pumps (PP), sewage pumps (PS) and controls (AR), equation 2 has been used, where M_O equals 2 for each unit. Thus:

$$K_{ZPS} = 1-(1-K_{ZPS1})^2 = 0.999999639$$

$$K_{PS} = 1-(1-K_{PS1})\cdot(1-K_{PS2}) = 0.99999932$$

$$K_{PP} = 1-(1-K_{PP1})\cdot(1-K_{PP2}) = 0.99999099$$

$$K_{AR} = 1-(1-0.99902850)(1\text{-}1) = 0.99902850$$

Thus the readiness coefficient for PPT subsystem obtained from equation (1) will be equal to:

$$K_{PPT} = K_{PZS}\cdot K_{PS}\cdot K_{PP}\cdot K_{AR}\cdot K_F = 0.99998994$$

3. UP Subsystem
In calculating the readiness coefficient for the pipeline system (UP) equation (2) has been used. Taking into consideration that readiness coefficient values for individual pipeline sections (PT) converge to 1, the readiness coefficient for UP subsystem will also converge to 1, i.e.

$K_{UP} \rightarrow 1.0$

B. Calculations for the second level of specificity – the analysis of subsystem states and obtaining readiness coefficient for the vacuum sewerage system.

Using the readiness coefficient values for the three subsystems, i.e. the network subsystem (SP), vacuum-sewage pumping station (PPT) and pipeline system (UP), calculated above, one uses equation (1) (gate AND), where M_A equals 3:

$$K_{SKP} = K_{SP}\cdot K_{PPT}\cdot K_{UP} = 0.99905775$$

The value of the readiness coefficient obtained in this way expresses the probability of the event such that the system under consideration is capable to convey at any time the predicted amount of wastewater from the given area into the wastewater treatment plant under the necessary vacuum level in the pipelines and under specific conditions of system operation. This value may be interpreted e.g. as an annual system operability. In this case it means that the system will perform its functions for about 364.66 days a year in total, and will be inoperable for about 0.34 days a year (about 8 hours).

6 CONCLUSION

The method presented above allows an evaluation of the quality of wastewater collection services for a given settlement unit or its part. Obtaining a numerical value for the reliability of a vacuum sewerage system allows, among other things, the analysis of different sewerage systems of this kind. However, the investigation of these systems, which is only possible on-site, remains the basis of reliability evaluation. Only such an investigation is the source of values for coefficients describing the reliability of individual elements of a vacuum sewerage system and consequently the whole system.

The development of plausible reliability criteria for vacuum sewerage systems is necessary to obtain quality standards for such systems.

ACKNOWLEDGEMENTS

The authors are also grateful to reviewers for their constructive comments.

This paper has been written based on the study conducted under the research programme no. 4 T07E 003 27 funded by KBN (The State Committee for Scientific Research)

REFERENCES

Amari, S.V. & Akers, J.B. 2004. Reliability analysis of large fault trees using the Vesely failure rate. *Annual Reliability and Maintainability Symposium. 2004 Proceedings.* (IEEE Cat. No.04CH37506C), *pp 391–6*

ATV-A116P Directive. 1992. Special Sewerage Systems, Vacuum Sewerage Systems – Pressure Sewerage Systems.

Dippold, W. & Jedlitschka, J. 1976. Vakuumsytstem— Druckentwässerung, *Wasser und Boden 28. no. 5, pp 100–104.*

Ermolin, Y.A. 2001. Estination of raw sewage discharge resulting from sewer network failures. *Urban Water 3. pp.131–136*

Ermolin, Y.A., Zats, L.I. & Kajisa, T. 2002. Hydraulic reliability index for sewage pumping stations. *Urban Water 4, pp.301–306*

Han, X-T., Yin, X-G & Zhang, Z. 2004. Application of fault tree analysis method in reliability analysis of substation communication system. *Power System Technology, v 28, no. 1. pp 56–59*

Jang-Soo, L. & Sung-Deok, C. 2005. Fault tree construction of hybrid system requirements using qualitative formal method. *Reliability Engineering & System Safety, v. 87, no 1. pp 121–31*

Kalenik M. & Ćwiek A. 2005. Badania eksploatacyjne sieci kanalizacji podciśnieniowej w systemie AIRVAC. (Examination of the vacuum sewerage network in operation in the AIRVAC system), *Gas, Water and Sanitary Technique, no. 3. pp. 26–30 (in Polish).*

Kwietniewski M., Roman M., Kłoss–Trębaczkiewicz H. 1993. *Niezawodność wodociągów i kanalizacji, (Reliability of Water Supply and Sewerage Systems).* ARKADY. Warsaw (in Polish).

Liudong, X. 2004. Maintenance-oriented fault tree analysis of component importance. *Annual Reliability and Maintainability Symposium. Proceedings.* (IEEE Cat. No.04CH37506C), *pp 534–539*

Mays L. 1996. Review of reliability analysis of water distribution systems. In Tickle K.S., Goulter I.C., Xu C., Wasimi S.A. (Eds.). *Proceedings of the seventh IAHR International symposium, Stochastic Hydraulics '96 (pp. 53–62). Machay, Queens-Land, Australia*

Mays, L., Lansey, K.E. & Cullinare, M.J. 1989. *Reliability analysis of water distribution systems.* New York ASCE

Narret, R. 1988. Vacuum sewers: Construction and operating experience. *Proceedings of the 15th Annual Water Resource Conference, Critical Water Issues and Computer Applications, Norfolk, Virginia, June 1–3, ASCE, pp 352–354*

Olszewski, W. & Franek, D. 2002. Kanalizacja grawitacyjno–podciśnieniowa (Gravitational-vacuum sewerage system), *Rynek instalacyjny no. 5, pp 32–35*

PN-EN 1091: 2002. *Systemy zewnętrzne kanalizacji podciśnieniowej. (External Vacuum Sewerage Systems)*

PN-IEC 1025 1994. *Analiza drzewa niezdatnści (FAT) (Fault Tree Analysis)*

Surendran, S. & Tanyimboh T.T. & Tabesh, M. 2005. Peaking demand factor-based reliability analysis of water distribution systems. *Advances in Engineering Software, Volume 36, Issues 11–12, pp 789–796*

Tanyimboh, T.T. & Burd, R. & Burrows R. & Tabesh M. 1999. Modelling and reliability analysis of water distribution systems *Water Science and Technology, Volume 39, Issue 4, pp 249–255*

Tchórzewska-Cieślak B., Rak J. 2006. Analysis of risk connected with water supply system operating by means of the logical trees method. *Jurnal of Konbin. Instytut Techniczny Wojsk Lotniczycht.*1, z.1, *pp 315–322*

Thomas, J.H., Hoffman, D. & Singh B. 1998. *Handbook of vacuum science and technology.* Academic Press

Wieczysty, A. 1990. *Niezawodność systemów wodociągowych i kanalizacyjnych, (Reliability of Water Supply and Sewerage Systems)* Wydawnictwo Politechniki Krakowskiej, (in Polish).

Yan-Gang, Z. & Ono, T. 2001. Moment methods for structural reliability. *Structural Safety 23 pp.47–45*

Environmental Engineering – Pawłowski, Dudzińska & Pawłowski (eds)
© 2007 Taylor & Francis Group, London, ISBN13 978-0-415-40818-9

The influence of velocity gradient on removal of ammonia nitrogen from water

Alina Pruss & Marian Błażejewski
Institute of Environmental Engineering; Poznan University of Technology, Poznan, Poland

ABSTRACT: The study was carried out on the physical model of rapid filter, filled with activated biological filtration bed taken from water treatment plant in Poznan. The water was prepared from chlorine-free tap water by addition of a standard solution of NH_4Cl. Three filtration cycles with filtration rates 5.0, 7.5 and 10.0 m/h were carried out and analysed. During all filtration cycles as the filtration time was increased the ammonia nitrogen was removed with the consumption of oxygen lower than stoichiometric for the nitrification process. This indicated that besides nitrification, heterotrophic bacteria also assimilated the ammonia nitrogen. The analysis showed that the biofilm formed on the bed had been periodically removed causing changes in the concentration of organic nitrogen in filtrated water. The time required to reach the peak concentration of organic nitrogen in filtered water depended on the velocity gradient during the filtration process.

Keywords: Velocity gradient, biofilm, nitrification, assimilation.

1 INTRODUCTION

The three nitrogen conversion pathways traditionally used for the removal of ammonia-nitrogen in aquaculture systems are photoautotrophic removal by algae, autotrophic bacterial conversion of ammonia-nitrogen to nitrate-nitrogen, and heterotrophic bacterial conversion of ammonia-nitrogen directly to microbial biomass. Heterotrophic bacterial growth is stimulated through the addition of organic substrate. At high carbon to nitrogen (C/N) feed ratios, heterotrophic bacteria will assimilate ammonia-nitrogen directly into cellular protein (Ebeling et al., 2006).

The studies carried out on physical models of the filters demonstrated that during the filtration only part of removed ammonia-nitrogen had been transformed into nitrate-nitrogen (Bray & Olańczuk-Neyman, 2001, 2003, Błażejewski & Pruss, 2000). The rest was assimilated and cumulated in the cells of nitrifying bacteria and heterotrophic bacteria forming biofilm on the particles of the filtration bed (Pruss, 2006). The ammonia nitrogen was removed with the consumption of oxygen lower than stoichiometric consumption for the nitrification process. It also indicates that not only nitrification bacteria, but also heterotrophic bacteria assimilate the ammonia and participate in the process of biofilm building, by using a soluble organic carbon (Lipponen et al., 2004, Summerfelt et al., 2003). Autotrophic nitrification bacteria and oxidising or relatively oxidising heterotrophic bacteria compete for dissolved oxygen and space in the inner layers of the biofilm (Bouver & Crowe, 1988, Vandenabelle et al., 1995, Larrea L. et al., 2004). Investigations carried out by Stief & et al. have shown that a thin oxic surface layer is a site of initially high rates of nitrification, i.e. O_2, NH_4^+, and NO_2^- consumption, and NO_3^- production. These data imply factors other than the frequently suggested competition between nitrifiers and heterotrophs for NH_4^+, NO_2^- or O_2 as causes for the loss of nitrification activity (Stief et al., 2004).

The exponential growth in the number of bacteria caused increasing thickness of the biofilm (Liu & Capdeville 1996). Consequently, the nitrification process depended on the processes taking place on the way from the border between the aqueous phase and biofilm to the surface of the filtration bed particles. Heterotrophic bacteria are less firmly attached to the biofilm than nitrifying bacteria (Oleńczuk-Neyman & Bray, 1998, 2000; Lazarova et al., 1992), therefore it can be expected that the shear stress will cause the composition of biofilm organisms to vary during the filtration cycle as the porosity of the bed decreases and velocity gradient of the water flow increases.

2 METHODS AND MATERIAL

The study was carried out on the physical model of rapid filter, filled with activated biological filtration bed taken from the rapid filters of the water treatment plant in Poznań.

The filtration column was made from methyl polymethacrylate with internal diameter 8.4 cm and 200 cm height. The filter consisted of a drainage system, a 5 cm gravel structural layer and a 110 cm sand filtration layer. In order to sustain constant temperature the filtration column was placed in a water bath, fed with water flow parallel to the filtration flow.

The water fed to the test system was prepared from chlorine-free tap water, which flowed through the float valve to the preliminary tank, where it was aerated and heated to a constant temperature of 18°C. The concentration of the ammonia nitrogen in the tested water was adjusted with a standard solution of NH_4Cl, which had been added proportionally to the required nitrogen load to the tube between preliminary tank and the filter. The average concentration of phosphorus in the tap water was approx. 0.25 mg P_2O_5/dm^3, what was considered sufficient for biological processes in the filtration bed.

Filtration was carried out with the pressure approx. 7 m water head. Head losses were determined during the filtration cycle every 24 h by measurement of the flow rate and corresponding filtration velocity. The filter was rinsed after each filtration cycle with a counter-current flow of tap water, with intensity securing 50% expansion of the bed.

Water was sampled upstream and downstream of the filter and tested for ammonia nitrogen, nitrite, nitrate, total and organic nitrogen. All tests were carried out according to Polish Standards.

2.1 *Methodology of determination of biofilm thickness*

In order to quantify growth of the biofilm the changes of the porosity of the filtration bed were measured, assuming that it determines the percentage of free space between bed particles in total volume of the bed. Decreased amount of free space in the bed might be caused by both growth of the biofilm and in some degree by removed fractions of the biofilm. In order to simplify the study, it was assumed that decreased porosity of the bed was caused only by growth of the biofilm on the particles of the filtration bed.

On the basis of the time dependence of the porosity of the bed the clogging ratio (S) was calculated. It was calculated as a difference between initial porosity and porosity after the filtration and was determined by the following formula:

$$S = m_0 - m_x \tag{1}$$

with:

S – clogging ratio [%]

m_0 – initial porosity of the bed [%]

m_x – porosity of the bed after a certain time of the filtration [%]

Before the filtration cycle and after rinsing of the filter the initial porosity (m_0) of the bed was determined. Knowing the volume of the bed ($V_{bed} = 5.54\,dm^3$), it was filled with water and the volume of water was measured. It was assumed that it reflected the volume of pores ($V_{pores} = 2.20\,dm^3$). Than the initial porosity of the bed was calculated:

$$m_0 = \frac{V_{pores}}{V_{bed}} * 100\% \tag{2}$$

$$m_0 = \frac{2.2}{5.54} * 100\% = 39.7\%$$

During the filtration cycles, the flow rate of water changed as a result of growth of the biofilm or shearing of the fraction of the film. By transforming the formula to the real flow water velocity in the pores of the bed, the porosity of the bed during filtration was determined i.e. the changing throughout filtration volume of the pores between particles of the bed was calculated.

$$m = \frac{Q}{F * V} \tag{3}$$

m – the so-called surface porosity, equal to the quotient of the surface of the pores at a certain cross section and the whole surface of the bed in the cross section (dimensionless value). In order to simplify the study it was assumed that the surface porosity is equal to the volume porosity;

Q – water flow rate in the filtration bed [cm^3/s];

F – surface of the cross section of the bed (brut surface – total surface of particles and pores channels in right cross section to the direction of the flow) [cm^2].

V – velocity of the flow in pores channels [cm/s]

Using the empirically determined initial porosity of the bed, the flow rate and bed surface, the initial velocity of flow in the pores channels was calculated on the basis of the transformed formula 3. Furthermore, it was assumed that during the whole filtration cycle that the velocity had remained constant.

Based on the sieve analysis, the effective diameter d_e of the bed particles was calculated. Assuming a spherical shape of the particles the volume of standard particles was calculated on the basis of calculated effective diameter.

Further analysis was carried out for the filtration layer of thickness equal to the effective diameter d_e of bed particles. The hypothetical volume of the particles in the layer was determined on the basis of the volume of the layer and porosity of the bed. Then, the number of particles in the analysed layer was determined on the basis of the calculated volume of a single particle.

The growth of the biofilm on the bed particles caused increasing diameter of particles and decreased the size of the pores channels. As a result, the porosity of the analysed layer changed. The difference of the porosity and constant number of particles of analysed layer was used for the calculation of changing throughout the time diameter of the bed particles. The difference between the calculated and effective diameters equalled the thickness of the biofilm.

The average gradient of velocity of liquid as a function of filtration rate, grain size distribution, porosity and density of filtration bed was calculated using following formula:

$$G = \frac{1-\varepsilon}{\varepsilon} * \frac{V}{d} * \sqrt{\frac{\varsigma_m - \varsigma}{\varsigma} * K} \qquad (4)$$

G – gradient of velocity of liquid [s^{-1}],
ε – porosity,
V – filtration velocity [mm/s],
d – grain size distribution of the bed [mm],
ζ_m – density of the bed [g/cm^3] = 2,65 g/cm^3 for sand,
ζ – density of water [g/cm^3] = 1,0 g/cm^3,
K – Kozeny – Carman constant = 70.

3 RESULTS AND DISCUSSION

Three filtration cycles with velocities 5.0, 7.5 and 10.0 m/h were carried out and analysed. Results are shown in the following Tables 1–3.

The results of the study (Tabs 1–3) show that the velocity gradient in the filtration bed at the beginning of the cycle for each filtration velocity was respectively 51.9; 77.9 and 103.8 s^{-1}. Those values changed during the filtration time as a result of growth of the biofilm on the particles of the filtration bed.

The velocity gradient increased up to the specific maximum and then decreased for each filtration cycle

Table 1. Results during the filtration with initial velocity 5 m/h.

Time	Days	0	6	7	13	14	15	20	27
$V_{filtration}$	m/h	5.00	4.98	4.98	4.75	5.09	4.75	4.87	4.98
N-NH_4^+ – inflow	mg/dm^3	–	2.840	3.320	0.800	3.500	1.200	2.230	2.040
N-NH_4^+ – outflow	mg/dm^3	–	1.080	0.810	0.760	1.350	0.480	0.730	0.410
Removed N-NH_4^+	mg/dm^3		1.76	2.51	0.04	2.15	0.72	1.5	1.63
Removed N-NH_4^+	%		62	76	5	61	60	67	80
N-NO_2 – inflow	mg/dm^3		0.150	0.200	0.045	0.289	0.013	0.184	0.042
N-NO_2 – outflow	mg/dm^3	–	0.300	0.440	0.140	0.073	0.031	0.149	0.067
N-NO_3 – inflow	mg/dm^3		0.055	0.172	0.000	0.099	0.165	0.154	0.337
N-NO_3 – outflow	mg/dm^3	–	0.327	0.788	0.099	0.117	0.150	0.238	0.656
Total N – outflow	mg/dm^3	–	4.300	3.300	1.600	1.900	1.500	1.700	2.300
Organic N – outflow	mg/dm^3	–	2.593	1.262	0.601	0.360	0.839	0.583	1.168
$D_{particles}$	mm	1.087	1.113	1.113	1.122	1.108	1.122	1.118	1.113
Porosity	%	40	35.6	35.6	33.9	36.4	34.0	34.8	35.6
Thickness of biofilm	mm	0.000	0.013	0.013	0.018	0.011	0.018	0.015	0.013
Gradient	s^{-1}	51.9	68.5	68.5	73.0	66.4	73.0	70.7	68.5

Table 2. Results during the filtration with initial velocity 7.5 m/h.

Time	Days	0	2	7	9	15	17	21
$V_{filtration}$	m/h	7.5	7.15	6.93	6.50	7.47	7.36	7.36
N-NH_4^+ – inflow	mg/dm^3	–	2.000	2.000	2.000	2.000	2.000	2.000
N-NH_4^+ – outflow	mg/dm^3	–	0.186	0.000	0.395	0.326	0.232	0.484
Removed N-NH_4^+	mg/dm^3		1.81	2.0	1.605	1.674	1.768	1.516
Removed N-NH_4^+	%		91	100	80	84	88	76
N-NO_2 – inflow	mg/dm^3		0.001	0.001	0.001	0.001	0.001	0.001
N-NO_2 – outflow	mg/dm^3	–	0.006	0.005	0.001	0.001	0.000	0.001
N-NO_3 – inflow	mg/dm^3		0.162	0.194	0.356	0.000	0.373	0.292
N-NO_3 – outflow	mg/dm^3	–	1.312	0.940	1.215	0.907	0.697	0.956
Total N – outflow	mg/dm^3	–	2.504	2.778	2.820	2.421	2.230	2.122
Organic N – outflow	mg/dm^3	–	1.000	–	1.209	1.187	1.301	0.681
$D_{particles}$	mm	1.087	1.103	1.109	1.122	1.093	1.096	1.096
Porosity	%	40	37.4	36.2	34.0	39.1	38.5	38.5
Thickness of biofilm	mm	0.000	0.008	0.011	0.018	0.003	0.004	0.004
Gradient	s^{-1}	77.9	87.3	91.6	99.5	82.0	83.9	83.9

Table 3. Results during the filtration with initial velocity 10 m/h.

Time	Days	0	23	25	30	31	32	36
$V_{filtration}$	m/h	10.0	10.11	10.11	9.89	9.77	9.77	10.13
$N\text{-}NH_4^+$ – inflow	mg/dm^3	–	2.000	2.000	2.000	2.000	2.000	2.000
$N\text{-}NH_4^+$ – outflow	mg/dm^3	–	0.319	0.319	0.256	0.527	0.417	0.663
Removed $N\text{-}NH_4^+$	mg/dm^3	–	1.681	1.681	1.744	1.473	1.583	1.337
Removed $N\text{-}NH_4^+$	%	–	84	84	87	74	79	67
$N\text{-}NO_2$ – inflow	mg/dm^3	–	0.000	0.000	0.000	0.000	0.000	0.065
$N\text{-}NO_2$ – outflow	mg/dm^3	–	0.000	0.000	0.000	0.004	0.000	1.442
$N\text{-}NO_3$ – inflow	mg/dm^3	–	0.154	0.154	0.231	0.198	0.213	0.238
$N\text{-}NO_3$ – outflow	mg/dm^3	–	1.000	1.000	0.568	0.660	0.521	0.656
Total N-outflow	mg/dm^3	–	2.6	2.6	2.6	2.5	1.6	3.0
Organic N-outflow	mg/dm^3	–	1.281	1.281	1.776	1.309	0.662	0.239
$D_{particles}$	mm	1.087	1.094	1.094	1.100	1.102	1.102	1.094
Porosity	%	40	38.8	38.8	37.9	37.4	37.4	38.8
Thickness of biofilm	mm	0.000	0.004	0.004	0.007	0.008	0.008	0.003
Gradient	s^{-1}	103.8	113	113	117	119.4	119.4	113.3

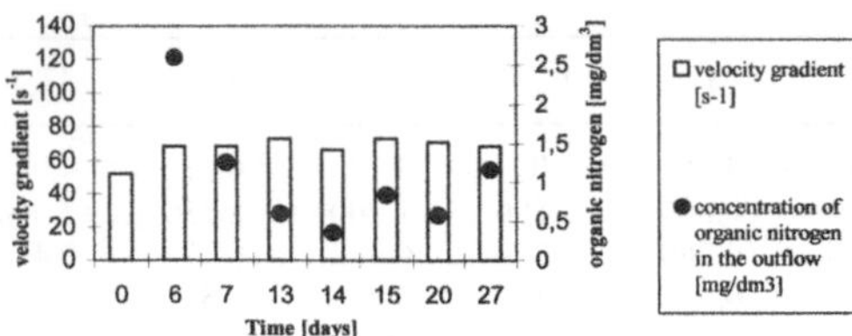

Figure 1. Velocity gradient and concentration of the organic nitrogen in the effluent during filtration with initial velocity 5 m/h.

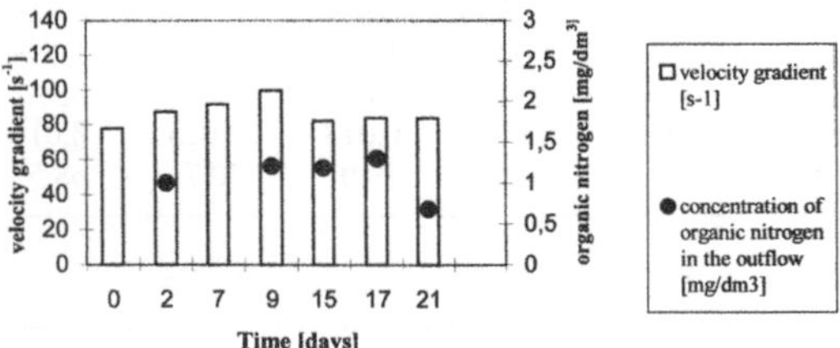

Figure 2. Velocity gradient and concentration of the organic nitrogen in the effluent during filtration with initial velocity 7,5 m/h.

(Figs 1–3). For the filtration rate 5 m/h two peaks occurred on the 13th and 15th day (Fig. 1). Those peaks were certainly caused by growth of the biofilm on the filtration bed and periodical removal of the shorn biofilm.

Further analysis of the results presented in tables and in the Figures 1–3 indicates that during the filtration organic nitrogen appeared in the effluent. At the beginning of the process the concentration of the organic nitrogen in the effluent was low. It increased during the filtration process, reached a peak value and

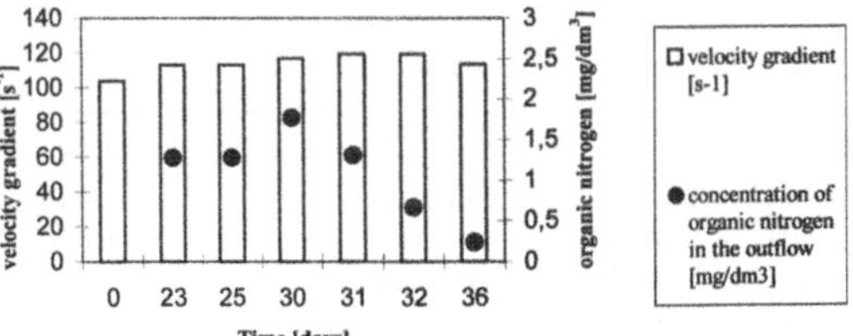

Figure 3. Velocity gradient and concentration of the organic nitrogen in the effluent during filtration with initial velocity 10 m/h.

decreased. The maximum of the concentration of organic nitrogen in the effluent for the filtration rate 7.5 and 10 m/h was preceded by shearing of the biofilm as a result of the high velocity gradient.

For the filtration with low velocity gradient, filtration rate 5 m/h, the maximum concentration of organic nitrogen in the effluent occurred after a shorter time of operation. It was probably caused by the growth of biofilm with large participation of heterotrophic bacteria less resilient to the shear stress.

In order to quantify the above phenomena, the input and output loads of total, mineral and organic nitrogen were calculated for the whole filtration cycle. The calculations were carried out by integrating the surface under the curve of the diagram of function of concentration to the volume of filtrated water. The results are presented in the Table 4.

The mineral nitrogen in the effluent in the above table represents that part of the nitrogen, which was transformed to nitrate nitrogen. The contribution of mineral nitrogen was 31% of the input load for the filtration rate 5 m/h, 57% of the input load for the filtration rate 7.5 m/h and 76% of the input load for the filtration rate 10 m/h respectively. The rest of the

Table 4. Total, mineral and organic nitrogen load in the influent and effluent in the filtration of water with filtration rate 5 m/h for 27 days, 7.5 m/h for 21 days and 10 m/h for 36 days.

Filtration rate [m/h]	Sampling point –	Total nitrogen [g]	Mineral nitrogen [g]	Organic nitrogen [g]
5.0	Influent	20.43	20.43	0.00
5.0	Effluent	22.08	6.26	15.82
7.5	Influent	44.41	44.41	0.00
7.5	Effluent	43.35	25.24	18.11
10.0	Influent	40.49	40.49	0.00
10.0	Effluent	45.83	30.83	15.00

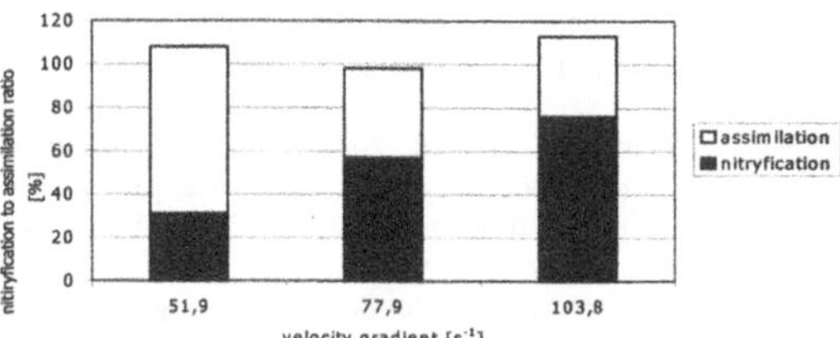

Figure 4. Nitrification to assimilation ratio depending on the initial velocity gradient.

Table 5. Nitrification to assimilation ratio as a function of initial filtration velocity gradient.

Cycle	$V_{filtration}$ [m/h]	Filtration velocity gradient [s^{-1}]	Nitrification to assimilation ratio
1	5	51.9	0.4:1.0
2	7.5	77.9	1.4:1.0
3	10	103.8	2.0:1.0

nitrogen was probably assimilated and incorporated in the bacteria cells. Some of them were carried away with the effluent while a part remained attached to the particles of the bed. That way the effluent carried away was ca. 77%, 41% and 37% respectively of the nitrogen as organic nitrogen (Fig. 4). The sludge and rinsing water contained approx. 0.1 g of total nitrogen, which was considered negligible from the point of view of total mass balance.

The nitrification to assimilation ratio, depending on the initial velocity gradient, was determined by correlating loads of mineral and organic nitrogen with filtration rate and velocity gradient for each filtration cycle. Results are presented in the Table 5.

The lowest contribution of the nitrification process to the nitrogen removal in the cycle with filtration rate 5 m/h indicates that the heterotrophic bacteria play a major part in the process. As the filtration rate

Table 6. Maximum concentration of organic nitrogen in the effluent depending on the filtration velocity gradient.

Cycle	$V_{filtration}$ [m/h]	Filtration velocity gradient [s^{-1}]	Concentration of organic nitrogen in the effluent [mg/dm^3]	Filtration time required to reach max concentration of organic nitrogen in the effluent [days]
1	5	68.5	2.593	6
2	7.5	83.9	1.301	17
3	10	117	1.776	30

increased the contribution of the nitrification process to the nitrogen removal increased and correspondingly, the significance of assimilation of ammonia nitrogen by heterotrophic bacteria decreased.

Table 6 shows that the time required to reach the maximum concentration of organic nitrogen in the effluent depends on the filtration velocity gradient.

The study demonstrated that the lower the filtration velocity gradient the quicker the growth of the biofilm up to the maximum thickness, and after reaching that the biofilm was shorn and removed by the effluent, which in turn contained a higher concentration of organic nitrogen. The lower growth of biofilm for the higher filtration velocity gradient was certainly caused by constant shearing of some of the heterotrophic bacteria in the most exposed places in the bed and slower growth of nitrifying bacteria. This conclusion was confirmed by the higher contribution of the nitrification process in ammonia nitrogen removal at higher filtration rates.

As a result, the biofilm growing on particles of the bed was more compact and reached the shearing thickness later.

4 CONCLUSIONS

Analysis of the influence of the velocity gradient on the nitrification process in the filtration bed revealed that biofilm growing on the particles of the bed were periodically shorn and removed with effluent. The effect was indicated by the changing concentration of organic nitrogen in the effluent during the filtration process. At the beginning of the filtration the concentration of organic nitrogen in the effluent was low, then it increased, reached a maximum value and then decreased. The time to reach the peak concentration of organic nitrogen in the effluent depended on the filtration velocity gradient.

As a consequence, an increased filtration velocity gradient influenced the nitrification and assimilation ratio increasing the nitrification weight from ratio

0.4:1.0 for gradient 51.9 s^{-1} to 1.4:1.0 for gradient 77.9 s^{-1} and 2.0:1,0 for gradient 103.8 s^{-1}.

The higher filtration velocity gradient caused the creation of a fine biofilm resilient to shear stress. The increased contribution of nitrification at higher velocity gradients confirmed the earlier theory of the higher shear stress resilience of *nitrifying* bacteria compared to *heterotrophic* bacteria. Therefore, shear stress resilience was a decisive factor in the different mechanisms of nitrogen removal. The shear stress resilience influenced also the periodical shearing of the biofilm and contamination of the effluent with organic nitrogen.

REFERENCES

Błażejewski M., Pruss A. 2000. Some problems of manganese removal in case of ammonium contents. Municipal and Rural Water Supply and Water Quality IV International Conference, Kraków, Poland, pp. 315–325

Bouwer Edward J., Crowe Patricia B. 1988. Biological Processes in Drinking Water Treatment // J. AWWA vol. 8

Bray R., Olańczuk-Neyman K. 2003. Treatment of groundwater containing high amounts of manganese and ammonia nitrogen using activated filtration beds. Environmental engineering Studies, Polish Research on the Way to the EU, Kluwer Academic/Plenum Publisher, N.York, USA, pp. 87–100

Bray R., Olańczuk-Neyman K. 2001. The influence of changes in groundwater composition on the efficiency of manganese and ammonia nitrogen removal on natural quartz sand filtering beds. Water Science and Technology: Water Supply Vol. 1, no. 2. pp. 91–98

Ebeling J.M., Timmons M.B., Bisogni J.J 2006. Engineering analysis of the stoichiometry of photoautotrophic, autotrophic, and heterotrophic removal of ammonia-nitrogen in aquaculture systems. Aquaculture vol. 257, pp. 346–358

Larrea L., Abad A., Gayarre J. 2004. Improving nitrogen removal in predenitrification-nitrification biofilters. Water Science & Technology. Vol. 48, no 11. pp 419–428, IWA Publishing

Lipponen M.T.T., Pertti J., Martikainen P.J., Ritva E., Vasara R.E., Servomaa K., Outi Zachevsa O., Kontroa M.H. 2004. Occurrence of nitrifiers and diversity of ammonia-oxidizing bacteria in developing drinking water biofilms. Water Research. Vol. 38, pp. 4424–4434

Liu Y., Capdeville B. 1996. Specific activity of nitrifying biofilm in water nitrification process. Wat. Res. Vol. 30, no. 7, pp. 1645–1650

Lazarova V., Capdeville B., Nikolov L. 1992. Biofilm performance of fluidized bed biofilm reactor for drinking water denitrification. Wat. Sci. Technol. Vol. 26 (3/4), pp. 555–566,

Olańczuk-Neyman K., Bray R. 1998. Biological processes in the manganese and ammonia nitrogen removal from groundwater, Municipal and Rural Water Supply and Water Quality III International Conference, Poznań, Poland

Olańczuk-Neyman K., Bray R. 2000. The role of physico-chemical and biological processes in manganese and ammonia removal from groundwater. Polish Journal of Env. Studies. Vol. 9, no 2, pp. 91–96

Pruss A. 2006. Heterotrophic bacteria during ammonia removal from water. VII-th International Conference and Water Quality, Poznań-Zakopane Poland

Stief P., Schramm A., Altmann D., Dirk de Beer. 2003. Temporal variation of nitrification rates in experimental freshwater sediments enriched with ammonia or nitrite. Microbiology Ecology. Vol. 46, pp. 63–71

Summerfelt S., Sharrer M.J. 2004. Design implication of carbon dioxide production within biofilters contained in recirculating salmonid culture systems. Aquacultural Engineering. Vol. 32, pp. 171–182

Vandenabeele J., De Beer D., Germinpre R, Van de Sande R., Verstraete W. 1995. Influence of nitrate on manganese removing microbial consortia from sand filters. //Wat. Res. Vol. 29, no 2, pp. 579–587

Environmental Engineering – Pawłowski, Dudzińska & Pawłowski (eds)
© 2007 Taylor & Francis Group, London, ISBN13 978-0-415-40818-9

Dimensioning of non-conventional storm overflows with the new method of throttling the outflow to the treatment plant

Andrzej Kotowski, Jerzy Wartalski & Patryk Wójtowicz
Wrocław University of Technology, Wrocław, Poland

ABSTRACT: A method for hydraulic dimensioning of a non-conventional storm overflow with a throttling system, which reduces the frequency and upgrades the quality of discharge from a combined sewer to the recipient during heavy rain, is presented. On the basis of experimental data, a mathematical model is proposed to describe the performance of a side weir with a high-elevated overflow edge and equalizing unit located behind the weir chamber. The paper presents the results of model studies, including the procedure, as well as an example, of hydraulic dimensioning of storm overflows with flow stilling chambers before and after the weir chamber and with high-elevated overflow edges. A new method of throttling the outflow to the wastewater treatment plant is proposed.

Keywords: Sewer system, Combined sewage, Modelling of flow, Overflow, Dimensioning.

1 INTRODUCTION

Volumetric separators of rainfall sewage (generally referred to as storm overflows) are used in combined sewage systems (Oliveto et al. 2004) mainly to protect a wastewater treatment plant against hydraulic overloading (and, consequently, against reduced treatment efficiency) during torrential rains. Another benefit of using these separators is that they allow the interceptor size to be reduced. When used in semi-separate and separate sewage systems, storm overflows are to discharge a certain portion of rainfall sewage to the recipient streams or directly to the environment. At maximal rate of sewage inflow (Q_{in}) the object of the weir is to split this discharge, in assumed proportions, into two streams: one (denoted by Q) entering, directly or indirectly the recipient, and the other (referred to as Q_o) passing to the wastewater treatment plant.

In hydraulic terms, the storm overflows, functioning as volumetric separators of rainfall sewage can be divided into two types (Kotowski 1998):

- those with a low weir edge (with no devices that throttle the outflow of the sewage to the treatment plant), located at a height equal to the normal depth of flow (H_n) in the inlet channel before the weir, at limit rate of flow (Q_{lim}), and
- those with a high weir edge (with throttling devices), enabling the use of sewer system retention before the overflow, both at limit (Q_{lim}) and maximal (Q_o) rates of flow, as well as the flattening of the hydrogram of sewage inflow to the overflow.

Because of the substantial inertia of the flowing sewage, the side weirs with a low overflow edge must be very long (their length not uncommonly reaching a dozen or so meters) (May et al. 2003). Furthermore, when the rate of sewage flow into the interceptor increases abruptly, the runoff to the treatment plant increases in an uncontrollable way. Side weirs with a low overflow edge and controlled (throttled) outflow rarely suffer from these shortcomings. The adopted conditions of weir operation can be maintained via the regulating units that are in use now, i.e. throttling pipes of appropriate length (l_r) and diameter (d_r), gates with adjustable openings, or hydrodynamic regulators of various type, with properly selected flow characteristics. The application of throttled sewage outflow facilitates the use of sewer system retention and reduces the frequency of overflow operation throughout the year, even at the limit rate of flow (Q_{lim}). Throttling the outflow (Q_o) to the treatment plant at maximal swollen inflow of the sewage ($Q_{in\ sw}$) to the unit upgrades the hydraulic efficiency of the side weir and thus enables the length of the overflow to be shortened, in most instances to several meters.

The paper presents examples of model test results, the procedure for (and an example of) hydraulic dimensioning of non-conventional storm overflows with high-weir edges (both single and double-sided) and stilling chambers after the side weir chamber. A new method of throttling the outflow to the treatment plant is proposed.

2 MATERIALS AND METHODS

2.1 *Criteria for the protection of the recipient against pollution*

When designing the volumetric separators of sewage flow, the storm weirs, it is necessary to include quantitative or qualitative criteria for protecting the recipient against pollution. They are expressed in terms of admissible annual frequency of storm flow discharge, or length of duration, admissible volume, or admissible concentration and load of pollutants, entering the recipient water body together with the sewage (ATV A128 1992). In Poland, the obligatory criterion in this regard (Order of the Ministry of Environment 2004) is the average annual number of storm sewage discharges derived from at least 10-year observations. As for the sewage dumps from the storm weirs located along combined sewer system into inland or coastal waters, the number of dumps must not exceed 10. As far as storm sewage dumps alone are concerned (from weirs along the storm sewage system), which enter lakes or other water reservoirs with constant inflow or outflow of surface water, the number of dumps must not exceed 5. When the design data required for the application of the foregoing criteria are lacking, the sewage from the combined system overflows can be discharged only if:

- the sewage system is discharging to a treatment plant of a Population Equivalent (PE) below 100 000, and
- the rate of inflow to the weir is at least 3 times as high as the average flow rate of the sewage under dry weather conditions, specified for the yearly average inflow to the treatment plant.

If direct analysis has shown a deterioration of water quality in the recipient contributed by the dumps from combined sewer systems, the maximal number of the dumps can be reduced.

It is conventional to express the quality criteria for the protection of the recipient water body against pollution in terms of the admissible pollutant concentration in the sewage discharged to the recipient. In Poland, rainfall and melting snow sewage outflow from industrial areas, city centres, roads and car-parks (area greater than 0.1 ha) (Order of the Ministry of Environment 2004) must be purified at a rate of at least 15 dm^3/s per 1 ha of impervious surface until the maximal content of total solids is 100 mg/dm^3 and the maximum concentration of petroleum derivatives amounts to 15 mg/dm^3. As for the outflow from an impervious surface of a fuel storage and distribution area, the required purification must be carried out at a flow rate greater than the one produced by a rainfall of a frequency of once a year and duration 15 minutes but not lower than that of the discharge induced by rainfall of a flow rate of 77 $dm^3/s{\cdot}ha$.

The criteria described above determine the choice of the limit flow rate (Q_{lim}), which influences the beginning of the dump of combined sewage into the recipient:

$$Q_{lim} = Q_s + Q_{in\,lim} + \sum Q_o \tag{1}$$

where Q_s is dry-weather rainfall sewage flow rate, dm^3/s; $Q_{in\,lim}$ stands for the limit rate of rainfall sewage flow from the direct drainage area, dm^3/s; and $\sum Q_o$ denotes sum of runoffs from upstream overflows, dm^3/s.

The rate of limit rainfall sewage flow ($Q_{in\,lim}$) for the combined sewer system can be determined by two methods. Using the definition of the initial dilution coefficient (n_{id}), we obtain:

$$Q_{in\,lim} = n_{id} Q_s \tag{2}$$

where $n_{id} \in <1, 10>$; for small and protected recipients: $n_{id} \geq 3$.

To protect the recipient water body it is essential that the pollutants from the drainage area are discharged to the treatment plant (during a dry weather period), and washed away from the reduced surface (ψA_{dr}) by runoff of flush intensity (q_{fi}):

$$Q_{in\,lim} = q_{fi} \psi A_{dr} \tag{2a}$$

In Poland, it is assumed (minimum) that $q_{fi} = 6\,dm^3/\,s{\cdot}ha$; in Germany it is anticipated that $q_{fi} = 7.5\text{–}15\,dm^3/s{\cdot}ha$, depending on the duration of flow in the interceptor ($q_{fi} = 7.5\,dm^3/s{\cdot}ha$ at a duration of flow $t_p \geq 120$ minutes and $q_{fi} = 15 \cdot 120/(t_p+120)$, when $t_p < 120$ minutes) (ATV A128 1992).

2.2 *Methods for the regulation of sewer structures*

Depending on the throttling method, the available structures for controlling the flow rate in sewer structures, like storm overflows, settling or retention tanks, can be divided into two groups:

- linear throttling devices, e.g. rectilinear sections of pipelines with an adequate diameter, length and wall roughness, referred to as throttling pipes,
- local throttling devices such as orifices, reducing pipes, gates and gate valves, hydrodynamic flow regulators, etc.

The proposed new flow rate control method involves pressure flow of the sewage through a properly selected system of elbows or bends made of plastics (e.g. PVC-U, or PP). In principle, it applies to the group of local throttling devices, which display the features of advanced structures (without reducing the internal area of the channel), but are much cheaper than the conventional (steel) throttling pipe. A major drawback of the throttling pipe is its considerable

Table 1. Loss coefficient $\zeta(n)$ of a throttling system built from plastic (n) elbows or (n) segmental bonds (βi)

System No	Description, scheme $\beta_i \cdot \beta_{sum}$	A	B	C	D
Version		R/d = 4.25	R/d = 2.25	R/d = 1.75	R/d = 1.00
		– system built from bends $\beta = 15°$	– system built from bends $\beta = 30°$	– system built from bends $\beta = 45°$	– system built from elbows $\beta = 90°$
		$\zeta/l_{ax}/(l_{piping})$	$\zeta/\ l_{ax}/\ (l_{piping})$	$\zeta/\ l_{ax}/\ (l_{piping})$	$\zeta/\ l_{ax}/\ (l_{piping})$
1	2	3	4	5	6
1	4 bends or elbows 90° (360°)	0.90 26.7d (17.0d)	1.0 14.1d (9.0d)	1.5 11.0d (7.0 d)	4.1 6.3d (4.0d)
2	4 bends 60° (240°)	0.65 17.8d (14.7d)	0.83 9.4d (7.8d)	–	–
3	4 bends 45° (180°)	0.47 13.3d (12.0d)	–	0.91 5.5d (4.9d)	–
4	4 bends 30° (120°)	0.30 8.9d (8.5d)	0.44 4.7d (4.5d)	–	–
5	8 bends or elbows 90° (720°)	1.9 53.4d (34.0d)	2.0 28.3d (18.0d)	3.0 22.0d (14.0d)	8.0 12.6d (8.0d)
6	8 bends 60° (480°)	1.4 35.6d (29.4d)	1.6 18.8d (15.6d)	–	–
7	8 bends 45° (360°)	1.0 26.7d (24.0d)	–	2.2 11.0d (9.9d)	–
8	8 bends 30° (240°)	0.65 17.8d (17.0d)	0.82 9.4d (9.0d)	–	–
9	12 bends 60° (720°)	2.1 53.4d (44.2d)	2.3 28.3d (23.4d)	–	–
10	12 bends 45° (540°)	1.5 40.1d (36.1d)	–	3.2 16.5d (14.8d)	–

length, which often reaches several dozen meters. Equivalent to the resistance of a throttling pipe of such considerable length, the hydraulic resistance of an appropriately selected throttling system occurs within several meters of axial length (or piping length), when the system consists of (n) elbows or within approximately a dozen meters when the system is made of (n) bends. The selected throttling system consists of sinusoidal waves made of elbows with relative radius of curvature $R/d \leq 1$ in the case of slightly polluted liquids. Bends with a curvature radius $R/d > 1$ are used for highly polluted liquids (raw sewage), both (elbows and bends) with the same diameter (d) as that of the throttling pipe. This becomes evident if we compare the ("local") head loss of such systems and the corresponding frictional head loss in the conventional throttling pipe:

$$\zeta_{(n)} \frac{\upsilon^2}{2g} = \lambda(\mathrm{Re}) \frac{l_e}{d} \frac{\upsilon^2}{2g}. \tag{3}$$

Hence, the equivalent (substitute) length of a rectilinear throttling pipe becomes

$$l_e = \frac{\zeta_{(n)}}{\lambda(\mathrm{Re})} d. \tag{4}$$

Since the local loss coefficient (Tab. 1) of the systems built from (n) elbows or (n) bends ($\zeta_{(n)}$) is many times higher than the friction factor of the rectilinear

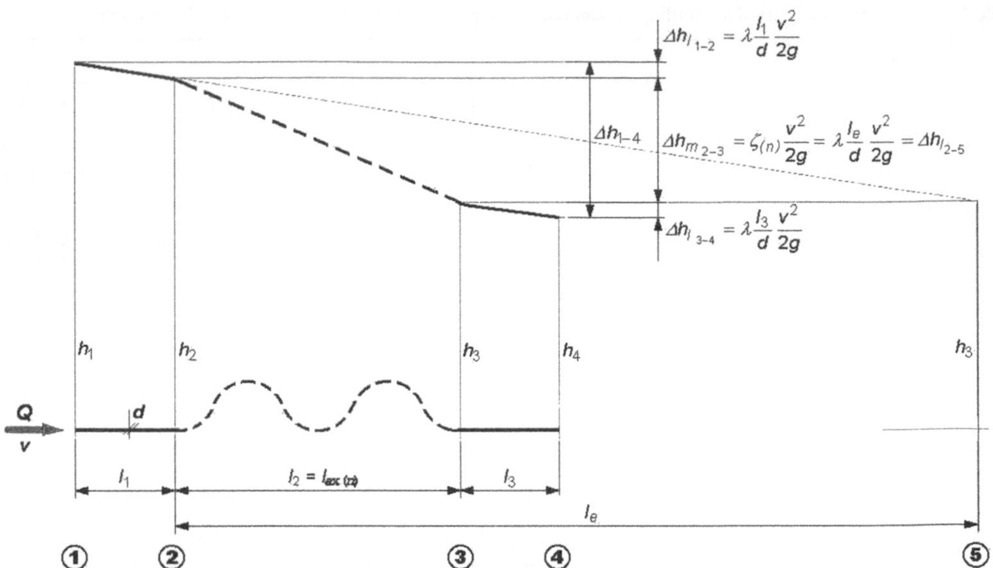

Figure 1. Operation idea for a throttling system consisting of sinusoidal waves built from (n) segmental bends (hi stands for energy line height in sections 1 to 5).

throttling pipe (λ(Re)), the equivalent length (l_e) of the throttling pipe must be many times the axial length ($l_{ax(n)}$) of the system made of (n) elbows or (n) bends ($l_e \gg l_{ax(n)}$).

Figure 1 depicts the operation idea for a system consisting of (n) segmental bends (axial length $l_2 = l_{ax(n)}$)– in the form of sinusoidal waves, built in a pipeline of diameter d, between straight connection pipes of length l_1 and l_3 – before and after the throttling system.

The example of a throttling system in Figure 1 consists of two sinusoidal waves built from eight alternatingly connected (segmental) bends n = 8 with central angles $\beta_I = \beta = 45°$ (e.g. made of PVC-U) and a radius of curvature R = 1.75d, (hence, of a length equal to the axial length of the system $l_{ax\,(8\times 45°)} = (n\beta_i^{\circ}/360°)2\pi R = 11d$) and yields a "local" loss $\zeta_{(8\times 45°)} = 2.2$ (Tab. 1: system 7, version C). With a Reynolds number of e.g. Re = 1 000 000, we obtain a friction coefficient λ(Re) = 0.012 (PVC-U pipeline (Kotowski 2005)) and an equivalent length of the rectilinear throttling pipe $l_e = (2.2/0{,}012)$ d = 183.3d. This is approximately 17 times the axial length ($l_{ax} = 11d$) of the throttling system depicted in Figure 1 and about 19 times the (real) length of the piping: $l_{piping} = n \sin\beta_i^{\circ}\, R = 9.8d$ (Tab. 1).

3 RESULTS: PROCEDURE AND EXAMPLE

3.1 *Initial assumptions*

The available methods for hydraulic dimensioning of side weirs with throttled outflow suffer from simplifications, such as the omission of variations in the liquid level ordinate along the length of the overflow and the assumption of a constant value for the weir discharge coefficient (depending on weir crest shape), by analogy with sharp-crested, non-submerged rectangular weirs. Furthermore, in the light of technological research (Saul et al. 1981, Luyckx et al. 1999) the widespread standardization of such structures – without flow stilling chambers after the side weir – is not a recommendable trend since the bottom wastes enter the recipient via overflow. It would be advisable to establish a new construction standard for the weirs under consideration and carry out relevant model tests (Kotowski 2001). It has been assumed that the shape of the cross-section for the overflow is identical to that of the stilling chamber – up to the level of the intersection axis. Above this level, the cross-sections are rectangular in shape (above D/2 the channel is circular, and above $2H_c/3$ the channel is egg-shaped, etc.), and have a width b = D. The stilling chamber length after the side weir is $l_s = 2b = 2D$. The computational scheme for the weir is shown in Figure 2.

The computational method proposed involved the following procedure:

- at the limiting rate of inflow (Q_{lim}) to the weir, an appropriate height of the overflow crest (p) is adopted, taking into account the hydraulic and operating conditions for the occurrence of (swollen up) subcritical flow in the vicinity of the weir, and thereafter the throttling element (with an appropriate value of the loss coefficient $\zeta_{(n)}$ consisting of (n)

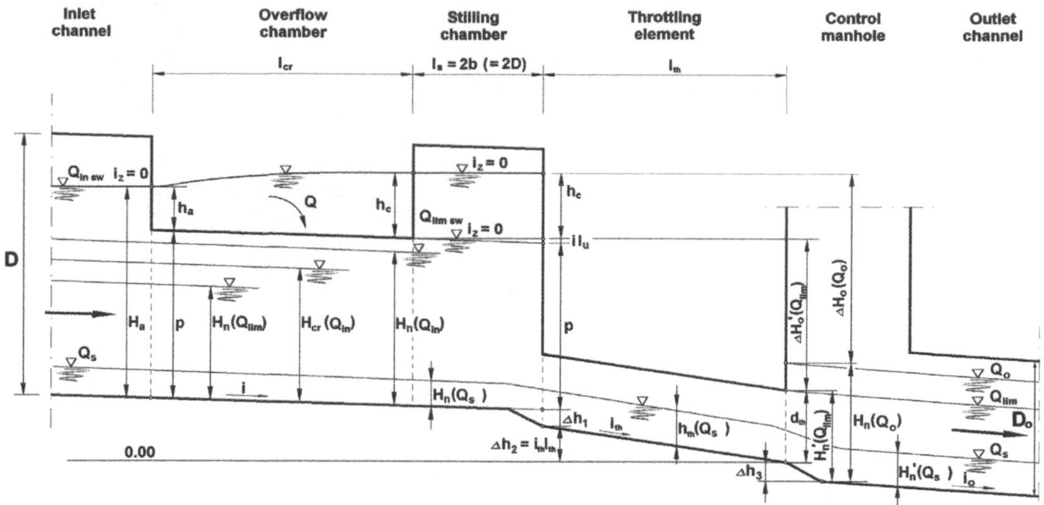

Figure 2. Computational scheme for a non-conventional side weir with throttled sewage outflow to the treatment plant.

segmental bends, connected in series, of an axial length $l_{ax} = l_{th}$) is selected;

- at the maximal rate of flow (Q_{in}) the desired flow division at the overflow is specified: for the assumed rate of outflow to the treatment plant $Q_o \in <1.1\ Q_{lim}; 1.2Q_{lim}>$, the hydraulic losses of the previously selected throttling system and the height of the liquid layer above the weir edge (h_c) at the overflow end are calculated – according to ΔH_o;
- for the outflow to recipient $Q = Q_{in} - Q_o$ and calculated height h_c, the necessary length of weir crest (l_{cr}) is iteratively determined, by discrete change in the height of liquid layer above the weir edge h_a – at the beginning of the weir.

3.2 *Input parameters of the weir*

Initial data:

- drainage area: $A_{dr} = 100$ ha,
- mean run-off coefficient: $\psi = 0.3$,
- initial dilution coefficient: $n_{id} = 3$,
- flush intensity: $q_{fi} = 15\ dm^3/s \cdot ha$,
- flow rate of municipal sewage: $Q_s = 150\ dm^3/s$,
- flow rate of rainfall sewage: $Q_{in\,max} = 2500\ dm^3/s$,
- terrain slope at the overflow location: $i_t = 1.0$ ‰.

Limiting rate of combined sewage inflow in terms of eq. (1) and eq. (2):

$$Q_{lim} = Q_s + n_{id} Q_s = 0.150 + 3 \cdot 0.150 = 0.600\ m^3/s$$

or by virtue of eq. (1) and eq. (2a):

$$Q_{lim} = Q_s + q_{fi} \psi A_{dr} = 0.150 + 0.015 \cdot 0.3 \cdot 100 = 0.600\ m^3/s$$

Effective (maximal) rate of inflow to the weir:

$$Q_r = Q_s + Q_{in\,max} = 0.150 + 2.500 = 2.650\ m^3/s$$

Inlet channel. For the effective rate of inflow $Q_{in} = 2.65\ m^3/s$ and the assumed bottom slope i = 1‰, a concrete channel of a diameter D = 1.80 m is selected. Calculations were carried out using the nomographs for the Manning equation (circular channels) at $n = 0.013\ s/m^{1/3}$. But use can also be made of plastic channels; then the nomographs for the Darcy-Weisbach, Colebrook-White and Bretting equations, at k = 0.4 mm, are applied.

Standard depth of flow:

$H_n(Q_s) = 0.25$ m ($\upsilon = 0.73$ m/s),
$H_n(Q_{lim}) = 0.49$ m ($\upsilon = 1.00$ m/s),
$H_n(Q_{in}) = 1.12$ m ($\upsilon = 1.60$ m/s).

Critical depth of flow: $H_{cr}(Q_{in}) = 0.77$ m ($\upsilon = 2.65$ m/s),

The weir crest. The height of the weir crest (p) should be assumed by including the following hydraulic conditions (Kotowski 1998):

$$p > H_n(Q_{lim}), \quad (5)$$

$$p > H_{cr}(Q_{in}), \quad (6)$$

$$p + h_a > H_n(Q_{in}), \quad (7)$$

In addition, consideration should be given to the following operating conditions:

$$p > 0.6D \text{ (or } H_c), \quad (8)$$

$\upsilon_{min}(Q_{lim\,sw}) \geq 0.30$ m/s, (9)

The initial assumption was: $p = 1.30$ m ($p = 0.72D$), and the condition (9) of minimal flow velocity in the overflow chamber was checked at the limiting flow rate, swelled to the height (p) of the weir crest:

$$\upsilon_{min}(Q_{limsw}) = Q_{lim} / \left[\frac{\pi D^2}{8} + \left(p - \frac{D}{2}\right)D\right] =$$

$$= 0.600 / \left[\frac{3.14 \cdot 1.8^2}{8} + (1.3 - 0.9)1.8\right] = 0.30 \text{ m/s}$$

If condition (9) is not satisfied, then the weir crest height p must be lowered.

3.3 *Calculated parameters of the weir at the rate of flow Q_s*

Throttling element. At the flow rate ($Q_s = 0.150$ m^3/s) of the municipal sewage, it is necessary to select the diameter of the throttling element (d_{th}) (consisting of n bends β_i, the axial length being $l_{ax(n\beta)} = l_{th}$) and assume the depth of flow $h_{th}(Q_s)$, taking into account the following restrictions:

$d_{th\,min} = 0.20$ m,

$h_{th}/d_{th} \leq 0.6$,

$\upsilon_s(Q_s) \geq 1.0$ m/s

If we assume that $d_{th} = 0.60$ m and $h_{th} = 0.30$ m, then the flow area of Q_s becomes $\pi d_{th}^2/8 = 3.14 \cdot 0.6^2/8 = 0{,}141$ m^2, and the sewage flow velocity in the throttling system: $\upsilon_s(Q_s) = 0.15/0.141 = 1.06$ m/s.

The requirement of water level compensation (at the inlet to the throttling system) at flow Q_s often necessitates the lowering of the throttling element bottom (at its beginning), in relation to the stilling chamber bottom (at its end), by the value of Δh_1 (Fig. 2):

$$\Delta h_1 = h_{th}(Q_s) - H_n(Q_s) = 0.30 - 0.25 = 0.05 \text{ m}$$

3.4 *Calculated parameters of the weir at flow rate Q_{lim}*

Choice of the throttling system. Using the Bernoulli equation, derived for the sections immediately before the inlet and just after the outlet of the throttling system (at flow rate Q_{lim}), we obtain the following hydraulic loss equation (Fig. 2):

$$il_s + p + \Delta h_1 + \Delta h_2 - d_{th} \equiv \Delta H_o^I(Q_{im}) = = \zeta_{in}\frac{\upsilon_{lim}^2}{2g} + \zeta_{(n\beta)}\frac{\upsilon_{lim}^2}{2g} + \zeta_{out}\frac{\upsilon_{lim}^2}{2g} \quad (10)$$

where υ_{lim} is average velocity of flow at Q_{lim}: $\upsilon_{lim} = 4Q_{lim}/(\pi\, d_{th}^2)$, m/s; ζ_{in} stands for the local loss coefficient at the inlet of the throttling system (ATV A111 1994): $\zeta_{in} = 0.45$; $\zeta_{(n\beta)}$ denotes the hydraulic ("local") loss coefficient for the throttling system consisting of (n) segmental bends (connected in series) with central angles β_i and axial length $l_o = l_{th}$ (Tab. 1), ζ_{out} is local loss coefficient at the outlet of the throttling system, assumed equal to the Coriolis coefficient: $\alpha = 1 + 2.93\lambda - 1.55\,\lambda^{3/2}$; for $\lambda \in <0.0017;0.031>$– $\alpha \in <1.05;1.08>$. Assumption for plastics (Kotowski et al., 2004): $\zeta_{out} = \alpha = 1.05$.

The required value of the loss coefficient $\zeta_{(n\beta)}$ should be calculated from the rearranged relation of (10), neglecting the component Δh_2 in the first approximation. Thus we have:

$$\zeta_{(n\beta)} > \frac{il_s + p + \Delta h_1 - d_{th} - (\zeta_{in} + \zeta_{out})\frac{8Q_{lim}^2}{g\pi^2 d_{th}^4}}{\frac{8Q_{lim}^2}{g\pi^2 d_{th}^4}} =$$

$$= \frac{0.001 \cdot 3.6 + 1.3 + 0.05 - 0.6 - 1.5\frac{8 \cdot 0.6^2}{9.81 \cdot 3.14^2 \cdot 0.6^4}}{\frac{8 \cdot 0.6^2}{9.81 \cdot 3.14^2 \cdot 0.6^4}} = 1.78$$

The throttling system characterized by the loss coefficient $\zeta_{(8\times45°)} = 2.2 > 1.78$ was selected from Table 1 (system no. 7, according to version C). The system consists of $n = 8$ bends connected in series, having central angles $\beta_i = 45°$ and a radius of curvature $R = 1.75d$, so the axial length is $l_{ax(8\times45°)} = (n\beta_i^o/360°)\,2\pi R = 11.0d = 6.6$ m (the piping length being: $l_{piping(8\times45°)} = n \cdot \sin\beta_i^o\, R = 9.9d = 6.0$ m).

For the above mentioned parameters of the throttling system, the difference in the invert height between the inlet and outlet of the system ($\Delta h_2 = i_{th} l_{th}$, Fig. 2), which is equal to the hydraulic head loss at the flow rate Q_s, can be calculated from the following equation:

$$\Delta h_2(Q_s) = \zeta_{(8x45°)}\upsilon_s^2/2g = 2.2 \cdot 1.06^2/19.62 = 0.13 \text{ m}.$$

Hence the invert slope of the throttling element (along the axial length) equals $i_{th} = \Delta h_2/l_{o(8\times45°)} = 0.13/6.6 = 0.020$ (the real slope along the piping length being: $0.13/6.0 = 0.022$).

The correction of the weir crest height (p) – from the rearrangement of equation (10) – includes the previously neglected component Δh_2 and takes the form:

$$p_{(1)} = (\zeta_{in} + \zeta_{(n\beta)} + \zeta_{out})\frac{8Q_{lim}^2}{g\pi^2 d_{th}^4} - il_s - \Delta h_1 - \Delta h_2 + d_{th} \quad (10a)$$

$$p_{(1)} = (0.45 + 2.2 + 1.05)\frac{8 \cdot 0.6^2}{9.81 \cdot 3.14^2 \cdot 0.6^4} \cdot$$
$$- 0.001 \cdot 3.6 - 0.05 - 0.13 + 0.6 = 1.27 \text{ m}$$

The conditions (5)–(9) must be verified for the new value of $p_{(1)}$ (i.e. $\upsilon_{min(1)}(Q_{lim\,sw}) = 0.31$ m/s). If the conditions are not met, we have to select (from Tab. 1) a throttling system with another value of the loss coefficient $\zeta_{(n\beta)}$, and then calculate a new value of the head $\Delta h_2(Q_s)$ and the height of the weir crest $p_{(i)}$ using eq. (10a).

Outlet channel (to the sewage treatment plant). For the flow rate $Q_{lim} = 0.6\ m^3/s$ and selected invert slope $i_o = 1‰$, the diameter of the outlet channel $D_o = 1.0$ m and the normal depth of flow $H_n^I(Q_{lim}) = 0.67$ m ($H_n^I(Q_s) = 0.30$ m) were selected. Mostly, in case of

$$H_n^I(Q_{lim}) > d_{th} \tag{11}$$

it is necessary to lower the inlet channel with respect to the assumed reference datum level by the value of (Fig. 2):

$$\Delta h_3 = H_n^I(Q_{lim}) - d_{th} = 0.67 - 0.60 = 0.07 \text{ m}$$

If $H_n^I\ (Q_{lim}) < d_{th}$, we have to correct the parameter: i_o or D_o.

3.5 *Calculated parameters of the weir at the flow rate Q_{in}*

Height of the liquid layer above the weir edge (h_c) at the end of the overflow. At maximal inflow to the weir $Q_{in} = Q_s + Q_{in\,max} = 2.65\ m^3/s$ we should assume a value for the rate of outflow Q_o through the throttling element which is by 10 to 20% higher than that of Q_{lim} (ATV A111 1994):

$$Q_o \in <1.1Q_{lim}; 1.2Q_{lim}> \tag{12}$$

For the assumed value of Q_o it is necessary to determine the normal depth of flow $H_n(Q_o)$ in the outlet channel, and then calculate the hydraulic head loss $\Delta H_o(Q_o)$ in the throttling system (by virtue of the modified equation (10)):

$$\Delta H_o(Q_o) = (\zeta_{in} + \zeta_{(n\beta)} + \zeta_{out})\frac{8Q_o^2}{g\pi^2 d_{th}^4} =$$
$$= (0.45 + 2.2 + 1.05)\frac{8 \cdot 0.69^2}{9.81 \cdot 3.14^2 \cdot 0.6^4} = 1.12 \text{ m}$$

assuming that $Q_o = 1.15\ Q_{lim} = 0.690\ m^3/s$. The normal depth of flow in the outlet channel is $H_n(Q_o) = 0.75$ m ($\upsilon_o = 1.05$ m/s).

The height of the liquid layer above the weir edge at the end of the weir should be calculated in terms of the following equation:

$$h_c = H_n(Q_o) + \Delta H_o(Q_o) - (i \cdot l_s + p + \Delta h_1 + \Delta h_2 + \Delta h_3); \tag{13}$$

hence:

$$h_c = 0.75 + 1.12 - (0.001 \cdot 3.6 + 1.27 + 0.05 + 0.13 + 0.07 = 0.35 \text{m})$$

Length of the weir crest (l_{cr}). The length of the weir crest (single-sided) must meet the condition $l_{cr} \leq 4D$. If $l_{cr} > 4D$ there is a need to apply a double-sided weir. The required length of the weir crest should be calculated iteratively, either by virtue of the dimensionless form of the differential equation of motion (Kotowski 2001), or in terms of the following equations:

$$l_{cr} = \frac{Q}{(2/3)\mu\sqrt{2g}\ h_m^{3/2}}, \tag{14}$$

where:

Q is rate of flow through the side weir: $Q = Q_{in} - Q_o$

μ is side weir discharge coefficient ($\mu \in <0.50; 0.60>$):

$$\mu = 0.64 - 0.052q_r + 0.0088L_0 + 0.035W_0 - 0.075Fr_0 - 0.065K_0 \tag{15}$$

q_r is coefficient of flow division in the weir ($q_r \in <0.5;1>$): $q_r = Q/Q_{in}$,

L_0 is relative length of the weir crest ($L_0 \in <1.8;5.1>$): $L_0 = l_{cr}/H_a$, where $H_a = p + h_a$,

W_0 is relative height of the liquid layer above the overflow edge in the initial part of the weir ($W_0 \in <0.13;0.35>$):

$$W_0 = h_a / H_a,$$

Fr_0 is the Froude number in the initial cross-section of the overflow chamber ($Fr_0 \in <0.1;0.5>$), where $Fr_0^2 = Q_{in}^2/(A_0^2 g H_a)$ by definition (Kotowski, 2001),

K_0 is shape factor of the channel bottom in the initial part of the overflow chamber ($K_0 \in <1;1.2>$):

$$K_0 = bH_a / A_0,$$

h_m is effective (weighted average) height of the liquid layer above the weir edge:

$$h_m = h_a + \frac{3}{5}(h_c - h_a) \tag{16}$$

and

$$h_c = h_a + 0.9\frac{\alpha_{in}\upsilon_a^2}{2g} \tag{17}$$

υ_a is velocity of flow in the initial part of the overflow chamber:

$$\upsilon_a = Q_{in} / A_0 \tag{18}$$

A_0 is cross-sectional area of flow in the initial part of the overflow chamber,

$$A_0(Q_{in\,sw}) = [\pi D^2/8 + (p + h_a - D/2)D] \tag{19}$$

α_{in} is kinetic energy (Coriolis) coefficient in the inlet channel before the weir: $\alpha_{in} = 1.15$ for cylindrical channels ($\alpha_{in} = 1.20$ for prismatic channels).

In the side weir, the sewage swells along the weir length. It is therefore necessary to assume such a value for the height of the liquid layer above the weir edge at the beginning of the overflow (h_a) that will be by several centimetres lower than the value of h_c ($1.05 \leq h_c/h_a \leq 1.4$). In the first approximation, we assumed that $h_{a(1)} = 0.27$ m, hence the cross-sectional area of flow in the initial part of the overflow chamber (by virtue of eq. (19)) takes the form:

$$A_{0(1)}(Q_{in\,sw}) = [3.14 \cdot 1.8^2/8 + (1.27 + 0.27 - 1.8/2) \cdot 1.8] = 2.42\ \text{m}^2$$

and the flow velocity in the initial part of the overflow chamber (in terms of eq. (18)) becomes:

$$\upsilon_{a(1)} = 2.65/2.42 = 1.10\ \text{m/s}.$$

The height of the liquid layer above the edge of the weir at its end (by virtue of eq. (17)) as:

$$h_{c(1)} = 0.27 + 0.9(1.15 \cdot 1.1^2/(2 \cdot 9.81)) = 0.33\ \text{m}$$

From the calculated hydraulic head loss in the adopted throttling system (at the flow rate $Q_o = 0.690\ \text{m}^3/\text{s}$) it follows that $h_c = 0.35$ m is higher than $h_{c(1)} = 0.33$ m from the first approximation (the difference being greater than 1 cm). Thus, in the second approximation it is necessary to adopt a new value for the height of the liquid above edge of the weir in its initial part (h_a), e.g. $h_{a(2)} = 0.29$ m. Finally, in the second approximation we shall have $A_{0(2)}(Q_{in\ sw}) = 2.46\ \text{m}^2$, $\upsilon_{a(2)} = 1.08$ m/s and $h_{c(2)} = 0.35$ m.

The effective value of the height of the overflowing liquid in the side weir (by virtue of eq. (16)) can be written as

$$h_m = 0.29 + 3/5(0.35 - 0.29) = 0.33\ \text{m}$$

Hence, assuming primarily that the value of the weir discharge coefficient is $\mu = 0.60$ for the stream $Q = Q_{in} - Q_o = 2.65 - 0.69 = 1.96\ \text{m}^3/\text{s}$, the initial length of the side crest of weir (eq. (14)), becomes

$$l_{cr(1)} = 1.96/((2/3) \cdot 0.60 \cdot (2 \cdot 9.81)^{0.5} \cdot 0.33^{1.5}) = 5.84\ \text{m}$$

With the preliminarily established length of the weir crest (lcr(1)), it is possible to calculate a value of the weir discharge coefficient, after having determined the dimensionless values of the factors q_r, L_0, W_0, Fr_0 and K_0 (similarity numbers (Kotowski 1998)):

$q_r = Q/Q_{in} = 1.96/2.65 = 0.74$ (condition $q_r \in <0.5;1>$);

$L_{0(1)} = l_{cr(1)}/H_a = 5.84/1.27 + 0.29 = 3.74$ (condition $L_0 \in <1.8;5.1>$);

$W_0 = h_a/H_a = 0.29/1.56 = 0.19$ (condition $W_0 \in <0.13;0.35>$);

$Fr_0 = Q_{in}/(A_0\sqrt{gH_a}) = 2.65/(2.46\sqrt{9.81 \cdot 1.56}) = 0.28$ (condition $Fr_0 \in <0.1;0.5>$);

$K_0 = bH_a/A_0 = 1.8 \cdot 1.56/2.46 = 1.14$ (condition $K_0 \in <1.0;1.2>$);

thus, the weir discharge coefficient calculated (by virtue of eq. (15)) for $l_{cr(1)} = 5.84$ m takes the form:

$$\mu_{(1)} = 0.64 - 0.052 \cdot 0.74 + 0.0088 \cdot 3.74 + 0.035 \cdot 0.19 - 0.075 \cdot 0.28 - 0.065 \cdot 1.14 = 0.54$$

and enables the correction of side edge length of the weir (in terms of eq. (14)):

$$l_{cr(2)} = 1.96/((2/3) \cdot 0.54 \cdot (2 \cdot 9.81)^{0.5} \cdot 0.33^{1.5}) = 6.48\ \text{m}$$

After successive approximation, for $L_{0(2)} = l_{cr(2)}/H_a = 6.48/(1.27 + 0.29) = 4.15$, we obtain: $\mu_{(2)} = 0.55$ and

$$l_{cr(3)} = 1.96/((2/3) \cdot 0.55 \cdot (2 \cdot 9.81)^{0.5} \cdot 0.33^{1.5}) = 6.37\ \text{m}$$

Since the difference in the length of the weir between $l_{cr(2)}$ and $l_{cr(3)}$ (from the last approximation) is comparatively small (amounting to 10 cm), the calculations can be thought of as being completed. Hence, we adopted the single-sided weir of the side edge length $l_{cr} = 6.4$ m (the condition $l_{cr} \leq 4D$ has been satisfied $6.4 \leq 7.2$).

Double-sided weir as an alternative. If a double-sided weir is to be constructed, which is characterized by a lower hydraulic efficiency as stated by Kotowski (1998), the length of the side edges will not be half the value defined in the foregoing example. For the initial approximations, however, it can be assumed that $l_{cr(1)double} = l_{cr}/2 = 3.2$ m, and calculate the value of the dimensionless coefficient which describes the relative length of the weir $L_{0(1)} = l_{cr(1)double}/H_a = 3.2/(1.27 + 0.29) = 2.05$ (the condition $L_0 \in <1.8;5.1>$ has been fulfilled); then the value of the factor $\mu_{(1)double} = 0.53$ can be computed (using eq. (15)). The length of the side edges can be determined as follows:

$$l_{cr\,double} = \frac{Q/2}{(2/3)\mu_{double}\sqrt{2g}\;h_m^{3/2}} \qquad (20)$$

hence:

$$l_{cr(2)double} = (1.96/2)/\left((2/3)\cdot 0.53\cdot(2\cdot 9.81)^{0.5}\cdot 0.33^{1.5}\right)$$
$$= 3.30 \text{ m}$$

In the second approximation the new value of the factor $L_{0(2)} = l_{cr(2)double}/H_a = 3.30/(1.27 + 0.29) = 2.12$ is calculated, and thereafter the coefficient $\mu_{(2)double} = 0.53$. Thus the length of the weir crest becomes $l_{cr(3)double} = 3.30$ m (by virtue of eq. (20)).

Finally the double-sided weir of the side edge length $l_{cr} = 3.3$ m (and total length 6.6 m) was selected.

Adopting the condition of overflow sharp-edged operation, the width of the weir crests was assumed to be $s \leq h_m/2 = 0.15$ m. The adopted diameter of the storm sewer (outlet to the recipient) is $D_b = 1.5$ m, the bottom slope being $i_b = 1.0‰$ and the normal depth of flow accounting to $H_n(Q) = 1.05 \text{ m} < p = 1.27$ m, according to the condition of free (non-submerged) weir operation.

4 CONCLUSIONS

The new construction standard and dimensioning method for side weirs (single and double-sided) with throttled outflow to the sewage treatment plant apply to structures:

- with high-elevated overflow crests (fulfilling the condtions: $p > H_n(Q_{lim})$; $p > H_{cr}(Q_{in})$; $p > 0.6D$ (or H_c); $p + h_a > H_n(Q_{in})$),
- with crests of practical shape (meeting the sharp-edged (of width $s \leq h_m/2$) and non-submerged operation condition (in outlet storm sewer – depth of flow: $H_n(Q) < p$),
- with a cylindrical cross-section of the overflow chamber (the same as the intersection of the inlet channel) up to the level of the horizontal axis (the so-called haunch of the channel), and rectangular (of a width $b = D$) above this level,
- with stilling chambers after the overflow (of the length $l_s = 2b = 2D$),
- with throttled sewage outflow, via in-series systems of segmental bends (in the form of sinusoidal waves), having a noticeably shorter piping length (compared to the equivalent length of the throttling pipe) and limiting the outflow rate (Q_o) of the sewage to the treatment plant (at maximal inflow to the weir: Q_{in}) to the predetermined value.

The improved standard of side storm weir construction involves a new method of throttling, more effective compared to the classical throttling pipe. It also formulates principles of dimensioning, which allow the frequency of overflow dumps to be limited (owing to the retention capabilities of the channels situated above the overflow), and the quality of the storm dumps to be notably improved, thus protecting the sewage treatment plant against hydraulic overload.

REFERENCES

ATV-Arbeitsblatt A128, 1992. Richtlinien für die Bemessung und Gestaltung von Regenentlastungsanlagen in Mischwasserkanälen. Hennef.

ATV-Arbeitsblatt A111, 1994. Richtlinien für hydraulische Dimensionierung und den Leistungsnachweis von Regenwasser-Entlastungsanlagen in Abwasserkanälen und -leitungen. Hennef.

Kotowski, A., 1998. Principles of the dimensioning of a non-conventional storm overflow. Oficyna Wydawnicza Politechniki Wrocławskiej, Wrocław.

Kotowski, A., 2001. Dimensionless equation for side-channel weirs. *Archives of Hydro-Engineering and Environmental Mechanics*, Vol. 48 (1), p. 97–113.

Kotowski, A., Wójtowicz, P., 2004. Analysis of research methods of isothermal liquid flows in plastic pipes. *Environment Protection Engineering*, Vol. 30 (3), p. 71–80.

Kotowski, A., 2005. Durchflusswiderstaende in Kniekruemmer- und Segmentkruemmer- Anordnungen aus Kunststoff. *Gas Wasserfach Wasser/Abwasser* Jg. 146, No. 2, pp. 134–140.

Luyckx, G., Vaes, G., Berlamont, J. 1999. Experimental investigation on the efficiency of a high side weir overflow. *Water Science Technology*, Vol. 39, No. 2, pp. 61–68.

May, R.W.P., Bromwich, B.C., Gasowski, Y., Rickard, C.E.. 2003. *Hydraulic design of side weirs*. Thomas Telford Ltd.

Oliveto, G., Fiorentino, M. 2004. Design and hydraulics of a combined storm overflow structure. J. Irrig. and Drain. Engrg. Volume 130, pp. 331–334.

Rozporządzenie Ministra Ochrony Środowiska Ministerstwo Ochrony Środowiska z dnia 8 lipca 2004 w sprawie warunków, jakie należy spełnić przy wprowadzaniu ścieków do wód lub do ziemi, oraz w sprawie substancji szczególnie szkodliwych dla środowiska wodnego. Dz. U. RP No 168, Position 1763, Warszawa. (Order of Ministry of Environment)

Saul, A.J., Delo, E.A., 1981. Performance of a high-side-weir storm-sewage chamber incorporating storage. *2nd Int. Conf. Urbana Storm Drainage. Urbana*, Illinois, USA, p. 110–119.

Environmental Engineering – Pawłowski, Dudzińska & Pawłowski (eds)
 ISBN13 978-0-415-40818-9

Separation of sewage flow in separators with internal by-pass of the coalescence chamber

Andrzej Kotowski, Henryk Szewczyk & Patryk Wójtowicz
Wrocław University of Technology, Wrocław, Poland

ABSTRACT: The purpose of the present study is to assess, in hydraulic terms, the functions of petroleum derivative separators integrated with the settling tank and with a lateral or central interior by-pass of the coalescence chamber. The hydraulic models of separator functioning were developed in two variants: with a free discharge of sewage from the settler chamber through a side overflow into the by-pass channel and with a throttled discharge of sewage from the coalescence chamber (via siphon elbows or outflow tubes with float valves of specific diameters) into the discharge channel. Calculations have not confirmed the occurrence of sewage flow distribution inside the unit (both under the rated and maximum flow rates) assumed by the manufacturer. Thus, the hydraulic modernization of the devices proposed by the authors is recommended.

Keywords: Wastewater system, separator, modelling of flow division, hydraulic reconstruction.

1 INTRODUCTION

The subject of the present paper is the hydraulic assessment of the functions performed by petroleum derivative separators integrated with the settling tank. The separators fulfil the European standards EN 858 (2003) with regard to the volume of the petroleum derivative discharges for separators with coalescence filters. The separators under study those with by-passes of the coalescence filters inside the devices – are used mainly for the pretreatment of rainfall sewage. The principle of operation for such devices involves two phenomena: gravitational division of solids and floatation of petroleum derivatives. The devices in question consist of three principal elements mounted inside a steel (horizontal) drum, namely:

- a settler chamber for trapping and collecting solids,
- a coalescence chamber (provided with an automatic discharge cut-off system at the siphon elbows to prevent the collected substances from entering the environment), designed for capturing and collecting petroleum derivatives,
- a by-pass inside the unit for relieving the coalescence chamber in case of torrential rains.

Such separators are used for pretreatment of rainfall sewage from road surfaces, car-parks, car handling yards, airports etc. The size of the unit is matched to the maximum volume of rainfall sewage inflow (Q_{max}), assuming that one fifth of this flow undergoes continuing treatment to remove petroleum derivatives and deposits, including the so-called first wave, i.e. highly polluted rainfall sewage washed away from the drainage surface and referred to in the literature as rated flow (Q_n). The remaining inlet flow (i.e. the four fifths of Q_{max}), after pretreatment for the removal of deposits, is sent directly to the by-pass channel (which goes past the coalescence chamber) to enter the outlet channel.

2 MATERIALS AND METHODS

Hydraulic calculations were carried out to verify the sewage flow separation assumed by the manufacturer inside the unit, i.e. at the inlet to the coalescence chamber (at Q_n), and the one discharged through the by-pass channel (at Q_{max}), with the suggestion for modifying the construction to enable a more precise division of flows (by changing the diameters and slopes of the channels, as well as the diameters of the siphons and thus the height and length of the weir crests in the by-pass channel). It has been assumed that when sewage inflow into the separator is greater than the rated flow (Q_n) and this includes the maximum flow (Q_{max}) the actual flow (Q_{storm}) through the coalescence chamber must not be smaller than Q_n and greater than $Q_{storm} = 1.15Q_n$ (so that $Q_{max}/Q_{storm} \leq 5$). The calculations refer primarily to a specific type of series of separators – with a lateral and central by-pass inside the device, i.e. to the following series of types: $Q_n \leq 0.080\,m^3/s$ (with side by-pass,

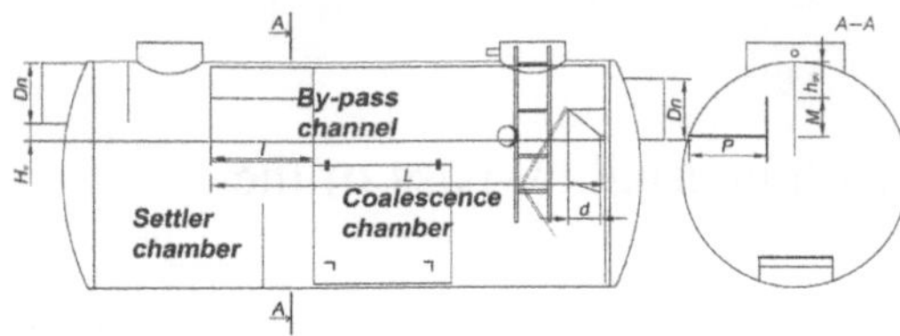

Figure 1. Separator 060AA with side (internal) by-pass channel.

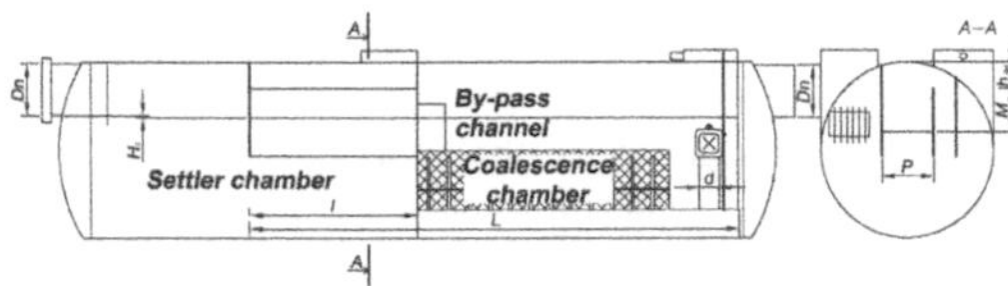

Figure 2. Separator 120AA with central by-pass channel.

Fig. 1) and $Q_n > 0.080\,m^3/s$ (with central by-pass, Fig. 2).

A typical separator with a side by-pass is designed for the rated flow $Q_n = 60\,dm^3/s$ and the maximum flow $Q_{max} = 5Q_n = 300\,dm^3/s$. The inner diameter of the inlet and outlet channel is $D_n = 0.50\,m$ with the difference in the ordinates of their bottom $H_0 = 5\,cm$ (structural variables are shown in Fig. 1). According to the manufacturer's specification, the device functions as follows:

- with a flow lower than or equal to the rated one Q_n ($60\,dm^3/s$), the rainfall sewage flows through the settler chamber and then through the coalescence chamber, thus being subject to full pretreatment (up to the extent desired); with the rate of flow equal to Q_n, the sewage in the settler chamber is swollen up to the ordinate of the side overflow threshold in the by-pass channel, i.e. up to the height M above the outlet channel, located at the bottom level of the by-pass channel; rated discharge from the coalescence chamber takes place through two siphoned elbows of diameter $d = 0.20\,m$;
- with the inlet flow rate Q_{max} ($300\,dm^3/s$), the sewage in the settler chamber is swollen above the side overflow threshold height M in the by-pass channel and is discharged at the rate $Q_{by\text{-}pass} = Q_{max} - Q_n$ ($240\,dm^3/s$) through the by-pass channel directly to the receiver and at the rate Q_n ($60\,dm^3/s$) through the coalescence chamber.

The by-pass channel, which has a total length L and a side overflow edge (inside the settler chamber) of a length l, is located on the side of the drum wall. The shape of the cross section area of this channel resembles an equilateral triangle (with a base P and a height $M + h_{(p)}$). The lengthwise slope of the by-pass channel is equal to zero.

A typical separator with a central by-pass channel is depicted in Figure 2 and has been designed for the rated flow $Q_n = 120\,dm^3/s$ and maximum flow $Q_{max} = 5Q_n = 600\,dm^3/s$. The internal diameter of the inlet and outlet is $D_n = 0.60\,m$, the difference in the ordinates of their bottom amounting to $H_0 = 5\,cm$. The cross-section area of the by-pass is rectangular in shape (with a width P and a height $M + h_{(p)}$). The lengthwise slope of the by-pass channel is equal to zero.

3 MODELLING OF SEWAGE SEPARATION IN THE DEVICES

3.1 *Inlet and outlet channels*

The flow rate q_v of the sewage in the inlet and outlet channels was calculated from the de Chézy formula (Yen 1992, Subhash 2001, Kotowski et al. 2005):

$$q_v = CA\sqrt{IR_h} \tag{1}$$

where C the (de Chézy) velocity coefficient was determined from the Manning-Strickler formula:

$$C = \frac{1}{n}\sqrt[6]{R_h} \tag{2}$$

where q_v is flow rate, m^3/s; A is cross-sectional flow area, m^2; I is channel bottom slope; R_h is hydraulic radius ($R_h = A/U$), m; U is wetted perimeter, m; and n is channel roughness coefficient: $n = 0.013\,s/m^{1/3}$.

For a given depth of flow in the channel and for a given flow rate, the bottom slope required was calculated from the formula:

$$I = \frac{n^2 q_v^{\,2}}{A^2 \sqrt[3]{R_h^{\,4}}} \tag{3}$$

Let us consider a 060AA type separator at $Q_n = 0.060\,m^3/s$. Assuming a full depth of flow in the inlet channel, $h = D_n = 0.50\,m$, at the maximum inlet flow of rainfall sewage, $q_v = Q_{max} = 5Q_n = 0.30\,m^3/s$, the required bottom slope in the inlet channel will be $I_{in} = 0.0066$ (The very high value of the slope indicates that the inlet channel diameter has been underestimated – the proper one for this rate of flow should be $D_n = 0.6\,m$, then $I_{in} = 0.0024$, which means that there is a 2.7-fold decrease in the slope).

With such a bottom slope in the inlet channel, the normal depth of flow (h) at the rated flow $q_v = Q_n = 0.060\,m^3/s$ may be determined from the equation:

$$\frac{nq_v}{\sqrt{I}} = A(h)R_h^{\,2/3}(h) \tag{4}$$

where $A(h)$ and $R_h(h)$ are functions of depth (h) in inlet and outlet channels.

Assuming that the depth of flow for the outlet channel is about $h = 0.9D = 0.45$ m (the highest hydraulic efficiency of the channel), then for $q_v = 0.30\,m^3/s$ we obtain $I_{out} = 0.0058$. (For a greater diameter, i.e. $D_n = 0.60$ m, we could obtain a bottom slope $I_{out} = 0.0021$, which means that there is a 2.8-fold decrease in the slope.)

Taking into account the location of the by-pass channel in the device, flow separation in the separators was modelled separately for two series of types:

- with a side by-pass channel (for type dimension 060AA: $Q_n = 60\,dm^3/s$),
- with a central by-pass channel (for type dimension 120AA: $Q_n = 120\,dm^3/s$).

3.2 *Coalescence chamber*

Discharge q_v through the coalescence chamber of the separator with a side by-pass (Fig. 1) under rated operating conditions (without involving the by-pass channel) was calculated from the equation (Kotowski et al., 2005):

$$q_v = n_s \frac{\pi d^2}{4} \sqrt{\frac{2g\Delta h}{\zeta + \frac{\lambda l_r}{d}}} \tag{5}$$

where n_s is the number of outlet siphons in the coalescence chamber ($n_s = 2$); d is the siphon pipe diameter, m; Δh is the difference between the levels of sewage in the settler chamber and in the outlet channel, m; ζ is the local loss coefficient in the siphon (inlet into the pipe, 90° elbow and outlet from the pipe: $\zeta = 2.5$); λ is the friction factor of the siphon pipe ($\lambda = 0.03$) of length l_r, and g is gravitational acceleration ($9.81\,m/s^2$).

It follows from the construction of the separators with a central by-pass (Fig. 2) that the rate of flow through the coalescence chamber should be calculated from the following equation (Kotowski et al. 2005):

$$q_v = \frac{\pi d^2}{4} \sqrt{\frac{2g\Delta h}{\zeta + \frac{\lambda l_r}{d} + \left(\frac{\pi d^2/4}{\mu_o A_o}\right)^2}} \tag{6}$$

where, $n_s = 1$ and $\zeta = 2$; μ denotes discharge coefficient of the opening between the settler chamber and coalescence chamber ($\mu = 0.6$); A_o stands for the cross-sectional area, m^2, of the opening (the notation for other quantities as in eq. 5).

In equations (5) and (6) the losses during flow through the coalescence cartridges were neglected, assuming that they are small compared to other losses, since the sewage velocities are very low.

The value of the height Δh was determined by assuming that the sewage may be swollen at most to the height of the overflow edge M on the side of the inlet to the separator. On the outlet side the level of the sewage determined by the depth of the outlet channel during discharge with at most rated flow ($q_v \le Q_n$). By virtue of equations (5) and (6) we can determine the position of the overflow edge in such a way that, for $q_v = Q_n$, discharge will take place only through the coalescence chamber of the separator. Only when the discharge exceeds the rated value, some portion of the sewage will flow to the by-pass channel.

The discharge of the sewage flowing through the coalescence chamber at $Q_{max} = 5Q_n$ was also calculated in terms of equations (5) and (6). The Δh value was determined by assuming that the inlet channel may be filled in full and the outlet channel may be filled up to $0.9D_n$ at most. In view of the fact that the outlet channel in the separators lies deeper than the inlet channel (by $H_0 = 5$ cm), the quantity Δh can take the value $H_0 + 0.1D_n$ at the most.

Substituting $q_v = Q_{max}$ into the relations of (5) and (6), it is possible to establish the actual rate of flow Q_{storm} through the coalescence chamber during torrential rainfalls (i.e. at $Q_{max} = 5Q_n$). This results in the following rate of discharge through the by-pass channel: $Q_{by\text{-}pass} = Q_{max} - Q_{storm}$. This may become be a basis for determining the length of the side overflow of the by-pass channel.

During the maximum rate of inflow to the separator, the levels at the inlet and outlet of the coalescence chamber differ from those observed during rated flow (Q_n). Now they are determined by the depths of flow in the inlet and outlet channels. This means that it is not possible to select the parameters for the hydraulic device in such a way that the discharge through the coalescence chamber of the separator would be the same both under rated(Q_n) and maximal (Q_{max}) flow conditions – ($Q_n \neq Q_{storm}$). What can be achieved is that the rates of flow obtained do not very much differ in their order of magnitude. This applies to all hydraulic devices without adjustable control units, such as valves or gates, which characterized by variable hydraulic resistance. From the viewpoint of environmental pollution control, the following technological requirement must be met: $Q_{max}/Q_{storm} \le 5$. In other words, it must be assumed that at least discharge Q_n should undergo pretreatment for the removal of petroleum derivatives, the rate of flow to the separator being $\ge Q_n$.

3.3 *By-pass channels*

The problem of calculating the discharge of sewage through the by-pass channel is much more complicated than the issues discussed previously. The liquid flows to this channel through side overflows which are usually examined when the mass of the liquid decreases towards the discharge (Kotowski 1998). The case under study is a different one: the liquid is collected from the environment and reaches the by-pass channel by flowing over the side overflow crest. Therefore, two parts should be distinguished in the by-pass channel: one with a side overflow and the other one with an open channel. The part with the side overflow is extremely complicated from the hydraulic point of view. If it is to be treated as a regular overflow, then the flow rate can be calculated as follows (White 2003):

$$q_v = \frac{2}{3}\mu\delta l h_{(p)}\sqrt{2gh_{(p)}} \tag{7}$$

where δ stands for the submergence coefficient of the overflow, at $\delta = 1$; μ denotes the discharge coefficient of the overflow; l is the length of the weir crest, m, and $h_{(p)}$ is the height of the liquid layer over the overflow edge, m.

During low flow, when the level of the sewage in the by-pass channel does not reach the level of the overflow crest (or only slightly exceeds it), the overflow is treated as non-submerged. But when with increasing rate of flow, the liquid level in the by-pass channel exceeds 30% of the height of the liquid layer over the overflow edge ($h_{(p)}$), the weir becomes submerged. Then the flow rate is calculated using equations 8a and 8b (Mostkov 1963):

$$\delta = 1.05\sqrt[3]{\frac{z}{h_{(p)}}} \tag{8a}$$

$$\mu = 0.607 + \frac{0.00452}{h_{(p)}} \tag{8b}$$

where z is the difference between the level of the liquid above the weir and that in the by-pass at the end of the overflow: $z/h_{(p)} < 0.7$ (when $z/h_{(p)} \geq 0.7$ then $\delta = 1$), m.

The value of z can be calculated, when the depth of flow in the by-pass channel at the end of the overflow is known. To achieve this, it is necessary to solve the differential equation of steady, non-uniform motion in the channel of a constant shape and horizontal bottom slope:

$$\frac{dh}{dx} = -\frac{\frac{U}{C^2A^3}q_V^{\;2}}{1-\frac{\alpha b}{gA^3}q_V^{\;2}} \tag{9}$$

where $h = h(x)$ is the depth of the channel at distance x from the origin of the coordinates, m; $\alpha = 1.3$ is the Coriolis coefficient (Kotowski 1998); $b = b(h)$ is the width of the liquid level at distance x, m; $A = A\,(h)$, $C = C(h)$, $U = U(h)$ are the functions of depth h in the by-pass channel.

Equation (9) should satisfy the initial condition that at the end of the by-pass channel ($x = L_c - l$) the depth must be equal to the level of the liquid in the outlet channel – for the same rate of flow as in the by-pass channel, while in the outlet channel flow rate equals to the total discharge through the unit.

It should be noted that before the overflow to the by-pass channel (Fig. 1 and 2) there is an inlet siphon which is the source of certain hydraulic losses. The discharge cross-sectional area of this siphon is 3/2 of the discharge cross-sectional area through the overflow, and the velocity head in this siphon for typical separators approaches 2 cm (Kotowski et al. 2005).

4 RESULTS AND DISCUSSION

The flow through the separators depends on many discrete or continuous parameters. The discrete ones are, e.g., the diameter of the channels or siphons, local losses and assigned characteristic discharges. The major continuous can be itemized as follows: length of the weir crest (l), height of the weir crest above the bottom of the by-pass channel (M) and slope of the outlet channel (I_{out}). Each of the parameters mentioned above may take values from the limited intervals, imposed by the design and operation of the separators with respect to hydraulics, etc.

The minimal bottom slope value of the inlet channel (I_{in}) was determined assuming that its full fill occurs at the rate of flow $q_v = 5Q_n$. The value of the bottom slope of the outlet channel (I_{out}) must be consistent with the calculated one (included in Table 1), as the depth of its

Table 1. Operating parameters of typical (modernized) separators.

Q_n (dm^3/s)	Q_{storm}/Q_n (Q_n) (–)	Q_{storm}/Q_n ($2Q_n$) (–)	Q_{storm}/Q_n ($3Q_n$) (–)	Q_{storm}/Q_n ($4Q_n$) (–)	Q_{storm}/Q_n ($5Q_n$) (–)
60	1.00	1.05	1.10	1.13	1.15
120	1.00	1.07	1.12	1.15	1.15

flow exerts a noticeable effect on the stream division of the liquid flowing to the unit – i.e. the part of the fluid flowing through the coalescence chamber (at a height of $1.15Q_n$) and the remaining part discharged to the by-pass channel. This slope was selected so as to fill the channel to the height $h_{od} \in 0.8; 0.9 > D_n$, preferring values closer to $0.9D_n$ at Q_{max}.

Parameters like the diameters of channels and outlet siphons were selected by discrete trials, focusing on the calculation of the continuous parameters, which were quantities {l, M, I_{out}}. It follows from fact that relations between these quantities can be described by equations of continuous variables, which form a set of three equations:

$$\{Q - q = \frac{2}{3}\mu \delta l h_{(p)} \sqrt{2gh_{(p)}} \tag{10a}$$

$$q = \sqrt{\frac{M + h_{(p)} + \zeta_b \left(\frac{Q-q}{1.5lh_p}\right)^2 \frac{1}{2g} - h_{out}}{\Delta h}} \tag{10b}$$

$$I_{out}\left(\frac{q}{\chi}, \frac{M - \Delta h}{D_n}\right) = I_{out}\left(Q, \frac{h_{out}}{D_n}\right)\} \tag{10c}$$

where Q is discretely assumed sewage inflow to the unit ($Q \in \{1, 2, 3, 4, 5\}Q_n$), m^3/s; q denotes rate of flow through the coalescence chamber ($Q_{storm} \in <Q_n; 1.15\ Q_n>$), m^3/s; ζ_b stands for local loss coefficient in the inlet siphon to the overflow situated at the by-pass channel: $\zeta_b = 1.0$; h_p indicates height of overflow opening in the vertical plane of the weir crest, m, and χ is overload coefficient of the coalescence chamber, assumed as $\chi \in <1; 1,15>$.

Assuming that $Q = 5Q_n$, $q = \chi Q_n$ and $h_{(p)} = h_p$, we can calculate values of quantities {l, M, h_{od}}by virtue of the set of nonlinear equations (10a, 10b, 10c). If the values calculated from the solution of the set of equations (10a, 10b, 10c) are not the expected ones, the correction of D_n or d value will be necessary. After having calculated the quantities {l, M, h_{od}} we have to determine the head loss in the inlet siphon to the overflow situated in the by-pass channel, in order to prevent the depth of flow in the inlet channel from exceeding D_n during flow $5Q_n$.

The final calculation stage is to verify whether or not during discharge through the separator at kQ_n (for $k = 1, \ldots, 5$) the rate of flow through the coalescence chamber exceeds the admissible value $1.15Q_n$. Calculations were carried out using the first two equations of the set (10a and 10b). Since the dimensions of the overflow and bottom slope of the outlet channel are known, the actual flow through the coalescence chamber can be calculated, as well as the height of the liquid layer above the overflow edge, required to check the type of flow through the overflow (submerged or non-submerged). The calculated are summarized in Table 1 (where the division of sewage flow Q_{storm}/Q_n is shown for modernized separators at flow Q_n, $2Q_n$, $3Q_n$, $4Q_n$ and $5Q_n$). Consequently, the assumed hydraulic correctness of the operation of the modernized units has been proved.

The calculated results (which verify the functioning of the former construction of the separators considered i.e. the division of sewage flow inside the units at discharge $5Q_n$) and the hydraulic reconstruction proposed in the paper (i.e. after modification of some design and operation parameters of typical separators) are shown in Table 2.

For example, before modernization, separator series of types 120AA – with a central by-pass channel ($Q_n = 120\ dm^3/s$), was characterized by the following parameters: $Q_{storm} = 0.85Q_n = 102\ dm^3/s$ at $Q_{max} = 600\ dm^3/s$, which is definitely too small a value ($Q_{storm} < Q_n$), and $Q_{by\text{-}pass} = 600 - 102 = 498\ dm^3/s$ and $Q_{max}/Q_{storm} = 5.9 > 5$, which is definitely too high a value (Tab. 2, row 2).

The hydraulic modernization of the separator (which consisted inter alia, of simultaneous increase in the inlet and outlet channel diameter from $D_n = 0.60$ m to $D_n = 0.80$ m and the siphon diameter respectively), brought about an over four-fold decrease of slopes required for these channels ($I_{in} = I_{out} = 2.2‰$)– as compared to the initial diameters, preserving, with the assumed accuracy, the division of sewage flow inside the unit: $Q_{storm} = 1.15Q_n = 138\ dm^3/s$

Table 2. Selected design and operation parameters of typical separators.

No. –	Q_n (dm^3/s)	D_n (m)	d (m)	I_{in} (min) (mm/m)	I_{out} (const) (mm/m)	Q_{storm} (dm^3/s)	$Q_{max} = 5Q_n$ (dm^3/s)	$Q_{by\text{-}pass}$ (8)-(7) (dm^3/s)	Q_{max}/Q_{storm} (8)/(7) (–)
1	2	3	4	5	6	7	8	9	10
1	60[1]	0.50	0.20	6.6	5.8	58	300	242	5.2
1a*		0.60	0.20	2.4	2.5	69		231	4.3
2	120[2]	0.60	0.40	9.5	8.4	102	600	498	5.9
2a*		0.80	0.45	2.2	2.2	138		462	4.3

* – modernized parameters.
[1] side by-pass; [2] central by-pass.

at $Q_{max} = 600\,dm^3/s$, then $Q_{max}/Q_{storm} = 4.3$ and $Q_{by\text{-}pass} = 600 - 138 = 462\,dm^3/s$ (Tab. 2, row 2a). The operation of the unit at moderate rates of flow, $Q \in \{1, 2, 3, 4, 5\}Q_n$ (Tab. 1, row 2) is described as $Q_{storm} \in <1.0; 1.15>Q_n$.

5 CONCLUSIONS

1. Hydraulic calculations have shown that a typical separator (a series of types with a side or central by-pass channel inside the unit) enables the rated flow capacity ($Q_{max} = 5Q_n$) to be attained when a specified volume of sewage flows through the by-pass ($Q_{by\text{-}pass} = Q_{max} - Q_{storm}$), provided that the slopes of the inlet and outlet channel bottoms are sufficiently large, as shown in Table 2 (rows 1 and 2). There is however no guarantee that the accuracy of the division of sewage inside the investigated separators, assumed by the manufacturer, will be sufficient, since $Q_{max}/Q_{storm} > 5$, because $Q_{storm} < Q_n$.
2. Hydraulic modernization has been proposed for the design of the separators with a side or central by-pass channel. The major changes included, increasing the diameter of the inlet and outlet channels, and the determination of the channel slopes (as well as the correction of the siphon diameters in the coalescence chambers and the height and length of the weir crest). The result was a 2.7-fold to 4.3-fold decrease in the minimal slopes required for the inlet and outlet channels at the assumed sewage flow division inside the units under consideration, with the adopted accuracy $Q_{storm} \in <Q_n; 1.15Q_n>$ (for examined rates of flow to the separators $Q \in \{1, 2, 3, 4, 5\}Q_n$), provided that their operation exerts no environmental impact, which means that $Q_{max}/Q_{storm} \leq 5$ is fulfilled.
3. The study has shown that further modernization is necessary. The other units of the two series of types considered need to be improved according to the methodology formulated in this paper.
4. It should be noted that the hydraulic modelling of sewage division in this study was mainly based on the data reported in the literature. Thus, to attain the actual distribution of streams in the units considered, it will be necessary to perform model testing.

REFERENCES

EN 858-1: 2002. Separator systems for light liquids (e.g. oil and petrol) – Part 1: Principles of product design, performance and testing, marking and quality control. CEN/TC 165. 89/106/EEC.

Kotowski A., 1998. *Principles of the dimensioning of a non-conventional storm overflow*. Oficyna Wydawnicza Politechniki Wrocławskiej, Wrocław.

Kotowski, A., Szewczyk, H., Pawlak, A., 2005. Hydraulic modelling of sewage flow in separators of petroleum distributors. *Environment Protection Engineering*, Vol. 31 (1).

Mostkov, M. A., 1963. *Applied Hydromechanics*. Gosudar. Energet. Izdatelstvo, Moskva – Leningrad.

Subhash C.J., 2001. *Open channel flow*. John Wiley & Sons. New York.

White, F.M., 2003. *Fluid Mechanics*. McGraw-Hill, Boston.

Yen, B.C., 1992. Dimensionally Homogeneous Manning's Formula. *Journal of Hydraulic Engineering*., Vol. 118, pp.1326–1332.

Water treatment and supply

Environmental Engineering – Pawłowski, Dudzińska & Pawłowski (eds)
© 2007 Taylor & Francis Group, London, ISBN13 978-0-415-40818-9

Oxidation of Fe(II) to Fe(III) by heterogeneous oxidant as a convenient process for iron removal from water

Romuald Bogoczek, Elżbieta Kociołek-Balawejder, Ewa Stanisławska & Agnieszka Żabska
Chair of Industrial Chemistry, University of Economics, Wrocław, Poland

ABSTRACT: Beads of a macroporous and macromolecular oxidant, polystyrene cross-linked by divinylbenzene containing pendant *N,N*-dichlorosulfonamide groups, was used to oxidize ferrous ions to ferric in dilute aqueous media. The redox polymer, a macromolecular analogue of Dichloramine T, was prepared starting from the sulfonate cation exchanger *Amberlyst* 15 by a three step transformation of its functional groups. The resulting copolymer contained 8.2 mequiv of active chlorine/g. The investigation was carried out using both batchwise and column methods and 0.005 and 0.02 M $FeSO_4$ solutions with different acidity. The oxidative ability amounted to ca 450 mg of Fe^{2+}/g of the copolymer. We found that the investigated oxidation reaction is favored by a low pH. The column effluent after its neutralization, precipitation, sedimentation and filtration was iron free.

Keywords: Fe(II) oxidation, iron removal from water, active chlorine support, macromolecular oxidant.

1 INTRODUCTION

Iron compounds appear in natural water in diverse concentrations, however they always influence the water's quality. Despite of their low toxicity they are undesirable in potable water. Iron in water, even in low concentration (at levels of 1.0 mg/dm^3) is markedly perceptible, it causes a deterioration of organoleptic properties it influences negatively the flavor, color and taste. It causes deposition of sludge in pipelines and heaters, it covers sanitations with sediments, it causes occurrences of patches at laundered fabrics. In water reservoirs it favors a development of algae, while present in water-supply-installations it favors the development of ferruginous bacteria causing the spongy rust. Natural processes are the source of iron compounds in water: for example the dissolution of minerals and also via sewage and industrial waste products and corrosive processes. Iron compounds are highly undesirable and must be carefully removed from water applied in a wide range of industries (pharmaceutical, fermentative- etc.), alimentary-, textile, cellulose-, paper-, dye-stuffs- and in paint and lacquer manufacture. In water used for manufacture of white pigments and of some sorts of paper the concentration of iron cannot exceed 0.001 mg Fe/dm^3. The admissible concentration of the iron in potable water is 5 mg Fe/dm^3.

In surface waters well saturated with oxygen, the concentration of soluble iron ions is generally not high, scarcely exceeding several mg/dm^3. In that water, predominates iron at its third degree of oxidation, which compounds being in the conditions existing in surface water, are difficult soluble.

The elimination of iron is essential in the purification of *underground* water. Such waters, saturated with carbon dioxide but low in oxygen, may be considerably iron contaminated, sometimes containing up to 100 mg Fe/dm^3. In these waters, ferrous (II) ions represented mainly by $Fe(HCO_3)_2$ and $FeSO_4$, dominate. These compounds are easily soluble in conditions actually existing in underground water.

The essence of the removal process of iron from water is the oxidation of its easily soluble bivalent form to the trivalent ferric form, which, as the hydroxide, is difficultly soluble. This is accomplished by the addition of oxidizing agents to the reaction medium followed by a quantitative removal of insoluble reaction products using sedimentation and filtration at the required pH. Oxygen provided by aeration can be used as the oxidant or one of the known oxidizing agents generally applied in the treatment of water, e.g. chlorine:

$$4Fe^{2+} + O_2 + 10H_2O \rightarrow 4Fe(OH)_3 + 8H^+ \quad (1)$$

$$2Fe^{2+} + Cl_2 + 6H_2O \rightarrow 2Fe(OH)_3 + 2Cl^- + 6H^+ \quad (2)$$

It is worthy of note that the reaction of chlorine with ferrous salts is a very convenient one, as it can be used for two different purposes: (i) to remove iron from water and (ii) to produce a coagulant for both water and sewage treatment (for example, in pickle liquor treatment, a by-product of steel mills):

$$6FeSO_4 \cdot 7H_2O + 3Cl_2 \rightarrow 2FeCl_3 + 2Fe_2(SO_4)_3 + 42H_2O \quad (3)$$

Chlorine reacts easy with ferrous ion and converts it to the ferric form. Depending upon the hydroxyl ion activity, the ferric chloride formed can quickly hydrolyze to ferric hydroxide. The latter precipitates as a reddish fluffy mass, depending on the concentration of ferric ion. The reaction can proceed over a wide pH range (4–10), but the optimum pH is above 7.0 (White 1999).

For iron elimination from water we attempted to use reactive polymers and the process of ionic exchange. The sorptive properties of cationites and anionites in relation to iron ions have been investigated many times (Gutsanu 1990). As these copolymers show a high affinity to multivalent metal ions, their regeneration is not easy. The elimination of iron from solutions by means of ion exchange is onerous and runs up against numerous limitations. If oxygen is present in the purified solution, colloidal and suspended forms of iron compounds can be precipitated inside the resin beads thus blocking the pores and the functional groups of the ionite. Periodic elimination of such foulants using solutions of especially well-chosen composition is essential. Adding a reducing agent to the deironized water before passing it over to the column can counteract this disadvantageous phenomenon.

It is valuable to add here that polymers containing iron ions in various forms (even precipitated inside the polymer's structure) have been recently suggested to be very desirable as adsorptive materials. One reason for their usefulness is in the elimination of arsenic from water. There is a large variety of materials including various resins based on iron and its derivatives as adsorbents for arsenic. The following iron-loaded polymers have been used as arsenic sorbents – cation exchangers in the Fe^{3+} form (Dambies 2004), polymeric/inorganic hybrid sorbents, e.g. uniformly and irreversibly dispersed submicron hydrated Fe oxide particles (Cumbal et al. 2003, DeMarco et al. 2003), iron-oxide-coated polymers prepared by adsorptive filtration (Katsoyiannis et al. 2002), ion exchangers prepared by polymerization using iron-oxide and iron-oxyhydroxide-containing mixtures and functionalizing the resultant polymers (Podszun et al. 2005).

As the oxidation process running from Fe(II) to Fe(III) – the main reaction in the removal of iron from water – is dependant on a number of factors, i.e. of the used reagent's concentration, in this investigation we attempted to use water insoluble, i.e. heterogeneous, oxidizing agents. This kind of procedure, i.e. the oxidation of ferrous ions in a column process, will make it possible to realize the reaction in a quantitative and efficient mode. The solid phase high oxidant concentration will have a favorable influence on the direction and the speed of the reaction.

The solid oxidant is a *N,N*-dichlorosulfonamide S/DVB resin (DCSR) comprising a macroporous matrix bearing $-SO_2NCl_2$ functional groups:

$$[P]\text{-}SO_2NCl_2 + 4Fe^{2+} + 2H^+ \rightarrow [P]\text{-}SO_2NH_2 + 4Fe^{3+} + 2Cl^- \quad (4)$$

[P] stands for the copolymer styrene/divinylbenzene, macroporous structure in bead form (ca 1 mm in diameter). These groups contain the actual oxidizing agent – chlorine with oxidation number +1 covalently bound to the macromolecular solid carrier. The concentration of active chlorine in this well water swollen copolymer attains ca $2.5\,M/dm^3$. We have shown its strong oxidizing activities for cyanides, thiocyanates, sulfides and nitrites in previous contributions (Kociołek-Balawejder 2000a, 2000b, 2002, Bogoczek et al. 2006).

Macromolecular, water-insoluble and chemically reactive materials are very useful reagents for the removal of trace concentration pollutants in large volumes of solution. The advantage of a solid phase oxidation is the potential to use a local large excess of oxidant to drive the reaction to completion.

The described procedure promises to be a highly efficient process for removing iron from water mainly because the more conventional low efficiency processes of aeration or oxidation with micromolecular reagents, are not involved. The oxidation proceeds only slowly when using traditional water-soluble agents, because of the low concentration of iron in the purified water. The application of water-insoluble heterogeneous oxidizing agents can considerably improve this process.

The *N,N*-dichlorosulfonamide copolymer used here is a high molecular weight equivalent of the known micromolecular oxidant, the Dichloramine T, which is used in analytical chemistry for the quantitative, titrimetric, determination of Fe^{2+} ions (Jacob et al. 1972, Gowda et al. 1981). In this paper, we show the oxidation of ferrous ions by means of a polymer-supported oxidizing reagent instead of gaseous chlorine. By the application of this macromolecular oxidant numerous new oxidation methods can be realized.

2 MATERIALS AND METHODS

2.1 *Reagents*

The copolymer bearing *N,N*-dichlorosulfonamide groups was prepared by the method previously described (Emerson et al. 1978, Bogoczek & Kociołek-Balawejder 1986, 1989). As the starting material, *Amberlyst* 15 (produced by Rohm and

Haas Co.), a commercially available sulfonate cation exchanger was used. *Amberlyst* 15 is a macroporous poly (S/20% DVB) resin that contained 4.7 mmol/g $-SO_3H$ groups in its air dried state (surface area: 45 m^2/g; average pore diameter: 25 nm). We transformed the sulfonic groups to chlorosulfonyl- and then to sulfonamide groups which incorporated chlorine after reaction with sodium hypochlorite in acetic acid medium. The product contained 2.05 mmol/g $-SO_2NCl_2$ groups (i.e. 4.10 mmol of active chlorine/g or 8.20 mequiv of active chlorine/g) as well as a small amount of sulfonic groups (0.60 mmol/g).

Analytical grade ferrous sulfate was used for the preparation of the aqueous solutions containing $FeSO_4$ alone or in a mixture with sulfuric acid. Solutions used in the batch regime experiments were: 0.02 M $FeSO_4$ (i.e. 1120 mg Fe^{2+}/dm^3 in: (1) water, (2) 0.001, (3) 0.01 or (4) 0.1 M H_2SO_4. The solutions used in the experiment carried out in a dynamic regime were: 0.005 M $FeSO_4$ (280 mg Fe^{2+}/dm^3) in: (1) water, (2) 0.005, (3) 0.01 or (4) 0.05 M H_2SO_4.

2.2 *Analytical methods*

The ferrous and ferric ions concentrations were determined by spectrophotometric methods (Spekol 1200, Analytic Jena, Germany). The Fe^{2+} concentration was determined using the formation of an orange complex compound with 1,10 phenanthroline monohydrate. The absorbency measurement was taken at 510 nm wavelength (Fries 1971). Fe^{3+} concentration was determined by the use of the red colored complexes formed by the reaction of Fe^{3+} with thiocyanates. The absorbency was determined at 480 nm. Chloride ions were estimated by argentometric titration using 0.01 M $AgNO_3$ with the Ag/AgCl/calomel electrodes system. The active chlorine content of the resin was determined by iodometry.

The redox titration of the DCSR was performed by the use of 0.02 M $FeSO_4$ in 0.01 M H_2SO_4. Into eleven separate samples of DCSR (0.24 g) of active chlorine (~2.0 mequiv) were introduced the following increasing volumes of 0.02 M $FeSO_4$, respectively: (1) 0, (2) 12.5 cm^3, (3) 25 cm^3, (4) 37.5 cm^3, (5) 50 cm^3, (6) 62.5 cm^3, (7) 75 cm^3, (8) 87.5 cm^3, (9) 100 cm^3, (10) 125 cm^3 and (11) 150 cm^3. To the first (1) sample of the copolymer only 0.01 M H_2SO_4 was added. The stated increasing solution volumes of $FeSO_4$ were needed to bring about the reduction, for example of: (1) 0%, (5) 50%, (9) 100% of the functional group's active chlorine. However, the last two samples contained (10) 125% and (11) 150% of the stoichiometric amount of ferrous ion. All of these samples, in closed vessels, were shaken at constant temperature (20°C). After 24 h, the electric potentials of the reaction media were measured by means of the platinum/calomel electrode pair and the pH values of the solutions by a glass/calomel couple. At the end, the content of ferrous ions in the post reaction solution was determined.

2.3 *Ferrous solution treatment*

In all studies carried out in the batch regime, at room temperature, a measured amount of the resin (ca 0.25 g) placed in a flask was shaken with 0.02 M $FeSO_4$ solution, employing media of different acidity:

(a) 200 cm^3, an 100% excess of ferrous ions relative to stoichiometry of Eq. (4).
(b) 50 cm^3, an 100% excess of active chlorine relative to stoichiometry of Eq. (4).

Time-dependent measurements of the Fe^{2+} and Fe^{3+} contents in solution were made. After the reaction end, the copolymeric reagent was separated from the reaction medium by filtration, was water-washed, and then washed with 50 cm^3 of 1 M H_2SO_4. In the effluent, the concentrations of the Fe^{2+} and Fe^{3+} ions were determined and finally the active chlorine content was analysed.

In the experiments carried out in the dynamic regime, a sample of DCSR (10.0 g in its dry state, 82 mequiv active chlorine content) was packed into a glass column (inner diameter ~1.15 cm; height of the package ~17.5 cm). 0.005 M $FeSO_4$ solutions of various acidity were passed through the column bed. The flow rate was 15 bed volumes per hour. Fractions (250 cm^3) were collected to estimate their composition in terms of pH and the ferrous, ferric and chloride ions. When the copolymer lost its oxidizing capacity the resin bed was washed with distilled water. Next 1 M H_2SO_4 passed the column and fractions of about 25 cm^3 were collected. They were analysed for the content of Fe^{2+} and Fe^{3+} ions. Finally, the exhausted copolymer was removed from the column, water-washed, air-dried, and subjected to analysis of active chlorine contents.

3 RESULTS AND DISCUSSION

Hitherto we used the *N,N*-dichlorosulfonamide copolymer as a heterogeneous oxidizing agent for removal of undesirable chemical compounds present as impurities in very dilute aqueous solutions. These compounds were in the form of anions. We transformed these admixtures into less harmful products (e.g. oxidizing sulfides to sulfates). In the present study, for the first time, the copolymer was used for oxidizing cations. This improves the deironing process during water treatment. The chemical stability of DCSR in aqueous media will play an important role in any new method based on its use. Earlier experiments showed that the DCSR looses its activity in strongly alkaline media. The reason is the loss of chlorine by dechlorination – both of the active chlorine atoms leave the resin phase easily and fast (although the first one more easily than the second). The highest stability of DCSR is in neutral and slightly acidic media. These are the pH conditions under which iron occurs in the environment, for example in the natural and industrial waters.

As was shown in the experimental part DCSR, besides the basic functional groups $-SO_2NCl_2$ with oxidative properties, contains a small amount of sulfonic groups with cation exchange properties. This implies that these residual groups are capable of binding iron cations (both the substrates and the products of the oxidation reaction). The cation exchange ability of the copolymer following the content of $-SO_3H$ groups is 0.6 mmol H^+/g. In relation to polyvalent ions of iron possibly 2–3 times lower by counting the number of mmol. Sulfonic groups show larger affinity to Fe^{3+} ions than to Fe^{2+} ions. The oxidizing capacity of DCSR in relation to Fe^{2+} calculated from Eq. (4) is 8.2 mmol Fe^{2+}/g. Thus, DCSR is a considerably stronger oxidizing agent than cation exchanger.

Initial investigations on the oxidation the Fe^{2+} ions by means of DCSR were carried out using the stationary (batch) method. With the aim of assessing the dependence of the oxidative ability of DCSR on the reaction conditions, samples of DCSR were treated with an excess of $FeSO_4$aq in media of different acidity. We examined the time dependence of the process (Fig. 1), and after 24 h we analyzed the post reaction solutions and the copolymers (Table 1).

In each of the four media the oxidation reaction took place – the Fe^{2+} ions concentration decreased and the initially absent Fe^{3+} ion concentration rose. The highest turnover of the reaction was in the sample having the largest acidity (0.02 M $FeSO_4$ in 0.1 M H_2SO_4). In this case the operating oxidative capacity of the copolymer was 7.2 mmol Fe^{2+}/g , and reached ca 88% of the theoretical oxidative capacity (following from the active chlorine content in the fresh copolymer).

After the reaction the pH value was 1.65, and the complete iron content (Fe^{2+} and Fe^{3+} ions) was present in the *solution* (i.e. the ions were not bound to the sulfonic groups because of the low pH in the reaction medium). In the residual three samples the progress of the oxidation reaction was lower.

In the case of the solution containing $FeSO_4$ alone (without addition of H_2SO_4) the operating oxidative capacity of the copolymer was the lowest, it was 4.08 mmol Fe^{2+}/g of the resin. The oxidized ions Fe^{3+} were partially present in the solution, and partially they were combined with the sulfonic groups and could be determined only after eluation from the copolymer with 1 M H_2SO_4 (the pH value in solution after the reaction was 2.95).

In the next part of the batchwise research the reagents were applied in the reciprocal proportions – of the samples contained an excess of the active chlorine with relation to $FeSO_4$.

We wished to examine how much the ferrous ions concentration in the solution at the given conditions decrease. As is indicated in Table 2, the most rapid reaction took place in the 0.1 M H_2SO_4 medium (like this was previously) – after 24 h the ion concentration

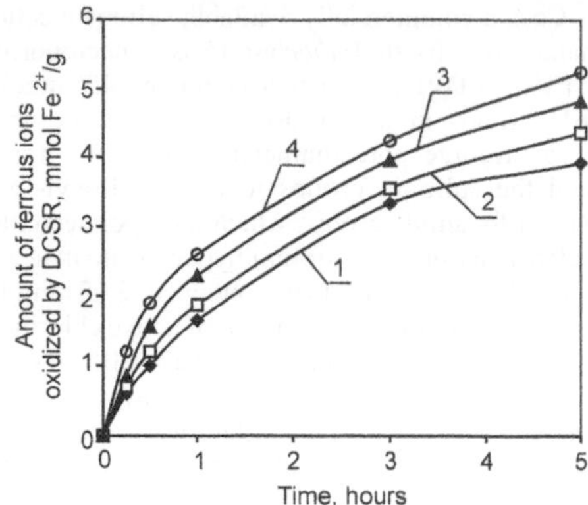

Figure 1. Removal of ferrous ions from aqueous solution in a batchwise reaction: 0.25 g DCSR + 200 cm^3 0.02 M $FeSO_4$ in (1) water, (2) 0.001 M H_2SO_4, (3) 0.01 M H_2SO_4, (4) 0.1 M H_2SO_4.

Table 1. Results of a 24 hr batchwise reaction between DCSR (0.25 g) and ferrous ions (200 cm^3 of 0.02 M $FeSO_4$) in media of different acidity, i.e. an 100% excess of ferrous ions in relation to stoichiometry of Eq. (4).

0.25 g of DCSR + 200 cm^3 of 0.02 M $FeSO_4$				
	Alone	In 0.001 M H_2SO_4	In 0.01 M H_2SO_4	In 0.1 M H_2SO_4
Fe^{2+} in solution, mmol				
Before reaction	4.0	4.0	4.0	4.0
After reaction	2.98	2.79	2.46	2.20
Fe^{3+} in solution, mmol				
Before reaction	0.0	0.0	0.0	0.0
After reaction	0.49	0.74	1.49	1.78
Iron eluted from resin				
Fe^{2+}, mmol	0.05	0.02	<0.01	<0.01
Fe^{3+}, mmol	0.51	0.44	0.04	0.01
pH				
Before reaction	3.58	3.09	2.43	1.68
After reaction	2.95	2.79	2.40	1.65
Cl^- in solution after reaction, mmol				
	0.60	0.72	0.76	0.80
Active chlorine in sample of resin, mequiv				
Before reaction	2.0	2.0	2.0	2.0
After reaction	0.60	0.46	0.34	0.24
Operating oxidizing capacity of DCSR, mmol Fe^{2+}/g				
	4.08	4.84	6.16	7.20

Table 2. Concentration drop of ferrous ions in solution during batchwise contact with a sample of DCSR, i.e. an 100% excess of active chlorine in relation to stoichiometry of Eq. (4).

0.25 g of DCSR + 50 cm^3 of 0.02 M $FeSO_4$				
Time (hours)	Alone	In 0.001 M H_2SO_4	In 0.01 M H_2SO_4	In 0.1 M H_2SO_4
Concentration of ferrous ions in solution, mg Fe^{2+}/dm^3				
0	1120	1120	1120	1120
1	780	750	700	670
3	540	470	430	400
5	440	350	300	270
24	230	160	130	110
Concentration of ferric ions in solution, mg Fe^{3+}/dm^3				
24	630	784	957	1010
Iron eluted from the copolymer, mg Fe^{3+}				
24	13.0	8.90	1.80	0.0
pH after reaction				
24	2.33	2.30	2.03	1.20
Active chlorine in a sample of DCSR, mequiv				
Before reaction	2.0	2.0	2.0	2.0
After reaction	0.91	0.90	0.89	0.88

of Fe^{2+} dropped from 1120 to 100 mg/dm^3, i.e. ca 90% of the Fe^{2+} ions contained in the initial solution were oxidized to Fe^{3+} (this was quantitatively determined in the solution). In the medium with the least acidity ($FeSO_4$ alone) the degree of oxidation was less, ca 80% of the Fe^{2+} ions contained in the initial solution was oxidized to Fe^{3+} ions (they were present partially in the post reaction medium and partially combined with the copolymer). The post reaction copolymers contained a considerable amount of the active chlorine.

An important experiment in the case of the investigation of the redox reaction is the potentiometric titration with respect to the platinum electrode. We titrated the DCSR (the oxidant) with 0.02 M $FeSO_4$ (the reductant) in 0.01 M H_2SO_4. Figure 2a represents the curve of the redox titration prepared as follows: the electrical potential value for each point we have read after 24 h of the beginning of the reaction run; each point was taken in a separate copolymer sample, while successive samples of the copolymer were treated with an increasing amount of the reducer (the pH value during the reaction run was ca 2.45).

Because the reagents represented different phases and the reaction ran too slowly an acknowledged special titration method was used. It results from the course of the redox titration curve that two stages can be distinguished in the investigated redox reaction. The first stage (the addition of 0–50% of the reducer) runs at a high redox potential (from +900 mV to +500 mV) and a quantitative oxidation of Fe^{2+} ions to Fe^{3+} ions

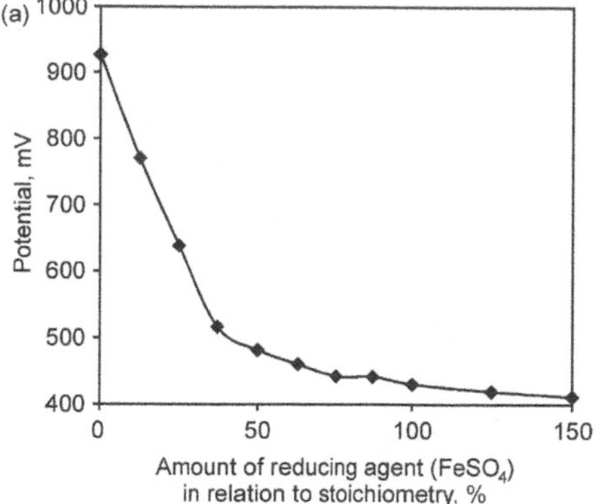

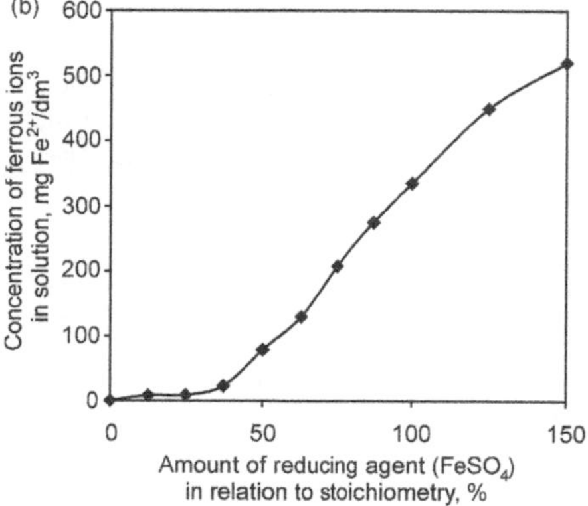

Figure 2. (a) Redox titration curve of DCSR (0.24 g) by 0.02 M $FeSO_4$ in 0.01 M H_2SO_4, (b) Concentration of ferrous ions in solution.

takes place (Fig. 2b). This results from the course of the following reaction:

$$[P]\text{-}SO_2NCl_2 + 2Fe^{2+} + H^+ \rightarrow [P]\text{-}SO_2NClH + 2Fe^{3+} + Cl^- \quad (5)$$

In the second stage (the addition of 50–100% of the reductant) the redox potential is lower and it remains approximately at a constant level of ca +450 mV. Since Fe^{2+} ions were detected in the reaction medium, the reaction (6) did not run quantitatively to its end. The reason is the lower oxidation power of the second chlorine atom in the *N*-monochlorosulfonamide group:

$$[P]\text{-}SO_2NClH + 2Fe^{2+} + H^+ \rightarrow [P]\text{-}SO_2NH_2 + 2Fe^{3+} + Cl^- \quad (6)$$

In that part of the research conducted with the dynamic method we used beds of DCSR with the volume of ca 17.5 cm^3, which contained ca 82 mequiv of

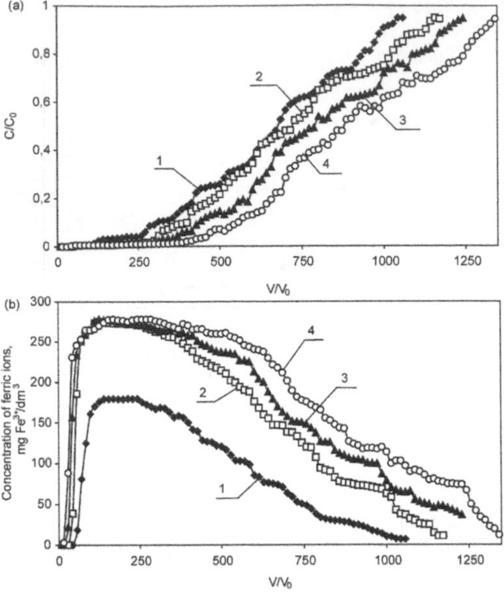

Figure 3. (a) Ferrous ions breakthrough curve for DCSR in the column process: influx 0.005 M $FeSO_4$ in (1) water, (2) 0.005 M H_2SO_4, (3) 0.01 M H_2SO_4, (4) 0.05 M H_2SO_4; flow rate 15 bed volumes/h. (b) Concentration of ferric ions in effluent.

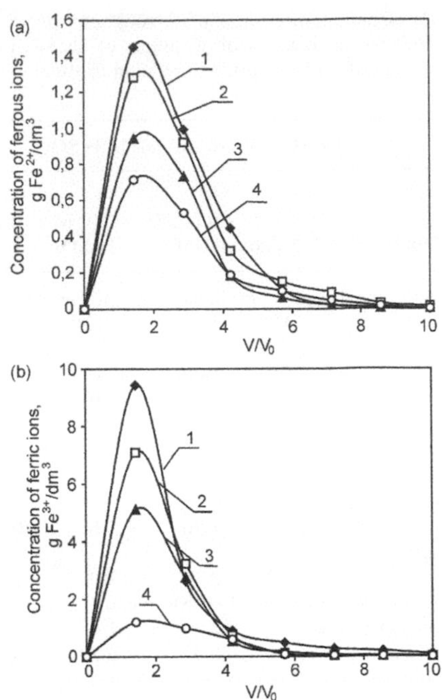

Figure 4. Elution curve of (a) ferrous ions, (b) ferric ions by 1 M H_2SO_4 from exhausted resin bed (number of the column process like in Fig. 3).

active chlorine. Different solutions of $FeSO_4$ having increasing acidity, containing ca 280 mg Fe^{2+}/dm^3 were allowed to pass through the column. We applied such a high iron concentration to enable us to quickly establish the operating capacity of the resin bed in the column. Because of a favorably active chlorine content in the copolymer, column runs were time-consuming. To exhaust the full oxidizing capacity of the copolymer 20 to 25 dm^3 influent had to be passed through the column.

In Figure 3a breakthrough curves for the influent Fe^{2+} ion concentration, whereas in Figure 3b the corresponding effluent Fe^{3+} ion concentration, observing the alternating acidity rising media were presented.

Of the four breakthrough curves presented in Figure 3a the fourth has the most advantageous form; it characterizes the oxidation of Fe^{2+} to Fe^{3+} ions in 0.05 M H_2SO_4. This medium represents the most acidic solution (the pH of the effluent was 1.95). Analyzing the initial samples of the effluent (up to 7 dm^3, $V/V_0 = 400$) only traces of ferrous ions (<1.0 mg Fe^{2+}/dm^3) were observed. Instead ferric ions (ca 275 mg Fe^{3+}/dm^3) and chlorides (ca 90 mg Cl^-/dm^3) were detected. With the progress of the column process the Fe^{2+} cations concentration increased whereas the Fe^{3+} and chloride concentrations dropped. When the C/C_0 value exceeded 0.95, i.e. after passing 24 dm^3 of the solution (V/V_0 ca 1350), the column process was terminated. On the basis of the breakthrough curve 4 the amount of ferrous ions oxidized during the process was estimated. It was 78.75 mol Fe^{2+} (at $C/C_0 = 0.5$, $V/V_0 = 900$, i.e. 900 bed volumes $\times$ 17.5 cm^3 $\times$ 0.005 M/dm^3). This is close to the active chlorine content of the resin bed.

The amount of iron missing from the balance was found by rinsing the exhausted resin bed with 1 M H_2SO_4 (Fig. 4a curve 4 and Fig. 4b curve 4). Fe^{2+} (0.71 mmol) ions and Fe^{3+} (1.27 mmol) ions were eluted. Thereafter, the column was emptied and the spent copolymer was dried and weighed. Its mass was 8.0 g. It did not contain any active chlorine.

When the influent was used alone 0.005 M $FeSO_4$, i.e. without additional acid, it was susceptible to oxidation with atmospheric oxygen. The tested solutions after their preparation were used as soon as possible (i.e. within a few hours). The treated solution loaded on the column had a pH value 3.85. The first effluent fractions were the most acidic and had a pH value of ca 3.05. In the next fractions the pH increased steadily, until, at the end of the process, the pH of the effluent and influent were equal in value. The course of breakthrough curve was less favorable than in the case of the previously discussed (Fig. 3a, curve 1). The calculated operating oxidizing capacity of the column bed was ca 62.0 mmol Fe^{2+} ($C/C_0 = 0.5$, $V/V_0 = \sim 700$). From a comparison of the Figure 3a curve 1 and the Figure 3b, curve 1 it follows that the Fe^{2+} ions decrease was not compatible with the Fe^{3+} ions increase in the effluent (this is to say that in the column process more Fe^{2+} ions disappeared than Fe^{3+} ions arrived). If the pH in the medium was ca 3.0 a partial precipitation of ferric compounds occurred.

During this column process the resin bed changed its appearance. The previous cream-colored beads of the copolymer, took on an intensive gray-green color. After termination of the column oxidation process and after washing the resin bed with 1 M H_2SO_4 the gray-green coloration of the copolymer disappeared (the first fractions of the effluent had an intensive yellow color). This time the quantity of the iron retained by the resin bed was considerable (Fig. 4a, curve 1 and Fig. 4b, curve 1) – 1.35 mmol Fe^{2+} and 6.53 mmol Fe^{3+} ions were eluted. The spent copolymer had a mass of 8.4 g. It contained a little of the active chlorine, ca 5.0 mequiv.

To remove iron from the eluate, the fractions below $C/C_0 < 0.02$ (they contained 5 mg Fe^{2+}/dm^3 and 275 mg Fe^{3+}/dm^3) were neutralized with 1 M NaOH up to pH 6.5. During the neutralization a red-brown precipitate of $Fe(OH)_3$ was observed. The liquor above the precipitate was clear and colorless.

4 CONCLUSIONS

The copolymer S/DVB which contain active chlorine in the $-SO_2NCl_2$ groups is an effective oxidant of ferrous ions present in water, transforming them into ferric ions which are removable considerably more easily. *N,N*-dichlorosulfonamide oxidation reactions are favored by a low reaction pH in contrast to the known aeration process used in water treatment. Both active chlorine atoms have a powerful but differentiated oxidative power, the first chlorine is much stronger than the second. The investigated reaction has a very favorable stoichiometry – 1 mequiv of the active chlorine oxidizes 1 mmol ferrous ions to ferric ions, thus the oxidative capability of the copolymer in relation to the iron is very high and amounts to ca 450 mg Fe^{2+}/g. The oxidation reaction of ferrous ions, by means of the water insoluble heterogeneous reagent, proceeds favorably as can be seen from the large flow intensities applied in the column experiments. At a flux intensity of 15 bed volumes/hour a ca 100 – fold lowering of the concentration of ferrous ions occurred. Ferric ions formed in the treated solution were easily and quantitatively removed by precipitation of $Fe(OH)_3$ after neutralization of the solution. If the pH value of the reaction medium was >2.0, a part of the iron remained in the copolymer (bound by sulfonic groups or precipitated in the resin bed). It has been shown that the iron remaining in the copolymer can be easily eluted by means of 1 M H_2SO_4 and that a higher part in the eluate belongs to the Fe^{3+} than to the Fe^{2+} ions. This method of oxidation of ferrous ions with polymer-supported active chlorine is especially useful and effective for the treatment of low concentrated solutions. Thus, it provides iron free solutions, otherwise difficult to obtain by traditional methods using low molecular weight homogenous water-soluble oxidants.

REFERENCES

Bogoczek, R. & Kociołek-Balawejder, E. 1986. *N*-Mono-halogeno- and *N,N*-dihalogeno (styrene-co-divinylbenzene)sulfonamide. *Polym. Commun.* 27(9): 286–288.

Bogoczek, R. & Kociołek-Balawejder, E. 1989. Studies on a Macromolecular Dichloroamine – the *N,N*-Dichloro Poly(Styrene-co-Divinylbenzene)sulfonamide. *Angew. Makromol. Chem.* 169(2774): 119–135.

Bogoczek, R., Kociołek-Balawejder, E. & Stanisławska, E. 2006. A macromolecular oxidant, the *N,N*-dichlorosulfonamide for removal of residual nitrites from aqueous media. *React. Funct. Polym.* 66(6): 609–617.

Cumbal, L., Greenleaf, J., Leun, D. & SenGupta, A.K. 2003. Polymer supported inorganic nanoparticles: characterization and environmental applications. *React. Funct. Polym.* 54(1–3): 167–180.

Dambies, L. 2004. Existing and prospective sorption technologies for the removal of arsenic in water. *Sep. Sci. Technol.* 39(3): 603–627.

DeMarco, M.J., SenGupta, A.K. & Greenleaf, J.E. 2003. Arsenic removal using polymeric/inorganic hybrid sorbent. *Water Res.* 37(1): 164–176.

Emerson, D.W., Shea, D.T. & Sorensen, E.M. 1978. Functionally modified poly(styrene-divinylbenzene). Preparation, characterization, and bacterial action. *Ind. Eng. Chem. Prod. Res. Dev.* 17(3): 269–274.

Fries, J. 1971. *Spurenanalyse. Erprobte photometrische Methoden.* Darmstadt: E. Merck.

Gowda, H.S., Shakunthala, R. & Subrahmanya, U. 1981. Redox indicators in titrations with dichloramine-T. *J. Indian Chem. Soc.* 58(6): 567–570.

Gutsanu, V.L. 1990. Sorption and state of iron in ion exchangers. *Khimiya I Tekhnologiya Vody* 12(12): 1074–1094 (in Russian).

Jacob, T.J. & Nair, C.G.R. 1972. Dichloramine-T as a new oxidimetric titrant in nonaqueous and partially aqueous media. *Talanta* 19(3): 347–351.

Katsoyiannis, I.A. & Zouboulis, A.I. 2002. Removal of arsenic from contaminated water sources by sorption onto iron-oxide-coated polymeric materials. *Water Res.* 36(2): 5141–5155.

Kociołek-Balawejder, E. 2000a. A macromolecular *N,N*-dichlorosulfonamide as oxidant for cyanides. *Eur. Polym. J.* 36(2): 295–302.

Kociołek-Balawejder, E. 2000b. A macromolecular *N,N*-dichlorosulfonamide as oxidant for thiocyanates. *Eur. Polym. J.* 36(6): 1137–1143.
Kociołek-Balawejder, E. 2002. A macromolecular *N,N*-dichlorosulfonamide as oxidant for residual sulfides. *Eur. Polym. J.* 38(5): 953–959.
Podszun, W., Schleger, A., Klipper, R., Seidel, R. & Herrmann, U. 2005. Process for the preparation of iron-oxide- and /or iron-oxyhydroxide-containing ion exchangers. US 0038130 A1.
White, G.C. 1999. *Handbook of Chlorination and Alternative Disinfectants*. New York: Wiley.

Environmental Engineering – Pawłowski, Dudzińska & Pawłowski (eds)
© 2007 Taylor & Francis Group, London, ISBN13 978-0-415-40818-9

Chlorine decay and disinfection by-products in water distribution systems

Beata Kowalska, Dariusz Kowalski & Anna Musz
Department of Environmental Protection Engineering, Lublin University of Technology, Lublin, Poland

ABSTRACT: Presented work contains the literature review of chlorination as a one of frequently used water disinfection process. The main aspect of the paper is a presentation of chlorine characteristics in comparison with other disinfectants, review of various models of chlorine decay and disinfection by-product formation, major of which is trihalomethane (THM). All included formulas can help to predict chlorine decay in extensive and complex distribution systems which can lead to limitation of water-quality deterioration process. The following article is the first part of a wide research on description of contaminants distribution and connected with them secondary pollution process in water distribution system.

Keywords: Chlorine decay, chlorine demand, disinfection by-products (THMs).

1 INTRODUCTION

Chlorine is the most popular disinfectant used in drinking water distribution systems and performs several important functions. It must achieve an adequate inactivation of microorganisms before the treated water reaches the first consumer, and be large enough to ensure an adequate residual at the periphery of the distribution system. Although the disinfection protects the drinking water from the pollution, side reactions can cause health risks due to the formation disinfection by-products (DBPs), major of which are trihalomethanes (THM). In order to help predict chlorine decay, biofilm growth and substrate utilisation, numerical models have been developed.

Typically, source water is clarified and disinfected at a water treatment plant, before it is discharged into a drinking water distribution system (Kowal 2003). Disinfection is the most important stage in the treatment of drinking water and chlorine (or other disinfectant) is applied in the clearwell, the final stage of treatment (Abdullah et al. 2003, Boccelli et al. 2003, Sohn et al. 2004). Water treatment process was to produce water of acceptable quality complying with established Polish standards. Despite this, there are still failures in estimating the quality of water especially at the periphery of the distribution system.

In recent years there has been a decrease of water consumption in Poland, and as a consequence of this, lower hydraulic velocity is evident in the distribution pipelines (Winn-Jung et al. 1999). This situation may change the functioning of the existing systems and the water quality inside them (Kłos-Trębaczkiewicz et al. 2000, Kuś et al. 2003). Such systems function as biological and chemical reactors and with the long detention time, impact on the transported water. This can lead to the water quality deterioration in water that reaches the consumer (Kłos-Trębaczkiewicz et al. 2000, Kuś et al. 1999, Roman et al. 2001). Such deterioration is known as "secondary pollution" and it is presently being studied by many researcher (Dharmarajah et al. 1991, Garcia-Villanova et al. 1997, Kowal 2003, Świderska-Broż 2003, Zacheus et al. 2001). The following is a brief review of recent studies undertaken to describe chlorine decay and disinfection by-products formation in distributed water reported in the literature. It includes the first part of a wide study on secondary pollution in the water distribution system of the city of Lublin.

This will be discussed in the next paper described.

2 DISINFECTANTS USED IN DRINKING WATER

A high-quality disinfectant should have a strong bactericidal effect and a high stability in aqueous environment to ensure to inactivate of microorganisms and inhibit their regrowth in the network (Haas 1999). It is also essential to maintain the optimum disinfectant concentration within the whole distribution system (Świderska-Bróż 2001). In order to achieve all of the above, the following are desirable (Hill et al. 1999, Kowal 2003)

- maintain a high level of water treatment that ensures inactivation and/or removal of microorganisms. The

Table 1. Standard oxidation potential of selected disinfectants and their reactions in water (Haas 1999).

Compound	Symbol	Potential [V]	Reaction
Ozone	O_3	2.07	$O_3 + 2e^- + 2H^+ \leftrightarrow O_2 + H_2O$
Chlorine dioxide	ClO_2	1.91	$ClO_2 + 5e^- + 2H_2O \leftrightarrow Cl^- + 4OH^-$
		0.95	$ClO_2 + e^- \leftrightarrow ClO_2^{-*}$
Chlorine	Cl_2	1.36	$Cl_2 + 2e^- \leftrightarrow 2Cl^-$
Bromine	Br_2	1.09	$Br_2 + 2e^- \leftrightarrow 2Br^-$
Iodine	I_2	0.54	$I_2 + 2e^- \leftrightarrow 2I^-$

* This reaction occurs frequently in water.

process is also effective in the removal of the organic carbon and nitrogen compounds,
- treatment plants and reservoirs are maintained in a good condition,
- periodical cleaning of the distribution network,
- to protect against water contamination during repair of damage or failure of the network system,
- to ensure that distributed drinking water meets microbiological quality standards at all times,

to maintain a disinfectant residual within the distribution system for an adequate time i.e it is maintained to reach the customer tap.

To evaluate the bactericidal effect of a disinfectant, it is important to know its ability to damage the cellular membranes of microorganisms or its ability to diffusion through the membranes and disrupt metabolism, biosynthesis and cell growth (Haas 1999). The higher the oxidation potential of the disinfectant, the easier it oxidizes the organic matter. The standard potentials of some disinfectants and their reactions in water are shown in table 1.

Disinfectant stability can be described by its lack of reactivity with constituents other those that lead to destruction or inactivation of microorganisms, and its effective bactericidal time (Kłos-Trębaczkiewicz et al. 2000). The bactericidal effectiveness of disinfectants changes according to the order shown in table 1, though their stability in the water is reverse to this (Hoff et al. 1981)

$O_3 < Cl_2 < ClO_2 <$ CHLORAMINES

2.1 *Chlorine demand of bulk water*

Rossman et al. (Rossman et al. 2001) defined bulk chlorine decay based on reactions with dissolved and suspended matter in the water, mostly natural organic matter (NOM), and thus ignored those reactions with compounds attached to or derived from pipe materials. For most waters, the reactions of chlorine with NOM make up the majority of the chlorine demand. Chlorine also reacts with various inorganic compounds, for example, with ammonia to form different species of chloramines (Jadas-Hecart et al. 1992).

In distribution water, a low number of suspended particles can be introduced into the system because of incomplete removal of particles from raw water during water treatment, external contamination in sources or pipes etc. (Gauthier et al. 1999). They can be found in different amounts especially in a large distribution systems and they can differ throughout the system. Suspended particles generate the following problems for distributed water quality: they can carry bacteria fixed on their surface which may protect them from disinfection and they contribute to the formation of loose deposits in sources and pipework, which are resuspended into the water phase when a change occurs in the hydraulic properties of the system (direction, velocity, water hammer etc.) (Gauthier et al. 1999, Lu et al. 1999).

Disinfectant decay in the bulk water may be separated from wall influenced decay by carrying out chlorine decay experiments on the treated water under controlled conditions in the laboratory (Hua et al. 1999). The bulk decay was observed to have an inverse relationship with the initial chlorine concentration (Hallam et al. 2002). Kinetic rate equations describing the decay of chlorine are further discussed later in this paper.

2.2 *Chlorine demand of biofilm*

It is important in predicting chlorine decay to understand the relation between biofilm (fixed biomass) formation and its chlorine demand. Deposits covering the pipe walls impede the contact of chlorine with biofilm, which consists of microorganisms like bacteria, fungi, and their metabolites (Donlan et al. 1986, Świderska-Bróż 2003, Van der Wende et al. 1990). The optimum conditions for growing the biomass of biofilm are at temperature above 15°C and with no-flow of water in the network (Schmidt et al. 2002, Ndiongue et al. 2005). However, LeChevallier observed that biofilm growth was possible at temperatures above 5°C (LeChevallier et al. 1987).

In describing the chlorine demand of biofilms in water distribution systems, Wen Lu *et al.* (Lu et al. 1999) pointed out that the main parameters influencing biomass formation was temperature and natural organic matter (NOM). The fraction of NOM available to biofilm as organic carbon is defined as biodegradable dissolved organic carbon (BDOC). The BDOC depends on substrate type and environmental conditions at the time DOC entered to water phase. The BDOC, consisting primarily of humic substances, amino acids and carbohydrates reacts with disinfectant (Kaplan et al. 1993, Malcolm 1991). For example, amino acids, are substrates that have high demand for chlorine and result in the formation of high amounts

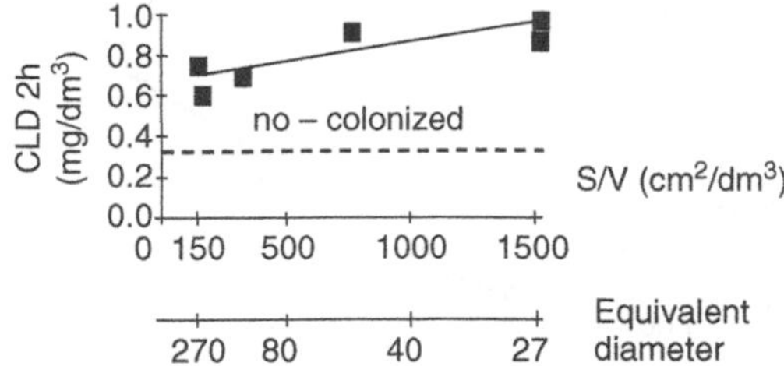

Figure 1. Variation of initial chlorine demand (CLD) as a function of S/V (at water temperature 18°C) (Lu et al. 1999).

of biomass whereas carbohydrates have a low chlorine demand and result in the least biofilm biomass formation in the presence of chlorine (Butterfield et al. 2002). Butterfield (Butterfield et al. 2002) and Winn-Jung (Winn-Jung et al. 1999) have shown that the presence of free chlorine results in less accumulation of biomass within biofilm compared to non-chlorinated water with the same quality parameters.

In practice, chlorine doses are frequently not sufficient to remove all biomass, but sufficient to not allow regrowth of microorganisms (Clement et al. 1991). Adequate chlorine level in the water and its contact time (the time measured from the point of disinfectant application to the first consumer) depends on the treated water quality. Under Polish drinking water standards, a sufficient chlorine dose to prevent biomass formation, equals $0.3 \div 0.5$ free chlorine, or higher if it occurs as chloramines. Too high levels of free chlorine cause odours and deterioration of water quality (Suffet et al. 1996) which is one of the most frequent causes of consumer complaint.

Chlorine decay depends on the surface/volume (S/V) ratio, which is an important factor to consider in modelling chlorine decay in the distribution system. This dependence increases when the ratio S/V varies from 150 to 1500 cm^2/dm^3 (Lu et al. 1999). Wen Lu et al. found a linear relation between initial chlorine demand (in 2 hours) of biomass and the S/V ratio (Fig. 1). When the S/V increases, that is when the pipe diameter decreases, the chlorine demand also increases. It means that for small pipe diameters biomass formation should be controlled to avoid rapid chlorine consumption (Lu et al. 1999).

2.3 *Chlorine demand of pipe wall*

Wen Lu et al. (Lu et al. 1999) pointed out that chlorine decay due to the pipe wall is also affected by the ratio surface/volume, (S/V); the higher the S/V ratio, the more the inner pipe surface is in contact with chlorine. Wen Lu et al. (Lu et al. 1999) showed that chlorine consumption due to new plastic pipes (PVC and PE) is negligible compared to the bulk water chlorine demand. Kiéne and Lévi (Kiéné et al. 1996) reported that in aged grey cast iron pipes, chlorine is principally consumed by the material corroded and deposits. Chlorine demands from different kinds of pipes are shown in Figure 2.

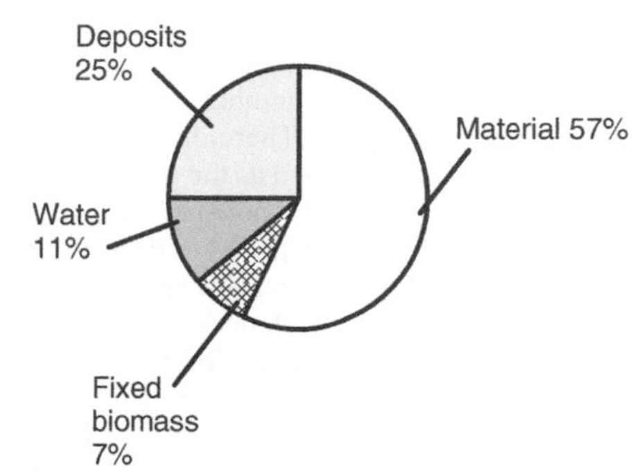

Figure 2. Chlorine demand (2 hours) in networks made of (a) PVC and PE pipes and (b) grey cast iron pipes of a diameter of 250 mm (Lu et al. 1999).

As shown in Figure 2, total chlorine consumption (2 hours) in plastic pipes is over two times lower than for cast iron pipes. In cast iron pipes, the chlorine demand of biomass is negligible, and chlorine is principally consumed by the wall material, deposits and the bulk water components.

In summary, it can be stated that the main factors influencing chlorine consumption are as follows

- reaction with organic and inorganic chemicals in the bulk aqueous phase,
- reaction with biofilm at the pipe wall,
- consumption by the corrosion process of the pipe wall,
- mass transport of chlorine and other reactants between the bulk flow and the pipe wall S/V ratio and pipe material (Lu et al. 1999).

3 MODELLING OF CHLORINE DECAY IN WATER DISTRIBUTION SYSTEMS

Chlorine disinfection requires that a minimum disinfectant level be maintained in all parts of a distribution system. Therefore it is important that the factors that influence chlorine decay be identified and that models that can reliably predict chlorine residual levels in treated and distributed water be developed. In order to achieve a balance between sufficient chlorination to ensure bacteriology quality, it is necessary to understand the mechanism of chlorine decay in water distribution systems and the factors affecting it. Chlorine disappears due to its reactions with compounds present in water, which in most cases is unknown. For these reasons, most kinetic models describing chlorine decay have been established empirically or semi-empirically. The key issue of using decay equations is the determination of the decay constants, which can very with the quality of the source water, the water temperature, the Reynolds number and the material properties of the water pipes. Therefore, the total decay constant (k) is often expressed as the decay due to the chlorine demand of the pipe (known as the wall decay constant (k_w) and that due to quality of the water itself, known as the bulk decay constant (k_b) (AWWARF 1996, Beatty et al. 1996). Bulk decay may be isolated from wall decay by carrying out chlorine decay experiments on the source water under controlled conditions in the laboratory.

A number of models have been developed to predict chlorine decay in drinking water (Boccelli et al. 2003, Clark 1998, Clakr et al. 1994, Feben et al. 1951, Gallard et al. 2002, Sohna et al. 2004). Some of these models are described below.

3.1 *First order chlorine decay model*

The most popular model is a first order decay model in which the chlorine concentration is assumed to decay exponentially (Boccelli et al. 2003, Clark et al. 1994, Hua et al. 1999)

$$C = C_0 e^{-kt} \tag{1}$$

where C is the concentration at time t in $mg{\cdot}dm^{-3}$, C_0 is initial chlorine concentration in $mg{\cdot}dm^{-3}$, k is the decay rate in min^{-1}, t is time in min.

The decay constant k in the model is often considered as the bulk decay constant k_b (due to the reaction in the bulk water) and the wall decay constant k_w (due to the reaction with biofilm at the pipe wall or with the pipe wall material itself).

Fang Hua et al. (Hua et al. 1999) reported the effects of water quality parameters on the bulk decay constant of free chlorine, in different water samples. They also found an empirical relationship between initial chlorine concentration and the bulk decay constant (k_b) for a fixed temperature for three types of water. These are represented as follows,

$$k_b \approx \frac{0.018}{C_0} - 0.024 \tag{2}$$

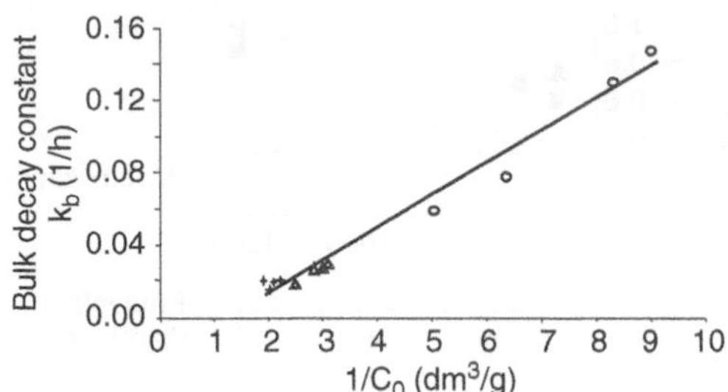

Figure 3. The relationship between the bulk decay constant (measured at 9°C) and the initial chlorine concentration, Δ – final water; ° – tap water (water from a tap at the laboratory without re-chlorination, sourced from the same treatment plant after flowing through the distribution pipes), * – re-chlorination water (water from the same tap, which was re-chlorinated with sodium hypohlorite to an initial concentration of about 0.5 $mg{\cdot}dm^3$) (Hua et al. 1999).

Table 2. Effect of pipe material on wall chlorine decay constant k_w (Hallam et al. 2002).

Pipe material	CI	SI	DICL	MDPE	PVC
Wall chlorine decay constant k_w (h^{-1})	0.67	0.33	0.13	0.05	0.09

CI – cast iron, SI – spun iron, DICL – cement-lined ductile iron, MDPE – medium density polyethylene, PVC – polyvinyl chloride.

As can be seen from Figure 3 the decay constant k_b is inversely proportional to the initial concentration of chlorine, C_0.

Hallam et al. (Hallam et al. 2002) conducted the experiment on different kinds of pipes in order to find an effect of pipe material on the wall chlorine decay constant k_w in the first-order decay equation. They separated pipes into relatively reactive pipes (CI – cast iron and SI – spun iron pipes) and relatively unreactive pipes (MDPE – medium density polyethylene, PVC – polyvinyl chloride and DICL – cement-lined ductile iron pipes).

Their study aimed to examine a wide range of pipe types with more repeated wall decay determinations for *in situ* conditions. Results of their study are shown in table 2.

From the results shown in table 2, the pipe reactivity, as determined from the wall decay constants, are as follows CI < SI < DICL < PVC < MDPE. Unlined cast iron pipes have decay rates between 4 and 100 times greater than lined or plastic pipes.

3.2 *Second-order chlorine decay model*

Clark (Clark 1998) developed a second-order chlorine decay model, based on the concept of competing reacting substances. He investigated two component model accounts for both disinfectant and a fictitious reactant via the hypothetical irreversible reaction (Clark 1998)

$$aA + bB \rightarrow pP \qquad (3)$$

where A is the chlorine component, B is an unknown fictitious reactive component, P is the disinfectant by-product component, and a, b and p are stoichiometric reaction coefficients.

Clark (Clark 1998) assumed the reaction rate to be first-order with respect to A and B, and second-order overall,

$$\frac{dC_A}{dt} = -k_A C_A C_B, \qquad \frac{dC_B}{dt} = -k_B C_A C_B \qquad (4)$$

where C and k are the concentration and decay rate coefficient for the disinfectant (subscript A) and reactive component (subscript B), respectively. The analytical solution to (4) is

$$C_A(t) = \frac{C_{A,0} - aC_{B,0}/b}{1-(aC_{B,0}/bC_{A,0})} \cdot \frac{1}{\exp[-(bC_{A,0}/aC_{B,0}-1)k_A C_{B,0} t]} \qquad (5)$$

where $C_{A,0}$ and $C_{B,0}$ are the initial concentrations of the disinfectant and unknown fictitious reactive component at $t = 0$, and the ratio a/b and k_A are the parameters that must be estimated from experimental data. Since the reactive component is unknown, the initial concentration $C_{B,0}$ must also be incorporated in the estimation process.

Jedas-Hecart et al. (Jadas-Hecart et al. 1992) and Ventresque et al. (Ventresque et al. 1990) divided chlorine decay into two phases – an initial phase of immediate consumption during the first 4 hours and the second phase of slower consumption after the first 4 hours, known as the long term demand. Because the initial decay is rapid, Dharmarajah and Patania (Dharmarajah et al. 1991) suggested a second order decay equation for the first phase and the first order decay equation for the second phase.

3.3 *Combined first and second order model*

Fang Hua et al. (Hua et al. 1999) proposed a semi-empirical combined first and second order model, which provides a good description of chlorine decay in the first (initial) and the second stages.

This model was derived from the equation

$$\frac{dC}{dt} = -k_1 C - k_2 C^2 \qquad (6)$$

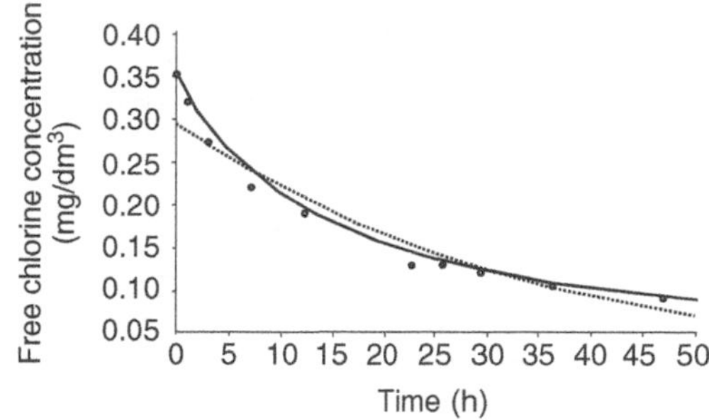

Figure 4. Comparison of model predictions and experimental data for water from a treatment plant: — combined first and second order model, $k_1 = -0.004\,h^{-1}$, $k_2 = 0.0199\,dm^{-3}\cdot mg^{-1}\cdot h^{-1}$, $r^2 = 0.992$, ... first order model, $k_b = 0.029\,h^{-1}$, $r^2 = 0.926$, ° – experimental data (Hua et al. 1999).

After integration (4) we have

$$\frac{1}{C} + \frac{k_2}{k_1} = \left(\frac{1}{C_0} + \frac{k_2}{k_1}\right) e^{k_1 t} \qquad (7)$$

where the decay constants k_1 and k_2 are the functions of the overall decay constant k and were determined by deriving the best fit of equation (7) with the experimental data. The comparison of the first order model and this combined model using experimental data for water from a treatment plant at a fixed temperature 9°C is shown in Figure 4.

It can be seen that the combined first and second order model provide a better description of chlorine decay than the first order model especially in the initial (rapid) stage.

In summary, the modeling of chlorine decay is very difficult because of complexity of water quality and the changes in the distribution system. But such modeling is important because disinfection processes change the composition and characteristics of the water distributed to the consumer. Mathematical models for predicting the decay of chlorine help to predict the formation of disinfectant by-products which can cause risk to health.

4 FORMATION OF DISINFECTION BY-PRODUCTS

The interaction between disinfectant and precursor materials in source water results in the formation of disinfection by-products (DBPs). Chlorination was reported to form trihalomethanes (THMs), haloacetic acids (HAAs), haloacetonitriles (HANs), chlorophenols anh chloroketones (Abdullah et al. 2003, Adin et al. 1991, Clark et al. 1998, Clark et al. 2001, Rodriguez et al. 2001, Urs von Gunten et al. 2001,

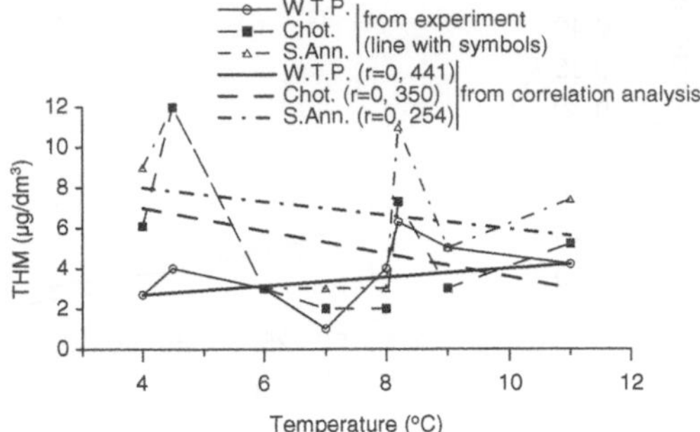

Figure 5. Formation of THM in different temperature (El-Dib et al. 1995).

Waller et al. 1998, Zbieć et al. 1999). Special attention has been paid to the concentration of trihalomethanes because of their potential carcinogenic effects (Boccelli et al. 2003, Bull et al. 1991). THMs (mainly chloroform $CHCl_3$) which result from the reaction of chlorine with naturally occurring organic matter, principally humic and fulvic acids, represent between 5 and 20% of the chlorinated products formed during the chlorination process (Abd–El.-Shafy et al. 1998).

The focus on the occurrence of DBPs in drinking water has increased in recent years and concentrates on the attempting to characterize the nature of DBPs and the conditions that govern their formation in drinking water. Much research has been invested in developing of mathematical models describing the decay of chlorine (or other disinfectants) for predicting the formation of DBPs themselves (Abdullah et al. 2003, Boccelli et al. 2003, Clark et al. 1994, Cozzolino et al. 2005, Elshorbagy et al. 2000, Garcia-Villanova et al. 1997, Nokes et al. 1999). The main precursor compounds of THM are humic acids, chlorophyll-a, metabolites of aqueous compounds, aliphatic hydroxy acids, mono-, di-, tri-carboxyacids, aromatic carboxyacids (Morris et al. 1975).

Many factors can affect THM formation, such as organic matter, pH value, temperature, contact time, dose and concentration of chlorine etc. (Adin et al. 1991, Gang et al. 2003, Garcia-Villanova et al. 1997, Zbieć et al. 1999).

Abd El.-Shafy and Grünwald (Abd El.-Shafy et al. 1998) found an increase in THM formation due to the increase of temperature for different sampling points, as can be seen in Figure 5. It was confirmed by other researches (El-Dib et al. 1995) but the correlation between temperature and THM is not clear because of the small change of this parameter during the study.

A good correlation was found between THM and chloroform ($CHCl_3$) concentrations and THM and $CHCl_3$ creation (Fig. 6). The increase of THM and $CHCl_3$ concentrations at different residence time was formulated as exponential functions of the residence

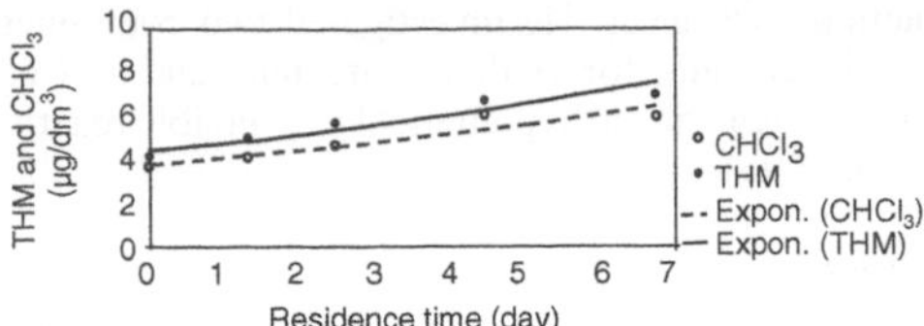

Figure 6. THM and $CHCl_3$ concentration at different residence times (El-Dib et al. 1995).

time t (or first order increase) as the following (Abd El.-Shafy et al. 1998)

$$THM_t = THM_0 \exp(kt) \tag{8}$$

$$CHCl_{3t} = CHCl_{30} \exp(kt) \tag{9}$$

where k is the coefficient of first order increase, THM_t and $CHCl_{3t}$ are the concentrations of THM and $CHCl_3$ at time, t, and THM_0 and $CHCl_{30}$ refer to their initial concentrations. These formulae can be used to predict the concentration of trihalomethane and chloroform in the network after the determination of the coefficient k.

Clark (Clark et al. 1998) noted that THMs are formed as a linear function of the chlorine demand

$$THM(t) = T\,x(t) + M \tag{10}$$

where x(t) is the total chlorine demand at time t, T is a parameter relating THM formation to chlorine demand, and M is the THM concentration present at $t = 0$ ($M > 0$ if the model is applied to waters that have natural background concentrations and/or were previously chlorinated).

Boccelli et al. (Boccelli et al. 2003) suggested that, because of non-zero value of M, the relationship between THM formation and chlorine demand during the initial rapid chlorine decay phase may not be linear. Therefore, the THM model parameters can be determined directly from the experimental data, or by estimating them through a coupled chlorine decay and THM formation model.

Abdullah et al. (Abdullah et al. 2003) established some relationships between THM formation and different environmental conditions. THM formation was higher with increasing soluble humic material content in natural occurring water. The rate of THM formation is equal to that of the total organic carbon (TOC) consumption. In general, the rate of THM production increases with pH. Abdullah et al. (Abdullah et al. 2003) also determined the effect of chlorination dosage on the production of THM. They observed a weak but definite relationship. A similar correlation was found between THM formation and the distance from treatment plant. From the effect of residue chlorine on the production of THM, a small relationship was found between the two factors. As THM concentration increases, the concentration of HOCl (formed

Table 3. Relationship between formation of THM and the factors group of water treatment plant and distribution system in conditions of Tampin and Sabak Bernam district (Abdullah et al. 2003).

THM formation within dependent variable	Tampin ($n = 74$) Pearson coeficient, r	Sabak Bernam ($n = 117$) Pearson coeficient, r
TOC	0.380	0.478
pH	0.362	0.215
Chlorine dose	0.233	0.505
Distance from treatment plant	0.353	0.205
Chlorine residue	−0.311	−0.134

by chlorine with water) was found to decrease. These results of simple regression analyses are shown in table 3.

5 CONCLUSIONS

Disinfection is the most important process at a water treatment plant, to reduce risk of infectious disease from water consumption. However, the interaction between chemical disinfectants and precursor materials in source water results in the formation of disinfection by-products (DBPs), themselves pose a health risk. It is necessary to understand the mechanism of chlorine decay (or other disinfectant decay) in the water distribution system and the factors affecting it.

It is difficult to predict chlorine decay in extensive and complex distribution systems because it reacts with the compounds present in bulk water, which in most cases remain unknown. It also reacts with the biofilm at the pipe wall or with the pipe material.

Much research has been invested in attempting to characterize the nature of reactions associating the chlorination of drinking water in distribution systems. One aspect of the research is the development of mathematical models for predicting the decay of chlorine and for predicting the formation of DBPs themselves. Knowledge of actual relationships between kinetic parameters provides a better understanding of the true effects of free chlorine on the formation of DBPs.

The large number of variables limits extensive applications of the chlorine decay models. For these reasons it is impossible to find universal model predicting chlorine decay or formation disinfection by-products. Generally mathematical models are established empirically or semi-empirically. Each distribution system should be considered individually, according to its different environmental conditions. Studies on the disinfection of drinking water and the formation of disinfection by-products are still the aim of many researchers and are central to managing water quality by the water treatment industry.

REFERENCES

Abd El.-Shafy, M., Grünwald, A., Macek, L. 1998. Changing water quality in drinking water pipelines. *In Proceedings of the Master Plans for Water Utilities, June 17–18,* Prague, Czech Republic.

Abdullah, Md.P., Yew, C.H., bin Ramli, M.S. 2003. Formation modeling and validation of trihalomethanes (THM) in Malaysian drinking water: a case study in the districts of Tampin, Negeri Sembilan and Sabah Bernam, Selangor, Malaysia. *Water Research 37: 4637–4644.*

Adin, A., Katzhendler, J., Alkaslassy, D., Rav, A.C. 1991. Trihalomethanes formation in chlorinated drinking water: a kinetic model. *Water Research 25: 797–805.*

AWWARF: *Characteristics and modelling chlorine decay in distribution system,* USA: AWWA 1996.

Axworthy, D.H., Karney, B.W. 1996. Modeling Low Velocity/High Dispersion Flow in Water Distribution Systems. *Journal of Water Resources Planning and Management, May/June: 218–221.*

Beatty, R., Bliss, P.J., Vintage, D.C. 1996. Analysis of factor influencing chlorine decay in pipe distribution systems. *J. AWWA 16: 159–165.*

Boccelli, D.L., Tryby, M.E., Uber, J.G., Summers, R.S. 2003. A reactive species model for chlorine decay and THM formation under rechlorination conditions. *Water Research 37: 2654–2666.*

Bull, R.J., Kopfler, R.C. 1991. Health effects of disinfectants and disinfection by-products. Denver, CO: *American Water Works Association Research Fundation.*

Butterfield, P.W., Camper, A.K., Ellis, B.D., Jones, W.L. 2002. Chlorination of model drinking water biofilm: implications for growth and organic carbon removal. *Water Research 36: 4391–4405.*

Clark, R.M. 1998. Chlorine demand and TTHM formation kinetics: a second-order model. *J Environ Eng ASCE 124(1): 16–24.*

Clark, R.M., Sivaganesan, M. 1998. Predicting chlorine residuals and the formation of TTHMs in drinking water. *Journal of Environmental Engineering Division of the ASCE 124(12): 1203–1210.*

Clark, R.M., et al. 1994. Managing water quality in distribution systems: Simulating TTHM and chlorine residual propagation. *Journal of Water Supply, Research and Technology – Aqua 43(4): 182–191.*

Clark, R.M., Thurnau, R.C., Sivaganesan, M., Ringhand, P. 2001. Predicting the formation of chlorined and brominated by-products. *Journal of Environmental Engineering Division of the ASCE 127(6): 495–501.*

Clement, J.A., et al. 1991. The disinfectant residual dilemma. *Journal AWWA 1: 24–30.*

Cozzolino, L., Pianese, D., Pirozzi, F. 2005. Control of DBPs in water distribution systems through optimal chlorine dosage and disinfection station allocation. *Desalination 176: 113–125.*

Dharmarajah, H., Patania, N. 1991. Empirical modeling of chlorine and chloramine residual. *AWWA Proceedings: Water Quality for the New Decade. Annual Conference, Philadelphia, June 1991, PA: 569–577.*

Donlan, R.M., Pipes, W.O. 1986. Pipewall biofilm in drinking water mains. *Proc. AWWA Water Qual. Tech. Conf. Portland, OR, Vol. 14: 637–660.*

El-Dib, M., Ali, R. 1995. THMs formation during chlorination of Raw Nile River water. J.R. *J. Wat. Res., Egypt 29(1): 375–378.*

Elshorbagy W.E., Abu-Qdais, H., Elsheamy, M.K. 2000. Simulation of THM species in water distribution system. *Water Research 34(13): 3491–3439.*

Feben D., Taras, M.J. 1951. Studies on chlorine demand constants. *Journal of the American Water Works Association 43(11): 922–932.*

Gallard H., Urs von Gunten. 2002. Chlorination of natural organic matter: kinetic of chlorination. *Water Research 36: 65–74.*

Gang, D., Clevenger, T.E., Benerji, S.K. 2003. Relationshop of chlorine decay and THMs formation to NOM size. *Journal of Hazardous Materials A96: 1–12.*

Garcia-Villanova, R.J., Garcia, C., Gomez, J.A. et al. 1997. Formation evolution and modeling of trihalomethanes in the drinking water of a tawn: I. At the municipal treatment utilities. *J. Wat. Res.41: 251–255.*

Gauthier, V., Gérard, B., Portal, J-M., Block, J-C., Gatel, D. 1999. Organic matter as loose deposits in a drinking water distribution system. *Water Research 33(4): 1014–1026.*

Haas, C.N. 1999. Benefits of using a disinfectant residual. *Journal AWWA, 1: 65–69.*

Haas, C.N. 1990. Water Quality and Treatment – Disinfection. A Handbook of Community Water Supplies, 4d ed, *American Water Works Association*, Mc Graw-Hill, Inc., New York 1990.

Hallam, N.B., West, J.R., Forster, C.F., Powell, J.C. , Spencer. 2002. The decay of chlorine associated with the pipe wall in water disinfection systems. *Water Research 36: 3479–3488.*

Hill, J.Y., et al. 1999. The effect of water treatment on biological stability of potable water. *Water Research 11(33): 2287.*

Hoff, J.C., Geldreich, E.E. 1981/1. Comparison of the biocidal efficiency of alternative disinfectants. *JAWWA: 40.*

Hua, F., West, J.R., Barker, R.A., Forster, C.F. 1999. Modelling of chlorine decay in municipal. *Water Research 33(12): 2735–2746.*

Jadas-Hecart, A., El Morer, A., Stitou, M., Bouillot, P., Legube, P. 1992. Modelisation de la Demande en Chlore D'une Eau Traitee. *Water Research 26(8): 1073.*

Kaplan, L.A., Newbold, J.D. 1993. Biogeochemistry of dissolved organic carbon entering streams. In: Ford TE (ed.), *Aquatic microbiology, an ecological approach.* Oxford: Blackwell Scientific Publications: 139–165.

Kiéné, L., Lu, W., Levi, Y. 1996. Relative importance of phenomena responsible of the chlorine consumption in drinking water dystribution systems. *Proc. of WQTC AWWA, MA, 17–21 Nov, Boston.*

Kłoss-Trębaczkiewicz, H., Osuch-Pajdźińska, E., Roman, M. 2000. Decrease In water consumption in Polish towns and cities: its causes and effects. *Journal of Water Supply: Research and Technology SRT – AQUA.*

Kowal, A.L. 2003. Przyczyny i zapobieganie zmianom jakości wody w systemach wodociągowych. *Ochrona Środowiska 4: 3–6.*

Kuś, K., Grajper, P., Ścieranka, G., Wyczańska-Kokot, J., Zakrzewska, A. 2003. Wpływ spadku zużycia wody w miastach zaopatrywanych przez wodociąg grupowy GPW w Katowicach na jakość wody w systemie dystrybucji. *Ochrona Środowiska 3: 29–34.*

LeChevallier, M.W., Babcock, T.M., Lee, R.G. 1987. Examination and characterization of distribution system biofilms. *Apll. Environ. Microbiol. 53(12): 2714–2724.*

Lu, W. 1995. Etude des phénomènes responsables de la connsommation du chlore en réseau de distribution systems. *Thesis of Paris 7 University: 232.*

Lu, W., Kiéné, L., Lévi, Y. 1999. Chlorine demand of biofilms in water distribution systems. *Water Research 33(3): 827–835.*

Malcolm, R.L. 1991. Factors to be considered in the isolation and characterization of aquatic humic substances. In: Boren H, Allard B. (ed.), *Humic substances in the aquatic and terrestrial environment.* London: Wiley.

Morris, J.C., McKay, G. 1975. Formation of halogenated organics by chlorination of water supplies. US EPA, Washington.

Ndiongue, S., Huck, P.M., Slawson, R.M. 2005. Effect of temperature and biodegradable organic matter on control of biofilms by free chlorine in a model drinking water distribution system. *Water Research 39: 953–964.*

Nokes, C.J., Fenton, E, Randall, C.J. 1999. Modeling the formation of brominated trihalomethanes in chlorinated drinking waters. *Water Research 33(17): 3557–3568.*

Rodriguez, M.J., Seradel, J.B. 2001. Spatial and temporal evolution of trihalomethanes In three water distribution systems. *Water Research 6: 1572–1586.*

Roman, M., Kłoss-Trębaczkiewicz, H, Osuch-Pajdźińska, E., Kałużna, M., Mikulska, E. 2001. Zmiany zużycia wody w miastach polskich w latach 1987–1998. *Ochrona Środowiska 3(82): 3–6.*

Rossman, L.S., Brown, R.A., Singer, P.C., Nuckols, J.R. 2001. DBP formation kinetics in a simulated distribution system. *Water Research 14: 3483–3489.*

Schmidt, W., et al. 2002. Biofilmbildung en modifizierten SiO_2 – Schutzichten. *Vom Wasser, B. 98: 177–192.*

Sohn, J., Amy, G., Cho, J., Lee, Y., Yoon, Y. 2004. Disinfectant decay and disinfection by-products formation model development: chlorination and ozonation by-products. *Water Research 38: 2461–2478.*

Suffet, H., Corado, A.,Chou, D., McGuire, M.J., Butterworth, S. 1996. Taste and odour survey. *Journal AWWA 4: 168–180.*

Świderska-Bróż, M. 2001. Niepożądane zmiany jakości wody podczas jej oczyszczania i dystrybucji. *Inżynieria i Ochrona Środowiska 4(3–4): 283–300.*

Świderska-Bróż, M. 2003. Skutki braku stabilności biologicznej wody. *Ochrona Środowiska 4: 7–12.*

Świderska-Bróż, M. 1999. Wybrane problemy w oczyszczaniu wody do picia i na potrzeby gospodarcze. *Ochrona Środowiska 3(74): 7–12.*

Urs von Gunten, Driedger, A., Gallard, H., Salhi, E. 2001. By-products formation during drinking water disinfection: a tool to assess disinfection efficiency? *Water Research 2001 (8):2095–2099.*

Van der Wende, E., Characklis, W.G., Smith, D.B. 1989. Biofilms and bacterial drinking water quality. *Water Research 23(10): 1313–1322.*

Ventresque, C., Bablon, G., Legube, B., Jadas-Hecart, A., Dore, M. 1990. Development of chlorine demand kinetics in drinking water treatment plant. *Water chlorination:*

Chemistry, environmental impact and health effects, R.L. Jolley, et al., eds., Vol. 6, Lewis Publications, Inc., Chelsea, MI: 715–728.

Waller, K., Swan, S.H., DeLorenze, G., Hopkins, B. 1998. Trihalomethanes in drinking water and spontaneous abortion. *Epidemiology 9(2): 134–140.*

Winn-Jung, H., Hsuan-Hsien, Y. 1999. Reaction of chlorine with NOM adsorbed on PAC. *Water Research 1: 65.*

Zacheus, O.M., Lehtola, M.J., Korhonen, L.K., Martikainen, P.J. 2001. Soft deposits, the key site for microbial growth in drinking water distribution networks. *Water Research 7: 1757–1765.*

Zbieć, E., Dojlido, J.R. 1999. Uboczne produkty dezynfekcji wody. *Ochrona Środowiska 3(74): 37–44.*

Environmental Engineering – Pawłowski, Dudzińska & Pawłowski (eds)
© 2007 Taylor & Francis Group, London, ISBN13 978-0-415-40818-9

Comparing the efficiency of dissolved organic fractions removal by coagulation and PAC-aided coagulation in the water treatment

Małgorzata Szlachta & Wojciech Adamski
Wrocław University of Technology, Poland

ABSTRACT: The efficiency of coagulation as a unit process in removing dissolved organic pollutants during water treatment was compared with the efficiency of coagulation combined with adsorption on powdered activated carbon. The paper describes the conditions of adsorption of the dissolved organic compounds from raw riverine water and the same water after coagulation and sedimentation.

Keywords: Coagulation, sorption, powdered activated carbon, natural organic matter.

1 INTRODUCTION

Owing to the specific nature of some pollutants that have health implications and are difficult to remove from polluted water, conventional treatment methods, such as coagulation with rapid filtration, may not ensure the water quality required under Polish regulations. The coagulation process provides mainly the removal of colloidal and insoluble organic water pollutants, but fails to remove dissolved fractions such as micropollutants, to the levels postulated in relevant standards.

The efficiency of organic matter removal by coagulation can be enhanced by conducting the process over the optimal pH range, selecting an appropriate coagulant and its dose, as well as establishing the duration and velocity gradient (G) for rapid mix and flocculation. However, even if the process runs under optimal conditions, some of the dissolved organic pollutants remain resistant to removal by coagulation (Exall et al. 2000).

Some information about the character and susceptibility of the dissolved organic matter to removal by coagulation can be drawn from the SUVA (Specific Ultraviolet Absorbance) index, expressed as the ratio of absorbance at the wavelength of 254 nm (UV, m^{-1}) to the concentration of dissolved organic carbon (DOC, mg/dm^3). For natural water with high SUVA values (e.g. $\geq 4\,dm^3/mg{\cdot}m^{-1}$), the extent of dissolved organic matter removal by coagulation is comparatively high and reaches 80%. High SUVA values indicate that aromatic hydrophobic high-molecular weight compounds are dominant in the water. SUVA values equal to, or lower than, $3\,dm^3/mg{\cdot}m^{-1}$ for example suggest the dominance of primarily non-humic hydrophilic and low-molecular weight fractions displaying efficiencies of removal by coagulation lower than 30% (EPA 2005; Karanfil et al. 2002). Thus, to efficiently decrease the concentration of the dissolved fraction, it is necessary to make use of a highly effective water treatment technology such as adsorption by porous adsorbents, e.g. powdered activated carbon (PAC).

The benefit of combining PAC adsorption with coagulation, manifests itself not only in the reduction of the coagulant dose or extension of the optimal coagulation pH range, but also in the significant improvement in floc sedimentation properties, which was confirmed by the investigations reported on in this paper. The PAC added to the water in the course of the coagulation process promoted the formation of large and heavy flocs, which settled very well. What is more, owing to its very good sorptive properties, the PAC significantly raised the extent of dissolved organic fraction removal by coagulation.

A detailed analysis of the sorptive properties of the PAC, as well as the identification of the mechanisms and phenomena involved in coagulation-PAC adsorption allows the process to be optimised at the stage of design and in existing water treatment plants.

This paper presents the results of bench-scale research on the influence of PAC on raw water coagulation, as well as the conditions of the adsorption process. Consideration is given to the problem of how PAC contributes to the removal of dissolved organic pollutants during coagulation-PAC adsorption.

Table 1. Selected physicochemical parameters of the Odra River water.

Parameter	Unit	Range
pH	–	7.68–8.15
Tu	NTU	6.30–28.90
C	gPt/m^3	13.82–26.54
COD_p	gO_2/m^3	4.60–8.20
TOC	gC/m^3	3.76–5.35
DOC	gC/m^3	3.17–4.55
UV	m^{-1}	9.40–11.00

2 MATERIALS AND METHODS

The research was divided into stages. At the initial stage the optimal parameters of the coagulation process were determined. Then the kinetic and static parameters were established for the adsorption of dissolved organic compounds (in terms of DOC concentration) from raw water and water after coagulation and sedimentation. The final stage included tests of PAC-aided coagulation.

Samples of raw water, water after conventional coagulation and PAC-aided coagulation were analysed for the following water quality parameters, pH (Multi-Parameter Instrument 340i, WTW); chemical oxygen demand – COD_p (permanganate method); turbidity – Tu (turbidimeter 2100N IS, Hach); colour – C and absorbance at 254 nm – UV (spectrophotometer UV-1202, Shimadzu); total organic carbon – TOC and dissolved organic carbon – DOC (5050 TOC analyser, Shimadzu). DOC was measured in the PAC adsorption tests.

2.1 *Raw water*

The tests involved samples of Odra River water collected from November 2004 to May 2005. The parameters of the water are summarised in Table 1. The SUVA value of the water ranged from 3.07 to 2.58, water temperature after the stabilisation in ambient air being $20 \pm 2°C$.

2.2 *Powdered activated carbon*

Use was made of the Norit SA Super powdered activated carbon produced by steam activation of dedicated vegetable raw materials. According to the specification supplied by the manufacturer (Norit Datasheet), the PAC's

- apparent density (tamped) is $250\,kg/m^3$,
- total surface area is S_{BET} $1150\,m^2/g$,
- particle size $d_{50} = 10\,\mu m$,
- ash content amounts to 10 mass %, and
- pH is alkaline.

The Norit SA Super was chosen because of its very high adsorptive capacity, which is suitable for the removal of both low- and high-molecular weight dissolved organic pollutants.

2.3 *Conventional coagulation*

The experimental method used in this study was based on the jar-test procedure (Velp Scientifica Jar-tester). In order to determine the optimal parameters of the process, the following was assumed: 3-minute rapid mix at $G = 180\,s^{-1}$ and 30-minute flocculation at $G = 10\,s^{-1}$.

Velocity gradients were calculated for $2\,dm^3$ samples at 20°C according to the formula (Bouyer et al. 2005):

$$G = \sqrt{\frac{P}{\eta \cdot V}},\ s^{-1} \tag{1}$$

where P is power, W; V is volume of solution, m^3, and η is dynamic viscosity, kg/m·s.

The study was carried out at adjusted water pH, which ranged from 5.5 to 7.5 (pH adjustment involving 0.1 M HCl and NaOH solution). Based on the measured physicochemical parameters (Table 1), further studies were performed with the pH for which the efficiency of the process was the highest. The coagulant used in the experiments was alum, $Al_2(SO_4)_3 \cdot 18H_2O$. Its doses were established on the basis of the following alum amounts 1.2, 1.0, 0.8, 0.6 and 0.4 D_{Al}.

The optimal contact time and velocity gradient for rapid mix fell in the Camp number range of 5040 to 32400 and those for flocculation ranged from 18000 to 64800.

After coagulation the samples were allowed to settle for two hours.

2.4 *Kinetic and static parameters of adsorption*

The kinetic and static PAC adsorption tests were carried out with raw water and water after coagulation-sedimentation. Use was made of the conventional jar-test.

The selected velocity gradient guaranteed the maintenance of good intermixing between water and PAC at low and high concentration. The G-values examined ranged from 19 to $84\,s^{-1}$. PAC was used in suspension with distilled water, the PAC doses varying from 1 to $50\,g/m^3$. On the basis of preliminary tests results it was assumed that 5-hour adsorption was sufficient to attain the adsorption equilibrium under conditions used in this study. Water samples were collected hourly, filtered through a $0.45\,\mu m$ filter paper and analysed for DOC.

Thereafter, the following parameters of the Langmuir adsorption isotherm were determined: the equilibrium concentration (C_E), estimated using the kinetic curves that describe the adsorbate concentration variations in the sample contacting with the selected doses of PAC at the fifth hour of the process, and the adsorptive capacity (x) corresponding to the equilibrium concentration. The data obtained were approximated by the Langmuir isotherm equation (Faust 1998):

$$x = \frac{x_m \cdot b \cdot C_E}{1 + b \cdot C_E}, \qquad (2)$$

where x_m stands for monolayer capacity, g/kg; C_E denotes the adsorbate equilibrium concentration, g/m^3, and b is the Langmuir isotherm parameter, m^3/g.

2.5 *Coagulation + PAC*

The jar-test method was used to examine the PAC-aided coagulation process. The batching sequence of carbon and alum was investigated using the optimal process parameters established in conventional coagulation tests.

PAC doses (which varied from 25 to 50 g/m^3) were added 30 minutes before the addition of alum, at the time of alum addition, after rapid mix and during flocculation. On the basis of experimental data, as well as kinetic and static adsorption tests, we decided to further the experiments with the PAC dose of 40 g/m^3.

The next stage of the study was to compare the efficiency of coagulation and PAC adsorption with that of conventional coagulation. The process involved a 3-minute rapid mix at $G = 180\,s^{-1}$, 30-minute flocculation at $G = 10\,s^{-1}$, an alum dose of 0.8 D_{Al}, pH = 6.0, and such a batching sequence of PAC and alum that yielded the best results. After coagulation the samples were allowed to settle for two hours.

3 RESULTS AND DISCUSSION

3.1 *Conventional coagulation*

At the initial stage, the tests were performed with a constant coagulant dose,

$D_{Al} = 2.5\,gAl/m^3$, and a pH ranging from 5.5 to 7.5. The highest extent of turbidity removal, 89–86%, was achieved at pH 6.0–6.5, respectively, whereas the efficiency of coloured matter removal decreased with rising pH. The removal of organic matter (expressed as TOC and COD_p) showed the following pattern: with the pH of 5.5–6.5, the removal of TOC and COD_p was 36–31% and 46–53%, respectively. Increasing the pH to >6.5 brought about a substantial drop in coagulation efficiency. The results show that the coagulation efficiencies obtained with the pH adjusted to 5.5 and 6.0 are comparable. However, considering the fact that pre-acidification intensifies the corrosive action of the water after the process, we decided to continue our study with the pH = 6.0.

The next experimental series aimed at establishing the alum dose for coagulation at the optimal pH.

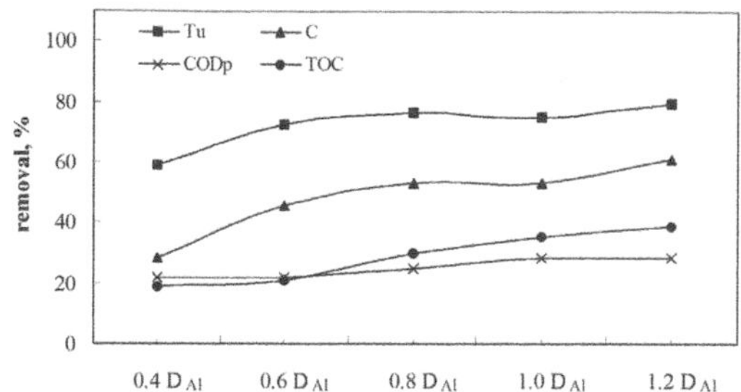

Figure 1. Efficiency of various alum doses in turbidity, colour, COD_p and TOC removal.

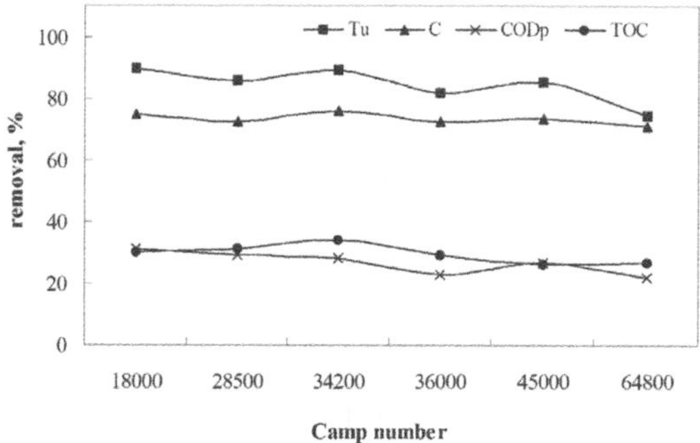

Figure 2. Removal of turbidity, colour, COD_p and TOC related to Camp number.

The doses applied in the test included 1.2, 1.0, 0.8, 0.6 and 0.4 D_{Al}, where

$D_{Al} = 2.5\,gAl/m^3$. Generally, the extent of coloured and organic matter removal, as well as that of the pollutants promoting turbidity, was found to increase with increasing alum dose (Fig. 1), but it turned out that the dose equal to 0.8 D_{Al} was sufficiently effective.

In the following stage the optimal velocity gradients and duration for rapid mix and flocculation were determined. Coagulation was run with adjusted pH (pH = 6.0) and an alum dose of 0.8 D_{Al} ($D_{Al} = 2.4\,gAl/m^3$). Applying 1-, 2- and 3-minute rapid mix, we varied the gradient values from 84 to $180\,s^{-1}$. The duration of flocculation was 20, 25 and 30 minutes, whereas the velocity gradient ranged between 10 and $54\,s^{-1}$. In the case of rapid mix, the most effective intermix of alum with treated water was achieved at T = 3 min and $G = 180\,s^{-1}$.

Satisfactory treatment effects were achieved when flocculation was carried out over the Camp number range of 18,000 to 34,200. When the Camp number was greater than 35,000, i.e. when the G-value was higher than $30\,s^{-1}$, the treatment effects were poorer due to the damage of the floc structure (Fig. 2). Thus, the optimal process parameters were established i.e. T = 30 min. and $G = 10\,s^{-1}$, with which the flocs

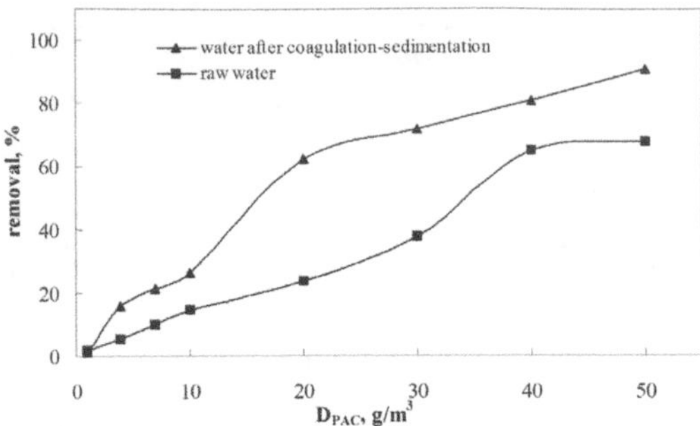

Figure 3. Removal of DOC by adsorption related to PAC dose (after 5 hours of the process).

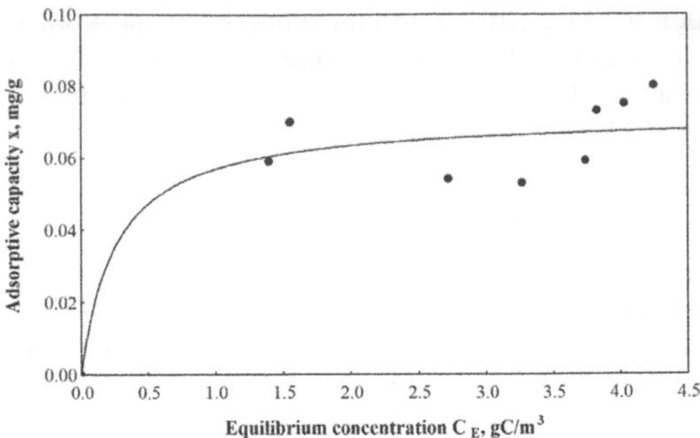

Figure 4. Adsorption isotherm for DOC: raw water.

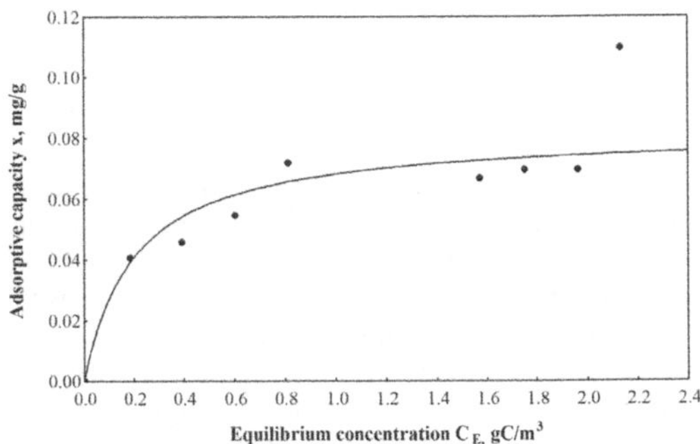

Figure 5. Adsorption isotherm for DOC: water after coagulation and sedimentation.

did not settle and their structure did not undergo damage.

3.2 *Kinetic and static parameters of adsorption*

Further jar-tests were performed to determine the kinetic and static parameters of adsorption for dissolved organic fractions. The velocity gradient of mixing PAC with water amounted to $5\,s^{-1}$ and provided a uniform adsorbate concentration in the sample volume. The PAC dose applied ranged from 1 to 50 g/m^3.

The experiments revealed that the higher the dose of the adsorbent, the greater the extent of dissolved organic carbon adsorption. Moreover, the notable differences in DOC removal between particular PAC doses imply that the adsorptive capacity of the carbon was effectively utilised. The most effective was the adsorption of DOC after coagulation and sedimentation (Fig. 3). For example, in the fifth hour of the adsorption process with $D_{PAC} = 40\,g/m^3$, the efficiency of DOC removal from water after coagulation-sedimentation and from raw water totalled 81% and 65%, respectively. This finding can be attributed to the higher initial DOC concentration in raw water, as well as to the presence of other adsorbates (e.g. colloidal and insoluble organic water pollutants) which could compete with dissolved fractions.

The stronger affinity of the PAC for the dissolved organic matter persisting after coagulation-sedimentation is reflected in the isotherm parameter b. With adsorption of the dissolved fractions from raw water, the value of b was lower, amounting to 3.83 m^3/g; with adsorption from water treated by coagulation and sedimentation the b-value equalled 4.97 m^3/g. The value of monolayer capacity, x_m, followed a similar pattern: it was higher for water after coagulation, $x_m = 0.082$ mg/g and lower for raw water, $x_m = 0.068$ mg/g.

The shape of the isotherms obtained (Figs. 4 and 5) indicates that the sorptive properties of the PAC are very good with respect to the organic pollutants referred to as DOC.

4 COAGULATION + PAC

Comparisons show that the extent of DOC and UV removal was comparable when PAC and alum were dosed simultaneously and when PAC was added 30 minutes before alum. (Fig. 6). Thus, when PAC was added prior to alum ($D_{PAC} = 50\,g/m^3$), the DOC concentration was reduced from 3.29 to 0.98 gC/m^3 and UV dropped from 9.82 to $1.6\,m^{-1}$. Simultaneous alum and PAC batching decreased the DOC concentration and UV from the same initial level to 0.83 gC/m^3 and $1.56\,m^{-1}$, respectively.

A comparatively high extent of DOC and UV removal was achieved when PAC was dosed after the rapid mix (Fig. 6), this , however, can be attributed to retention of the pollutants on the filter paper. It is worth noting that in this case turbidity, TOC and COD_p removal was, on the average, by 23%, 24% and >50% lower, respectively, than when PAC and alum were dosed simultaneous or when PAC was added 30 minutes before alum. Similar trends were observed when PAC was batched during flocculation. In that particular case, however, the poorer treatment effects are due to insufficient time of contact between the PAC and the treated water, as well as to the blocking of the active centres by the flocs. As for the PAC-aided coagulation,

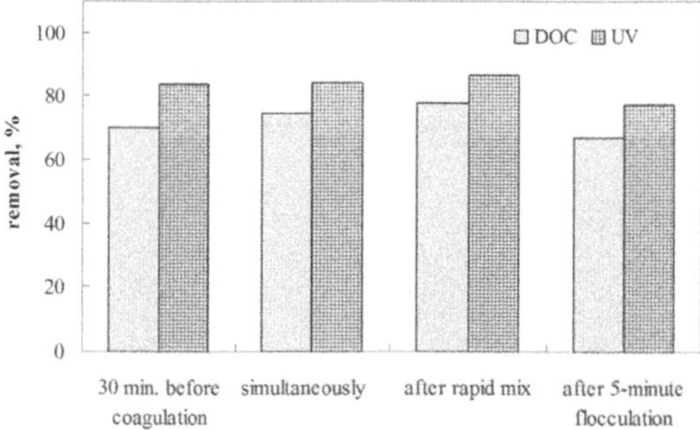

Figure 6. Removal of DOC and UV by PAC-aided coagulation related to the method of PAC batching.

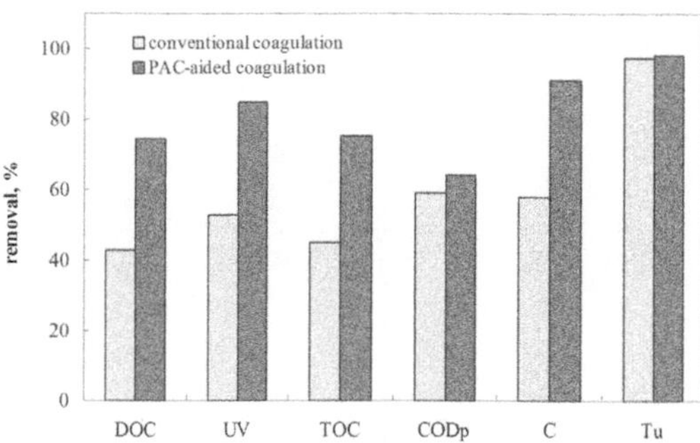

Figure 7. Comparison of removal of selected pollutants obtained by conventional and PAC-aided coagulation.

Table 2. Values of selected water quality parameters measured after conventional and PAC-aided coagulation.

Parameter	Raw water	Conventional coagulation	PAC-aided coagulation
DOC, gC/m^3	3.29–4.04	2.04–2.10	0.84–0.91
UV, m^{-1}	9.82–10.42	4.60–4.92	1.28–1.60
TOC, gC/m^3	3.85–4.39	2.14–2.49	0.89–1.05
COD_p, gO_2/m^3	5.10–6.80	2.50–2.60	2.00–2.20
Colour, gPt/m^3	8.83–11.37	4.07–4.37	0.93–0.94
Turbidity, NTU	17.00–28.90	0.36–0.88	0.25–0.35

the addition of the adsorbent after rapid mix or during flocculation does not seem to be a good idea. It is, however, advantageous to add the activated carbon prior to alum or simultaneously with alum.

Further tests were run with optimal coagulation parameters, using simultaneous dosage of PAC ($D_{PAC} = 40\,g/m^3$) and coagulant ($0.8\,D_{Al} = 2.1\,g\,Al/m^3$). The results revealed that the removal efficiency of coagulation as a unit process with respect to the dissolved organic fractions was low as compared to PAC-aided coagulation, amounting, on the average, to 43% and 53% for DOC and UV, respectively (Fig. 7). This implies that the water under test contained certain amounts of organic pollutants which were resistant to removal by coagulation. The remarkable removal of dissolved matter (DOC, 75%; UV, 85%) attained after PAC adsorption substantiates the advantage of combining coagulation and adsorption, especially when high water quality is required.

There was also considerable TOC and residual colour removal, which can be attributed to the high efficiency of dissolved organic matter removal (Table 2).

5 CONCLUSIONS

The study enabled the following coagulation parameters for the optimal removal of dissolved organic compounds to be established: pH = 6.0; alum dose $0.8\,D_{Al}$; duration of rapid mix, 3 min; duration of flocculation, 30 min; velocity gradient of rapid mix, $180\,s^{-1}$; velocity gradient of flocculation, $10\,s^{-1}$.

The Langmuir isotherm very well described the adsorption of dissolved organic fractions from the water under test and the powdered activated carbon used in the study showed a strong affinity for the adsorbates expressed as DOC.

It is advisable to combine coagulation and PAC adsorption, as this results in enhanced removal of the dissolved organic pollutants. The integration of these processes yields a 75% removal of DOC, while coagulation as a unit process of the water treatment train brings about only 43% removal efficiency. When applying PAC-aided coagulation, it is legitimate to batch the PAC either before or simultaneously with the addition of alum.

ACKNOWLEDGMENTS

The study was supported by Grant 3 T09D 02628 from the Ministry of Science and Higher Education.

REFERENCES

Bouyer D., Coufort C.e., Line A. & Do-Quang Z. 2005. Experimental Analysis of floc size distributions in a 1-L jar under different hydrodynamics and physicochemical conditions. *Journal of Colloid and Interface Science*, 292:413–428.

Environmental Protection Agency, February 2005. Determination of Total Organic Carbon and Specific UV Absorbance at 254 nm in Source Water and Drinking Water. EPA 600-R-05-055.

Exall K.N. & VanLoon G.W. 2000. Using coagulants to remove organic matter. *AWWA*, 92:11:93.

Faust S.D. & Aly O.M. 1998. *Chemistry of Water treatment*, 2nd edition. CRC Press LLC, USA.

Karanfil T., Schlautman M.A. & Erdogan I. 2002. Survey of DOC and UV Measurement Practices with Implications for SUVA Determination. *AWWA*, 94:12:68.

Norit Datasheet: Manufacturer's Specification for Powdered Activated Carbon Norit SA Super.

Environmental Engineering – Pawłowski, Dudzińska & Pawłowski (eds)
 ISBN13 978-0-415-40818-9

Application of a Finnish model to predict mutagenicity of drinking water – model research

Agnieszka Trusz & Teodora M. Traczewska
Institute of Environmental Protection Engineering, Wroclaw University of Technology, Wroclaw, Poland

ABSTRACT: To verify the correlation between a Finnish model to predict the mutagenicity of drinking water from surface sources disinfected with chlorine and water mutagenicity determined by Ames' test, model research was conducted. Mutagenicity tests were carried out for different TOC and ammonia concentrations. Disinfection was carried out with chlorine on the basis of the chlorine demand curve. Generally, the mutagenicity rate (MR) increased with the increase of the TOC and ammonia content. The highest MR growth with the increase of concentration rate was observed for the water with the highest analyzed TOC and ammonia parameters. The observed model water mutagenicity determined in the Ames assay showed a low correlation with the water mutagenicity as obtained by the Finnish model for the prediction of the mutagenicity of drinking water.

Keywords: Chlorine, disinfection, chlorination by-products, mutagenicity, Ames test, Finnish model.

1 INTRODUCTION

Chlorine disinfection of surface waters containing natural and anthropogenic organic compounds, results in the formation of numerous by-products (Nissinen et al. 2002), some of which might possess biological activity (Sujbert et al. 2005), including mutagenicity and carcinogenicity, and many of which are still not well researched (WHO 2004).

Epidemiological studies have suggested the increased risk of cancer in some geographic areas where chlorinated surface waters are used as water sources for water supply systems (Murphy et al. 1990). For example, research carried out from the mid-seventies of the last century in Finland, revealed a relationship between mutagenic activity of chlorinated water and morbidity of some carcinoma varieties – mainly kidney and bladder cancer (Koivusalo et al. 1994, Koivusalo et al. 1998). That relationship was considered by International Agency for Research on Cancer as evidence (IARC 1991).

The presence of chlorination by-products in drinking water then poses a danger to human health and even life, so it is undesirable and should be controlled by adequate methods.

Using current analytical resources there is no possibility of routine examination of the full spectrum of micro pollutants present in drinking water and thus also of its chlorination by-products (Liu et al. 1999, Guzzellaa et al. 2005). This situation has aroused great interest in biological methods of assaying the water consumer's health risk – for example, the Ames test (Ohea et al. 2004).

The Ames test is probably the most widely applied mutagenicity test and has a high sensitivity. This procedure was developed in the mid 1970's and, to date has been used in the assessment of many thousands of chemical compounds. This application is recommended by Standard Methods for Examination of Water and Wastewater in the assessment of drinking water quality (Eaton et al. 1995).

The Ames test procedure is complicated and time-consuming. It also needs qualified staff. This less-than-ideal situation has led to attempts at correlating the Ames test results with standard parameters used for characterizing drinking waters. Some of the first such attempts were carried out by Vartiainen et al. in 1988 and were the basis for the creation of a mathematical model for the prediction of drinking water mutagenicity. That model was based on the three main chemical water quality parameters: total organic carbon (TOC), ammonia concentration (NH_3) and chlorine dose (Cl_2) and is confirmed by Ames test with the use of *Salmonella typhimurium* TA 100 strain, without metabolic activation.

The results of the Finnish research encouraged an attempt at verification under US conditions, where a very high correlation ($r = 0.96$) between the predicted and observed mutagenicity level was obtained (Schenck et al. 1998). It confirmed the high usefulness of Vartiainen's model and suggested the possibility of its still wider usage in other countries.

It also inspired an attempt at using the Vartiainen model under Polish conditions.

The model research of the correlation of the three basic parameters characteristic of purified water (TOC, NH_3 concentration and Cl_2 dose) with water mutagenicity was carried out. The established composition of that model water eliminated the influence of its other components, which could affect the observed mutagenic activity.

2 MATERIALS AND METHODS

2.1 *The Vartiainen model*

The form of Vartiainen's model proposed for the prediction of drinking water mutagenicity is expressed by the following equation:

$$f = A * (1 - e^{k*c}) \quad (1)$$

where f = mutagenicity of drinking water – expressed as revertants per liter equivalents as determined in *Salmonella typhimurium* strain TA 100 without metabolic activation A and k = constants: 4000 and 0.054, respectively c = is variable defined as:

$$c = [TOC] * [Cl_2] * (1 \; NH_3)^2 \quad (2)$$

where [TOC] = total organic carbon concentration, $[NH_3]$ = ammonia concentration $[Cl_2]$ = chlorine disinfections dose; mg/dm^3.

If two chlorination steps in water treatment technology are used then:

$$f = f_R + f_D \quad (3)$$

where R = refers to the parameters of the raw water D = refers to the parameters of the treated water.

2.2 *Preparation of model water*

The model water was prepared from:

- tap water from Wroclaw Municipal Water Treatment Plant "Mokry Dwor", which is supplied by surface water from the Olawa River. The treatment technology consisted of coagulation, filtration on sand filters and chlorination. The water was additionally purified in a laboratory on filter with WD-extra activated carbon.
- water rich in natural, dissolved humus compounds from the "Batorow" Peat-Bog.

The two water sources were mixed in appropriate ratios to achieve TOC concentrations: 2.0, 4.0 and 6.0 g C/m^3. To these waters with known TOC content, ammonium chloride solution was added to achieve the appropriate concentration of ammonia: 0.1; 0.5; 1.0 gNH_3/m^3.

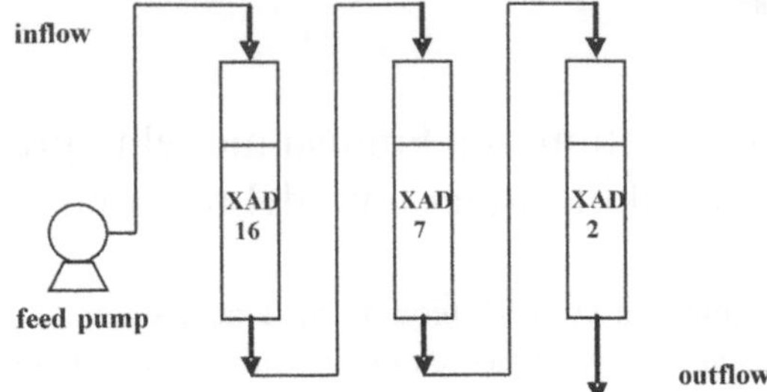

Figure 1. The schematic diagram of laboratory scale adsorption system.

The prepared model water was disinfected with chlorine on the basis of the chlorine demand curve (Eaton et al. 1995). The water contact time with the disinfectant was set at 16 hours, corresponding to the estimated average time that the water was in the supply system.

2.3 *Procedure of isolation, concentration and extraction of organic compounds from model waters*

In order to isolate and concentrate the organic compounds from the analyzed model water, adsorption on Amberlite XAD resins was used. Before the adsorption process the resins were purified and activated in a Soxhlet extractor in three eight hour extraction steps with ethyl acetate, acetone and methanol.

Following preliminary studies adsorption was carried out in a serial flow 3 glass column system. In each column there was a resin with different adsorption ability. The treated water flows through resins in experimentally established sequence: XAD16, XAD7 and XAD2.The system was fed by a peristaltic pomp. A schematic diagram of the laboratory scale adsorption system is presented in Figure 1. In all tests the water sample (10 dm^3) was filtered through the resins.

Organic compounds adsorbed on resins were subsequently extracted with acetone.

2.4 *Procedure of bioindicative tests*

The assessment of mutagenic and carcinogenic activity of the organic fractions extracted from the model water was carried out using the Ames test procedure, in accordance with Ames et al. 1983, for water samples of different concentration: 100, 200, 300, 400, 500, 600, 800 and 1000 times concentrated. It was equivalent to 10 to 100 cm^3 of analyzed model water, respectively.

The bioindicative tests of the model water disinfected with chlorine were performed with the use of one strain of tested *Salmonella typhimurium* TA 100 bacteria with and without the addition a metabolic activator, i.e. microsomal fraction from homogenate of the rat liver (S9-mix), used in the transformation of promutagens into mutagens.

Each test was carried out on five sets of samples. Before the beginning of the proper research, genetic markers of those strains, their spontaneous reversion and their reaction to 100% mutagens were checked each time.

The results obtained by the Ames test were the basis for the calculation of the mutagenicity rate MR (Mathur et al. 2005). Each case in which the MR value was greater than or equal to 2 (i.e. when the induced reversion was at least double that of the spontaneous reversion) was considered as a positive result for the particular model water sample. The mutagenicity of drinking water (as revertants per liter) determined by the Ames test was calculated using the ordinary last squares method.

All calculation was conducted by OriginPro 7,5 – statistical computer program software. These calculations were the basis for the comparison with the Vartiainen model.

3 RESULTS AND DISCUSSION

The Ames test results performed both with and without the addition of a metabolic activator, allowed for the assessment of the potential mutagenic activities of micro-pollutants that were present in the model water. That activity refers to the mutagens as well as to the promutagens.

For all model water samples containing compounds of mutagenic character, the results of the Ames tests carried out with added metabolic activator, showed MR < 2.That means, monooxygenase activity in microsomal fractions could detoxicate these compounds. At the same time the negative results of the Ames test proved the absence of promutagens in the test waters.

The results of the Ames tests (Figs 2–4) without the metabolic activation showed that organic compounds present in the tested waters caused mutation in *Salmonella typhimurium* TA100 strain which in turn resulted in the increase of induced revertants colony number, in relation to the spontaneous revertants. The analysed sample with the highest analysed pollutant indicator concentration: TOC – 6.0 gC/m^3 and ammonia – 1.0 gNH$_3$/m^3, respectively showed the highest MR value of 4.85.

Model water having TOC's in a range 4.0–6.0 gC/m^3, irrespective of the ammonia concentration range from 0.1 to 1.0 gNH$_3$/m^3, showed potentially mutagenic properties. The value of mutagenicity rate (MR) for these waters increased with the TOC and ammonia content increase (Figs 2, 3). However, waters with the TOC contents of 2.0 gC/m^3, irrespective of the ammonia concentration, showed no mutagenic features (Fig. 4).

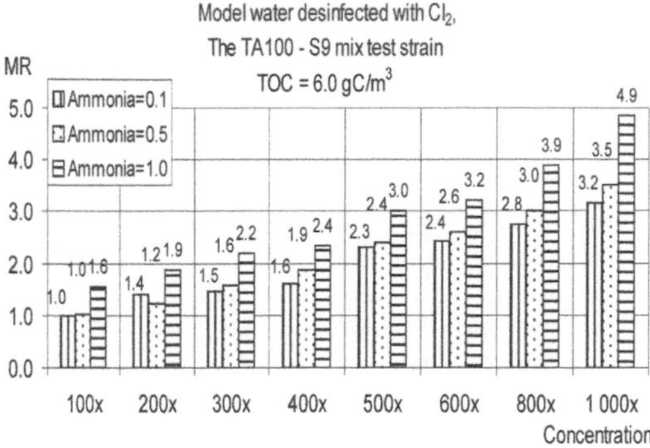

Figure 2. The mutagenicity rate for chlorinated model water containing TOC on 6.0 gC/m^3 level and different ammonia concentrations.

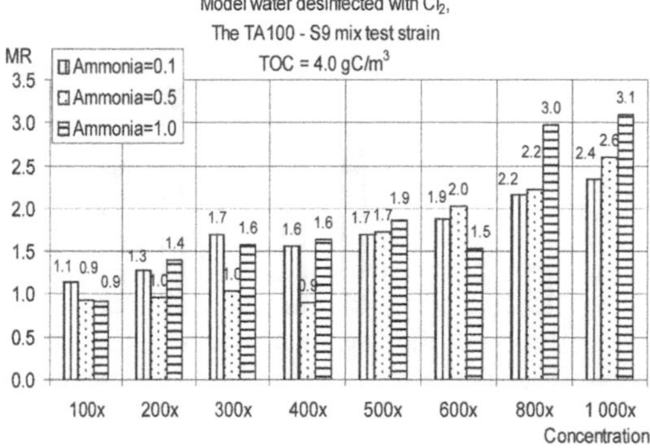

Figure 3. The mutagenicity rate for chlorinated model water containing TOC on 4.0 gC/m^3 level and different ammonia concentrations.

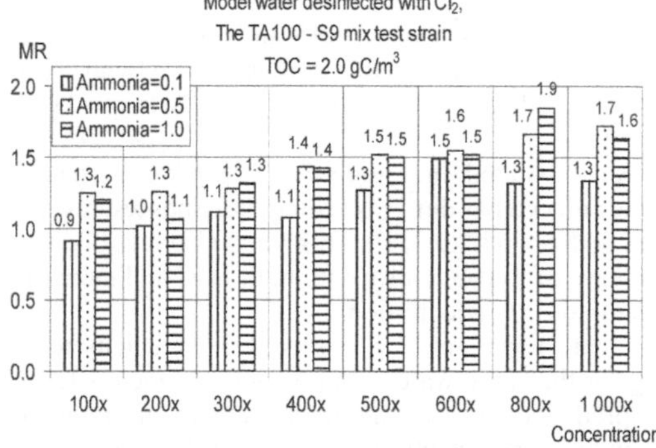

Figure 4. The mutagenicity rate for chlorinated model water containing TOC on 2.0 gC/m^3 level and different ammonia concentrations.

On the basis of the dose–response curves the mutagenicity of each water sample as the amount of revertants induced per dm^3 was calculated. The average value of spontaneous revertants, calculated using

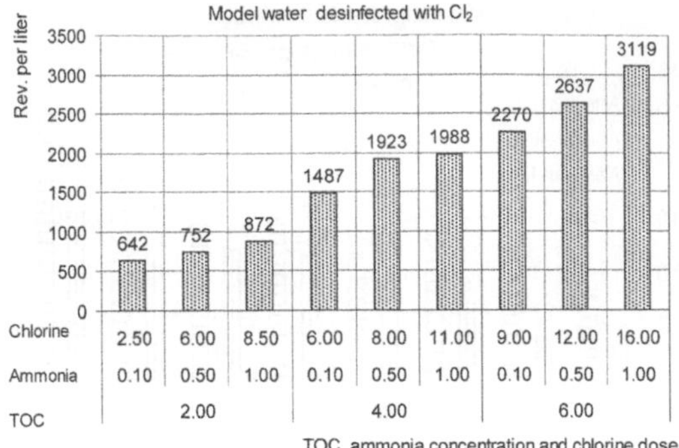

Figure 5. Revertants per liter equivalent observed in the Ames test with the use of Salmonella typhimurium TA100 strain without metabolic activity.

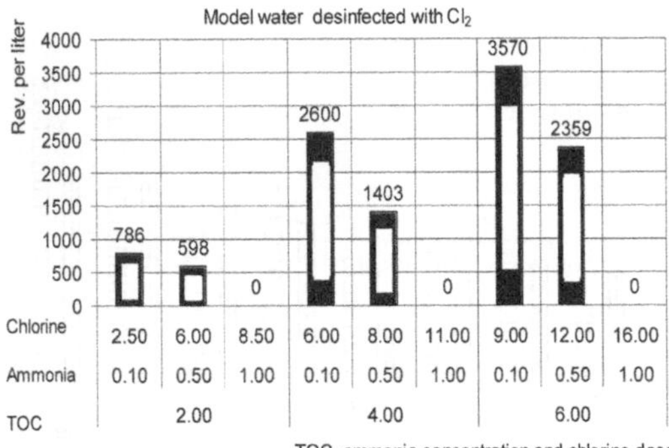

Figure 6. Revertants per liter equivalent as predicted by the Vartiainen model.

this method, was between 115 and 135. The average number of revertants in the positive control (with use a sodium azide as a 100% mutagen) was between 2800 and 2950.

As illustrated in Figure 5 the number of revertants induced per dm^3, according the Ames test results ranged from 642 to 3119. This result also revealed the increase of mutagenicity with both TOC and increase in ammonia concentration.

Figure 6 presents the results of mutagenicity testing for the same water composition parameters level as predicted by the Vartiainen model. The number of induced revertants was in the similar range (598 to 3570) and its increase was correlated with the TOC concentration increase and ammonia concentration decrease. For the highest examined ammonia concentration (1.0 gNH_3/m^3), irrespective of the TOC content, it showed the value of 0.

The model water mutagenicity determined in the Ames assay showed a low correlation with the water mutagenicity obtained using the Finnish model for predicting drinking water mutagenicity. The mutagenicity correlation rate for analyzed model water was 0.23.

4 CONCLUSIONS

If surface water is subject to chlorination treatment, formation of a large number of chlorinated by-products with potentially mutagenic and carcinogenic characteristics occurs. It is not possible to isolate and examine the full spectrum of micro-pollutants present in water with conventional analytical methods. The application of drinking water genotoxicity monitoring using the Ames test requires qualified staff and assurance of appropriate conditions to perform it, resulting in considerable costs. For this reason it would be propitious to apply, after prior adjustment to Polish conditions, the Finnish model for predicting drinking water mutagenicity.

This allows, on the basis of the main chemical parameters of water (e.g. TOC, ammonia concentration, chlorine dose) the potential mutagenicity of purified water to be estimated. However, on the basis of the Ames test results, the mutagenicity of the test model water demonstrated only a weak correlation with the mutagenicity predicted with the Finnish model. Clearly, further research and analyses are needed in order to verify the Finnish model under Polish conditions or to justify its modification to include the specifics of the country.

ACKNOWLEDGEMENTS

The research was carried out within Grant (Project No 7T09D 027 21) financed by the Committee for Scientific Research of Poland.

REFERENCES

Eaton, A.D., Clesceri, L.S., Greenberg, A.E. 1995. Standard methods for the Examination of Water and Wastewater – 19th Edition. American Public Health Association, Washington, ISBN 0-87553-223-3.

Guzzellaa, L., Monarcab, S., Zanic, C., Ferettic, D., Zerbinic, I., Buschinid, A., Polid, P., Rossid, C., Richardsone, S.D. 2004. In vitro potential genotoxic effects of surface drinking water treated with chlorine and alternative disinfectants. *Mutation Research* 564: 179–193.

IARC Monograph of the evaluation of carcinogenic risk to humans Chlorinated drinking water; chlorination by-products; some other halogenated compounds, Cobalt and Cobalt Compounds. 1991, volume 52.

Koivusalo, M., Jaakkola, J., Vartiainen, T. 1994. Drinking water mutagenicity in past exposure assessment of the studies on drinking water and cancer. Application and evaluation in Finland. *Environmental Research* 64: 90–101.

Koivusalo, M.T. 1998. Drinking water mutagenicity and cancer. Publications of the National Public Health Institute, Division of Environmental Health, Finland.

Liu, Q., Jiao, Q.C., Huang, X.M., Jiang, J.P., Cui, S.Q., Yao, G.H., Jiang, Z.R., Zhao, H.K., Wang, N.Y. 1999. Genotoxicity of Drinking Water from Chao Lake. *Environmental Research* A 80: 127–131.

Maron, D.M., Ames, B.N. 1983. Revised method for Salmonella mutagenicity test. *Mutation Research* 113: 173–215.

Mathur, N., Bhatnagar, P., Bakre, P. 2005. Assessing mutagenicity of textile dyes from Pali (Rajasthan) using Ames bioassay. *Applied Ecology and Environmental Research* 4(1): 111–118.

Murphy, P.A., Craun, G.F. 1990. A review of recent epidemiologic studies reporting associations between drinking water disinfection and cancer. Water Chlorination, *Chemistry*, Environmental Impact and Health Effects. Vol.6, Chap.29. Lewis Publishers, Chelsea.

Nissinen, T.K., Miettinen, I.T., Martikainen, P.J., Vartiainen, T. 2002. Disinfection by-products in Finnish drinking waters. *Chemosphere* 48: 9–20.

Ohea, T., Watanabeb, T., Wakabayashic, K. 2004. Mutagens in surface waters: a review. *Mutation Research* 567: 109–149.

Schenck, K.M., Wymer, L.J., Lykins, B.W., Clark, R.M. 1998. Application of a Finnish mutagenicity model to drinking waters in the U.S. *Chemosphere* 37(3): 451–464.

Sujbert, L., Rácz, G., Szende, B., Schröder, H.C., Müller, W.E.G., Török d, G. 2006. Genotoxic potential of by-products in drinking water in relation to water disinfection: Survey of pre-ozonated and post-chlorinated drinking water by Ames-test. *Toxicology* 219: 106–112.

Vartiainen, T., Liimatainen, A. 1988. Relation between drinking water mutagenicity and water quality parameters. *Chemosphere* 17(1): 189–202.

World Health Organization, 2004. Guidelines for third edition drinking-water Quality, vol. 1, Geneva.

Environmental Engineering – Pawłowski, Dudzińska & Pawłowski (eds)
© 2007 Taylor & Francis Group, London, ISBN13 978-0-415-40818-9

Estrogenic micropollutants in the aqueous environment: sources, dispersion and removal

Michał Bodzek & Mariusz Dudziak
Faculty of Environmental and Energy Engineering, Silesian University of Technology, Gliwice, Poland

ABSTRACT: Estrogens are female sex hormones. They are mainly extracted with human and animal urine and have always been been present in the environment. Normally, the concentrations of these compounds are very low in an aquatic environment but still sufficient to exert a harmful effect on the endocrine system functions in organisms exposed to estrogens. The present paper includes data that enable the initial risk of estrogen contamination to be assessed for selected waters of Poland. Samples were extracted by solid-phase extraction C_{18} cartridges, followed by gas chromatography/mass spetrometry of the trimethylsilyl derivatives. This method was applied to the monitoring of estrogens removal from water stream. A conventional water treatment methods for removal trace amounts of estrogens were developed and compared with membrane techniques.

Keywords: Estrogens, occurrence, determination and removal.

1 INTRODUCTION

The Endocrine disrupting (ED) compounds are defined as "external substances which adversely affect the health of living organisms and their offspring which brings about changes in endocrine functions". ED compounds form a class of substances defined by the biological effect they cause and not their chemical nature (López de Alda & Barceló 2001). A wide range of pollutants are classified as ED: pesticides, PAHs, phthalates, alkylphenols, synthetic steroids and natural products, such as phytoestrogens.

Estrogens and progestogens are of special interest because of their high estrogenic activity and wide application. People use them not only as contraceptives, but also for therapeutic purposes or to treat a number of hormonal diseases or cancer. They are also used in veterinary medicine for animal breeding.

Natural sex hormones, excreted in urine, have always been present in the environment. It was not until recently that phenomena such as the feminisation or hermaphroditism in water organisms were observed (Diaz-Cruz et al. 2003). They are attributed to the biological activity of chemical pollutants, notably natural and synthetic estrogens identified in the water environment.

The development of analytical techniques to determine micropollutants at a level of pg/dm^3-ng/dm^3 enabled not only the identification of particular compounds but also research on their distribution in different media. These developments are particularly important in the light of hypotheses which indicate that estrogens probably possess biological activity even well below their current limits of detection (Siemiński 2001).

The scope of our interest covers free estrogens significant in terms of environmental protection, such as estrone (E1), estradiol (E2), estriol (E3), ethynylestradiol (EE2), mestranol (MeEE2) and diethylstilbestrol (DES).

Estrogens are mostly excreted in urine as free forms or conjugated with sulfuric (sulfates) and glucuronic (glucuronides) acids in the liver. The fate of synthetic ethynylestradiol – from the moment it was taken as medication (26 μg/d) – to its excretion from the body is depicted in Figure 1. The figure indicates that only 43% (11.2 μg/d) of a therapeutic dose is metabolized in a woman's body while the remaining part is excreted 6 μg/d of the amount has the form of free ethynylestradiol, the remaining part being conjugated ethynylestradiol and its metabolites.

In the water environment, the conjugated forms-biologically inactive – are transformed into free forms which have specific estrogenic properties. It is observed primarily in the special ecosystem of wastewater treatment plants where some researchers found lower concentrations of estrogens in the influent than effluent. This transformation is usually induced by micro-organisms present in the biocenosis of activated sludge and the enzymes produced by them (β-glucuronides and arylsulfates enzymes). The investigations showed that approx. 80% of estradiol assayed in the conjugated form undergoes a change into free estrogens i.e. estradiol and estrone in the environmental matrix during initial 20–30 h. After 50 h, 10–20% of the amount is not degraded (Danish Environmental Protection Agency 2003).

Figure 2 depicts the pathway of estrogens follow from their excretion from the organism to the natural environment. Schowanek and Webb (2000) claim that both industrial (pharmaceutical) and municipal wastewater constitute the primary source of estrogens in the water environment. The wastewater treated in WWTPs accounts for 70% of estrogens while untreated wastewater discharged accidentally for 30%. Due to the hydrophobic nature of estrogens they can be adsorbed on the particles of sedimenting inorganic and organic suspensions. Thus in the case of a rapid increase of river water flow, a part of riverbed deposit might be transferred to the main stream and micropollutants will undergo desorption. The riverbed deposit may, in such circumstances, therefore act as secondary source of estrogens.

The number of studies addressing the analysis of estrogens in the environment is very limited, Table 1. Estrogenic steroids have been detected in effluents of sewage treatment plants (STPs) at concentrations ranging up to 82.1 ng/l. E2 concentrations in river waters from Germany, Italy and the Netherlands ranged up to 5.5 ng/l. The very few data available on their occurrence indicates their presence in river sediments at concentrations up to 22.8 ng/g and somewhat higher in sludge (up to 64 ng/g). In one of the studies conducted with drinking water samples originating from southern Germany, of the four estrogens, the average concentrations of the estrogenic steroids E1, aE2, bE2 and EE2 in drinking water samples were 0.40, 0.30, 0.70 and 0.35 respectively.

The compounds from this group are not eliminated completely during wastewater treatment, despite their trace concentrations. The 14-h hydraulic contact time with activated sludge used in European countries removes 85% of 17β-estradiol, estriol and 17α-ethynylestradiol. Estrone is removed to a lesser extent (Kuster et al. 2004). Sorption or biodegradation processes, under aerobic conditions, play a major role in the elimination of estrogens from the aquatic environment. However, this needs further investigation.

More advanced and effective techniques, such as UV radiation and ozonation are not used very often

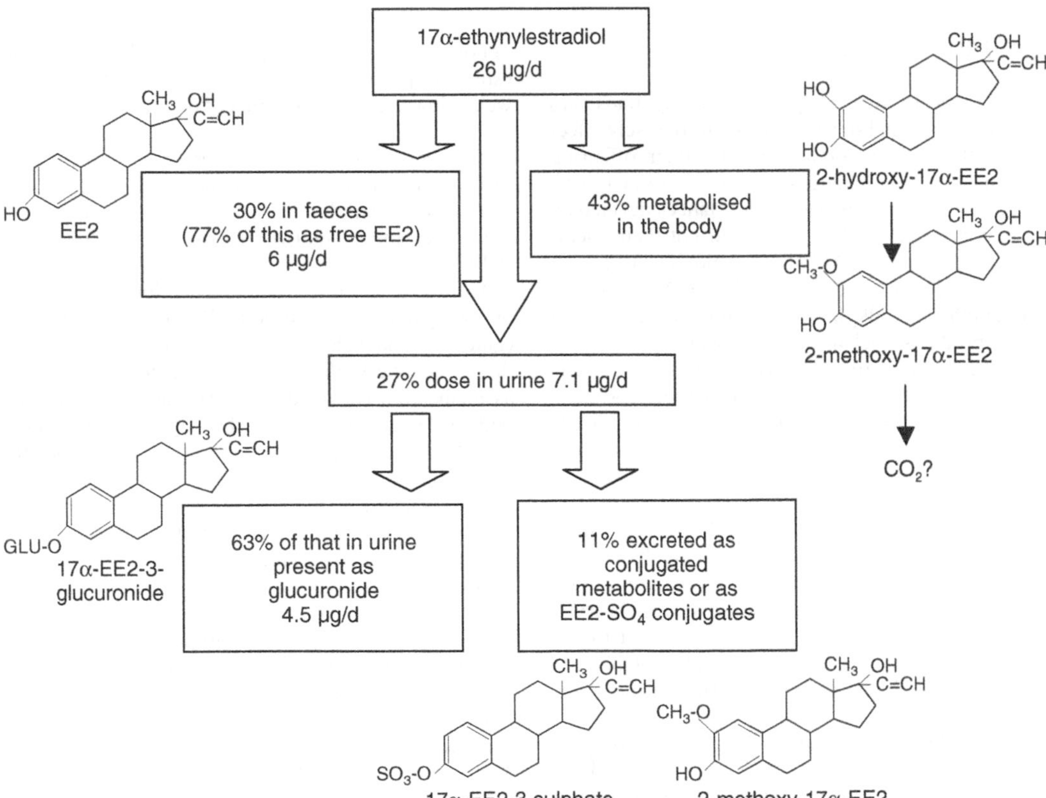

Figure 1. Fate and excretion of 17α-ethynylestradiol in the body (Johnson & Williams 2004).

since they are extremely costly (Kuster et al. 2004). The question of whether these more advanced techniques should be employed becomes important because of the scarcity of knowledge on the toxicity of their by-products. Estrogens lose their biological activity, but this is not synonymous with their becoming harmless either in the environment or in drinking water.

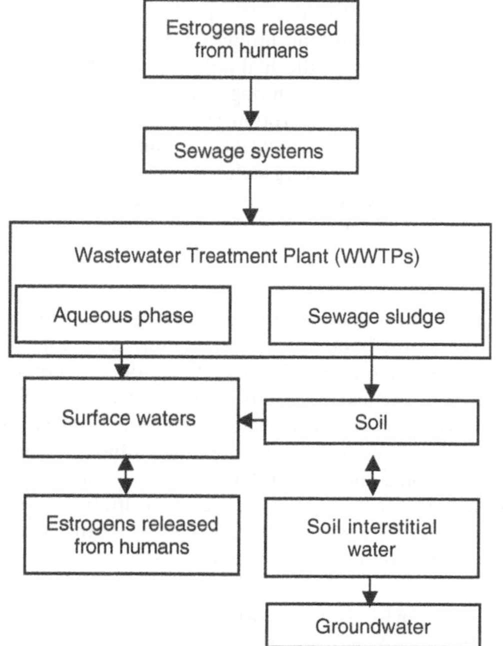

Figure 2. Pathways of estrogens in the environment (Danish Environmental Protection Agency 2003).

In our research, we compared the effectiveness of coagulation and sorption on activated sludge which are predominant processes in natural water treatment and frequently used to improve the effect of wastewater treatment. However, there is no reference data in literature on estrogen removal by conventional water treatment processes.

Our investigations are aimed at the determination of estrogen concentrations in water fluxes resulting from water treatment, and the assessment of the effectiveness of pressure driven membrane operations in the elimination of the micropollutants from water fluxes compared to the conventional techniques.

Our earlier investigations into PAHs and phthalates indicate that membrane operations might be an efficient tool to achieve high retention of organic micropollutants from waters. They may be used as single processes or in hybrid systems as an additional treatment of water flux (Bodzek et al. 2004).

2 MATERIALS AND METHODS

2.1 *Description of the estrogens analysis by SPE – GC/MS (ion trap)*

Estrogens were extracted from a water sample in columns filled with octadecylsilane (C_{18}, J.T. Baker). Prior to GC-MS analysis, the extracted estrogens underwent silylation with a mixture of

Table 1. Occurrence of estrogens in environmental samples.

Matrix (location)	Compounds detected	Detected concentration	Reference
Effluents of WWTPs			
Italy	E1, E2, E3, EE2	0.44–82.1 ng/dm^3	Baronti et al. 2000
Netherlands	E1, E2, EE2	0.4–47 ng/dm^3	Belfroid et al. 1999
Germany	E1, E2, EE2	LOD-70 ng/dm^3	Ternes et al. 1999
UK	E1, E2	1.4–76 ng/dm^3	Desbrow et al. 1998
Activated sludge			
Germany	E1, E2, EE2	<4–37 ng/g	Kuster & Barceló 2004
Israel	"Estrogen"	19–64 ng/g	
River			
Germany	E1, E2, EE2	0.1–5.1 ng/dm^3	Kuch & Ballschmiter 2001
Italy	E1, E2, EE2	0.04–1.5 ng/dm^3	Baronti et al. 2000
Netherlands	E1, E2, EE2	0.1–5.5 ng/dm^3	Belfroid et al. 1999
River sediment			
Germany	E1, E2, EE2	<0.2–2 ng/g	
Spain	E1, E3, EE2, DES	0.05–22.8 ng/g	Kuster et al. 2004
UK	E1, E3, DES	<0.04–0.388 ng/g	
Drinking water			
Germany	E1, E2, aE2, EE2	0.30–0.70 ng/dm^3	Kuch & Ballschmiter 2001

E1, estrone; E2, 17β-estradiol; aE2, 17α-estradiol; E3, estriol; EE2, 17α-ethynylestradiol; DES, diethylstilbestrol; MeEE2, mestranol; WWTPs, Wastewater Treatment Plant; LOD, limit of detection.

Table 2. Parameters of estrogen analysis by GC-MS.

Gas chromatograph	Mass spectrometer			
Injector temperature: 290°C	Temperature of ion source: 290°C			
Sample: 3 μl	Temperature of ion trap: 180°C			
Oven	SIM Diagnostic Ions of Derivative, m/z			
Program: 150°C (2 min)	DES	412	384	217
12°C/min to 240°C	E1	342	257	244
3°C/min to 290°C	E2	416	326	285
(10 min)	EE2	425	286	232
	MeEE2	367	227	173
	E3	504	414	387

Derivatization procedure: The derivatization mixture containing MSTFA/TMIS/DTE was added to SPE extract (after evaporation) and heated at 60°C for 30 min.

MSTFA/TMIS/DTE reagents in the ratio 1000:4:2 (v/v/w Sigma-Aldrich). The resulting trimethylsilil derivatives were analyzed gas chromatographically (GC-MS Varian Saturn 2100T ion trap equipped with CP-Sil8 column). The qualitative identification of estrogens was carried out by comparing the retention times of the peaks of chromatographic standards and samples, and their mass spectra. The quantitative analysis was made using the method of internal standard (mirex, Sigma-Aldrich) which employed Single Ion Monitoring (SIM). Gas chromatographic and mass spectrometric conditions are presented in Table 2.

2.2 *Studies concerning the estrogen removal by coagulation and sorption on activated carbon*

The tests were carried out for model waters prepared from deionized water which contained humic acids of constant concentration of 10 mgTOC/dm^3 and particular standards of estrogens in the amount of 1 μg/dm^3. The waters were treated using four consecutive series of separate coagulation and sorption. The water samples were examined for estrogens directly after the preparation and treatment process by the analytical procedure SPE-GC/MS.

The coagulants applied were iron sulphate PIX-113 and polyaluminium chloride PAX-18 manufactured by Kemipol Sp. z o.o. (Poland). Detailed characteristics of the coagulants are presented in Bodzek and Dudziak (2006). The preliminary tests determined the effect of coagulant dose on the removal of organic compounds which constituted the water matrix (HAs, 10 mgTOC/dm^3) at the conventional phases of the process i.e. coagulation (stirring rate 110 s^{-1} for 1 min.), flocculation (stirring rate 25 s^{-1} for 15 min.) and sedimentation (20 min.). Further tests used the following optimum coagulant doses (D_{op}): PAX-18 5.4 mg Al/dm^3 and PIX-113 12.2 mg Fe/dm^3.

The sorption in the cross-flow mode was carried out on carbon columns (5 cm in diameter and 25 cm in height) filled with granulated activated carbon manufactured by GRYFSKAND Sp. z o.o., Hajnówka, Poland, at a constant inflow of working solution into the column at the hydraulic loading from 5 to 15 $m^3/m^2 \cdot h$. The flow through the bed was of gravitational nature. The sorption in the batch mode was investigated using conical flasks and involved 4-h shaking of the working solutions with 25-mg portions of powdered activated carbon manufactured by GRYFSKAND (Poland).

2.3 *Membrane operations*

Further investigations prompted us to focus on the elimination of estrogens using pressure driven membrane operations, notably nanofiltration. Membrane filtration was examined, using a laboratory installation whose main element was a high pressure SEPA CF-HP module produced by Osmonics Inc. Its structure enabled the use of the cross-flow mode. The tests employed a flat polyamide nanofiltration membrane, type DS5DK SEPA CF (Osmonics Inc.), cut-off 150–300 Da. Nanofiltration was investigated as follows:

- preliminary filtration which involved the conditioning of the membrane using deionised water, in the range of transmembrane pressure of 1.0–3.0 MPa for 5 hours,
- basic filtration of the solutions which contained estrogens carried out for 3 hours at 2.0 MPa.

Constant process parameters were maintained during membrane filtration: flow rate under the membrane surface was 0.75 m/s and temperature 20 ± 2°C.

Particular estrogens whose constant concentration was 1 μg/dm^3 were added to deionized water containing humic acids (10 mgTOC/dm^3). The waters were treated in four consecutive series using a single process of nanofiltration and a two-step hybrid process i.e. 1° coagulation with polyaluminium chloride (PAX-18) and 2° nanofiltration. Water samples were tested shortly after preparation and the filtrate was examined when it reached the volume of 1 dm^3 (assumed sample volume for estrogen assays), discarding the filtrate collected during initial 30 minutes.

3 RESULTS AND DISCUSSION

3.1 *Recoveries and blanks of analytical protocol*

Table 3 provides data which demonstrate the quantitative assays carried out by SPE-GC/MS. The water solutions prepared from estrogen standards of 10 ng/dm^3 and 100 ng/dm^3 concentrations were analyzed several

Table 3. Recovery of estrogens spiked into Milli-Q water (n = 4 replicates) (Luks-Betlej & Dudziak 2004).

Spiked level, ng/dm^3	DES	E1	E2	EE2	MeEE2
10 (RSD %)	99 ± 2.4 (2.4)	103 ± 2.4 (2.1)	98.5 ± 8 (8)	96.8 ± 15 (24)	105 ± 15 (30)
100 (RSD %)	95 ± 9 (9.6)	106 ± 4 (2.4)	97.9 ± 3.8 (3.9)	101 ± 7.4 (7.3)	99.4 ± 1.3 (1.3)
LOQ	0.5	1.0	1.0	0.5	0.5

Solid Phase Extraction (Bakerbond octadecyl C_{18} 1000 mg 6 ml): Conditioning: 7 ml acetonitrile, 5 ml methanol and 7 ml H_2O (pH = 3.0).
Concentration: 1 l samples (pH = 3.0).
Elution: 2 × 5 ml acetonitrile.

times, following the whole procedure. The results obtained provided a basis for the calculations of estrogens recovery (%). The mean values of that parameter ranged from 95% to 106% for diethylstilbestrol in the solutions of higher concentrations of standards. The reproducibility of that technique expressed as a relative standard deviation (RSD) amounted to 1.3–9.6% for solutions containing 100 ng/dm^3 of estrogens and 2.1–30% for solutions containing 10 ng/dm^3 of estrogens. The calculations allowed for losses during the assays. For control purposes, a constant amount of cholesterol acetate standard (2000 ng) was introduced into SPE column before filtering the water sample with estrogens and its concentration was assayed after complete analysis.

3.2 *Concentration levels of estrogens in water in Poland*

The analytical procedure described in experimental enabled the separation of estrogen mixtures and their quantitative determination in selected waters in Poland, Table 4.

The waters of the Vistula and Odra rivers showed traces of estrone whose concentration was over 1 ng/dm^3. Estradiol concentration in the Odra river reached 3.2 ng/dm^3 and its presence was also confirmed in the Vistula river. Synthetic ethynylestradiol occurred in negligible amounts in the sampling stations described or close to the detection limit for this analytical procedure. The water samples taken at a water treatment plant and customers' houses (Gliwice) did not reveal any signs of estrogens.

None of the samples assayed revealed the two forms of synthetic estrogens i.e. non-steroid diethylstilbestrol or steroid mestranol.

3.3 *Removal of estrogens in conventional water treatment processes*

The results shown in Figure 3 confirm the high efficiency of sorption (100% removal of particular estrogens) both on powdered activated carbon (PAC) and granulated activated carbon (GAC) in removing free estrogens from the aqueous environment.

Table 4. Concentration levels of estrogens in samples taken from locations in Poland.

	Concentration, ng/dm^3				
Location	DES	E1	E2	EE2	MeEE2
Vistula River, Kraków	nd	1.3	<LOQ	<LOQ	nd
Odra River, Kędzierzyn Koźle	nd	<LOQ	1.3	<LOQ	nd
Odra tributary Gliwice Canal, Kędzierzyn Koźle	nd	1.0	3.2	1.1	nd
Wisla Czarne Water Treatment Plant	nd	nd	nd	nd	nd
Drinking Water (Gliwice)	nd	nd	nd	nd	nd

LOQ, limit of quantification; nd, not detected.

Coagulation proved to be the least efficient. Estrogen removal depended on the coagulant applied. Much higher values were observed for polyaluminium chloride (PAX-18), from 17% for estrone (E1) to 40% for diethylstilbestrol (DES).

3.4 *Removal of estrogens by nanofiltrations and hybrid process*

Figure 4 gives the results obtained for the removal of particular estrogens during nanofiltrations and hybrid process.

The retention coefficients for estrogens during single nanofiltration ranged from 63% (E1) to 100%

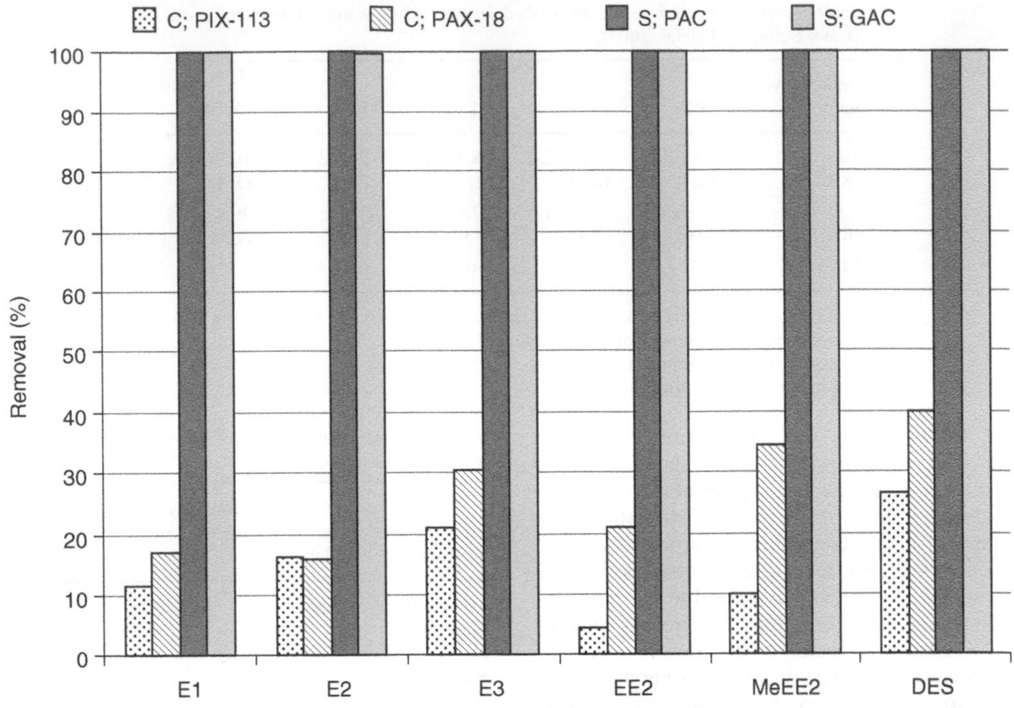

Figure 3. Retention efficiency of estrogens in conventional water treatment processes (C – coagulation, S – sorption).

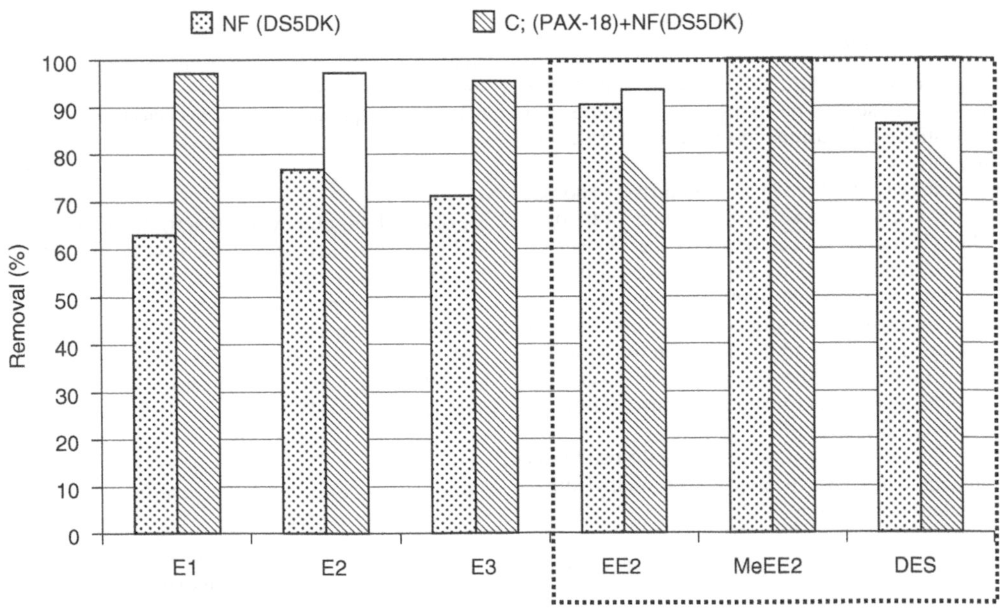

Figure 4. Removal efficiency of estrogens in nanofiltration (NF) and hybrid process (C + NF).

(MeEE2). Much higher values were observed for the retention of synthetic estrogens (dotted line) which were characterized by lower solubility in water and higher log K_{ow}.

Nanofiltration, as used in the second phase of water treatment, after coagulation–flocculation and sedimentation with aluminium salts (D_{op} of PAX-18 5.4 mg Al/l), turned out to be an efficient technique for additional water treatment with respect to estrogens removal. The retention coefficients increased to over 90% for all examined compounds. The removal of mestranol reached 100%, both during single nanofiltration and the hybrid process.

Publications discussing the application of membrane operations to estrogen removal (Agenson et al. 2002, Bellona et al. 2004, Nghiem et al. 2004 and Schäfer et al. 2003) point to a number of factors and phenomena typical of membrane filtration which affect the removal of the compounds during nanofiltration. Some of them are as follows:

- physicochemical properties of estrogens (log K_{ow}) and their morphology (Stokes' radius),
- reciprocal activity of the membrane and removed particles (hydrophobicity of membrane material and estrogen particles and/or polarity), which affects adsorption,
- pH of the flux treated by membrane filtration and the presence of salt,
- the mode of operation ("cross-flow" or "dead-end").

The great magnitude of estrogen elimination obtained during nanofiltration prompts us to expand our research into the issues presented herein. It might result in a better understanding of membrane operations and support this technique as an alternative method for purifying water streams by the removal of the micropollutants.

4 CONCLUSIONS

An analytical procedure for the determination of six estrogens by SPE and GC-MS has been developed. Only the presence of three estrogenic micropollutants (estrone, estradiol and ethynylestradiol) in concentrations up to 3.2 ng/l (estradiol) was noticed in samples taken from locations in Poland. The concentrations of estrogens in Poland depended on the source of natural water examined, but were consistent with the ranges of concentrations described in other publications.

The processes in question removed estrogens from the water stream in the following increasing order of efficiency: $C_{Fe} < C_{Al} < NF < C_{Al} + NF < S_{GAC} = S_{PAC}$. The lowest efficiency was observed during coagulation while sorption on activated carbon produced the best effects.

ACKNOWLEDGEMENTS

This work was performed with the financial support from the Polish State Committee of Scientific Research (KBN) under grant number 3 T09D 040 27.

REFERENCES

Agenson, K.O.Oh., Kikuta, J.I., Urase, T. 2002. Rejection mechanisms of plastic additives and natural hormones in drinking water treatment by nanofiltration, *Proceedings of the Conference "Membranes in drinking water and industrial water production"*, Mülheim an der Ruhr, Germany, 37a: 323–331.

Baronti, C., Curini, R.D., Ascenzo, G., Di Corcia, A., Gentili, A., Samperi, R. 2000. Monitoring natural and synthetic estrogens at activated treatment plants and in receiving river water. *Environ Sci. & Technol.* 34: 5059–66.

Belfroid, A.C., Van Der Horst, A., Vethaak, A.D., Schäfer, A.I., Rijs, G.B., Wegener, J., Cofino, W.P. 1999. Analysis and occurrence of estrogenic hormones and their glucuronides in surface water and waste water in the Netherlands. *Sci. Total Environ.* 225: 101–8.

Bellona, Ch., Drewes, J.E., Xu, P., Amy, G. 2004. Factors affecting the rejection of organic solutes during NF/RO treatment – a literature review. *Water Research* 38: 2795–809.

Bodzek, M., Konieczny, K., Dudziak, M. 2004. Biologically active micropollutants of natural waters. In M.R. Dudzińska & M. Pawłowska (eds), *Pathways of pollutants and mitigation strategies of their impact on the ecosystems*, Monographs of Environmental Engineering Committee of Polish Academy of Science PAN 27: 310–33.

Bodzek, M. & Dudziak, M. 2006. Removal of natural estrogens and synthetic compounds considered as endocrine disrupting substances (EDs) by coagulation and nanofiltration. *Polish Journal of Environmental Studies* 15(1): 35–40.

Danish Environmental Protection Agency. 2003. Evaluation of analytical chemical methods for detection of estrogens in the environment. *Working Report* no. 44.

Desbrow, C., Routledge, E.J., Brighty, G.C., Sumpter, J.P., Waldock, M. 1998. Identification of estrogenic chemicals in STW effluent: 1. Chemical fractionation and in vitro biological screening. *Environ Sci. & Technol.* 32: 1549–58.

Díaz-Cruz, M.S., López De Alda, M.J., Barceló, D. 2003. Determination of estrogens and progestogens by mass spectrometric techniques (GC/MS, LC/MS and LC/MS/MS). *Journal of Mass Spectrometry* 38: 917–23.

Johnson, A.C. & Williams, R.J. 2004. A model to estimate influent and effluent concentration of estradiol, estrone and ethynylestradiol at sewage treatment plants. *Environ. Sci. & Technol.* 38: 3649–58.

Kuch, H.M. & Ballschmitter, K. 2001. Determination of endocrine-disrupting phenolic compounds and estrogens in surface and drinking water by HRGC–(NCI)–MS in the picogram per liter range. *Environ Sci. & Technol.* 35: 3201–6.

Kuster, M., López De Alda, M.J., Barceló, D. 2004. Analysis and distribution of estrogens and progestagens in sewage sludge, soils and sediments. *Trends in Analytical Chemistry* 23(10–11): 790–8.

López De Alda, M.J. & Barceló, D. 2001. Review of analytical methods for the determination of estrogens and progestogens in waste water. *Fresenius J. Anal. Chem.* 371: 437–47.

Luks–Betlej, K. & Dudziak, M. 2004. Determination of natural and synthetic estrogens in water by solid phase extraction and gas chromatography–mass spectrometry. *Sixth International Symposium on Advances in Extraction Technologies*, Sept 6–8 2004, Germany–Lipsk.

Nghiem, L.D., Manis, A., Soldenhoff, K., Schäfer, A.I. 2004. Estrogenic hormone removal from wastewater using NF/RO membranes. *Journal of Membrane Science* 242: 37–45.

Schäfer, A.I., Nghiem, L.D., Wait, T.D. 2003. Removal of the natural hormone estrone from aqueous solutions using nanofiltration and reverse osmosis. *Environ. Sci. & Technol.* 37: 182–8.

Schowanek, D. & Webb, S. 2000. Examples of exposure assessment simulation for pharmaceuticals in river basins with the GREAT-ER 1.0 system. *Proceedings KVIV Seminar "Pharmaceuticals in the Environment"*, March 9, Brussels.

Siemiński, M. 2001. Risk of health by environment, Polish Scientific Publishers PWN, Warsaw (in Polish).

Ternes, T.A., Mueller, J., Stumpf, M., Haberer, K., Wilken, R.D., Servos, M. 1999. Behaviour and occurrence of estrogens in municipal sewage treatment plants—I. Investigations in Germany, Canada and Brazil. *Sci. Total Environ.* 225: 81–90.

Environmental Engineering – Pawłowski, Dudzińska & Pawłowski (eds)
© 2007 Taylor & Francis Group, London, ISBN13 978-0-415-40818-9

Removal of arsenic from underground waters on modified clinoptilolite

Anna M. Anielak*, Renata Świderska-Dąbrowska & Artur Majewski
Technical University of Koszalin, Department of Water and Wastewater Technology, Koszalin, Poland

ABSTRACT: Arsenic removal from underground waters is a subject of study in many research centers. Most often studies concern arsenic sorption on iron oxides, hematite and carbonates. Here we report on the use of manganese modified clinoptilolite (Mn-Clin) for arsenic elimination. Mn-Clin also successfully removes iron and manganese, which often occur in underground waters in high concentrations. The study showed that it also has catalytic and oxidizing properties. Oxidation of As(III) to As(V) occurs in the pH range from 4 to 9. Its efficiency significantly depends on water composition. Iron ions increased arsenic (V) removal in the presence of Mn-Clin probably due to adsorption of $H_2AsO_4^-$ or $Na_2AsO_4^-$, $NaAsO_4^2$, AsO_4^{3-} ions on hydroxy-complexes of Fe(III) resulting from Fe(II) oxidation. In the presence of fulvic acids, less arsenic is eliminated, while manganese ions do not affect arsenic removal.

Keywords: Arsenic removal, sorption, modified clinoptilolite, underground water.

1 INTRODUCTION

In many regions of the world arsenic is a great problem in underground water treatment. Its over-concentration has been proven locally in the waters of Bangladesh, Taiwan, Mongolia, China, Japan, Poland, Hungary, Germany, Belgium, Argentina, Mexico, USA and others. The World Health Organisation has defined the acceptable concentration of arsenic in drinking water (due to its high toxicity) to be 10 μg/L. Arsenic may exist in eight oxidation states (between +5 and −3) but arsenates As(V) and the more toxic arsenite As(III) are mainly found in water. In natural conditions organic matter reduces As(V) to As(III). That is why both forms need to be eliminated from water.

To remove arsenic, processes such as coagulation, absorption, ion exchange, reverse osmosis, etc. are used. A number of authors have shown that ferric oxides, hematite, goethite and carbonates (Farquar et al., 2002; Redman et al., 2002) are good adsorbents for the removal of arsenic. The high affinity of As(III) and As(V) towards hematite was proven and between As(III), hematite and humic substances. The adsorption of As(III) decreases in the presence of humic substances and oxidation to As(V) takes place (Redman et al., 2002). Some other studies have shown that arsenic removal can be achieved by adsorption on manganese oxide crystals with the general formula $Na^{+}_{0.333}\,(Mn^{4+}_{0.722}\,Mn^{3+}_{0.222}\,Mn^{2+}_{0.055})\,O_2$ (Manning et al., 2002). At the same time, the adsorption and biodegradation of As(V) in soil and the relation between the biodegradation process and the content of iron and pH value in soil, was found. In alkaline soils, biodegradation of As(V) is more effective (Yang et al., 2002). Physical and chemical processes applied to the removal of arsenic require a developed technological system and those processes are not always efficient. Hence, it was decided to carry out experiments on the removal of arsenic compounds from water in a filtration process using modified zeolite. The latter can also be used to remove manganese and iron cations, (which often occur in water containing arsenates and arsenites). Modified zeolite granules are covered with MnO_2 oxides which show catalytic and adsorptive properties. Batch and column studies carried out for more than 30 days brought good results. The results presented here include only batch studies.

2 MATERIALS AND METHODS

Research into the adsorption process of As(III) ions on modified zeolite was carried out in a static system, which means that adsorption was performed in a bath.

Firstly, the kinetics of adsorption of As(III) was tested (from an Na_3AsO_3 solution; 1 mg As/L; pH 8.4) on modified zeolite added to a solution in the amount of 10 g/L. Samples were mixed in closed conical flasks on a laboratory shaker at 200 r.p.m. After an appropriate mixing period (from 10 minutes to 24 hours) the samples were centrifuged for 15 minutes at a speed of 14000 r.p.m. In a clear solution free from adsorbent particles, the concentration of arsenic was determined

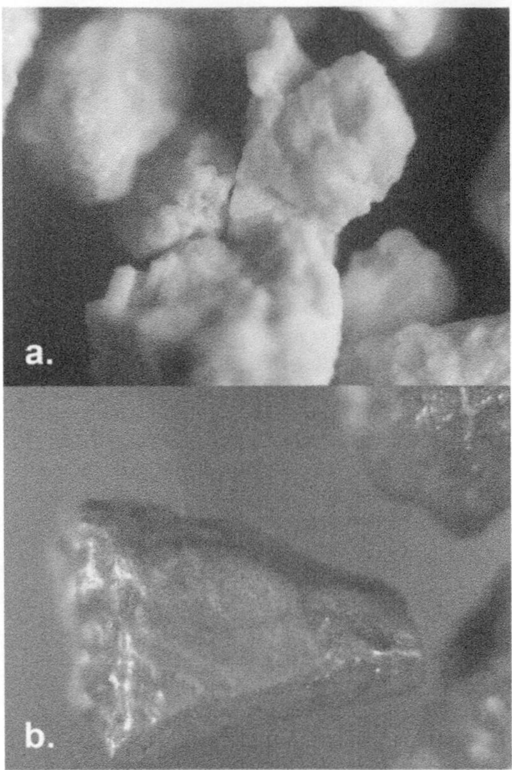

Figure 1. Photograph of zeolite: a) natural clinoptilolite, b) clinoptilolite modified with manganese.

with silver diethyldithiocarbamate according to Polish Standards (PN-88 C-04594/01).

Next, for the assumed contact time of 60 minutes (required for achieving equilibrium) research into the adsorption process of As(III) from Na_3AsO_3 solution with concentration of 1 mg As/L and pH 5.8, with different dose of sorbent (2–30 g/L) and fixed zeolite dose $C_z = 10$ g/L and variable pH (within pH 3 to 9) was carried out. The pH was corrected with 0.1 M NaOH or 0.1 M HNO_3.

In the next stage, the influence of Fe(II), Mn(II) ions and of fulvic acids on As(III) sorption on zeolite were tested. For this purpose, into an arsenic solution (1 mg/L) $FeSO_4 \cdot 7H_2O$ or $MnSO_4 \cdot H_2O$, or fulvic acids (FA) were added. FA were extracted from peat in accordance with the methods specified in previous works (Anielak et al., 2003; http://www.ihss.gatech.edu, 2005). For each substance its influence (with a constant dose of zeolite $C_z = 10$ g/L and pH 7.0) as well as the influence of solution pH (with a constant dose of zeolite $C_z = 10$ g/L and a concentration of tested reagent) on the level of removal of arsenic ions, was determined.

The adsorption process of arsenic ions was carried out on clinoptilolite, modified with manganese ions (http://www.klinosorb.pl, 2005). The surface of zeolite, micro and macropores was covered with manganese oxides, particularly MnO_2 and hydrated $MnO(OH)_2$. Zeolite, modified using a method developed in our own laboratories, was characterized by brown and black colours and a metallic sheen (Fig. 1). The electrokinetic potential of zeolite in a wide range of pH changes was negative, showing not only catalytic properties but also an affinity towards cations. The grain size of zeolite was between 0.5 and 1 mm. Natural zeolite, from which modified zeolite was obtained, included 55% of clinoptilolite, 26% of ash and volcanic glass, 6% of quartz and 13% of montmorillonite. The modified zeolite contained about 11% of Mn; its density was 1.74 g/cm^3.

3 RESULTS AND DISCUSSION

Kinetic studies of arsenic adsorption on modified zeolite were the basis for specifying the adsorption time for other experiments within which the equilibrium concentration of sorbate was obtained. The results presented in (Fig. 2) show that the state of equilibrium of adsorption is obtained after 30 minutes.

For the next tests, the contact time of sorbate with sorbent was 60 minutes. After that time a solution of initial concentration $C_o = 1$ mg As/L achieved equilibrium at concentration $C_r = 600\,\mu$g As/L. So the removal of ions at a constant dose of zeolite $C_z = 10$ g/L was $C_o - C_r = 400\,\mu$g/L.

An important factor in the adsorption process is the zeolite dose (Fig. 3). With a constant initial concentration of sorbate amounting to 1 mg As/L, the increase of zeolite concentration from 0 to 30 g/L and constant time of sorption, assumed as the time to reach equilibrium, the removal of As(III) ions increased from 0 to 800 μg/L. This proves that the adsorbent selected shows a high affinity towards As(III).

The amount of adsorbed arsenic calculated on 1 g of zeolite with constant contact time is shown in (Fig. 4). The obtained dependency can be approximated to the function:

$$(C_o - C_r)/m = 15\log10(C_r) - 5.61 \qquad (1)$$

which allows calculation of the amount of removed arsenic per 1 g of sorbent, depending on the value of the equilibrium concentration of sorbate after the adsorption process. The experiments showed that with the change of equilibrium concentration between 100 and 900 μg/L, with a constant zeolite dose of 10 g/L and an initial concentration of arsenic at 1 mg/L, the maximum removal of arsenic per 1 g of sorbent was 40 μg/L.

The results presented in (Fig. 5) show that the adsorption process of arsenite is associated with a decrease in solution pH (protons freed to solution), which proves that oxidation of As(III) to As(V)

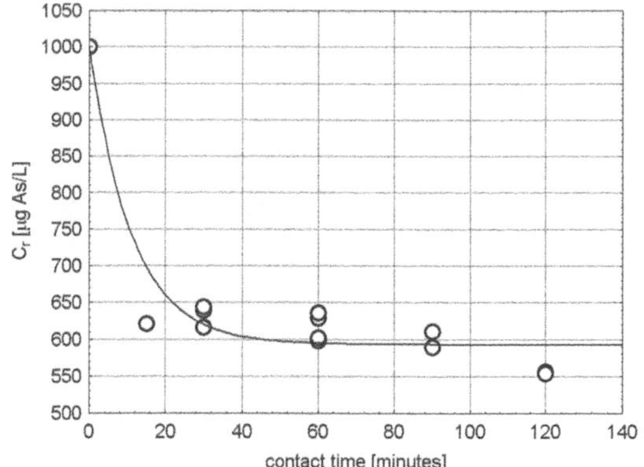

Figure 2. Kinetics of As(III) adsorption process from the solution with an initial concentration $C_0 = 1$ mg As/L and pH value 8.4 on modified zeolite. Dose of zeolite 10 g/L.

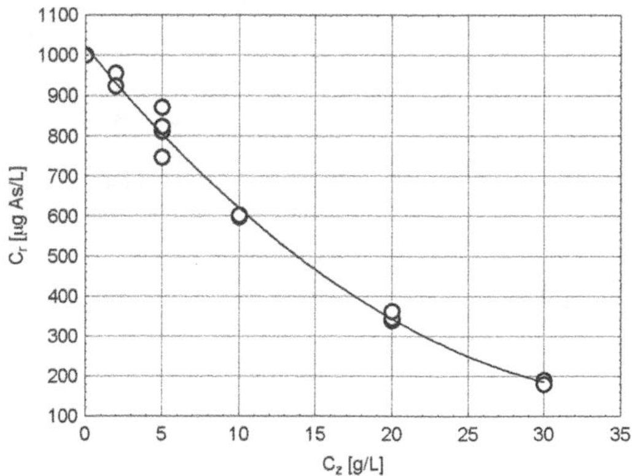

Figure 3. Influence of zeolite dose on the As(III) adsorption process. Initial concentration of solution $C_0 = 1$ mg As/L.

takes place with simultaneous reduction of Mn(IV) to Mn(III) according to the following:

$$2Mn^{IV}O_2 + Na_3As^{III}O_3 + H_2O = Na_3As^{V}O_4 + 2Mn^{III}OOH = Na_3As^{V}O_4 + 2Mn^{III}OO^- + H^+ \quad (2)$$

The surface of modified zeolite in water will be hydrated and manganese dioxide will be partly hydrated. The oxidation of As(III) to As(V) and reduction of Mn(IV) to Mn(III) will be as follows:

$$2Mn^{IV}O(OH)_2 + Na_3As^{III}O_3 = Na_3As^{V}O_4 + 2Mn^{III}OOH + H_2O = Na_3As^{V}O_4 + 2Mn^{III}OO^- + H_3O^+ \quad (3)$$

This dependence is confirmed by the test results presented in Fig. 5, where it is shown that, within the range of pH values between 4 to 9 (within which the effective adsorption of arsenic takes place),

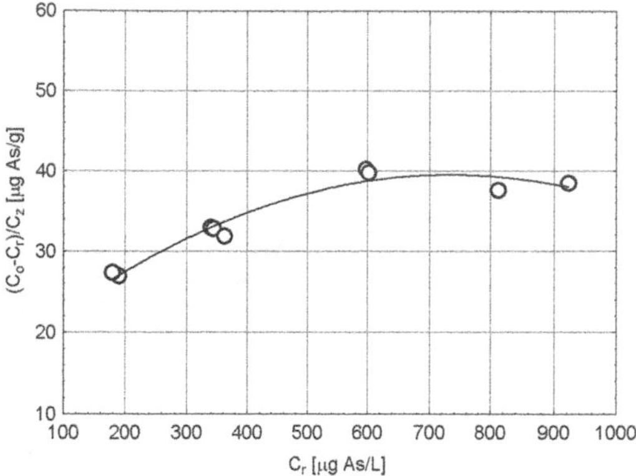

Figure 4. The amount of removed arsenic as a function of equilibrium concentration of sorbate in the adsorption process on modified clinoptilolite. Dose of zeolite 10 g/L.

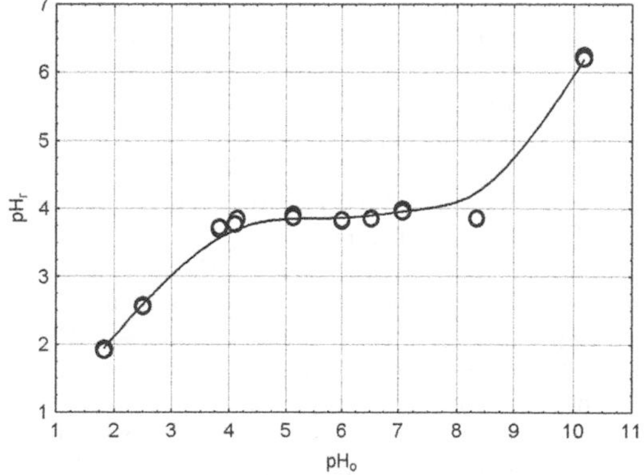

Figure 5. Influence of pH_0 value (initial) solution on pH_r value (equilibrium) obtained, in order to maintain the equilibrium of the adsorption process on modified zeolite of arsenic with an initial concentration of 1 mg As/L. Dose of modified zeolite 10 g/L.

an increase in solution pH after the adsorption process does not take place. Clearly the mixture shows buffering properties.

At the same time, at $pH < 4$, no change of pH value in the adsorption process occurs ($pH_0 = pH_r$). For instance, the solution pH before adsorption is 2 and after the adsorption process is also 2. This means that, within the given pH range, the reaction between substrates does not take place. At an initial $pH > 9$ a proportional increase in the pH_r value is noticed (in the solution after the adsorption process). Hydroxyl ions, liberated into the system, are neutralized by freed protons, the solution is weakly acidic (ca. 4).

The normal potential of redox reaction:

$$H_3AsO_3 + H_2O = H_3AsO_4 + 2H^+ + 2e^- \quad (4)$$

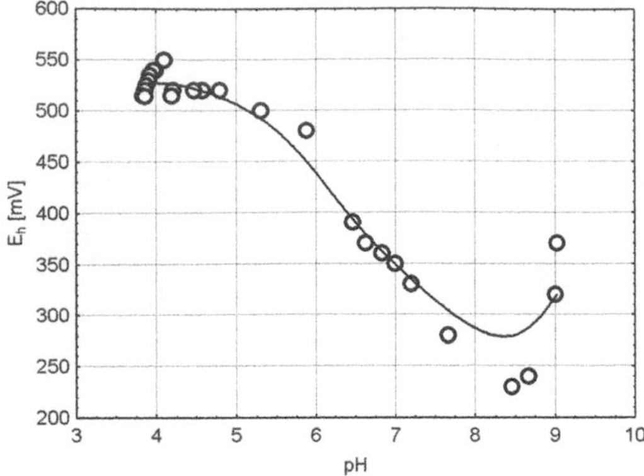

Figure 6. Dependency of redox potential from pH solution for configuration: Mn-Clin – As(III).

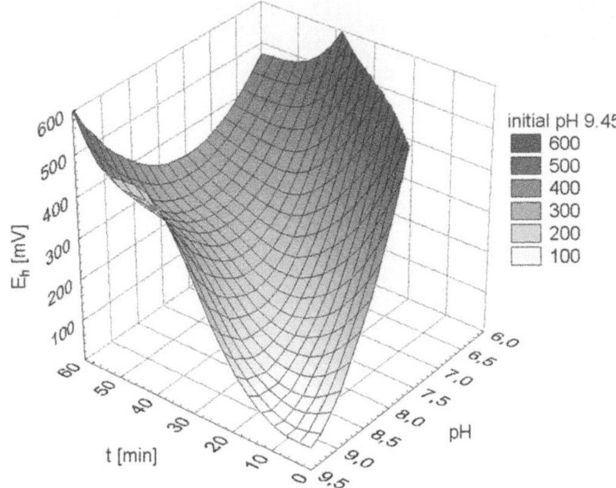

Figure 7. Influence of time for the change in redox potential of solution with the concentration 1 mg As/L and pH 9.45.

measured in relation to normal hydrogen electrode at 298 K, is 570 mV.

The redox potential of an Na_3AsO_3 solution of 1 mg As/L, in the presence of 10 g/L of modified zeolite, changes with time and at the same time a change of solution pH value takes place.

The results showed that the highest values of redox potential for the tested configuration were obtained at a pH of ca. 4 (Fig. 6). This means that, for the tested configuration, oxidation of As(III) to As(V) takes place at pH of 4, which explains why, within the range of the initial pH_0 of the solution changes between 4 and 9, in every test the final $pH_r \approx 4$ was obtained.

Those remarks are confirmed by the results presented in (Fig. 7). The Na_3AsO_3 solution with a concentration of 1 mg As/L and a pH value of 9.45 with 10 g/L of modified zeolite, changed its pH. The value of redox potential rose along with a decrease of pH, i.e. with increasing H^+ concentration.

In underground waters, apart from arsenic, manganese and iron are also often found and together with infiltrating waters, fulvic acids may also enter underground waters. For this reason it is very important to obtain information regarding the influence of manganese, iron and fulvic acids on the adsorption process of arsenic on zeolite modified with manganese.

Results of tests carried out with a constant concentration of arsenic (1 mg/L) and variable concentrations of Mn(II), Fe(II) and fulvic acids are presented in Figs. 8 and 9. Analysis shows that removal of arsenic increases along with an increase as a function of iron ions concentration in the solution. So, it can be assumed that iron ions (II), which oxidize quickly in the presence of oxygen to Fe(III), create flocs of hydroxides and hydroxy-complexes $Fe(OH)_3^0$, $Fe(OH)_2^+$, $Fe(OH)^{2+}$. These iron species participate in the adsorption of As(V), which occur as $H_2AsO_4^-$ or $Na_2AsO_4^-$, $NaAsO_4^{2-}$, AsO_4^{3-}. Electronegative ions are easily adsorbed by electropositive

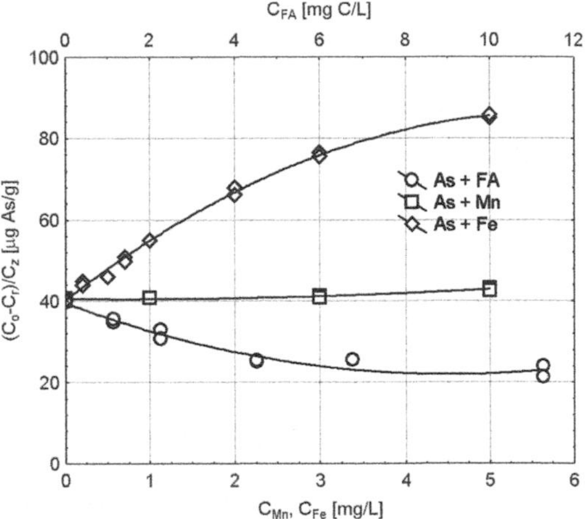

Figure 8. Influence of iron, manganese and fulvic acids concentration on the adsorption process of arsenic on modified zeolite. The dose of zeolite 10 g/L, initial concentration of arsenic $C_o = 1$ mg As/L, Cr – equilibrium concentration, contact time 60 minutes.

hydroxy-complexes of Fe(III). The presence of Mn(II) in the solution has no influence on the adsorption process. Mn(II) ions are dissolved in water and they are resistant to oxidation, and so the possibility of a change in their valence state is low . That is why divalent manganese cannot participate in the process of arsenic removal. It is a neutral factor for the process, and may even show reduction properties.

The influence of humic substances on the sorption process is different to those presented earlier. Fulvic acids added to a tested arsenic solution have dominant negative functional groups, mainly carboxyl and hydroxyl. Carboxyl groups dissociate in a slightly acid medium (at pH > 5), and hydroxyl groups in a slightly alkaline medium (at pH > 8). Hence, we carried out the

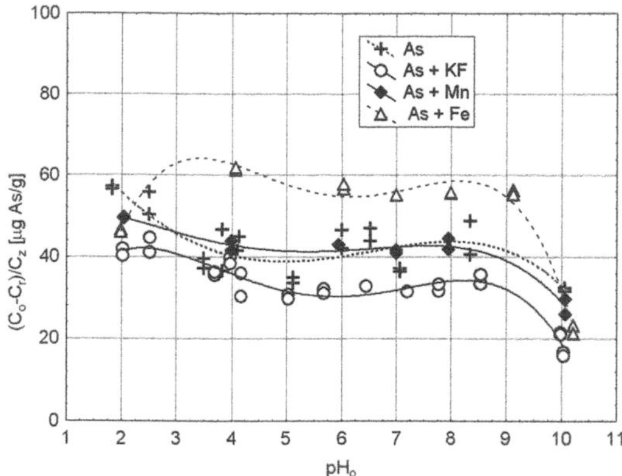

Figure 9. Influence of pH_0 arsenic solution with initial concentration $C_0 = 1$ mg As/L on the process of its sorption on modified zeolite in the presence of fulvic acids 4.51 mg C/L, Mn(II) 1 mg/L and Fe(II) 1 mg/L. The dose of zeolite 10 g/L, C_r – equilibrium concentration of arsenic.

work on effective adsorption processes over a range of pH values.

The electronegative character of fulvic acids does not create adequate conditions for creating complexes with arsenates As(V). The decreasing effectiveness of arsenic adsorption, together with the increase in fulvic acids concentration (from 0 to 10 mg C/L) shows that the increase of negative charge in solution makes adsorption more difficult and increases the dispersion of As(III) and As(V) ions. It also imparts a reducing character on the solution. With the increase of FA concentration from 0 to 10 mg C/L a decrease of removed arsenic by 19 µg/g was noticed.

Within the range of pH < 4 and pH > 9, a diminution of the adsorption process is noticed. Within the values of pH from 4 to 9 the amount of adsorbed arsenic per unit of sorbent mass is constant, independent of the pH (Fig. 9). This dependence has been obtained in every test, i.e. during removal of As(III) in the presence of fulvic acids, Mn(II) and Fe(II) cations. However, in the presence of iron, the reduction of arsenic was at its highest and amounted to 60 µg As per g of zeolite on average, but in the presence of manganese it was identical to that from arsenic water solution and amounted to 41 µg As/g; in the presence of fulvic acids, it took on its smallest value of 35–37 µg As/g.

4 CONCLUSIONS

The following conclusions are drawn from this study:

- Mn-Clin shows catalytic and oxidizing properties, and in its presence oxidation of As(III) to As(V) takes place.
- Zeolite modified by manganese oxides is a good sorbent for arsenic removal from water.
- Oxidation of As(III) and adsorption processes occur within the range pH from 4 to 9.
- The effectiveness of arsenic adsorption on modified zeolite depends on the adsorbent amount.
- The highest redox potential (550 mV) in the As(III) oxidation process, in the presence of modified zeolite, was obtained at a pH of approx. 4.
- Removal of arsenic from water in the presence of Mn-Clin depends on the composition of the test solution.
- Along with the increase of iron concentration, arsenic removal in the solution increases. Due to the structure of the products arising in the hydrolysis and oxidation process, it may be assumed that adsorption of H2AsO4- or Na2AsO4-, NaAsO42-, AsO43-ions, which arise in the process of As(III) oxidation to As(V), takes place on Fe(III) hydroxycomplexes obtained as a result of Fe(II) oxidation.
- Humic substances, in particular fulvic acids, make the removal of arsenic from arsenic solution more difficult.
- Mn(II) ions in the solution have no influence on the process of arsenic removal from water.

REFERENCES

Anielak A.M. & Majewski A., 2003, Physico-Chemical Properties of Fulvic Acids. In: Polish Research on the way to the UE, *Environmental Engineering Studies*, Kluwer Academic/Plenum Publisher, New York, pp. 421–429.

Farquhar M.L., Charnock J.M., Livens F.R. & Vaughan D.J., 2002. Mechanisms of Arsenic Uptake from Aqueous Solution by Interaction with Goethite, Lepidocrocite, Mackinawite, and Pyrite: An X-ray Absorption Spectroscopy Study. *Environmental Science Technology* 36 (8), 1757–1762.

http://www.ihss.gatech.edu/ 2005.

http://www.klinosorb.pl Anielak A.M.: Naturalne zeolity minerałami ery ekologicznej. 2005.

Manning B.A., Fendorf S.E., Bostick B. & Suarez D.L., 2002. Arsenic(III) Oxidation and Arsenic(V) Adsorption Reactions on Synthetic Birnessite. *Environmental Science Technology* 36 (5), 976–981.

Redman A.D., Macalady D.L. & Ahmann D., 2002. Natural Organic Matter Affects Arsenic Speciation and Sorption onto Hematite. *Environmental Science Technology* 36 (13), 2889–2896.

Yang J., Barnett M.O., Jardine P.M., Basta N.T. & Casteel S.W., 2002. Adsorption, Sequestration, and Bioaccessibility of As(V) in Soils. *Environmental Science Technology* 36 (21), 4562–4569.

Yuan T., Hu J.Y., Ong S.L., Luo Q.F. & Jun W. Ng., 2002. Arsenic Removal from Household Drinking Water by Adsorption. *Journal of Environmental Science and Health* 37, 1721–1736.

Environmental Engineering – Pawłowski, Dudzińska & Pawłowski (eds)
 ISBN13 978-0-415-40818-9

Changes of groundwater quality in the vicinity of a municipal landfill site

Izabela Anna Tałałaj & Lech Dzienis
Bialystok Technical University, Bialystok, Poland

ABSTRACT: This paper describes the effect of landfill pollution expansion on water quality. The Hryniewicze landfill – which is in the south-eastern part of Podlasie – was chosen as the object of investigation. The landfill can be assumed to be a typical object for the central and northern part of Poland including the Polish Depression Region. The scale and extent of pollution spread in groundwater was observed in seven piezometers and four house wells situated nearest to the analysed landfill. Analysis of variance (ANOVA) was used as the basic tool for the assessment of essential differences of pollution index values to be tested between individual investigating points. For statistically essential results, the Tukey test was carried out. Therefore, similar variability areas of investigated pollution indexes as well as those of similar values could be identified. The results show that detailed statistical factor analysis can be useful for assessment of the effect of landfill pollution expansion on water quality.

Keywords: Pollution, groundwater, landfill, wastes.

1 INTRODUCTION

Shallow location of an aquiferous layer, mainly groundwater, creates good conditions for penetration of impurities originating from different kind of waste disposal on land (Rowe & Booker 2000). Adverse effects of landfill on groundwater quality can be limited by its location on sites of suitable hydro-geological properties as well as construction of adequate safety devices (Bogchi 2004). Nevertheless, even well protected landfill sites are still dangerous for underground waters (Haq 2003) (Pujori & Deshpande 2005).

This paper describes and evaluates changes of groundwater quality in the vicinity of a landfill site.

Changes of groundwater quality next to the Hryniewicze landfill site were investigated. This landfill site is located in the south-eastern part of the Podlaskie Voivodeship in the area of Wysoczyzna Podlasie Mesoregion.

The free groundwater-table is separated from ground surface by an aeration zone and lies outside landfill site 0.95 to 5.4 m deep, i.e. on datum approx. 139.0 m to 142.0 m above sea level. The landfill site is undermined from western side by groundwater that is flowing from beneath the landfill in three waterways in the north-eastern, south-eastern and eastern directions. Some elevations existing between these three waterways prevent this groundwater from flowing in any other direction. Between Niewodnica and Horodnianka Rivers there is a watershed line constituting a barrier to this flow. Sands being the first aquiferous layer have variable strata thickness (0.7 to 12.0 m), uniform granulation ($\phi = 0.05\,mm \div 0.20\,mm$) and water-permeability factor of the order of magnitude $k = 10^{-4}\,m/s \div 10^{-5}\,m/s$. For these parameters and low hydraulic gradient values, so typical of watershed and water-head zones, groundwater flow is of 10 to 50 m/year (Fig. 1).

Farm buildings next to the landfill site are located on the eastern and south-eastern side without any forest complex between them. House well (S11) nearest to the landfill is approx. 400 m away. The sanitary protection zone around the landfill is 300 m.

2 METHODS AND MATERIALS

The scale and extent of pollution spread in the groundwater was investigated in seven piezometers and four house wells situated nearest to the analyzed landfill (Fig. 1). Observation points around the municipal landfill site were located in three zones, for three different water flow times. Since the operation time of the landfill was long enough (17 years) and because we were attempting to determine pollutant extent, most investigation points were carried out in zone III.

- Zone I was determined by the 200 days' water-flow time (30 m wide). There are points P2, P5 situated 20 and 30 m away from the landfill.
- Zone II corresponding to 2 years' water-flow time (50–100 m). It is point P3 situated 90 m away from the landfill.
- Zone III corresponding to more than 2 years' water-flow. It incorporates the nearest house wells. There

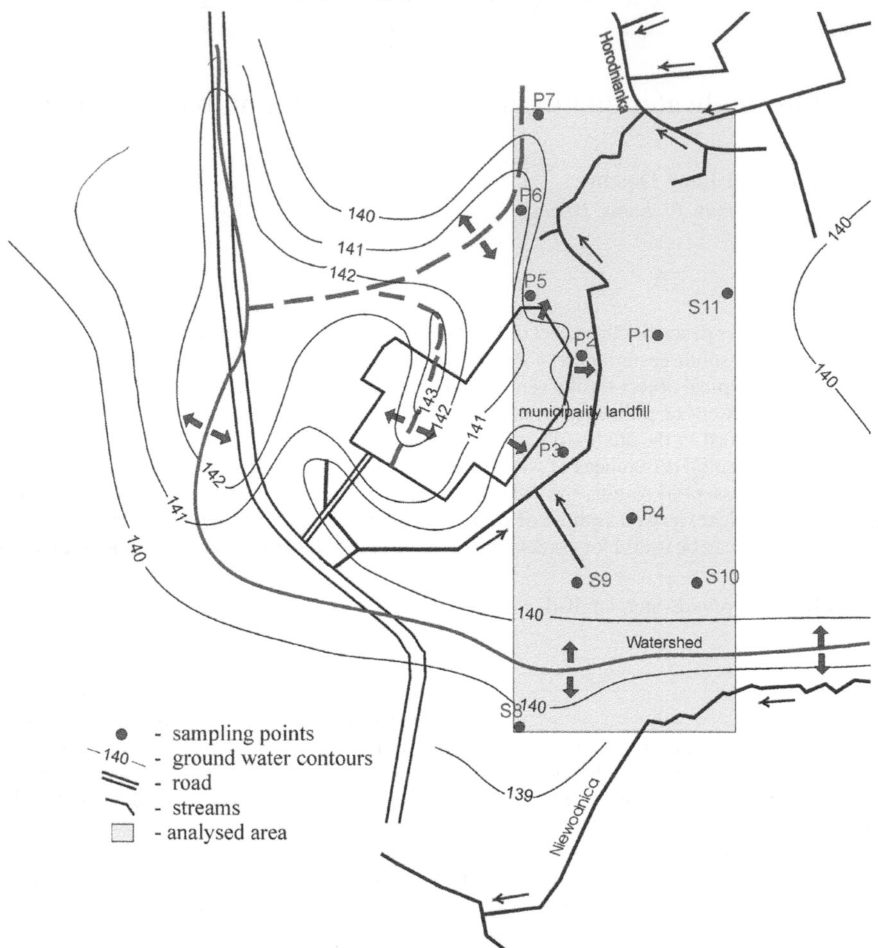

Figure 1. Groundwater contours next to the landfill site and groundwater sampling places.

are points S11, S9, S10 situated 400 m, 500 m, and 680 m away from the landfill and piezometers P7, P1, and P4 situated 650 m, 260 m, and 600 m away from the landfill respectively.

During the investigation the following parameters were determined:

- general suspension, boron, general hardness and hardness converted to calcium and magnesium, cyanides, iron, phosphates, $ChZT_{Cr}$, ammonia nitrogen, nitrate nitrogen, nitrite nitrogen, electrolytic conductivity, cadmium, copper, nickel, reaction, and temperature.

Determinations were carried out according to the Polish Standards and results obtained were the mean value of three determinations carried out simultaneously.

The analysis of variance (ANOVA) with the sampling point as a constant was used as the basic tool for the assessment of essential differences of pollution index values to be tested between individual investigating points. For statistically essential results the Tukey test was carried out (Gibbon & Coleman 2001). Therefore, similar variability areas of investigated pollution indexes as well as those of similar values could be identified. On the basis of our analysis, points of particularly polluted groundwater could be located and an internal uniform subset distinguished in the mean value set.

3 RESULTS AND DISCUSSION

The analysis of variance results show that the sampling point location and essential index differences differ as follows: electrolytic conductivity, nitrate nitrogen, ammonia nitrogen, iron compounds, cyanides,

Table 1. Analysis of variance – groundwater quality changes depend on water sampling point.

Index	SS btw. groups	df group	MS btw. groups	SS resid.	df resid.	MS resid.	F	p
Reaction/pH	11.796	10	1.179	45.71559	103	0.443	2.657	0.056
Temperature	53.946	10	5.394	2884.321	110	26.22	0.205	0.995
Conductivity	4,909,719	10	490,971	1,967,315	110	17,884	27.45	0,000
Nitrite nitrogen	0.002	10	0.000	0.05436	110	0.0004	0.573	0.832
Nitrate nitrogen	1642.3	10	164.2	2649.86	110	24.08	6.817	0.000
Ammonia nitrogen	123.63	10	12.36	355.935	110	3.235	3.820	0.000
$ChZT_{Cr}$	658,460	10	65,846	273,529	110	24,866	2.648	0.066
Phosphates	13.3311	10	1.333	251.568	110	2.286	0.582	0.824
Iron	7.3211	10	0.732	21.2064	110	0.192	3.797	0.000
Free cyanides	0.0008	10	0.000	0.00319	110	0.000	3.092	0.001
Hardness Ca	286,012	10	28,601	158,106	110	14,373	1.989	0.641
Hardness Mg	135,221	10	13,522	157,941	110	14,358	0.941	0.498
Hardness, general	417,819	10	41,781	245,875	110	22,352	1.869	0.057
Chlorides	120,009	10	12,000	212,156	110	1928	6.222	0.000
Boron	37.241	10	3.724	136.401	110	1.350	2.757	0.004
Suspended matter	721,843	10	72,184	379,688	110	34,517	2.091	0.030
Cadmium	0.0031	10	0.000	0.00991	110	0.0001	1.996	0.548
Copper	0.0132	10	0.001	0.06227	110	0.0010	1.323	0.238
Nickel	0.0117	10	0.001	0.04336	110	0.0006	1.685	0.104

SS btw. groups – sum square between groups; df group – degrees of freedom between groups; MS btw. groups – mean square between groups; SS resid. – sum square inside groups (residual); df resid. – degree of freedom inside groups (residual); MS resid. – mean sum square inside groups; F – F test value; p – probability level.

chlorides, boron and suspended matter. For the results see Table 1.

Tukey's test results relating to electrolytic conductivity (Fig. 2a) show large conductivity differences between the test points. Data analysis allows of the formation of three internally uniform groups, statistically different. I: S8, S11; II: P2, P3; III: P5, P6. Group I is characterized by the highest electrolytic conductivity values (666.25 μS/cm ÷ 832.86 μS/cm), group II has high conductivity (585.00 μS/cm ÷ 719.00 μS/cm) and group III – the lowest one (585.00 μS/cm ÷ 719.00 μS/cm). Polluted water is in the neighborhood of the landfill site (P2 and P3 points) as well as close to house wells (S8 and S11).

For nitrate nitrogen, three groups of high uniformity inside the groups can be distinguished (Fig. 2b): I: S9, S10, S8; II: S10, S8, S11; III: P1, P2, P3, P4, P5, P6. Group I (3.6933 mg/dm^3 ÷ 7.6406 mg/dm^3) and group II (7.6406 mg/dm^3 ÷ 16.071 mg/dm^3) are characterized by a high concentration of nitrate nitrogen compared to the lowest value (1.335 mg/dm^3 ÷ 8.3889 mg/dm^3) of group III. A source of strong pollution with nitrate nitrogen was found close to house well S11. Water in other house wells (S10, S8, S11, S9) is more polluted with nitrates (V) than water originating from the area next to the landfill (P1, P2, P3, P4, P5, P6). Lower nitrate (V) concentrations in water close to the landfill can be the result of domination of other nitrogen forms, i.e. ammonia and nitrate nitrogen originating from the municipal landfill site.

Figure 2c shows the diversity of the average ammonia nitrogen concentration. As a result of Tukey's test two uniform groups were separated: I: S9, S11, S10, P1; II: S11, P1, P2, P3, P4, P5, P6, P7. Group I is characterized by a low ammonia nitrogen value (0.9275 mg/dm^3 ÷ 1.7243 mg/dm^3) and group II – by high ones (1.7243 mg/dm^3 ÷ 4.2770 mg/dm^3). The highest groundwater pollution with ammonia nitrogen was observed close to the landfill site – P2 piezometer – but the lowest one has water from S9 and S10 dug wells. The results confirm previous observations that groundwater close to a municipal landfill site is characterized by a higher percentage of ammonia nitrogen than its other forms ($N_{NO_2^-}$, $N_{NO_3^-}$). Higher $N_{NH_4^+}$ concentration results from flow-in of fresh organic pollutants as well as from domination of oxygen-free conditions in the area closest to a landfill site.

Figure 2g shows the diversity of the average concentration of iron compounds in groundwater. Here we can distinguish uniform groups as follows: I: S9, S11, S10, P1; II: P3, P4, P5, P6, P7. Group I is characterized by lower concentration values (0.07 mg/dm^3 ÷ 0.46 mg/dm^3), and group II – higher ones (0.30 mg/dm^3 ÷ 0.79 mg/dm^3). The point located nearest the landfill site (piezometer P2), had the highest concentration of iron compounds. The increase of iron compound concentration in water next to the landfill is prompted by the considerable content of organic

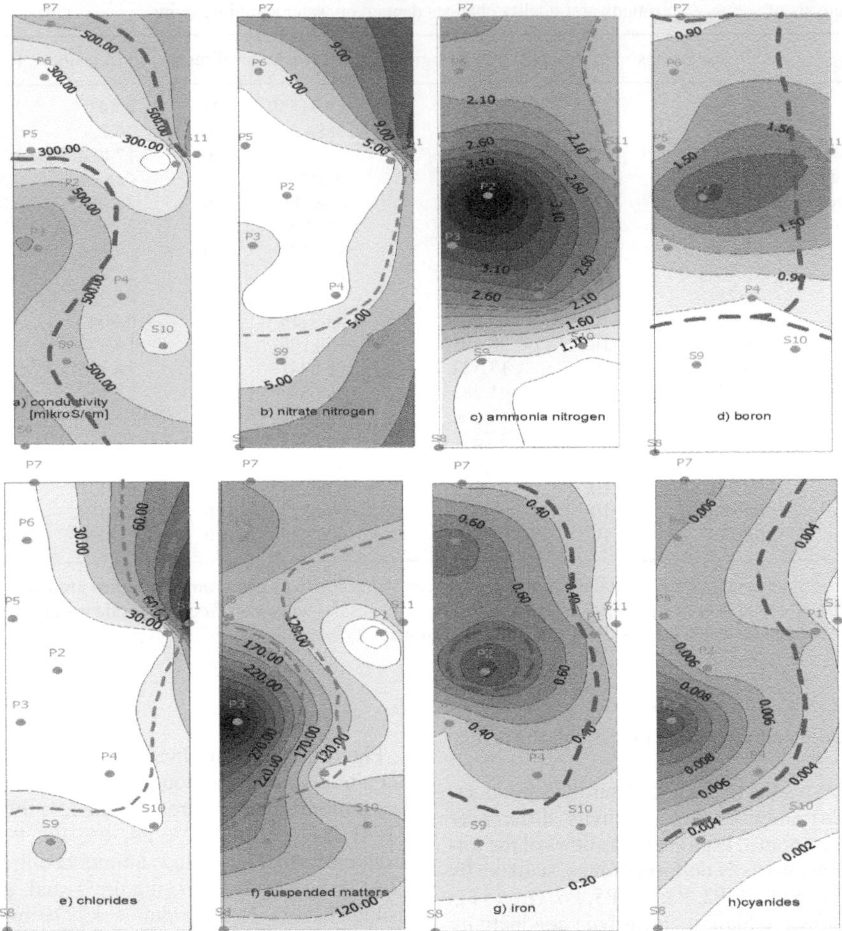

Figure 2. Concentration distribution [mg/dm^3] for selected groundwater pollution indexes 1998–2001.

compounds originating from leachate because together with iron they can create many complex forms. The presence of aggressive carbon dioxide can be another source of iron, inducing iron bicarbonates to migrate into water.

Figure 2h shows Tukey's test results for cyanides. Based on the data two uniform statistic groups were distinguished. I: S9, S11, S10, P1; II: S11, P1, P2, P3, P4, P5, P6, P7. The highest cyanide concentration was observed in the area closest to the landfill site, i.e. for point P3. Group I contains places of lower cyanide concentration (0.0020 mg/dm^3 ÷ 0.0055 mg/dm^3) than for group II (0.0051 mg/dm^3 ÷ 0.0122 mg/dm^3). The test results show that the source of groundwater pollution with cyanides could be leachates from the analyzed landfill. According to Appelo (Appelo 1994) polluted groundwater contains nitrogen in form of ammonia nitrogen and cyanides. This regularity is confirmed by the relationships achieved for $N_{NH_4^+}$ and CN^- as the highest concentration of these ions was observed in piezometers at the landfill.

Figure 2e shows variation of chloride concentration. On the basis of the results two uniform groups can be distinguished: I: S10, S8, S9; II: S11, P1, P2, P3, P4, P5, P6, P7. Points belonging to Group I (house wells) are characterized by high chloride ion concentration (1.5 mg/dm^3 ÷ 33.3 mg/dm^3). Group II is an area of the lowest concentration (4.9 mg/dm^3 ÷ 14.17 mg/dm^3). Point S11 is of highest concentration. The data show strong water pollution with chloride ions in the house wells being investigated. Chloride ion concentration in house well water exceeds values observed in the water at the municipal landfill site.

Figure 2d shows results of Tukey's test for boron. The data confirms that two uniform average groups were achieved: I: S10, S8, S9, S11, P1; II: S11, P1, P2, P3, P4, P5, P6, P7. Group I is characterized by lower boron concentration value

($0.36\,mg/dm^3 \div 1.85\,mg/dm^3$) while group II has higher ones ($0.61\,mg/dm^3 \div 2.26\,mg/dm^3$). Increased concentration values are typical of area close to a municipal landfill site. According to Appelo (Appelo 1994) boron is present in different kinds of waste from where it can migrate into leachates and penetrate groundwater.

Figure 2f shows changes of general suspension values for successive investigated points. For suspended matters three uniform groups were distinguished: I: S10, S8, S9, S11, P1; II: P2, P3; III: P4, P5, P6, P7. A group of points next to landfill P2 and P3 ($202.0\,mg/dm^3 \div 392.0\,mg/dm^3$) and group I – $105.4\,mg/dm^3 \div 202.0\,mg/dm^3$) are characterized by the highest values. The lowest suspension values have piezometers P4, P5, P6, P7 of group III ($83.0\,mg/dm^3 \div 193.3\,mg/dm^3$).

4 CONCLUSIONS

In the groundwater close to the solid waste landfill site higher concentrations of ammonia nitrogen, iron compounds, cyanides and boron (Fig. 2c, d, g, h) were observed. Boron, iron compound, cyanide and ammonia nitrogen contents in house well water were on the average twofold lower than for piezometer water close to the landfill site.

Moving away from the landfill in the direction of groundwater flow, values of the above mentioned groundwater components decreased. The distribution of component concentrations (Fig. 2) shows that a range of groundwater pollution reaches piezometer P1. Quantity changes observed indicate that self-cleaning proceeds in the groundwater.

Investigations carried out show that this communal landfill site is not the only strong source of pollution but also farm buildings close to the investigated house wells. On many occasions V. concentrations of the investigated house wells parameters were higher than those for groundwater next to the landfill site. This applies to electrolytic conductivity, nitrate nitrogen and chlorides (Fig. 2a, b, e). The highest values of these indices were observed in S11 well water. Slightly lower concentrations of the above mentioned variables were found in groundwater close to the landfill site and the lowest ones – in piezometers P5 and P6 (Fig. 2a, b, e). Such a concentration distribution indicates that leachate migration from the landfill is limited to the northern direction where points P1 and P6 are located. Probably it is enhanced by a new dumping site, situated on the north side of the landfill, having sealed ground and leachate drainage sloping in the SE direction.

Data analysis carried out shows that piezometers P2 and P3 situated nearest to the landfill site are under the influence of leachate pollution. Higher boron and cyanide concentrations reach points P1 and P4 situated 260 m and 400 away from the landfill site respectively. Compared to data from the literature this distance is not long. This means that self-cleaning proceeds relatively fast. Undoubtedly, fine sand with water permeability from 10^{-4} m/s to 10^{-5} m/s and uniform granulation ($\phi = 0.005\,mm \div 0.200\,mm$) as well as low hydraulic gradient have an effect on it. These conditions are averse to pollution filtration and sorption processes and reduce groundwater flow, thus limiting pollutant dissemination.

The investigations enabled separate zones of similar average values for individual variables on the investigated area to be discerned. Statistical methods used for the analysis enabled the estimation of the propagation range of particular pollution sources as well as to determine (for electrolytic conductivity) the approximate borderline between interaction of pollution from different sources.

ACKNOWLEDGMENT

This work was supported by the university internal grant W/IIŚ/34/03.

REFERENCES

Appelo, C.A.J. & Postma, D. 1994. *Geochemistry, groundwater and pollution*. Rotterdam: Balkema.

Bogchi, A. 2004. *Design of Landfills and Integrated Solid Waste Management*. New York: John Wiley and Sons, Ltd.

Gibbons, D. & Coleman, D. 2001. *Statistical Methods for Detection and Quantification of Environmental Contamination*. New York: John Wiley and Sons, Ltd.

Haq, I. 2003. Environmental Impact Assessment Study: leaching of chemical contaminants from a municipal dump site Hastsal, Delhi. *International Journal of Environmental Studies* 60: 363–377.

Pujori, P. & Deshpande, V. 2005. Source apportionment of groundwater pollution around landfill site in Nagpur, India. *Environmental Monitoring and Assessment* III (1–3): 43–54.

Rowe, R.K. & Booker, J.R. 2000. Modelling and Application in Geomechanics. In Zaman et al. (eds), *A practical modelling technique for assessing potential contaminant impact due to landfills*: 493–504. New York: John Wiley and Sons, Ltd.

Environmental Engineering – Pawłowski, Dudzińska & Pawłowski (eds)
© 2007 Taylor & Francis Group, London, ISBN13 978-0-415-40818-9

Impact of leachate from an unsealed municipal landfill site on surface and groundwater quality

Czesława Rosik-Dulewska
Instytut Podstaw Inżynierii Środowiska, Zabrze, Poland

Urszula Karwaczynska & Tomasz Ciesielczuk
Opole University, Department of Land Protection, Opole, Poland

ABSTRACT: In this paper the results of physico-chemical investigations of leachate, surface and ground water in an unsealed municipal landfill site area are presented. The research objective was the investigation of the quality of leachate and the degree of pollution of the Odra River surface waters by compounds transported with underground water flowing from the direction of the landfill. No clear (statistically significant) impact of leachate on the Odra River waters was observed, however the ground waters in this area were heavily polluted.

Keywords: Leachates, surface waters, chemical composition, landfill, municipal wastes.

1 INTRODUCTION

The main method of municipal waste disposal in Poland and worldwide is storage in landfills (Sang 2005, Scott et al. 2005). Wastes, which are delivered to a landfill site, are levelled, compressed and covered by an inactive insulating material layer every day. This procedure hinders, for example water infiltration, the access of insects to waste. It also prevents the dispersion of fetid odours and causes a landfill to look not inattractive.

These precautions amount to minimal sanitary conditions. Bacteria, endospores or other pathogens often infect municipal wastes. The organic matter they contain is a good feeding ground for animals, which may spread diseases. Covering of waste with mineral waste limits these hazards.

A municipal waste landfill is a specific bioreactor where deposited wastes are digested in the presence of atmospheric and microbiological agents. These changes can occur in aerobic or anaerobic conditions. Organic compounds such as proteins, fats and polysaccharides have a very widely differing susceptibility to biodegradation. Moreover, there are some non-biodegradable materials such as plastic and rubber. The inorganic fraction of wastes does not undergo biodegradation, however it does affect its course.

The initial decomposition process of organic materials results from aerobic microorganisms (bacteria and fungi). Aerobic reactions lead to production of water, carbon dioxide, energy production and residues of oxyacids in their highest oxidation states (Golimowski et al. 2001, Scott et al. 2005). Microbiological processes of waste decomposition lead also to release of hazardous compounds, which migrate beyond the landfill area with the leachates. Leachates can fill free cells in different landfill levels in perched water form or stored in the lower part of the landfill bowl. From they it can migrate through the aeration level to the water layer and thus disperse over long distances. To ensure elimination of groundwater pollution the bowl of landfill should be fully sealed.

In this work, the impact of leachate from an unsealed municipal landfill, previously used for marl mining, on the pollution levels of surface and groundwater present in the area of the landfill has been investigated.

2 MATERIALS AND METHODS

Research was carried out on the unsealed municipal landfill "Grundman" in Opole. This landfill is situated in an old post-marl working and was used from 1945 to 1998.

The landfill is situated in the central part of the city, between the railway lines Opole-Katowice and Opole-Czestochowa. About 500 meters from the landfill border there are allotment gardens within the built-up area (Fig. 1).

In the landfill area there are two layers of Cretaceous water level: Cenomanian and Turonian. The Turonian

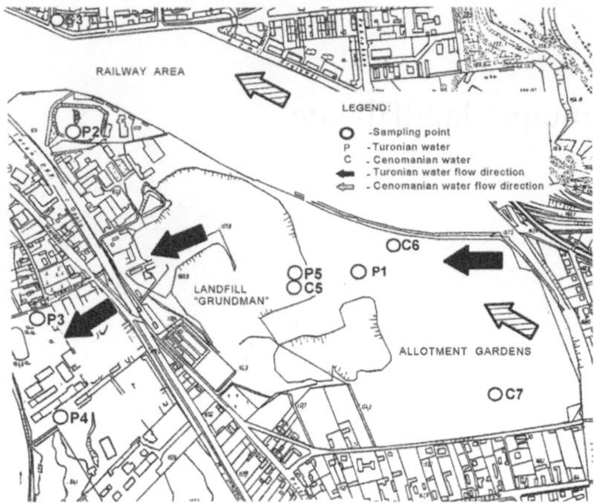

Figure 1. Study area with marked sampling points.

level is of a creviced character. The water-bearing rocks are marls and Turonian marly limestones, at a level of 35 meters under ground. Because of rock variability Turonian profile crevasses have often only small ranges and are filled with clay material. This is the reason why water circulation is limited. The Turonian free water-table is found 1.5–2.5 meters under ground. The amount of water flow depends on rock fissuring and atmospheric precipitation.

The filtration factor for marly limestone is 4.2×10^{-5} m/s. Water migration in the lower Turonian parts is complicated because it is comprised of clay marls which have small and short crevasses. Additionally, the crevasses are full of clay. For these rocks the filtration factor ranges from 1.7×10^{-7} to 4.2×10^{-9} m/s.

The Turonian level is supplied by atmospheric precipitation on the whole outcrop area through the rock mantle and infiltration from Quarternary rocks. From the hydrogeological data it is known that Turonian waters move in the direction of the Odra River.

The Cenomanian rocks form the second water-bearing layer. The filtration time through this 7–9 meter thick marl is 21–38 years. The water table of the Cenomanian layer is perched. The perched level is formed from Turonian rocks, and the bottom clay of Keuper (red marl). The water-bearing layer is comprised of differently grained Cenomanian sandstones with a filtration factor of $8.4 \times 10^{-4} - 2.6 \times 10^{-5}$ m/s. They are at the depth of 35–65 meters from the ground surface. The water table at the Cenomanian level is at a depth of 15–17 meters under ground level. This water layer is supplied through the outcrops of the Opole Basin. The Opole climate is affected by the Sudety mountains. It has features of oceanic climate. After the Schmuck, district the Opole area is one of warmest areas in Poland with small temperature fluctuations. Precipitation is highest in July and August (two or three times higher than in February or March). The area is well ventilated with winds blowing mostly from the NWW direction.

A post-marl working without insulation was used as a waste landfill from 1945. Wastes were deposited from the depth of 25 meters over an area of ca. 15 hectares. During exploitation, the composition of waste was not under control.

Researches into wastes were begun there in 1975. The capacity of wastes storage at this time was about 130–140 thousand cubic meters per year. In the years 1990–95 about 180–225 thousand cubic meters of wastes were deposited. In 1997, in the eastern and north-eastern parts, wastes from the great flood were deposited.

The reclamation of the landfill site (larger basin) was started in 1996 according to a project encompassing: formation of landfill profile, removal of leachates and gases, putting on a 30 cm ground layer, mechanical services, fertilizing and liming, sieving of grass seed. Tree and bushes planting completes this project.

During exploitation, the leachates (O-1 and O-2) from the area near the allotment gardens were collected (leachates were found in the depression in 1995 only). Leachate from the well (O-3) (in 1995 recognized as an "old" leachate) was collected. This leachate reappeared in 1997 when, on the landfill, different wastes from big flood were stored. In this case the leachate was recognized as "new" leachate.

Surface water sampling points (from the Odra River) were located 200 m in front of the landfill (K-1) and 500 m after the landfill (K-2).

Underground water sampling points were located in the inflow and downflow directions from the landfill (west and south-west – Turonian, north-west – Cenomanian waters). A reference point for Turonian water was a piezometer P-1 located on the waters inflow (to the east from the landfill), about 200 m from it. The location of other piezometers for this layer was as follows: piezometer P-2 – in the direction of the downflow of Turonian waters, 260 m to the south-west from the landfill, piezometer P-3 – in the direction of the downflow of Turonian waters, 220 m to the west from the landfill, piezometer P-4 – in the direction of the downflow of Turonian waters, 270 m to the south-west from the landfill, piezometer P-5 – on the landfill "Grundman" – waters from this point come from under the landfill basin. For the Cenomanian water layer, C-7 piezometer was a reference point located on the inflow of Cenomanian water to the landfill, about 22 m south-west from it. The location of other piezometers was as follows: C-3 – in the direction of the downflow of Cenomanian waters, 800 m to the south-west from the landfill, C-5 – near piezometer P-5 – water from this point comes from under landfill basin, C-6 – in the direction of the down flow of Cenomanian waters,

200 m to the east from the landfill. Before sampling the water was removed from piezometer. The samples were taken to glass and PP bottles and fixed. The samples were analyzed according to PN methodology. All photometric determinations were made on a spectrophotometer PHILIPS PU8620. Heavy metals content determinations were made on ASA PU 9100X UNICAM-PHILIPS. Statistical analyses of results were made using Excel for each point and the averages as well as differences between them were calculated. The T-Student distribution was calculated with significance level of 0.05.

3 RESULTS AND DISCUSSION

Chemical composition and physical properties depend on the quality and quantity of deposited wastes, climate conditions, landfill age, type of wastes deposited and landfill exploitation. But the migration of pollution from landfill depends on the geological environment, which can facilitate the flow of groundwater and dissolved substances (sand levels or high hydraulic gradients) or can inhibit movement of substances from the landfill (low ground permeability low hydraulic gradients) (Baker and Curry 2004).

At the beginning of the research in 1995 the landfill was still working, so leachate samples were collected in an old part of it and from freshly deposited wastes (Table 1). In the analyzed "new leachates" higher COD levels were detected (maximum about 55 g O_2/dm^3) than BOD_5 (about 7 g O_2/dm^3) which shows that mostly wastes unsuitable for microbial decomposition were deposited – higher values than reported by Oygard et al. (2004). Oxidizability and conductivity in analyzed leachates are comparable to samples from other landfills from the Netherlands, Italy, Germany USA and Norway [1,3,8], but the total chloroorganic pesticides content, HCH (from 0.42 to 1.0 $\mu g/dm^3$) and HCB (from 1.84 to 1.89 $\mu g/dm^3$) suggest that these compound can be detected also in groundwater.

In the group of mineral indicators, nitrogen ions, chlorides and sulphates are worth special attention. The concentration of chloride ions in analyzed leachates fluctuates from 106 mg/dm^3 (leachate from wastes deposited a couple of years ago) to 660 mg/dm^3 (leachate from freshly deposited wastes). These values are lower than cited by Breukelen (2003) and Scott (2005), but the sulphate ions determined are comparable to theirs. Literature sources report that the ammonium ion concentration in landfill leachates is several dozens to several hundreds of times higher than the nitrate content. This phenomenon was found also in this work. In new leachates the ammonium ion concentration was 643 times higher than nitrate ions content and in "old" leachates – 70 times higher. The ammonium content is comparable to this parameter value measured in leachates from Canadian (Manning and Hutcheon 2004) and Netherlands (Breukelen et al. 2003) landfills.

From the literature it is known that heavy metal concentrations in leachates are not so high. It is possible that with increasing age of the landfill, metals become bound by organic and mineral matter.

In our analyzed leachates higher contents of lead, cadmium and copper (Table 1) than average values from Polish, German, Norwegian and USA landfills (Al.-Yaqout and Hamoda 2003) Oygard et al. 2004) were detected. The heavy metal content in leachates from this landfill can be ranked as follows: Zn > Pb > Cu > Ni > Cr > Cd.

In our researches only minimal deterioration of the Odra River water was detected after landfill leachate inflow (K-2). Pollution is mainly restricted to the following species: ammonium, nitrite, chloride, chloroorganic pesticides and Coli Index. Statistical analysis did not confirm significant differences between sampling points K-1 and K-2.

The composition of groundwater is modified by mixing with infiltrating waters. On this basis the concentrations of ions and compounds can fluctuate within high limits. With infiltrating waters not only heavy metals and organic compounds but also bacteriological pollution flow contribute to the groundwater level.

Holistic analysis of groundwater quality and the impact of pollution sources with all factors taken into consideration is a very complicated task.

On the GRUNDMAN landfill two groundwater levels were detected: the Turonian and the Cenomanian. A stabilized water table for Cenomanian waters was detected 14 meters and for Turonian waters 2–2.5 meters under ground.

Of all analyzed parameters, the most important were: conductivity, oxidizability BOD_5, nitrogen ions, and chlorides, sulphates and, of the heavy metals: cadmium, lead, copper, chromium and zinc.

The lowest (Class I according to the Polish Groundwater Guidelines Directive) and stable long term conductivity was detected only in piezometer P-5 (max. 412 $\mu S/cm$), located on landfill datums. Slightly higher values of this parameter were noted in samples from piezometer P-1 located on inflow (from 484 $\mu S/cm$ – April 2002 to 1100 $\mu S/cm$ – August 1995 – II class according to the Directive). Waters from piezometers located on runoff have very high conductivity values (4480 $\mu S/cm$ – March 1995 for P-2 and 4620 $\mu S/cm$ – March for P-3) which classify them as V quality class (water from P-4 – 2930 $\mu S/cm$ – May 2002), IV quality class according to Directive. EC values were ten times higher than those reported by Breukelen (2003).

In Cenomanian waters inflowing into the landfill area (C-7 max. 890 $\mu S/cm$) the conductivity

Table 1. Results of physico-chemical and biological parameters analysis of landfill leachates.

			Leachates											
No.	Parameters	Units	O-1 22.03.95	O-2 22.03.95	O-3 22.03.95	O-1 25.05.95	O-2 25.05.95	O-3 25.05.95	O-1 12.07.95	O-2 12.07.95	O-3 12.09.95	O-3 10.10.97	O-3 20.01.1998	O-3 8.03.1998
1.	Reaction	pH	6.61	6.87	6.61	6.65	6.64	6.80	6.55	6.75	4.66	7.07	7.03	6.99
2.	Conductivity	μS/cm	4040	3910	1370	5110	4820	2510	5230	4750	1160	5960	6220	6480
3.	BOD_5	mgO_2/dm^3	400	260	1800	730	680	6400	510	226	4200	500	43	82
4.	COD_{KMnO_4}	mgO_2/dm^3	275	476	2729	240	384	7392	255	180	4200	558	368	199
5.	$COD_{K_2Cr_2O_7}$	mgO_2/dm^3	883	864	13616	804	739	54785	941	941	15232	4358	1680	67
6.	Ammonium nit.	mgN_{NH_4}/dm^3	9.8	8.6	1.4	n.o.	n.o.	n.o.	n.o.	n.o.	n.o.	256.8	239.5	352.3
7.	Nitrate nitrogen	mgN_{NO_3}/dm^3	n.o.	n.o.	0.02	n.o.	n.o.	n.o.	n.o.	n.o.	n.o.	0.40	3.22	4.25
8.	Chlorides	$mgCl/dm^3$	n.o.	n.o.	106.0	n.o.	n.o.	n.o.	n.o.	n.o.	n.o.	662.7	533.6	10.8
9.	Sulphates	$mgSO_4/dm^3$	n.o.	n.o.	218.0	n.o.	n.o.	n.o.	n.o.	n.o.	n.o.	24.2	1.97	44.8
10	Zinc	$mgZn/dm^3$	0.730	0.470	19.20	0.356	0.928	34.00	1.290	1.830	21.00	3.700	1.304	1.216
11.	Chromium	$mgCr/dm^3$	0.01	0.010	0.61	0.004	0.004	0.860	0.120	0.080	0.670	0.170	0.040	0.020
12.	Cadmium	$mgCd/dm^3$	0.02	0.020	0.20	0.000	0.000	0.220	0.040	0.060	0.160	0.090	0.004	0.004
13.	Copper	$mgCu/dm^3$	0.01	0.040	4.120	0.028	0.116	5.640	0.140	0.130	3.606	0.040	0.016	0.008
14.	Nickel	$mgNi/dm^3$	0.18	0.200	1.10	0.052	0.068	1.200	0.090	0.090	0.805	0.040	0.006	0.032
15.	Lead	$mgPb/dm^3$	0.05	0.030	3.710	0.048	0.168	7.520	0.110	0.200	4.110	0.060	0.006	0.020
16	Phenols	mg/dm^3	0.696	0.154	5.220	0.022	0.029	5.123	0.167	0.094	1.94	n.o.	n.o.	n.o.
17	Cyanide	mg/dm^3	0.000	0.000	0.060	0.000	0.000	0.000	0.042	0.012	0.092	n.o.	n.o.	n.o.
18	DDT + metabolites	$\mu g/dm^3$	0.0	0.0	0.0	0.0	0.0	n.o.	0.53	0.0	0.0	n.o.	n.o.	n.o.
19	HCH	$\mu g/dm^3$	0.0	0.0	0.81	0.0	1.0	n.o.	0.0	0.42	0.59	n.o.	n.o.	n.o.
20	HCB	$\mu g/dm^3$	0.0	0.0	0.0	0.0	0.0	n.o.	0.0	1.84	1.89	n.o.	n.o.	n.o.
21	Metoxychlor	$\mu g/dm^3$	0.0	0.0	0.0	0.0	0.0	n.o.	0.0	0.0	0.0	n.o.	n.o.	n.o.
22	Σ Cl-org pesticides	$\mu g/dm^3$	0.0	0.0	0.81	0.0	1.0	n.o.	0.53	2.26	2.48	n.o.	n.o.	n.o.
23	Coli Index		1000000	1000000	10000	n.o.	n.o.	n.o.	1000000	1000000	1000	n.o.	n.o.	10000
24	Coli Index fecal type		100000	100000	10000	10000000	10000000	100000	1000000	100000	0	n.o.	n.o.	520

n.o. – not determinated.

Table 2. VOC and PAH's content in surface and groundwater.

	VOC [$\mu g/dm^3$]		PAH [$\mu g/dm^3$]															
Smpl	Dich	Tri	Dip	Tet	Naph	Acy	Acn	Flu	Phe	Anth	Flu	Pyr	Chr	BaA	B(b,k)F	BaP	D(ah)A I(123)P	B(ghi)P
P-1	<0.01	n.d.	n.d.	<0.01	n.d.	n.d.	n.d.	0.001	0.311	1.604	0.425	0.032	0.077	8.259	n.d.	0.049	n.d.	n.d.
P-2	n.d.	n.d.	n.d.	<0.01	n.d.	n.d.	0.207	n.d.	n.d.	0.058	0.054	n.d.	8.040	n.d.	0.666	n.d.	n.d.	n.d.
P-3	<0.01	<0.01	n.d.	<0.01	n.d.	n.d.	n.d.	0.016	0.085	0.098	n.d.	n.d.	0.969	3.692	n.d.	n.d.	n.d.	n.d.
P-4	n.d.	n.d.	<0.01	<0.01	n.d.	n.d.	n.d.	0.048	0.139	n.d.	n.d.	n.d.	3.924	n.d.	1.172	0.003	n.d.	n.d.
P-5	n.d.	<0.01	<0.01	<0.01	n.d.	n.d.	n.d.	0.083	0.073	0.140	0.041	n.d.	7.229	n.d.	1.009	0.065	n.d.	n.d.
C-5	n.d.	n.d.	<0.01	<0.01	n.d.	n.d.	n.d.	0.076	0.034	0.266	0.178	n.d.	n.d.	7.061	0.914	0.013	n.d.	n.d.
K-2	<0.01	n.d.	<0.01	<0.01	n.d.	n.d.	0.011	n.d.	0.159	0.415	0.192	n.d.	9.604	n.d.	1.307	0.105	n.d.	n.d.

n.d. – not detected, Dich – 1,2-dichloroethane, Tri – trichloroethene, Dip – 1,2-dichloropropane, Tet – 1,1,2,2-tetrachloroethane.

values confirmed their high quality. Statistical analysis showed significant differences between the inflowing and remaining waters.

Similar oxidizability determinations (an important factor for characterizing organic pollution in waters) showed differences of values for Turonian (P-1 max. 5.2 mg O_2/dm^3) and Cenomanian (C-7 max. 5.62 mg O_2/dm^3) water levels in their inflow, in comparison to the remainder of the water samples (P-2 max. 63.2 mg O_2/dm^3, P-3 29 mg O_2/dm^3, C-3 16.3 mg O_2/dm^3, C-6 29.4 mg O_2/dm^3). Similar values to those determined in water inflow were also observed in samples from piezometer P-5 – 6.56 mg O_2/dm^3 (Turonian) and C-5 – 5.5 mg O_2/dm^3 (Cenomanian) located on landfill datums. This means that pollution from landfill spreads with Turonian waters (flowing at the depth of 2.5–14 m) into the whole analyzed area. Values of BOD_5 (P-2 max. 66 mg O_2/dm^3, P-3 20.4 mg O_2/dm^3) determined taken from piezometers in Turonian waters show that these waters are polluted. Although some differences occurred between the samples, these differences were not of statistical significance. An increase in the BOD_5 value was found in Turonian waters, however in Cenomanian waters this phenomenon was rare. Also, in this case, low values (0.2–3.0 mg O_2/dm^3) in samples from the piezometer C-5 located under the landfill, were confirmed. The greatest fluctuations of BOD_5 values were found in piezometer C-3 (min. 0.8 mg O_2/dm^3, max. 16.8 mg O_2/dm^3), located 800 meters from the landfill in the direction of water runoff. In Cenomanian waters in the runoff the following items were also detected: chloroorganic pesticides (C-3 max. 14.82 $\mu g/dm^3$, C-6 max. 1.15 $\mu g/dm^3$), metoxychlor (C-3 max. 1.56 $\mu g/dm^3$, C-6 max. 0.20 $\mu g/dm^3$), DDT and metabolites (C-3 max. 14.82 $\mu g/dm^3$, C-6 max. 1.15 $\mu g/dm^3$), HCH (C-3 max. 0.065 $\mu g/dm^3$, C-6 max. 0.116 $\mu g/dm^3$), HCB (C-3 max. 0.109 $\mu g/dm^3$, C-6 max. 0.068 $\mu g/dm^3$).

The VOC and PAH's content (Table 2) in waters of both levels also showed that ground waters are contaminated with landfill pollutants.

Higher values of ammonium in Turonian waters in the runoff direction (P-2 max. 144.1 mg NH_4/dm^3, P-3 max. 314.8 mg NH_4/dm^3, P-4 max. 9.1 mg NH_4/dm^3) were noted. Statistically important differences were noted between nitrogen content in inflow (P-1) waters and those from out of the landfill (P-3) piezometers in the runoff direction (P-3/P-4 and P-5). The mobility of nitrite ions leads to their high concentrations in Turonian waters (P-5 0.342 mg NO_2/dm^3) and Cenomanian waters (C-5 0.757 mg NO_2/dm^3) taken from the landfill area and from the most distant piezometer (C-3 0.181 mg NO_2/dm^3).

In Turonian waters collected in the runoff direction, the highest values of nitrite (P-2 1.481 mg NO_2/dm^3, P-3 2.204 mg NO_2/dm^3, P-4 0.724 mg NO_2/dm^3) were observed. This provides evidence for self-purification of waters (ammonium oxidation). The higher concentrations of this form of nitrogen were also observed in Cenomanian waters in the runoff direction (C-3 0.181 mg NO_2/dm^3, C-5 0.757 mg NO_2/dm^3, C-6 0.286 mg NO_2/dm^3).

Nitrate nitrogen content points to I class (according to Directive) of Cenomanian waters (from all piezometers) and of Turonian waters from (P-1) in the inflow. The remaining Turonian waters were classified as IV class (P-2 max. 89 mg NO_3/dm^3 – March and May 1996) and V class (P-3 max. 124 mg NO_3/dm^3 – May 1996 and P-4 max. 151 mg NO_3/dm^3 – May 1996).

Chloride ions in Turonian waters reached values of up to 930 mg/dm^3 (P-3, P-4). The lowest values were detected in control points of Turonian waters (P-1 min. 11 mg/dm^3) and also in piezometers localized in the landfill (P-5 max. 57 mg/dm^3 and C-5 max. 31 mg/dm^3). Chlorides in waters from other piezometers from the Turonian layer reached higher values and these samples were classified as V class. Cenomanian waters, due to their chloride content (C-3 317.7 mg/dm^3) were classified as IV class. The chloride content in analyzed water were comparable to samples taken in the UK (Baker and Curry 2004).

Sulphates classifying water as V quality class were found in water from the C-3 (max. 843 mg/dm^3) piezometer only. At other points, the concentration of sulphates is high (from 160 to 289.6 mg/dm^3) which is probably a result of their elution from sedimentary rocks.

Cadmium concentration in the first stage of research, allowed the classification of the ground waters as useful (V class: P-2 max. 0.0400 mg/dm^3, P-3 max. 0.030 mg/dm^3, P-4 max. 0.032 mg/dm^3, C-3, C-5 and C-6 max. 0.024). In following years the quality of water rose to II class (about 0.002 mg/dm^3). Every water sample taken from a control point (except for small fluctuations) was purer. But in this case, statistically significant differences were not found.

A similar phenomenon was noted in the case of zinc, copper, chromium and nickel concentrations. The lead content was the reason for qualifying all Turonian waters from years 1995 and 1996 as class V. In following years lead concentration dropped and waters were re-qualified as classes III and II. Also in this case (however concentrations in control points were clearly lower) statistically significant differences were not found.

The physico-chemical properties of ground waters fluctuated in the course of the experiment. Only in 1998 were both water levels (Turonian and Cenomanian) cleaner in comparison to earlier years. Probably the high precipitation which occurred in 1997 ("thousand-year flood") improved the water quality.

The sanitary quality of the investigated waters was low. Only in the inflow direction (piezometers P-1 and

C-7) at the beginning of the work (to July 1995) could waters be classified as drinkable. In the following periods these waters were classified as not suitable for drinking. Samples from other Turonian points were classified as out-of-class because they exceeded at least one of the determined parameters (counts of psychro- or mesophilic bacteria). Bacteriological factors (except fluctuations) in samples show on decrease of groundwater quality in analyzed time period. The worst quality of Cenomanian waters was observed in the C-3 (the runoff direction) and C-5 (landfill located) piezometers. At the C-3 point the bacteriological water quality steadily deteriorated and in samples taken in May 1997 *Streptoccoccus* Sp. and *Proteus vulgaris* were detected.

Like Turonian, the Cenomanian waters showed deterioration of bacteriological indices. This could be a result of infiltration of water from the Turonian to the Cenomanian layer through the marl rocks, which so far were considered to provide an insulating hydraulic contact for water from both layers.

4 CONCLUSION

Although the geological features of the landfill area appear to fully insulate it from the environment, pollutants in Cenomanian layer were detected which point to their hydraulic contact with polluted Turonian waters. This is a result of rocks fissuring in the landfill area. The results of researches show that:

Landfill leachates have contact especially with Turonian waters. This phenomenon shows statistically significant differences (between inflow and the rest of waters) in oxidizability, conductivity, BOD_5, ammonium nitrogen, chlorides, zinc, copper, nickel and bacteriological factors. Also in Cenomanian waters this phenomenon shows statistically significant differences (between inflow and the rest of waters) in conductivity, ammonium nitrogen, chlorides and bacteriological indexes. The most polluted are waters from Turonian and Cenomanian piezometers located close to the landfill area in the runoff direction (P-3 and C-3). But, in fact, the whole area with piezometers to distant points in the runoff direction is under the impact of landfill leachates – P-4 located 270 meters from landfill (Turonian) and C-3 located 800 meters from the landfill (Cenomanian waters). Due to the lack of piezometers at greater distances it is impossible to determine the actual area under landfill influence.

Analysis of heavy metal concentrations shows that mostly cadmium in 1995 and 1996, qualify these ground waters as useless, but in the following period (1997–98) the metals content (zinc, copper, chromium and nickel) qualify waters as class III and cadmium content as class Ib. In each sample, inflow waters were much cleaner than remaining waters, however statistically significant differences were not found. In Turonian and Cenomanian waters, high counts of psychro- and mesophilic bacteria were found (especially in March 1997) which disqualifies these waters as drinking waters.

REFERENCES

Al-Yaqout A.F., Hamoda M.F. (2003) Evaluation of landfill leachate in arid climate – a case study. *Environmental International* 29: 593–600.

Baker A., Curry M. (2004) Fluoroescence of leachates from three contrasting landfills. *Water Research* 38: 2605–2613.

Breukelen van B.M., Roling W.F.M., Groen J., Griffioen J., Verseveld van W. (2003) Biogeochemistry and isotope geochemistry of a landfill leachate plume. *Contaminant Hydrology* 65: 245–268.

Golimowski J., Koda E., Mamelka D. (2001) Water monitoring in reclaimed municipal landfill, Materials from VII Scientific–Technical *Conference on Municipal Wastes Management, 14-17.05.2001 Koszalin-Kołobrzeg*: 87–102.

Manning A.C., Hutcheon I.E. (2004) Distribution and mineralogical controls on ammonium in deep groundwaters. *Applied Geochemistry* 19: 1495–1503.

Oygard J.K., Mage A., Gjengedal E. (2004) Estimation of the mass-balance of selected metals in four sanitary landfills in Western Norway, with emphasis on the heavy metal content of the deposited waste and the leachate. *Water Research* 38: 2851–2858.

Sang N., Li G., Xin X. (2005) Municipal landfill leachate induces cytogenetic damage in root tips of Hordeum vulgare. *Ecotoxicology and Environmental Safety* 63: 469–473.

Scott J., Beydoun D., Amal R., Low G., Cattle J. (2005) Landfill management, Leachate generation and leach testing of solid wastes in Australia and Overseas. *Critical Reviews in Environmental Science and Technology* 35: 239–332.

Environmental Engineering – Pawłowski, Dudzińska & Pawłowski (eds)
© 2007 Taylor & Francis Group, London, ISBN13 978-0-415-40818-9

The influence of the local pipeline leak on water hammer properties

Apoloniusz Kodura
Warsaw University of Technology, Environmental Engineering Faculty, Institute of Water Supply and Water Engineering, Warsaw

Katarzyna Weinerowska
Gdansk University of Technology, Faculty of Civil and Environmental Engineering, Department of Hydraulics and Hydrology, Gdansk

ABSTRACT: In the paper the results of the experiments and numerical analysis of water hammer phenomenon are presented. The measurements were carried out for the 36 meter long pipeline made of MDPE, in which the local leak was modelled. For that purpose, in the certain distance from the downstream valve the equipment for local leak modelling was installed. The analysis of the obtained pressure characteristics was carried out and specific parameters were indicated: wave period, amplitude of pressure increase, number of oscillations, phenomenon duration time. The comparison of parameters for obtained characteristics in different leak conditions led to indicate the values that could have the influence on dissipation of water hammer energy in local leak conditions. The possibility and efficiency of numerical simulation of the water hammer phenomenon in pipeline with local leak were discussed. The conformity between calculated and observed (measured) pressure characteristics was analysed.

Keywords: Pressure wave, water hammer, leak, pipelines, Preissmann scheme.

1 INTRODUCTION

The analysis of unsteady flow phenomena in pipelines under pressure has a big importance for the sake of designing and use of water-pipe networks. The aim of numerous research works is to recognize and characterize the factors influencing the nature and run of these phenomena and to create a mathematical model enabling to simulate the considered effects in sufficiently accurate way (Chaudry 1987, Pezziga 2002, Ramos et al. 2004, Zhao & Ghidaoui 2003).

In the case of real water-pipe network, there are many additional factors affecting energy dissipation in unsteady flow. One of them is occurrence of the local leak in the pipeline. Several attempts to recognize and describe from mathematical point of view the case of unsteady flow in the pipeline with local leak, have been presented in literature (e.g. Covas 2002). In the papers the special emphasis is usually put on the question of the consistence of calculated and observed during experiments pressure characteristics, but the analysis is usually focused on a few initial pressure amplitudes. The biggest increases of pressure which are then observed, are particularly important in regard to the loads to which the pipeline is submitted during the water hammer phenomenon. However, considering only a few first amplitudes is not sufficient for the proper description of the energy dissipation in unsteady flow. The water hammer phenomenon in the pipeline is damped in consequence of intensive dissipation of energy, much stronger than in the case of steady flow. For the correctness of mathematical description of the phenomenon, it is advisable to obtain the consistence not only for initial pressure amplitudes, but for all the pressure characteristics during the phenomenon run as well.

The main purpose of the work was to analyze the influence of the local leak in pipeline on the chosen parameters characterizing water hammer phenomenon, and to indicate those characteristics, which show the highest sensitivity to the occurance and intensiveness of outflow from the leak. Additionally, the possibility of mathematical modeling of the considered relations was analyzed. Both elements of the work are of the big importance for the problems of leak localization or for analysis of the local leak influence on the water hammer phenomenon run in the case of more complicated pipeline networks.

In the paper the experimental analysis of water hammer in a pipeline with local leak is described and the results of numerical simulations of the phenomenon for the chosen experimental cases are presented.

2 MATERIALS AND METHODS

2.1 *Experimental analysis*

Experiments were carried out on the model schematically presented in Figure 1. The main element is the MDPE pipeline, of the length L equal to 36 m, extrinsic diameter D equal to 50 mm and the wall thickness e equal to 4.6 mm (1). In the distance of 6.08 m from the valve (3) a system enabling local leak modeling is installed. The equipment consists of the short ½″ polypropylene pipeline with a valve enabling regulation of the leak outflow discharge (2). The water hammer pressure characteristics were measured by tensiometer indicators (5), located 0.39 m and 6.13 m, respectively, from the valve (3) and recorded in computer's memory. Time of valve closing was measured by precise electronic stop-watch connected to the valve (4).

The experiment was carried out for three cases:

- simple positive water hammer for the pipeline with no local leak, which was the basis for estimation of the local leak influence on unsteady flow parameters – series Z,
- water hammer in pipeline with a leak to the overpressure reservoir (with overpressure lower than in main pipe under steady flow conditions) – series X,
- water hammer in a pipeline with a local leak with free outflow (to atmospherical pressure, with the possibility of sucking in air in negative phase) – series Y.

During the experiment the values of steady flow discharges (total discharge in pipeline Q_t, discharge from the leak Q_l and discharge in the outflow cross-section Q_o) and unsteady flow pressure characteristics were measured. In next step the analysis of registered pressure characteristics was carried out, with estimation of the characteristic parameters, such as: water hammer wave period, pressure amplitude, number of oscillations, duration of the phenomenon. The purpose of this analysis was to indicate the factors that may influence the energy dissipation in water hammer in pipeline with a local leak. For considered analysis it was assumed that by the term of local leak one can

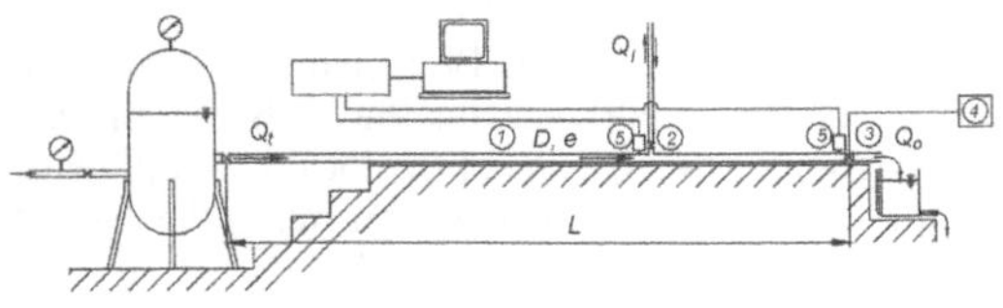

Figure 1. Scheme of experimental model.

understand the value of Q_l which is not higher than 15% of total discharge. The analysis of the obtained results was referred to the percentage ratio of the leak discharge Q_l to total discharge Q_t.

2.2 *Numerical simulation of the phenomenon*

Numerous analyses of the possibility of numerical modeling of the water hammer proved, that the mathematical description of the phenomenon running in plastic pipelines is much more difficult than in the case of metal pipes, as many additional effects appear, that modify the 'pure' water hammer phenomenon run. One of such effects is viscoelastic nature of pipe wall deformation (Pezzinga & Scandura 1995, Brunone et al. 2000, Covas et al. 2002, Covas et al. 2004, Covas et al. 2005). As a consequence, the clasical mathematical description of the phenomenon in the form of equations (Chaudhry 1979, Streeter & Wylie 1979, and others):

$$\frac{\partial H}{\partial x}+\frac{1}{gA}\frac{\partial Q}{\partial t}+R_o Q|Q|=0 \text{ where } R_o=\frac{8}{g\pi^2}\frac{\lambda}{D_i^5} \quad (1a)$$

$$\frac{\partial H}{\partial t}+\frac{a^2}{gA}\frac{\partial Q}{\partial x}=0 \quad (1b)$$

where Q = a rate of discharge; H = water head, g = acceleration due to gravity; a = wave velocity; A = cross-section area; D_i = internal pipe diameter; and λ = the linear friction factor, is often replaced by more complicated description, taking into account the unsteady nature of flow resistance and viscoelastic character of pipe wall deformations. Such approach is required especially in those problems, where the accuracy of the phenomenon reconstruction by the model is of essential importance, e.g. in the problem of leak localization. However, there is a group of problems, for which the simplified approach mentioned above is acceptable, as it leads to sufficiently accurate results and makes the analysis much easier.

For simple equations in the text always use superscript and subscript (select Font in the Format menu). Do not use the equation editor between text on same line.

For the needs of analyses presented in the paper, because of its preliminary character, the simplified description of the phenomenon, in the form of equations (1a,b), was assumed as a mathematical model. To solve the problem, the Preissmann's scheme (Cunge 1980) was applied as numerical method. Computational time step was matched in a way enabling to obtain Courant number Ca close to unity. For the presented calculation $\Delta x = 0.375$ m (what gives 97 computational sections of main pipeline) and $\Delta t = 0.00097$ s were imposed. As it is known, computations carried out with Ca $\neq 1$ suffer from introducing artificial diffusion of numerical nature, not observed on physical model. However, the lack of numerical diffusion may lead to unphysical oscillations, resulting from numerical dispersion. In considered calculations unphysical oscillations, that could arise for Ca values close to unity were reduced by appropriate choice of the value of θ parameter in Preissmann scheme, higher than 0.5. For each of the considered scenarios, in the first step of calculations the values of λ in steady flow for the main pipeline and for the section representing the leak were estimated (on the basis of the measured value of discharge and the values of pressure recorded by tensiometer indicators and imposed at the outflow of the leak). Then, on the basis of the recorded pressure characteristics, preliminary values of wave velocity a were estimated, which were then corrected in the further phase of calculations. Eventually, the main calculations were carried out for each considered scenario. In calculations the possibility of wave velocity changes due to density variations caused by pressure changes was taken into account, and changes of the friction factors due to velocity variations were allowed. However, the analysis proved that these changes do not have any significant influence on the phenomenon run and thus they may be neglected.

3 RESULTS AND DISCUSSION

3.1 *Experimental analysis*

In the Figure 2 the example of pressure characteristics observed for water hammer in pipeline with no leak (series Z) and with local leak (series X and Y) is presented. The common feature of the three cases is the comparable period of the wave, and thus the comparable wave velocity a. The increase of the energy dissipation in the case of local leak, manifesting in duration of the phenomenon, can be vividly seen. Similar relations were observed for different values of total discharge in steady flow.

The analysis of all registered pressure characteristics confirms comparable values of wave period T for all the considered cases. The average value of T for all registered characteristics is equal to 0.1763 s. This value refers to the comparable for all the cases value of wave velocity a equal to 408 m/s (Fig. 3). In comparison with literature data, the obtained wave velocity is consistent with typical values for polyethylene pipes PE 80 of a length less than 40 m (Mitosek & Chorzelski 2003). No influence of the gas presence in water to the wave velocity was observed. According to other researchers' experiences, even a very little amount of gas in water (about 1% of volume share) causes significant decrease of wave velocity, even down to the values lower than 20% of wave velocity for the case of lack of gas (Fig. 4) (Mitosek 2002).

As the decrease of wave velocity was not observed in analyzed cases, it can be suspected as the most probable explanation, that the amount of air sucked in to the main pipeline was too little (in Fig. 2 the disturbance of pressure characteristics can be seen for two

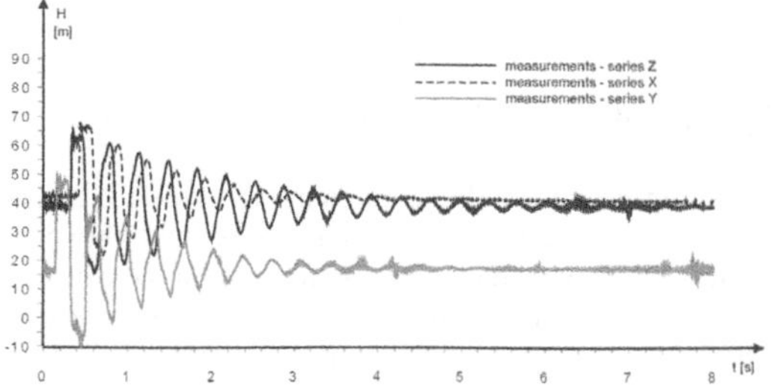

Figure 2. Example of the observed pressure characteristics for the case of total discharge $Q_t \approx 0.75$ l/s and leak discharge $Q_l \approx 0.08$ l/s (series X) and $Q_l \approx 0.05$ l/s (series Y).

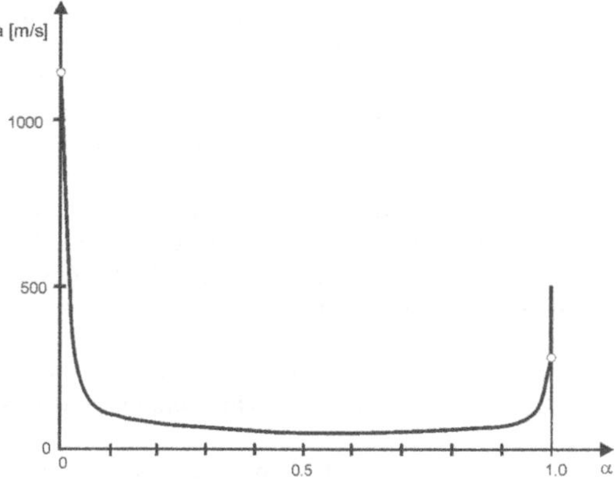

Figure 3. Influence of the gas amount in control volume of the stream α on wave velocity a.

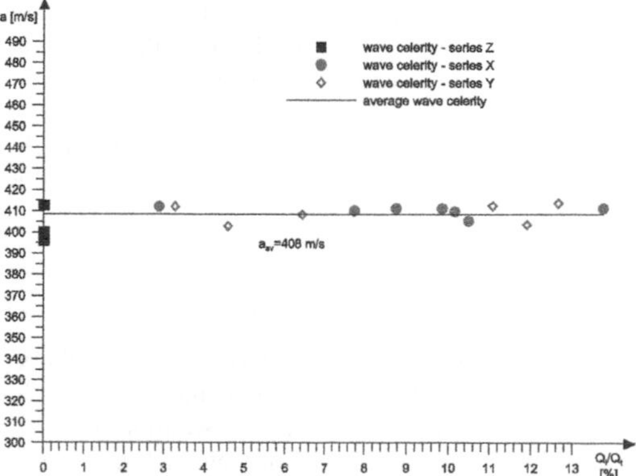

Figure 4. Relation between wave velocity a and percentage ratio of leak discharge to total discharge in the pipeline.

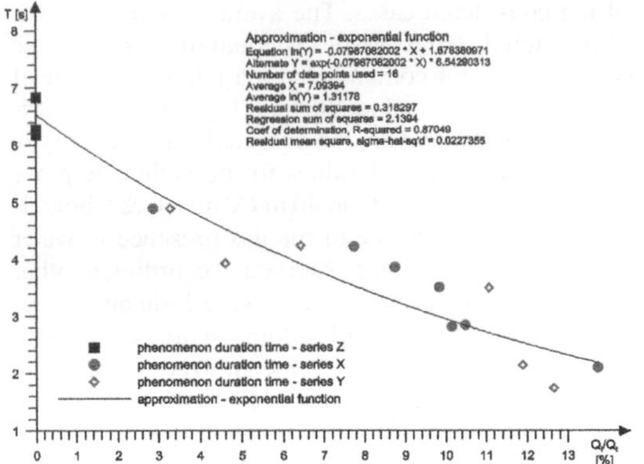

Figure 5. Relation between water hammer duration time T and percentage ratio of leak discharge to total discharge in the pipeline.

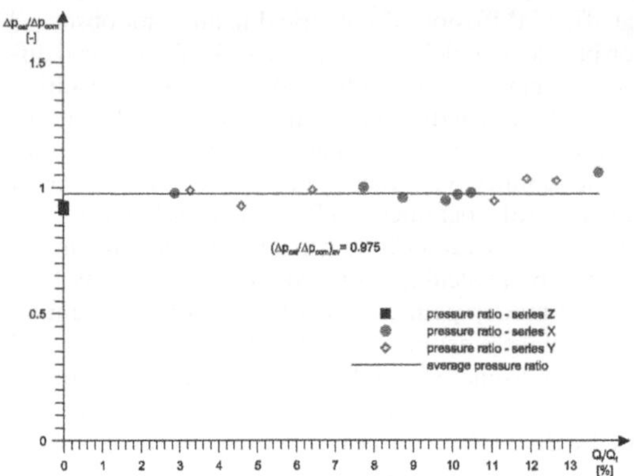

Figure 6. Relation between observed and calculated maximal pressure amplitudes ratio and percentage ratio of leak discharge to total discharge in the pipeline.

initial amplitudes only), which was probably the result of the length of the additional pipe imitating the leak.

The local leak in pipeline significantly increases energy dissipation in water hammer. The relation between duration time of the phenomenon and the percentage share of the leak discharge Q_l in total discharge Q_t is presented in Figure 5. As 'duration time of the phenomenon' such time was assumed, after which the following increases of the pressure were less than 5% of the first amplitude. The bigger ratio Q_l/Q_t was, the shorter duration time of the phenomenon was observed. During the experiment, no influence of the kind of the leak (to the overpressure reservoir – series X, or free outflow – series Y) to the duration time was observed.

Next step of the analysis was the comparison of the maximal pressure amplitudes. Observed values of maximal pressure amplitudes were referred to the theoretical ones calculated according to Zukowski formula: $\Delta p_{max} = \rho a v_0$, where ρ = density of liquid; a = wave velocity; v_0 = mean velocity of steady flow.

The calculations were carried out for the velocities in the inflow cross-section. As the experiment proved, the ratio of the calculated maximal pressure amplitude to the observed one is the value independent from the rate of leak discharge (Fig. 6). Average obtained value of $\Delta p_{max\,cal}/\Delta p_{max\,com}$ is equal to 0.975.

3.2 *Numerical simulation analysis*

As a consequence of the assumptions presented above, the numerical solution of water hammer for the observed cases was obtained. The solution was presented in the form of pressure characteristics and compared with those obtained during experiment. The calculated characteristics show high accordance to the observed ones. The example of the results obtained

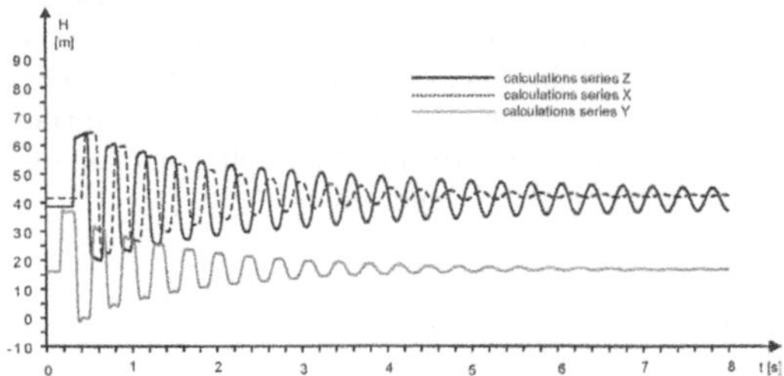

Figure 7. Example of the calculated pressure characteristics for the case of total discharge $Q_t \approx 0.75$ l/s and leak discharge $Q_l \approx 0.08$ l/s (series X) and $Q_l \approx 0.05$ l/s (series Y).

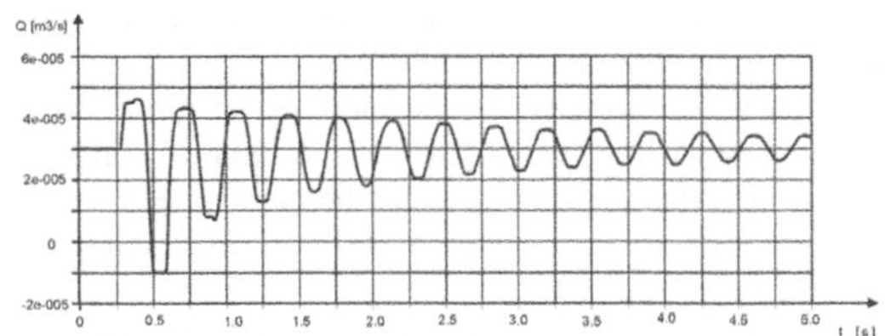

Figure 8. Example of the calculated pressure characteristics for the case of total discharge $Q_t \approx 0.75$ l/s and leak discharge $Q_l \approx 0.08$ l/s (series X) and $Q_l \approx 0.05$ l/s (series Y).

by calculations for the case presented in Figure 2 are shown in Figure 7.

In Figure 8 the calculated characteristic of the discharge from the leak for the case presented above is shown. The vivid changes, both in values and the direction of the discharge in the first negative phase of water hammer are seen. Such situation testifies the inflow to the main pipeline through the leak, what causes additional energy dissipation in unsteady flow.

The presented numerical analyses confirm the previous conclusions, drawn at the basis of experiments

Table 1. Experimental results specification.

No.	Q_o l/s	v_o m/s	Q_l l/s	Q_l/Q_o %	Q_t l/s	Q_l/Q_t %	t_c s	$\Delta p_{max\ obs}$ MPa	$\Delta p_{max\ cal}$ MPa	$\frac{\Delta p_{max\ cal}}{\Delta p_{max\ com}}$ –	p Mpa	T s	a m/s	n –	t_d s
Z2	0.935	0.72	0	0.0	0.935	0.0	0.037	0.3060	0.283	0.9259	0.0050	0.182	396.1	17	6.18
Z3	0.784	0.60	0	0.0	0.784	0.0	0.044	0.2630	0.240	0.9123	0.0080	0.180	400	19	6.84
Z4	0.744	0.57	0	0.0	0.744	0.0	0.024	0.2550	0.235	0.9209	0.0133	0.174	412.7	18	6.28
X2a	0.935	0.72	0.03	3.0	0.963	2.9	0.027	0.3100	0.303	0.9790	0.1500	0.175	412.3	14	4.89
X2b	0.922	0.70	0.09	9.6	1.010	8.7	0.045	0.3306	0.318	0.9611	0.1553	0.175	411.4	11	3.85
X2c	0.936	0.72	0.11	11.7	1.046	10.5	0.045	0.3306	0.324	0.9813	0.1713	0.178	405.6	8	2.84
X2d	0.936	0.72	0.15	15.9	1.085	13.7	0.045	0.3213	0.341	1.0623	0.1633	0.175	411.4	6	2.1
X3a	0.778	0.60	0.07	8.4	0.843	7.7	0.045	0.2640	0.265	1.0024	0.1510	0.175	410.5	12	4.21
X3b	0.735	0.56	0.08	10.9	0.816	9.8	0.045	0.2700	0.257	0.9503	0.1510	0.175	411.4	10	3.5
X3c	0.773	0.59	0.09	11.3	0.860	10.1	0.027	0.2769	0.270	0.9737	0.1490	0.176	410	8	2.81
X3d	0.738	0.56	0.16	21.7	0.899	17.8	0.025	0.2620	0.284	1.0833	0.1693	0.174	413.1	7	2.44
Y2a	0.933	0.71	0.03	3.4	0.965	3.3	0.033	0.3070	0.304	0.9904	0.0100	0.175	412.3	14	4.89
Y2b	0.961	0.74	0.07	6.9	1.027	6.4	0.044	0.3233	0.321	0.9923	0.0133	0.176	408.5	12	4.23
Y2c	0.990	0.76	0.12	12.5	1.113	11.1	0.028	0.3700	0.351	0.9494	0.0147	0.175	412.6	10	3.49
Y2d	0.967	0.74	0.05	4.8	1.014	4.6	0.040	0.3367	0.312	0.9279	0.0067	0.179	403.1	11	3.93
Y2e	1.042	0.80	0.14	13.5	1.182	11.9	0.034	0.3520	0.365	1.0369	0.0133	0.178	403.7	6	2.14
Y2f	0.987	0.75	0.07	7.2	1.058	6.7	0.034	0.3300	0.326	0.9882	0.0127	0.179	403.2	7	2.5
Y2g	0.664	0.51	0.05	6.9	0.710	6.4	0.031	0.3310	0.220	0.6652	0.0120	0.178	405.6	12	4.26
Y2h	0.951	0.73	0.14	14.5	1.089	12.7	0.045	0.3350	0.344	1.0283	0.0130	0.174	413.8	5	1.74

Q_o = steady flow discharge in the outflow cross-section; v_o = steady flow mean velocity in the outflow cross-section; Q_l = steady flow discharge from the leak; Q_l/Q_o = discharge ratio; Q_t = total steady flow discharge in pipeline; Q_l/Q_t = discharge ratio; t_c = time of closing the valve in the outflow cross-section; $\Delta p_{max\ obs}$ = maximum observed pressure amplitude – 1st indicator; $\Delta p_{max\ cal}$ = maximum calculated pressure amplitude – 1st indicator; $\Delta p_{max\ cal}/\Delta p_{max\ com}$ = pressure amplitude ratio; p = overpressure – 3rd indicator; T = wave period; a = wave velocity; n = number of oscillations; t_d = duration time of the phenomenon.

on the physical model. In all considered cases the maximal consistence between calculations and observations was achieved for the constant value of wave velocity, equal to 408 m/s, what corresponds to the average value of period of oscillations T equal to 0.17 s. The difficulties in obtaining full consistence between measurements and calculations during all duration of the phenomenon are the result among other things of the simplified mathematical description of water hammer in applied model. Some of the authors try to reduce these discrepancies by imposing some coefficient increasing the value of friction factor λ (e.g. Wichowski 2002). However because of the complicated nature of the phenomenon such approach is not always successful.

In spite of the problems mentioned above and despite the relatively simple mathematical model taken into account the developed numerical analyses showed high accordance with the results of the experiments which suggests that the effective modeling of the water hammer in a pipeline with local leak is possible and the factors influencing the phenomenon run are properly represented. It also seems that by applying more detailed mathematical description of the phenomenon it will by possible to increase the consistence between calculations and observations in the future research.

4 CONCLUSIONS

During the experiments carried out for unsteady flow in PE 80 pipeline with a leak the following effects were observed:

- lack of the influence of the ratio Q_l/Q_t (discharge from local leak to total discharge in the pipeline) to the values of period of oscillations and – in a consequence – to the value of wave velocity if the outflow to the overpressure reservoir from the leak was imposed; when the free outflow from a leak was considered the change in pressure characteristics only for the two first amplitudes were observed. what indicates very little amount of air sucked in to the pipeline;
- no influence of the rate of discharge from local leak on the maximal value of pressure for the considered range of percentage ratio between discharge from the leak and total discharge in the pipeline;
- consistence between observed values of maximal pressure in first amplitude and corresponding values calculated according to Zukowski's formula (except the difference in series Y for the presented example), irrespective of the rate of discharge from the leak;
- significant influence of the rate of the discharge from the leak on the vivid decrease of duration of

the water hammer phenomenon. what suggests the possibility of utilization of this fact to the pipeline leaktightness assessment especially that the duration time decreases with the increase of the outflow from the leak.

The carried out numerical analyses confirmed the possibility of numerical modeling of the phenomenon of water hammer in the pipeline with a leak despite relatively simple mathematical description applied in the model. The results of the computations show high consistence with observations and thanks to this they confirm the conclusions drawn on the basis of physical experiment. Thus mathematical modeling may be an effective tool enabling more detailed analysis of the phenomenon. However it seems advisable to develop the model in order to take into account the factors particularly important for the case of plastic pipelines. such as viscoelastic behavior of the material. The results of calculations carried out for the two models could be compared then and the conclusions connected with the influence of different factors to the solution could be drawn. However. it is necessary to analyze the influence of numerical parameters to the accuracy of the solution and thus it is important to present and control their values in each case as they may very strongly influence the results and lead to totally wrong interpretations.

REFERENCES

Brunone. B. & Karney. B.W. & Mecarelli. M. & Ferrante. M. 2000. Velocity Profiles and Unsteady Pipe Friction in Transient Flow. *Journ. of Water Res. Planning and Management.* Vol.126. No.4. 236–244.

Chaudhry. M.H. 1979. Applied Hydraulic Transients. Van Nostrand Reinhold Company (eds.). New York.

Chaudry. M.H. 1987. Applied Hydraulic Transients. 2nd ed. Van Nostrand Reinhold Company. Litton Educational Publishing Inc. New York.

Covas. D. & Stoinanov. I. & Graham. N. & Maksimović. C. & Ramos. H. & Butler. D. 2002. Inverse transient analysis for leak detection and calibration – a case study in a polyethylene pipe. in: *Fifth International Conference on Hydroinformatics.* Cardiff. UK. 1154–1159.

Covas. D. & Stoinanov. I. & Mano. J. & Ramos. H. & Graham. N. & Maksimović. C. 2004. The dynamic effect of pipe-wall viscoelasticity in hydraulic transients. Part I-experimental analysis and creep characterization. *Journ. of Hydr. Res.* Vol.42. No.5. 516–530.

Covas. D. & Stoinanov. I. & Mano. J. & Ramos. H. & Graham. N. & Maksimović. C. 2005. The dynamic effect of pipe-wall viscoelasticity in hydraulic transients. Part II model development, calibration and verification. *Journ. of Hydr. Res.* Vol.43. No.1. 56–70.

Cunge. J.A. & Holly. F.M. Jr. & Verwey. A. 1980. Practical Aspects of Computational River Hydraulics. Pitman Publishing Limited. Vol.3. London.

Mitosek. M. 2001. *Mechanika płynów w inżynierii i ochronie środowiska (Mechanics of Fluids in Engineering and Environmental Protection.* in Polish). PWN. Warszawa.

Mitosek. M. & Chorzelski. M. 2003. Influence of Visco-Elasticity on Pressure Wave Velocity in Polyethylene MDPE Pipe. *Archives of Hydro-Engineering and Environmental Mechanics.* Vol.50 (2003). No.2. 127–140.

Pezzinga. G. & Scandura. P. 1995. Unsteady Flow in Installations with Polymeric Additional Pipe. *Journ. of Hydr. Eng.* Vol.121. No.11. 802–811.

Pezzinga. G. 2002. Unsteady Flow in Hydraulic Networks with Polymeric Additional Pip. *Journ. of Hydr. Eng.* Vol.128. No.2. 238–244.

Ramos. H. & Covas. D. & Borga. A. & Loureino. D. 2004. Surge damping analysis in pipe systems: modeling and experiments. *Journ. of Hydr. Res.* Vol.42. No.4. 413–425.

Streeter. V.L. & Wylie. E.B. 1979. *Hydraulic transients.* New York. McGraw-Hill Book Co.

Wichowski R. 2002. *Wybrane zagadnienia przepływów nieustalonych w sieci wodociagowej pierścieniowej (Chosen problems of unsteady flow in looped water-supply pipeline network.* in Polish). Monograph No.27. Publ. of Gdansk University of Technology.

Zhao. M. & Ghidaoui. S. 2003. Efficient Quasi-Two Dimensional Model for Water Hammer Problems. *Journ. of Hydr. Eng.* Vol.129. No.12. 1007–1013.

Environmental Engineering – Pawłowski, Dudzińska & Pawłowski (eds)

Treatment of waters polluted with organic substances

Krystyna Konieczny
Silesian University of Technology, Gliwice, Poland

ABSTRACT: The paper presents the results of studies on the treatment of natural waters having high contents of organic compounds. The research involved the unit processes coagulation, ultrafiltration, and the hybrid process coagulation-sedimentation + ultrafiltration and in-line coagulation ultrafiltration. In the membrane filtration process we applied the 0.03 μm PVDF immersed hollow – fiber membranes. The volumetric permeate flux was constant (5.7×10^{-05} $m^3/m^2 \cdot s$). Simulated water containing humic acids in amount of 20 and 30 mg/l. was used in the investigations. Four coagulants were tested for which optimal process parameters were determined experimentally. We tested the effectiveness of coagulation and ultrafiltration and that of the hybrid system coagulation-membrane filtration, based on measurements of membrane yield (permeate flux) and physicochemical analysis of raw water and permeates. The application of the hybrid system combining coagulation and ultrafiltration offers better removal of organic pollution (100% for TOC).

Keywords: Ultrafiltration, coagulation, PVDF membrane, water treatment, hybrid process.

1 INTRODUCTION

Water is the most common substance present on the earth and plays a significant role in the physiological and cultural life of humankind. It is the main component of plant and animal organisms. Apart from satisfying human necessities, it is also of great importance in many branches of economic activity – it is used, among others, in industry, agriculture, for fire-fighting or sanitary purposes. Water, as the basic component of the environment is, at present, very polluted and is continually subjected to further pollution. Therefore, the water environment should be taken care of to provide people with good quality drinking water and ensure appropriate quality standards for water for industrial needs.

The requirements concerning the quality of drinking water were specified in the Directive 98/83/EC (Roman 2001), according to which all necessary measures must be undertaken to ensure that drinking water for people is healthy and clean. High quality drinking water can be achieved by using membrane processes combined with conventional processes in the water treatment technology. Microfiltration as a single process is not so commonly applied due to the problems it brings about, and in particular the fouling effected by natural organic matter (NOM) present in surface waters. Therefore, we suggest adding coagulant before the membrane filtration, not only to improve the effectiveness in removing natural organic substances, but also to reduce fouling, especially of the irreversible kind (Choi 2004, Konieczny et al. 2005). With the application of the hybrid process coagulation-MF/UF, organic or inorganic pollution can be removed more effectively than in unit processes applied in environmental protection (Guigui et al. 2002, Strathmann 1998). Also hybrid processes are introduced, combining e.g. ultrafiltration and coagulation or ultrafiltration with adsorption on activated carbon. The application of activated carbon is indispensable in the treatment of surface waters which contain the precursors of chlorinated organic compounds (Simpson et al. 1998, Konieczny et al. 2002).

A number of the waters taken in for treatment (particularly surface waters) are characterized by variable physicochemical composition, since they may have direct contact with solid, liquid and gaseous wastes. The most important and most relevant anthropogenic pollution present in natural waters involve the following: polycyclic aromatic hydrocarbons, surface active substances, chlorinated organic compounds, pesticides, oil derivatives, heavy metals, nitrogen compounds and other dissolved substances. Efficient coagulation reduces turbidity, coloring, indices of organic pollution including the precursors of disinfection by-products as well as chemical oxidation. It removes much micropollution and colloidal substances including humic ones, which can not be retained by sedimentation tanks or filters due to their small size. Coagulation ensures also a high removal level of heavy metals and it lowers the potential for the formation of trihalomethanes (THM). It also contributes to the reduction of organic micropollution absorbed on colloidal particles, and also, through the removal of organic carbon, it limits the demand of water for disinfection agents such as chlorine or ozone (Nawrocki et al. 2000).

The most commonly applied coagulants in the processes of water treatment are ferric salts (e.g. ferric chloride(III), ferric sulphate(II/III), PIX) and aluminum salts (e.g. aluminum chloride, aluminum sulphate, PAX), which, via hydrolysis, form positively charged combinations (hydrocomplexes) which efficiently destabilize the colloidal pollution present in treated water and perfectly precipitate pollution present there.

The application of PAX as compared to traditional inorganic coagulants has many relevant advantages: it requires smaller doses, facilitates the formation of floccules, ensures better treatment level of water or sewage, generates less sludge, allows coagulation in a wider range of pH to be carried out, allows easy and precise batching, whereas PIX binds sulphides, eliminating at the same time unpleasant odor and corrosion risks.

The addition of coagulant to water in the membrane treatment brings about better quality of permeate, and in the microfiltration process it offers other advantages, confirmed by the research of M.R. Wiesner (Fig. 1) (Mallevialle J., Odendaal P.E., Wiesner M.R., 1996). Applying coagulation-membrane filtration we improve the quality of permeate flux as a result of the following factors:

1. Lowering the penetration of substances generating fouling to the inside of the membrane, since after the addition of coagulant, the substances responsible for the phenomenon of fouling undergo complexing, aggregation or sorption. Colloids form larger aggregates which are retained on the surface of the membrane. Natural organic matter is regarded as the main factor effecting the drop of permeate flux, and its complexation brings about advantages by reducing this drop.
2. Change of the character of the substance deposited on membrane surface. The formation of larger complexes by colloidal particles lowers the resistance of cake – the higher the number of particles, the lower the resistance.
3. Better transport conditions of particles. The forces which cause particles to float above the membrane surface by adding coagulant to water result in the formation of larger particles which are deposited to lower degree on the membrane. The upward forces increase with the size of particles. They may reduce the deposition of particles on the membrane surface, and during backflashing they can remove the particles deposited on the membrane more effectively (Fig. 1).

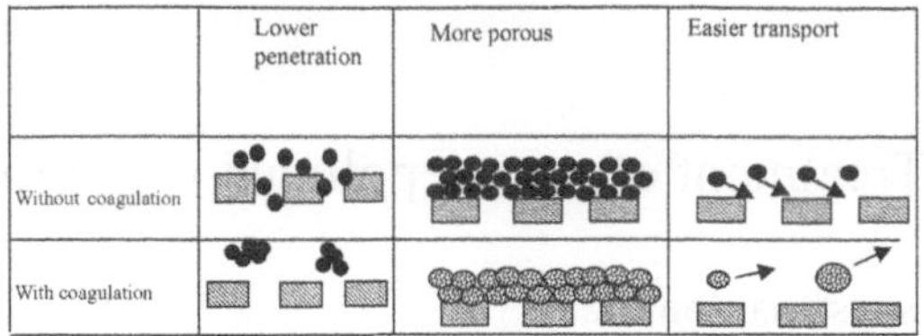

Figure 1. Mechanisms improving the quality of permeate flux with the application of pre-coagulation [Mallevialle J., Odendaal P.E., Wiesner M.R., 1996].

Moulini et al. (Moulin et al. 1991, Schlichter et al. 2003) carried out studies in the hybrid system coagulation + ultrafiltration and coagulation/oxidation + ultrafiltration. They found that the addition of coagulant (aluminium polychlorosulphate) considerably improves the removal of organic substances and allows the acquisition of higher permeate flux. The combination of coagulation and oxidation resulted in 90% removal of organic substance. The studies were carried out with the dose of coagulant of 120 mg/dm^3 and chlorine of 6 mg/dm^3. The application of the above hybrid processes contributes to lower fouling of membranes and to the acquisition of higher permeate flux and its better quality (Schlichter et al. 2003). The main objective of the work was to assess the efficiency of hybrid processes, i.e. combined unit processes: coagulation/sedimentation + membrane ultrafiltration and in-line coagulation ultrafiltration in the treatment of waters with raised content of humic acids. The currently binding standards were accepted as assessment criteria, i.e. The Directive of Health Ministry dated 19 November 2002 on the requirements involving the quality of drinking water (Dz.U. Nr 203, poz. 1718). Another objective was to find the most efficient and economical hybrid process.

2 MATERIALS AND METHODS

In the studies carried out we applied the testing unit Zee Weed 10 (Zenon Systems). The Zee Weed water treatment technology is a membrane process of very low energy consumption which is based on an ultrafiltration module with the flow from the outside to the filtration fibers immersed directly in the water subjected to treatment. The membrane module consists of a large number of membrane fibers fixed between two heads (the characteristic of membranes is presented in Table 1). This kind of membrane is classified as porous. The length of the module is 692 mm and its width is 110 mm. In the middle part of the upper head there is a joint for process air which, via an appropriate pipe, moves from the upper part of the module to the bottom part, and there, in the form of small bubbles,

Table 1. Characteristics of membrane modules ZW10.

Membranes type	Hollow-fiber, hydrophilic
Nominal area, m^2	0.93
Diameter of the fiber, μm	0.1
Cutoff, nm	30
Length of the module, mm	692
Width of the module, mm	109.5
Membrane material	Polivinyl fluoride
Transmembrane pressure	Max. 62 kPa at 40°C

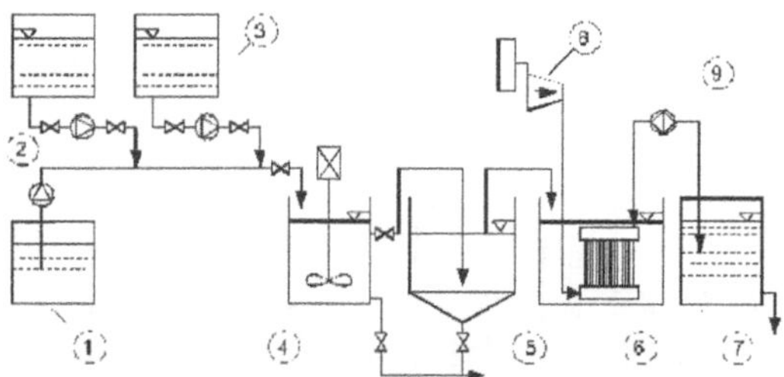

Figure 2. Schematic diagram of the coagulation and ultrafiltration. (1) – water tank; (2) – coagulant tank; (3) – pH correcting solution tank; (4) – stirring cell; (5) – sedimentation tank; (6) – ultrafiltration pilot unit; (7) – permeation tank; (8) – blower; (9) – two-way pump.

is dissipated and goes upwards putting the membrane into motion during its operation. Filtration fibers are built in such a way that the outward structure is covered by a polymer film with pores of nominal diameter of 0.03 μm; the absolute magnitude of the pores is 0.2 μm. This means that all microbiological pollution larger than 0.2 μm (including therefore also Giardia and Cryptosporidium), suspension and oxidized compounds of iron and manganese have no access to the flux of treated water.

The in-line coagulation process with ultrafiltration was realised as presented on the diagram above (Fig. 2). Table 1 presents the characteristic parameters of the applied module with immersed membranes ZW10.

The simulated water used in the laboratory tests was prepared by dissolving the powdered humic acid SH manufactured by Aldrich in deionized water, in amounts of 20 and 30 mg/dm^3, which corresponded accordingly to 7 and 10 mg of the total organic carbon TOC/dm^3. The studies involved five basic stages which were carried out as follows:

- the first stage of the investigation involved the selection of the optimal technological parameters of the coagulation process. We defined the influence of the kind and amount of coagulant and the influence of the environment's reaction on the run and efficiency of coagulation. Four coagulants were tested in the studies: ferric chloride ($FeCl_3 \cdot 6H_2O$), aluminium sulphate ($Al_2(SO_4)_3 \cdot 18H_2O$), ALF (aqueous solution of ferric chloride and aluminium sulphate) and PAX-16 (aqueous solution of polyaluminum chloride). The coagulation was carried out for four doses of coagulant, within the range 1÷7.2 mg Me/dm^3 with three reactions for ferric chloride and aluminium sulphate and two reactions for ALF and PAX-16. The reaction range was pH 5.5÷8.8. The correction of pH was carried out using the 1% solution of NaOH and HCl. The details involving the coagulation process have been presented previously (Konieczny et al. 2005).
- the second stage involved the testing of membranes through the filtration of deionized water under constant process conditions for 5 hours in order to determine the dependence of volumetric flux of deionized water (J_w) on time for the investigated membrane.
- the third stage was carried out in the hybrid system coagulation/sedimentation + water ultrafiltration after coagulation, to assess the transport and separation properties of the investigated membranes after pretreatment of water in the coagulation process. After the preparation of simulated water, reaction correction, coagulant addition, a fast coagulation was carried out over one minute (fast mixing t = 1 min) and slow stirring (t = 30 min) and then the sedimentation process (for a further 30 min). The water treated in this way was subjected to membrane filtration lasting eight hours.
- the fourth stage involved fast coagulation lasting 1 minute and then the feed was passed directly (in-line) to the process tank in which the ultrafiltration membranes were immersed, and the measurements were carried out for eight hours.
- the fifth stage covered the filtration of simulated water in the unit process of membrane filtration (for 8 hours), to assess transport properties of ultrafiltration membrane, based on the obtained value of the volumetric permeate flux (J_v), depending on the duration of the process and the efficiency of the process realized in that way.

The constant parameters of the process were retained: Q = 20 dm^3/h, pressure p = 0.02 MPa, temperature T = 293 ± 1 K. The analysis involving the content of chemical compounds was carried out in simulated raw water, in the water after coagulation and in permeates taken every 60 minutes of the filtration process for all investigated coagulants. Membrane filtration during the third, fourth and fifth stages was carried out in the open system, i.e. the permeate was not returned to the feed tank; the water after coagulation (third stage) and the simulated water (fourth and fifth stages) was passed to the tank in the amount equal to the sampled permeate.

The efficiency of the unit process (coagulation, ultrafiltration) and hybrid processes (coagulation/sedimentation + UF and in-line coagulation + UF) was assessed based on the physicochemical analysis of simulated water, water after coagulation or ultrafiltration, as well as on the parameters from two realized methods of hybrid process. The analysis covered the following:

- determination of organic carbon TOC, turbidity, absorbance at wavelength 254 nm, conductivity, residual iron in the water after coagulation, residual aluminum in the water after coagulation. Also other indices of water loading were analyzed.

3 RESULTS AND DISCUSSION

3.1 *Selection of optimal coagulation parameters*

Before the investigation studies on membrane filtration were begun, optimal parameters of coagulation process were determined experimentally using a vessel test, i.e. the dose of reagent and the pH of the aqueous environment. The obtained results are presented in Figs 3 and 4 as well as in table 2 (Konieczny et al. 2005).

Two types of simulated waters were subjected to coagulation: water with the concentration of organic compounds at the level of 7 and 10 mg of TOC/dm^3. The efficiency of the applied reagents was assessed based on the percentage removal of organic compounds determined as the content of total organic carbon (TOC) and as the absorbance at wavelength 254 nm (Figs 3 and 4). Also the applied dose which ensured respective unloading of water was accounted for. The lowest optimal dose was determined for the aluminum coagulant PAX-16, but due to low removal of TOC for both modal waters, this coagulant was not used in further experiments. Higher doses were determined for the remaining three coagulants, with different solution pH values. The efficiency of organic substance removal was high – 57–98% in the case of absorbance and 61–91% for TOC depending on the type of coagulant.

3.2 *Yield of membrane filtration process – transport properties of immersed membranes*

In the second stage of the investigation studies involved the conditioning of membranes and the determination of the dependence of volumetric flux of deionized water (J_w) on time (5 h-long filtration) as well as the occurrence of fouling in the module ZW 10. The tests were carried out using mains water for three hours. The yield of the filtration process did not decrease. The said membrane was characterized by constant flux at the level of $J_v = 5.65 \cdot 10^{-5}$ m^3/m^{2}s in all experiments, that is, both in the unit filtration process and in the hybrid systems (coagulation/sedimentation + UF and in-line coagulation +UF).

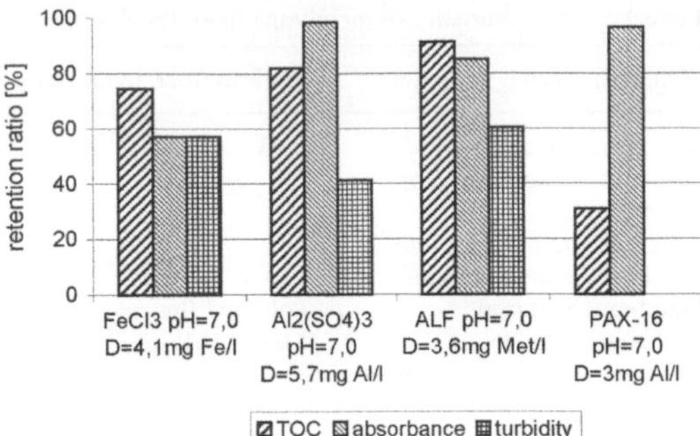

Figure 3. Efficiency of coagulants in the removal of pollution from water (TOC concentration 7 mgC$_{org}$/dm^3).

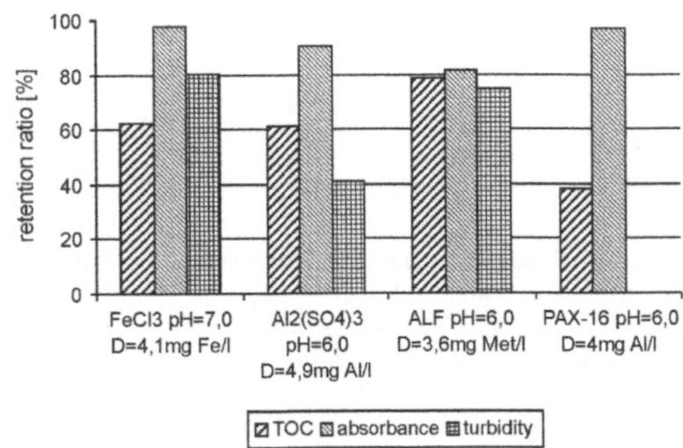

Figure 4. Efficiency of coagulants in the removal of pollution from water (TOC concentration 10 mgC$_{org}$/dm^3).

Table 2. Optimal parameters of the applied coagulants with the content of organic substances measured as TOC – 10 mg/dm^3.

Coagulant	Reaction pH	Dose
$FeCl_3$	7.0	4.1 mg Fe/dm^3
$Fe_2(SO_4)_3$ (PIX-113)	7.5	4.0 mg Fe/dm^3
$Al_2(SO_4)_3$	7.5	4.1 mg Al/dm^3
PAX-25	7.5	3.6 mg/dm^3

3.3 *Removal efficiency of organic compounds in the hybrid system coagulation/sedimentation + UF process*

Tables 3–5 present the values of physicochemical parameters for the investigated concentration of humic

Table 3. Average values of physicochemical parameters of water subjected to treatment.

Parameter	No. of measurements, n	Feed Average	Feed σ*	Permeate Average	Permeate σ
PH	6	7.0		7.9	
Conductivity, mS/cm	6	3.080		2.300–3.010	
Turbidity, NTU	6	8.02	0.01	0.14	0.01
Total iron, mg/dm^3 Fe	6	0.66	0.01	0.10	0.02
Oxidizability with $KMnO_4$,mg O_2/dm^3	6	5.52	0.08	3.13	0.05
Absorbance $UV_{\lambda 254}$, cm^{-1}	6	0.329	0.002	0.081	0.001

* σ – standard deviation.

Table 4. Average values of physicochemical parameters of water subjected to treatment with the application of coagulant 1 ($FeCl_3$) and coagulant 2 (PIX-113) (coagulation/sedimentation/ultrafiltration).

Parameter	No. of measurements, n	Feed 1		Coagulation 1		Permeate 1		Feed 2		Coagulation 2		Permeate 2	
		av*	σ*	av	σ	av	σ	av	σ	av	σ	av	σ
pH	6	7.0–7.1		7.0–7.2		7.7–7.9		7.0		6.3		7.5	
Conductivity, mS/cm	6	2.76–2.86		2.78–2.89		2.77–2.89		1.07–1.10		1.10–1.17		1.15–1.17	
Turbidity, NTU	6	6.5	0.10	7.63	0.70	0.11	0.06	7.78	0.01	3.25	0.55	0.21	0.02
Total iron, $mgFe/dm^3$	6	0.21	0.02	1.15	0.73	0.03	0.02	0.32	0.01	3.84	0.08	0.07	0.01
Oxidizability with $KMnO_4$, mgO_2/dm^3	6	4.47	0.37	4.63	0.44	2.37	0.46	6.07	0.08	3.10	0.00	3.00	0.00
Absorbance $UV_{\lambda=254}$	6	0.37	0.003	0.40	0.007	0.054	0.011	0.42	0.012	0.145	0.008	0.035	0.002

* av – average, * σ – standard deviation.

Table 5. Average values of physicochemical parameters of water subjected to treatment with the application of coagulant 3 ($Al_2(SO_4)_3$ and coagulant 4 (PAX-25) (coagulation/sedimentation/ultrafiltration).

Parameter	No. of measurements, n	Feed 3		Coagulation 3		Permeate 3		Feed 4		Coagulation 4		Permeate 4	
		av*	σ*	av	σ	av	σ	av	σ	av	σ	av	σ
pH	6	7.5		7.4–7.7		7.9–8.0		7.5		7.4–7.6		7.8–8.0	
Conductivity, mS/cm	6	2.67–2.74		2.64–2.82		2.78–2.83		2.70		2.75–2.78		2.76–2.78	
Turbidity, NTU	6	6.63	0.04	7.71	1.18	0.10	0.03	6.87	0.00	9.13	1.36	0.11	0.04
Total iron, $mgFe/dm^3$	6	–	–	–	–	–	–	0.21	0.00	0.56	0.12	0.00	0.00
Oxidizability with $KMnO_4$, mgO_2/dm^3	6	4.10	0.00	4.60	0.00	2.15	0.08	4.50	0.00	4.43	0.08	1.83	0.15
Absorbance $UV_{\lambda=254}$	6	0.417	0.006	0.45	0.032	0.148	0.013	0.415	0.001	0.478	0.015	0.118	0.013
Aluminum, $mgAl/dm^3$	6	0.66	0.02	0.18	0.01	0.09	0.01	0.68	0.02	0.27	0.03	0.14	0.01

* av – average, * σ – standard deviation.

Table 6. Average values of physicochemical parameters of water subjected to treatment with the application of coagulant 1 ($FeCl_3$) and coagulant 2 (PIX-113) (in-line coagulation/ultrafiltration).

Parameter	No. of measurements, n	Feed 1		Coagulation 1		Permeate 1		Feed 2		Coagulation 2		Permeate 2	
		av*	σ*	av	σ	av	σ	av	σ	av	σ	av	σ
pH	6	7.0–7.2		7.7–7.9		7.7–7.9		7.1		7.9		7.9	
Conductivity, mS/cm	6	2.6–2.65		2.55–2.67		2.61–2.74		1.36		1.34–1.35		1.31–1.32	
Turbidity, NTU	6	7.15	0.34	12.16	0.27	0.15	0.08	7.7	0.01	15.53	0.24	0.17	0.02
Total iron, $mgFe/dm^3$	6	0.30	0.01	0.71	0.09	0.01	0.01	0.45	0.01	1.70	0.10	0.00	0.01
Oxidizability with $KMnO_4$, mgO_2/dm^3	6	5.30	0.26	5.92	0.73	2.30	0.21	6.15	0.09	6.48	0.61	3.09	0.08
Absorbance $UV_{\lambda=254}$	6	0.389	0.003	0.745	0.023	0.130	0.005	0.441	0.001	0.712	0.054	0.117	0.006

* av – average, * σ – standard deviation.

Table 7. Average values of physicochemical parameters of water subjected to treatment with the application of coagulant 3 (($Al_2(SO_4)_3$) and coagulant 4 (PAX-25) (in-line coagulation/ultrafiltration).

Parameter	No. of measurements, n	Feed 3		Coagulation 3		Permeate 3		Feed 4		Coagulation 4		Permeate 4	
		av*	σ*	av	σ	av	σ	av	σ	av	σ	av	σ
pH	6	7.6		8.0–8.1		8.0–8.1		7.5		7.9–8.0		7.9–8.0	
Conductivity, mS/cm	6	2.70		2.71–2.78		2.73–2.82		2.62		2.6–2.71		2.61–2.78	
Turbidity, NTU	6	7.12	0.02	13.4	1.89	0.16	0.05	7.23	0.00	13.6	0.31	0.12	0.01
Total iron, $mgFe/dm^3$	6	–	–	–	–	–	–	0.26	0.00	0.67	0.03	0.01	0.01
Oxidizability with $KMnO_4$, mgO_2/dm^3	6	4.70	0.00	5.78	0.67	2.215	0.12	4.70	0.00	6.72	0.04	2.28	0.13
Absorbance $UV_{\lambda=254}$	6	0.439	0.004	0.818	0.046	0.127	0.028	0.43	0.00	0.849	0.02	0.155	0.003
Aluminum, $mgAl/dm^3$	6	0.66	0.02	0.19	0.02	0.011	0.01	0.68	0.02	0.37	0.02	0.16	0.01

* av – average, * σ – standard deviation.

acids in simulated water (10 $mgTOC/dm^3$) and in the hybrid system coagulation/sedimentation + UF.

3.4 *Removal efficiency of organic compounds in hybrid systems using in-line coagulation/ UF process*

The obtained results from the process of in-line coagulation + UF for the applied coagulants are presented in Tables 6 and 7 with the values of standard deviation σ. Based on the average concentration of parameters of the substances present in the investigated water, the retention coefficients R were calculated using the formula $R = 1 - C_n/C_p$ (where C_n – feed concentration, C_p- permeate concentration). Table 8 represents the retention coefficients for ultrafiltration process together with retention coefficients of physicochemical parameters for two hybrid processes, i.e. coagulation, sedimentation/UF and in-line coagulation/UF. It can be observed from table 8 that the treatment process on ultrafiltration membranes is efficient, especially with the removal of substances causing turbidity and with the removal of total iron.

This process can be additionally supported by coagulation, which results in the rise of retention coefficient R, in particular with respect to organic substances determined as UV_{254} and oxidizability with $KMnO_4$.

Table 8. Retention coefficients R for the applied processes and coagulants.

Parameter	Feed/ permeate	PIX-113		$FeCl_3$		$Al_2(SO_4)_3$		PAX-25	
		f/c/p*)	in-line	f/c/p*)	in-line	f/c/p*)	in-line	f/c/p*)	in-line
		Retention coefficients R, %							
Turbidity	98.3	97.3	97.8	98.4	97.9	98.4	97.8	98.4	98.4
Total iron	84.8	79.3	99.3	87.0	97.8	–	–	100.0	96.8
Oxidizablity with $KMnO_4$	42.9	50.6	49.7	7.0	56.6	47.6	52.1	59.3	51.4
Absorbance UV_{254}	75.1	91.6	69.5	85.4	66.5	64.4	71.1	71.6	64.0
Aluminum	–	–	–	–	–	87.0	84.0	79.4	77.2

* f/c/p – feed/coagulation/permeate.

The application of the coagulation process does not result in lower turbidity compared to the process carried out in simulated water passed directly to the ultrafiltration membrane. In cases when two types of hybrid processes and four coagulants are applied, the turbidity of the investigated water remains at approximately the same level.

The application of coagulation has a considerable effect on the removal of organic substances from water, determined as absorbance UV_{254}. The high removal efficiency of those substances was exhibited by the coagulant PIX-113 in the process coagulation, sedimentation/ultrafiltration. The process in-line/ultrafiltration involving the removal of organic compounds determined as absorbance UV_{254} turned out to be less effective than the ultrafiltration process without the application of coagulants.

The organic substances determined as oxidizability with $KMnO_4$ are more efficiently removed from water in hybrid processes. PAX-25 turned out to be the best reagent in the removal of organic substances in the process of coagulation, sedimentation/ultrafiltration, and coagulant $FeCl_3$ in the process in-line/ultrafiltration. The substances determined as oxidizability with $KMnO_4$ are removed from water to a smaller degree than the organic substances determined as absorbance UV_{254}. This can result from the different measurement ranges of both methods. The oxidizability with $KMnO_4$ covers also the determination of certain inorganic substances which can be simultaneously oxidized under the standard conditions.

The application of both ferric and aluminum coagulants will be accompanied by a rise of these elements in water, and therefore the content of aluminum and iron in water was assessed in the investigation. The application of the process coagulation/sedimentation + UF for the reagent PAX-25 is characterized by 100% removal efficiency of iron from water. The highest efficiency, of removal of aluminum from water was exhibited by the process coagulation/sedimentation + UF with the application of $Al_2(SO_4)_3$.

4 CONCLUSIONS

The following conclusions can be drawn when analyzing the results involving the unit process (ultrafiltration):

1. high efficiency of turbidity removal from water was found,
2. low removal efficiency of organic substances determined as oxidizability with $KMnO_4$ from water was found.

With regard to the combined process of coagulation and ultrafiltration, the following conclusions can be presented:

1. The application of the hybrid system results in higher removal efficiency of organic substances from water,
2. With respect to the applied iron coagulants ($FeCl_3$, PIX-113), the coagulant PIX-113 is more efficient, and the process coagulation, sedimentation/ultrafiltration is more effective than the process in-line/ultrafiltration,
3. The aluminum coagulant PAX-25 is more efficient than aluminum sulphate in the process of coagulation, sedimentation/ultrafiltration, while the opposite is the case with in-line/ultrafiltration,
4. The metals (Fe, Al) being the part of the applied coagulants were efficiently removed to the level required by the Directive of Health Ministry (dated 19 November 2002) only after the application of hybrid processes.

ACKNOWLEDGEMENTS

The financial support of the Polish Committee of Science, Grant 3 T09D 018 26, is greatly appreciated.

REFERENCES

Choi K.Y., Dempsey B.A., 2004, *In-line coagulation with low-pressure membrane filtration*, Water Research, 38, 4271–4281

Choksuchart P., Héran M., Grasmick A., 2002, *Ultrafiltration enhanced by coagulation in an immersed membrane system*, Desalination, 145, 265–272

Guigui C., Rouch J.C., Durand-Bourlier L., Bonnelye V., Aptel P., 2002, *Impact of coagulation conditions on the in-line coagulation/UF process for drinking water production*, Desalination, 147, 95–100

Konieczny K., Bodzek M., Rajca M., 2005, *Coagulation – MF system for water treatment using ceramic membranes*, International Membrane Conference "PERMEA 2005", Poland, (Desalination 2006 – in press)

Konieczny K., Klomfas G., 2002, *Using activated carbon to improve natural water treatment by porous membranes*, Desalination, 147, 109–116

Mallevialle J., Odendaal P.E., Wiesner M.R., 1996, *Water Treatment Membrane Processes*, McGraw-Hill, New York–San Francisco–Washington

Moulin C., Bourbigot M.M., Tazi-Pain A., Bourdon F., 1991, *Design and performance of membrane filtration installations: Capacity and product quality for drinking water applications*, Environmental Technology, 12 (10), 841–858

Nawrocki J., Biłozor S., 2000, Uzdatnianie wody, Procesy chemiczne i biologiczne, Państwowe Wydawnictwo Naukowe, Warszawa – Poznań (in Polish)

Roman M., 2001, Jakość wody do picia w przepisach Unii Europejskiej i w przepisach polskich. Monografie PZITS, Seria: Wodociągi i Kanalizacja Nr 9, Wyd. Polskie Zrzeszenie Inżynierów i Techników Sanitarnych, Warszawa

Schlichter B., Mavrov V., Chmiel H., 2003, *Study of hybrid process combining ozonation and membrane filtration – filtration model solutions*, Desalination, 156, 257–265

Simpson K.L., Hayes K.P., 1998, Drinking water disinfection by-products. An Australian perspective. Water Research, 32, 1522–1528

Strathmann H., 1998, *Economical evaluation of the membrane technology*, in "Future Industrial Prospects of Membrane Processes" (Cecille L., Tousaint J.-C., Eds) Elsevier Applied Science, London–New York

Environmental Engineering – Pawłowski, Dudzińska & Pawłowski (eds)
© 2007 Taylor & Francis Group, London, ISBN13 978-0-415-40818-9

Reliability analysis of variant solutions for water pumping stations

Jarosław Bajer
Institute of Water Supply and Environmental Protection, The Cracow University of Technology, Kraków, Poland

ABSTRACT: A method for analysing the reliability of water pumping stations of various technical structures is presented in this paper. The results of reliability investigations for several dozens of solutions commonly used by designers for connecting pumps and pipelines in water pumping stations with a total number of pump sets up to 5 (6 in some cases) are presented. Only systems of one or two main delivery manifolds were considered. To choose the best solutions an optimising task was constructed. The solution of this task enabled tabular presentation of so-called compromise structures of water pumping stations (in sense of the Pareto postulate). It was recognised for the structures under analysis, that lower or comparable costs (investment expenditures in specified units) have higher reliability indices (stationary availability factor, mean time between failures, mean repair time). In addition, general recommendations were established for accepting most reliable designs for water pumping stations.

Keywords: Water pumping station, design, reliability, optimisation.

1 INTRODUCTION

So far, despite recommendations presented in numerous papers (Novohatnij 1972, Ilin 1979, Valcur 1980, Cullinane 1985, Mays et al. 1986, Ilin 1987, Mays 1989, Duan & Mays 1990), only reliability issues are considered intuitively or insufficiently in designing water pumping stations (PoW). Unreliable or non-optimal solutions are frequently the consequence. This applies primarily to the so-called technical structures of such facilities being a general conception of its design solution. The PoW technical structure is established by its technical design which comprises connection diagram for pump sets (AP), layout of suction and delivery pipelines and the number and arrangement of basic control and safety fixtures (gate valves, check valves), to which a specified emergency structure of "n from M" pump sets is assigned (where n – number of basic APs, $M = n + m$ – number of installed APs, m – number of emergency APs).

Since there are no uniform recommendations based on measurable (quantitative) reliability assessment for the purpose of selecting appropriate PoW technical solutions under given design conditions, the author decided to undertake appropriate research studies, based on the ideas initiated by Novohatnij (1972). The reliability theory methods allow an assessment of reliability of PoW function, enabling comparison of various variants based on accepted criteria and the choice of the best solution. The aim of this study was to analyse an effect of technical structures of water pumping stations on their reliability, thus enabling general design recommendations for accepting the most reliable designs for specified reserve structures of pump sets. The study was undertaken to enable quantitative comparison of the methods used for increasing PoW (setting a number of reserve pump sets, use of ring delivery manifolds and doubled gate valves), while indicating the most efficient solution. A catalogue of technical structures recommended due to its higher reliability and economic efficiency is to be made.

This paper presents the results of studies based on several dozens of common solutions for connecting pumps and pipelines in water pumping stations with one (Table 1) or two (Table 2) main delivery collectors (GKT) and of various pump set reserve structures. The author is not aware of past research regarding PoW, which were conducted in such a wide range and for similar purposes, however the issues concerning reliable operation of these objects were the focus of different studies. Usually the researchers did not cover details of these objects; their analyses were mainly focused on the operation of pumping units. Such an approach is demonstrated in research studies conducted on modeling and optimising the operation of water distribution systems (Ertin et al. 2001, Shinstine et al. 2002, Baranowski et al. 2004, Farmani et al. 2004, Farmani et al. 2005, Lippai 2005). Issues of water pumping stations reliability were usually treated in examining the impact of these objects, specifically pumping units and their hydraulic (pressure

Table 1. Specification of analysed solutions for pump and pipeline connections in water pumping stations (PoW) with a single main delivery collector.

NUMBER OF PUMP SETS (M = n + m) IN WATER PUMPING STATION				
M = 2	M = 3	M = 4	M = 5	M = 6*
2A	3A	4A	5A	6A
2B	3B	4B	5B	6B
		4Ba	5Ba	6Ba
		4Bb	5Bb	6Bb
2Ba	3Ba	4Bc	5Bc	6Bc
ZT, KZ, AP [P]–[S], ZS } O	substitute designation, including: pump set (pump - P, engine - S) - AP with fixtures (suction damper - ZS, check valve - KZ, delivery damper - ZT)			
2C	3C	4C	5C	6C
2D	3D	4D	5D	6D
2Da	3Da	4Da	5Da	6Da
Designation:	3Db	4Db	5Db	6Db
			5Dc	

┼ - collector shut-off damper
n - number of basic pump sets
m - number of reserve pump sets
e.q. 3B - denotes a structure containing 3 pump sets and of delivery side layout of B type
* - the selected reservation structures of "n from M" type were analysed ("3 from 6" and "4 from 6")

Table 2. Specification of analysed solutions for pump and pipeline connections in water pumping stations (PoW) with two main delivery collectors.

NUMBER OF PUMP SETS (M = n + m) IN WATER PUMPING STATION				
M = 2	M = 3	M = 4	M = 5	M = 6*
2E	3E	4E	5E	6E
		4Ea	5Ea	6Ea
			5Eb	6Eb
			5Ec	
2F	3F	4F		6Fb
		4Fa	5Fb	
2G	3G	4G		
ZT, KZ, AP [P]–[S], ZS	Apply to H type structures	4Ga	5Ga	6Ga
		4Gb	5Gb	6Gb
2H	3H	4H	5H	6H

Designation: ┼ - collector shut-off damper
e.q. 3H - denotes a structure containing 3 pump sets and of delivery side layout of H type
n - number of basic pump sets, m - number of reserve pump sets
* - the selected reservation structures of "n from M" type were analysed ("3 from 6" and "4 from 6")

and capacity), economic (pumping costs) and reliability (failures) parameters on the reliable and optimal operations of these systems.

2 ASSUMPTIONS AND DOMAIN OF STUDY

The scope of this study was established by the following assumptions:

(1) Water pumping stations having the following basic (working) pump sets $n = 1, 2, 3, 4$ and reserve pump sets $m = 1, 2, 3, 4\,(5)$ were considered, while the total number of pump sets $M = n + m$ does not exceed five (except some cases where $M = 6$). It is assumed that all APs are of the same type and identical operational parameters and are connected in parallel and that the reserve ones are loaded.

(2) The PoW under investigation are of different technical structures (number of APs and reserve pumps ("n from M") as well as delivery network design). It was assumed that the solution of suction side is repeatable (independent control of suction pipelines equipped with shut-off dampers, while the number of pipelines corresponds to the number of pumps).

(3) Only basic fixtures are taken into account, i.e. suction and delivery shut-off dampers (mounted on suction and delivery pipelines, respectively) and check valves. The reliability of these components is determined with estimators of mean time between failures (Tp) and mean repair time (Tn). When considering reliability homogeneity for fixtures, one may assume the same values of parameters Tp and Tn ($Tp = 70{,}000$ [hr], $Tn = 8$ [hr]), based on operation analysis, while for pump sets $Tp = 3700$ [hr], $Tn = 40$ [hr], respectively.

(4) The assumed efficiency criterion for water pumping stations is based on the necessity to maintain efficiency $Q = Qn$, where Qn – nominal efficiency.

(5) Systems possessing one or two delivery collectors are considered, while assuming that to meet the adopted efficiency criterion for a PoW with two GKTs it is sufficient that only one of them is used. No pipeline failures are analysed.

(6) Estimated investment expenditures Kn expressed in a specified unit are taken into account.

3 A METHOD FOR COMPUTING RELIABILITY OF WATER PUMPING STATIONS

To conduct the reliability study for various PoW technical structures, the method of minimum unserviceability cross-sections (MPN) was selected in a version modified by Gumiński (1977). The method, closely connected with the reliability binary model (Barlow & Proschan 1965, Reinschke 1974, Henley & Gandhi 1975, Bondy & Murty 1980, Hwang et al. 1981, Tan 2003), belongs to the so-called two-parameter methods that are most suitable for describing renewable systems. In the MPN method the basic properties of system reliability, namely failure-free operation, repairability and availability, are expressed by mean time between failures (Tp), mean repair time (Tn) and, related to these both quantities, the stationary availability factor (K_s), correspondingly.

The method employed, as one of the methods based on structural and functional analysis, requires identification, i.e. so-called reliability structure for the system under examination. Such an approach consists of an analysis of possible flow routes (e.g. current or water in the case of PoWs) laid when its function is performed. The reliability structure determines a relationship between reliability states of system components (1 – operative, or 0 – defective) and a reliability state of the system. This relationship is represented with zero-one flow matrices constructed for individual cases and used as the basis for determining so-called unserviceability cross-sections containing one, two, three or more elements. Any one-element unserviceability cross-section is a single element that repeats in all flow routes. When such elements are present, they form a set of one-element unserviceability cross-sections. Two-element unserviceability cross-sections contain pairs of elements which break all flow routes if both become defective concurrently, thus causing failure of the system. The elements that form one-element cross-sections are excluded from analysis. Similarly, one can define unserviceability cross-sections containing three, four or more elements, depending on the required computational accuracy. The number and type of the minimal system unserviceability cross-sections containing the minimum number of elements depend on the technical structure of the system and, for accepted efficiency criterion and known reliability indices of elements, decide on the values of the computed system parameters.

For a water pumping station performing the specified function (task), defined by the accepted efficiency criterion $Q = Qn$ (see paragraph 4), the number of possible flow routes, thus also the number and type of minimal unserviceability cross-sections depend on the number of working and reserve pump sets, the number and layout of pipelines, dampers, check valves and connection structure of these components.

After determining the minimal unserviceability cross-sections it is possible to conduct a reliability study for the system by employing formulae used in the two-parameter reliability assessment method for systems of serial and parallel structure (Sozański 1982). The equations were derived based on the assumption that the state of each system element is described with an independent stationary normal and ergodic stochastic process and the random variable distributions of mean time between failures (Tp) and repair time (Tn) are exponential.

In addition, in the method of minimal unserviceability cross-sections it is assumed also that the disjoint rule applies to events of minimal cross-sections (Gumiński 1977, Sozański 1982). According to this rule, failure of the system may occur only if a single unserviceability cross-section occurs (concurrent failure of two or more MPNs is excluded). This enables the description of the so-called system failure rate (Λ_s), as a sum of failure rates for individual minimal unserviceability cross-sections containing one, two, three or N elements in general:

$$\Lambda_s = \sum_{i=1}^{M_1} \lambda_i^{(1)} + \sum_{j=1}^{M_2} \lambda_j^{(2)} + \sum_{l=1}^{M_3} \lambda_l^{(3)} + \ldots + \sum_{\alpha=1}^{M_0} \lambda_\alpha^{(N)} \qquad (1)$$

where: $\lambda_i^{(1)}$ = failure rate for i-th one-element unserviceability cross-section ($\lambda_i^{(1)} = \lambda$, $\lambda = 1/Tp$ – element failure rate, Tp – mean time between failures); $\lambda_j^{(2)}$ = failure rate for j-th two-element cross section; $\lambda_l^{(3)}$ = failure rate for l-th three-element unserviceability cross-section; $\lambda_\alpha^{(N)}$ = failure rate α-th N-element unserviceability cross-section; and M_1, M_2, M_3, M_0 = number of unserviceability cross-sections, one-, two-, three- and N-element, respectively.

The values of $\lambda_i^{(1)}$ (λ) apply to single elements that form one-elements unserviceability cross-sections of the system. They are computed from statistical analysis or taken from the literature, while the values of $\lambda_j^{(2)}, \lambda_l^{(3)}, \ldots, \lambda_\alpha^{(N)}$ are computed from formulae (as for parallel structures) (Gumiński 1977). For example, when reducing considerations to one- and two-element unserviceability cross-sections, the failure rate for the latter can be written as follows:

$$\lambda_j^{(2)} = \frac{\lambda_{j1} \cdot \lambda_{j2} \cdot (Tn_{j1} + Tn_{j2})}{1 + \lambda_{j1} \cdot Tn_{j1} + \lambda_{j2} \cdot Tn_{j2}} \cong \lambda_{j1} \cdot \lambda_{j2} \cdot (Tn_{j1} + Tn_{j2}) \qquad (2)$$

where: $\lambda_j^{(2)}$ = as in expression (1) above; λ_{j1}, λ_{j2} = failure rates for the first and the second element in j-th two-element unserviceability cross-section; and Tn_{j1}, Tn_{j2} = mean repair times the first and the second element in j-th two-element unserviceability cross-section.

The values of $Tn_i^{(1)}$ and $\lambda_i^{(1)}$, related to one-element cross-sections are calculated from operating data, while for two-element cross-sections the mean repair time for two-element parallel structure is computed from the following formula (Sozański 1982):

$$Tn_j^{(2)} = \frac{Tn_{j1} \cdot Tn_{j2}}{Tn_{j1} + Tn_{j2}} \tag{3}$$

where: Tn_{j1}, Tn_{j2} – as in formula (2).

The mean repair time for the system is:

$$Tn_s \cong \frac{\sum_{i=1}^{M_1} \lambda_i^{(1)} \cdot Tn_i^{(1)} + \sum_{j=1}^{M_2} \lambda_j^{(2)} \cdot Tn_j^{(2)}}{\Lambda_s} \tag{4}$$

where: $Tn_i^{(1)}$, $Tn_j^{(2)}$ = mean repair time for i-th one-element and j-th two-element unserviceability cross-sections, respectively.

Knowing Λ_s and Tn_s it is possible to compute the stationary availability factor (K_s) from formula (5). This index indicates the probability that the system remains ready for performing assigned tasks (i.e. state of efficiency) (Rao 1992):

$$K_s = \frac{Tp_s}{Tp_s + Tn_s} \tag{5}$$

where the error-free running time Tp_s for the system is defined as the inverse of the failure rate:

$$Tp_s = 1/\Lambda_s \tag{6}$$

To determine reliability indices for water pumping station technical structures the method presented above was generalised and a dedicated computer program was developed. The program SZPN (Ability Paths Unserviceability Cross-sections) was used for computational purposes.

4 COMPARATIVE ANALYSIS OF WATER PUMPING STATION TECHNICAL STRUCTURES

To obtain an objective assessment for the analysed water pumping station technical structures (Tables 1–2) being a set of alternative decisions and to choose an optimal variant, the complex assessment was carried out by using economic and reliability criteria.

Some difficulties in simultaneous optimisation of appropriate criteria in a specified variant forced further using a polyoptimisation set of so-called compromise solutions (most effective solutions). The compromise solutions cannot be comparable with each other without additional selection criteria. The final selection of optimal solution depends on availability of additional requirements for significance of individual quality criteria (technological, economic, reliability).

Since it is practically impossible to establish every quality criterion that could be set in detail in specific design project only, the assessment of various PoW technical structures was constrained to comparison in the terms of achieved reliability indices and estimated investment expenditures (in specified units), while aiming at construction of a set of compromise structures.

As a consequence of necessary simplifications related to estimation of investment expenditures according to a rule described below, the comparability of the analysed technical structures was limited to those which enabled the "cheaper-more expensive" relations to be established. Thus, it was assumed that only structures of identical number and type of working APs could be compared. This results in division of the set of possible alternative decisions X into subsets X^i hereinafter referred to as division groups. Due to difficulties in relative assessment, also some complicated systems (ring or with two independent delivery pipelines connected to each pump) were excluded from comparison with simple systems (all others).

The following general rule was accepted to assign estimated investment expenditures (in specified units) to individual variants:

(1) basic expenditures for each variant to be calculated as follows: $100 \times (n+m)$; where n, m = the number of working and reserve pump sets, respectively;
(2) basic expenditures shall be increased by costs of additional fixtures (dampers on delivery collectors) calculated from the following formula: $10 \times lz$; where lz = the number of additional fixtures compared to the least equipped variant of pumping station belonging to given division group.

To assess the analysed technical structures of water pumping stations, thus also to establish compromise sets for specified division groups, an optimisation task (ZO) was constructed containing L partial tasks (ZO^i) to be solved individually for each division group X^i:

$$ZO = \{ZO^1, \ldots, ZO^i, \ldots, ZO^L\} = \{ZO^i\}^{i=1,\ldots,L} \tag{7}$$

where: L = number of division group; and i = partial task number.

Since it was necessary to find extremes for more than one function of decision variables, it was decided to formulate a partial multicriteria optimisation task (ZO^i) in the form:

$$\mathrm{ZO}^i : \text{to find } \mathrm{K}^i = \{\min \phi_e(x_j^i), e = 1, \ldots, r \ \forall \ \max \phi_e(x_j^i), e = r+1, \ldots, r+s; x_j^i \in X^i; j = 1, \ldots, l^i\} \tag{8}$$

where: K^i = set of compromises for i-th partial task, i.e. a set of the best and already incomparable technical structures within a specified division group; x_j^i = decision variables considered as PoW technical structures belonging to i-th selected division group; X^i = set of possible decision variants for i-th task understood a set of all technical structures forming i-th division group ($X^i = \{x_1^i, \ldots, x_j^i, \ldots, x_{l^i}^i\}$, card $X^i = l^i$, l^i – number of elements in i-th division group); $\phi_e(x_j^i)$ = objective function indicating proportionally to e as a measure of losses resulting from selection of decision x_j^i from the set of possible decisions X^i and then minimised, or treated as a measure of benefits, and thus maximised; e = criterion number ($e = 1, \ldots, r+s$); r = number of criteria being a measure of losses (in ZO^i $r = 2$ and there are: estimated investment expenditures – Kn and mean repair time for specified PoW – Tn_s); and s = number of criteria being a measure of benefits (in ZO^i there are: mean time between failures of specified PoW Tp_s and availability factor K_s).

Individual partial optimisation tasks (ZO^i) were solved numerically by using a modified form of the Pareto postulate (Szpindor & Piotrowski 1986) allowing two decisions x_i and x_j to be classified (better or worse) based on objective functions defined in equation (8). This postulate in a simplified notation, not taking into consideration the indexation of partial tasks, is expressed by the following condition:

$$\begin{aligned} \phi_e(x_i) \le \phi_e(x_j) \quad & \forall \quad e = 1, \ldots, r \\ \phi_e(x_i) \ge \phi_e(x_j) \quad & \forall \quad e = r+1, \ldots, r+s \end{aligned} \tag{9}$$

If simultaneously at least one of the $(r+s)$ inequalities is satisfied sharply, it might be stated, that the decision x_i is "better" than decision x_j ($x_i \succ x_j$). The above mentioned means, that in order to get a compromise decision, the following condition must be fulfilled: in the given possible solution set there is no such x_j decision, for which the substitution of x_0 for x_i the set of equations (9) would be fulfilled. This is exactly the modified version of the Pareto condition, whose fulfillment determines the necessary condition of polioptimality. The application of the Pareto condition eliminates the non-compromising variants, not classifying however, the remaining ones.

5 DISCUSSION

According to the adopted methodology (formal notation of ZO and method of solution) for each division group the sets of compromise variants were chosen in the sense of Pareto (Tables 3–4), to group those schematic solution of delivery side of pumping stations that are to be finally selected by the designer only.

For better clarity, all determined sets of compromise structures were divided into subgroups corresponding to the specified number of reserve pump sets. Because of obvious relationship between PoW reliability indices (Tp_s, Tn_s, K_s) and actual values of reliability parameters (Tp, Tn), its elements listed along with technical structures of water pumping stations presented in tables can be used only for understanding the differences in reliability levels achieved by individual structures.

A review of selected most effective PoW technical structures, while maintaining the design nominal efficiency Qn, leads to the following detailed conclusions:

I. For structures containing only one delivery collector (Table 3):

(1) In the case of simplest structures (without dampers on deliver collectors) the highest reliability regardless of the number of basic APs is achieved for two reserve pump sets. Thus, increasing the number of pump sets serves no purpose.

(2) For structures with lateral GKT and the number of reserve APs not less than 2 ($m \ge 2$), the highest reliability parameters are reached by installing a collector damper between the connections of n-th and $n+1$-th AP, counting in the direction opposite to the GKT lateral.

(3) In structures with front main delivery collector and the number of reserve pump sets not less than 2 ($m \ge 2$), it is favourable to install two collector dampers located symmetrically on both side of the GKT junction, while its proper location depends on the "n from M" reservation schedule. Appropriately locating these dampers should allow operation of the required number of AP, while enabling the remaining ones to be shut off.

(4) By using the number of reserve pump sets $m \ge 2$ it is possible to stabilise the mean shutdown time of water pumping station that correspond practically (minute differences only) with the value of this parameter accepted for fixtures. This indicates a slight effect of mean repair time of pump sets on the PoW expected down-time.

II. For structures with two delivery collectors, while admitting emergency shutdown of one of them (Table 4):

(1) The increased number of reserve pump sets considerably increases PoW reliability.

(2) Doubled collector dampers improves reliability of solutions with a single basic pump set. Such configuration becomes ineffective for other solutions ($n = 2$; $m = 1, 2$ and $n = 3$; $m = 1$).

Table 3. Compromise technical structures (according to *ZO*) of water pumping stations (PoW) with a single main delivery collector.

Number of basic pump sets (AP)	NUMBER OF RESERVE PUMP SETS (AP)						
	m = 1		m = 2				m = 3
n = 1	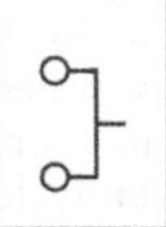	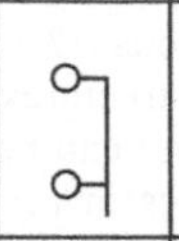	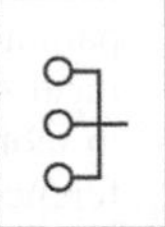	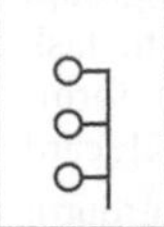	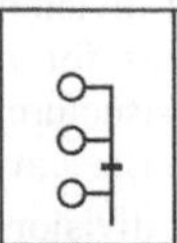		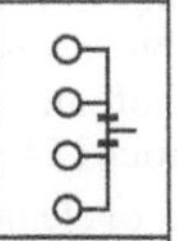
	2A	2C	3A	3C	3D		4B
Tps [hr]	28,541	28,541	23,276	23,276	34,414		34,970
Tns [hr]	9.9	9.9	8.0	8.0	8.0		8.0
Ks	0.999652	0.999652	0.999656	0.999656	0.999768		0.999771
n = 2	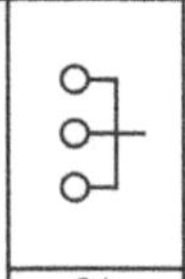	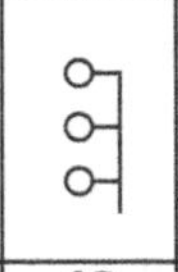	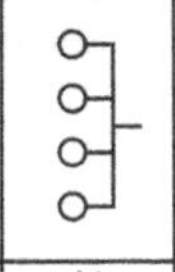	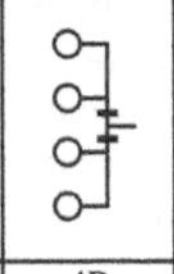	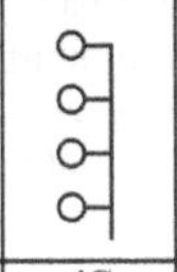	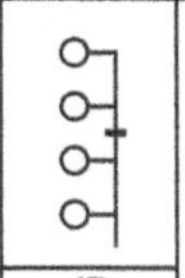	
	3A	3C	4A	4B	4C	4D	
Tps [hr]	16,063	16,063	17,371	32,960	17,371	22,706	
Tns [hr]	11.3	11.3	8.0	8.0	8.0	8.0	
Ks	0.999300	0.999300	0.999537	0.999757	0.999537	0.999647	
n = 3	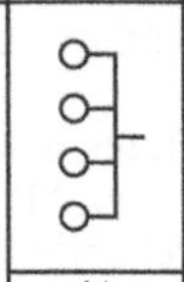	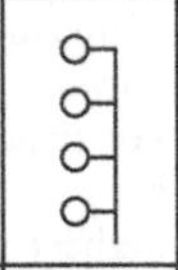	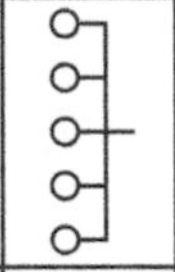	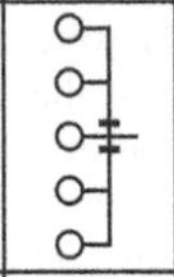	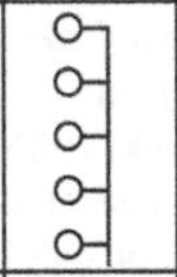	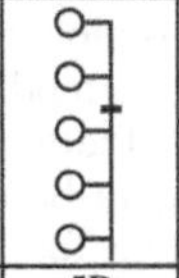	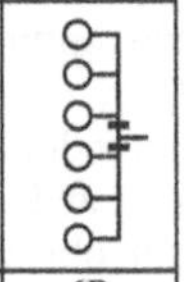
	4A	4C	5A	5B	5C	5D	6B
Tps [hr]	10,423	10,423	13,795	21,641	13,795	16,852	31,232
Tns [hr]	12.2	12.2	8.1	8.0	8.1	8.0	7.8
Ks	0.998828	0.998828	0.999416	0.999629	0.999416	0.999523	0.999750
n = 4	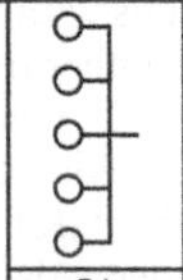	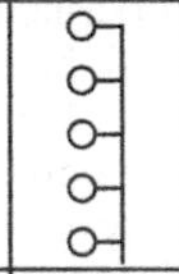	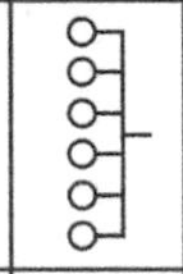		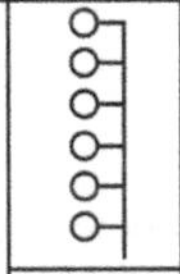	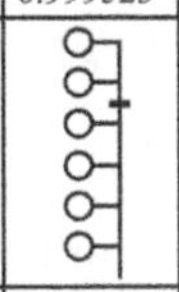	
	5A	5C	6A		6C	6D	
Tps [hr]	7348	7348	11,385		11,385	13,322	
Tns [hr]	13.0	13.0	8.1		8.1	8.1	
Ks	0.998238	0.998238	0.999289		0.999289	0.999393	

(3) In PoW with a single basic pump set ($n = 1$) and one or two reserve pump sets ($m = 1$ or $m = 2$) the use of complicated solutions for the delivery side such as independent supply of delivery collectors by each of pump (H type structures) or circulation collectors (G type ring structures) – is purposeless. For other solutions such structures achieve the highest reliability indices.

(4) By using the number of reserve pump sets $m > 2$ one may expect that the mean downtime of designed PoW should not exceed that of fixtures.

6 SUMMARY AND CONCLUDING REMARKS

The results of the reliability study presented in this paper and obtained for water pumping stations of various technical structures enabled a so-called catalogue of compromise structures to be developed (in the sense of the Pareto postulate) to group those solutions among the analysed ones that are recommended for design purposes as the most advantageous for reliability reasons. This, along with detailed recommendations resulting from this study and referring to an effect of number of reserve pumps and number and layout of shut-off dampers on delivery collectors on reliability indices achieved by various PoW technical structures, can assist designers in selecting the most adequate solution under given design conditions. The above apply only to comparison of PoWs of identical pump sets and similar technical schemes. Otherwise, it is necessary to consider actual investment expenditures and operating costs related to alternative solutions under investigation, when the selection of an optimal solution requires more detailed

Table 4. Compromise technical structures (according to *ZO*) of water pumping stations (PoW) with two main delivery collectors (while allowing shutdown of one of them).

Number of basic pump sets (AP)	NUMBER OF RESERVE PUMP SETS (AP)									
	m = 1			m = 2			m = 3			m = 4
n = 1	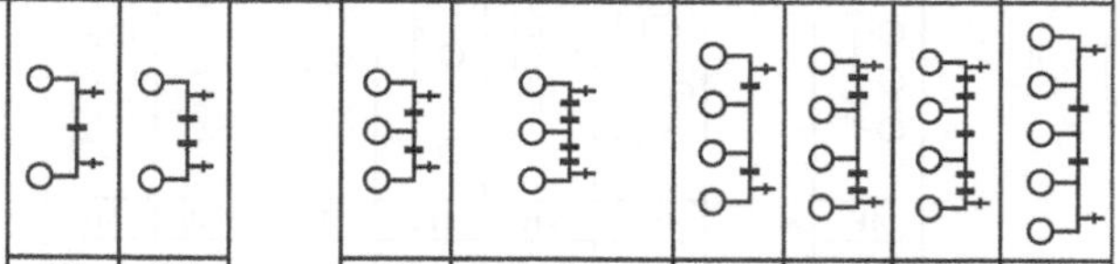									
	2E	2F		3E	3F		4E	4F	4Fa	5Ea
Tps [hr] Tns [hr] Ks	46,459 11.1 0.999761	130,978 16.6 0.999873		~1,900,000 7.6 0.999996	~6,300,000 9.8 0.999998		~2,400,000 6.4 0.999997	~22,500,000 4.6 0.999997	~29,000,000 4.3 0.999998	~17,500,000 4.1 0.999999
n = 2	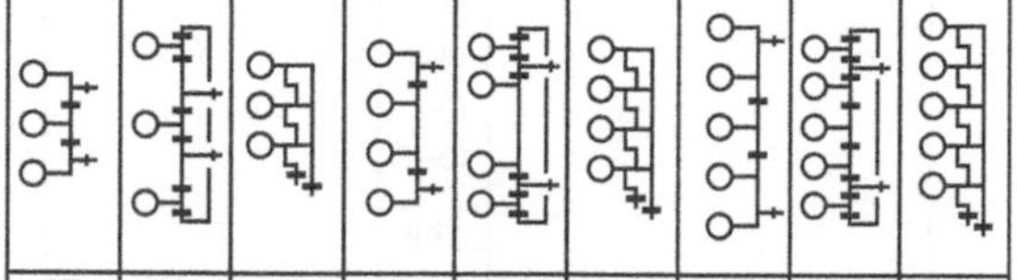									
	3E	3G	3H	4E	4Ga	4H	5Ea	5Gb	5H	
Tps [hr] Tns [hr] Ks	15,676 11.1 0.999290	43,597 16.6 0.999619	43,597 16.6 0.999619	17,358 8.0 0.999538	176,597 7.6 0.999984	~1,700,000 10.1 0.999994	~1,100,000 6.3 0.999995	~11,900,000 4.0 0.999996	~7,800,000 4.3 0.999994	
				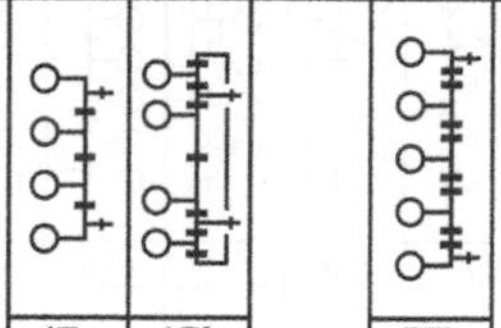						
				4Ea	4Gb		5Fb			
			Tps [hr] Tns [hr] Ks	61,313 8.0 0.999870	~1,000,000 8.8 0.999992		~6,600,000 4.8 0.999993			

analysis. Such an analysis should be based on economic effectiveness, while taking into account also the measurable reliability factor. The concept of such analysis that leads to determination of expected losses resulting from failure of the facility that become the basis for searching for the best solution among all under consideration, will be described scientifically by the author.

In summary, the following general concluding remarks can be made:

(1) The number of working and reserve pump sets and layout of pipelines connected to the pumps as well as arrangement of fixtures should not be based on the designer's intuition and experience, as such a approach may lead to choice of faulty or overprotected solutions. It is advisable to choose solutions that meet the required reliability level deciding on the significance of the facility (e.g. related to purpose) and may be accepted from reliability study or available literature.

(2) The reliability of a water pumping station depends in the first place on its technical structure, while in many cases the appropriate number and layout of fixtures have similar advantageous effects on PoW reliability parameters as those of increased number of reserve pump sets.

(3) An increased number of reserve pump sets can lead to adverse effects, i.e. to decreasing reliability indices of pumping stations. This applies primarily to PoW technical structures with a single main delivery collector without fixtures, where installation of two reserve pump sets is an optimal solution, regardless of the number of working pump sets.

Table 4. (*Continued*)

Number of basic pump sets (AP)	NUMBER OF RESERVE PUMP SETS (AP)									
	m = 1				m = 2				m = 3	m = 4
n = 3	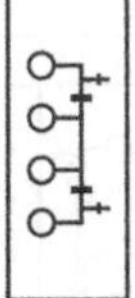		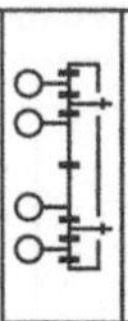	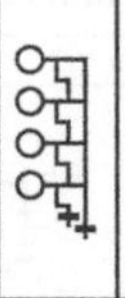	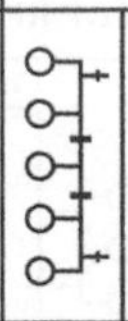	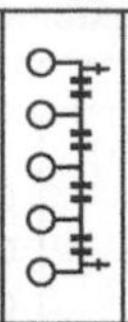				
	4E		4Gb	4H	5Ea	5Fb	5Gb	5H		
T_{ps} [hr] T_{ns} [hr] K_s	10,180 12.1 0.998813		16,967 14.7 0.999132	21,810 16.6 0.999239	21,113 8.0 0.999622	209,404 7.7 0.999963	396,298 8.7 0.999978	711,614 10.5 0.999985		
					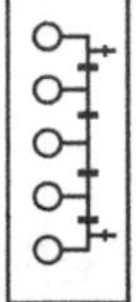					
					5Eb					
				T_{ps} [hr] T_{ns} [hr] K_s	21596 8,0 0,999629					
n = 4	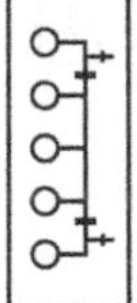				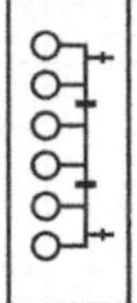	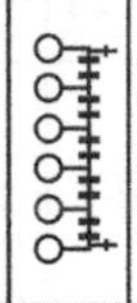				
	5E	5Fb	5Gb	5H	6Ea	6Fb	6Gb	6H		
T_{ps} [hr] T_{ns} [hr] K_s	7187 12.8 0.998218	13,098 16.6 0.998732	9825 14.5 0.998524	13,089 16.6 0.998732	15,662 8.0 0.999487	203,050 8.9 0.999956	202,540 8.8 0.999956	365,574 10.7 0.999971		

(4) It seems to be purposeful to continue present study to include much wider representation of possible solutions, thus enabling a wider base of compromise solutions to be obtained.

REFERENCES

Baranowski, T.M., Walski, T.M., Wu, Z.Y., Makowski, R. & Hartell, W. 2004. Trading off reliability and cost in optimal water distribution system design. In P. Bizier, P. DeBarry (eds), *World water and environmental resources congress and related symposia; World water and environmental resources congress 2003, Philadelphia, Pennsylvania, 23–26 June 2003*. New York: ASCE.

Barlow, R.E. & Proschan, F. 1965. *Mathematical theory of reliability*. New York: Wiley.

Bondy, J.A. & Murty, U.S.R. 1980. *Graph theory in applications*. New York: North-Holland.

Cullinane, M.J. & Jr. 1985. Reliability evaluation of water distribution systems components, *U.S. Army Engineering, Waterways Experiment Station*: 353–358. New York: ASC.

Duan, N. & Mays, L.W. 1990. Reliability analysis of pumping systems. *Journal of Hydraulic Engineering* 116(2): 230–238.

Ertin, E., Dean, N., Moore, M. & Priddy, K. 2001. Dynamic optimization for water distribution systems. In K.L. Priddy, P.E. Keller, P.J. Angeline (eds), *Applications and science of computational intelligence IV; Proceedings of the SPIE conference, Orlando, 17–18 April*.

Farmani, R., Savic, D.A. & Walters, G.A. 2004. The simultaneous multi-objective of anytown pipe rehabilitation, tank sizing, tank siting and pump operation schedules. In G. Sehlke, D.F. Hayes, D.K. Stevens (eds.), *Critical transitions in water and environmental resources management; World water and environmental resources congress 2004, Salt Lake City, Utah, June 27–July 1*. New York: ASCE.

Farmani, R., Walters, G.A. & Savic, D.A. 2005. Trade-off between total cost and reliability for anytown water distribution network. *Journal of Water Resources Planning and Management* 131(3): 161–171.

Gumiński, J. 1977. The method of calculation of indicators of discreteness of admission of network nodes distributive about classic arrangements. *Journal of Power Industry Installations* 12257.

Henley, E.J. & Gandhi, S.L. 1975. Process reliability analysis. *American Institute of Chemical Engineering Journal* 21(4): 677–686.

Hwang, C.L., Tillman, F.A. & Lee, M.H. 1981. *System reliability evaluation techniques for complex/large systems – a review*. Institute of Electrical and Electronics Engineers Transaction on Reliability R-30 (5).

Ilin, Yu. A. 1979. Optimum borders of blocking of pumper devices. *Journal of Water Supply and Sanitary Technology* 11.

Ilin,Yu. A. 1987. *Calculation the reliability of water delivery.* Москва: Стройиэдат.

Lippai, I. 2005. Water system design by optimization: Colorado Springs utilities case studies. In C. Vipulananandan & R. Ortega (eds), *Optimizing pipeline design, operations, and maintenance in today's economy; The pipeline division specialty conference, Houston, Texas 21–24 August 2005.* New York: ASCE.

Mays, L.W. (ed.) 1989. *Reliability analysis of water distribution systems*. New York: ASCE.

Mays, L.W., Duan, N. & Su, Y.-C. 1986. *Modeling reliability in water distribution network design*. University of Texas, Austin.

Novohatnij, W.G. 1972. Reliability analysis of pumping stations delivery and resolution of water. *Journal of Water Supply and Sanitary Technology* 8: 1–6.

Valcur, H.C. & Jr. 1980. Pumping station reliability. How and how much. *Journal of the American Water Works Association* 4: 187–191.

Rao, S.S. 1992. *Reliability-based design*. New York: McGraw-Hill Inc.

Reinschke, K. 1974. *Zuverlässigkeit von Systemen*. Berlin: VEB Verlag Technik. (in German)

Shinstine, D.S., Ahmed, I. & Lansey, K.E. 2002. Reliability/availability analysis of municipal water distribution networks: Case studies. *Journal of Water Resources Planning and Management* 128(2): 140–151.

Sozański, J. 1982. Reliability of electric energization. Warszawa: Wydawnictwa Naukowo-Techniczne.

Szpindor, A. & Piotrowski, J. 1986. Water economics. Warszawa: Państwowe Wydawnictwo Naukowe.

Tan, Z. 2003. Minima cut sets of s-t networks with k-out-of-n nodes. *Reliability Engineering and System Safety* 82: 49–54.

Solid waste and sludge treatment

Environmental Engineering – Pawłowski, Dudzińska & Pawłowski (eds)
© 2007 Taylor & Francis Group, London, ISBN13 978-0-415-40818-9

The improvement and evaluation methods of settling properties of activated sludge

Stanisław Cytawa, Renata Tomczak-Wandzel & Krystyna Mędrzycka
Department of Fats and Detergent Technology, Chemical Faculty, Gdańsk University of Technology, Poland

ABSTRACT: In many biological treatment plants problems with settling of activated sludge have recently become very common. Poor settling properties of activated sludge result mostly from massive growth of filamentous bacteria. The aim of this research was to determine to what extent it is possible to improve sludge settling properties, based on control of process parameters in the SBR system. The parameters under consideration were temperature and oxygen levels. Focus was also on simplifying the method of evaluating sludge settling. Comparison of various methods of evaluating the properties of settling sludge shows that the direct measurement of the real sludge sedimentation time for four meters in the reactor is much more precise than laboratory tests of the sludge volume index (SVI) and it helps in making a decision on changing the aeration conditions.

Keywords: Sludge settling properties, sludge bulking, SBR system, SVI index.

1 INTRODUCTION

Biological treatment of wastewater is commonly accompanied by a problem with settling of the activated sludge after treatment and, due to higher COD and turbidity, the purified wastewater can not fulfil the quality requirements. Besides, due to difficulties in thickening and dewatering of the sludge huge expenses are involved. This is caused by uncontrolled development of bacteria, which are characterized by a high surface to volume ratio, making the separation of the treated wastewater from activated sludge bacteria very difficult. This phenomenon is known as "bulking" of the sludge. Usually, sludge bulking occurs when its SVI index is >150 ml/g (Albertson 1991) and when excessive growth of filamentous bacteria takes place. The poor settling of the sludge is a severe problem in many treatment plants all over the world. Thus, within recent years many scientific centres have participated in research on the causes of bulking and its control. It is very important from the technological, environmental and economic standpoint.

One method to improve the settling efficiency is to increase the flocks density by using weighting material e.g. bentonite, talc etc (Albertson 1991; Liu 2003). There is also work suggesting mechanical breaking of filaments (Albertson 1991). Disintegration of filamentous bacteria can be also achieved by using lime, chlorine, ozone or synthetic polymer in amounts adjusted so that other bacteria inside the flocks are not damaged (Juang 2005; Martins et al. 2004; Saayman et al. 1996). However, these methods can be applied only in emergency, due to their high costs. The most reasonable alternative is suppressing filamentous growth by kinetic or metabolic selection. Bacteria selection can be obtained by changing conditions during the process as well as by changing the substrate availability.

Research on metabolic selection has developed since the method of biological nutrient removal was made possible by the option of operating under unaerated or aerated conditions (Gaval & Pernelle 2003; Holmstrom et al. 1996; Ingvorsen et al. 2003; Martins et al. 2003; Wanner et al. 1987). However, in spite of the knowledge of the metabolic selection mechanisms, foaming and bulking of sludge still remains the main problem in wastewater practice. Most operators try to resolve it based on their experience. In experiments carried out by Gabb (Gabb et al. 1989) it was found that in a fully aerated, fully anoxic and intermittently aerated reactor the filaments of low F/M type did not proliferate in fully aerobic and fully anoxic systems, but proliferated in intermittent aeration systems. However, they did not establish *which* factors in the intermittent aeration procedure promote filament proliferation.

Continuing their research Casey (Casey et al. 1993) have stated that the formation of intracellular NO affects the oxygen uptake rate, and thus, it affects the rate of proliferation of floc-forming microorganisms, while not affecting the filamentous bacteria growth, as they do not accumulate NO. However, this explanation relates only to the systems with N removal due to nitrification-denitrification. Besides, they have stated that biodegradability of COD, and thus, substrate availability, is one of the crucial factors affecting the rate of microorganism proliferation.

In Poland most of the treatment plants suffer from bulking. The case study discussed in this paper (the Swarzewo Treatment Plant) is no exception; such problems appear there seasonally.

The aim of this research was to determine to what extent it is possible to improve the sludge settling properties in this treatment plant, based on the control of process parameters. We also focused on simplification of the method of sludge settling evaluation.

2 MATERIALS AND METHODS

The treatment plant in Swarzewo treats wastewater from coastal cities and villages (Puck, Władysławowo, Swarzewo, Gnieżdżewo, Rzucewo, Białdzikowo, Chłapowo, Chałupy, Żelistrzewo Smolno i Mrzezino). The process is performed in an activated sludge chamber, operating in the SBR (Sequencing Batch Reactor) mode. Each bioreactor is 4000 m^3 and has a diameter of 30 m. Decantation of treated wastewater proceeds through the point-decanters and, due to its high flow rate (1000 m^3/h), the sludge flocks are entrained with the wastewater stream. The duration of the sedimentation phase and clarity of the out-stream depends to a great extent on the rate of sludge settling to a safe level in a chamber, ensuring the wastewater outflow through a decanter without carrying out the flocks.

An increase of the SVI index up to 100 ml/g requires the prolongation of the sedimentation and decantation phase to >400 min., while it is only 120 min. when SVI is close to 40 ml/g. Some sedimentation and decantation periods last even longer than 600 min.

The improvement of settling properties was achieved when, the aeration conditions and the mode of feeding of the bioreactor with wastewater were changed based on the analysis of particular stage of the process. Besides, various methods of measuring the settling properties of the sludge were used and compared. Attention was mostly on the SVI index and on the settleability. The latter was based on direct measurement of the time of sedimentation in the bioreactor to the level of 4 m above the bottom.

3 RESULTS AND DISCUSSION

In Figure 1 values of three parameters are presented: the adjusted by the operator time of sedimentation, the adjusted time of sedimentation and decantation phases and sludge settling time. These data relate to the period May 2004–January 2005.

As can be seen, in the period of fast sedimentation of the sludge, the clarification time was below 120 min., while in the periods of worse sedimentation

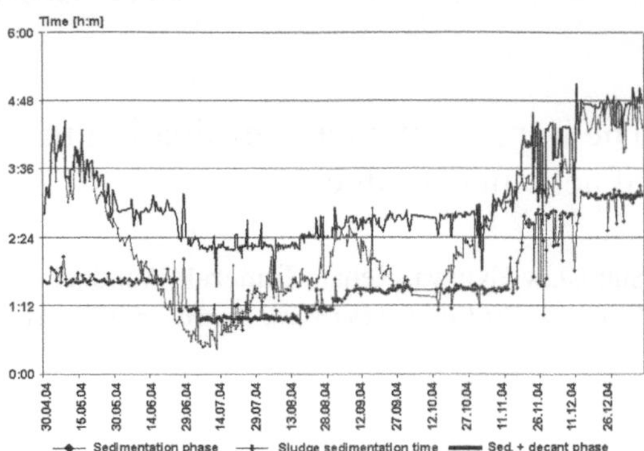

Figure 1. The adjusted time of sedimentation, total time of sedimentation and decantation and measured sludge settling time.

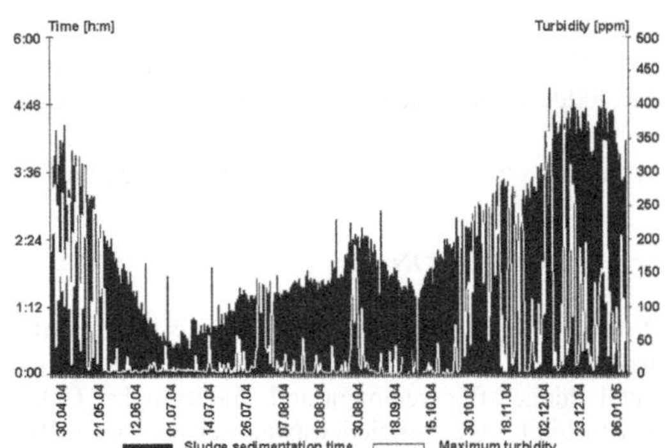

Figure 2. The sludge sedimentation time and outflow turbidity within the period May 2004–January 2005.

this time increased to 290 min. What is more, in spite of the long settling time the flocks were entrained into the decanter, especially in the final phase of decantation. The use of effective decanters with high outflow rate (required in case of high capacity reactors) causes disturbances in the decantation of the purified wastewater.

In Figure 2 the sludge settling rate is compared with turbidity of wastewater at the outflow (maximal value registered in each cycle).

It is clearly visible that there is a correlation between both parameters. In the period of good settling ability (June–July 2005, Figs. 1–2), despite the shortened sedimentation and decantation times, the maximal turbidity of wastewater was very low (<10 ppm, Fig. 2). On the other hand, in the periods of poor settling ability (May 2004, November–December 2005) turbidity was high (200–350 ppm), the purification effect. As is evident from the above data, both, the time of settling and turbidity measurements can be very useful to evaluate the settling properties of the sludge. However,

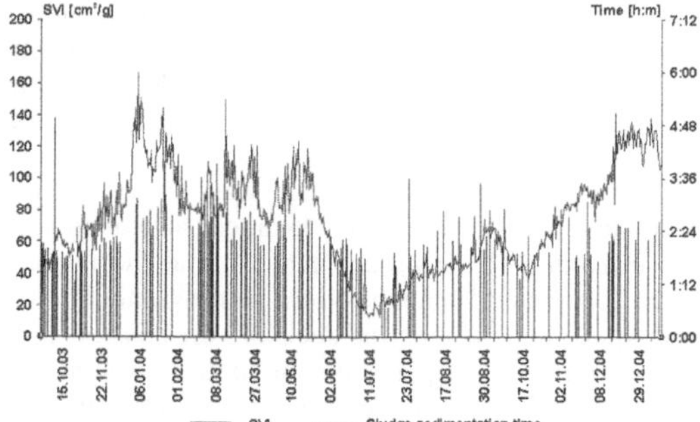

Figure 3. The time of sludge sedimentation (line) and sludge volume index (bars) within the period of September 2003–January 2005.

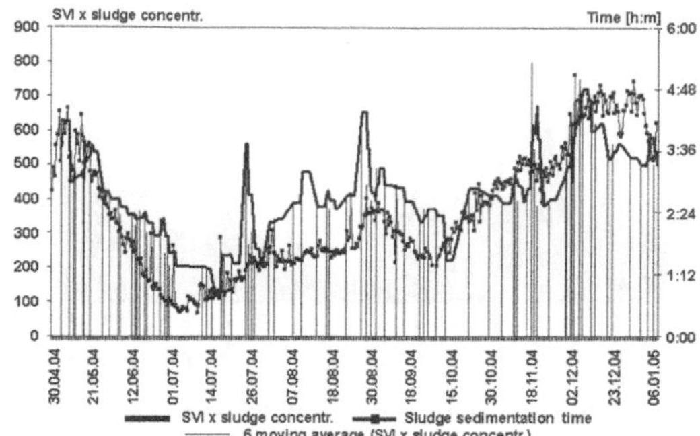

Figure 4. The rate of sludge sedimentation (thin line with points), SVIxAS$_c$ (bars) and 6 moving average of SVIxAS$_c$ (thick line).

the comparison of settleability and SVI results leads us to interesting conclusions.

In Figure 3 the results obtained within the period from September 2003 to January 2005 are shown.

It can be seen that both methods give comparable results; the increase in SVI index relates to the increase of settling time. However, using the settling time measurement is much more advantageous because of its simplicity, automatic registration on-line and the ability to perform the measurements under reproducible conditions, always at the end of the aeration period. SVI determinations performed on samples taken from the reactor require much more time and are very often affected by continuing denitrification, which can falsify the results. Thus, the method based on the rate of settling is more useful because of its simplicity and also because direct observations in the bioreactor are possible.

In Figure 4 another attempt at finding a correlation between SVI and settling time, is shown. If one multipies the SVI value by the activated sludge concentration in the reactor (AS$_c$), then the obtained values (bars) are related to the sedimentation rate (thin line with points). Further, if one calculates the "six point moving average value" of SVI × AS$_c$ then the obtained line (thick line) is similar to that of settling time values (thin line, Fig. 4).

The range of variability of settling time was from 27 min. to 360 min., while the SVI index ranged from 18 to 137 cm^3/g. The very good repeatability of neighbouring measurements of settling time, which ensure evaluation of the sludge settleability tendency much faster than when SVI was used, should be noted. Besides, the effect of sludge thickening in the Imhoff cylinder on the SVI results is clearly visible. In the case of "heavy" sludge the settling rate is slowed and SVI values are overstated compared with the results of real settleability. In the case of "loose" sludge, the opposite tendency is observed (the SVI values are underrated).

The lowered accuracy and great scattering of neighbouring measurements of SVI makes the rapid interpretation of sludge sedimentation changes resulting from often changing settings of technological parameters, impossible. In the case when settings are chosen on the basis of registered settling time, the amelioration of sludge settleability is achieved rather fast (even after days the improvement is visible).

The above observations prove that the usefulness of measurements of real time sludge settling is higher than that of the SVI index, also because of simplification of the method of sludge settling evaluation in the treatment plant. Besides, it has been stated that the SVI value depends on the real time of sampling during the consecutive stages of the process in SBR reactor.

Within a very long period (few years) the settling properties in sedimentation chamber were monitored and the results were compared with some process parameters, e.g. temperature, activated sludge concentration and its loading with BOD and others. In Figure 5 the profile of sludge settling rate is compared with the air and wastewater temperature for selected period, September 2003–January 2005.

It has been confirmed that temperature (especially that of the wastewater) is one of key parameters affecting the settling properties, which is a common observation. During the autumn months the settling time constantly increases and is the highest during winter. During winter season the lowest temperature in the bioreactor can even reach 6°C, which stimulates the growth of filamentous bacteria. At wastewater temperatures below 15°C the continuous development of those bacteria took place, which leads to progressive worsening of the SVI up to 100 cm^3/g and prolongation of settling time up to 290 min. In the late spring, after temperature rise, the improvement of settling properties can be observed and when temperature drops in the fall it worsens again (Fig. 5).

However, the temperature is not the only parameter affecting the settling properties of the sludge. The excessive growth of filamentous bacteria depends strongly on oxic conditions during then process and

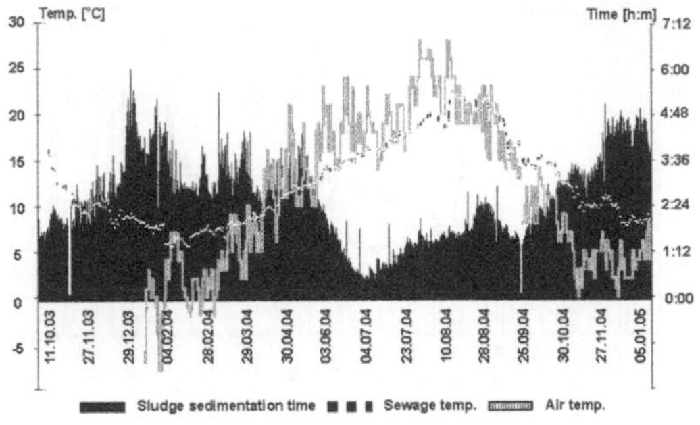

Figure 5. Sludge sedimentation rate, temperature of air and temperature of wastewater measured from September 2003 till January 2005.

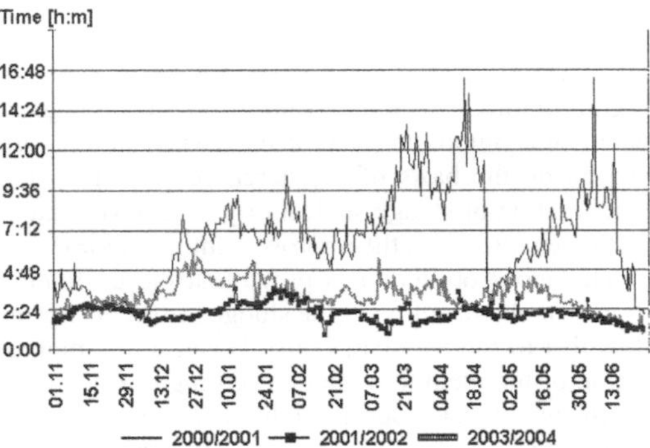

Figure 6. The rate of settling time of the sludge in three winter seasons (years 2000/2001, 2001/2002, 2003/2004).

on substrate availability. During winter seasons a very rapid aggravation of SVI can be observed, when aeration is switched on during filling the bioreactor with wastewater. Even a short period of aeration of inflowing wastewater, repeated within only 5 cycles is sufficient to prolong the settling time lastingly. Such a situation was observed within a few years.

In Figure 6 the settling times of the sludge in three winter seasons are presented. It should be mentioned that only in winter 2000/2001 the wastewater was aerated during inflow to bioreactor; in this case the settling time was the longest (>15 hours).

Simultaneously, the measured SVI value was the highest (>200 cm^3/g; the results not presented here).

The same kind of the sludge aeration during the raw sewage inflow performed in the summer results in decreasing the sludge sedimentation, however the tendencies are reasonably slow. This can be observed from 10th July to 31st August 2004 (Fig. 1). A gradual extension of the sludge sedimentation time from 28 to 156 minutes could also be noticed when simultanously SVI was altered from 35 cm^3/g to 65 cm^3/g.

Another change in aeration performed on the 1st September also resulted in shortening the sludge sedimentation time – the change was based on eliminating the aeration during the sewage inflow to the reactor. However, it lasted only till the sewage temperature fell to 15°C (Fig. 5). After that time, gradual deterioration in sludge sedimentation time was observed, this means the extension of time sedimentation and the growth of the turbidity of purified sewage was observed. The automatic registration helps to track several items: the influence of the proper aeration and the impact of all parameters' changes on the settling rate. Therefore, quick conclusions and decisions are possible and the sensitivity of a particular biological system can be seen. Stabilized changing tendences were visible even only four days after launching the project. Also, microscopy reveals the increased level of filamentous bacteria. What is crucial is the growth of diverse bacteria existing in various seasons. *Microthrix parvicella* prevails mainly in the winter, whereas numerous colonies i.e. 041 type and for instance *Nostocoida* sp. are in the summer as well as type 092 and 021N.

4 CONCLUSIONS

The study compares various methods of monitoring the properties of sedimenting sludge, including the sludge volume index (SVI) and the indirect measurement of the settling rate for four meters in the reactor. It has been stated that the measure of the real sludge sedimentation time – as a method – helps in making a decision. The method is much more precise than testing the index in the laboratory.

Contrary to index measurements, measuring the sedimentation time of the real sludge, always at the very end of the process, is continuous via an online probe. The index varies during the treatment process making measurements at definite times difficult, if not impossible.

The comparison between various aeration conditions and various inflows of raw sewage helps to visualise the impact of these changes on the development of filamentous bacteria. The maximum deterioration of the SVI in the SBR system can be reached easily, confirming that the mechanism of development of various bacteria is identical in both the batch and continuous flow reactors.

The improvement of the sedimentation lasts longer than the deterioration. Moreover, there is a good correlation between the temperature and the process.

It was also found that the rate of improvement of the settling properties of the sludge depends very strongly on detention time in anaerobic and aerobic zones and is different at different temperatures (different seasons of the year). Using the suggested method of control, excellent results of sedimentation were achieved (SVI equals 40 and even less).

On the basis of the obtained results we conclude that in the SBR-type treatment plant it is possible to choose the operating parameters which guarantee a long term elimination of undesirable bacteria in the activated sludge and to sustain very good sludge sedimentation parameters.

REFERENCES

Albertson, A. 1991. Bulking sludge Control – Progress, Practise and Problems. *Wat. Sci. Tech.* 23: 835–846.

Casey, T.G., Wentzel, M.C., Ekama, G.A., Loewenthal, R.E., Marais, G.R. 1993. An hypothesis for the causes and control of anoxic-aerobic filament bulking in nutrient removal activated sludge systems. In *First International Specialized Conference on Microorganisms in Activated Sludge and Biofilm Processes, Paris, September 1993*: 169–178.

Liu, Y. 2003. Chemically reduced excess sludge production in the activated sludge process. *Chemosphere* 50: 1–7.

Martins, A.M.P., Pagilla, K., Heijnen, J.J., van Loosdrecht, M.C.M. 2004. Filamentous bulking sludge – critical review. *Wat. Res.* 38: 793–817.

Saayman, G.B., Schutte, C.F., van Leeuwen, J. 1996. The effect of chemical bulking control on biological nutrient removal in full scale activated sludge treatment plant. *Wat. Sci. Tech.* 34: 275–282.

Gabb, D.M.D., Still, D.A., Ekama, G.A., Jenkins, D., Marais, G.R. 1991. The selector effect on filamentous bulking in long sludge age activated sludge systems. *Wat. Sci. Tech.* 23: 867–877.

Holmstrom, H., Bosander, J., Dahlberg, A.-C., Dillner-Westlund, A., Flyborg, L., Jokinen, K. 1996. Severe bulking and foaming at Himmerfjarden WWTP. *Wat. Sci. Tech.* 33: 127–135.

Wanner, J., Kucman, K., Ottova, V., Grau, P. 1987. Effect of anaerobic conditions on activated sludge filamentous bulking in laboratory systems. *Wat. Res.* 21: 1541–1546.

Martins, A.M.P., Heijnen, J.J., van Loosdrecht, M.C.M. 2003. Effect of feeding pattern and storage on the sludge settleability under aerobic conditions. *Wat. Res.* 37: 2555–2570.

Ingvorsen, K., Nielsen, M.Y., Joulian, C. 2003. Kinetics of bacterial sulfate reduction in an activated sludge plant. *FEMS Microbiology Ecology* 46: 129–137.

Gaval, G. & Pernelle, J.-J. 2003. Impact of the repetition of oxygen deficiencies on the filamentous bacteria proliferation in activated sludge. *Wat. Res.* 37: 1991–2001.

Juang, D.-F. 2005. Effects of synthetic polymer on the filamentous bacteria in activated sludge. *Bioresource Technology* 96: 31–40.

Environmental Engineering – Pawłowski, Dudzińska & Pawłowski (eds)
© 2007 Taylor & Francis Group, London, ISBN13 978-0-415-40818-9

The influence of lime, water-glass and clay addition on sealing properties of waste rock from Bogdanka

Martyna Wiśniewska & Witold Stępniewski
Chair of Land Surface Protection Engineering, Lublin University of Technology, Lublin, Poland

ABSTRACT: A landfill should be safe for the environment. Bottom and top liners play an important role in controlling migration of leachates in waste disposal. Compacted clay liners are widely used due to their cost effectiveness, large attenuative capacity and durability. Clay liners reduce the rate of contaminant migration by their low permeability and their sorption capacity. Though compacted mineral liners possess many advantages such as low permeability and a large capacity of attenuation, they have high shrinkage and swelling potential causing problems of instability. The aim of this study was to examine the effects of lime, water-glass and clay addition on the compactability and water permeability of mineral waste from coal mining which, after modification, is a potential sealing material for the construction of mineral landfill liners.

Keywords: Waste management, landfill, mineral sealing, coal mining wastes, hydraulic permeability.

1 INTRODUCTION

Long-term preservation of a landfill liner impermeable to water is the basis for protection of soil and ground waters from contamination with landfill leakages. The compacted mineral layers, characterized by a low water permeability, prevent vertical leachate migration. They are commonly used in constructing bottom, slope and top sealing in case of landfills already being exploited or designed, as well as in protecting old objects. Some experts (e.g. Holzlöhner et al. 1995, Met et al. 2005) consider a low value of the hydraulic conductivity as the main feature determining the usefulness of a material for mineral sealing construction. The published results widely describe the issues of selecting optimal compaction parameters for the mineral layers (Seymour & Peacock 1994, Łuczak-Wilamowska 2002, Met et al. 2005) and the possibilities of reducing their permeability by using mineral and organic additives (Grantham & Robinson 1989, Pingpank et al. 1998). Numerous publications (Quigley et al. 1988, Cancelli et al. 1994, Quigley & Fernandez 1994, Rowe 1994) present studies on the influence of organic and inorganic leachates on the water permeability of clay material and describe the problem of selecting a method of filtration rate measurement *in situ* and under laboratory conditions. During constructing the mineral liners two variables determining their water permeability are of importance, namely water content during compaction and the resultant degree of compaction. Determination of optimal water content and performance of compaction with a heavy compactor in several layers enables obtaining sufficiently low water permeability. Depending on the material, it is possible to reduce the coefficient of water permeability k to a value in the range 1×10^{-7}–1×10^{-9} m s^{-1}, which is required in most countries (Holzlöhner et al. 1995). The desired effect may be obtained by using additives such as bentonite, cement, amorphous silica or water-glass (Prashanth et al. 2001, Tay et al. 2001). The addition of clay material improves the grain composition and the ion exchange capacity of the mixture. Sodium water-glass (Na_2SiO_3), due to its sealing properties, is used in artificial linings and in strengthening building ground (the petrification process). The effect of sealing is a result of silica gel peptization. The silica penetrates the pores of the material and seals them. Additionally, introduction of sodium cations increases the amount of molecular water in the soil and causes its peptization, which leads to a reduction of water permeability. According to Belouschek & Kügler (1990), water-glass is practically resistant to all chemicals, apart from hydrofluoric acid and strong alkalis, such as sodium hydroxide. Another amendment which may added to the mineral layer during its processing is quicklime (CaO) which enables a reduction of water content of the material to the optimum value required during forming of the layer (Junge & Horn 2001, Horn & Stępniewski 2004). Decreasing the water content takes place both directly (by its binding in the reaction of $Ca(OH)_2$ formation) and indirectly during temperature rise (an exothermic reaction) and water evaporation.

The following materials are used in constructing of mineral liners:

- natural mineral materials with adequate properties,
- natural mineral materials with the appropriate amendments,
- waste materials.

While searching for materials meeting the requirements for sealing clay liners, more and more attention is paid to waste mineral materials (e.g. coal mining wastes, fly ashes). Waste rock stemming from hard coal mining is one of the industrial wastes which are troublesome for the environment mainly due to its huge mass, and to a smaller extent, due to its harmfulness and toxicity. Owing to a large content of clay minerals (about 80%), these wastes may be useful in the formation of sealing layers in landfills.

'Waste rock' is a wide concept, most often defined as rock which is considered useless with reference to the mined raw mineral (Jaroszewski et al. 1985, Haller & Włoszek 1998). Individual types of wastes which are released in the process of mining and recovery differ in petrographic, mineralogical and chemical composition. The particle size distribution and the proportion of mineral components change with time. The main component of these rocks are clay minerals with a particle size <0.002 mm: kaolinite – about 30%, illite – about 26%, chlorite – about 4% and some amounts of montmorillonite. The content of quartz reaches as much as 28–30%. Also such minerals as siderite, pyrite and coal are present in small amounts (Borys et al. 2004, Stępniewski et al. 2004). Research concerning the waste rock from Bogdanka S.A. Coal Mine (Skarżyńska et al. 1997, Haller & Włoszek 1998), confirms the possibility of its use in road construction and in the production of brick. It is widely used in the reclamation of degraded lands.

A special method of waste rock utilization, due to low water permeability, is its application in water engineering (Borys et al. 2004). The permeability of mine waste depends on its composition, on its susceptibility to weathering processes, on the water content during compaction and degree of compaction obtained. The processes of mechanical weathering (disintegration) of rocks proceed very fast in the initial period (the first year) of coal waste storage. As a result of fast weathering, the share of the clay fraction the material increases (Sullivan & Sobek 1982, Twardowska et al. 1988), which leads to a reduction of water permeability and stabilisation of filtration conditions. Further changes proceed more slowly. The properties of mineral liner depend not only on the characteristics of the selected material but also on the conditions of its construction. The mineral layer fulfills the sealing function well, if it is characterized by: low enough water permeability, resistance to crack formation, sufficiently high mechanical resistance as well as by hydromechanical and chemical stability in time. It was shown that part of them, if not all, depend on the water content during formation (Horn & Stępniewski 2004, Wysocka et al. 2004).

The aim of the paper was to study the influence of the amendments applied on compactability and the possibility to reduce the water permeability of coal mine waste rock likely to be used for mineral landfill construction. The materials under study were 1 – the original waste rock from Bogdanka coal mine, 2 – the waste rock amended with water-glass and quicklime, and 3 – the waste rock amended with natural clay.

2 MATERIALS AND METHODS

The basic material under was waste rock sampled from a hard coal mine in Bogdanka. For use in the research it was ground into particles <4 mm. As additives, natural clay from Markowicze and quicklime applied together with sodium water-glass (density 1.38 $Mg\,m^{-3}$, 35% $Na_2O + SiO_2$) were used. Markowicze clay was characterized by 32.5% fraction <0.002 mm, 30% of 0.01–0.002 mm fraction, 36% of 0.06–0.01 mm fraction and 1.5% of the fraction >0.06 mm. It contained 30–40% montmorillonite, 12–16% of illite, 11–15% of chlorite and 5–8 of kaolinite. Other components of this material are: quartz 28–30%, calcite 4–6%, dolomite 2–3% and spars 2–3%.

On the basis of laboratory compaction tests, the optimal water content (W_{opt}) for compaction (Proctor optimum water content) was determined and the compaction curves were determined for all the materials under study. The studies were carried out on the following materials:

1. waste rock from Bogdanka (East Poland) coal mine – the material was denoted with an 'S' symbol,
2. the waste rock with the addition of 2% (by mass) of quicklime and 6% by mass of water-glass – the material denoted as 'SWS',
3. the waste rock with 30% (by mass) amendment of Markowicze clay – the treatment denoted as 'SI'.

For all the materials under study the compaction was performed at the following five water contents: $W_{opt} - 4\%$; $W_{opt} - 2\%$; W_{opt}; $W_{opt} + 2\%$; $W_{opt} + 4\%$ (all by mass basis). Measurements of the coefficient of water permeability k were performed on the compacted materials.

Determining the optimal water content for compaction of the material was carried out in the Proctor apparatus with a rammer. The material was compacted in a small cylinder with the capacity of 1 dm^3 and weight of 1278 g. According to the previously selected

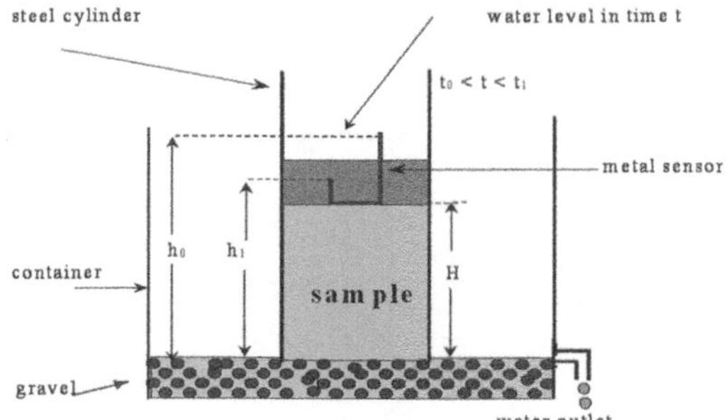

Figure 1. The principle of hydraulic conductivity determination with the falling-head method (Niedźwiecki et al. 2003).

method the following parameters of the test were applied:

- the type of a rammer – light 2.5 kg
- the number of inserted layers – 3
- the number of strokes – 25
- the height of drop of the rammer – 32 cm
- the unitary compaction energy – 0.6 kJ dm^{-3}

The coefficient of water permeability k was determined with the falling-head method (Hartge & Horn 1992, Niedźwiecki et al. 2003), in ten replications for each material. The determination consisted in a measurement of the level of a column of water falling in a time interval ($t_1 - t_0$) in a soil sample saturated with water (Fig. 1).

The values of the coefficient of water permeability k were calculated by the following formula:

$$k = \frac{H}{t_1 - t_0} \log_e \left(\frac{h_0}{h_1} \right) \qquad (1)$$

where k = coefficient of water permeability [m s^{-1}]; t_0 = time of the beginning of measurement [s]; t_1 = time of the end of measurement [s]; h_0 = water level at time t_0 [mm]; h_1 = water level at time t_1 [mm]; and H = height of the sample [mm].

3 RESULTS AND DISCUSSION

On the basis of the compaction test the optimal water content for compaction was determined. The values of bulk density for the untreated material are presented in Figure 2. The optimal water content, at which the maximal compaction of waste rock 1.88 Mg m^{-3} was obtained, is 10.5%. The compaction of waste rock on the left side of the Proctor's curve gives higher bulk densities compared to that on the right (wetter) side.

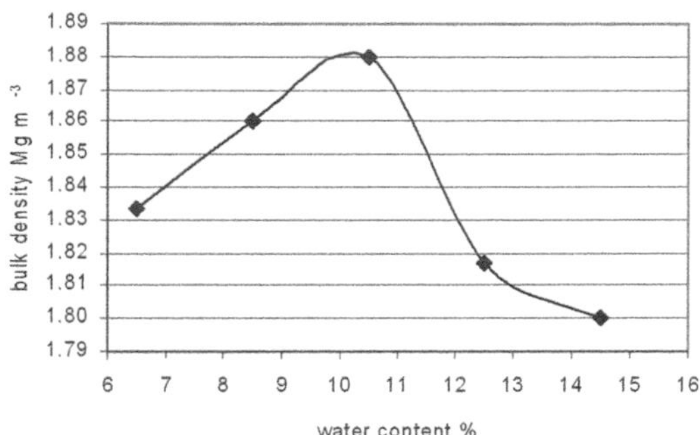

Figure 2. The compaction curve for waste rock (S).

Table 1. $\rho_d/W_{opt}/k$ characteristics of the basic and modified material.

Type of material	Bulk density (ρ_d [Mg m^{-3}])	Optimal water content (W_{opt} [%])	Coefficient of water permeability (k [m s^{-1}])
S	1.88	10.5	9.71E−07
SWS	1.93	16.8	1E−10
SI	1.74	15.0	1.33E−07

S – waste rock.
SWS – waste rock + quick-lime (2% by mass) + water-glass (6% by mass).
SI – waste rock + clay (30% by mass).

The values of optimal water content for compaction and the maximal bulk densities for individual materials are presented, together with the corresponding values of water permeability coefficient k, in Table 1. The simultaneous addition of water-glass and quick-lime caused an increase of optimum water content by 6.3% and an increase in Proctor's bulk density to 1.93 Mg m^{-3}, as well as a decrease of the coefficient of water permeability k from 9.7×10^{-7} m s^{-1} to as low a value as 10^{-10} m s^{-1}. The latter value fulfills the requirements of all the European regulations. The addition of the 30% of Markowicze clay shifted the Proctor optimum water content to the right by 4.5% and lowered the Proctor's bulk density to 1.74 Mg m^{-3} with simultaneous decrease of the water permeability coefficient to 1.33×10^{-7} m s^{-1}.

The addition of CaO causes a shift of the Proctor's optimum of the material to the right. This effect is described in the literature (Junge & Horn 2001, Horn & Stępniewski 2004). Moreover, the Ca^{2+} cation, being divalent, reduces shrinking and swelling, but also increases coagulation and improves the structure of the soil – thus it may increase permeability. In the results obtained for the SWS material the impact of water content on compactability and water permeability is

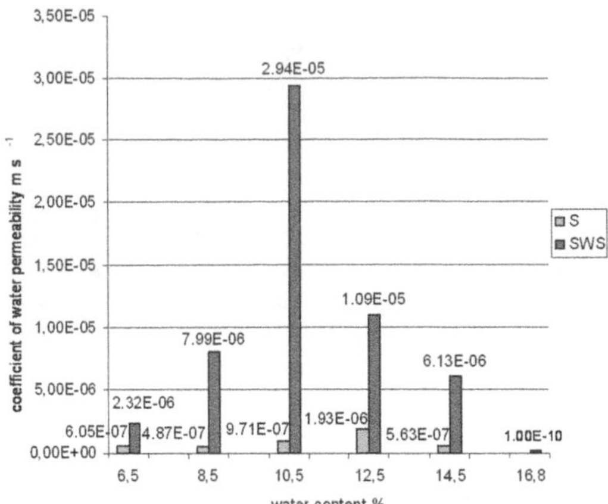

Figure 3. Influence of lime and water-glass addition on the coefficient of water permeability of waste rock (S – waste rock, SWS – waste rock with the addition of quicklime and water-glass)*.
*At water content 16.8% coefficient of water permeability was determined for SWS material.

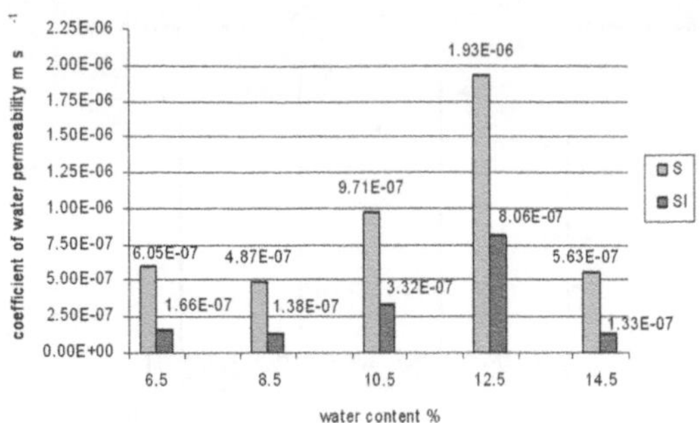

Figure 4. Influence of the 30% clay addition on coefficient of water permeability of waste rock (S – waste rock, SI – waste rock with the clay addition).

marked. A negative influence of lime on the permeability was observed for the SWS material, compacted at a water content value lower than the optimal one (Fig. 3).

At water content lower than 14% the stage of chemical binding of water by CaO during $Ca(OH)_2$ formation is clearly visible. In fact, the water bound by quicklime, reduces water content during compaction. The obtained compaction is lower and water permeability higher in comparison to the basic material. The coefficient of water permeability reaches a mean value of $2.94 \times 10^{-5}\,m\,s^{-1}$. With the increase of water content of the material during compaction (up to 16.8%), the effect of sealing the pores with silica gel may be clearly observed. Water-glass cements the pores and forms a structure of very low permeability. Introduction of sodium cations increases the amount of molecular water in the soil and results in its peptization, which leads to a reduction of soil permeability. The bulk density reaches the value of $1.93\,Mg\,m^{-3}$. At such high compaction the coefficient of water permeability is lower than $10^{-10}\,m\,s^{-1}$ (Tab. 1). The SWS material was characterised by the highest compaction and much lower permeability than the other tested materials. A possibility of water-glass application in case of mineral linings is confirmed in the literature. Pingpank et al. (1998), researched the sealing of cracks formed in the mineral layer under the influence of drying and shrinking. The basic sealing mixture on the basis of kaolin was modified with the addition of water-glass (5–8% by mass) and Portland cement. The treatments modified with water-glass showed only a clear advantage over other materials.

The addition of clay increases the contents of the clay fraction in the basic material, i.e. waste rock. Significant changes in the coefficient of water permeability reported by Benson et al. (1994) are explained by the authors as a result of changes in the mineral composition of the material and conditions of its compacting. The coefficient of water permeability should therefore be considered as a function of geotechnical and mineralogical properties of the material. It decreases with the increase of the clay fraction content. For the SI treatment with a higher content of the clay fraction, despite lower bulk density ($1.74\,Mg\,m^{-3}$), the coefficient of water permeability was lowered by as much as one order of magnitude (Fig. 4). Similar dependencies are reported by Daniel (1987), Benson et al. (1994) and Met et al. (2005).

Under landfill conditions one additional factor influencing water permeability of the mineral liner are leachates from the waste body. The Na^+ cations, present in the leachates from municipal wastes, are exchanged for Ca^{2+} cations present in the clay material. This may lead to the decrease of water permeability (Met et al. 2005).

4 CONCLUSIONS

The results of the studies indicate that waste rock is a potential material for building mineral liners for landfills. The analysis of waste rock compaction in the experiment lead to the following conclusions:

1. It was confirmed that the water content during compaction of materials has a decisive influence on the bulk density and water permeability obtained.
2. The coefficient of water permeability of the original waste rock compacted at different water contents was highest for the water content $W_{opt} + 2\%$, and lowest for compaction at $W_{opt} - 2\%$.
3. The addition of quicklime and of water-glass to the coal mine waste rock enables (during compaction at

Proctor's optimum water content) the coefficient of water permeability of order of 10^{-10} m s^{-1} required for a landfill, to be obtained.

4. Addition of 30% of Markowicze clay to the mine waste rock, despite worse compaction parameters, causes a decrease in the coefficient of water permeability by one order of magnitude.

ACKNOWLEDGEMENT

This work was supported by PB 3 T09D 068 26 State Committee for Scientific Research Grant.

REFERENCES

Belouschek, P. & Kügler, J.U. 1990. Labortechnische Untersuchungen zur Rißbildung sowie zur Rißsicherung von mineralischen Dichtsystemen für Zwischen- und Deckeldichtungen. In: *Deponie. Ablagerung von Abfällen 4. Berlin: EF-Verlag für Energie und Umwelttechnik* 261–278.

Benson, C., Zhai, H. & Wang, X. 1994. Estimating the hydraulic conductivity of compacted clay liners. *Journal of Geotechnical Engineering*, ASCE 120: 366–387.

Borys, M. & Filipowicz, P. 2004. Possible utilization of mining wastes from the Lubelskie Coal Basin for hydrotechnical structures. *Polish Journal of Environmental Studies* Vol. 13, Suppl. III: 146–148.

Cancelli, A., Cossu, R., Malpei, F. & Offredi, A. 1994. Effects of leachate on the Permeability of Sand-Bentonite Mixtures. In: *T.H. Christensen, R. Cossu, R. Stegmann, Landfilling of Waste: Barriers*. London: 259–293.

Daniel, D. 1987. Earthen liners for land disposal facilities. In: *Geotechnical Practice for Waste Disposal '87*, GSP No. 13, ASCE: 21–39.

Grantham, G. & Robinson, M. 1989. Instrumentation and monitoring of a bentonite landfill liner. *Waste Management and Research* 6: 125–139.

Haller, C.M. & Włoszek, J. 1998. Nowe możliwości wykorzystania odpadów powęglowych. *Wiadomości Górnicze* 7–8: 277–280.

Hartge, K.H. & Horn, R. 1992. Die physikalishe Untersuchung von Böden. *Enke Verlag*, Stuttgard: 116–121.

Holzlöhner, U., August, H., Meggyes, T. & Brune, M. 1995. Landfill Liner Systems. *A State of the Art Report*, Penshaw Press, University of Newcastle.

Horn, R. & Stępniewski, W. 2004. Modification of mineral liner to improve its long-term stability. *International Agrophysics* 18: 317–323.

Jaroszewski, W., Marks, L. & Radomski, A. 1985. Słownik Geologii dynamicznej. *Wydawnictwa Geologiczne*, Warszawa.

Junge, T. & Horn, R. 2001. Branntkalk zur Trocknung von Substraten für den Bau von Deponieabdichtungen – Möglichkeiten und Probleme. *Wasser und Boden* 53: 4–7.

Łuczak-Wilamowska, B. 2002. Neogene clays from Poland as mineral sealing barriers for landfills: experimental study. *Applied Clay Science* 21: 33–43.

Met, I., Akgün, H. & Türkmenoğlu, A.G. 2005. Environmental geological and geotechnical investigations related to the potential use of Ankara clay as a compacted landfill liner material. Turkey. *Environmental Geology* 47: 225–236.

Niedźwiecki, J., Czyż, E.A. & Dexter, A.R. 2003. Effect of particle size distribution, organic matter and bulk density on saturated hydraulic conductivity of arable soils. Edited by A.R. Dexter & E.A. Czyż, *International Workshop 'Soil Physical Quality'* 2–4 Oct. 2003: 61–66.

Pingpank, H., Baumgartl, T. & Horn, R. 1998. Verbesserung mineralischer Deponieabdichtungen durch Austrocknung und Verfüllung der entstandenen Schrumpfrisse. *Kulturtechnik und Landentwicklung* 39: 6–11.

Prashanth, J.P., Sivapullaiah, P.V. & Sridharan, A. 2001. Pozzolanic fly ash as a hydraulic barrier in landfills. *Engineering Geology* 60: 245–252.

Quigley, R.M. & Fernandez, F. 1994. Effect of Organic Liquids on the Hydraulic Conductivity of Natural Clays. In: *T.H. Christensen, R. Cossu, R. Stegmann, Landfilling of Waste: Barriers*, E & FN Spon, London: 203–218, ISBN 0419159908.

Quigley, R.M., Fernandez, F. & Row, R.K. 1988. Clayey barrier assessment for impoundment of domestic waste leachate (Southern Ontario) including clay-leachate compatibility by hydraulic conductivity testing. *Canadian Geotechnical Journal* 25: 574–581.

Rowe, R.K. 1994. Diffusive Transport of Pollutants through Clay Liners. In: *T.H. Christensen, R. Cossu, R. Stegmann, Landfilling of Waste: Barriers*, London, E & FN Spon, London: 219–245, ISBN 0419159908.

Seymour, K.J. & Peacock, A.J. 1994. Quality Control of Clay Liners. In: *T.H. Christensen, R. Cossu, R. Stegmann, Landfilling of Waste: Barriers*, London, E & FN Spon, London: 69–79, ISBN 0419159908.

Skarżyńska, K.M. 1997. Odpady powęglowe i ich zastosowanie w inżynierii lądowej i wodnej. *Agriculture University Press,* Kraków.

Stępniewski, W., Wysocka, A., Rożej, A., Węgorek, T., Wiśniewska, M., Kotowska, U. & Nosalewicz, M. 2004. Utlenianie metanu w warunkach biologicznej rekultywacji składowisk odpadów komunalnych z wykorzystaniem przywęglowej skały płonnej. *Monografie Komitetu Inżynierii Środowiska PAN*, Vol. 19, ISBN 83-89293-40-4.

Sullivan, P.J. & Sobek, A.A. 1982. Laboratory weathering studies of coal refuse. *Minerals and Environment* 4: 9–18.

Tay, Y.Y., Stewart, D.I. & Cousens, T.W. 2001. Shrinkage and desiccation cracking in bentonite – sand landfill liners. *Engineering Geology* 60: 263–274.

Twardowska, I., Szczepańska, J. & Witczak, St. 1988. Wpływ odpadów górnictwa węgla kamiennego na środowisko wodne. Ocena zagrożenia, prognozowanie, zapobieganie. *Prace i studia* 35, PAN.

Wysocka, A., Stępniewski, W. & Horn, R. 2004. Effect of liming of a clay material on its hydraulic conductivity and shrinkage potential. *Polish Journal of Environmental Studies* 13, Suppl. 3: 301–304.

Environmental Engineering – Pawłowski, Dudzińska & Pawłowski (eds)
© 2007 Taylor & Francis Group, London, ISBN13 978-0-415-40818-9

A concept for the optimisation of the hydraulic transport of sludge

Zbyslaw Dymaczewski, Joanna Jez-Walkowiak & Marek M. Sozanski
Institute of Environmental Engineering, Poznan University of Technology, Poznan, Poland

ABSTRACT: Vast amounts of sludge are produced during water and wastewater treatment processes. One of the important issues is its hydraulic transport. Thickened sludge often has non-Newtonian properties, which appear under a certain value of the water content. In the paper an optimisation concept of hydraulic sludge transport, based on minimization of energy losses of flow of sludge solid phase, is presented. The main factor affecting the optimisation of sludge transport in pipelines is its water content. This is a result of its influence on rheological parameters, as well as on the flow rate. The proposal of such a formulated problem solution is presented in this paper.

Keywords: Hydraulic transport, non-Newtonian fluids, optimisation, sludge.

1 INTRODUCTION

One of the most intricate problems, especially for large water and wastewater treatment plants, is the disposal of sludge produced during treatment processes. Finding the proper solution for this problem in the design phase is one of the conditions to guarantee the technical efficiency of waterworks to be constructed. The dominant role of water content in sludge is a reason why gravital and mechanical thickening are common processes applied in sludge disposal systems. There is deviation from Newtonian properties of thickened sludge during its flow (Dentel 1997, Slatter 1997). Omitting this fact leads to calculation errors causing severe exploitation problems in the entire sludge disposal system. Establishing the proper hydraulic parameters for sludge pipe transport is crucial for the project conception. A method for the optimal design of thickened and partly dewatered sludge hydraulic transport has not yet been established (Slatter 2004). One of the reasons is the lack of rational criteria allowing an optimum design of pipe systems for sludge transport as a part of the entire sludge disposal system. This issue is a subject of this paper.

2 AIM AND CONCEPTION

Difficulties in optimising technologies applied for water and wastewater treatment plants are related to existence of local optimisation criteria for unit processes as well as their influence and correlation with general optimisation criterion. The basic requirement for general optimisation criterion is obtaining the maximum economic efficiency. There are ongoing research on formulating and applying technological criteria for unit processes and devices (Slatter 2004).

The aim of the research problem is establishing an optimum state for sludge pipe transport in sludge disposal and thickening systems according to accepted criterion of minimization energy loss value. The goal of optimisation is establishing a value of water content of sludge guarantying the lowest energy loss during its pipe transport, fulfilling the desired mass capacity of the solid phase.

The idea of the research task results from the formulated thesis:

Because each change of water content always results in change of volume, described by the strictly increasing function and change of internal friction resistance by strictly decreasing function, so there is always such water content value, which minimizes energy loss value of the sludge motion in tube.

The presented problem concerns typical sludge disposal installations in waterworks systems, with thickening process included (Vesilind 2004, AWWA/ASCE 2005). An example scheme is shown in Figure 1. Gravity and mechanical thickening is applied to change water content to the optimum value, minimizing the head loss value of its pipe transport. So the thickening devices are controlling object of sludge disposal systems.

Establishing the optimisation criterion and its solution should be based on the assumption of interaction between parameters of pipe sludge transport and effects of gravital and mechanical thickening. The

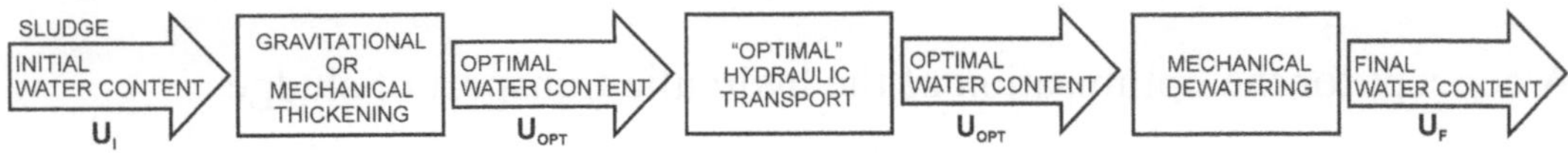

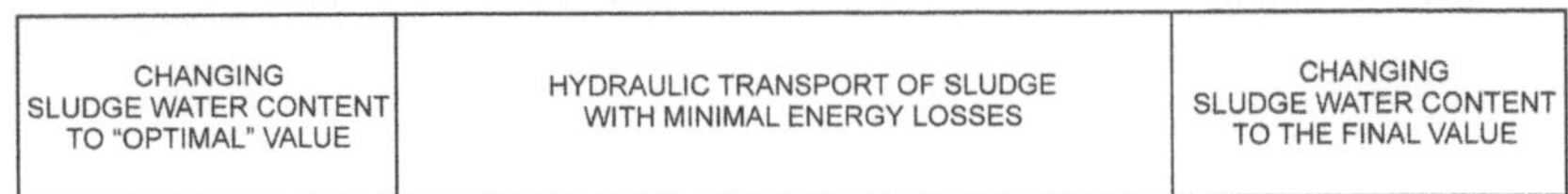

Figure 1. Conception of solution for optimal hydraulic sludge transport.

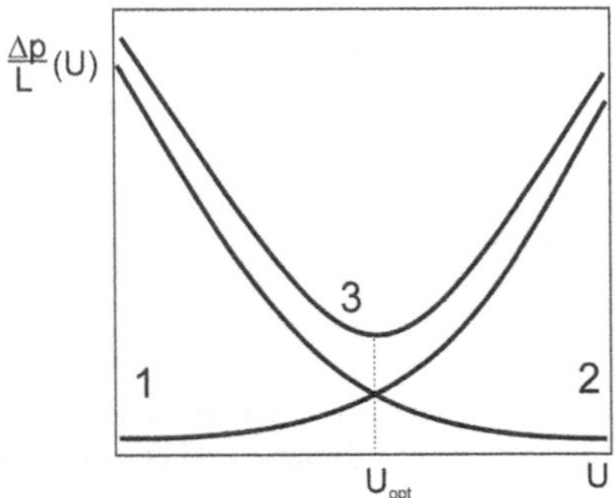

Figure 2. The influence of water content (U) on unit pressure loss ($\Delta p/L$) of hydraulic sludge transport. 1 – relationship describing the influence of sludges flow rate (v) on hydraulic pressure head loss: $(\Delta p/L)_1 = f_1[v(U), \ldots]$. 2 – relationship describing the influence of rheological parameters on hydraulic pressure head loss: $(\Delta p/L)_2 = f_2[\eta_{pl}(U), \tau_0(U), \ldots]$. 3 – resultant relationship describing the influence of sludges flow rate and rheological parameters on hydraulic pressure head loss: $(\Delta p/L)_3 = f_3[v(U), \eta_{pl}(U), \tau_0(U), \ldots]$.

pipes connect objects and devices for solid phase separation, what makes the assumption rationally acceptable.

The change of thickening process efficiency leads to exactly described changes of water content and volume of sludge causing the change of sludge flow rate in pressure pipe. It means that with a given mass capacity of solids the increase or decrease of water content will respectively increase or decrease the volume capacity and the flow rate (Figure 2, curve 1). Water and sewage sludges with concentrations higher than comprimation point, are positioned among non-Newtonian liquids and their characteristics cannot be described with equations applied for water and wastewater flows. These sludges are characterized by complex internal structure with features of homogenous systems, which makes it possible to apply method of phenomenological rheology for determination of flow characteristics.

Hydro-mechanical properties of sludges, as non-Newtonian liquids, are described by flow models substituting rheological parameters for viscosity (Slatter 2004, Spinosa and Wichmann 2004). For rheological properties of sludge, diverse rheological models can be used. The choice of a certain model is based on experimental work (Slatter 1997, Sozanski et al. 1997). For this paper discussion of the two-parameter Bingham model was selected.

$$\tau - \tau_0 = \eta_{pl} \frac{dv}{dr} \tag{1}$$

Parameters included in this model: yield stress – τ_0 and plastic viscosity – η_{pl} are dependent on sludge water content, determining the value of pressure hydraulic loss of pipe transport (Figure 2, curve 2).

3 METHODS

The goal of the presented problem is to find the optimum water content of sludge minimizing the energy loss of pipe sludge transport,

$$\frac{\Delta p}{L}\left(U_{opt}\right) = \min \frac{\Delta p}{L}(U) \tag{2}$$

with a given capacity of solid phase (sp) particles (3).

$$\frac{dM_{sp}}{dt} = m_{sp} \tag{3}$$

The relation to be found: $\frac{\Delta p}{L}(U)$ can be derived from a general Darcy-Weisbach formula:

$$\frac{\Delta p}{L} = \lambda\left(Re_{gen}, \frac{k}{D}\right) \cdot \frac{v^2 \cdot \rho_{sl}}{2D} \tag{4}$$

where: D – pipe diameter, ρ_{sl} – density of sludge, λ – friction coefficient.

The exact form of this formula is a function of rheological parameters of sludge and a value of general Reynolds number (Re_{gen}). Non-Newtonian properties appear in sludge below the certain value of water content so-called critical. The value of critical water content depends on the properties of sludge flocs structure. For water and wastewater treatment sludges, the critical value of water content (U_{KR}) is in the range of 94–98%.

For sludges with Newtonian properties, the tubes for hydraulic sludge transport should be designed applying rules for water transport design, but in place of water viscosity and density, appropriate values for the sludge should be used. In the range of water content lower than critical, sludge belongs to the non-Newtonian fluids, with properties described by the rheological model developed by experimental research.

The value of general Reynolds number (Re_{gen}) calculated according to the Bingham model (1) is described by the equation:

$$Re_{gen} = \frac{v \cdot D \cdot \rho_{sl}}{\left(\eta_{pl} + \frac{D \cdot \tau_0}{6 \cdot v}\right)} \tag{5}$$

The range of laminar flow of sludges is specified by the following criterion (6):

$$Re_{gen} \leq 2320 \tag{6}$$

The pipe system can be precisely calculated for laminar flow transport of thickened sludge (Govier and Aziz 1972) by applying the following formula 7:

$$\lambda = \frac{64}{Re_{gen}} \tag{7}$$

combining the general Reynolds number and friction coefficient (λ) from the Darcy-Weisbach formula.

Rheological parameters also influence the pipe sludge transport in the zone of hydraulically smooth flow. In that case friction coefficient λ is empirically estimated and the results are approximated using Re number. For practical aspects very useful is the Blasius formula, expanded by Thomas for liquids of Bingham properties (Govier and Aziz 1972).

4 RESULTS AND DISCUSSION

The equation describing energy loss in pipe sludge transport can be generally formulated, depending on the chosen rheological model, as a function consisting of four or five internal functions and a variable pipe diameter:

$$\frac{\Delta p}{L} = f_1\left(D, v(U), \eta_{pl}(U), \tau_0(U), \rho_{os}(U)\right) \tag{8}$$

The water content of sludge is treated as an independent continuous variable and the pipe diameter is considered as a constant. The flow rate is a function of water content according to the following equation (Dymaczewski et al. 2005):

$$v(U) = v_{KR} \cdot \frac{1 - U_{KR}}{1 - U} \tag{9}$$

To solve the optimisation problem it is necessary to establish the remaining internal functions in equation 8. Conversion of these parameters into their functions of water content means that these parameters are no longer independent non-controllable variables and become dependent controllable variables. Control of rheological parameters is possible due to precise change of water content of sludge during the thickening process.

The formulas of the internal functions $\eta_{pl}(U)$, $\tau_0(U)$ are obtained as from research on values of rheological parameters of sludges with various water contents at the same temperature and further approximation of the results obtained. Establishing a formula describing the influence of water content on the density of sludge also needs experimental measurements.

To solve an optimisation problem it is necessary at first to conduct a rheological experiment supplemented with research on sludge structure, granulometry parameters, chemical composition, water content and temperature. The goal of the rheological experiment is to simulate the flow of sludge in a viscosimeter over a broad range of deformation rates and shear stresses. The change of water content, temperature and deformation rate of sludge allows its behaviour at different stages of flow to be observed.

The relationships between values of rheological parameters and water content or solid phase of sludge are usually described as exponential or potential functions as seen following equations:

$$\eta_{pl} = \eta \cdot \alpha \cdot \exp\left[\beta\left(U_{KR} - U\right)\right] \tag{10}$$

$$\tau_0 = \chi \cdot \exp\left[\varepsilon\left(U_{KR} - U\right)\right] \tag{11}$$

where: α, β, χ, ε are experimentally determined constants and depend on the kind and temperature of the sludge.

The solution of the optimising problem is based on determining the optimum value of water content of sludge that minimizes the value of the function described by equation 2 supplemented with equations 3–11. These relations consist of potential and exponential expressions. The rational and exponential functions as well as their derivatives are continuous, so the presented criterion of optimisation and established limitations are continuous and differentiable in the range of established values of their variables and parameters. Continuity of the accepted criterion of optimisation and its definity in a closed and bounded set are, according to the Weierstrass statement, conditions for the existence of solution to the optimising problem. The forms of the optimisation criterion and its limitations classify the problem as a non-linear programming issue. Further analysis can be done according numerical methods and be presented in graphical form. Based on differential calculus analysis it can be proved that both the objective function and the established limitations are convex functions. In such a case, the local optimum is also a global optimum. This fact allows further simplification of the calculations. The recognition of the convex goal function and accepted limitation makes it easier to solve an optimising problem. It should be pointed that it is impossible to find, in a finite number of algebraic operations, the analytical solution describing the exact value of U_{opt} as an expression consisting of the coefficients of the criterion equation.

5 CONCLUSIONS

As a conclusion it should be pointed that water content of sludge is one of the basic factors, which makes the optimisation of sludge pipe transport possible. It is a result of the influence of water content on flow rate in the pipe as well as on the rheological parameters of sludge. This issue reflects the proposition of the paper. Practically, such an optimisation is only possible in large-scale sludge disposal systems, where it is possible to change the water content of sludge in gravital and mechanical thickening. The criterion of optimising sludge transport and a method for solving this problem are presented in this paper, pointing to the indirect influence of rheological parameters and pipe diameter on the value of the optimum water content. Water content is a variable, which enables intentional control of the values of rheological parameters.

Temperature and composition of sludge are non-controllable variables, which can be only directly measured. The general method of finding the optimum water content of sludge should allow solving this issue for cases in which the occurrence of laminar as well as turbulent flow is possible. Then the energy loss of pipe transport should be estimated according to the Darcy-Weisbach formula, determining λ and Re according to a properly chosen rheological model. Experimentally establishing a rheological model is a primary issue for optimising sludge pipe transport. Calculations for turbulent flow require experimental values of the friction coefficient – λ. The subject of optimisation of sludge pipe transport has not yet been analysed in the literature.

The main conclusion is that there is an optimum value of water content of sludge that minimizes energy loss during sludge pipe transport. The value of the optimum water content of sludge grows with the value of rheological parameters as well as with the pipe diameter.

ACKNOWLEDGEMENTS

The authors acknowledge the financial support of the Polish Ministry of Education and Science (project 4 T07E 085 26).

REFERENCES

AWWA/ASCE 2005. Water Treatment Plant Design, 4th edition, McGraw-Hill

Dentel S. 1997. Evaluation and role of rheological properties in sludge management, *Wat.Sci.&Technol.* Vol 36 No 11 pp 1–8.

Dymaczewski Z. Jez-Walkowiak J. Sozanski M.M. 2005. Optimisation of sewage sludge hydraulic transport – problem's formulation and solving conception. *Proc. IWA spec. conf. Nutrient Management in Wastewater Treatment Processes and Recycle Streams, Krakow, 19–21 September 2005*: 1047–1051.

Govier G.W. i Aziz K. 1972. The Flow of Complex Mixtures in Pipes, New York, Van Nostrand Reinhold Company.

Slatter P.T. 1997. The rheological characterisation of sludges. *Wat.Sci.&Technol.* Vol 36 No 11 pp 9–18.

Slatter P.T. 2004. The hydraulic transportation of thickened sludge. *Water SA* vol 30 no 5 spec. issue pp 66–68.

Sozanski M. Kempa E. Grocholski K. Bien J. 1997. The rheological experiment in sludge properties research. *Wat.Sci.&Technol.* Vol 36 No 11 pp 69–78.

Spinosa L. Wichmann K. 2004. Sludge characterization: the role of physical consistency. *Wat.Sci.&Technol.* Vol 49 No 10 pp 59–65.

Vesilind P.A. 2004. Wastewater Treatment Plant Design, IWA Publishing, WEF.

Environmental Engineering – Pawłowski, Dudzińska & Pawłowski (eds)
 ISBN13 978-0-415-40818-9

The activated sludge process simulation – a preparation stage for Poznan WWTP

Zbyslaw Dymaczewski
Institute of Environmental Engineering, Poznan University of Technology, Poznan, Poland

ABSTRACT: Application of computer simulation of the activated sludge process was presented for the example of Poznan WWTP. The presented preparation stages include the model selection, data collection, plant model construction and calibration. As a result, a calibrated model, ready for solving specific tasks, was obtained. The ASM 2d model was chosen as the best for complex C, N and P removal. The pilot research enabled collecting a suitable data set, necessary for model construction and calibration. Selected model parameters were determined. The applied calibration method consisted of combining simulations of the steady state pilot plant model and kinetic experiments performed in batch reactors. To achieve the best possible match of simulations with collected data, the model parameters were adjusted. Using the calibrated model of the pilot plant, a model of the full scale WWTP was built. That model was applied for the plant design verification and for drawing up initial operational guidelines.

Keywords: Wastewater treatment, activated sludge, computer simulation, ASM 2d, calibration.

1 INTRODUCTION

In the last two decades a significant increase of interest in the mathematical modelling of wastewater treatment processes was observed. The dynamic development of personal computers has had an impact on applying simulators as practical tools for engineers and scientists. Better and better knowledge of processes of pollutant transformations taking place during biological wastewater treatment, allows improvement of mathematical models. Because of newly implemented regulations, wastewater treatment systems must meet more and more stringent quality requirements, especially in the field of nutrient removal. Characteristic features of new systems are process complexity, different recirculations, different functions of individual reactors, dependency of treatment effectiveness on C:N:P ratios and maintaining the process variables within a small range of optimal values. Applying simulation programs began setting new design, operation and management standards for WWTPs in developed countries.

The development of modern wastewater treatment technologies in Poland can be dated as beginning in the nineties. It was largely spurred on by the introduction, in 1991, of the decree of Minister of Environment Protection, Natural Resources and Forestry, where, for the first time, the requirement of nutrient (nitrogen and phosphorus) removal was set. The new regulations, adapted to EU standards (decree from 29.11.2002, valid from 1.01.2003 till 30.06.2004 and now in force decree from 8.07.2004) as a matter of fact decreased the requirements for some smaller WWTPs but for most of them the requirements increased, especially regarding nitrogen removal.

Some of limitations of a wider usage of mathematical modelling of wastewater treatment processes in Poland are relatively poor experience, high complexity of the models, necessity of widening the range of analytical analyses and difficulties of correct calibration.

The practical application of computer simulation is presented using, as example, the pilot plant for Poznan Central WWTP. The steps of the simulation preparation stage (Fig. 1) include the mathematical model selection, appropriate data selection, building the plant model and its calibration. The calibrated model can be used to suggest best practices for the plant operation and to solve exploitation problems.

2 THE PILOT RESEARCH AND AIM OF SIMULATIONS

Regulations introduced in 1991 caused the need of modernisation of the Poznan Central WWTP. At that time, the plant consisted only of mechanical wastewater treatment and primary sludge stabilisation process

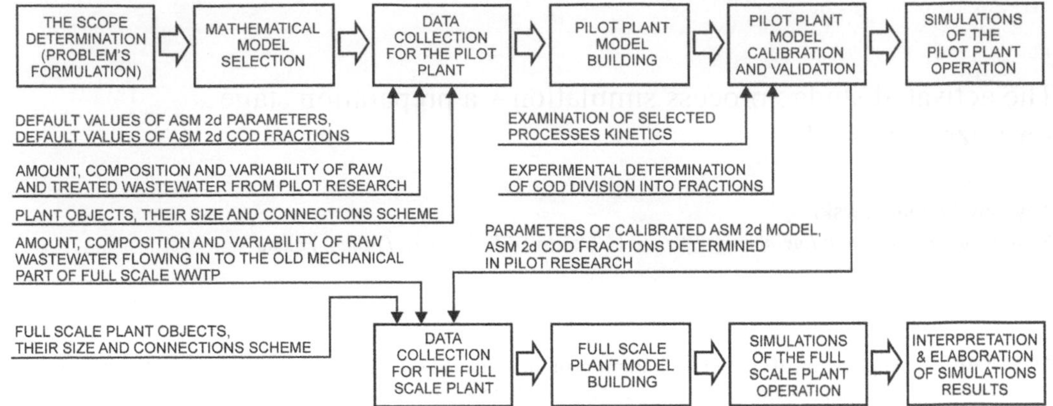

Figure 1. Stages of simulation preparation for the Poznan WWTP project.

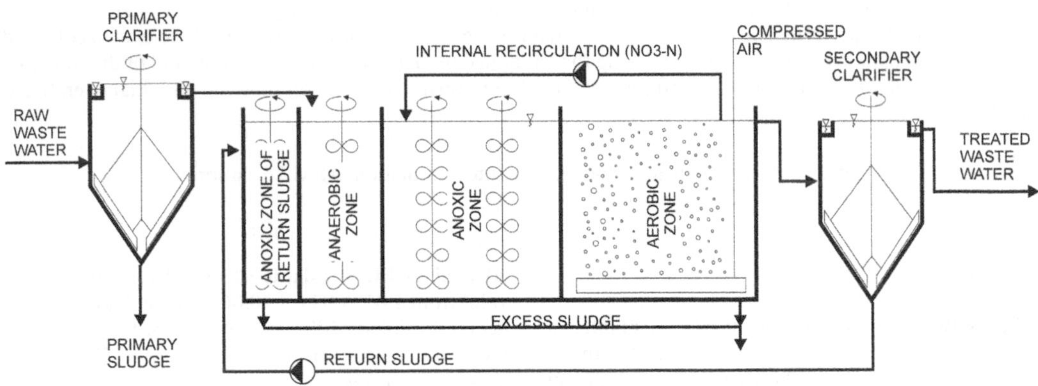

Figure 2. A technological scheme of the pilot plant for the Poznan WWTP.

utilities and had to be extended by modern biological organics and nutrient removal system. As in Poland there were no experience in designing and operation of such systems it was decided to carry out pre-design pilot research at Poznan WWTP. The technological start-up of the pilot plant took place in 1996.

During pre-design research, the pilot plant was operated under varied technological conditions such as the sludge retention time, F/M ratio, internal and external recirculations and different proportions of anoxic and aerobic zones within the biological reactor (Dymaczewski et al. 1999). A technological scheme of the pilot plant is presented in Figure 2. The collected experimental data enabled building a model of the pilot plant and its calibration. Based on the calibrated model of the pilot plant, a full-scale plant model was developed and series of simulations were performed.

The first goal of the simulation analysis was to check the plant efficiency considering the assumed design values: amount and composition of wastewater, volumes of tanks, recirculations and technological parameters. The second goal was to find the optimal strategy of plant operation in various conditions.

3 MATHEMATICAL MODEL OF THE PROCESS

The currently applied models vary in complexity and number of processes taken into account. Among the models currently widely used in the world are the following:

- ASM 1 model (IWA 2000), which takes into account carbon removal, nitrification and denitrification processes;
- ASM 2 model and its improved version ASM 2d (IWA 2000), which takes into account carbon removal, nitrification and denitrification processes, biological and chemical phosphorus removal;
- ASM 3 model (IWA 2000), which takes into account carbon removal, nitrification and denitrification

processes and has a structure allowing addition of new processes;

- General model (Barker & Dold 1997), similar to ASM 1, with added phosphorus removal module.

The complexity level of the model depends on how many processes are included. The complexity of a model causes an increase in the number of kinetic and stoichiometric parameters. The wastewater characteristic is also more complex, so it is very important to choose the adequate model. The chosen model should include processes present on the given WWTP and important for the planned simulation analysis. Another important issue is to check a model's limitations and its range of applicability.

For the simulation of Poznan WWTP, the most complex ASM model was selected – ASM 2d. The model in its original version does not predict conditions in the biological reactor well when a relatively high amount of VFA is in the influent and not all VFA are utilised in the anaerobic zone – these limitations were also noticed by the model's authors (IWA 2000). Because of this, a small modification to the model was introduced to correct this problem. The modified equation is shown below (1):

$$\rho_{10} = q_{PHA} \cdot \frac{S_A}{K_A + S_A} \cdot \frac{K_{O2}}{K_{O2} + S_{O2}} \cdot \frac{K_{NO3}}{K_{NO3} + S_{NO3}} \cdot \frac{S_{ALK}}{K_{ALK} + S_{ALK}} \cdot \frac{X_{PP}/X_{PAO}}{K_{PP} + X_{PP}/X_{PAO}} \cdot X_{PAO} \quad (1)$$

where: K_A, K_{O_2}, K_{NO_3}, K_{ALK}, K_{PP} are the appropriate kinetic constants, S_{O_2}, S_{NO_3}, X_{PP}, X_{PAO}, S_{ALK} are the concentrations of dissolved oxygen, nitrate, polyphosphate, PAO and alkalinity respectively.

The modification includes adding the appropriate switching functions ($K_{O_2}/(K_{O_2} + S_{O_2})$ and $K_{NO_3}/(K_{NO_3} + S_{NO_3})$), to the original rate (ρ_{10}) expression for PHA accumulation.

The ASM 2d model was selected for two reasons. Firstly, the wastewater treatment plant was designed for integrated C, N and P removal. Secondly, it was important in the simulation analysis to determine the efficiency of removal of all these pollutants (C, N and P) under various plant operation conditions.

4 DATA FOR THE PILOT PLANT MODEL

The main problem occurring when simulating real objects is to collect sufficient data to calibrate and verify the model. This problem, in many cases, leads to decreasing the number of independent data, conditioning their selection from existing empirical data and existence of clear correlations with predicted variables. Such behaviour however, increases the random nature of the prediction. The wastewater characteristic, adapted to the model specification, is very extended and the model parameters and coefficients result from specially designed experiments. Adequate research is difficult and expensive.

Determination, which of many parameters have significant influence on simulation results and characterization of that influence are important in decision meaning during the phase of planning the analytical analyses for mathematical modelling needs. Optimising the research scope to the necessary minimum gives economical savings and enables the labour demand of the whole project to be decreased.

The pilot research created an excellent possibility for performing extended tests. The result of pilot research was a rich dataset, which included data on the daily and seasonal diversity of pollutant concentrations in raw and treated wastewater and appropriate technological parameters of the pilot plant. Investigation of the diurnal variation of raw wastewater composition was carried out for 1.5 months. Test series consisted of 10 measurements with 12 samples taken every 2 hours. The average values were assumed as representative for the whole period. The averaged daily variations of COD, suspended solids, total nitrogen and total phosphorus are given in Table 1.

Three series of tests were done to evaluate COD fractions of ASM 2d model. The resulting data are presented in Table 2.

It was assumed that X_A, X_{PAO} and X_{PHA} fractions are equal or close to zero and together with X_H fraction make ca. 10% of total COD. Their values were taken by decreasing the X_S fraction. The other fractions were indicated analytically. Measurement methods were based on works of Roeleveld and Kruit (1998) and Baetens (2000). The details are given in Table 3.

Several kinetic tests were conducted in batch reactors (nitrification, denitrification, release and uptake of phosphorus). The obtained values were used for calculation of the processes rates. Results of tests are presented in Table 4.

Calculated rates of the individual processes are generally in the range of values given in the literature (Kristensen et al. 1992, Kujawa-Roeleveld 2000, Brdjanovic 1998, Baetens 2000).

5 CONSTRUCTION AND CALIBRATION OF THE PILOT PLANT MODEL

During pilot plant model construction, a flow sheet was built (Fig. 3). It covers all objects and their connections, flows of wastewater and sludges, which are to be taken into account in simulations. Based on collected data, the size of each object and technological conditions were determined.

Table 1. Diurnal variation of raw wastewater composition (individual values referred to the average = 1).

Hours	COD	Suspended solids	Total nitrogen	Total phosphorus
0.00 ÷ 2.00	1.07	0.99	1.03	1.07
2.00 ÷ 4.00	0.94	0.86	0.92	0.97
4.00 ÷ 6.00	0.82	0.72	0.87	0.81
6.00 ÷ 8.00	0.78	0.70	0.74	0.72
8.00 ÷ 10.00	0.90	0.66	0.91	0.86
10.00 ÷ 12.00	1.02	1.27	1.08	1.01
12.00 ÷ 14.00	1.04	1.36	1.19	1.05
14.00 ÷ 16.00	1.08	1.25	1.10	1.09
16.00 ÷ 18.00	1.08	1.12	1.04	1.14
18.00 ÷ 20.00	1.05	0.95	1.10	1.09
20.00 ÷ 22.00	1.13	1.08	1.04	1.11
22.00 ÷ 24.00	1.09	1.05	0.97	1.06

Table 2. Division of COD into fractions in the WWTP influent (values as shares of total COD).

Parameter	Period 1	Period 2	Period 3	ASM2d*
Share of S_F fraction	0.167	0.192	0.150	0.115
Share of VFA (S_A fraction)	0.162	0.152	0.277	0.078
Share of S_I fraction	0.080	0.065	0.074	0.115
Share of X_S fraction	0.371	0.388	0.409	0.481
Share of X_I fraction	0.120	0.103	0.040	0.096
Share of X_H fraction	0.093	0.093	0.043	0.115
Share of X_A fraction	0.005	0.005	0.005	0
Share of X_{PAO} fraction	0.001	0.001	0.001	0
Share of X_{PHA} fraction	0.001	0.001	0.001	0

* Fractions were calculated based on "typical" wastewater composition provided for ASM 2d model (IWA 2000).

Table 3. The determination of COD fractions for the ASM 2d model applied in pilot research at Poznan WWTP.

Measured parameter	method of parameter measurement	Equation parameter	Model
Influent			
COD	Acc. to Polish Norm	$COD = S_A + S_F + S_I + X_I + X_S$	X_I
BOD_5	Acc. to Polish Norm	$BCOD = S_A + S_F + X_S$	X_S
COD_{filtr}	Coagulation $ZnSO_4$ + filtration	$COD_{filtr} = S_A + S_F + S_I$	S_F
VFA	Chromatographic	$S_A = COD_{VFA}$	S_A
N_{Kj}	Acc. to Polish Norm	$S_{NH_4} = N_{Kj} - (i_{NSI} * S_I + i_{NSF} * S_F + \cdots\cdots)$	Calibr. $i_{N,?}$
P_{og}	Acc. to Polish Norm	$S_{PO_4} = P_{og} - (i_{PSI} * S_I + i_{PSF} * S_F + \cdots\cdots)$	Calibr. $i_{P,?}$
$N\text{-}NH_4$	Acc. to Polish Norm		S_{NH_4}
$N\text{-}NO_3$	Acc. to Polish Norm		S_{NO_3}
$P\text{-}PO_4$	Acc. to Polish Norm		S_{PO_4}
Effluent			
COD_{filtr}	Coagulation $ZnSO_4$ + filtration	$S_I = COD_{filtr-effl}$	S_I

To proceed with the calibration process, three periods were selected from the whole pilot study. Characteristics of raw and treated wastewater measurements were used. In the first step, default values of stoichiometric and kinetic parameters of ASM 2d model were used (IWA 2000).

The aim of calibration was to achieve the best possible match of simulations with analytical data and kinetic tests. The following contaminants were taken into account: COD, BOD_5, suspended solids, different forms of nitrogen (organic N, ammonia, nitrates + nitrites), total phosphorus and soluble

Table 4. Rates of nitrification, denitrification, release and uptake of phosphorus obtained in kinetic tests.

Test	Unit	Process rate Period 1	Period 2	Period 3
Ammonia utilisation rate (AUR)	mg N/g SS·h	2.2	1.8 ÷ 4.5	2.3 ÷ 2.6
Nitrate utilisation rate (NUR)	mg N/g SS·h	–	2.3 ÷ 2.4	1.8 ÷ 5.5
Phosphorus release rate (PRR)	mg P/g SS·h	–	7.5 ÷ 17.2	15.4 ÷ 25.4
Phosphorus uptake rate (PUR) aerobic conditions	mg P/g SS·h	–	33.0 ÷ 34.4	24.2 ÷ 70.6
Phosphorus uptake rate (PUR) anoxic conditions	mg P/g SS·h	–	–	7.5 ÷ 12.9

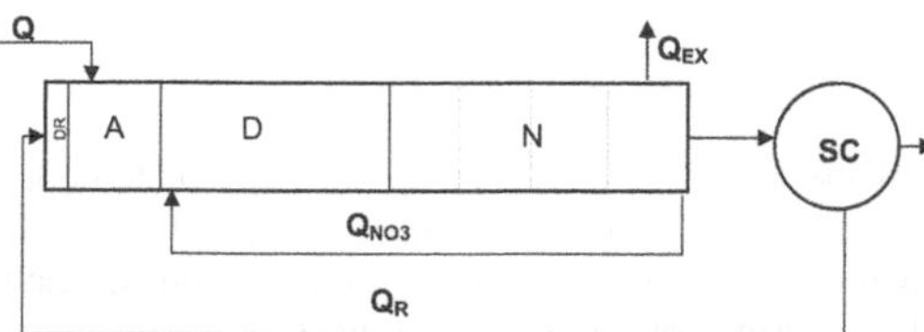

Figure 3. Flow sheet of Poznan pilot plant, used for computer simulations. DR – return sludge denitrification zone, A – anaerobic zone, D – denitrification zone, N – nitrification zone, SC – secondary clarifier, Q – influent, Q_{NO_3} – internal recirculation, Q_R – return sludge, Q_{EX} – excesss sludge.

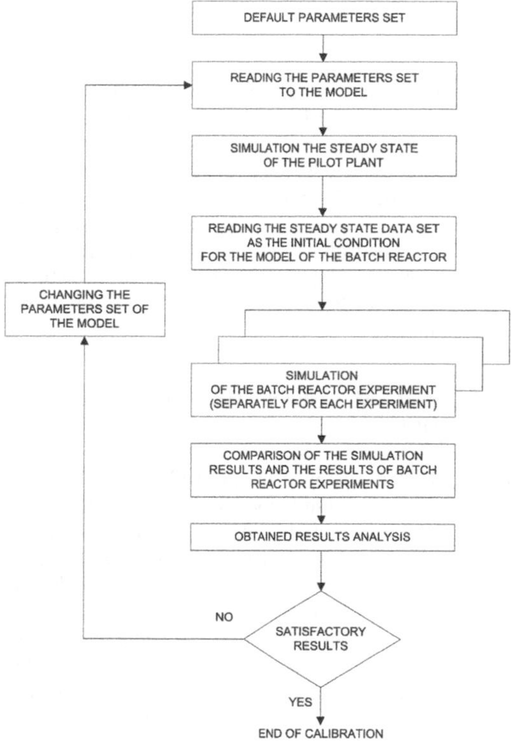

Figure 4. Second step of calibration procedure for the Poznan pilot plant.

orthophosphate. Also the sludge concentration in the biological reactor, at a certain SRT was considered.

Calibration was conducted in two steps. In the first step suspended solids, COD and organic nitrogen in the effluent, sludge concentration in the reactor and organic fraction of N and P in raw sewage were adjusted. None of the kinetic parameters was changed at this step. At first, the value of suspended solids in the effluent was calibrated by changing the efficiency of secondary clarifier. Next, COD and BOD_5 of the effluent as well as the sludge concentration in the biological reactor were adjusted by changing proportions of previously analytically determined COD fractions in the influent. At the end of this step, stoichiometric coefficients "$i_{N,?}$; $i_{P,?}$" for recalculating COD fractions to corresponding N and P fractions were iteratively adjusted.

In the second step of the calibration procedure the results of kinetic tests were used. The tests included kinetics of ammonia utilisation rate (AUR), nitrate utilisation rate (NUR), phosphorus release rate (PRR) and phosphorus uptake rate (PUR). The tests were performed in laboratory batch reactors filled with activated sludge from the biological reactor. Model parameters were changed iteratively (with results analysis after each iteration).

Each batch experiment simulation started from initial loading conditions. These conditions corresponded with the steady state of the pilot plant, for the actual set of kinetic and stoichiometric parameters. Next, simulations of the selected batch experiment were conducted. Simulation results were compared to experimental data. The next step was changing the value of the selected model parameter, simulation of the steady state of the pilot plant (with the modified model parameters set), loading new initial conditions etc. The scheme of the calibration procedure is given in Figure 4. The calibration procedure was applied independently to two research periods. For the

Table 5. Comparison of the results from the start-up period with the simulations for Poznan WWTP, on the basis of design criteria and the actual legislative regulations.

Parameter	Unit	WWTP start-up	Simulation results	Design criteria	Actual Polish limits
BOD_5	g O_2/m^3	6.7	3.04	15	15
COD	g O_2/m^3	52.1	38.83	150	125
Suspended solids	g/m^3	11.5	8.24	50	35
Ammonium nitrogen	g N/m^3	1.7	1.41	6	–
Organic nitrogen	g N/m^3	4.2	3.13	–	–
Nitrite and nitrate	g N/m^3	4.9	6.15	–	–
Total nitrogen	g N/m^3	11	10.7	30	10
Total phosphorus	g P/m^3	0.49	0.53	1.5	1.0
Orthophosphate	g P/m^3	–	0.11	–	–
MLSS in biological reactor	g/m^3	3900	4062	3500–4000	–

first period eight kinetic parameters were changed ($\eta_{NO_3} = 0.8$; $\eta_{fe} = 0.1$ for hydrolysis, $K_{fe} = 20$ for heterotrophs, $q_{PHA} = 7$; $q_{PP} = 5$; $K_{MAX} = 0.2$ for PAOs and $\mu_{AUT} = 0.8$; $K_{NH_4} = 0.2$ for nitrifiers). For the second period, two stoichiometric ($Y_{PO_4} = 0.2$; $Y_{PHA} = 0.15$) and ten kinetic parameters ($K_h = 5$; $\eta_{fe} = 0.1$ for hydrolysis, $K_{O_2} = 1$; $K_F = 2.9$; $K_A = 2.9$ for heterotrophs, $q_{PHA} = 8$; $q_{PP} = 5$; $\eta_{NO_3} = 0.35$; $K_{MAX} = 0.36$ for PAOs and $\mu_{AUT} = 0.8$ for nitrifiers) were changed.

6 PRACTICAL USE OF THE CALIBRATED MODEL

The calibrated model of the pilot plant was the base for building the simulation model of the technical scale WWTP. Using this model, the design of the Poznan WWTP was evaluated and the initial operational conditions were worked out. The conducted simulations included a couple of tasks, such as testing the dependence of treatment efficiency on the sludge age, or from internal recirculation in winter and in summer time. Some of the simulation results for the full scale WWTP are presented in the literature (Dymaczewski & Szetela 2005).

Data collected during the WWTP start-up period (Jaroszynski 2003), were compared with simulation results. The selected results from the start-up period, design criteria and the actual legislative regulations are summarised in Table 5.

Analysing data from Table 5, one can conclude that Poznan WWTP meets the design criteria and the actual stringent regulations (except for a small excedance of N-total). Comparison of the simulation and operational data shows high convergence.

The general conclusion is that the full scale WWTP model built on the basis of pilot research data and a pilot plant calibrated model is correctly calibrated and its predictions satisfactorily reflect the full scale WWTP operation in different conditions. Such model is a good tool for process optimisation and elucidating procedures for different situations (e.g. unexpected plant overload or removing an object for maintenance or repair).

7 CONCLUSION

The dynamic development of personal computers has had an important influence on the increased interest in the mathematical modelling of wastewater treatment processes. Computer simulators began setting new standards in designing, operation and management of wastewater treatment plants in highly-developed countries. Some attempts at practical use of computer simulations in that field are also carried out in Poland. Poznan WWTP can be taken as an example. Preparing the plant model for simulation experiments is a complex and laborious task. It includes mathematical model selection, appropriate data collection, model building and its calibration.

For the simulation of the Poznan WWTP, the ASM 2d model, which includes organics removal, nitrification, denitrification and phosphorus removal, was selected. The model in its original version did not predict the conditions in the biological reactor particularly well. The reason was the relatively high amount of VFA in raw wastewater. A small modification to the model was made to correct this problem. The data set necessary for model building, its calibration and conducting simulations was divided into three groups: information about the object, extended wastewater characteristics and determination of kinetic and stoichiometric model parameters. All the data were collected during pre-design pilot research. To utilise the model for quantitative analysis and solving operational problems, its calibration and validation was necessary. Using the model, WWTP design was verified and initial operational guidelines were drawn up. The simulation results confirmed the correctness of the design assumptions. The results from the start-up of the newly constructed WWTP and its correlation with the model predictions confirm the high practical usefulness and credibility of the correctly calibrated model.

ACKNOWLEDGEMENTS

The authors acknowledge the financial support of Polish Ministry of Education and Science (project 4 T07E 088 28).

REFERENCES

Baetens D., 2000. Enhanced Biological Phosphorus Removal: Modelling and Experimental Design. Ph.D. Thesis, Ghent University, Belgium.

Barker P.S., Dold P.L., 1997. General Model for biological nutrient removal activated sludge systems: model presentation. *Water Environ. Research*, 69 (5), pp. 969–984.

Brdjanovic D., 1998. Modeling biological phosphorus removal in activated sludge systems. Ph.D. Thesis, Delft University of Technology, The Netherlands.

Choi E-H., Klapwijk B., Mels A., Brouwer H., 2005. Evaluation of wastewater characterization methods. *Wat. Sci. Technol.* 52, No.10–11, pp. 61–68.

Dymaczewski Z., Gruszecka K., Jaroszynski T., 1999. The pilot plant research for the central wastewater treatment plant in Poznań. *Proc. VIII-th IAWQ Conference on design, operation and economics of large wastewater treatment plants. Budapest University of Technology, Budapest*, pp. 431–438.

Dymaczewski Z., Szetela R., 2005. Computer simulation for Poznan Wastewater Treatment Plant based on pilot plant study. *Proc. IWA spec. conf. Nutrient Management in Wastewater Treatment Processes and Recycle Streams, Krakow, 19–21 September 2005*: 1381–1386.

IWA Task Group on Mathematical Modelling for Design and Operation of Biological Wastewater Treatment, 2000. Activated Sludge Models ASM1, ASM2, ASM2d and ASM3. *IWA Scientific and Technical Report No 9*, IWA Publishing, London.

Jaroszynski T., 2003. Scale-up of activated sludge process with nutrient removal – practical experiences in the Poznań WWTP. *Proc. 9th IAWQ Conference on Design, Operation and Economics of Large Wastewater Treatment Plants. Praha, Czech Republic*, pp. 195–198.

Kristensen G.H., Jorgensen P.E., Henze M., 1992. Characterization of functional microorganism groups and substrate in activated sludge and wastewater by AUR, NUR and OUR. *Wat. Sci. Technol.* 26, No.6, pp. 43–57.

Kujawa-Roeleveld K., 2000. Estimation of denitrification potential with respiration based techniques. Ph.D. Thesis, Wageningen University, The Netherlands.

Roeleveld P.J., Kruit J., 1998. Guidelines for wastewater characterization in the Netherlands. *Koresp. Abwasser* 45, No.3, 1998, pp. 465–468 (English translation).

Roeleveld P.J., Loosdrecht M.C.M., 2002. Experience with guidelines for wastewater characterisation in The Netherlands. *Wat. Sci. Technol.* 45, No.6, pp. 77–87.

Environmental Engineering – Pawłowski, Dudzińska & Pawłowski (eds)

Impact of anionic surfactants on activated sludge properties

Ewa Liwarska-Bizukojc & Maciej Urbaniak
Department of Environmental Engineering, Technical University of Lodz, Lodz, Poland

ABSTRACT: The results of the biodegradation of three selected anionic surfactants in a continuous flow activated sludge system are presented. The experiments were carried out in lab-scale aeration chambers at dilution rates varying from 0.0256 to 0.25 h^{-1}. The main aim of this work was to estimate the impact of anionic surfactants of different chemical structure on the biochemical activity and the settling properties of activated sludge. Among three tested anionic surfactants sodium alkylbenzene sulphonate, belonging to the linear alkylbenzene sulphonates (LAS), had the strongest influence on the activated sludge properties. The inhibition concentration (IC50) towards activated sludge microorganisms, determined with the dehydrogenase test, was the lowest for this surfactant and equal to 25.6 mg L^{-1}. This anionic surfactant also inhibited the growth of freely suspended bacteria. Moreover, the obtained results indicated that the presence of anionic surfactants in the influent did not significantly diminish the settling properties of sludge, however, it could induce a higher turbidity in the effluent.

Keywords: Activated sludge, anionic surfactants, dehydrogenase activity, inhibition concentration.

1 INTRODUCTION

The consumption of the synthetic surfactants remains at a very high level. For example, in Western Europe about 2 million tons (excluding soaps) were used in 2000; 41% of them were anionic surfactants (Petrovic & Barceló 2004). The majority of anionic surfactants in use are linear alkylbenzene sulponates (LAS) representing more than 41%, the second most important group are alkylpolyoxyethylene sulphates (AES), which comprise 31% of all anionic surfactants. Finally, the third group is alkyl sulphates (AS), whose contribution is estimated at 9%.

After use synthetic surfactants are discharged into the municipal sewer systems and reach wastewater treatment plants (WWTPs). There they are removed by a combination of biodegradation and sorption processes. The average concentration of synthetic surfactants varies from 10 to 20 mg L^{-1} in domestic wastewater, but exceeds 300 mg L^{-1} in some industrial wastewater (Prats et al. 1997, Fauser et al. 2003). WWTPs are regarded as central in the flow analysis of hazardous chemicals that are used in households and industry (Fauser et al. 2003). Therefore, the experiments performed within this study simulated large-scale wastewater treatment processes.

Many studies have been performed in order to investigate removal of the synthetic surfactant from the aquatic and terrestrial environment. These investigations comprised screening tests and other experiments carried out in laboratories as well as others conducted in the full-scale wastewater treatment plants (Prats et al. 1997, Zhang et al. 1999, Fauser et al. 2003, Liwarska-Bizukojc et al. 2005). Literature data concerning removal of synthetic surfactants from domestic wastewater in WWTPs, show that high degrees of removal, exceeding usually 95%, are achieved (Fauser et al. 2003). At the same time the biodegradation of surfactants at higher concentrations is still not well documented (Zhang et al. 1999). What is more important, there is a shortage of data concerning the influence of synthetic surfactants on the quantity and quality of activated sludge microorganisms as well as sludge flocs (Proksová et al. 1998, Warne & Schifko 1999, Contreras et al. 2004, Liwarska-Bizukojc & Bizukojc 2005). The results of experiments carried out so far indicated that anionic surfactants could change the morphological properties of activated sludge flocs and deteriorate the biochemical activity of microorganisms, even at low concentration 2.5–25 mg L^{-1} (Liwarska-Bizukojc & Bizukojc 2005).

In this work we examine more closely the impact of anionic surfactants on activated sludge biomass changes. Three anionics, which belong to the most common groups of anionic surfactants described above, were selected. The experiments were conducted in a lab-scale continuous flow system at similar concentration as those found in industrial wastewater. The main purpose of this study was to compare the effect of three different anionic surfactants on biomass activity and quantity. Also the settling properties of activated sludge and degree of removal of anionics at higher concentrations were investigated.

2 MATERIALS AND METHODS

2.1 *Substrate and inocula*

The influent was synthetic wastewater prepared according to Polish Norm PN-72/C-04550. It contained one of the tested anionic surfactants at concentration 250 mg MBAS L^{-1}, which remained constant during the experiments and was frequently controlled in all experiments. Systematics and chemical formulae of the investigated surfactants are presented in Table 1. Two of them (A2 and A3) are commercial products provided by a Polish detergent manufacturer, whereas sodium dodecyl sulphate (A1) was purchased as a PA grade compound from POCh (Poland).

The characteristics of the influents containing surfactant and pure synthetic wastewater is shown in Table 2. The content of total solids was similar in all influents (excluding the control run) and varied from 910 to 1180 mg L^{-1}. A BOD_5 to COD ratio above 0.4 indicated that all substrates were easily biodegradable. The highest value of this quotient 0.607 was obtained for the synthetic wastewater without surfactant, whereas the lowest value 0.411 characterised sodium alkylbenzene sulphonate run (A2 surfactant run).

The activated sludge taken from the aeration chamber at the Wastewater Treatment Plant (WWTP) in Zgierz (Poland) was used as an inoculum. Inoculation was always performed within one hour of sampling the activated sludge from the WWTP. The aeration chamber operated under the following conditions: sludge loading rate from 2.1 to 2.9 g BOD g $VSS^{-1}day^{-1}$, sludge retention time from 19 to 22 days, hydraulic retention time about 16 days. The total solids (TS) content in activated sludge varied from 3.51 to 5.4 g L^{-1}, total suspended solids (TSS) were between 3.0 and 4.4 g L^{-1} and volatile suspended solids (VSS) were between 2.12 and 3.45 g L^{-1}. The sludges used had good settling properties and their sludge volume index (SVI) remained in the range from 39 to 69 ml g^{-1} TS. The dehydrogenase activity of all activated sludges was similar – from 13.4 to 25.9 mg TF $L^{-1}h^{-1}$.

2.2 *Experimental setup and process conditions*

The experiments were conducted in the continuous flow system without recirculation and in shake flasks. The description of the shake flasks culture was given by Liwarska-Bizukojc & Bizukojc (2005). The main item of the continuous experimental setup was the model of aeration chambers coupled with clarifiers to simulate the large-scale biological wastewater treatment plant (Fig. 1). Each chamber had a total working

Table 1. Chemical names and structural formulae of tested anionic surfactants.

Code	Chemical name	Group of surfactants	Structural formulae
A1	Sodium dodecyl sulphate (SDS)	Alkyl sulphate (AS)	$CH_3—(CH_2)_{11}—O—SO_3Na$
A2	Sodium alkylbenzene sulphonate	Linear alkylbenzene sulphonate (LAS)	$H_3C—(CH_2)_{8÷11}—CH_2—C_6H_4—SO_3Na$
A3	Sodium alkyltrioxyethylene sulphate	Alkyl polyoxyethylene sulphate (AES)	$H_3C—(CH_2)_{11÷13}—O—(CH_2CH_2O)_3—SO_3Na$

Table 2. Physicochemical properties of the influent (synthetic wastewater with/without proper surfactant).

No.	Code of surfactant/run	pH (-)	COD (mg O_2 L^{-1})	BOD_5/COD (-)	TS (mg L^{-1})	VS (mg L^{-1})	Turbidity (FTU)
1	A1 (surfactant run)	7.5	910	0.588	900	600	34
2	A2 (surfactant run)	7.6	1180	0.411	990	730	83
3	A3 (surfactant run)	7.5	1130	0.469	1003	730	14
4	Synthetic wastewater (control run)	7.3	420	0.607	380	80	10

volume of 7.8 L. The aerated part comprised 5.6 L^{-1} of it. Dissolved oxygen concentration, temperature and pH were controlled during the experimental runs. The aeration flow rate was constant and identical in each experimental run (0.321 vvm). The experiments were conducted at ambient temperature ($22 \pm 1°C$). Substrate was added to aeration chambers at different dilution rates from 0.0256 to 0.250 h^{-1}. The steady state in the chamber was assumed when at least five residence times had completed.

2.3 *Analytical techniques*

The concentration of anionic surfactants was determined by the methylene blue method (APHA-AWWA-WPCF 1995) and expressed as Methylene Blue Active Substances (MBAS). Anionics form ion pairs with methylene blue that are extracted by chloroform and determined spectrophotometrically at 652 nm.

Chemical Oxygen Demand (COD) was measured by the standard dichromate method (APHA-AWWA-WPCF 1995). Turbidity was measured using HACH spectrophotometer DR/2000 at 450 nm according to method no. 750 (Instrument Manual HACH Co. 1991/96). Total suspended solids (TSS) and volatile suspended solids (VSS) were measured gravimetrically. The sludge volume index (SVI) was also determined. Dehydrogenase activity was determined according to Miksch (1985) using 2,3,5 triphenyltetrazolium chloride (TTC). This test was also used in order to estimate the inhibition concentration of anionic surfactants towards activated sludge microorganisms. All inhibition tests were performed using the same sample of activated sludge. Additionally, microscopic observations of activated sludge microorganisms according to Eikelboom & van Buijsen (1999) were performed.

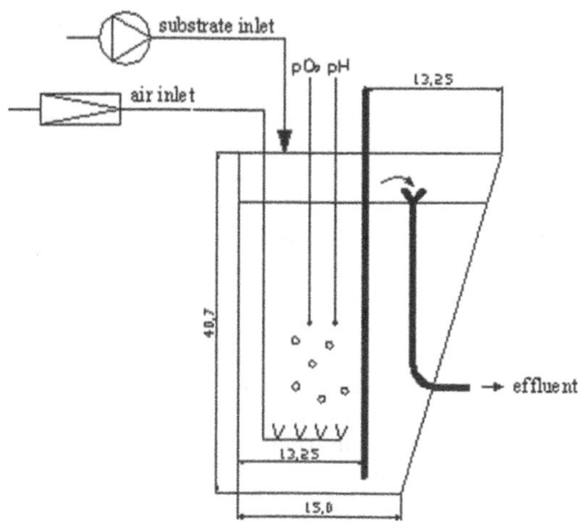

Figure 1. Cross section of aeration chamber.

3 RESULTS AND DISCUSSION

Activated sludge flocs exposed to anionic surfactants were subjected to saponification. As a result of this process sludge flocs diminished and the content of total solids decreased (Liwarska-Bizukojc & Bizukojc 2005). The biomass quantity expressed as volatile suspended solids also decreased with the increase of dilution rate in all surfactant runs (Fig. 2). On the contrary, volatile suspended solids in the control run increased from about 1 g L^{-1} at the lowest dilution rate (0.0282 h^{-1}) to 3–3.5 g L^{-1} at all higher dilution rates. In surfactant runs at the shortest retention time (4–5 h) the concentration of volatile suspended solids was lowest and remained in the range between 0.1 and 0.3 g L^{-1}, dependent on the surfactant added (Fig. 2). This means that the VSS was at least ten times lower in surfactant runs than in the control run. It must be also noted that for A2 and A3 biomass concentration, expressed as the VSS, decreased gradually with dilution rate, whereas for A1 it started decreasing at dilution rates above 0.08 h^{-1}.

The changes of dehydrogenase activity of activated sludge, which was used as an indicator of biomass quality, are similar to the changes in volatile suspended solids. Dehydrogenase activity decreased with the increase of dilution rate in all surfactant runs (Fig. 3). Comparing the results of TTC tests for the shortest residence times, the lowest values of dehydrogenase activity were observed for sodium alkylbenzene sulphonate (A2). Furthermore, the changes of dehydrogenase activity and volatile suspended solids in the whole studied range of dilution rate indicated that A2 had the strongest influence on the activated sludge biomass.

In order to establish the inhibition concentrations (IC50) towards activated sludge microorganisms TTC test were applied. In Table 3 the results of inhibition concentration IC50 estimation are presented.

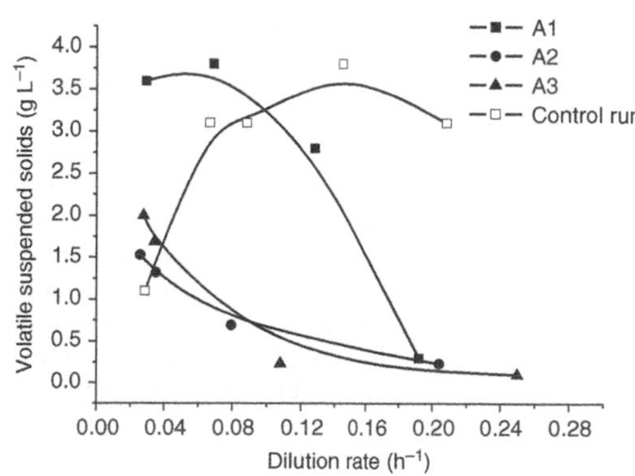

Figure 2. Volatile suspended solids vs. dilution rate.

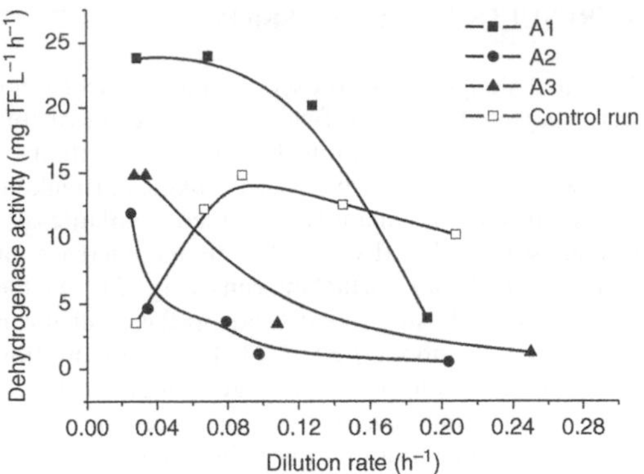

Figure 3. Dehydrogenase activity vs. dilution rate.

Table 3. The estimation of inhibition concentration IC 50 and maximum specific growth rate. Maximum specific growth rate (μ_{max}) was calculated on the basis of turbidity results.

No.	Code of anionic surfactant	IC 50 (mg L^{-1})	μ_{max} (h^{-1})
1	A1	99.6	0.172 ± 0.061
2	A2	25.6	0.054 ± 0.012
3	A3	42.1	0.43 ± 0.27
4	control	–	0.177 ± 0.009

The lowest IC 50, equal to 25.6 mg L^{-1}, was obtained for A2. It confirmed that linear alkylbenzenesulphonates were stronger inhibitors of dehydrogenase activity of activated sludge microorganisms than the other anionic surfactants studied. At the same time, it must be noted that all estimated IC 50 values are higher than the average concentration of anionics in domestic wastewater. Therefore, the presence of anionics at low concentrations in the influent to WWTPs should not significantly deteriorate the biodegradation processes.

Anionic surfactants can diminish oxygen transfer to wastewater (Wagner & Pöpel 1996). As a result, the metabolism of aerobic bacteria can be inhibited. Taking this into account, the concentration of dissolved oxygen in wastewater suspension was continuously measured and controlled. The changes of dissolved oxygen concentration with dilution rate are shown in Figure 4. Although the dissolved oxygen concentration decreased gradually with the increase of dilution rate, the amount of oxygen remained at the level above 2 mg L^{-1} in all runs (surfactant and control) i.e. sufficient for proper aerobic wastewater treatment processes.

In parallel to the gravimetric and biochemical analyses of activated sludge biomass, microscopic observations were also performed. They indicated that in

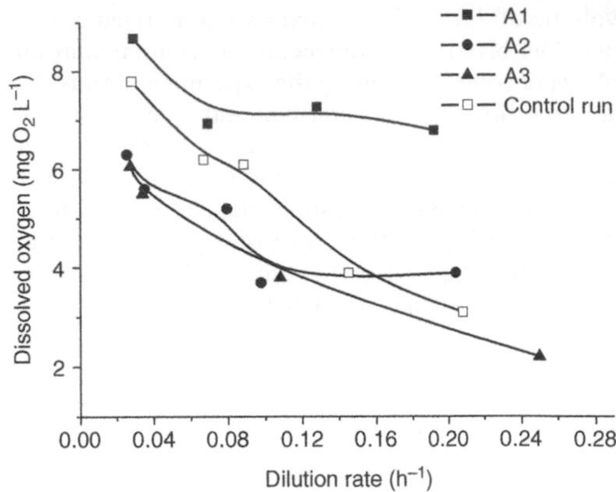

Figure 4. Changes of dissolved oxygen concentration in wastewater with dilution rate.

the surfactant runs the number of freely suspended bacteria in the liquid phase was higher than in the control run. Moreover, it was clearly seen that the number of freely suspended bacteria increased with the dilution rate in surfactant runs. This phenomenon is typical at higher loading rate and/or when toxic substances are present in wastewater (Eikelboom & van Buijsen 1983, Eckenfelder & Mustermann 1995). The images of freely suspended bacteria taken in surfactant and control run at the highest dilution rate (about 0.2 h^{-1}) are shown in Figure 5. Additionally, it was observed that the biomass diversity also changed with the increase of dilution rate. At the higher dilution rates used in surfactant runs rod bacteria dominated. Also the number of cocci was significant, while at the lowest dilution rate of surfactant runs, in the control run and in the inoculum, *Spirochetae* were mainly observed.

The increase of freely suspended bacteria was confirmed by the increase of supernatant turbidity (Fig. 6).

In all surfactant runs the turbidity determined in liquid phase increased with the dilution rate, whereas in the control run its increase was only slight. Generally, it remained at similar level, between 14 and 25 units. In A1 and A3 surfactant runs turbidity increased above 100 FTU at dilution rates above 0.012 h^{-1}. Additionally, the results of turbidity measurements in shake flask culture for all surfactants studied are presented in Figure 7. Both the results achieved in continuous flow systems as well as the results of batch experiments shown that turbidity well reflected the growth of freely suspended bacteria.

In order to ascertain whether turbidity arises from the dissolved surfactants, the additional measurement of turbidity in aqueous solutions of surfactants in the concentration range 5–5000 mg L^{-1} was carried out. The tests revealed that below 250 mg L^{-1} none

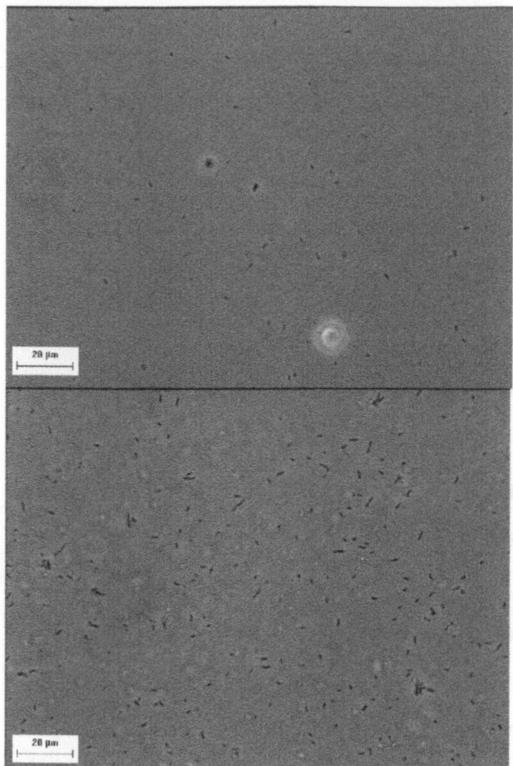

Figure 5. Freely suspended bacteria in the control run (upper image) and in the surfactant run (lower image) at the highest dilution rate (about 0.25 h^{-1}) – greyscale images made in phase contrast at magnification 1000.

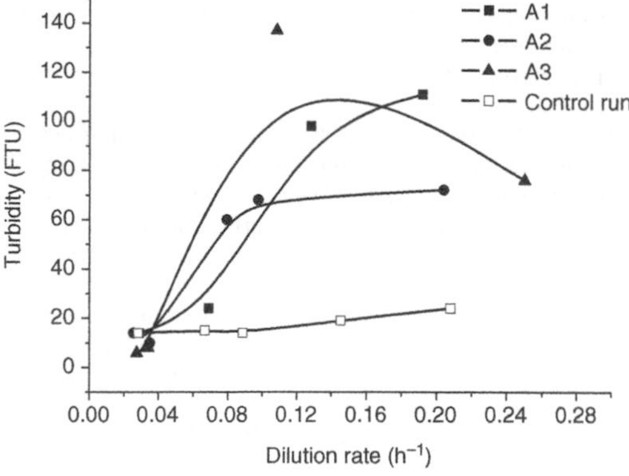

Figure 6. Changes in the turbidity of effluent with dilution rate.

of tested surfactants showed turbidity above 5 FTU. Therefore, the freely suspended bacteria and liberated extracellular polymeric substances seem to be the only source of turbidity.

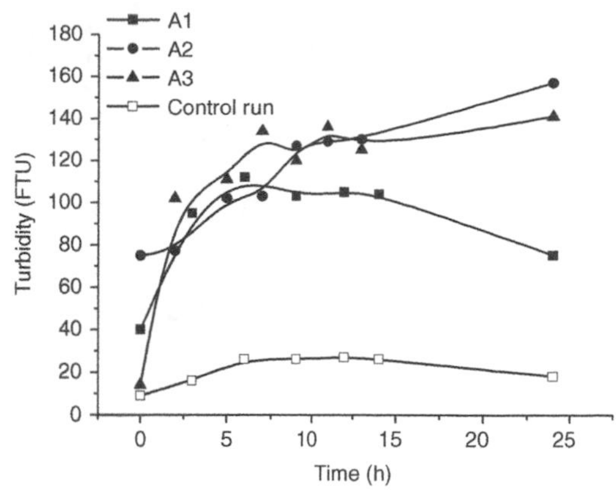

Figure 7. Changes in the turbidity of effluent in shake flask culture.

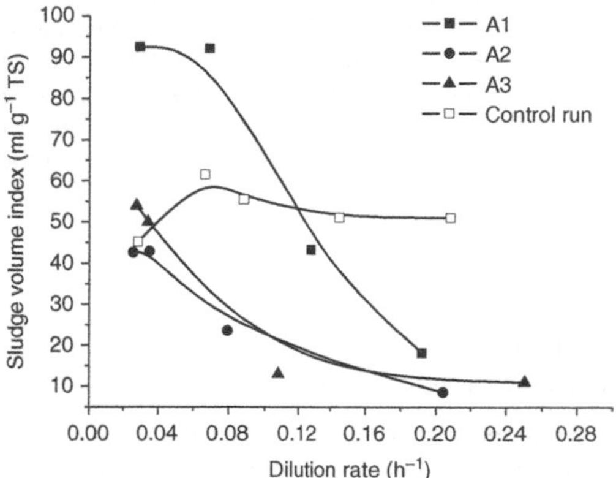

Figure 8. Sludge volume index vs. dilution rate.

On the basis of turbidity measurements made in the shake flask culture, the maximum specific growth rate was determined (Table 3). Here the lowest maximum specific growth rate was obtained in the A2 run. This means that sodium alkylbenzene sulphonate inhibited the growth of freely suspended bacteria, confirming again that it was the most toxic of all anionics tested.

The changes of sludge volume index (SVI) are presented in Figure 8. SVI decreased with the increase of dilution rate in all surfactant runs, while in the control run SVI increased from 45 at $D = 0.0282\,h^{-1}$ to 61 ml g^{-1} TS at $D = 0.0667\,h^{-1}$ and remained at the level 51–55 ml g^{-1} TS. The low values of SVI at higher dilution rates in surfactant runs were caused by the saponification of flocs. Generally, anionic surfactants at high concentration did not worsen the settling properties of activated sludge, however, they did induce the increase of turbidity in supernatant. This turbidity certainly deteriorated the quality of effluent.

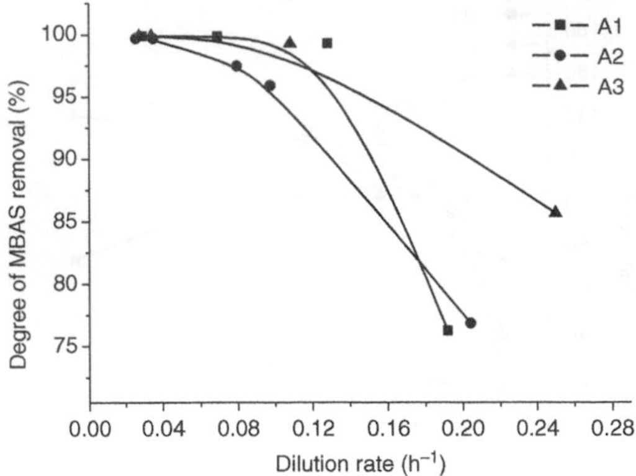

Figure 9. Degree of MBAS removal vs. dilution rate.

Last but not least, the changes in the degree of removal of anionic surfactants are shown in Figure 9. It is very evident that the degree of removal was generally high, but that it also decreased with the increase of dilution rate. At lower dilution rates it varied from 95 to 99% in all surfactant runs. Only at the highest dilution rate did it decrease to 75–80% for two of the surfactants tested (A1 and A2).

4 CONCLUSIONS

On the basis of the obtained results the following conclusions can be drawn:

1. All anionic surfactants studied caused the saponification of sludge flocs. As a result, the content of total and volatile solids decreased with the increase in dilution rate.
2. Anionic surfactants inhibited the dehydrogenase activity of activated sludge microorganisms. The lowest value of IC50, 25.6 mg L^{-1} was obtained for sodium alkylbenzene sulphonate. Results of continuous flow experiments confirmed that sodium alkylbenzene sulphonate had the strongest impact on respiration activity.
3. Anionic surfactants did not significantly deteriorate the settling properties of sludge. Nevertheless, they induced turbidity in the effluent due to the development of freely suspended bacteria. The number of freely suspended bacteria was higher in surfactant runs compared to control runs. This was especially evident at higher dilution rates (above 0.08 h^{-1}) when the organic volumetric load was high.

REFERENCES

APHA-AWWA-WPCF 1995. Standard Methods for the Examination of Water and Wastewater, 19th edition. Washington DC: APHA-AWWA-WPCF.

Contreras, E.M., Gianuzzi, L., Zaritzky, N.E. 2004. Use of image analysis in the study of competition between filamentous and non-filamentous bacteria *Water Research* 38: 2621–2630.

Eckenfelder, W.W. & Mustermann, J.L. 1995. *Activated sludge treatment of industrial wastewater*. Lancaster-Basel: Technomic Publishing Company.

Eikelboom, D.H. & van Buijsen, H.J.J. 1983. *Handbuch für die mikrobiologische Schlammuntersuchung*. Munich: F Hirthammer Verlag GmbH.

Fauser, P., Vikelsøe, J., Sorensen, P.B., Carlsen, L. 2003. Phthalates, nonylphenols and LAS in an alternately operated wastewater treatment plant – fate modelling based on measured concentrations in wastewater and sludge *Water Research*. 37: 1288–95.

Instrument Manual HACH Company (1991/96) Loveland, Colorado, USA.

Liwarska-Bizukojc, E. & Bizukojc, M. 2005. Digital image analysis to estimate the influence of sodium dodecyl sulphate on activated sludge flocs *Process Biochemistry* 40: 2067–2072.

Liwarska-Bizukojc, E., Miksch, K., Malachowska-Jutsz, A., Kalka, J. 2005. Acute toxicity and genotoxicity of five selected anionic and nonionic surfactants *Chemosphere* 58: 1249–1253.

Miksch, K. 1985. Auswahl einer optimalen Methodik für die Aktivitätsbestimmung des Belebtschlammes mit Hilfe des TTC-Testes (in German) *Vom Wasser* 64: 187–198.

Prats, D., Ruiz, F., Vazquez, B., Rodriguez-Pastor, M. 1997. Removal of anionic and nonionic surfactants in a wastewater treatment plant with anaerobic digestion. A comparative study *Water Research* 31: 1925–1930.

Proksová, M., Vrbanovà, A., Sládcková, D., Gregorovà, D., Augustin, J. 1998. Dialkyl sulfoccinate toxicity towards Commonas terrigena N3H *Journal of Trace Microprobe Techniques* 16: 475–480.

Polish Norm: Water and Wastewater PN–72/C-04550 (in Polish).

Petrovic, M. & Barceló, D. 2004. Fate and removal of surfactants and related compounds in wastewater and sludges. *The Handbook of Environmental Chemistry* 5(1): 1–28.

Wagner, M. & Pöpel, H.J. 1996. Surface active agents and their influence on oxygen transfer *Water Science and Technology* 34: 249–256.

Warne, M.St.J. & Schifko, A.D. 1999. Toxicity of laundry detergent components to a freshwater cladoceran and their contribution to detergent toxicity *Ecotoxicogy and Environmental Safety* 44: 196–206.

Zhang, C., Valsaraj, K.T., Constant, W.D., Roy, D. 1999. Aerobic biodegradation kinetics of four anionic and nonionic surfactants at sub- and supra-critical micelle concentrations (CMCs) *Water Research* 33: 115–124.

Environmental Engineering – Pawłowski, Dudzińska & Pawłowski (eds)
© 2007 Taylor & Francis Group, London, ISBN13 978-0-415-40818-9

Influence of aggressive liquids on hydraulic conductivity of hardening slurries with the addition of different fluidal fly ashes

Paweł Falaciński & Zbigniew Kledyński
Warsaw University of Technology, Institute of Water Supply and Hydraulic Engineering, Warsaw, Poland

ABSTRACT: This article presents a testing methodology and results for the basic characteristic of hardening slurries, with the addition of non-activated and mechanically activated fluidal fly ash from hard coal and from brown coal, used in cut-off walls, i.e. hydraulic conductivity in the conditions of persistent, filtration over several months with liquids chemically aggressive to cement binders.

Keywords: Hardening slurries, fluidal fly ash, hydraulic conductivity, bentonite, cut-off walls.

1 INTRODUCTION

The changing standards of environmental protection result in the need for modernisation of the installations used for combustion of solid fuels. Increasingly, the power industry in Poland is deciding to replace conventional furnaces with furnaces with a circular fluidal layer.

A lower combustion temperature and the addition of a desulphurisation sorbent produce a difference between fluidal and conventional fly ashes. This limits the possible applications of fluidal ashes in civil engineering (Havlica et al., 2004).

In the search for new ways of utilisation of fluidal ashes, research was undertaken to examine the possibilities of adding them to hardening slurries.

A hardening slurry is a mixture of water (the major constituent by volume), bentonite and cement, which in the liquid state has tixotropic characteristics (important in the case of liquids expanding trenches during excavation), and, after binding, characteristics of a construction material for cut-off walls, i.e. adequately durable, strong and waterproof. Since various additives and admixtures can be added to hardening slurries and thus modify their technological or functional properties, it is possible to design various formulae depending on the destination of the material.

Combustion wastes, including fluidal ashes, are rather special additives used in hardening slurries. They are mineral materials with binding characteristics. Therefore, they can act not only as fillers but also as binders (Falaciński et al., 2005; Giergiczny, 2005).

The complicated nature of the chemical changes occurring in mineral binding materials and the characteristics of fluidal ashes call for a deeper insight into their application in hardening slurries. A special challenge is to recognise the properties of hardening slurries with the addition of fluidal ashes functioning in a chemically aggressive water-ground environment, which is the case of cut-off walls in hydraulic structures (e.g. waste dumps or facilities protecting against pollution of underground water intakes) (Kledyński, 2004; Fratalocchi, Pasqualini, 1998; Jefferis, 1990).

2 DETERMINING FORMULAE FOR HARDENING SLURRIES

The following components were used for hardening slurries: sodium bentonite "Dywonit S", cement CEM I 32,5R "Ożarów", and non-activated and mechanically activated fluidal fly ash from hard coal (Katowice Thermal-electric Power Station) and from brown coal (Turów Power Plant). Formulae determined and well-tried in earlier projects were used in the research. Table 1 shows the compositions of the tested hardening slurries.

Mechanical activation of fly ash is a physical process running in specially constructed mechanical activators (Patent No. 180380). The process itself does not require the application of chemical reagents.

Mechanical activation involves a supply of mechanical energy to the fly ash in a way that does not bring about a breakage of ash particles but induces a change in their internal energy and pore structure, as well as the creation of "active centres" on the particle surface which has been renewed by friction and collision of the particles.

Table 1. Formulae of hardening slurries.

No.	Component	Amount of component [kg] according to formulae signed as: PKN	PKA	PBN	PBA
1	Tap water	1000	1000	1000	1000
2	Bentonite "Dywonit S"	40	40	30	30
3	Non-activated fluidal fly-ash from hard coal	323	–	–	–
4	Mechanically activated fluidal fly-ash from hard coal	–	323	–	–
5	Non-activated fluidal fly-ash from brown coal	–	–	326	–
6	Mechanically activated fluidal fly-ash from brown coal	–	–	–	326
7	Cement CEM I 32,5R Ożarów	163	163	170	170

PKN – hardening slurries with non-activated fluidal fly-ash from hard coal; PKA – hardening slurries with mechanically activated fluidal fly-ash from hard coal; PBN – hardening slurries with non-activated fluidal fly-ash from brown coal; PBA – hardening slurries with mechanically activated fluidal fly-ash from brown coal.

3 TESTING OF PARAMETERS OF LIQUID SLURRIES

Tests were performed to determine the following parameters of liquid slurries: density (ρ_{pc}), viscosity ratio (L), and 24 h water setting (O_d).

The density test and the viscosity ratio test (using Marsh's funnel) were performed according to the Polish standard BN-90/1785-01. The 24 h water setting test can be described as determining the percentage share of spontaneously separating water in 1 dm^3 of liquid slurry after one day of it standing in a measuring cylinder.

Test results for properties of liquid slurries are shown in Table 2.

4 PREPARATION AND STORING OF SAMPLES FOR HYDRAULIC CONDUCTIVITY RESEARCH

Hardening slurry test cylinders were prepared in PVC moulds of 8 cm in diameter and 8 cm in height. Before the slurry set, the samples were kept under a foil covering in the laboratory. After 3–4 days the samples were submerged under water. The water temperature was +18°C ± 2°C. The samples stayed under water until the moment of measurement. Leak tightness of the contact of the sample and the mould was assured by crimping of the internal wall of the mould and an additional silicon seal.

Table 2. Properties of liquid slurries.

No.	Parameter	Results of research of liquid slurries' properties signed as: PKN	PKA	PBN	PBA
1	Density of fluid slurry ρ_{pc} [g/cm^3]	1.29	1.29	1.29	1.29
2	Viscosity ratio L [s]	41	42	43	55
3	24 h water setting O_{sd} [%]	3.0	2.0	2.0	2.0

Abbreviations shown in Table 1.

5 PREPARATION AND STORING OF SAMPLES FOR HYDRAULIC CONDUCTIVITY RESEARCH

Hardening slurry test cylinders were prepared in PVC moulds of 8 cm in diameter and 8 cm in height. Before the slurry set, the samples were kept under a foil covering in the laboratory. After 3–4 days the samples were submerged under water. The water temperature was +18°C ± 2°C. The samples stayed under water until the moment of measurement. Leak tightness of the contact of the sample and the mould was assured by crimping of the internal wall of the mould and an additional silicon seal.

6 HYDRAULIC CONDUCTIVITY TESTS OF HARDENING SLURRIES

6.1 *Measuring apparatus*

Hydraulic conductivity tests of hardening slurries exposed to the action of water and aggressive liquids were carried out in specially constructed chemically resistant plastic apparatus (PVC and plexiglass). A section of the apparatus with a sample of hardening slurry is presented in Figure 1.

From among typical environments aggressive towards binding cement three were selected: leaching – distilled water; general acid – 0.5% solution of nitric acid; and sulphates – 1% solution of sodium sulphate.

6.2 *Methodology of measurement and calculation of results*

The hydraulic conductivity of hardening slurries is quite low (similar to that of cohesive soils) and so the time needed to obtain the balance of supply and outflow of water from the sample is quite long. In such

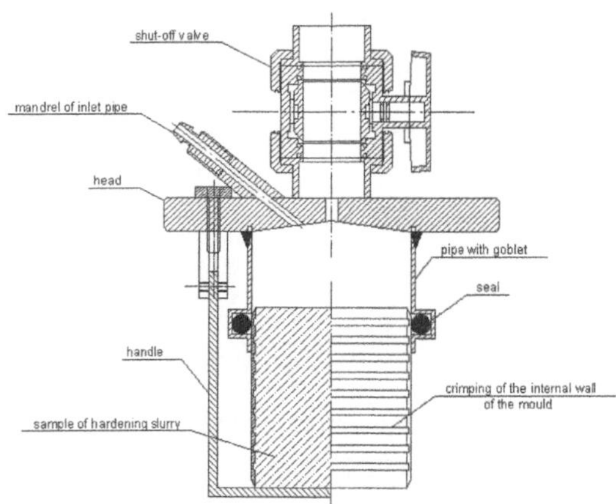

Figure 1. Apparatus for testing hydraulic conductivity with a sample of hardening slurry.

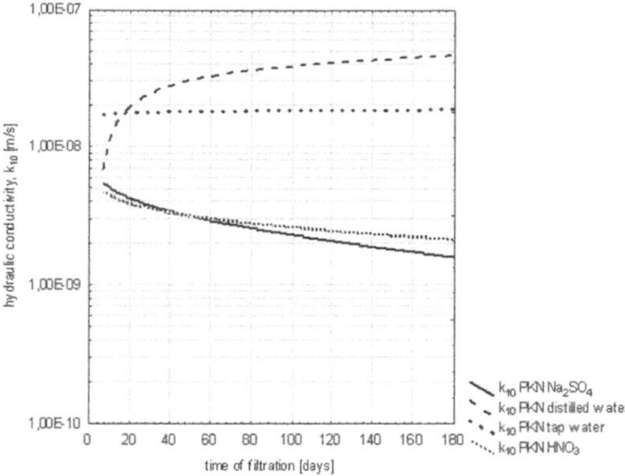

Figure 2. Hydraulic conductivity of hardening slurry with addition of non-activated fluidal ash from hard coal (PKN) as a function of time and type of filtrating liquid.

cases, conductivity tests are performed with a variable hydraulic gradient. This method consists of determining the values of water pressure (h_1, h_2, etc.) in the supply tube of cross-section area a, in established times (t_1, t_2, etc.) during the liquid's flow through the sample of length (height) L and cross-section area A. In this case the hydraulic conductivity is calculated with the following formula:

$$k_T = \frac{a \cdot L}{A \cdot t} \ln\left(\frac{h_1}{h_2}\right) \quad (1)$$

k_T – hydraulic conductivity in temperature T [m/s]; a – cross-section area of the supplying tube [m^2]; L – length (height) of the sample [m]; A – cross-section area of the sample [m^2]; t – time between pressure measurements (h_1, h_2), $t = t_2 - t_1$ [s]; $h_{1,2}$ – values of water pressures at times t_1 and t_2 [m].

The main advantage of this testing method is the possibility it offers of measuring small water flows and forcing high water pressures.

The action of the filtering media (tap water, distilled water and aggressive water solutions) on the tested sample was of gravitational nature. The measurements were performed with a decreasing initial hydraulic gradient.

The sample was placed in the apparatus (Fig. 1) and liquid poured over it up to the level which forced the maximum hydraulic gradient equal to 45 (hydraulic gradient is the quotient of water pressure measured in cm and height of investigated sample in cm). Over each measuring tube there was a container linked to it with an elastic pipe. As the level of liquid fell, the container was filled up in a way which protected the pipe from getting air into it and limited changes of hydraulic gradient between the measurements of hydraulic conductivity.

Once a week six measurements were taken on a half-hourly basis and the decreasing water level in the tube was recorded. A final measurement was taken after another hour. After the measurements were completed, the pipe and the container were filled up (including de-aerating), and the initial (maximum) hydraulic gradient was established. The samples were left in this state until the day of the next measurement, filling up the liquid in the containers when needed.

The hardened slurries were exposed to the action of the filtering liquids for 5 months. By the start of the research program, the samples had matured for 65 days. The range of hydraulic gradients acting on the samples was from 20 to 45, and gradients lower than 45 were only acting on the days of the hydraulic conductivity measurements (once a week) for no longer than 4 hours.

The hydraulic conductivity calculated with the formula no. 1 does not take account of the influence of temperature of the filtering liquids. The k_T values obtained during the tests (at temperature T) were recalculated into k_{10} values corresponding to the temperature of $+10°$C. The following formula (Pisarczyk, 1998) was used:

$$k_{10} = \frac{k_T}{0,7 + 0,03T}$$

The test results are shown in time function in Figures 2 and 3 (hardening slurries with addition of fluidal fly ash from hard coal), and Figures 4 and 5 (hardening slurries with addition of fluidal fly ash from brown coal). The diagrams show trends (matching lines) for the series of individual hardening slurries (formulae) exposed to filtering tap water, distilled water, and the solution of nitric acid and sodium sulphate. For each day of the hydraulic conductivity test, four repeated measurements were used with a gradually decreasing initial gradient (the first and the last measurement

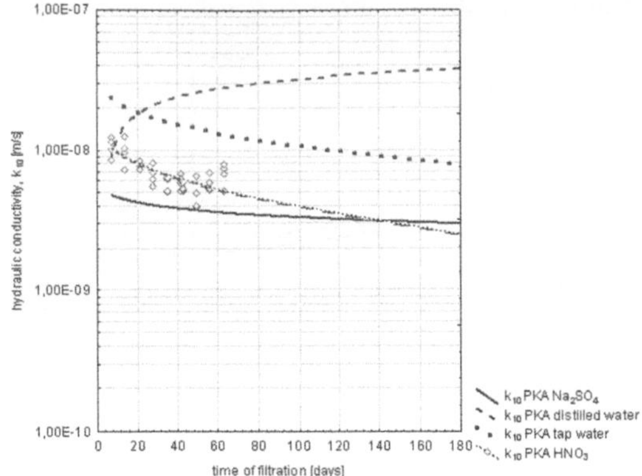

Figure 3. Hydraulic conductivity of hardening slurry with addition of mechanically activated fluidal ash from hard coal (PKA) as a function of time and type of filtrating liquid.

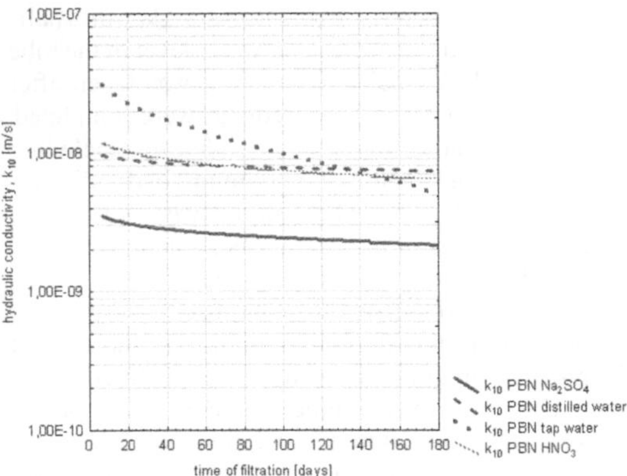

Figure 4. Hydraulic conductivity of hardening slurry with addition of non-activated fluidal ash from brown coal (PBN) as a function of time and type of filtrating liquid.

in the series were rejected). Earlier, the influence of hydraulic gradient on the result of conductivity measurements had been tested – in the applied range of gradients there was no relationship between these results. After about 60 days of testing the sample of slurry with activated ash from hard coal exposed to the action of acid, de-sealed in the area of contact with the mould so the test was discontinued. In Figure 3 the test results of this sample are shown as experimental points and a trend curve extrapolated until the end of the analysed period.

7 ANALYSIS OF THE OBTAINED RESULTS

Comparing the results of the long lasting filtering action of liquids chemically aggressive to cement binder and of tap water on slurries with fluidal fly ash, differences in the materials' response depending on

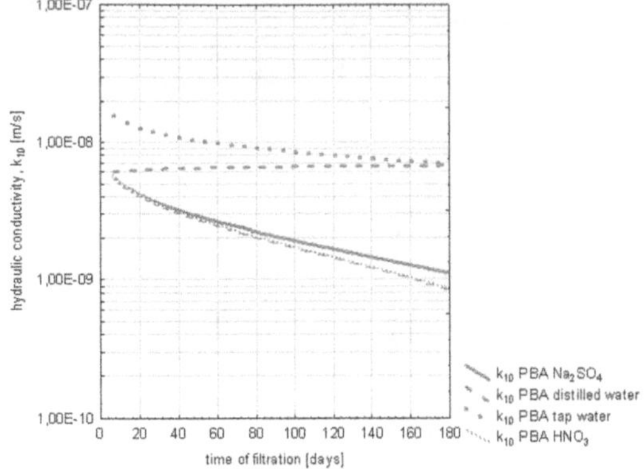

Figure 5. Hydraulic conductivity of hardening slurry with addition of mechanically activated fluidal ash from brown coal (PBA) as a function of time and type of filtrating liquid.

the kind of aggressive action and the type of ash used (hard and brown, non-activated and activated) can be easily noted.

Distilled water caused de-sealing of the slurry with addition of ash from hard coal by leaching some of its ingredients (probably calcium compounds) (Figs. 2 and 3). It caused an increase in hydraulic conductivity k_{10} from ca. $7.0 \cdot 10^{-9}$ m/s to ca. $4.5 \cdot 10^{-8}$ m/s in slurries with non-activated ash, and from ca. $1.0 \cdot 10^{-8}$ m/s to ca. $4.0 \cdot 10^{-8}$ m/s in slurries with activated ash. The differences in the final values of conductivity are lower than the measurement uncertainty for this quantity (about 15% of the measured value).

The values of conductivity of the slurries exposed to acid and sulphate aggression were fundamentally different – they decreased during the testing and stabilised at the level of ca. $(2 \div 3) \cdot 10^{-9}$ m/s. Larger differences in hydraulic conductivity were observed in the initial testing phase between the slurry with activated ash and that with non-activated ash, both exposed to the action of nitric acid. Perhaps the effect of lack of tight sealing between the mould and the slurry with activated ash was also revealed here.

It is worth noting that the values of hydraulic conductivity of the slurries exposed to acid and sulphate aggressions are similar and change over the testing period in a similar way (especially in the case of the slurry with non-activated ash). This is in spite of differences between the chemical processes of sealing of the structure of the materials. With sulphate aggression, complex hydrated sulphate salts are formed (Kledyński, 2004), and in the case of acid aggression it is possible to reseal the material with amorphous products (gel) of C-S-H phase decay in acid environment (Kledyński, 2000).

The most intensive changes in hydraulic conductivity of slurries occur in each case at the beginning of the exposure. The investigated quantities stabilise with time.

In order to compare changes in the response of the materials exposed to various aggressive liquids, analogous tests on slurries with tap water actually used as mixing water were held. The samples exposed to long-lasting filtrating action of potable water show stable values of hydraulic conductivity (ca. $2.0 \cdot 10^{-8}$ m/s) in the case of addition of non-activated ashes to the slurry, or a decrease in conductivity from $2.5 \cdot 10^{-8}$ m/s to ca. $8.0 \cdot 10^{-9}$ m/s in the case of activated ashes. At this point it is difficult to explain the causes of the differences in the response of the slurries. The hydraulic conductivity of the slurries compared to tap water does not deteriorate with time (within the scope covered by the testing), and the obtained values are intermediate between the values of their conductivity in the presence of distilled water, and solutions of acid and sulphate salt.

Analysing the hydraulic conductivity of hardening slurries with addition of fluidal ashes from brown coal (non-activated (PBN) – Fig. 4, and activated (PBA) – Fig. 5), one notices a number of similarities to the slurries with addition of ashes from hard coal, as well as one fundamental difference. The latter concerns the filtration of distilled water. While in the slurries with ashes from hard coal it caused de-sealing of the material and an increase of hydraulic conductivity, in the slurries with ashes from brown coal such changes did not occur. In the case of addition of non-activated ash, there is a small drop of hydraulic conductivity from ca. $1 \cdot 10^{-8}$ m/s to ca. $7.5 \cdot 10^{-9}$ m/s, and in the case of activated ash the conductivity is practically stable ($6 \div 7 \cdot 10^{-9}$ m/s).

A decrease of hydraulic conductivity of slurries with addition of ash from brown coal is noticeable with the action of nitric acid solution and sodium sulphate. In the case of non-activated ash (Fig. 4) and the action of nitric acid, the values of hydraulic conductivity k_{10} change within the limits of ca. $1.5 \cdot 10^{-8}$ m/s and ca. $6.5 \cdot 10^{-9}$ m/s, and in the case of sodium sulphate between ca. $3.5 \cdot 10^{-9}$ m/s and ca. $1.5 \cdot 10^{-9}$ m/s. More significant changes in conductivity (sealing up) occur with the filtration of solutions of nitric acid and sodium sulphate through the slurries with addition of activated ash (Fig. 5). The values of the hydraulic conductivity of these slurries were similar at the beginning of the testing (ca. $5.5 \cdot 10^{-9}$ m/s). At the end they fell to ca. $1 \cdot 10^{-9}$ m/s in the case of filtration with sodium sulphate solution and to ca. $8.5 \cdot 10^{-10}$ m/s in the case of filtration with nitric acid solution.

The samples of slurries exposed to the action of tap water showed a tendency to seal up under the influence of the filtration. A decrease of hydraulic conductivity appeared with both activated and non-activated ashes. In the case of addition of non-activated ash, the conductivity decreased from ca. $3 \cdot 10^{-8}$ m/s to ca. $5 \cdot 10^{-9}$ m/s, and, in the case of activated ash, from ca. $1.5 \cdot 10^{-8}$ m/s to ca. $7 \cdot 10^{-9}$ m/s.

When analysing the response of slurries, depending on the sort of added ash (from hard or brown coal), it is necessary to pay attention to the different effects of distilled water filtration. There was an unquestionable de-sealing of the structure of the slurry with ash from hard coal (non-activated and activated), and virtual invariability of hydraulic conductivity of the slurries with addition of ash from brown coal (non-activated and activated).

The other solutions (nitric acid and sodium sulphate) acted in similar ways on each slurry, despite quantitative differences and the type of ash or its activation.

The values of hydraulic conductivity as well as the direction of their changes in the case of action of tap water are similar for the slurries PKA, PBN and PBA. The slurry PKN shows practically stable conductivity in relation to this medium.

8 SUMMARY

The results presented above of the research on slurries with addition of fluidal ashes show the diverse and – above all – the small influence of mechanical activation of ashes on the hydraulic conductivity of slurries.

In acid or sulphate salts environments chemically aggressive towards cement binders, there occurs a tendency of lowered conductivity caused by sealing up of the structure of the material as a result of the chemical changes occurring between aggressive substances and the ingredients of hardened slurries.

In the case of long lasting filtration of distilled water (leaching aggression) through the slurries with addition of fluidal ashes from hard coal (PKN and PKA), the samples de-sealed which was proven by the gradually increasing value of the material's hydraulic conductivity (almost by one order of magnitude during ca. 180 days). This was probably caused by leaching of some of the ingredients from the structure of the material.

The slurries with addition of ashes from brown coal (non-activated and activated) are far more stable in this interaction. The values of hydraulic conductivity were stabilised at one level during the whole testing period (PBA) or showed insignificant tendencies to seal up the slurry structure (PBN), which was demonstrated by a small drop in k_{10}.

Against this background, the slurries exposed to potable water filtration showed intermediate or slightly higher values of conductivity.

The results of hydraulic conductivity tests confirm the possible application of fluidal ashes from hard and brown coal, non-activated or mechanically activated, as additives to hardening slurries for the construction of leak proof cut-off walls, durable also in acidic and sulphate-ionic water environments chemically aggressive towards cement binders, and in the case of the slurries with addition of ashes from brown coal, also in a soft water environment.

REFERENCES

Falaciński P., Garbulewski K., Kledyński Z., Skutnik Z., Ziarkowska K., 2005. *Fluidised Fly-Ash Cement-Bentonite Cut-off Walls in Flood Protection*. Archives of Hydro-Engineering and Environmental Mechanics, vol. 52, no. 1, pp. 7–20.

Fratalocchi E., Pasqualini E.,1998. *Permeability over time of cement-bentonite slurry walls*. Proc. of the III International Congress on Environmental Geotechnics, Lisbon, Portugal, vol. 2, pp. 509–514.

Giergiczny Z., 2005. *Effect of fly ash from different sources on the properties of hardened cement composites. Silicates Industriets*. Ceramic Science and Technology, vol. 70, no. 3–4, pp. 35–40.

Havlica J., Oder I., Brandstetr J., Mikulikova R., Walther D., 2005. *Cementious materials based on fluidised bed coal combustions ashes*. Advances in Cement Research, vol. 16, no. 2, pp. 61–67.

Jefferis S.A., 1990. *Cut-off walls: methods, materials and specifications*. Proc. of the International Conference on Polluted and Marginal Land, London, pp. 117–125.

Kledyński Z., 2004. *Influence of Fly Ashes on Hardening Slurries Resistance to Sulphate Attack*. Archives of Hydro-Engineering and Environmental Mechanics, vol. 51, no. 2, pp. 119–133.

Polish standard BN-90/1785-01. *Drilling slurry. Testing methods in field conditions*.

Patent No. 180380. *The Procedure and System for Producing the Bindings Material from Boiler Incineration Ashes, Particularly from Fluidal Furnace*. Patent Office of Poland (in Polish).

Pisarczyk S., 1998. *Soil Mechanics*. Publishing House of Warsaw University of Technology, Poland, Warsaw.

Environmental Engineering – Pawłowski, Dudzińska & Pawłowski (eds)
© 2007 Taylor & Francis Group, London, ISBN13 978-0-415-40818-9

Leaching of PCBs from sewage sludge after biochemical stabilization processes

Agata Rosińska, Marta Janosz-Rajczyk & Jacek Płoszaj
Faculty of Environmental Engineering and Protection, Department of Chemistry, Water and Wastewater Technology, Czestochowa University of Technology, Czestochowa, Poland

ABSTRACT: The leaching of PCBs (28, 52, 101, 118, 132, 153, 180) from sewage sludge (before and after biochemical stabilization processes) was conducted. The dewatered digested sludge, raw sludge and fermenting sludge from the municipal wastewater treatment plant were used as materials. The process of PCBs extraction from sludge was being observed by three sprinkling columns with fixed doses of water what stimulated extraction of contamination by rainfall. Dynamics of leaching for chosen PCBs congeners from sewage sludge was determined by examining changes of PCBs in sewage sludge before and after the extraction and in effluents.

Keywords: Polychlorinated biphenyls, methane fermentation, aerobic stabilization, lysimeter column.

1 INTRODUCTION

Polychlorinated biphenyls (PCBs) are a group of micropollutants which, when introduced together with e.g. sewage sludge into the environment in amounts exceeding maximum admissible concentration, may be dangerous both for the environment and the health or life of higher organisms (Berset & Holzer 1996, Ericson 1999).

When analysing the fertilizing properties of sewage sludge, its further disposal in terms of agricultural use and the nature should be taken into consideration. Processes of sewage sludge stabilization should be performed so as to enable the reuse of the nutritive substances contained in a sludge, macro- and microelements, while preventing secondary pollution (Lazzari et al. 2000, Mangas et al. 2002, Mantis et al. 2005, Pereira & Kuch 2005, Working Document on Sludge 2000). Micropollutants, introduced together with the sludge and adsorbed on the finest suspension may be released during sludge disposal and/or be washed out by rain water.

It should be mentioned that effective processes which prevent the harmful influence of sewage sludge on the environment and sewage sludge disposal may generate problems of an organizational and technological nature. Therefore, work on enhancement of the biochemical stabilization process should contribute to acceleration of the process of chemical decomposition of compounds. It is therefore reasonable to perform investigations on the influence of the intensification of methane fermentation or aerobic sludge stabilization on the changes in micropollutant concentrations in stabilized sludge while taking its reuse for nature and/or agriculture into consideration (Katsoyiannis & Samara 2005, Oleszek-Kudlak et al. 2005).

2 MATERIAL AND METHODS

2.1 *Material*

Sewage sludge from the Sewage-Treatment Plant "Warta" S. A. in Czestochowa, Poland was used in the study. It is a municipal wastewater-treatment plant which uses the activated sludge method for biological removal of carbon and nitrogen compounds and the chemical precipitation method (PIX coagulating agent) in order to remove phosphorus.

The materials under investigation were raw sludge collected from preliminary sedimentation tank thickened, excess, digested sewage sludge and dewatered sludge after the digestion process, dewatered sludge after aerobic stabilization. The digested sludge was sampled only once from the bottom of a covered digester. The samples for the excess sludge were taken before and after mechanical thickening. They were heterogeneous material samples; in order to homogenise and separate the mechanical contamination the sludges where strained through a 3 mm mesh.

To initiate the methane fermentation process, the raw sludge was inoculated with the digested sludge in a volume ratio of 1:2.

In order to obtain sludges with the appropriate hydration, the proper excess sludge was reached as a result of mixing, with the 3:2 volume proportion of the sludge sampled before and after thickening.

During the aerobic stabilization process, the mixed sludge was used; it was obtained a by mixing the raw and proper excess sludge with the volume proportion of 1:2.

Previous investigations (Janosz-Rajczyk et al. 2005) showed that thermal conditioning accelerates the hydrolysis of organic substrate. Therefore, before the methane fermentation process, raw sludge as well as initial and excess sludge mixture was conditioned before aerobic stabilization.

2.2 *Methane fermentation of sludge*

For the methane fermentation process the following sludge mixtures were used:

- raw and digested sludge (as a reference),
- raw sludge after thermal conditioning and digested sludge.

The process of methane fermentation was conducted in glass bioreactors. The bioreactors were filled with sludge mixtures, and placed in the incubator at $35 \pm 1°C$. The incubation took 18 days.

2.3 *Aerobic stabilization*

For the process of aerobic stabilization the following sludge mixtures were used:

- raw and proper excess sludge, after thermal conditioning,
- raw and proper excess sludge, without thermal conditioning (reference).

The process took 22 days and was performed at ambient temperature in chambers with a working volume of 14 litres. The sludges were aerated by means of membrane pumps with the output of 60 dm^3 of air per hour, keeping the oxygen in excess in the bioreactors.

2.4 *Leaching dynamics of PCBs after the processes of biological stabilization for sewage sludge*

The investigations on leaching of sewage sludge were performed in plexiglass lysimeters (diameter 103 mm). Into these columns, the sludges (thickened in a test-tube centrifuge) were placed after the processes of biochemical stabilization. The height of filling with sludge after methane fermentation was ca. 20 cm, the wet sludge mass was 1570 g. The thickness of the sludge layer after stabilization was ca. 17 cm, the wet sludge mass amounted to 1110 g. Lysimeter sludges were sprayed with distilled water (200 cm^3 per day). The amount of eluates was measured every day and stored in dark-glass bottles under the lysimeter columns. PCB analysis was performed after gathering of 1300 cm^3 of the eluate. The leaching dynamics was completed after collection of eight eluates. Before and after performing a leaching, quantitative and qualitative changes of PCBs were determined.

2.5 *Determination of PCBs for the sewage sludge and eluates*

After initial activities (centrifuging, drying, grinding), the sludge was subjected to sonification using dichloromethane as a solvent. Then, the sample was centrifuged and the extract was filtered in BAKERBOND spe Silica Gel (SiOH) columns. Next, the extraction of PCBs to the solid phase was performed by means of BAKERBOND spe octadecyl C_{18} type. Finally, PCBs were washed from the column sorbent using dichloromethane. The resulting eluate was evaporated to 1 cm^3.

The eluate with the volume of 500 cm^3 was centrifuged and the fluid from above the sludge passed through the previously conditioned columns of BAKERBOND spe octadecyl C_{18} type. PCBs were washed from the column sorbent using dichloromethane. The resulting eluate was evaporated down to 1 cm^3.

The qualitative and quantitative analysis of chosen congeners of PCB was performed by means of CGC-MS.

Determination of PCBs was performed by comparison of retention times for a standard mixture and the determined substances and the comparison of their mass spectra. The quantitative analysis consisted in comparison of the surface of peaks for chosen characteristic ions and the appropriate peaks for the investigated sample.

For the quantitative and qualitative determination of PCBs employed here, the recovery values for the analysed congeners are provided elsewhere (Sułkowski & Rosińska 1996).

3 RESULTS AND DISCUSSION

3.1 *Concentration of PCB in eluates during leaching dynamics of sewage sludge after methane fermentation*

In the investigated eluates of the sludge after methane fermentation process, only tri- and tetrachlorobiphenyl with codes 28 and 52 were determined, while the PCB 52 congener was determined for eluates from 1st to 5th and PCB 28 for all the received eluates. The highest PCB 28 concentration was found for the second eluate (15.59 ng/dm^3) and the sixth eluate (19.18 ng/dm^3), while the lowest – in the fourth, the seventh and the eighth eluate which amounted to, respectively: 1.14, 1.58, and 1.47 ng/dm^3. The highest concentration of PCB 52 (2.20 ng/dm^3) was found for the fourth eluate, the lowest for the second one (0.23 ng/dm^3) (Tab. 1).

For the sludge after methane stabilization but before the leaching process, seven PCB congeners were determined; their total concentration amounted to 117760 ng/kg d.m. The highest, 50% participation was found for trichlorobiphenyl PCB 28 (64680 ng/kg d.m.). After leaching, the total concentration of seven PCB congeners amounted to 36460 ng/kg d.m. It was observed that for the analysed PCBs, the most intensive decrease in concentration was found for the trichlorobiphenyl with code 28 (84.16%) and hexachlorobiphenyl PCB 153 (71.46%) (Fig. 1).

Comparing the concentration of leached congeners and the concentration of PCBs in the sludges before and after leaching, it was found that analysed PCB are leached only at a low level – this proves their strong adsorption on solids, which is especially durable when the investigated matrix is rich in organic substances.

Concentrations of PCBs in the investigated sludges were lower than anticipated from their aqueous solubility, which is connected with their hydrophobic nature. Water solubility of PCBs decreases with the increase in number of chlorine atoms in the biphenyl fragment. Among the investigated PCBs, the best solubility in water was observed for PCB 28. Two low-chlorinated PCB congeners with codes 28 and 52 were washed out of the investigated sludge. Such a washing out of low-chlorinated PCBs results from the fact that the received eluates may have contained e.g. detergents which facilitated migration of PCBs to the water phase.

3.2 *Concentration of PCBs in eluates during leaching dynamics of sewage sludge after aerobic stabilization*

In each eluate of the sludge after aerobic stabilization, a gradual increase in the quantity of the determined congeners of PCB was observed. In the

Table 1. Concentration of PCBs in eluates during leaching process of sewage sludge after methane fermentation.

Congener PCB	Concentration of PCBs in eluates, ng/dm^3							
	1	2	3	4	5	6	7	8
PCB 28	9.46	15.36	10.00	1.14	8.04	19.18	1.58	1.47
PCB 52	0.48	0.23	0.77	2.20	0.61	<0.01	<0.01	<0.01
PCB 101	<0.01	<0.01	<0.01	<0.01	<0.01	<0.01	<0.01	<0.01
PCB 118	<0.01	<0.01	<0.01	<0.01	<0.01	<0.01	<0.01	<0.01
PCB 138	<0.01	<0.01	<0.01	<0.01	<0.01	<0.01	<0.01	<0.01
PCB 153	<0.01	<0.01	<0.01	<0.01	<0.01	<0.01	<0.01	<0.01
PCB 180	<0.01	<0.01	<0.01	<0.01	<0.01	<0.01	<0.01	<0.01
Σ PCB	9.94	15.59	10.77	3.34	8.65	19.18	1.58	1.47

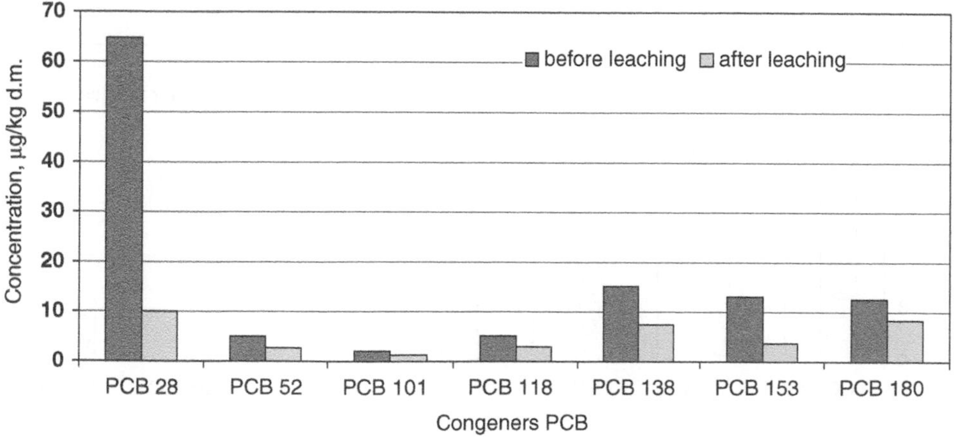

Figure 1. Concentration of PCBs in sewage sludge (after methane fermentation) before and after leaching.

eluates No 1 and 2, lower chlorinated biphenyls PCB 28 and 52 were determined, their total concentrations being respectively 2.32 and 0.32 ng/dm^3. In eluate No 3, PCB 101 was additionally determined and total concentration of the determined congeners amounted to 2.41 ng/dm^3. For the fourth eluate, only PCB 138 was not determined; in the fifth eluate, six PCB isomers were observed (total concentration 4.38 ng/dm^3); in the sixth and the seventh eluates, PCB 28, 52, 101, and 118 were determined, while total concentrations were respectively 4.65 and 4.57 ng/dm^3. In the last, the eighth eluate, all analysed PCBs were determined and the highest total concentration at a level of 28.81 ng/dm^3 was observed (Tab. 2). This fact could have been caused by the change in sorption ability of the sludge in relation to PCBs, which may have been a result of aerobic environment interaction with the sludge particles.

For the sludge after aerobic stabilization before the process of leaching, seven PCB congeners were determined; their total concentration amounted to 62570 ng/kg d.m. The highest, 87% participation was found for tri-, tetra- and pentachlorobiphenyls with codes 28, 52, 101, and 118 whose total concentration amounted to 54220 ng/kg d.m. After the process of leaching, only the above mentioned congeners with a total concentration of 22040 ng/kg d.m. were determined. The highest leaching degree was observed for PCB 153 (4.2%), PCB 180 (3.9%) and for PCB 138 (3.8%), while the lowest one was for PCB 52 (1.0%) as well as PCB 101 and 118 (1.1%) (Fig. 2).

Considering the leaching conditions (light, temperature, matrix rich in organic and mineral compounds) it seems safe to suggest that PCBs undergo degradation due to the presence of the microorganisms in the sludges and/or chemical and physical factors. Strikingly the most intensive decrease was in the concentration of low-chlorinated biphenyls: tri- and tetrachlorobiphenyls i.e. PCB 28 and PCB 52. According to the literature (Sierra et al. 2003, Vrana et al. 1998), these low-chlorinated biphenyls are subject

Table 2. Concentration of PCBs in eluates during leaching process of sewage sludge after aerobic stabilization.

Congener PCB	Concentration of PCBs in eluates, ng/dm^3							
	1	2	3	4	5	6	7	8
PCB 28	2.32	<0.01	1.59	2.31	1.30	3.01	3.25	4.02
PCB 52	<0.01	0.32	0.44	1.78	0.64	0.43	0.38	3.44
PCB 101	<0.01	<0.01	0.38	2.02	0.62	0.76	0.57	4.06
PCB 118	<0.01	<0.01	<0.01	2.03	0.60	0.45	0.37	3.96
PCB 138	<0.01	<0.01	<0.01	<0.01	0.46	<0.01	<0.01	4.72
PCB 153	<0.01	<0.01	<0.01	1.71	0.76	<0.01	<0.01	3.54
PCB 180	<0.01	<0.01	<0.01	3.36	<0.01	<0.01	<0.01	5.07
Σ PCB	2.32	0.32	2.41	13.21	4.38	4.65	4.57	28.81

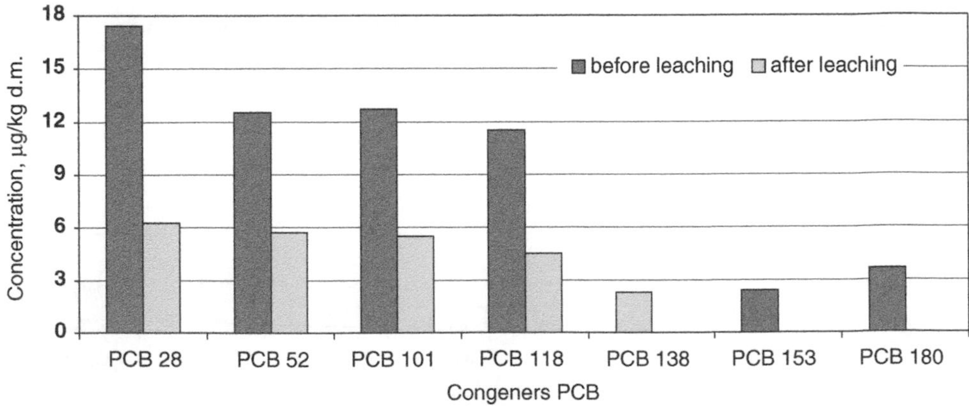

Figure 2. Concentration of PCBs in sewage sludge (after aerobic stabilization) before and after leaching.

to biodegradation under aerobic conditions. Apparently, during aerobic biodegradation, microorganisms attack the PCBs containing 1–5 chlorine atoms in the biphenyl skeleton. The end products are CO_2, H_2O and Cl^-.

4 CONCLUSIONS

During the process of leaching of PCBs out of the sludge both after methane fermentation and aerobic stabilization it was found that PCBs were leached only to a low degree. This result is not entirely unexpected owing to their known strong adsorption on solid molecules.

After the process of methane fermentation, the low-chlorinated PCBs with codes 28 and 52 were washed out of the investigated sludge, which may result from their comparatively better solubility in water as compared to the rest of the analysed congeners (solubility of the PCB in water decreases with the increase in the number of atoms of chlorine in the biphenyl molecule).

During the leaching of PCBs from the sludge after the process of oxygen stabilization, a gradual increase in the number of determined PCBs was observed in the eluates. For example, in the last eluate seven congeners were determined. The ability to leach the PCBs from the sewage sludges proves that dissolution of the PCBs in cosolvents (such as, for example, surface active substances) must have taken place.

The differences in PCB concentration in sludge (after the process of anaerobic and aerobic stabilization) before and after leaching and the quantity of leached congeners proved that during exposition of the sludge some alteration (physical and biochemical) and/or of strong sorption on a matrix molecules occurred (Dercova et al. 1999, Sierra et al. 2003, Triska et al. 2004).

Because of the observed leaching of PCBs from the investigated sludge, waste dumps can be a source of PCBs entering the soil and water environment.

ACKNOWLEDGEMENTS

This research was supported by the State Committee for Scientific Research, Poland within the Research Projects numbered KBN 7T09C03420, 3T09D07627 and BS-402–02/2004.

REFERENCES

Berset, J.D. & Holzer, R. 1996. Determination of coplanar and ortho substituted PCBs in some sewage sludges of Switzerland using HRGC/ECD and HRGC/MSD. *Chemosphere*. 32, 12, 2317–2333.

Dercová, K., Vrana, B., Baláž, Š. 1999. A kinetic distribution model of evaporation, biosorption and biodegradation of polychlorinated biphenyls (PCBs) in the suspension of *Pseudomonas stutzeri*. *Chemosphere*. 38, 6, 1391–1400.

Ericson, M.D. 1997. Analytical Chemistry of PCBs. second edition, CRC, Lewis Publishers Boca Raton, New York.

Janosz-Rajczyk, M., Rosińska, A., Płoszaj, J. 2005. Effect of conditioning of sewage sludge on the determination of PCBs. *32nd International Conference of SSCHE*. Tatranske Matliare, Slovakia.

Katsoyiannis, A. & Samara, C. 2005. Persistent organic pollutants (POPs) in the conventional activated sludge treatment process: fate and mass balance. *Environmental Research*. 97, 245–257.

Lazzari, L., Sperni, L., Bertin, P., Pavoni, B. 2000. Correlation between inorganic (heavy metals) and organic (PCBs and PAHs) micropollutant concentrations during sewage sludge composting processes. *Chemosphere*. 41, 427–435.

Mangas, E., Vaquero, M.T., Comellas, L., Broto-Puig, F. 2002. Analysis and fate of aliphatic hydrocarbons, linear alkylbenzenes, polychlorinated biphenyls and polycyclic aromatic hydrocarbons in sewage sludge-amended soils. *Chemosphere*. 36, 1, 61–72.

Mantis, I., Voutsa, D., Samara, C. 2005. Assessment of the environmental hazard from municipal and industrial wastewater treatment sludge by employing chemical and biological methods. *Ecotoxicology and Environmental Safety*. 62, 397–407.

Oleszak-Kudlak, S., Grabda, M., Czaplicka, M., Rosik-Dulewska, Cz., Shibata, E., Nakamura, T. 2005. Fate of PCDD/PCDF during mechanical-biological sludge treatment. *Chemosphere*. 61, 389–397.

Pereira, M. & Kuch, B. 2005. Heavy metals, PCDD/F and PCB in sewage sludge samples from two wastewater treatment facilities in Rio de Janeiro State, Brazil. *Chemosphere*. 60, 844–853.

Sułkowski, W. & Rosińska, A. 1999. Comparison of the efficiency of extraction methods for polychlorinated biphenyls from environmental wastes. *J. Chromatogr. A*. 845, 349–355.

Sierra, I., Valera, J.L., Marina, M.L., Laborda, F. 2003. Study of the biodegradation process of polychlorinated biphenyls in liquid medium and soil by a new isolated aerobic bacterium (*Janibacter* sp.). *Chemosphere*. 53, 609–618.

Třiska, J., Kuncová, G., Macková, M., Nováková, H., Paasivirta, J., Lahtiperä, M., Vrchotová, N. 2004. Isolation and identification of intermediates from biodegradation of low chlorinated biphenyls. *Chemosphere*. 54, 725–733.

Working document on sludge – 3rd draft, 2000. ENV.E.3/LM, Brussels.

Vrana, B., Tandlich, R., Balaz, S., Dercova, K. 1998. Aerobic biodegradation of polychlorinated biphenyls by bacteria. *Biologia*. 53/3, 251–266.

Environmental Engineering – Pawłowski, Dudzińska & Pawłowski (eds)
© 2007 Taylor & Francis Group, London, ISBN13 978-0-415-40818-9

Solid waste vitrification using a direct current plasma arc

January Bień, Witold Białczak & Katarzyna Wystalska
Institute of Environmental Engineering, Czestochowa University of Technology, Czestochowa, Poland

ABSTRACT: The storage of wastes (ashes, municipal wastes, sewage sludge) in landfill sites is limited by the available area of the country. Thermal methods are a possible way of reducing the volume of waste. Plasma technologies offer flexible methods to treat the various wastes. In this paper the authors report their own experiments on the treatment of wastes (slag from an incineration plant, sewage sludge) in a plasma reactor. The results show that the solid product of the plasma process (glassy-slag) is environmentally benign.

Keywords: Waste utilization, vitrification, plasma.

1 INTRODUCTION

In environmental applications plasma technology has generally been used as a non-thermal (non-equilibrium) process – mainly for gas conditioning, as well as a thermal (equilibrium) plasma. Wastes to be submitted to plasma utilization processes may be solid, liquid or gaseous. Solid waste is usually input by means of a feeder, while for liquid or gaseous waste an "in-flight" system is used, i.e. simultaneous feeding of plasma gas and waste. Initially, the plasma process was proposed for radioactive waste. However, in recent years, its use with other groups of wastes (medical waste, ashes left after a combustion process (Cheng et al. 2002), municipal waste (Beck 2003), carbonaceous waste (Nishikawa et al. 2004) and hazardous waste (Kavouras et al. 2003)) has been shown to be a feasible alternative.

At the present time, the most popular way of using plasma technology is for fly ash and bottom ash. Ashes generated by municipal waste incineration may contain significant amounts of PAH, PCB, chlorobenzenes, chlorophenols and chlorinated dioxins and benzenofurans. Polar mutagenic organic chemicals can also be found (Reijnders 2005).

Vitrification process may be conducted to obtain a glassy slag (Kavouras et al. 2003) or glass ceramics (Cheng & Chen 2003).

Until now, application of plasma technology for waste reduction in Poland has been limited to scientific investigations (Bień et al. 2003, Kołaciński & Cedzyńska 1998). In some countries, operating industrial plasma plants, designed for the complex utilization of the wastes, do exist. One of the plasma utilization systems is PEM™ – Plasma Enhanced Melter™ Process Hearth (Fig. 1). The system is used for hazardous, medical, radioactive, industrial and municipal waste. Such systems are in operation in Richland, USA (Allied Technology Group, Inc.'s Mixed Waste Treatment Fac. (APET)) for medical and hospital waste utilization.

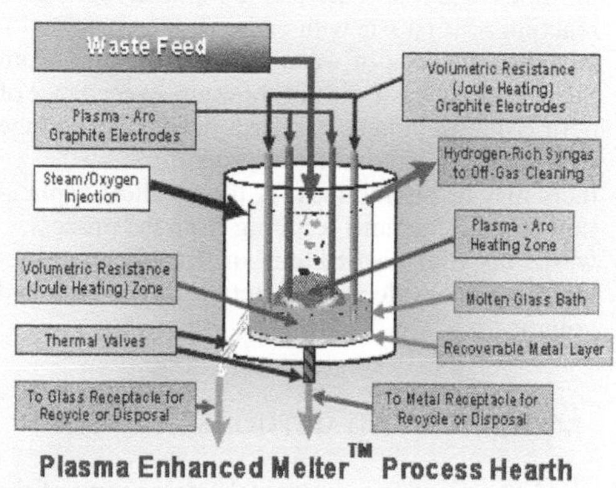

Figure 1. Diagram of the Plasma Enhanced Melter™ Process Hearth System.

Plasma reactors are frequently used for vitrification of the ashes obtained after the waste combustion process. Results of leaching tests of ashes show that such waste might be hazardous to the environment. The vitrification process allows for a significant reduction of the volume of hazardous components leaking to the soil (Haugsten & Gustavson 2002). However, a glassy slag derived from a mineral phase is much less troublesome to handle than solid waste containing significant amounts of organic substances.

The applications of the plasma technologies to hazardous and toxic waste are justified because of the possibility of conducting the process in such way that the received products are not hazardous for the environment (Cedzyńska & Kołaciński 1998, 1999, Inaba et al. 1999, Kordylewski & Robak 2002). The plasma processes might make it easier to adapt waste management in accord legislative environmental protection norms of the European Union.

Initial tests (Bień et al. 2003) on waste utilization in a plasma reactor, carried out by the authors, allowed the following conclusions:

- it will be possible to solidify the tested waste in the vitrification process carried out in the plasma reactor (even including significant amounts of organic substances),
- the waste vitrification can be conducted without the necessity of adding any fixing substances or substances to lower the melting temperature,
- in some cases it is beneficial to input oxygen in the reaction zone (waste with carbon),
- chemical analysis of water extracts sampled from the converted waste did not show any exceedance of permissible pollutants standards defined for sewage sludge injected into water or ground,
- there may be possibilities for the practical application of solid fractions resulting from the process,
- waste disposal carried out in the plasma reactor causes considerable reduction in its mass and volume.

2 MATERIALS AND METHODS

At the Institute of Environmental Engineering of the Czestochowa University of Technology a prototype plasma DC reactor for waste utilization (including those of a hazardous nature) has been constructed (Fig. 2). The reactor is a hermetically closed device, and capable of being operated underpressure as well as at excess pressure (up to 0,05 MPa). Hence, it is possible to carry out the process in conditions of a controlled atmosphere. The power supply is adjusted to 150 kW, with a continuous regulation of current intensity in the range between 50 and 1300 A.

A plasma burner has been constructed to move vertically in order to adjust the length of the arc to the maximum distance from the crucible up 0,35 m. On the lower electrode (anode) ceramic or graphite crucibles (10 dm^3) can be placed. Argon is used as the plasma creating gas. Additional gas (for example O_2) can also be input to the zone of reaction Exchanging the crucibles takes place through a charging window.

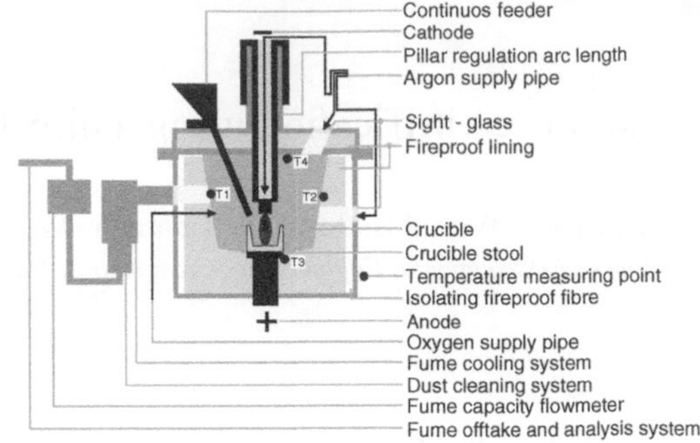

Figure 2. The scheme of the plasma reactor.

The reactor is equipped with sight-hole (quartz), which allows visible observation of the reaction zone and optical measurements (e.g. temperature measurement with a pyrometer).

The plasma reactor is equipped with a computer system for control and recording. The system is based on the PCLab Card. The measuring advices are connected to PC converters, which are used to modify measured signals and to isolate the PC. The measurement signals at the entrance of the card were attached via a multi-vein cable and a clamping terminal.

The temperatures are displayed on the monitor's screen, whereas other physical quantities are simultaneously monitored on a control desk, with an exclusion of the combustion gas flow that is visible directly on a metal rotameter. Currently some actions have been continued to design a automatic control system of the plasma reactor, and in this way to enable more precise operation of the utilization process.

The experiments using plasma, presented in this paper, were carried out in conditions of periodical feeding with material, and with the use of small graphite crucibles. Each time 50 g of the material was input into the crucible. During the experiments the following parameters of the reactor were changed: current intensity of the electrical arc, oxygen flow, and the duration of the process. Constant parameters of the process were: the intensity of flow of the plasma gas (20 dm^3/min) and the length of the plasma arc (0,15 m).

In this paper the authors compare the stability of the vitrificates produced in the plasma reactor. The substrates were obtained by two methods of the thermal utilization of medical waste:

- conventional thermal pyrolysis – temperature 800°C;
- thermal pyrolysis at the lower temperature – 400°C.

In the first method the medical waste were subject to conversion in a waste utilization plant. The process

Table 1. Characteristics of waste before plasma modification.

Indicator (%)	Slag from the medical waste	Char from the sewage sludge
Total humidity	3,10	1,10
Flammable parts	40,21	44,60
Inflammable parts	56,69	54,30
Total carbon	22,00	36,00

Table 2. Content of oxides in the mineral part of the wastes (as % of total of waste mass).

Component	Slag from the medical waste – % weight	Char from the sewage sludge – % weight
SiO_2	15,90	21,60
Al_2O_3	4,69	2,50
CaO	12,24	9,50
MgO	2,24	2,30
Fe_2O_3	6,25	5,60

is conducted in two stages: the first – in a pyrolysis chamber, at about 800°C – the thermal decomposition of the waste takes place; in the second stage the formed gas goes into an after-burning chamber. The after-burning process takes place at about 1200°C. The post-combustion gas obtained in such conditions, after cooling to 250°C are subject to a three-phase wet purifying system. The system consists of a Venturi washer, a wet ventilator and an absorption column with an expanded superficial filling. Before going to the atmosphere the gas flows also through a dropper (taking out the water). The process heat is used for warming plant buildings. The reduction of the waste mass during this thermal process of utilization is more than 90% (the solid product – slag).

In the second method sewage sludge was subjected to pyrolysis at 400°C for 30 minutes. During the process three fractions were obtained: solid (80%), liquid (3%), and gaseous (17%).

Medical waste and char received after the pyrolysis of sewage sludge was subjected to a physico-chemical analysis, where the following parameters were determined: total humidity of waste, content of the combustible parts, total carbon, content of oxides, content of heavy metals. The substrates of the plasma process were subjected to a leaching test carried out in accordance with PN-Z-15009. In the water extract some selected indicators as well as the heavy metal concentration were determined. The results of the tests are shown in tables 1, 2, 3.

3 RESULTS

As a result of the plasma modification of the waste (slag and char), conducted with the parameters stated above, the vitrificators showed in fig. 3 were obtained.

Plasma treatment of waste resulted in considerable reduction of its mass and volume. For the vitrificator (slag from medical waste – conventional method) reduction of weight was between 45 and 71% and for vitrificator from char between 67 and 97%. A leaching test was carried out on the resulting slag. The analysis is shown in table 3.

4 DISCUSSION

Waste treatment in our plasma reactor allowed vitrificator from the converted waste to be obtained. In the presented experiments the degree of transformation of the substrate to glassy-slag depended on the amount of oxygen in the reaction area. With the volume flow of oxygen 1 dm^3/h, no vitrificator was obtained at all. Consecutive tests of the plasma modification with $[O_2] = 3–9\ dm^3/min$ allowed vitrificator to form. Its mass compared to the total mass of the test material was in the range between 35 and 100%.

Thus, application of plasma technology allows the mass of the modified waste to be reduced to a level of 50–90%.

The analysis of the water extract for the tested waste before and after the plasma utilization showed the reduction of pH, COD, ammonia nitrogen and chlorides up to the zero level. The metal concentration in the water extract is low, and it does not go above the level determined by the Ministry of Environment in its legislative act on hazardous substances.

The main aim of the plasma treatment of wastes was to decrease the quantity of heavy metals in the water extract. Volatile metals in the leakage were practically absent. It may suggested that they metals have vaporized to the gas phase. Their vaporization temperatures are: Pb – 1749°C, Cd – 765°C, Zn – 907°C.

In future work, we intend carrying out a complete mass-balance. Unfortunately this was impossible till now. The content of the residual metals in the solid products stemming from the plasma process is significant.

The leaching test on waste carried out before and after the plasma modification showed a reduction of leaching to the level of 0,7%.

The optimal parameters for the treatment of waste in our plasma reactor, allowing the vitrificator to be obtained are as follows: O_2 flowrate 7 dm^3/min; current intensity 350 and 400 A, respectively in the case of slag; whereas in the case of char (product of the pyrolysis at $T = 400°C$): 9 dm^3/min, with the current intensity $300 \div 350$ A.

Table 3. Leaching test: analysis.

Parameter	Slag from the medical waste	Vitrificator received from the slag	Char from the sewage sludge	Vitrificator from the char
pH COD, mg O_2/dm^3	9,2	6,8	7,6	6,5
Ammonia nitrogen, mg N_{NH_4}/dm^3	66	0,0	280,0	0,0
Chlorides, mg Cl/dm^3	2,15	0,0	220,0	1,0
Metals (mg/dm^3)	690	Not detected	11,0	0,0
Fe	0,4617	0,1036	0,671	1,980
Zn	14,6500	0,0687	0,071	0,012
Ni	0,0986	0,0101	0,073	<0,005
Co	0,0029	0,0016	0,010	<0,010
Cd	0,5869	0,0024	<0,010	<0,010
Cu	0,1426	0,0469	0,010	0,018
Cr	0,1207	0,1017	0,060	0,018
Pb	0,0928	0,0041	0,015	<0,010

Figure 3. Vitrificators obtained from the slag (left) and char (right).

5 CONCLUSIONS

The application of plasma technology enables considerable reduction of waste mass and volume. The obtained glassy-slag is environmentally harmless. Taking this into account possibilities for its useful utilization should be sought.

Currently we are conducting experiments involving the continuous feed of substrate to the plasma reactor. We hope to show the results in the near future.

REFERENCES

Beck R.W., 2003, Inc, City of Honolulu; Review of plasma arc gasification and vitrification technology for waste disposal - Final Report, 1–33.

Bień J.B., Bień J.D., Białczak W., Wystalska K., 2003, Organic and Mineral Waste Solidification by Arc Plasma Technology, *Inżynieria i Ochrona Środowiska*, t.6, nr 3–4, 469–477.

Cedzyńska K., Kołaciński Z., 1998, Plazmowa destrukcja ciekłych i gazowych odpadów chloroorganicznych, *V Jubileuszowa Konferencja Naukowo – Techniczna Termiczna utylizacja odpadów – wymiana doświadczeń i poglądów*, Poznań, 183–187.

Cedzyńska K., Kołaciński Z., 1999, Vitrification of hospital waste incinerator residues. *International symposium on plasma chemistry*, Prague, Czech Republic, 2435–2439.

Cheng T.W., Chu J.P., Tzeng C.C., Chen Y.S., 2002, Treatment and recycling of incinerated ash using thermal plasma technology, *Waste Managment* 22, 485–492.

Cheng T.W., Chen Y.S., 2003, On formation of $CaO–Al_2O_3–SiO_2$ glass–ceramics by vitrification of incineratior fly ash, *Chemosphere* 51, 817–824.

Haugsten K.E., Gustavson B., 2000, Environmental properties of vitrified fly ash from hazardous and municipal waste incineration, *Waste Menegment* 20, 167–176.

Inba T., Nagano M., Endo M., 1999, Investigation of Plasma Treatment for Hazardous Wastes Such As Fly Ash and Asbestos, *Electrical Engineering in Japan*, vol.126, No.3, 73–81.

Kavouras P., Kaimakamis G., Ioannidis Th. A., Kehagias Th., Komninou Ph., Kokkou S., Pavlidou E., Antonopoulos I., Sofoniou M., Zouboulis A., Hadjiantoniou C.P., Nouet G., Prakouras A., Karakostas Th., 2003, Vitrification of lead-rich solid ashes from incineration of hazardous industrial wastes, *Waste Management* 23, 361–371.

Kołaciński Z., Cedzyńska K., 1998, Technologia plazmowej witryfikacji – alternatywą dla składowisk odpadów specjalnych. *Giełda technologiczna "Nauka dla środowiska" RACE*, Katowice, 43–47.

Kordylewski W., Robak Ł., 2002, Witryfikacja odpadów i popiołów, *Gospodarka Paliwami i Energią* 7, 18–21.

Nishikawa H., Ibe M., Tanaka M., Ushio M., Takemoto T., Tanaka K., Tanahashi N., Ito T., 2004, A treatment of carbonaceous wastes using thermal plasma with steam, *Vacuum* 2004, 589–593.

Reijnders, 2005, Disposal, uses and treatment of combustion ashes: a review, *Resources, Conservation and Recycling* 2005, 313–336.

Polska Norma PN-Z-15009: 1997, Solid wastes. Preparation of water extract.

Environmental Engineering – Pawłowski, Dudzińska & Pawłowski (eds)
© 2007 Taylor & Francis Group, London, ISBN13 978-0-415-40818-9

Application of sewage sludge as a component of alternative fuel

Małgorzata Wzorek & Leon Troniewski
Department of Processing Engineering, Technical University of Opole, Poland

ABSTRACT: In the paper, it is suggested to make use of sewage sludge as a component of an alternative fuel which can be used in the process of clinker-burning. The results of experiments on the transformation of sewage sludge into an alternative fuel called "PBS fuel" whose properties meet the requirements of the cement industry, are described. The results of work concerning the composition of that fuel and a method for its production, are also presented in the paper. The physico-chemical properties of the produced fuel were analysed taking its energy characteristics and mechanical properties into particular consideration. The results show that the new fuel possesses the properties necessary to be utilized in the process of cement clinker-burning as a substitute for hard coal.

Keywords: Sewage sludge, alternative fuel, cement rotary kiln.

1 INTRODUCTION

Sewage sludge disposal is one of the most difficult problems to be solved in waste management in Poland and the world.

Its special properties, i.e. substantial moisture, biological harmfulness and heavy metal content, sewage sludge cause difficult technical problems in processing. It is also significant that the amount of sludge is growing by the year while, at the same time, more and more rigorous legal conditions are being introduced governing the dumping and agricultural use of such types of waste.

Thus, the current need is to seek and introduce such methods of sludge management which are favourable for the environment.

Due to the specific character of cement manufacturing, the cement industry offers advantageous conditions for the utilization of different types of waste as substitutes for raw material, fuel and cement additives. The interest of the cement industry in acquiring alternative sources of raw materials and energy results from its striving both to improve the effectiveness of production and to lower its production costs.

The process of cement clinker-burning is characterized by a particularly great demand for thermal energy. An increase in prices of traditional fuels is increasingly bringing about more frequent use of alternative fuels instead of conventional ones. In this way, different types of waste can be used as waste-free with good ecological and economic effects (Oss et al. 2003, Kaantee et al. 2004, Pipilikaki et al. 2005).

In the countries of the European Union, in cement plants associated in the CEMBUREAU (the European Cement Association), alternative fuels supply about 20–30% of thermal energy in the process of clinker-burning (Mokrzycki et al. 2003a).

In Poland, the combustion of alternative fuels as a source of thermal energy is applied to commercial production in a very limited range. In 2003 only about 6.5% of heat was obtained from alternative fuels in the process of clinker-burning.

In recent years, however, many Polish cement plants have begun to make use of a wide range of alternative fuels, including the combustion of waste.

The co-combustion of waste in the process of cement clinker-burning is very beneficial from the ecological point of view, but it cannot influence negatively the process of clinker-burning, and at the same time, the properties of the final product (Trezza et al. 2005, Kikuchi 2001). Therefore, the fuels in use must be characterized by definite physico-chemical properties.

On the basis of many years of experience of cement plants, associated in the CEMBUREAU, the qualitative parameters which must characterize fuels to be used as substitutes for the conventional fuels, have been determined. They are (Mokrzycki et al. 2003b):

- the mean calorific value 12,000 kJ/kg,
- moisture $< 30\%$,
- mineral matter $< 60\%$,
- sulphur $< 2.5\%$,
- chlorine $< 0.3\%$,
- PCB + PCT < 50 ppm,
- heavy metals < 2500 ppm, including mercury content < 19 ppm, cadmium content, thallium content < 100 ppm.

The stability and homogeneity of the chemical composition as well as the physical properties of fuel are other important parameters which should be taken into account. The physical properties determine its receptivity to such operations as transport, storage and batching to the kiln.

Sewage sludge is of no interest to the cement industry as the potential alternative fuel on account of its variable composition, considerable moisture (about 70–80%), which, at the same time, brings about both the reduction of its calorific value and a greasy sludge consistency.

Thus, the authors suggest applying sewage sludge as a component of an alternative fuel meeting the requirements of cement industry. It will be known as *PBS fuel*.

2 MATERIALS AND METHODS

The tests carried out were aimed at working out a technology of fuel production, selecting its composition and assessing the physico-chemical properties of the produced fuel as well as paying special attention to its energy characteristics and physical properties.

In the experimental tests, use was made of sewage sludge coming from a municipal wastewater-treatment plant operated on the basis of the mechanical and biological treatment system. Because of the low calorific value of sludge in its dry mass, an additional component, in the form of coal slime, was applied to increase its calorific value. Moreover, quicklime and coal shale were used to as binding additives. The properties of fuel components are presented in Table 1.

As a result of our experiments, a method of production of the so-called PBS fuel was worked out. It consists of subjecting sewage sludge and other fuel components to a granulation process in a drum granulator specially designed for this purpose.

The drum granulator is equipped with a batching system which determines the size of the granulated product.

The drum granulator is equipped with a batching system which determines the size of the granulated product. As a result of the granulation process, products with the sizes of 20 and 35 mm were obtained.

The granulation process was carried out for many proportions, but on account of the technical conditions of fuel production in a granular form as well as the energy characteristics and desired physical properties of the fuel, it was found that the proportions of components should be maintained in the ranges of (53–62)% of sewage sludge mass, (35–37)% of coal slime and (3–10)% of the binding additive.

After the granulation process, the granulated product was subjected to seasoning, during which it obtained the appropriate physical properties. The fuel in the form of granules of 20 mm in diameter is presented in Figure 1.

Table 1. The properties of sewage sludge and coal slime.

Parameter	Unit	Sewage sludge	Coal slime
Moisture	%	75.53	6.46
Volatile matter	%	44.20	33.72
Ash	%	55.80	26.18
Sulphur	%	0.12	0.73
Chlorine	%	0.09	–
Calorific value	kJ/kg (dry mass)	9,000	19,630

Figure 1. The fuel in the form of granules of 20 mm in diameter.

Table 2. Compositions of the PBS fuel.

Composition	Sewage sludge (% of mass)	Coal slime (% of mass)	Additives: Kind of additive	Additives: % of mass
A	60	34	Quicklime	6
B	60	34	Quicklime	3
			Coal shale	3
C	53	37	Quicklime	10

The energy characteristics and physical properties were analyzed and compared for the compositions are presented in Table 2.

3 RESULTS AND DISCUSSION

3.1 *Energy characteristics of the PBS fuel*

Analyses of the energy characteristics have shown that the produced fuel is characterized by a higher calorific value than the minimum heating value (12 MJ/kg), required for alternative fuels by the cement industry, and the contents of the limited harmful components, i.e. chloride, heavy metals and PCB are much lower than their permissible limits. The properties of the PBS

Table 3. The properties of the PBS fuel.

Parameter	Unit	PBS fuel Composition A	Composition B	Composition C	Hard coal
Calorific value	kJ/kg	13.44	14.80	14.20	16.70–32.70
Moisture	%	11.20	10.92	10.52	1.5–20
Volatile matter	%	31.50	33.50	32.10	2.5–40
Ash	%	32.41	29.15	31.71	3–40
Elementary composition:					
Carbon C	%	36.90	40.96	38.60	65–85
Hydrogen H	%	3.47	3.18	3.22	4.7–5.9
Oxygen O*	%	14.02	14.25	14.15	9–25
Nitrogen N	%	1.30	0.90	1.12	0.02–0.40
Sulphur S	%	0.64	0.58	0.62	0.5–3.5
Chlorine Cl	%	0.06	0.06	0.06	0.02–0.40
Content of heavy metals:	mg/kg of dry mass				
Cu		58	46	30	38.6–45.3
Ni		<25	<25	26	30.5–33.5
Cd		<12.5	<12.5	<12.5	0.58–0.62
Cr		183	<181	162	35.4–39.1
Mn		146	166	180	51.0–70
Zn		250	274	202	2824–3970
Fe		10.59	11.50	11.66	No data
Hg		–	–	–	0.16–1.29
PCB	ppm	<0.2	<0.2	<0.2	No data

Table 4. Chemical composition of ash from the PBS fuel and raw materials for clinker production.

Parameter	Ash from PBS fuel (% of mass)	Blast furnace slag (% of mass)	Lime stone (% of mass)	Siderite (% of mass)
SiO_2	26.16	38.10	7.20	32.20
Al_2O_3	4.82	6.90	2.40	12.20
Fe_2O_3	12.20	2.10	1.10	27.60
CaO	35.45	43.50	48.60	2.60
MgO	2.60	7.10	0.40	2.40
SO_3	8.08	0.21	0.60	3.20
Compounds not marked	10.61	0.96	0.50	0.20

The remainder is ignition loss (it is not marked in table).

fuel as compared with the properties of hard coal are given in Table 3.

The PBS fuel contains a considerable amount of ash. The ash, coming from the combustion of the fuel, has a similar chemical composition to blast furnace slag, which is an iron-bearing additive, and it contains the same oxides that are contained in raw materials used for the production of clinker. It can partly perform a function of the so-called "low raw material" in the set of raw materials, and it must be taken into account while blending.

The oxide composition of ash coming from the PBS fuel and some exemplary components of raw materials are presented in Table 4.

In order to estimate the combustion process of the fuel obtained from sewage sludge, a test of its behavior at a high temperature was carried out. The analyses of the DTA and TG were made in the atmospheric air in a range of temperatures of 25–1,400°C. Figure 2 shows the DTA and TG diagram for the PBS fuel (composition C).

In a range of temperatures of up to 130°C an intensive loss of mass occurred resulting from the evaporation of water. The process of combustion in a range of temperatures of 700°C showed two maximum exothermic peaks at temperatures of about 300°C and 380°C.

At a temperature of ca. 840°C, a small endothermic effect, probably connected with the decomposition of $CaCO_3$, could be observed. The total loss of mass occurred at a temperature of about 1,250°C.

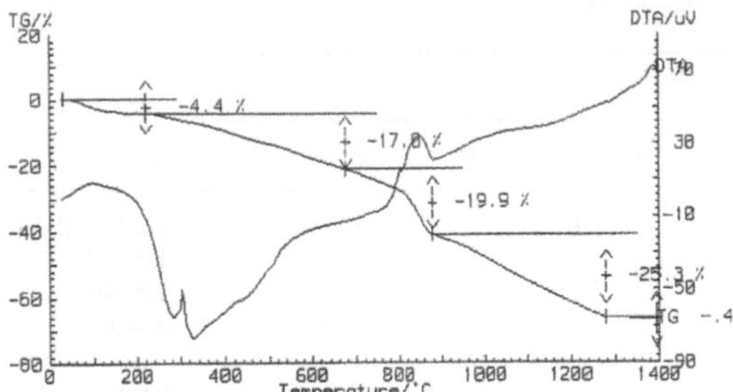

Figure 2. The DTA–TG diagram for the PBS fuel.

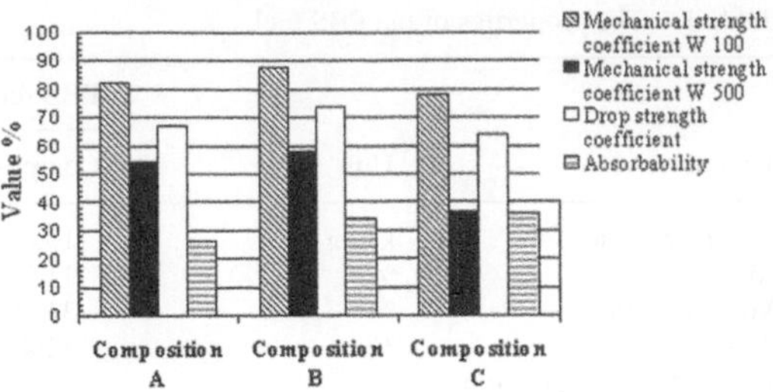

Figure 3. The physical properties of the PBS fuel with granules of 35 mm.

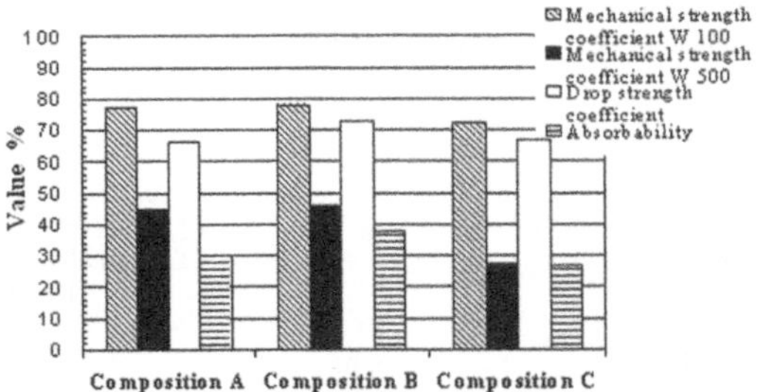

Figure 4. The physical properties of the PBS fuel with granules of 20 mm.

3.2 *Physical properties of the PBS fuel*

The physical properties of the granulated product are essential for its further transport, storage and batching to the kiln. All the compositions of the fuel with granules of 35 and 20 mm and the content of moisture of 10% were tested. The fuel used in the tests, was previously subjected to seasoning for a period of eight days.

The experimental tests were performed according to Polish standards for formed fuels. The mechanical strength, drop strength and fuel absorbability of samples were recorded.

The mechanical strength was determined with a drum method (according to the standard PN-G-04650) and the mechanical strength coefficient was defined after 100 revolutions (W_{100}) and after 500 revolutions (W_{500}) of the drum. This coefficient characterizes the fuel resistance to disintegration and abrasion inside the drum.

Still another tested parameter was the strength of fuel to its drop strength (according to the standard PN-G-04651). The test consisted in dropping fuel, under the standard conditions, from the height of 1.5 m to a hard surface, screening and sieving it and then weighing the residue. The obtained results determine the drop strength coefficient for the fuel.

In the case of fuel storage, it is also essential to know its ability to absorb water, i.e. its absorbability under pre-determined conditions.

The method for determining absorbability consists in defining the amount of water absorbed by the fuel under conditions determined according to the standard PN-G-04652.

A comparison of the results of testing is presented in Figures 3 and 4.

From the analyses carried out it results that the fuel with granules of 35 mm is characterized by better mechanical properties. Besides, a beneficial influence of a coal shale additive (composition B) on the mechanical strength of the fuel, was noted. A relatively high coefficient of the mechanical strength (W 100) suggests that the fuel can be transported short distances. All the compositions of the PBS fuel are characterized by a lower coefficient of the drop strength compared with the coefficient of the mechanical strength after 100 revolutions of the drum.

The mechanical properties of the PBS fuel indicate that the fuel can be subjected to the mechanical operations connected with transport, loading, unloading and batching to the kiln.

The determination of absorbability has shown that granulated products are characterized by a high ability to absorb water, hence the fuel ought to be protected from precipitations during storage.

In order to determine the environmental impact when the fuel is stored, the produced granulates were subjected to tests on harmful substance elution. To this end, aqueous eluates were prepared (according to Polish standard PN-Z-15009), and the values of the basic pollution indices were determined by making use of general procedures for testing waters and sewage.

The results of determining impurity elution were compared with the indices of the maximum values of the permissible pollution concentrations both in the inland and surface waters of the 3rd class of purity and with those for sewage (according to the Order of Polish Ministry of Environmental Protection, Natural Resources and Forestry of 5th November, Official Journal of Low No.116, item 503). They are presented in Table 5.

Table 5. The results of analyses of water extracts from the PBS fuel.

Parameter	Unit	PBS fuel Composition A	Composition B	Composition C	The maximum Permissible pollution concentrations for the 3rd class of the purity of water	The maximum Permissible pollution concentrations for sewage discharges into waters and soil
pH	–	12.41	11.13	12.48	6.5–9.0	6.5–9.0
Sulphides	$MgSO_4/dm^3$	26.14	71.00	19.34	250.00	500.00
Chlorides	mg Cl^-/dm^3	175.14	157.41	133.00	400.00	1000.00
Nitrite nitrogen	mg N_{NO_3}/dm^3	3.45	3.14	2.89	15.00	–
Nitrate nitrogen	mg N_{NO_2}/dm^3	0.34	0.29	0.27	0.06	30.00
Heavy metals:						
Cu	mg Cu/dm^3	1.25	1.25	1.00	0.005	0.50
Ni	mg Ni/dm^3	0.25	0.22	0.16	1.00	2.00
Cd	mg Cd/dm^3	<0.025	<0.025	<0.025	0.10	0.20
Cr	mg Cr/dm^3	<0.025	<0.025	<0.025	0.05	0.70
Mn	mg Mn/dm^3	0.01	0.20	0.01	0.80	1.00
Zn	mg Zn/dm^3	0.12	0.08	0.08	0.20	2.00
Fe	mg Fe/dm^3	0.10	0.64	0.08	2.00	10.00
Na	mg Na/dm^3	101.00	104.00	103.00	150.00	800.00
K	mg K/dm^3	144.00	164.00	108.00	15.00	80.00

The results show that the permissible values were exceeded only for the pH of the water extract and the contents of heavy metals, i.e. potassium (K) and copper (Cu). Much lower values of other parameters than those for indices of pollution concentrations in the surface water of the 3rd class of purity were observed.

On the basis of the analyses of harmful component elution, it can be concluded that, in spite of the fact that permissible pollution concentrations were exceeded only for three parameters, fuel should be stored on the grounds protected from penetration of harmful components into the soil. The fuel ought to be stored in a roofed area with a concrete foundation. Only then is the fuel fully protected from water penetration and potential infiltration of harmful components into the soil.

3.3 *The proposed method for introduction of the PBS fuel into a cement kiln*

The energy characteristics and physical properties of the produced fuel suggest that this type of fuel can be introduced into the chamber located between the cyclone preheater and the rotary kiln or into the pre-calciner burner installed in a state-of-the art modern kiln installation. These are the places where the kiln can be charged with low-energy fuels showing a high moisture content and likely to contain a larger amount of mineral matter.

The thermal and material simulation calculations were made for a kiln working on the basis of a dry method with the efficiency of 90.2 Mg/h. They allowed the determination of the range of a substitution of hard coal by the PBS fuel within 5–20% of the total heat consumption. However, 10% coal substitution is safe for technical and technological reasons.

4 CONCLUSIONS

Both energy costs and environmental protection standards have made the cement industry evaluate to what extent conventional fuels can be replaced by alternative ones. The process of clinker-burning is well-suited for the use of various alternative fuels. Conventional fuels, replaced by alternative ones, are beneficial to the cement industry for ecological and economic reasons. As a result of our experiments, the following conclusions can be drawn:

- Sewage sludge can be a component of alternative fuel with the appropriate energy characteristics and physical properties meeting the requirements of the cement industry.
- The method for the production for fuel from sludge is not complicated and it does not require either much investment or running expenses.
- The suggested method of sewage sludge management can be interesting from the economic point of view for wastewater-treatment plants and the cement industry.
- The thermal use of sewage sludge in clinker burning can solve the problem of sewage sludge

management in a safe way for those regions having cement plants.

- The physical composition and form of the produced PBS fuel make it able to be used in other processes involving industrial fuel combustion, e.g., in the process of co-combustion with coal in existing power plants and heat generating plants.

REFERENCES

Kaantee U., Zevenhoven R. & Backman R. 2004. Cement manufacturing using alternative fuels and advantages of process modeling. *Fuel Processing Technology*. 85: 293–301.

Kikuchi R. 2001. Recycling of municipal solid waste for cement production: pilot-scale test for transforming incineration ash of solid waste into cement clinker. *Resources Conservation and Recycling*. 31: 137–147.

Mokrzycki E., Uliasz-Bocheńczyk A. & Sarna M. 2003a. Use of alternative fuels for the cement industry. *Applied Energy*. 74: 95–100.

Mokrzycki E., Uliasz-Bocheńczyk A. & Sarna M. 2003b. Use of alternative fuels in the Polish cement industry. *Applied Energy*. 74: 101–111.

Oss H.G. & Padovani A.C. 2003. Cement Manufacture and the Environment. Part II: Environmental Challenges and Opportunities. *Journal of Industrial Ecology*. 71(1): 93–126.

Pipilikaki P., Katsioti M. & Papageorgiou D. 2005. Use of tire derived fuel in clinker burning. *Cement & Concrete Composites*. 27: 843–847.

Trezza M.A. & Scian A.N. 2005. Waste fuels: their effect on Portland cement clinker. *Cement and Concrete Research*. 35: 438–444.

Environmental Engineering – Pawłowski, Dudzińska & Pawłowski (eds)
© 2007 Taylor & Francis Group, London, ISBN13 978-0-415-40818-9

Detection and quantification of viable helminth ova in sludge from sewage treatment plants in the Podlaskie district of Poland

Andrzej Butarewicz
Chair of Sanitary Biology and Biotechnology, Białystok Technical University, Białystok, Poland

ABSTRACT: Municipal wastewater usually contains four major types of human pathogenic organisms: bacteria, viruses, protozoa and helminth. Most of them are found in sewage sludge. Polish regulations on agricultural re-use of sludge limit viable helminth ova to: *Ascaris* sp., *Trichuris* sp. and *Toxocara* sp. (ATT) and *Salmonella* genus bacteria. In recent years, various detection procedures for viable helminth ova in sewage sludge have been developed. This paper shows the results of viable helminth ova and *Salmonella* genus bacteria in sludge from sewage treatment plants located in the Podlaskie District. ATT detection and quantification was done by the Tulane Method and Polish Standards. It was shown that *Ascaris* ova was the most frequently detected in the sludge (87–92%). The presence of the other two helminth depended on the method used: *Trichuris* 4–7%, *Toxocara* 4–5%. *Salmonella* genus bacteria were detected in 13.3% of sewage sludge samples.

Keywords: Sewage sludge, Salmonella, helminths, egg viability.

1 INTRODUCTION

Sewage sludge is an inevitable by-product in the process of municipal and industrial sewage treatment. Most countries have seen an increase in the quantity of the sludge generated by sewage plants. The quantity of generated sludge is directly influenced by the ever-improving sewage treatment technology as well as legal regulations relating to the treated sewage quality.

All countries (Poland included) are concerned with the problem of sewage sludge management. Estimates presented by European Environmental Agency showed a continuous increase in sewage sludge, from 7.2 million tonnes of dry matter in 1998 to 9.4 million tonnes of dry matter in 2005 (EEA, 2001). Most of the available reports on the quantity of sewage sludge are published with a considerable delay and official EEA figures do not include the new member states, which joined the European Union in 2004. Figure 1 shows the total quantity of sludge produced in selected EU member countries.

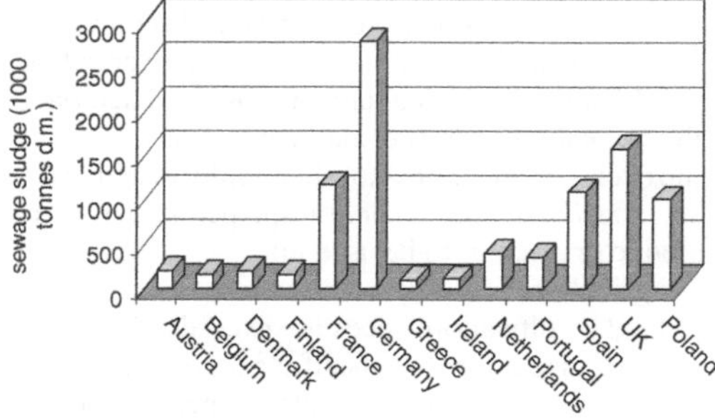

Figure 1. Predicted quantity of sewage sludge in selected EU member countries in 2005 (according to EEA, 2001 and Polish Ministry of Environment).

Of the new EU member states, Poland is the worst 'offender', producing the largest quantity of sewage sludge. The quantity of sludge produced in 2003 exceeded 1 million tonnes of dry matter, half of which came from industrial sewage treatment plants. According to the Polish Ministry of the Environment, in the next few years there will be a further increase in the quantity of sludge (including municipal sludge) from 0.45 million in 2005 to 0.7 million tonnes in 2014. The expected increase in sludge quantity should encourage developing effective processing methods and choosing the best and most effective disposal methods both in terms of economics and environment protection.

Apart from its desirable soil-forming properties, sewage sludge is a habitat of microfauna and microflora made up of pathogenic bacteria, viruses, parasitic worms, fungi and protozoans. Among the pathogens of epidemiological relevance, *Ascaris* eggs are the most resistant organisms (Capizzi-Banas et al., 2004). It is very important to take into account the presence of these parasite ova in the sludge, as they are extremely resistant to most treatment processes and have higher survival rate (5–7 years in sludge or soil) than other pathogens. They are carried into a sewage plant by sewage containing human and animal excrements, which consequently results in their accumulation in large quantities in sewage sludge.

Due to potential hazards caused by pathogenic organisms, microbiological and helminthological tests must be done, the results of which are important criteria in the sanitary assessment of sewage sludge.

The large number of pathogenic organisms which can be deposed of in sewage sludge does not allow determining all the species which are considered to be dangerous from a sanitary point of view. Many countries, including Poland, have made a list of indicator organisms which must be determined when examining the sewage sludge from municipal sewage plants. They include bacteria of the *Salmonella* genus and viable ova of human and animal helminth of *Ascaris*, *Trichuris* and *Toxocara* genera (ATT).

The 2002 Polish Ministry of the Environment Directive on municipal sewage sludge defined requirements which must be met in order to utilize municipal sewage sludge, the amount of sludge to be spread on agricultural land and the range, frequency and referential methods of examinations of municipal sewage sludge and the land on which it may be spread. Under this directive, municipal sewage sludge can be used if:

- its heavy metal contents do not exceed set limits;
- in 100 g of sludge to be used in agriculture or in land recultivation for agricultural purposes no bacteria of the *Salmonella* genus could be isolated;
- the overall number of viable helminth ova of *Ascaris* sp., *Trichuris* sp., *Toxocara* sp. per kilogram of sludge dry matter to be used in:
  - agriculture – is zero
  - land recultivation – does not exceed 300
  - land adaptation for specific purposes resulting from waste management plans, special management plans or decisions on construction requirements and land management – does not exceed 300
  - compost plants cultivation – does not exceed 300
  - non-consumption plant cultivation and fodder production – does not exceed 300.

2 MATERIALS AND METHODS

The aim of the research carried out in 2004–2005 was to estimate the degree of parasitological and microbiological contamination of sewage sludge produced by municipal sewage treatment plants. Raw, pre-fixing sludge, fixed sludge after pressure desiccation, and sludge dewatered by the Draimad® system or stored in drying beds were examined. Some of the examined sludge was sanitized by treating it with quicklime.

The scope of work covered in each of the examined sludges is as follows:

- determining the presence of bacteria of *Salmonella* genus
- defining the number of viable helminth ova of *Ascaris*, *Trichuris* and *Toxocara* genera by helminthological tests.

The analyses of *Salmonella* genus, isolated from the examined samples, were carried out on selectively-differentiating medium, with the use of the initial multiplication technique on the broth medium with sodium selenite, according to Leifson (Merck). The bacteria were isolated on SS and Rambach selectively-differentiating medium (Merck). API 20E tests were used to identify the strains.

The parasitological tests were carried out in accordance with Polish Standard PN-Z-19000-4 and the Tulane Method developed in the US (Bowman et al., 2003). The latter is used to detect parasite eggs in samples of waste and sewage sludge. Parasitological tests were done on dry (5%) sewage sludge samples: This is equivalent to 30–50 g of sludge wet matter and follows US EPA recommendations (1999). In order to unify the sample matter in both methods determination of selected helminth eggs in 50 g of sludge wet matter was carried out. Solutions of $NaNO_3$, $MgSO_4$ and $ZnSO_4$ of density 1.20 were used for flotation. Samples were tested twice. The extract from sludge parasite eggs were incubated in a wet chamber with formalin solution or in a test-tube, in 0.1 N H_2SO_4 at 26°C for 21 days, then the ova viability was determined on a filter using a microscope. Subsequently, the ATT index per kilo of dry matter was calculated.

Dry matter was determined according to Polish Standard PN-EN 12880.

The following formula was used to calculate the ATT index:

$$N = \frac{\text{number of embryonated ova per kilo of sludge dry matter}}{\text{sludge dry matter in \%}} \times 100\ \%,$$

where: N – number of embryonated ova per kilogram of sludge dry matter = ATT index.

3 RESULTS AND DISCUSION

In the recent years, a number of new or modified methods have been developed to detect parasite eggs in human and animal excrements, as well as in soil, sewage and sewage sludge (Ayres and Mara, 1996; EPA, 1999; Nelson and Darby, 2001; Bowman et al., 2003; Caballero-Hernandez et al., 2004). Most of the methods are based on the processes of eggs floatation and sedimentation. Polish legal regulations also recommend using these research procedures.

Table 1. Parasitological microbiological study results of sludge from selected municipal sewage treatment plants in Podlaskie District.

	Viable helminth ova concentration per kg of dry matter of sludge								
	Method I conforming with PN-Z-19000				Method II "Tulane"				Detection of genus
Nr	*Ascaris*	*Trichuris*	*Toxocara*	**ATT**	*Ascaris*	*Trichuris*	*Toxocara*	**ATT**	*Salmonella*
1	1040	80	230	1350	1360	0	120	1480	Present
2	0	0	0	0	120	0	0	120	Not detected
3	196	0	0	196	236	0	90	326	Not detected
4	0	0	0	0	0	0	0	0	Not detected
5	ns*	ns	ns	ns	130	0	0	130	Not detected
6	350	0	0	350	270	0	0	270	Not detected
7	742	0	90	832	620	96	0	716	Not detected
8	651	0	0	651	780	0	0	780	Not detected
9	430	120	0	550	560	120	0	680	Not detected
10	3320	340	116	3776	4200	180	120	4500	Present
11	0	0	0	0	0	0	0	0	Not detected
12	678	212	0	890	912	0	0	912	Present
13	0	0	0	0	124	0	0	124	Not detected
14	120	0	0	120	0	0	0	0	Not detected
15	ns	ns	ns	ns	450	120	0	570	Not detected
16	660	68	180	908	750	0	152	902	Not detected
17	416	120	120	656	378	0	80	458	Not detected
18	0	0	0	0	0	0	0	0	Not detected
19	360	0	0	360	450	0	0	450	Present
20	168	0	0	168	244	0	0	244	Not detected
21	ns	ns	ns	ns	620	0	0	620	Not detected
22	560	0	0	560	510	0	0	510	Not detected
23	5543	780	0	6323	6020	340	0	6360	Present
24	0	0	0	0	0	0	0	0	Not detected
25	140	0	0	140	160	0	0	160	Not detected
26	ns	ns	ns	ns	320	0	180	500	Not detected
27	0	0	0	0	130	0	0	130	Not detected
28	540	0	178	718	630	0	92	722	Present
29	0	0	0	0	0	0	0	0	Not detected
30	ns	ns	ns	ns	120	0	0	120	Not detected

* ns – not studied, because it was not possible using a method conforming with PN-Z-19000-4.

In this study two parasite ova determining techniques were used. Due to the lack of a special method to examine sewage sludge, the Polish Standard PN-Z-19000-4 of 2001 was used on *Ascaris lumbricoides* and *Trichuris trichura* helminth ova detection in soil samples. In Poland, the technique based on the flotation process is used by laboratories accredited in routine tests of sewage sludge. The other technique used in this research was the one developed at Tulane University, USA (Bowman et al., 2003).

Table 1 shows the results of parasites and *Salmonella* genus bacteria in sewage sludge in the Podlaskie District. Under our country's regulations the determination of *Salmonella* rods in sewage sludge is of qualitative nature only, not quantitative as is the case in many other countries, e.g. France, Italy, the RSA or the USA (Gantzer et al., 2001; Austin, 2001; EPA, 1999). On the other hand, in quantitative tests the requirements for the number of *Salmonella* vary from country to country and relate to different values of sludge dry matter, which should be standardized in the future.

In this study *Salmonella* genus bacteria was noted in a small number (13.3%) of tested samples. Similar results (16.1% positive) were obtained in 2001, when the sludge from the Podlaskie District sewage treatment plants was examined.

The parasitological study results shown in Table 1 indicate a large number of ova detected in the tested sludge. The most frequently detected ova were *Ascaris* genus, which depending on the method used, ranged from 87.8% to 92.0% (Fig. 2). In the case of the other two genera the total number of embryonated ova did not exceed 12.2% and 8%, respectively.

Similar results were recorded in the study conducted by Nelson et al. (2004). More than 85% of

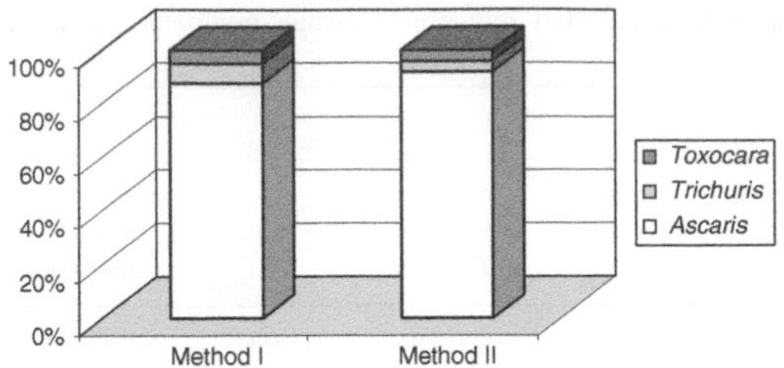

Figure 2. Percentage of parasite ova detected by the two diagnostic methods.

helminth eggs isolated from the sludge were *Ascaris* sp.; the remaining eggs were *Trichuris*, *Hymenolepis*, and *Toxocara* sp.

The study conducted by Schwartzbrod and Banas (2003) showed different results. This study was performed on liquid sludge samples from twenty wastewater treatment plants located in the north west of France. 78.6% of the samples contained helminth eggs belonging to the cestodes (6.1%) and nematodes (93.9%). Most of the detected nematode eggs were viable. The distribution, according to genera, indicated a high prevalence of *Toxocara* eggs (77.4%) followed by *Capillaria* (13.2%), *Trichuris* (8.1%) and *Ascaris* (only 1.3%).

Both diagnostic methods used in this parasitological study proved effective in sewage sludge testing, but each of them has its advantages and disadvantages. The one which complies with the Polish Standard is simpler and less time-consuming. It involves ova floating in $NaNO_3$ saturated solution, then collecting the floated material along with parasite ova from the surface layer. This should be repeated three times. Unfortunately, when using this method, it often happened that particles, which later disturbed the sludge microscope inspection, also floated to the surface. In addition, some of the examined sludge contained chemical compounds which reacted with the NaOH solution used, making, in some cases, examination impossible because of the lack of separation of the solid phase from the fluid on stirring. The Tulane procedure applied in such a case proved to be more effective. After reacting with Tween 80 solution and allowing solid particles to sediment, the sludge was filtered through two 1.27 mm and 0.51 mm size sieves. Then, after floating with $MgSO_4$ or $ZnSO_4$ solution, the floated fluid along with parasite ova was run through a third 0.06 mm size sieve, which to a large extent, solved the problem of larger particles which made further microscopical inspection difficult. This procedure is considerably longer and requires a great deal of precision. Reduction of the floating solution density, which is necessary for the ova sedimentation phase, is a very important stage of the study. While centrifuging, the ova settle and the supernatant should be removed. Supernatant aspiration must be performed very carefully, as the slightest turbulence could cause the ova to float and be accidentally removed along with the supernatant. A few minute delay is recommended, as according to Stoke's law, *Ascaris lumbriocoides* eggs sediment at 20 mm/min in 20°C water, whereas *Trichuris trichura* eggs descend at 16 mm/min (Ayres and Mara, 1996).

It is difficult to determine which ova are embryonated. The method is in accordance with Polish Standard PN-Z-19000-4 which makes it possible to determine the presence of helminth ova on a filter directly it has been applied. It can lead, however, to mistakes in determining the ova viability. That is why in the home study ova were incubated on a drain placed in a wet chamber so as to determine the phase of their growth. The ova incubation gave a better result when 0.1 N H_2SO_4 was used than with formalin. In the case of the other method, the determination of ATT ova viability was not performed until 21 days after incubation in a test-tube containing 0.1 N H_2SO_4.

4 CONCLUSION

Parasitological contamination of municipal sewage sludge is an issue that concerns many sewage treatment plants in the Podlaskie District. The studied sludge contained not only embryonated ova of *Ascaris* genus, but also *Toxocara* and *Trichuris* genera ova. *Salmonella* genus rods were found in a small number of the studied samples. Utilization of such sludge, without its sanitation, is impossible.

During helminthological examination, the study technique must be adapted to each particular sludge due to the large variety of sludge density and chemical composition. Parasitologial diagnostics conforming to Polish Standard PN-Z-19000-4 proved to be difficult, not to say impossible, when testing sludge to which coagulants or dehydrating agents had been added. The 'Tulane Method' developed in the USA proved more efficient.

The ever improving study methods do not guarantee that all embryonated parasite ova can be detected. That is why every precaution must be taken when using sludge for agricultural purposes.

In the years to come it is imperative that EU Member States develop common sewage sludge quality standards. The task falls under the responsibility of The European Agency for Environmental Protection.

ACKNOWLEDGEMENTS

The presented study was supported by Białystok Technical University (project No S/IIŚ/22/06).

REFERENCES

Austin, A. 2001. Health aspects of ecological sanitation. Internet Dialogue on Ecological Sanitation. http://www.ias.unu.edu/proceedings/icibs/ecosan/abstract.html

Ayres, R.M. & Mara, D.D. 1996. Analysis of wastewater for use in agriculture: a laboratory manual of parasitological and bacteriological techniques, WHO, Geneva, ISBN 92 4 154484 8.

Bowman, D.D., Little, M.D. & Reimers, R.S. 2003. Precision and accuracy of an assay for detecting *Ascaris* eggs in various biosolid matrices. *Wat. Res.* 37: 2063–2072.

Caballero-Hernandez, A.I., Castrejon-Pineda, F., Martinez-Gamba, R., Angeles-Campos, S., Perez-Rojas, M. & Buntinx S.E. 2004. Survival and viability of *Ascaris suum* and *Oesophagostomum dentatum* in ensiled swine faeces. *Bioresource Technology* 94: 137–142.

Capizzi-Banas, S., Deloge, M., Remy, M. & Schwartzbrod, J. 2004. Liming as an advanced treatment for sludge sanitation: helminth eggs elimination – *Ascaris* eggs as a model. *Wat Res*. 38: 3251–3258.

European Environmental Agency. 2001. Generation and treatment of sewage sludge. http://themes.eea.eu.int/Environmental_issues/waste/indicators/sewage/index_html

Gantzer, C., Gaspard, P., Galvez, L., Huyard, A., Dumouthier, N. & Schwartzbrod, J. 2001. Monitoring of bacterial and parasitological contamination during various treatment of sludge. *Wat. Res.* 35: 3763–3770.

Nelson, K.L. & Darby, J. 2001. Inactivation of viable *Ascaris* eggs by reagents during enumeration. *Appl. Environ. Microbiol.* 67: 5453–5459.

Nelson, K.L., Jimenez Cisneros, B., Tchobanoglous, G. & Darby, J.L. 2004. Sludge accumulation, characteristics, and pathogen inactivation in four primary waste stabilization ponds in central Mexico. *Wat. Res.* 38: 111–127.

PN-Z-19000-4. 2001. Soil quality-Assessment of the soil sanitary conditions – Detection of eggs of the intestinal parasites *Ascaris lumbricoides* i *Trichuris trichiura*.

PN-EN 12880. 2004. Characterization of sludges – Determination of dry residue and water content.

Schwartzbrod, J. & Banas, S. 2003. Parasite contamination of liquid sludge from urban wastewater treatment plants. *Water Science & Technology* 47 (3): 167–166.

U.S. Environmental Protection Agency. 1999. Environmental Regulations and Technology: Control of Pathogens and Vector Attraction in Sewage Sludge. EPA/625/R-92/013. Revised edition. U.S. EPA. Washington, D.C.

Environmental Engineering – Pawłowski, Dudzińska & Pawłowski (eds)
© 2007 Taylor & Francis Group, London, ISBN13 978-0-415-40818-9

Evaluation of the possibility of green waste composting – Lublin case study

Małgorzata Pawłowska
Lublin University of Technology, Faculty of Environmental Engineering, Lublin, Poland

Justyna Kurzak & Renata Orłowska
Lublin University of Technology, Faculty of Environmental Engineering, Lublin, Poland

ABSTRACT: Variations of chemical and physicochemical proprieties (pH, heavy metal contents, organic matter content, hydrolytic acidity, cation exchange capacity, base saturation, sum of the base exchangeable cations, buffer capacity) during composting are characterized. The compost was produced mainly from the green waste (mostly leaves collected in the city of Lublin, Poland). The heavy metal contents (Cr, Cu, Mn, Pb, Zn, Ni) did not exceed the values permitted for organic and mineral-organic fertilizers. Only the cadmium concentration exceeded permissible concentrations. The organic matter content during 6 months of composting decreased from 92.2% to 51.85%. The pH value of raw compost measured in H_2O (actual acidity) ranged between 8.22 and 8.92, and the pH values of dried material ranged between 7.82 and 8.41. Hydrolytic acidity was very low. CEC values were very high during all the time of composting and they oscillated around the value 80 cmol(+)/kg_{dmc}. The compost possessed a high buffer capacity, larger with respect to acids than to bases.

Keywords: Compost, sorptive properties, heavy metals, acidity, organic matter content.

1 INTRODUCTION

Green waste arising from city lawns, parks and gardens in Lublin are in most cases burnt at the location at which they are collected. This is specially burdensome during autumn when the amount of leaves is particularly voluminous. The low incineration temperature of what is often wet greenery leads to emission of hazardous pollutants. Hence, in same parts of Lublin city during the autumn months a strong and irritating odour of smoke is particularly evident.

Recently, one of the companies responsible for collection of garbage (Municipal Organic Waste Composting Plant administered by PW. KOM-EKO Sp. z o.o. in Lublin) has started to collect green waste, with the intention of setting up a central composting station. However, a serious obstacle is that much of the greenery collected in the city may contain elevated concentration of heavy metals and other impurities which may affect the process of composting and the quality of the compost produced.

The aim of this paper is to evaluate compost quality on the basis of some chemical and physicochemical parameters and to determine the changes in these parameters during composting.

2 MATERIAL AND METHODS

The composting material consists of: organic waste from green areas (grass, leaves, wood chips), market and food processing industry waste (outdated frozen foods). Leaves made up about 90% of this material.

The windrow composting method was applied. Organic waste was formed into rows of long piles and aerated by turning the pile periodically by mechanical means (compost turner TAK III-Morawetz). The windrows have the dimensions 3 m (width of base), 1.6 m (height) × 30 m (length).

The first composting phase lasts up to 12 weeks. During this time the temperature inside the composted biomass is monitored. When the temperature drops below 60°C the composting biomass is aerated by turning the pile. If, during this operation, an increase of temperature is not observed the second phase of composting is begun. It is in this phase that the temperature ranges between 40 and 60°C; intensive mineralization of organic matter is observed in this stage. The phase goes on for 4 to 8 weeks. The next stage is the compost maturation lasting up to 12 weeks, when a slowing of the biochemical processes and a decrease in the compost temperature to the ambient level occurs. All phases proceeded in an open area.

Samples of compost were taken four times during each half year (in 42, 63, 91, and 149 day of composting). The exact dates of sampling were as follows: 22 Nov., 13 Dec., 10 Jan. and 9 Mar. The production of this kind of compost is very dependent on season and shows a pronounced autumn-winter character. The composted biomass consisted mainly of leaves collected from the autumn of 2005. The composting process proceeded slowly because of the local climatic conditions (low temperatures, high and frequent precipitation).

Compost was dried at 105°C, ground in a ball mill, sieved by 0.5 mm mesh (according to Polish standard PN-Z-15011-3:2001).

The heavy metals concentration was determined by inductively coupled plasma-atomic emission spectrometry (ICP-AES), JY 238 ULTRACE, Jobin Yvon HORIBA.

The organic matter content was determined by the mass loss during roasting at 500°C according to Polish standard *PZ-A-15011-3:2001. Values were calculated based on compost dry mass.

Compost pH and exchangeable acidity (pH_{KCl}) were measured potentiometrically (pH-meter ELMETRON CX 551 with calomel electrode). Samples of the compost were mixed with distilled water (or a solution of 1 M KCl) in a ratio of 1:4 w/v.

Hydrolytic acidity (A_h) was determined by the modified Kappen method. Acidic ions were displaced from the sorptive complex using 1 M sodium acetate. The content of released ions was determined by titration, using 0.1 M NaOH in the presence of phenolphthalein.

Determination of the sum of the base exchangeable cations (BEC), was carried out by the Kappen method. Basic (alkaline) ions were displaced from the sorptive complex by treatment with 0.1 M HCl. The excess of acid was back titrated by 0.1 M NaOH in the presence of phenolphthalein. The sum of base absorption was calculated from the difference between volume of HCl used for the displacement of cations and that remaining in solution.

The total cation exchange capacity (CEC) and base saturation (BS) values were calculated from the equations:

$$CEC = A_h + BEC$$
$$BS = BEC/CEC \cdot 100\%$$

The buffer capacity of the compost was determined conventionally by addition of increasing quantities of 0.1 M HCl or 0.1 M NaOH solutions to compost samples. The buffer curve for the compost was compared with the model curve for a quartz sand.

3 RESULTS AND DISCUSSION

3.1 *Characteristics of organic matter content*

The activity of biomass degrading microorganisms leads to the decrease in organic matter content during composting. The loss of organic matter observed over a six month period was about 40% (from 92.2% in November to 51.85% in March).

3.2 *Characteristics of heavy metals concentrations*

A comparison of the results of our investigation with the permissible values is presented in Table 1. Concentrations of heavy metals (Cr, Cu, Ni, Pb, Zn) were below the permitted values for organic and mineral-organic fertilizers (Dz.U. Nr 236, poz. 2369). Cadmium is an exception because its concentration, ranging from 1.53 to 7.39 mg/kg_{dm}, exceeded the permissible value equal to 3 mg/kg_{dm}. This causes a problem in the practical use of the compost as it requires additional examination of composted substrates. Elimination of problematic waste or its dilution by addition of the less polluted waste would allow the improvement of the compost quality. The possible sources of cadmium in city green waste are vehicle fumes, attrition of car tyres, combustion of plastic, superphosphate fertilizers.

3.3 *Characteristics of pH value and hydrolytic acidity of compost*

The actual acidity of the examined compost was high. The pH value of raw compost measured in H_2O ranged from 8.22 to 8.92, and the pH values of dried material ranged from 7.82 to 8.41 (Tab. 2). As expected the exchangeable acidity showed lower values (from pH 7.22 to 7.96) since the acidic ions form a sorptive complex which is displaced by the K^+ ions.

$CaCO_3$ is prevalent in the rocks of the Lublin area and therefore the compost was rich in it. The

Table 1. Permissible concentration of heavy metals in organic and mineral-organic fertilizers and the concentration determined in compost.

Elements	Permissible concentration mg/kg_{dm}	Determined concentration mg/kg_{dm}
Cd	3	1.53–7.39
Cr	100	8.61–14.31
Cu	400	15.18–18.97
Hg	2	<LOD
Ni	30	5.05–6.75
Pb	100	29.02–83.24
Zn	1500	149.87–205.05

LOD – limit of detection, 4.45 mg/kg_{dm}.

concentrations of $CaCO_3$ determined by the Scheibler method ranged from 2.0 to 6.2%. Small pellets of this compound were observed in the composting material. The pH values for composts reported in the literature are lower. According to Smith et al. (2001) the pH of the moisture contained in garden waste compost ranged from 5.93 to 6.11, pH in $CaCl_2$ ranged from 5.8 to 5.97; pH in H_2O of compost produced from a mixture of garden and market waste (vegetable and fruit from municipal market) ranged from 6.58 to 6.79 and pH values in $CaCl_2$ for this material were 6.36–6.48. Usually, between 3 and 5 weeks of composting, a decrease of pH caused by intensive mineralization is observed. After that time the pH value increases. The lowering of pH values could be also induced by the aeration. This leads to acceleration of the degradation of organic matter and production of organic acids. The decrease of actual acidity is observed on the 63rd day of composting. After this time the pH value increases by about 8% after the 149th day, in comparison to the lowest value observed on 63rd day. Hydrolytic acidity was very low and it decreased during composting.

3.4 *Characteristics of sorptive properties of compost*

The values of specific parameters of sorption capacity are given in Table 3.

CEC values were very high during composting and they oscillated around the value 80 cmol(+)/kg_{dmc}. For comparison, the CEC value of garden waste compost ranged from 62.0 to 75.2 cmol/kg_{dmc} and of compost produced from a mixture of garden and market waste (vegetable and fruit from municipal market) ranged between 68.5 and 75.0 cmol/kg_{dmc}. The increase in the CEC values were caused by the accessibility of oxygen (Smith et al. 2001). Composts examined by Smith & Hughes (2004), of the same kind as those described above and composts from municipal waste (Drozd & Licznar 2004) had lower cation exchange capacities (oscillated in the range of 40–78 cmol/kg_{dmc}) than those examined by us. The latest had the CEC value lower by about 25% (the range 54.5–62.0 cmol(+)/kg_{dmc}). In contrast to these results Benito et al. (2003) reported higher CEC values, ranging from 99.9 to 149.2 cmol(+)/kg_{dmc} for pruning waste compost.

For comparison, the cation exchange capacity of humic substances ranged between 150 and 250 cmol(+)/kg (Gleboznawstwo 1999), of peat earth of coastal zone of Dąbie lake 15.2 cmol(+)/kg (BEC – 8.8 cmol(+)/kg, BS –57.9%) (Sammel 2004).

During the composting the carboxyl and phenolic groups, which contribute greatly to the CEC are formed, so an increase of CEC during composting should be expected (Lax et al. 1986, cited by Smith & Hughes 2004). Smith & Hughes (2004) observed an increase of CEC value of about 50% over 350 days, in the case of compost produced from market and garden refuse with biomass turning. Also Drozd & Licznar (2004), observed the tendency of increasing CEC with composting time. In contrast to this observation, Benito et al. (2003) did not observe any specific trend during composting of pruning waste (Tab. 4). In our examinations, variations of the CEC value between the 42 and 149 day were minimal (increase of about 1.6%).

Table 2. The actual and potential acidity of the green waste compost measured in different time of composting.

Day (week)	Actual acidity pH_{H_2O} Raw compost	Dried compost	Exchangeable acidity pH_{KCl}	Hydrolytic acidity cmol(+)/ kg_{dmc}
42 (6)	8.90	8.09	7.22	2.2
63 (9)	8.22	8.03	7.22	1.8
91 (13)	8.41	7.82	7.26	0.4
149 (22)	8.92	8.41	7.96	0.4

Table 3. Changes of sorptive properties of compost during composting.

Day (week)	BEC [cmol(+)/kg_{dmc}]	CEC [cmol(+)/kg_{dmc}]	BS [%]
42 (6)	78.4	80.6	97.3
63 (9)	78.2	80.0	97.8
91 (13)	79.8	80.2	99.5
149 (22)	81.2	81.6	99.5

dmc – dray mass of compost.

Table 4. Changes of CEC during composting of pruning waste compost and municipal waste compost, market and garden refuse.

Reference	Time (days)	CEC [cmol(+)/ kg_{dwc}]
Benito et al. 2003	33	149.2 ± 35.8
	81	99.9 ± 28.9
	120	188.4 ± 10.6
	145	117.9 ± 46.4
	169	92.0 ± 22.4
	190	100.7 ± 6.3
Drozd & Licznar 2004	54	54.5
	68	55.7
	95	56.5
	143	62.0
Smith & Hughes 2004	42	52
	83	65
	118	70
	153	72
	251	64
	335	78

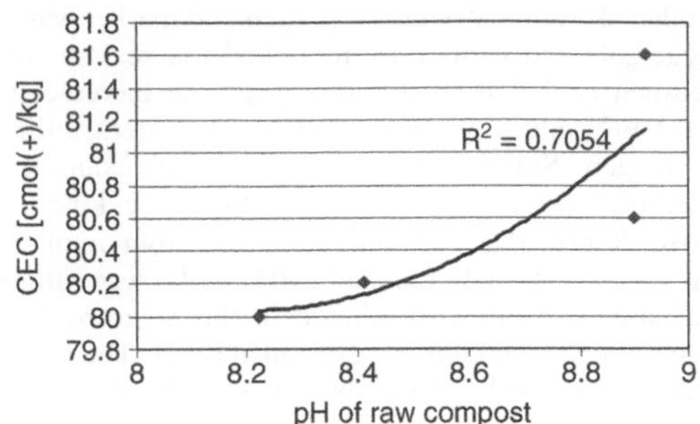

Figure 1. Correlation of pH value and cation exchange capacity of the examined compost.

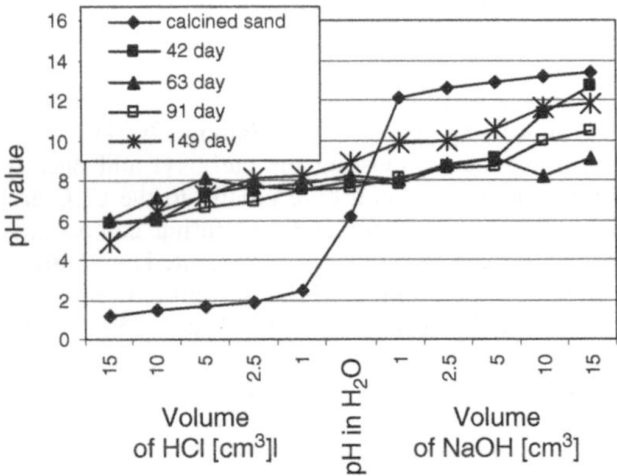

Figure 2. Buffer curves of green waste compost determined during composting.

Changes of sorptive capacity were strongly correlated with the pH of the compost. An increased pH causes an increased sorptive capacity of the compost. This is connected with high content of organic matter, in which dissociation of exchangeable groups depends on pH. Dissociation increases at higher pH.

The influence of pH value on cation exchange capacity is shown in Figure 1. The best fit curve describing the correlation was a polynomial. The correlation coefficient R^2 was 0.7054.

The total sorptive capacity of the compost is the sum of basic and acidic cations present. The sum of the base exchangeable cations changed insignificantly during composting. The base saturation of sorptive complex was about 97%. This means that the content of acidic ions in sorptive complex was very low.

3.5 *Characteristics of compost buffer capacity*

Changes in the compost buffer capacity can be seen in Figure 2. The buffer capacity for acids is higher than for bases. Therefore, application of such a compost to acidic soil would enhance its ability to counter the pH change caused by acidification.

4 CONCLUSIONS

Transformations of organic matter during composting lead to an increase in the sorptive properties of compost. The high value of the sum of base exchangeable cations and the base saturation value demonstrate the high content of cations available for plants. The examined compost has high buffer capacity, larger with respect to acids than to bases.

Addition of the compost to soil, especially those acidic in nature, improved the conditions of plant growth, provided mineral substances, increased the sorptive properties and enhanced the soil's ability to buffer pH changes caused by acidification (acid rain, acid fertilizers).

REFERENCES

Benito, M., Masaguer, A., Moliner, A., Arrigo, N. & Palma, R. 2003. Chemical and microbiological parameters for the characterization of the stability and maturity of pruning waste compost. *Biol. Fertil., Soils* 37: 184–189.

Iglesias-Jiménez, E., Poveda, E., Sánchez-Martín, M.J. & Sáanchez-Camazano, M. 1997. Effect of the Nature of Exogenous Organic Matter on Pesticide Sorption by the Soil. *Arch. Environ. Contam. Toxicol.* 33: 117–124.

Jakobsen, S.T. 1995. Aerobic decomposition of organic waste 2. Value of compost as a fertilizer. *Resources, Conservation and Recycling* 13, 57–71.

Jędrczak, A. & Haziak, K. Określenie wymagań dla kompostowania i innych metod biologicznego przetwarzania odpadów, Zielona Góra 2005.

Lax, A., Roig, A. & Costa, F. 1986. A method for determining the cation exchange capacity of organic materials. *Plant Soil* 94: 349–355.

Myśków, W., Rolnicze znaczenie próchnicy oraz sposoby regulowania jej ilości w glebie. IUNG, Puławy 1984: 48–49.

Pit, R., Lantripe, J., Harrison, R., Henry, Ch.R. & Hue, D. Infiltration through Disturbed Urban Soils and Compost-Amended Soil Effects on Runoff Quality and Quantity, Research Report. 1999, http://www.epa.gov/ednnrmrl/publications/reports/epa600r00016/beginningsection.pdf (page visited Sept, 21, 2006).

Sammel, A. 2004. Właściwości fizyczne i chemiczne gleb murszastych strefy brzegowej jeziora Dąbie. *Roczniki Gleboznawcze* LV(1), SGGW Warszawa 2004, 209–215.

Smith, D.C., Beharee, V. & Hughes, J.C. 2001. The effects of compost produced by a simple composting procedure on the yields of Swiss chard (*Beta vulgaris* L. var. *flavascens*) and common bean (*Phaseolus vulgaris* L. var. *nanus*). *Scienta Horticulturae* 91: 393–406.

Smith, D. & Hughes, J. 2004. Changes in maturity indicators during the degradation of organic wastes subjected to simple composting procedures. *Biol. Fertil. Soils* 39: 280–286.

Rozporządzenie Ministra Rolnictwa i Rozwoju Wsi z dn. 19 października 2004 w sprawie wykonania niektórych przepisów ustawy i nawozach i nawożeniu (Dz. U. Nr 236, poz. 2369).

Polish Standard PZ-Z-15001l-3:2001. Kompost z odpadów komunalnych. Oznaczanie: pH, zawartości substancji organicznej, węgla organicznego, azotu, fosforu i potasu.

Environmental Engineering – Pawłowski, Dudzińska & Pawłowski (eds)
© 2007 Taylor & Francis Group, London, ISBN13 978-0-415-40818-9

Physico-chemical and microbiological characteristics of leachates from Polish municipal landfills

Anna Grabińska-Łoniewska, Andrzej Kulig, Elżbieta Pajor, Andrzej Skalmowski, Waldemar Rzemek & Mirosław Szyłak-Szydłowski
Institute of Environmental Engineering Systems, Warsaw University of Technology, Poland

ABSTRACT: Leachates collected on five Polish landfills were investigated for the presence of microbes, indicating their level of biological impurity, along with a broad range of chemical determinations. Physico-chemical analysis showed that values of their impurity indicators vary greatly during the year and depend on the size and the age of the dumping site. These indicators correlate partly with results of microbiological determinations. The latter have proved that heterotrophic psychrophilic bacteria, spore bacteria and microscopical fungi, constitute the saprophytic microflora of leachates. Apart from the total number of mesophilic bacteria and thermotolerant coliform bacteria, pathogenic bacteria *Listeria monocytogenes* and *Clostridium perfringens*, as well as saprogenic bacteria *Proteus vulgaris* and microscopical fungi belonging to the moulds, constitute indicators of the leachates degree of sanitary pollution. Leachates collected in the landfills do not contain pathogenic opportunistic bacteria *Campylobacter sp.* and *Yersinia sp.* and helminths *Ascaris lumbricoides*.

Keywords: Bacteria, fungi, heavy metals, landfills, leachates, organic matter, pH, salinity, solid waste.

1 INTRODUCTION

The environmental impact of chemical and microbiological pollution from landfill sites designed for waste other than inert and hazardous waste (referred to, in brief, as municipal solid waste – MSW) has not been well explored even though about 12 million tons of municipal waste is deposited in Polish landfills every year (Skalmowski 2001). Additionally, the morphological composition of wastes deposited in such landfills (organic waste content about 60%) provides a suitable substrate for the multiplication of saprophitic micro-organisms and a transfer of pathogenic organisms that enter them with wastes and faeces of animals living on landfill sites.

Research into the influence of municipal landfills on the environment, established over many years in Poland, focused on estimations of air pollution (Grygorczuk-Petersons 2004, Kulig 1995, Traczewska & Karpińska-Smulikowska 2000, Wieczorek 1998) and covered also the identification of impurities of the ground and groundwater pollution associated with landfills. They provide a reference for evaluating the degree of chemical and microbiological impurities of investigated environments and to estimate the range of unfavourable influences (Karwaczyńska et al. 2005, Kulig 2004, Litwin & Pawłowska 1979, Tałałaj & Dzienis 2005). Investigations of drain waters from municipal solid waste landfills are currently conducted in many countries of the world, but in most cases they are limited to determining the physico-chemical properties and toxicity of leachates (Aluko et al. 2003, Kalyuzhnyi et al. 2003, Khattabi et al. 2002, Ward et al. 2002). There is no detailed data concerning the degree of microbiological contamination of landfill leachates.

This study examines leachates, collected from several Polish landfills for municipal waste, for the presence of microorganisms indicative of biological pollution (bacteria *Campylobacter* sp., *Yersinia* sp., *Clostridium perfringens*, *Proteus vulgaris* and eggs of *Ascaris lumbricoides*), including the broad context of chemical and biological determinations.

2 MATERIALS AND METHODS

2.1 *Description of the facilities studied*

This study examines leachates from lined landfills for municipal waste in the towns of Otwock-Świerk, Dębe Wielkie, Wola Suchożerbska near Siedlce, Rokitno near Lublin and from an unlined landfill in Lipiny Stare near Wołomin. Parameters of the facilities are given in Table 1.

Table 1. Parameters of the landfills from which leachates were sampled.

Landfill location year of opening	Source of wastes	Landfill size (ha) and capacity (Mg)	Waste type	Type of liner waste storage technology
Lipiny Stare near Wołomin 1974	Towns: Wołomin, Zielonka, Marki, Ząbki, Kobyłka, Radzymin and city of Warsaw	4.5 ha (ca. 1260000 Mg)	Municipal with a high amount of biodegradable organic matter also sewage sludge in the 1980s and 1990s	Unlined landfill , storage technology as in Wola Suchożebrska
Wola Suchożebrska near Siedlce 1993	Poviat and town of Siedlce	current 6.2 ha target 12 ha (ca. 230 000 Mg)	Municipal with a high amount of biodegradable organic matter	2 mm PEHD geomembrane. Storage plot thickness 1.5–2 m, insulation layer thickness 0.15 m
Rokitno near Lublin 1994	City of Lublin	stage I – 20.59 ha stage II – 17.60 ha (ca. 1120000 Mg)	Municipal	Two geomembrane layers made of thick polyethylene separated by an insulation layer. Storage plot thickness 2 m Sand insulation layer placed on the plot
Dębe Wielkie 1996	Gmina (district) of Serock	1.2 ha (ca. 276000 Mg)	Municipal with a small amount of biodegradable organic matter	Plastpap filter medium, storage technology as in Wola Suchożebrska
Otwock – Świerk 1998	Town of Otwock Poviat of Otwock & Warsaw	current 3 ha target 14 ha (ca. 300000 Mg)	Municipal with a high amount of biodegradable organic matter	2 mm PEHD geomembrane, storage technology as in Wola Suchożebrska

2.2 *Sampling*

Samples of leachates from lined MSW landfills were collected from the pumping station to which they were delivered via a drainage system.

In Otwock-Świerk, the pumping station is located ca. 20 m from the western side of the landfill site. Leachates are delivered to collective wells via two pipelines: a collective leachate pipeline from under the waste deposit and a girdling pipeline that catches leachates from under the foot of the landfill and run-off waters. Leachates are pumped to retention tanks from the well.

Only leachates are routed to the pumping stations in the other facilities. The location of the pumping station in relation to the waste deposit is as follows: the pumping station is situated ca. 15 m from the waste embankment in the village of Dębe Wielkie, 25 m from the southern slope of the waste deposit in the MSW landfill in Wola Suchożebrska. In Rokitno, samples were collected from the pumping station located on the eastern slope of the waste deposit ca. 15 m from the escarpment of the waste embankment to which leachates are routed from a new landfill section as well as from the main pumping station located in the northern part of the landfill area to which leachates from the old recultivated section are also routed.

The landfill site in Rokitno is equipped with operational and control drainage. Operational drainage collects and delivers leachates from the landfill basin, and is located on the proper liner. The horizontal drainage system, collecting leachates from the bottom of the basin, is connected with the pumping station by a collective pipe. Leachates from the bottom of the basin are routed gravitationally to the pumping station operating in the automatic system, and next via a sanitary channel through retention tanks in the direction of the treatment plant in Rokitno. The control drainage system, used in the landfill in Rokitno, serves the purpose of tightness control of the operation cover that seals the bottom and the escarpment of the landfill.

Leachate samples from the unlined MSW landfill, located in Lipiny Stare, were collected from the collection ditch situated along the northern escarpment of the waste embankment. In rain-free periods, the ditch also collects leachates from the landfill bottom.

2.3 *Chemical determinations*

Untreated leachates were used for chemical determinations. pH was determined according to the standard PN-90/C-04540/01, electrolytic conductivity in $mS \cdot cm^{-1}$ according to PN-77/C-04542, total organic carbon (TOC) in w $mg\ C \cdot l^{-1}$ according to PN-EN 1484, copper content in $mg\ Cu \cdot l^{-1}$ according to PN-92/C-04570/01, zinc, lead and cadmium content in $mg \cdot l^{-1}$ and (total) chromium content in $mg\ Cr \cdot l^{-1}$ according to PN-87/C-04570/08, mercury content in $\mu g\ Hg \cdot l^{-1}$ according to PN-82/C-04570/03, sodium and potassium content in $mg \cdot l^{-1}$ according to PN-88/C-04953, lithium content in $mg\ Li \cdot l^{-1}$ according to PN-84/C-04575, biochemical oxygen demand (BOD_5) in $mg\ O_2 \cdot l^{-1}$ – using Sapromat, chemical oxygen demand (COD) in $mg\ O_2 \cdot l^{-1}$ according to PN-74/C-0457803, sulphate content in $mg\ SO_4^{2-} \cdot l^{-1}$ according to PN-79/C-04566 (HACH methods) and chloride content in $mg\ Cl^- \cdot l^{-1}$ according to PN-ISO 9297.

2.4 *Biological determinations*

Untreated leachate samples were used to determine the abundance of psychrophilic (PB), mesophilic (MB) and spore forming bacteria (SB), thermotolerant coliform bacteria (TCB), microscopic fungi (MF), *Proteus vulgaris* as well as *Cl. perfringens*. Bacteria belonging to the genera *Yersinia*, *Campylobacter* and *Listeria* were examined with 1 ml untreated samples (1 ml) and samples concentrated from 10 and 100 ml with the membrane filtration method.

The total number of PB and MB were determined in culture on MPA agar, at 26°C for 7 days and 37°C for 48 hours. SB were determined in the way similar to that used for PB, using for inoculation samples free from vegetative forms by heating at 80°C for 15 min. The capability of fermenting lactose with acid and gas production in 24–48 h at a temperature of 37°C ± 1°C was used as presumptive tests to determine TCB. All tubes giving positive results were subjected to a confirmatory procedure composed of a lactose fermentation test in liquid medium with brilliant green and indole production from tryptophane in culture at 44°C for 24–48 hours. *Yersinia* sp. bacteria were first enriched in Osmer liquid medium and then isolated on selective Schiman agar medium using incubation at 30°C. Colonies characterised by typical morphology (round, diameter 1–2 mm, dark red), non-cytochrome oxidase producing were re-examined using Enteroplast EPL21 (Krogulska & Maleszewska 1992).

The occurrence of *Campylobacter* sp. was initially determined on the Oxoid Brucella liquid enrichment medium supplemented in compounds lowering E_h and after incubation at 37°C during 4 hours, also with a mixture of antibiotics. After 20 hours incubation at 42°C, culture was transferred on Oxoid Brucella agar supplemented with 5% of blood and incubating 2 days at 42°C in anoxic conditions. Characteristic colonies (pink point, beige coloured) were subjected to a confirmatory procedure, using the PCR method (Or & Jordan 2003).

Methods recommended by Merck were used to determine bacteria of the genus *Listeria*. First, selective enrichment of bacteria was conducted using the two-step D.G.Al. on half-concentrated Fraser broth medium, incubating at 30°C for 18–24 hours, and next on the medium as above, undiluted, incubating at 37°C for 18–24 hours. Cultures showing growth and blackening of the medium were inoculated on selective Palcam agar and incubated at 37°C for 24 hours. Grey-black colonies, surrounded by a blackening area, were re-examined for confirmation, including cell morphology (gram-positive rods), catalase- and oxidase-activity, as well as other biochemical properties using the API-*Listeria* test.

Proteus vulgaris was initially determined on MPA agar at 37°C for 24–48 hours. Cultures showing creeping growth and characteristic cell morphology (gram-negative, very motile rods) were re-examined for confirmation using the API 20E test.

The abundance of *Clostridium perfringens* was determined according to PN-77/C-04615, and that of microscopic fungi on agar according to Martin, incubating at 26°C for 7 days (Strzelczyk 1968). *Ascaris lumbricoides* eggs were determined using the microscopic method according to the methods described by Grabińska-Łoniewska et al. (1999).

The abundance of psychrophilic bacteria, spore bacteria, mesophilic bacteria, *Cl. perfringens* and microscopic fungi was given in $cfu \cdot ml^{-1}$ of the leachate, thermotolerant coliform bacteria as NPL $100\ ml^{-1}$ of the leachate, *Proteus vulgaris*, *Yersinia* sp., *Campylobacter* sp. and *Listeria monocytogenes* as titre, and *A. lumbricoides* the abundance of eggs in 1 litre of the leachate.

3 RESULTS AND DISCUSSION

In the thirteen month long examinations of effluents collected in Otwock, a relatively small increase in their pH (from 7.2 to 7.7) was observed (Table 2). The examinations in the landfill sites in operation for a longer period of time (Table 1) show pH values that are only slightly lower: 7.0 and above in Rokitno (Table 3) and 6.8 and above in Lipiny Stare (Table 4).

Relatively small concentrations of sulphates in the leachates collected in Otwock (mean ca. 160 mg $SO_4^{2-} \cdot l^{-1}$) show that wastes containing, for instance, calcium sulphate (gypsum) that may be present in many industrial wastes does not occur in this landfill site. However, sulphate concentrations, determined in the leachates collected in Rokitno, Dębe Wielkie, and in Lipiny Stare, are higher and show that wastes

Table 2. Results of the chemical analyses of landfill leachates from Otwock.

No.	Parameter	Unit	Date of sampling and examination						
			24.10.2003*	24.02.2004	29.03.2004	24.05.2004	31.08.2004	11.10.2004	22.11.2004
1.	pH	—	7.2	7.2	7.6	7.6	7.6	7.7	7.2
2.	Electrolytic conductivity	$mS \cdot cm^{-1}$	19.3	12.8	15.1	18.4	20.4	19.9	7.4
3.	Total organic carbon (TOC)	$mg\ C \cdot dm^{-3}$	1170	1115	1219	1661	1433	1448	733
4.	Copper	$mg\ Cu \cdot dm^{-3}$	0.124	0.288	0.193	0.024	0.014	0.428	0.072
5.	Zinc	$mg\ Zn \cdot dm^{-3}$	0.955	1.216	3.450	0.096	0.139	6.943	1.653
6.	Lead	$mg\ Pb \cdot dm^{-3}$	0.212	0.328	0.164	0.049	0.076	0.526	0.094
7.	Cadmium	$mg\ Cd \cdot dm^{-3}$	0.006	0.017	0.031	0.004	0.003	0.049	0.012
8.	Chromium	$mg\ Cr \cdot dm^{-3}$	0.254	0.096	0.186	0.038	0.089	0.720	0.129
9.	Mercury	$mg\ Hg \cdot dm^{-3}$	0.002	0.002	0.006	0.002	0.002	0.006	0.004
10.	Sodium	$mg\ Na \cdot dm^{-3}$	2040	1325	1660	2020	2300	2190	1550
11.	Potassium	$mg\ K \cdot dm^{-3}$	1485	867	1210	1420	1560	1560	600
12.	Lithium	$mg\ Li \cdot dm^{-3}$	30	28	33	31	39	39	25
13.	BOD_5	$mg\ O_2 \cdot dm^{-3}$	676	232	255	710	514	540	801
14.	COD	$mg\ O_2 \cdot dm^{-3}$	3132	2530	3480	4116	4469	4479	1685
15.	BOD_5/COD ratio	—	0.219	0.092	0.073	0.172	0.115	0.121	0.475
16.	Sulphates	$mg\ SO_4^{2-} \cdot dm^{-3}$	350	58	211	145	75	119	135
17.	Chlorides	$mg\ Cl^- \cdot dm^{-3}$	2440	1640	2160	2580	2880	2880	1140
18.	SO_4^{2-}/Cl^- ratio	—	0.146	0.035	0.098	0.056	0.026	0.041	0.118

* The arithmetic mean of the results of two analysed samples of leachates

Table 3. Results of the chemical analyses of landfill leachates from Wola Suchożebrska, Rokitno and Dębe Wielkie.

Landfill location			Date of sampling and examination				
			Wola Suchożebrska		Rokitno	Dębe Wielkie	
No.	Parameter	Unit	11.10.2004	22.11.2004	24.10.2003	09.12.2003	24.02.2004
1.	pH	–	7.5	7.5	7.0	7.8	7.7
2.	Electrolytic conductivity	$mS \cdot cm^{-1}$	18.6	11.9	18.5	6.4	3.2
3.	Total organic carbon (TOC)	$mg\ C \cdot dm^{-3}$	469	628	5613	336	238
4.	Copper	$mg\ Cu \cdot dm^{-3}$	0.556	0.059	0.214	0.037	0.049
5.	Zinc	$mg\ Zn \cdot dm^{-3}$	4.832	1.328	1.86	2.870	0.205
6.	Lead	$mg\ Pb \cdot dm^{-3}$	0.612	0.086	0.395	0.269	0.031
7.	Cadmium	$mg\ Cd \cdot dm^{-3}$	0.036	0.009	0.011	0.011	0.007
8.	Chromium	$mg\ Cr \cdot dm^{-3}$	0.688	0.141	0.409	0.048	0.051
9.	Mercury	$mg\ Hg \cdot dm^{-3}$	0.006	0.005	0.004	0.004	0.001
10.	Sodium	$mg\ Na \cdot dm^{-3}$	1950	2585	1710	725	308
11.	Potassium	$mg\ K \cdot dm^{-3}$	1665	1765	1350	595	262
12.	Lithium	$mg\ Li \cdot dm^{-3}$	43	52	29	23	13
13.	BOD_5	$mg\ O_2 \cdot dm^{-3}$	271	235	3450	90	28
14.	COD	$mg\ O_2 \cdot dm^{-3}$	2200	1970	12338	826	454
15.	BOD_5/COD ratio	–	0.123	0.119	0.280	0.109	0.062
16.	Sulphates	$mg\ SO_4^{2-} \cdot dm^{-3}$	52	38	225	680	310
17.	Chlorides	$mg\ Cl^- \cdot dm^{-3}$	2280	2420	2010	1040	392
18.	SO_4^{2-}/Cl^- ratio	–	0.023	0.016	0.112	0.654	0.791

Table 4. Results of the chemical analyses of landfill leachates from Lipiny Stare.

			Date of sampling and examination		
No.	Parameter	Unit	29.03.2004	24.05.2004	31.08.2004
1.	pH	–	7.8	6.8	7.6
2.	Electrolytic conductivity	$mS \cdot cm^{-1}$	10.5	17.7	20.4
3.	Total organic carbon (TOC)	$mg\ C \cdot dm^{-3}$	250	679	1433
4.	Copper	$mg\ Cu \cdot dm^{-3}$	0.058	0.041	0.014
5.	Zinc	$mg\ Zn \cdot dm^{-3}$	1.805	0.532	0.139
6.	Lead	$mg\ Pb \cdot dm^{-3}$	0.072	0.117	0.076
7.	Cadmium	$mg\ Cd \cdot dm^{-3}$	0.021	0.008	0.003
8.	Chromium	$mg\ Cr \cdot dm^{-3}$	0.081	0.076	0.089
9.	Mercury	$mg\ Hg \cdot dm^{-3}$	0.004	0.002	0.002
10.	Sodium	$mg\ Na \cdot dm^{-3}$	1630	2920	2300
11.	Potassium	$mg\ K \cdot dm^{-3}$	775	1260	1560
12.	Lithium	$mg\ Li \cdot dm^{-3}$	25	30	39
13.	BOD_5	$mg\ O_2 \cdot dm^{-3}$	55	196	514
14.	COD	$mg\ O_2 \cdot dm^{-3}$	1360	2550	4469
15.	BOD_5/COD ratio	–	0.040	0.077	0.115
16.	Sulphates	$mg\ SO_4^{2-} \cdot dm^{-3}$	310	780	75
17.	Chlorides	$mg\ Cl^- \cdot dm^{-3}$	2260	3960	2880
18.	SO_4^{2-}/Cl^- ratio	–	0.137	0.197	0.026

containing gypsum are deposited in these wastes (Table 3 and 4).

The SO_4^{2-}/Cl^- ratio that ranged from 0.026 to 0.146 in the case of the leachates collected in Otwock (Table 2, Fig. 1) is also representative of the other landfill sites studied, with the exception of the facilities in Dębe Wielkie, where the ratio equalled ca. 0.72 (Table 3, Fig. 2 and 3).

In the case of the facilities studied, it is quite difficult to confirm the tendency of the SO_4^{2-}/Cl^- ratio to decrease together with the age of the landfill site (Szymański 1995). The results obtained are indicative

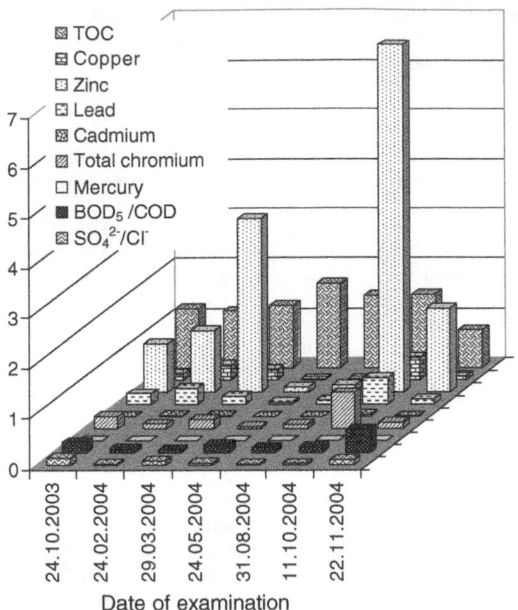

Figure 1. Results of the chemical examinations ($mg{\cdot}dm^{-3}$) of leachates collected in the landfill site in Otwock.

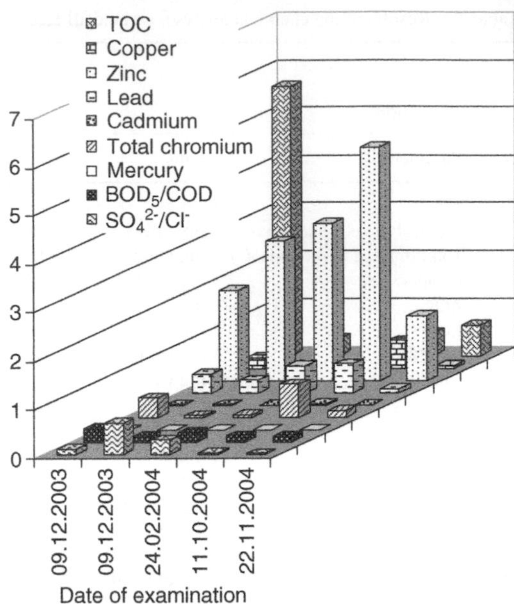

Figure 2. Results of the chemical examinations ($mg{\cdot}dm^{-3}$) of leachates from landfill site in Rokitno (1), Dębe Wielkie (2) and Wola Suchożebrska (3).

of far more complex relationships between individual indicators.

In leachates from Otwock, the greatest degree of leachate salinity occurred in October 2003 as well as in May, August and October 2004. The specific conductivities equalled 19.3, 18.4, 20.4, and 199 $mS{\cdot}cm^{-1}$, respectively; the chloride concentrations were 2440, 2580, 2880, and 2880 mg $Cl^{-}{\cdot}l^{-1}$, while the sulphate concentrations equalled 350, 145, 75, and 119 mg $SO_4^{2-}{\cdot}l^{-1}$. Leachate samples collected in this period were also characterised by a higher organic compound content, which is additionally confirmed by increased values of such indicators as TOC (Fig. 1), BOD_5 and COD (Table 2).

The composition of the leachate collected under the largest of the MSW landfill sites studied, i.e. that in Rokitno, differed significantly from that of the others, and was distinguished first of all by the content of easily degradable organic matter and organic matter difficult to degrade (Fig. 2). Inorganic compound content values were similar to concentrations determined in the leachates from the landfill in Wola Suchożebrska. Leachates collected under the landfill in Dębe Wielkie were characterised by significantly lower concentrations of organic matter. Apart from the facility in Rokitno (3450 mg $O_2{\cdot}l^{-1}$), BOD_5 values, lower than 800 mg $O_2{\cdot}l^{-1}$ in the leachates from Otwock and the other facilities, are indicative of very low amounts of easily degradable organic matter,

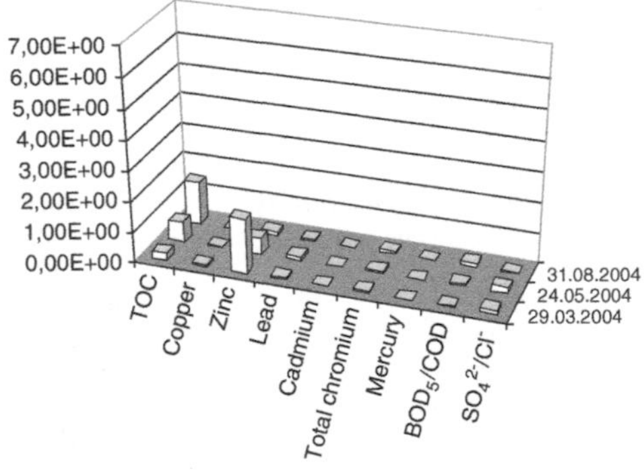

Figure 3. Results of the chemical examinations ($mg{\cdot}dm^{-3}$) of leachates collected from old landfill site in Lipiny Stare.

which is characteristic of landfills in the final stage of methane fermentation, that is methane genesis.

The examinations of the leachates collected in the landfill site in Lipiny Stare, conducted in March, May and August 2004, show an increase in the amount of organic and mineral matter (apart from heavy metals) in this period (Fig. 3). Very low initial BOD_5 values, characterising easily degradable organic matter content, also increased. A similar tendency was observed in the leachate samples collected in the landfill site in Otwock in the same period (Table 2).

COD values of the leachates in Otwock (from 1680 to 4480 mg $O_2 \cdot l^{-1}$) are representative also for the facilities in Wola Suchożebrska and Lipiny Stare. However in Rokitno a considerably higher COD value, equal to 12340 mg $O_2 \cdot l^{-1}$, was recorded. All these comparatively high values are situated in the lower values of the COD range, characteristic for the phase of the acid phase of methane fermentation or within the range characteristic for the final stages of these processes. On the other hand, the leachates collected from the facility in Dębe Wielkie were characterized by a very low COD value, below 1000 mg $O_2 \cdot l^{-1}$ (Table 3).

The BOD_5/COD ratio of the leachates collected in Otwock increased from 0.07 to 0.48 (Fig. 1). Its values for all the facilities examined fall within this range (Fig. 2 and 3). In the case of the oldest landfill site studied in Lipiny Stare, the mean value of the BOD_5/ChZT ratio was 0.08 and is consistent with literature data that show that the ratio is usually lower than 0.1 after ca. 20 year of landfill operation (Szymański 1995).

Characteristically, apart from the zinc concentration, which ranged from 0.1 to 6.9 mg $Zn \cdot l^{-1}$ in the samples collected in Otwock, all the leachate samples contained similar amounts of heavy metals (Fig. 1, 2 and 3). Metal concentrations in the leachates studied were small, indicating that the industrial wastes from which they were derived were not deposited in the landfill sites studied. The correlation indexes between metal concentrations are usually 0.8. This shows that the determined concentrations of metals such as Cu, Pb, Cd, Cr and Hg constitute a natural background of the wastes collected in the landfill sites.

Microbiological examinations of leachates collected in the four lined landfill sites showed that psychrophilic bacteria dominated among bacteria in the microflora. Their total number ranged between $27 \cdot 10^3$–$257 \cdot 10^5$ $cfu \cdot ml^{-1}$. The number of spore bacteria was between $2 \cdot 10^2$ and $464 \cdot 10^2$ $cfu \cdot ml^{-1}$. The participation of these bacteria in total PB ranged was between 0.004 and 2.86%. Microscopic fungi were the least numerous in the group of saprophytic microorganisms (2–590 $cfu \cdot ml^{-1}$). Bacteria indicative of faecal pollution also occurred in the leachates, i.e. MB growing at 37°C and TCB growing at 44°C. Their number ranged between $2 \cdot 10^3$–$297 \cdot 10^4$ $cfu \cdot ml^{-1}$ as well as 2300 up to > 240 000 in 100 ml, respectively. Pathogenic bacteria, *Listeria monocytogenes* and *Clostridium perfringens* (titre ranging between 10–10^{-5} as well as 1–10^{-2}, respectively) and saprogenic bacteria, *Proteus vulgaris* (titre from > 10^{-2}–10^{-5}), permanently determined in the leachates, as well as microscopic fungi belonging to moulds were characteristic for the leachates. The presence of *Cl. perfringens* in the leachates may also be indicative of a possible occurrence of parasitic protozoa in this environment. According to the WHO Guidelines, this assumption is justified by a similar susceptibility of spores of these bacteria and cysts of the protozoa to environmental factors. As a result of the studies conducted by Pajor (2004), the list of bacteria with sanitary importance, permanently occurring in the leachates discussed, was expanded to include species *Enterobacter cloaceae* and *Citrobacter braakii* belonging to the family *Enterobacteriaceae*. Other species of this family: *Citrobater cloaceae*, *C. freandii*, *Serratia fonticola*, *S. liquefaciens* and *Klebsiella pneumoniae*, occurred periodically in the microflora. The present study shows that pathogenic opportunistic bacteria, such as *Campylobacter* sp. and *Yersinia* sp. as well as eggs of *Ascaris lumbricoides*, did not occur in the leachates.

The problem of microbiological leachate pollution was also analysed in relation to the season and the time of the operation of the landfill sites on the basis of the results obtained for the lined facilities.

Examinations conducted for one year at the landfill site in Otwock (Table 5, Fig. 4) show that the greatest numbers of PB ($1879 \cdot 10^3$–$3500 \cdot 10^3$ $cfu \cdot ml^{-1}$) occurred in the autumn (October–November). The number of these bacteria in this period is slightly correlated with the BOD_5 of the leachates (the coefficient of Pearson's correlation = 0.6). Approximately 20–100 times lower values of the total number of PB than the numbers given above were detected in the leachates collected between late winter and the end of the summer. The percent of SB in the total number of PB ranged between 0.03 and 1.69%, but usually remained at a higher level starting from March to late autumn (0.96–1.69%). This may be indicative of an increase in the number of organic compounds more difficult to degrade in the leachates. The abundance of MB and MF showed similar seasonal correlations to that of PB. The maximum numbers recorded in the autumn were $91 \cdot 10^3$–$291 \cdot 10^3$ and 35–99 $cfu \cdot ml^{-1}$, respectively. 13 fungal species belonging to moulds, constituting the mycoflora of the leachates. *Trichoderma viride* Pers ex. Gray, *Gliocladium catenulatum* Gilm et Abbot as well as *Mariannea elegans* (Corda) Samson dominated in this group (19.3–25.7% of isolated strains).

The correlation between the abundance of MB and TCB occurred only in spring and summer, which are the seasons characterised by higher environment temperatures.

A differentiation in the abundance of pathogenic and saprophitic bacteria in the leachates depending on the season was also noticed. *Listeria monocytogenes* occurred most numerously in the leachates between February and March (titre ranging from 10^{-1} to 10^{-5}) as well as in November (titre 10^{-2}) when the temperature of the environment was the lowest. In May, August and October, their number decreased significantly, reaching the titre from 1 to 10. Autumn (October–November) was also characterised by an increased

Table 5. Results of microbiological analyses of landfill leachates from Otwock.

Date of examination	24.10.2003	24.02.2004	29.03.2004	24.05.2004	31.08.2004	11.10.2004	21.11.2004
Total number of bacteria cfu·ml^{-1} psychrophilic temp. 20°C, 72 hr	$3500 \cdot 10^3$	$234 \cdot 10^3$	$110 \cdot 10^3$	$27 \cdot 10^3$	$145 \cdot 10^3$	$1879 \cdot 10^3$	$2740 \cdot 10^3$
mesophilic temp. 37°C, 24 hr	$291 \cdot 10^3$	$16 \cdot 10^3$	$2 \cdot 10^3$	$23 \cdot 10^3$	$34 \cdot 10^3$	$250 \cdot 10^3$	$91 \cdot 10^3$
sporeforming temp. 28°C, 48 hr	$12 \cdot 10^2$	$2 \cdot 10^2$	$14 \cdot 10^2$	$4 \cdot 10^2$	$14 \cdot 10^2$	$290 \cdot 10^2$	$464 \cdot 10^2$
Percentage of sporeforming bacteria in total number of psychrophilic bacteria	0.03	0.08	1.27	1.45	0.96	1.54	1.69
Thermotolerant coliform bacteria MPN in 100 ml	2300	>240000	>2400	2400	>2400	>24000	>240000
Campylobacter sp titre	nf	nf	nf	nf	nf	nf	nf
Listeria monocytogenes titre	ni	10^{-1}	10^{-5}	10	1	1	10^{-2}
Yersinia sp. titre	ni	nf	nf	nf	nf	nf	nf
Clostridium perfringens cfu·ml^{-1} (or titre)	ni	(10^{-1})	(10^{-2})	(>1)	60 (10^{-1})	30 (10^{-1})	9 (10^{-1})
Proteus vulgaris titre	ni	ni	ni	>10^{-2}	ni	10^{-5}	10^{-5}
Microscopic fungi cfu·ml^{-1}	ni	14	2	0	35	85	99
Ascaris lumbricoides Number of eggs·l^{-1}	nf	nf	nf	nf	nf	nf	nf

Abbreviations: nf – not found, ni – not investigated

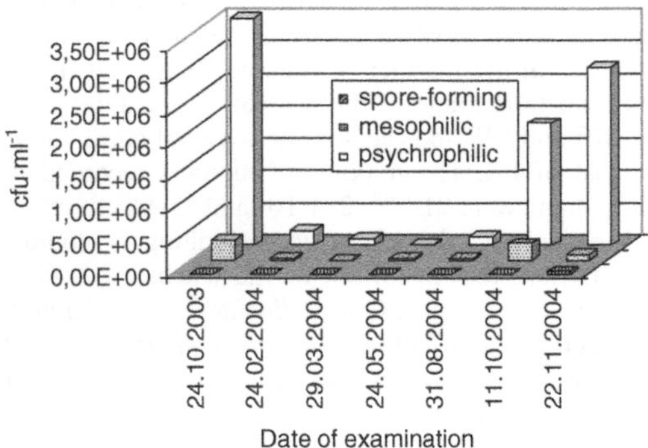

Figure 4. Changes in the abundance of psychrophilic, spore-forming and mesophilic bacteria in Otwock.

number of saprogenic bacteria *Proteus vulgaris* in the leachates (titre 10^{-5}). The number of *Clostridium perfringens* in the leachates studied was generally constant throughout the year, as demonstrated by the titre ranging >1 to 10^{-1}.

As the analysis of the findings given in Tables 5 and 6 shows, the abundance of individual groups of bacteria in the leachates collected from lined landfills with base insulation in a comparable period of time (months: October, November and December) did not depend on the time of the operation of the landfills (6–11 years).

Microbiological properties of the leachates collected in the unlined landfill site, located in Lipiny Stare (Table 7), did not differ significantly from those given above for the lined landfills. The total number of PB ranged between $41 \cdot 10^4$ and $193 \cdot 10^4$ cfu·ml^{-1}, and that of SB $12 \cdot 10^2$–$87 \cdot 10^2$ cfu·ml^{-1}. The percentage of occurrence of SB in the total number of PB was 0.06–2.12%. The abundance of MF was balanced throughout the study period, and was 50–70 cfu·ml^{-1}. 30 mould species and 2 undetermined yeast species constituted the mycoflora. The following species prevailed among moulds (12.4–27.5% of the total isolated strains): *Aspergillus fumigatus* Fres., *Penicillium expansum* Link et Gray, *Penicillium janthinellum* Biourge, *Penicillum verrucosum* Dierckx, *Trichoderma viride* Pers ex. Gray and *Diplodia mutila* (Fr.) Mont. The mycoflora of the leachates is discussed in greater depth in a study by Grabińska-Łoniewska et al. (2005). Considerable fluctuations in the abundance of MB and TCB in these leachates were recorded ($5 \cdot 10^3$–$1316 \cdot 10^3$ cfu·ml^{-1} and >2 400 cfu to >240 000 cfu 100 ml^{-1}, respectively), and no clear quantitative correlations between these bacteria can be established. Pathogenic bacteria *Listeria monocytogenes*

Table 6. Results of microbiological analyses of landfill leachates from Wola Suchożebrska, Rokitno and Dębe Wielkie.

Landfill location year of the opening	Wola Suchożebska 1993		Rokitno 1994	Dębe Wielkie 1996	
Date of examination	11.10.2004	21.10.2004	24.10.2003	09.12.2003	24.02.2004
Total number of bacteria $cfu{\cdot}ml^{-1}$					
psychrophilic temp. 20°C, 72 hr	$970{\cdot}10^3$	$168{\cdot}10^3$	$3890{\cdot}10^3$	$25700{\cdot}10^3$	$70{\cdot}10^3$
mesophilic temp. 37°C, 24 hr	$19{\cdot}10^3$	$3{\cdot}10^3$	$2970{\cdot}10^3$	$1150{\cdot}10^3$	$3{\cdot}10^3$
sporeforming temp. 28°C, 48 hr	$13{\cdot}10^2$	$6.4{\cdot}10^2$	ni	$453{\cdot}10^2$	$20{\cdot}10^2$
Percentage of sporeforming bacteria in total number of psychrophilic bacteria	0.13	0.004	ni	0.18	2.86
Thermotolerant coliform bacteria MPN in 100 ml	2400	>2400	6200	>2400	>2400
Campylobacter sp titre	nf	nf	nf	nf	nf
Listeria monocytogenes titre	10	1	ni	1	10^{-1}
Yersinia sp. titre	nf	nf	ni	nf	nf
Clostridium perfringens $cfu{\cdot}ml^{-1}$ (or titre)	20 (10^{-1})	4 (1)	700 (10^{-2})	5 (1)	(10^{-1})
Proteus vulgaris titre	10^{-5}	10^{-4}	ni	ni	ni
Microscopic fungi $cfu{\cdot}ml^{-1}$	20	99	ni	590	22
Ascaris lumbricoides number of $eggs{\cdot}l^{-1}$	nf	nf	ni	ni	nf

Abbreviations: nf – not found, ni – not investigated

(titre from 10 to $<10^{-1}$) and *Cl. perfringens* (titre from >1 to 10^{-2}) were permanent components of the microflora. Bacteria *Campylobacter* sp. and *Yersinia* sp. as well as eggs of *Ascaris lumbricoides* did not occur in the leachates.

4 CONCLUSIONS

The following general conclusions can be drawn from the examinations conducted:

- The results of physico-chemical investigations of leachate samples from MSW landfills show that the values of the examined pollution indicators (eg. BOD_5), partly correlate with results of microbiological determinations.
- Heterotrophic psychrophilic bacteria (PB), dominant in terms of the abundance, as well as less numerous spore bacteria and microscopic fungi, constitute the saprophytic microflora of the leachates collected in the lined landfill sites as well as from the unlined landfill. The number of these micro-organisms in the leachates increased in the autumn.
- Apart from the total number of mesophilic bacteria (MB) and the number of thermotolerant colifrom bacteria (TCB), pathogenic bacteria *Listeria monocytogenes* and *Clostridium perfringens* as well as saprogenic bacteria *Proteus vulgaris* and microscopic fungi belonging to moulds can be used as indicators of the degree of sanitary pollution of the leachates.
- The correlation between the abundance of MB and TCB occurred only in spring and summer that is, during seasons characterised by higher ambient temperatures. *Listeria monocytogenes* and *Proteus vulgaris* were most numerous in the leachates during periods characterised by low temperatures (Nov-Feb and Oct–Nov, respectively). An influence of season on the occurrence of *Clostridium perfringens* in the leachates was not observed.
- Leachates collected in the municipal waste landfills do not contain the pathogenic opportunistic bacteria *Campylobacter* sp. and *Yersinia* sp. or helminths, *Ascaris lumbricoides*.

Table 7. Results of microbiological analyses of landfill leachates from Lipiny Stare.

Date of examination	29.03.2004	24.05.2004	31.08.2004
Total number of bacteria cfu·ml^{-1}			
psychrophilic temp. 20°C, 72 hr	$104 \cdot 10^4$	$41 \cdot 10^4$	$193 \cdot 10^4$
mesophilic temp. 37°C, 24 hr	$5 \cdot 10^3$	$1316 \cdot 10^3$	$50 \cdot 10^3$
sporeforming temp. 28°C, 48 hr	$14 \cdot 10^2$	$87 \cdot 10^2$	$12 \cdot 10^2$
Percentage of sporeforming bacteria in total number of psychrophilic bacteria	0.13	2.12	0.06
Thermotolerant coliform bacteria MPN in 100 ml	>2400	>240000	>24000
Campylobacter sp titre	nf	nf	nf
Listeria monocytogenes titre	10^{-1}	$<10^{-1}$	10
Yersinia sp. titre	nf	nf	nf
Clostridium perfringens cfu·ml^{-1} (or titre)	100 (10^{-2})	(>1)	40 (10^{-1})
Proteus vulgaris titre	ni	10^{-2}	ni
Microscopic fungi cfu·ml^{-1}	66	50	70
Ascaris lumbricoides number of eggs·l^{-1}	nf	nf	nf

Abbreviations: nf – not found, ni – not investigated

ACKNOWLEDGEMENTS

The study was financed by the Ministry of Science and Informatics funds for years 2003–2006. We wish to thank Dr Dorota Korsak of the National Food and Nutrition Institute in Warsaw for her confirmatory determinations of *Campylobacter* sp.

REFERENCES

Aluko, O. O. et al. 2003. Characterization of leachates from a municipal solid waste landfill site in Ibadan, Nigeria. *J. of Environmental Health Research*. Vol. 2 Issue 1.

Grabińska-Łoniewska, A. 1999. Laboratory exercises of the general microbiology (in Polish), *The Publishing House WUT*, Warsaw, ISBN 83-7207-136-5.

Grabińska-Łoniewska, A. et al. 2005. Mycoflora of landfill leachates, *Acta Mycologica*, in press.

Grygorczuk-Petersons, E.H. 2004. Estimation of the influence of warehoused sewage sludge from the municipal sewage treatment plant on the composition of leachates from municipal dumping sites (in Polish). *Inżynieria Środowiska*. vol. 25, pp. 135–142.

Kalyuzhnyi, S. et al. 2003. Evaluation of the current status of operating and closed landfills in Russia, Finland and Ireland with regard to water pollution and methane emission. *Water Sci. Technol.*, 48(4), pp. 37–44.

Karwaczyńska, U. et al. 2005. Influence of leachates from unlined landfill of municipal waste on the quality of surface and underground waters (in Polish). *II Congress of the Engineering of the Environment*. Proceedings vol. 2, pp. 509–515. Monografie Komitetu Inżynierii Środowiska Polskiej Akademii Nauk. vol. 33. Lublin.

Khattabi, H. et al. 2002. Changes in the quality of landfill leachates from recent and aged municipal solid waste. *Waste Manag. Res.* Aug; 20(4), pp. 357–64.

Krogulska, B. & Maleszewska, J. 1992. Occurrence of the bacterium from the kind the Yersinia in surface waters (in Polish). *Roczn. PZH*, XLIII, pp. 295–300.

Kulig, A. 1995. Environmental impact assessment of municipal utilities. Proceedings of the Polish-British Conference in "*Environmental Engineering – British and Polish experience in linking education and research with industry*". pp. 253–268. Warsaw, 16–18 October 1995.

Kulig, A. 2004. Measurement and modelling based methods applicable in environmental impact assessment of municipal utilities (in Polish), *The Publishing House WUT*, Warsaw, ISBN 83-7207-518-2.

Litwin, B. & Pawłowska, L. 1979. Influence of dumping sites on the environment in the light of research of the Institute of Environment Shaping (in Polish). Materials of the Seminar PZITS No. 242. "Waste utilization with the method of sanitary landfills", Warsaw-Lublin, pp. 89–103.

Or, S.L.W. & Jordan, P.L. 2003. Evaluation of 11 PCR assays for species-level identification of Campylobacter jejuni and Campylobacter coli. *J. Clin. Microbiol.*, 41, pp. 330–336.

Pajor, E. 2004. Occurrence of the bacterium from the family Enterobacteriaceae, reckoned as opportunistic pathogen, in leachates from municipal landfills (in Polish). ISIŚ. Warsaw University of Technology, Warsaw.

Skalmowski, K. 2001. Municipal waste composting, models of technological solution (in Polish). Prace Naukowe Politechniki Warszawskiej. Inżynieria Środowiska, z 39.

Strzelczyk, A. 1968. Methods of soil fungi investigation (in Polish). *Roczn. glebozn.*, 19, pp. 405–424.
Szymański, K. 1995. Evaluation of the groundwater pollution (in Polish). The monograph of the Building and Environmental Engineering Faculty. No. 53, Koszalin, ISBN 83-86123-26-5.
Tałałaj, I.A. & Dzienis, L. 2005. Quality changes of the groundwater in the neighbourhood of the municipal landfill (in Polish). *II Congress of the Engineering of the Environment.* Proceedings vol. 2, pp. 777–785. Monografie Komitetu Inżynierii Środowiska Polskiej Akademii Nauk. vol. 33. Lublin.
Traczewska, T. M. & Karpińska-Smulikowska, I. 2000. Influence of the municipal landfill on the microbiological air quality (in Polish), *Ochrona Środowiska*, 77/2, pp. 35–38.
Ward, M.L. et al. 2002. Determining toxicity of leachates from Florida municipal solid waste landfills using a battery-of-tests approach. *Environ Toxicol.* 17(3), pp. 258–66.
Wieczorek, A. 1998. Emissions accompanied composting process (in Polish), *Ochrona powietrza i problemy odpadów*, 2, pp. 59–64.
WHO Guidelines for drinking-water quality, 2002, Second Ed. Add. Microbiological agents in drinking water, WHO Geneva, ISBN 83-906591-4-X.

Air pollution control

Environmental Engineering – Pawłowski, Dudzińska & Pawłowski (eds)

Modelling of heavy gas atmospheric dispersion in Poland

Maria T. Markiewicz
Institute of Environmental Engineering Systems, Technical University of Warsaw, Warsaw, Poland

ABSTRACT: Substances released to the atmosphere which have a density greater than the density of the atmospheric air are called heavy gases or dense gases. The dispersion of heavy gases is different from that encountered in the case of passive or positively buoyant gases. Specific models have been developed to describe it. The heavy gas dispersion models are important components of emergency response systems and valuable tools for environmental impact assessments and risk assessments in most of the West European countries and in the USA. In Poland the use of these models is at the early stage. In this article models of heavy gas dispersion in the atmosphere are reviewed and perspectives for their application in Poland are discussed.

Keywords: Heavy gases, dense gases, dispersion models, risk assessments.

1 INTRODUCTION

The use of hazardous toxic substances in modern industry is growing. Special legislation is initiated in order to prevent accidents and limit their consequences for man and the environment. These acts of legislation are introduced on the European Union (EU) as well as on international and national levels. The main acts concerning this subject at the international and European level are listed in Table 1.

The legislation on a national level undertaken in each of the European Union member states can differ slightly. However, they have to be harmonised with the international and European regulations. In Poland the main acts related to this topic include the Law of environment protection (Dz.U.2001.62.627) and the Acts on substances and chemical preparations (Dz.U.2001.11.84), road transport of hazardous materials (Dz.U.2002.199.167), transport of hazardous materials by rail (Dz.U.2004.97.962), Code of labour (Dz.U.1997.88.553), state fire brigades (Dz.U.1991.88.400), fire fighting protection (2002.147.1229), state inspectorate of environmental protection (Dz.1991.88.400), state labour inspectorate (Dz.U.2001.124.1362), state sanitary inspectorate (Dz.U.1998.90.575). These acts are followed by the Regulations of Ministries.

In some accidental releases heavy gas clouds are formed. These clouds have a density greater than that of the atmospheric air. Here gases are involved which either have a molecular weight higher than that of air or gases which "simulate" a high molecular weight due to one or combination of reasons. These reasons are as follows: the low release temperature, high storage pressure or chemical reactions of the released substance with the atmospheric water vapour.

Releases of heavy gases differ from conventional pollutant releases. The reasons are as follows:

- the volumes of released gas may be large due to the storage of material in a liquid phase;
- there are a large variety of types of emission sources and the specification of emission parameters is usually difficult due to their time dependence and very diversified source geometry;
- the releases may consist of different component mixtures of gases and liquids;
- there may be heat and/or mass transfer with the ground surface and the ambient air;
- the phase changes of the released material during the cloud formation typically take place.

The emission sources can be divided into classes based on different criteria:

- the release duration, which allows to distinguish instantaneous and continuous sources;
- release height, of which elevated and ground level releases are specified;
- characteristics of the released material encompassing the phase of released emissions, which includes gas phase, liquid phase and two-phase flows, and the composition of gases, including one-component and multi-component releases;
- characteristics of storage conditions, of which pressurised and non-pressurised releases (typically refrigerated) are specified;

Table 1. Main legislation acts initiated to prevent of accidents with the hazardous toxic substances and limit their consequences for man and the environment at the international and European level.

International level:
Guiding principles of OECD for chemical accident prevention (1992)
Convention on the transboundary effects of industrial accidents (1992)
Convention on prevention of major industrial accidents No 174 with the recommendation No 181(1993)
Convention of the long-range transport of air pollution (1979)
Convention on the international carriage of goods by rail with the uniform rules concerning the contract for international carriage of goods by rail (CIM) (1980)
Convention on the contract for the international carriage of goods by roads (1956)
Convention on the protection and use of transboundary watercourses and industrial lakes (1992)
Convention on the occupational safety and health and the working environment no 155 with the recommendation No 164 (1981)
Convention on civil liability for damage case during carriage of dangerous goods by road, rail and industrial navigation vessels (1989)
Convention on the protection of workers against occupational hazards in the working environment due to air pollution, noise and vibrations no 147 with recommendation No 156 (1977)
Convention concerning safety in the use of chemicals at work C170 with the recommendation R177 (1990)
Labour inspection convention C81 and recommendation (1947)

European level:
Directive 96/82/EC on control of major accident hazards involving dangerous substances (1996)
Directive 96/49/EC on the approximation of the laws of the member states with regard to the transport of dangerous goods by rail (1996)
Directive 94/55/EC on the approximation of the laws of the member states countries with regard to the transport of dangerous goods by roads (1994)
Directive 76/769/EEC on the transport of dangerous goods by vessels on inland waterways (1976)
European agreement concerning the international carriage of dangerous goods by inland waterways (ADN), (2000)
European agreement concerning international transport of dangerous goods by road (ADR), (1957)
European agreement concerning the work of crews of vehicles engaged in the international road transport (AETR), (2000)
Directive 89/391/EC on the introduction of measures to encourage improvements in the safety and health of workers at work with 18 individual directives (1989)

- and the source scenario, including low velocity jets, momentum jets, evaporation from the pools or explosions.

To determine a source emission rate for continuous releases or the volume of released emissions for instantaneous ones, different input data depending on the source type are needed. In general, the physical and chemical properties of the released material, the source geometry, the characteristics of the ground surface and mitigation measures should be supplied. The estimation of the source emission rate or the volume of released emissions is realised in an emission model.

The negative buoyancy of these clouds modifies their transport and dispersion in the atmospheric air in relation to the positively buoyant or passive clouds of pollution. The different behaviour of these dense clouds includes the gravity driven flow due to density gradients in the horizontal direction and turbulence dumping due to the stable stratification. Special models are being developed to perform the transport of heavy gas clouds in the atmospheric air. They are named heavy gas (or dense gas) dispersion models.

The law of environmental protection is the fundamental piece of law in Poland in relation to the dispersion modelling of accidental pollution releases to the atmosphere. According to the regulations of the Ministry of Economy describing in detail specific issues related to the development of the major accident prevention policy, safety reports, on-site and off-site emergency plans have to be prepared for industrial installations in which hazardous substances are processed, stored or transported in amounts above specified thresholds. In these documents, among many other issues, the evaluation of the atmospheric dispersion of the released substances, the distribution of the substance concentrations and the ranges of the forecast hazard zones have to be included.

In this article a classification of heavy gas releases and a classification of heavy gas dispersion models are given and each group of model is characterised. The current state and perspectives for the application of these models in Poland are shown. The need for future work is also discussed.

2 A CLASSIFICATION OF HEAVY GAS DISPERSION MODELS

The heavy gas dispersion models can be classified using different criteria. The most frequently used criteria are the following: the spatial scale, temporal scale, mathematical principles or its usage (Markiewicz 2006a). Based on this last criterion the heavy gas dispersion models are divided into three groups (Mercer et al. 1999, Markiewicz 2006b): phenomenological (empirical) models, intermediate (engineering) models and computational fluid dynamic (research) models.

In the first group of models the dispersion of heavy gas clouds released to the atmospheric close to the ground is described by a series of nomograms or simple correlations. They were created mainly in the eighties based on the results of field or laboratory measurements limited to flat terrain and neutral conditions. The centreline concentration of the heavy gas is calculated as a function of a downwind distance. The calculations are based on an assumption that the atmospheric stability does not influence on the heavy gas plume behaviour. Instantaneous and continuous releases are distinguished. The centreline concentrations are calculated in terms of the following parameters: the gravity constant, gas density relative to the ambient air, release volume or release rate, ambient wind velocity. Typical examples of these models are described in the Workbook on the dispersion of dense gases (Britter & McQuid 1988) and the German VDI Guidelines VDI 3783 (1990).

In the second group of models, five subgroups can be distinguished: box models, steady state plume models, generalised steady state plume models, one dimensional integral plume models and shallow layer models. These last ones are the most complex models among the intermediate models. They differ from other models of this group in their ability to treat the effects of topography and in the dimensionality.

The box models are used to describe instantaneous releases with grounded clouds. Mostly it is assumed that the pollution cloud forms a uniform cylinder. The cloud main variables such as its radius, mass and enthalpy are obtained by numerical integration of basic ordinary differential equations with respect to time. The basic equations represent the cloud horizontal spreading, mass and energy conservation. Other cloud variables such as concentration, height, volume, density and temperature are calculated from the main variables. The horizontal spreading influencing its radius is assessed using a front velocity. The exchange of the mass between the cloud and the atmospheric air taking place through a top and an edge of the cylinder is described by entrance velocities. The values of these parameters can differ for the top and the edge. They depend on the turbulence intensity, difference of densities between the cloud and the environment, transport velocity of the cloud. The cloud heating being the result of its contact with the ground and air is introduced straightforwardly. The cloud spreads in still air or is moved downwind at the wind velocity from the mid-height of the cloud. The concentration averaged over the box volume is calculated knowing the mass of the substance released and the box volume. The variation of concentration in the box can be later introduced by assuming empirical similarity profiles in the vertical or horizontal direction and dependence on the cloud dimensions. The box models based on this simple approach satisfactorily reproduce many aspects of field and laboratory experiments. Examples of box models include the DENZ-EDF model (Kaiser & Walker 1978), the GASTAR model (Britter 1990), the HEGABOX model in the HGSYSTEM computer package (Witlox 1994), the IIT Heavy Gas Model I (Mohan et al. 1995) and many other models described by Eidsvik (1980), Fay & Ranck (1982), Fay & Zemba (1985), van Ulden (1987), Webber (1993), Cleaver (1994), Kunsch & Fannelop (1995), Nielsen (1996), Kunsch & Webber (2000), Kumar et al. (2003). In general, in most of box models, a flat uniform terrain is assumed. In only some of these models are the dispersion on slopes (Britter 1989, 1990, Webber 1993, Nielsen 1994, Webber 2000, Kumar et al. 2003) or over fences and obstacles (Cleaver 1994) described. Some of the models of this group have been adapted to continuous releases of heavy gases on the ground. In these models the plume is simulated by a series of puffs of a rectangular shape.

The steady state plume models are used for continuous grounded releases. They are developed in a similar manner as the box models. All basic phenomena associated with the dense gas releases such as the horizontal spreading, exchange of mass between the plume and the surrounding air, plume heating are described by similar ordinary differential equations. However, here these equations are integrated with respect to the downwind distance, the main variables are the average mass and enthalpy fluxes through the plume cross section and the plume width. The plume moves downwind with the wind velocity. The plume cross section is assumed to be a rectangle and the rectangular or Gaussian profile is adjusted to this shape for concentrations. In case of rectangular profile the concentration averaged over the plume cross section is calculated knowing the mass flow rate of the substance and the volume flow rate of the plume. In case of Gaussian profile the concentration is allowed to vary in the cross section and is calculated directly from a Gaussian formula. The transition to the passive plume dispersion described by the same formula in the far field is natural. The model described by Fay & Zemba (1986) is an example of uniform plume models. The IIT heavy gas model (Mohan et al. 1994) is an example of a Gaussian plume model.

The generalised steady state plume models can be considered as an extension of the steady plume models in the sense that the spatial variation of concentrations and other parameters in the plume cross section do not need to follow Gaussian or rectangular profiles. Similarity profiles determined empirically are used to describe them. This allows us to model some of the physical processes more realistically. The concentration is expressed in terms of a centreline ground level concentration, vertical and horizontal dispersion parameters and width of the plume. These quantities are determined from a number of basic equations describing heavy gas mass conservation, air entrainment, horizontal crosswind gravity spreading and crosswind diffusion. In the far field these models make a smooth transition to passive gas dispersion. The HAGADAS model included into the HGSYSTEM computer package (Witlox 1994) and the DEGADIS model (Spicer & Havens 1984) are examples of generalised steady state plume models.

The one dimensional integral plume models are used to describe continuous, elevated releases. They are based on the integration of conservation equations of the mass, species, downwind and crosswind momentum and energy averaged over the plume cross section. These equations directly predict plume variables such as the concentration, velocity, density and temperature. In the steady state models the plume variables are evaluated as a function of the downwind distance. In more general time dependent models these variables are evaluated as a function of the downwind distance and time. The cross section of the plume is assumed to be a circle, ellipse or rectangle. Similarity profiles are used to describe the space variability of plume variables in the cross section of the plume while averaging plume variables over the plume cross section to simplify the equations so that they can be used to reintroduce spatial variability of these variables. The type of emission sources used is the basic difference in the identification of the general one dimensional integral model in relation to the one dimensional shallow layer models used for grounded releases. The elevated release may remain the passive elevated plume in the far field or may touch the ground still being sufficiently dense. The gravity, drag force of the ambient flow and momentum of the entrained air influence the plume path. The entrainment rate in these models is different from that for the models of grounded clouds. It mainly depends on the velocity shear between the elevated jet and the surrounding air. Examples of steady state integral plume models include the HMP model (Hoot et al 1973), the AEROPLUME and HFPLUME models in the HGSYSTEM package (Witlox & McFarlane 1994) and many other models described by Ooms (1974), Epstein et al. (1990), Muralidhar et al. (1995), Khan & Abassi (1999). The CLOUD model can follow the steady state or transient conditions (Banerjee et al. 1996). An up to date review of two phase jet dispersion models is given by Bricard & Friedel (1998).

The one or two dimensional shallow layer models are used for grounded releases. They are based on partial differential equations describing the principles of conservation of the mass, species, momentum and energy averaged over the cloud depth. This kind of averaging is convenient due to the geometry of the cloud. Its vertical dimension is small compared to its horizontal dimensions. The pollutant cloud behaviour is described using the variables changing in one or two dimensions in space and time. The top of the cloud is difficult to define in reality and a vertical concentration distribution is used to define it. Approximations of the shallow layer theory are adapted. They state that the pressure distribution is hydrostatic within the main body of the cloud and the dispensation is made only for the special processes at the leading edge. The exchange of the mass between the pollutant cloud and the atmospheric air is described by the entrainment velocity. The complex topography is introduced quite easily by adding some terms to the momentum equation. This is an advantage in comparison to the simpler models which, in general, are not suitable for complex topography. These models allow for a realistic description of the behaviour of the heavy gas clouds. The time to run them is somewhere between the time to run the simpler intermediate models and fluid dynamics models. The SLAB model is an example of one dimensional shallow layer models. The TWODEE model (Hankin & Britter 1999a, b, Hankin 2004a, b) and DISPLAY model (Wurtzs et al. 1996) are examples of the two dimensional shallow layer models.

The research models are three dimensional models, in which a full set of partial differential equations dependent on time and three space coordinates describing the principles of conservation of the mass, momentum, energy and substance, are solved. These models can be applied to any type of emission scenario, terrain or meteorological conditions. The description of the physical processes of the heavy gas dispersion is detailed and complete. Some of these models include a concentration fluctuation model. The models can be divided into groups depending on the turbulence description (models with K, K-epsilon or K-l type of closure or direct numerical simulation models without turbulence closure models), approximations introduced into equations (hydrostatic, anelastic or Boussinesq approximation) or solution methods of the equations (finite difference schemes, finite element methods, finite volume methods). The FEM3 model (Ermak et al. 1982), the MARIAH model, the HEAVY-GAS model (Deaves 1985), the ANDREA model (Wurtz et al. 1991, 1996), the MDPG model (Bellasio & Tamponi 1993), the MERCURE- GL model (Dujim 1994), the STD model (Ohba et al. 2004) and some other models described by Pereira & Chen (1995), Burman (1998), Sklavounos & Rigas (2004, 2006), Ohba (2006) are the examples of research models. The model of Ohba (2006) is a direct numerical simulation model which solves the fundamental equations with fine meshes and time steps without the assumptions of the turbulent closure model. It is expected that the group of scientific models will be extended in the near future by the development of the Large scale eddy simulation models suitable for heavy gas dispersion.

It is worth mentioning that the presented classification of the heavy gas dispersion models is an idealisation. Some models described in the literature seem not to fit to any of the categories. There are also computer packages incorporating several gas dispersion models and then each of models has to be treated independently in this classification.

The intermediate and empirical models are used in the West European countries and in the USA in the routine calculations. They are important components of emergency response systems and valuable tools for environmental impact assessment and risk assessment (Hepner & Finco 1995, Diatli et al. 2000, Bellasio & Bianconi 2005, Scenna & Cruz 2005). Input data needed by theses models are easy to obtain, their concept is usually simple and computing costs are low or reasonable. The fluid dynamic (research) models are usually used as a research tool to get to know better the heavy gas properties. However, although they are very attractive their usage is limited due to their high computational costs and the input data requirements.

Despite the fact that common work of many institutions mainly from the Western European countries and the USA has resulted in big advances in the area of heavy gas dispersion during the last decade there are still some problems, which need either solving or, at least, improvement. This includes the modelling of the dispersion of heavy gases in the following conditions: under low wind conditions, in stable or unstable conditions, within the chemical complex, in complex terrain. Special attention should be given to modelling the multi-component and multi-phase heavy gas releases (Britter 1998).

3 AN APPLICATION OF HEAVY GAS DISPERSION MODELS IN POLAND

None of the heavy gas dispersion models was developed in Poland. Up to now there has only been an attempt to adapt the models developed abroad to Polish conditions by introducing Polish language comments instead of the comments in the original language.

In Poland, the intermediate models developed abroad are applied for routine calculations. These are the simplest models of this group developed many years ago. The most popular are the ALOHA model and DEGADIS model. The intermediate models were applied in calculation of the atmospheric dispersion of hazardous materials in the big chemical Polish companies including the Nitrogen Works S.A. in Tarnow and Pulawy, the Chemical Works S.A. in Police, the Mazovian Refinery and Petrochemical Works S.A. in Plock or smaller companies such as the Waterwarks in Wieliszow and the Central Waterwarks in Warsaw.

Taking into account that in Poland application of heavy gas dispersion models is at the early stage much work has to be done. The main requirements are to: implement the more advanced models used abroad for regulatory purposes in our country, strengthen the research cooperation between Polish modellers with foreign institutions leading to the development of new models.

The implementation of models developed abroad for routine calculations should not be very difficult. However, some more detailed issues connected with this are worth discussing. They concern the availability of codes, availability of powerful computers and possibility of gathering a group of experts to train model users. As far as the first issue is concerned some model codes can be downloaded gratis from the internet, some other are available from institutions by establishing a cooperation and the rest can be bought. The second issue is not a limiting factor because the models used for regulatory purposes, even those developed during the last few years, need computers of only medium power. The third issue also has had a positive outcome. Assembling a group of experts to lead the work and train users will not be difficult. Some activities related to this have already been initiated. Workshops and training in this discipline were organised by the Centre of Excellence MANHAZ (Management of Health and Environmental Hazards). A Summer school was organised by the MANHAZ in cooperation with the Warsaw Technical University on 25 September, 2005.

Strengthening the cooperation between Polish scientists with foreign institutions seems promising. EU funded programs offer a splendid platform to support these activities and unite the efforts of different institutions related to heavy gas dispersion in order to extend knowledge and improve modelling methods.

Developing a methodology to evaluate atmospheric contamination due to heavy gas releases is complex. It should refer to such methodologies already applied in the Western European countries and the USA. However, special attention should be given to methods of meteorological data preparation for routine applications. The meteorological parameters used in Poland for the routine calculations are determined in slightly different manner to those in other countries. A stability classification scheme and wind profile determination are specific (Markiewicz 2006a).

4 SUMMARY AND CONCLUSIONS

Material related to the modelling of the transport and dispersion of heavy gases in the atmosphere is presented here in a condensed fashion. A number of different types of heavy gas dispersion models have been developed. These can be divided into three groups: empirical models, intermediate models and fluid dynamic models. The empirical and intermediate models are important components of emergency response systems and valuable tools for the assessment of environmental impact and risk.

Within the last decade, despite the considerable progress in heavy gas dispersion modelling there is still a need for further development. Models should be better able to describe the more realistic emission scenarios.

In Poland the use of heavy gas dispersion models is at an early stage. Work should be undertaken to:

- implement the more advanced intermediate models developed abroad and
- develop a methodology to evaluate atmospheric contamination due to accidental release of heavy gases.

Effort should be also directed towards strengthening cooperation between Polish experts and foreign institutions.

REFERENCES

Bartzis, J.G. 1991. ANDREA-HF: a three dimensional finite volume code for vapour cloud dispersion in complex terrain. CEC JRC Ispra Report EUR 13580 EN, Ispra: EC Joint Research Centre.

Bricard, P. & Friedel, L. 1998. Two-phase jet dispersion. *Journal of Hazardous Materials* 59: 287–310.

Bellasio, R. & Tamponi, M. 1993. MDPG: a new Eulerian 3D unsteady state model for heavy gas dispersion. *Atmospheric Environment* 28: 1633–1643.

Benerjee, S. et al. 1996. CLOUD: A vapour-aerosol dispersion model accounting for plume 3d motion and heat and mass transfer between phases. *Journal of Hazardous Materials* 46: 231–240.

Billeter, L. & Fannelop, T.K. 1996. Concentration measurements in dense isothermal gas clouds with different starting conditions. *Atmospheric Environment* 31: 755–771.

Britter, R.E. 1990. *GASTAR user manual*. Cambridge: Cambridge Environmental Research Centre.

Britter, R.E. 1989. Atmospheric dispersion of dense gases. *Annual Review of Fluid Mechanics* 21: 317–344.

Britter, R.E. 1998. *Recent research on the dispersion of hazardous materials*. University of Cambridge Report EC EUR 18198 EN, Brussels: Directorate of Science, Research and Development.

Britter, R.E. & McQuaid, J. 1988. *Workbook on the dispersion of dense gases*. HSE Contract Research Report No 17/1988. Sheffield: Health and Safety Directorate.

Bricard, P. & Friedel, L. 1998. Two phase jet dispersion. *Journal of Hazardous Materials* 59: 283–310.

Burman, J. 1998. An evaluation of topographic effects on neutral and heavy gas dispersion with CFD model. *Journal of Wind Engineering and Industrial Aerodynamics* 74–76: 515–325.

Cleaver, R.P. et al. 1994. Further development of a model for dense gas dispersion over real terrain. *Journal of Hazardous Materials* 40: 85–108.

Davies, J.K.W. & Hall, D.J. 1996. An analysis of some replicated wind tunnel experiments on instantaneously released heavy gas clouds dispersing over solid and cranellated fences. *Journal of Hazardous Materials* 49: 311–328.

Deaves, D.M. 1985. Application of advanced turbulence models in determining the structure and dispersion of heavy gas clouds. In G. Ooms & H., Tennekes (eds.) *Atmospheric Dispersion of Heavy Gases and Small Particles. Proc. intern. IUTAM symp.* Berlin: Springer Verlag.

Ditali, S. et al. 2000. Consequence analysis in LPG installations using an integrated computer package. *Journal of Hazardous Materials* 71: 159–197.

Duijm, N.J. 1994. *Research on the dispersion of two phase flashing releases- FLADIS*. Fladis TNO final report R94-451. Appeldorn: Institute of Environment and Energy Technology.

Eidsvik, K.J. 1980. A model for heavy gas dispersion in the atmosphere. *Atmospheric Environment* 14: 769–777.

Epstain, M. et al. 1990. A model of dilution of a forced two phase chemical plume in a horizontal wind. *Journal of Loss Prevention Process Industry* 3: 280–290.

Ermak, D.L. et al 1982. A comparison of dense gas dispersion model simulations with Burro Series LNG test results. *Journal of Hazardous Materials* 6: 129–160.

Fay, J.A. & Ranck, D.A. 1982. Comparison of experiments on dense gas cloud dispersion. *Atmospheric Environment*: 17: 239–248.

Fay, J.A. & Zemba, S.G. 1985. Dispersion of initially compact dense gas clouds. *Atmospheric Environment* 19: 1257–1261.

Fay, J.A. & Zemba, S.G. 1986. Integral model of dense gas plume dispersion. *Atmospheric Environment* 20: 1347–1354.

Hankin, R.K.S. & Britter, R.E. 1999a. TWODEE: the Health and Safety Laboratory's shallow layer model for heavy gas dispersion. Part I: Mathematical basis and physical assumptions. *Journal of Hazardous Materials* 66: 211–226.

Hankin, R.K.S. & Britter, R.E. 1999b. TWODEE: the Health and safety Laboratory's shallow layer model for heavy gas dispersion. Part II: Model validation. *Journal of Hazardous Materials* 66: 227–237.

Hankin, R.K.S. 2004a. Major hazard risk assessment over non-flat terrain. Part I: continuous releases. *Atmospheric Environment* 38: 695–705.

Hankin, R.K.S. 2004b. Major hazard risk assessment over non-flat terrain. Part II: instantaneous releases. *Atmospheric Environment* 38: 707–714.

Hanna, S.R. et al. 1993. Hazardous gas model evaluation with field observations. *Atmospheric Environment* 27A: 2265–2285.

Hepner, G.F. & Finco, M.V. 1995. Modelling dense gaseous contaminant pathways over complex terrain using a geographic information system. *Journal of Hazardous Materials* 47: 187–199.

Mercer, A. et al. 1999. *Heavy Gas Dispersion Expert Group, Model Evaluation Group Final Report*, Brussels: Directorate of Research, Science and Development.

Kaiser, G.D. & Walker, B.C. 1978. Releases of anhydrous ammonia from pressurised containers – the importance of denser than air mixtures. *Atmospheric Environment* 12: 2289–2300.

Khan, F.I. & Abassi, S.A. 1999. Modelling and simulation of heavy gas dispersion on the basis of modifications in plume path theory. *Journal of Hazardous Materials* A80: 15–30.

Kunsch, J.P. & Fannelop, T.K. 1995. Unsteady heat transfer effects on the spreading and dilution of dense cold clouds. *Journal of Hazardous Materials* 43: 169–193.

Kunsch, J.P. & Webber, D.M. 2000. Simple box model for dense gas dispersion in straight sloping channel. *Journal of Hazardous Material* A75: 29–46.

Kumar, A. et al. 2003. Study of the spread of a cold instantaneous heavy gas release with surface heat transfer and variable entrainment. *Journal of Hazardous Materials* B101: 157–177.

Markiewicz, M. 2006. Modelling of the air pollution dispersion. In *Modelling and techniques for health and environmental hazard assessment and management.* M. Borysiewicz & S. Potemski (eds.) Warsaw: Drukpol-Janusz Piatkowski.

Markiewicz, M. 2006. Modelling of the heavy gas dispersion. In *Modelling and techniques for health and environment hazard assessment and management.* M. Borysiewicz & S. Potemski (eds.) Warsaw: Drukpol-Janusz Piatkowski.

Mohan, M. et al. 1995. Development of dense gas dispersion model for emergency preparedness. *Atmospheric Environment* 29: 2075–2087.

Muralidhar, R. et al. 1995. A two-phase release model for quantifying risk reduction for modified HF alkylation catalysts. *Journal of Hazardous Materials* 44: 141–183.

Nielsen, M. 1996. Comment on: A model of the motion of a heavy gas cloud released on a uniform slope. *Journal of Hazardous Materials* 48: 251–258.

Ohba, R. et al. 2005. Validation of heavy and light gas dispersion models for the safety analysis of LNG tank. *Journal of Loss Prevention in the Process Industries* 17: 325–337.

Ooms, et al. 1974. The plume path of vent gases heavier then air. In C.H. Birschman (ed.) *Loss Prevention and Safety Promotion in the Process Industries. Proc. of the first intern. symp*. Haque. Amsterdam: Elsevier Press. 211–219.

Pereira, J.C.F. & Chen, X.Q. 1995. Numerical calculations of unsteady heavy gas dispersion. *Journal of Hazardous Materials* 46: 253–272.

Scenna, N.J. & Santa Cruz, A.S.M. 2005. Road risk due to transportation of chlorine in Rosario city. *Relability Engineering and System Safety* 90: 83–90.

Sklavounos, S. & Rigas, F. 2004. Validation of turbulence models in heavy gas dispersion over obstacles. *Journal of Hazardous Materials* A108: 9–20.

Sklavounos, S. & Rigas, F. 2006. Simulation of Coyote serial trials. Part I: CFD estimation of non-isotermal LNG releases and comparison with box model predictions. *Chemical Engineering Science* 61: 1434–1443.

Spicer, T.O. & Havens, J.A. 1985. Modelling the Phase I Thorney Island experiments. In McQuaid J. (ed), *Heavy gas dispersion trials at Thorney Island.* Proc. of symp., University of Sheffield, G.B., 3–5 April, 1984, Amsterdam: Elsevier.

Tickle, G.A. et al. 1992. An integral model for an elevated two phase jet in a cross flow. In Heat Transfer, *AICHE Symposium Series No.288, 88, 350–356.*

Van Ulden, A.P. 1987. The heavy gas mixing process in still air at Thorney Island and in the laboratory. *Journal of Hazardous Materials* 16: 411–425.

VDI Guidelines 1990. VDI 3783 Part II Environmental Meteorology, Dispersion of heavy gases (in German).

Webber, D.M. 1993. A model of the motion of a heavy gas cloud released on a uniform slope. *Journal of Hazardous Materials* 33: 101–122.

Wurtz, J. et al. 1996. A dense vapour dispersion code package for applications in the chemical and process industry. *Journal of Hazardous Materials* 46: 273–284.

Witlox, H.W.M. & McFarlane, K. 1994. Interfacing dispersion models in the HGSYSTEM hazard-assessment package. *Atmospheric Environment* 28: 2947–2962.

Witlox, H.W.M. 1994a. The HEGADAS model for ground level heavy gas dispersion. Part I. Steady state model. *Atmospheric Environment* 28: 2917–2932.

Witlox, H.W.M. 1994b. The HEGADAS model for ground level heavy gas dispersion. Part II. Time dependent model. *Atmospheric Environment* 28: 2933–2946.

Woodward, J.L. & Papadourakis, A. 1993. *Modelling and validation of the dispersing aerosol jet. Journal of Hazardous Materials* 46:185–207.

Environmental Engineering – Pawłowski, Dudzińska & Pawłowski (eds)
© 2007 Taylor & Francis Group, London, ISBN13 978-0-415-40818-9

Treating missing data at air monitoring stations

Szymon Hoffman
Department of Chemistry, Water and Wastewater Technology, Technology University of Częstochowa, Częstochowa, Poland

ABSTRACT: Data gathered continuously in air monitoring systems are never entirely complete. An assessment of the air quality, as is required by law, is handicapped by such a loss of data. These standards permit however, the use of modeling to recreate the missing data when the completeness of the monitoring issues is insufficient. In this paper, some computational methods are suggested to resolve the problem. The analysis was carried out in order to compare the methods of air pollutants concentration modelling. Some regression and forecasting methods based on artificial neural networks were compared. The analysed data set was built of hourly averages, gathered at two different air monitoring stations in Poland over a long-term period. It was shown that the application of regression models ensures relatively high precision of air pollution prediction. For short lookahead, time-series models are comparably precise.

Keywords: Missing data, air monitoring, modelling, artificial neural networks.

1 INTRODUCTION

The data gathered continuously in the air monitoring systems are never complete. An assessment of the air quality, that is legally required, is handicapped by this loss of data (Hauck et al. 1999). Over a whole year, the number of missing records usually equals several per cent but for some measured variables it may be considerably higher. Air quality standards permit the use of models in order to recreate the missing data when the completeness of the monitoring set is insufficient. Those regulations do not determine what modelling methods should be used. For this reason the air monitoring services do not exploit modelling as the preferred tool for completion of data.

The problem of missing data in environmental sciences has been solved in many ways, mainly using statistical methods (Tang et al. 1996, Bennis et al. 1997, Nosal et al. 2000). However, the wide usage of these methods in monitoring systems is doubtful. Recently, the use of artificial neural networks (ANNs) has been developed in environmental modelling. ANNs are very useful tools for modelling relationships whose mathematical formula is unknown. Because of their simplicity and accuracy, two types of ANN methods seem to be the most advantageous: (1) regression modelling (Gardner & Dorling 1998, Hadjiiski et al. 1999, Abdul-Wahab & Al Alawi 2002, Hoffman 2003), and (2) forecast modelling (Kolehmainen et al. 2001, Hoffman 2006). More advanced computational techniques have been tested as well, for example: genetic algorithms (Niska et al. 2004), Kohonen neural networks (Turias et al. 2006), hybrid fuzzy neural systems (Marabito et al. 2003), Bayesian neural networks (Wang et al. 2006).

In this paper, some computational methods are suggested to solve the problem of missing data modelling. The main objective of the analysis was to compare and verify the methods of air pollutants concentration modelling. The conception of autonomous models was assumed. In these models there is no need to exploit external data but only the data available in the monitoring systems are introduced as the model inputs. This concept makes it possible to formulate a universal algorithm of missing data treatment, practicable to use in any monitoring stations. The analysed data set was built of long-term hourly averages, gathered at two different air monitoring stations in Poland. Some regression and forecasting methods based on neural networks were compared. The value of the model error, resulting from divergences between the model output and the real concentration values, was assumed as the criterion in evaluating each model.

2 MATERIALS AND METHODS

Two long-term sets of hourly air monitoring data were used in the analysis:

- the data collected in the period 1994–1999 at the station SP-3 in Kędzierzyn-Koźle (South Poland),

D	H	WS	WD	T	P	SR	RH	O_3	NO	NO_2	SO_2	CO	PM
...	...	...	...	...	...	...	...	...	...	...	...	...	...
in 1	in 2	in 3	in 4	in 5	in 6	in 7	in 8	out	in 9	in 10	in 11	in 12	in 13
...	...	...	...	...	...	...	...	...	...	...	...	...	...

Figure 1. Modelling scheme employed in regression methods.

– the data collected in the period 1994–1997 at the station in Zabrze (South Poland).

These sets represent two different localizations. The station in Zabrze is situated in the center of a large industrial city, in the immediate vicinity of strong sources of air pollution emission, like a busy road (Wolności St.), a power plant, a coke plant and others. The station in Kędzierzyn-Koźle is located further from strong emission sources and represents the pollution background in the industrial region.

Both specified data sets were denoted respectively as Zabrze94/97 and KK94/99. The data sets comprised hourly averages of concentrations and meteorological observations. The following notation of the measured variables was used in this paper:

D – the day of measurement;
H – the hour of measurement;
T – temperature, °C;
SR – solar radiation, W/m^2;
RH – relative humidity, %;
WS – wind speed, m/s;
WD – wind direction, deg;
P – pressure, hPa
O_3 – ozone concentration, $\mu g/m^3$;
NO – NO concentration, $\mu g/m^3$;
NO_2 – NO_2 concentration, $\mu g/m^3$;
SO_2 – SO_2 concentration, $\mu g/m^3$;
CO – CO concentration, $\mu g/m^3$;
PM – PM_{10} concentration, $\mu g/m^3$.

Two different modelling methods were used in the assessment: regression and time series analysis. Artificial neural networks (ANN) were used as the tool to create all the models. The two types of models differ from each other on the applied data and computational algorithm used. A common feature of these modelling methods is exploitation of only the data available from the monitoring systems. This group of models may be called *autonomous models*. No external measurements are needed to create these models.

Regression methods exploit correlations between the chosen variable and other variables measured simultaneously. Model creation proceeded according to the rules presented in Figure 1. The modelled variable comprised the model output (signed as *out*) whereas the other variables (signed as *in*) were treated as the inputs. These models were almost independent of time because the variables introducing the day (D) and the hour (H) generally did not influence the modelling results. Also, during the computation the sequence of the cases is not important, so the models may be considered stationary. In the group of regression models two types of models were tested: (1) linear perceptrons, (2) non-linear perceptions, the latter with four neurons in a single hidden layer.

D	H	WS	WD	T	P	SR	RH	O_3	NO	NO_2	SO_2	CO	PM
...	...	...	...	...	...	...	...	...	...	...	...	...	...
...	...	...	...	...	...	...	...	in 3	...	...	...	...	...
...	...	...	...	...	...	...	...	in 2	...	...	...	...	...
...	...	...	...	...	...	...	...	in 1	...	...	...	...	...
...	...	...	...	...	...	...	...	...	...	...	...	...	...
...	...	...	...	...	...	...	...	...	...	...	...	...	...
...	...	...	...	...	...	...	...	out	...	...	...	...	...
...	...	...	...	...	...	...	...	...	...	...	...	...	...

Figure 2. Modelling scheme employed in time-series methods.

In this paper only results for the regression models including all input variables were considered. Such models reveal the limitations of approximation by regression methods, although more simplistic models might be equally precise in many cases.

In time-series analysis the source of knowledge of the modelled variable was the autocorrelation of this variable, resulting from the variable periodicity. The model output was the chosen pollutant concentration at the specified time (Figure 2). The inputs comprised concentrations of this pollutant at previous hours.

Generally, ANN time-series models are characterized by two parameters which influence the prediction quality: lookahead and steps. Lookahead specifies how far ahead of the last input case the modelled output is assigned. Steps specifies how many previous cases (inputs) should be taken to the network. In all models the constant number of steps (24), was assumed. The effect of changes of lookahead was analyzed. In the group of time-series models only the results for linear models were presented because non-linear models were not apparently any better.

The accuracy of different obtained models was compared. The value of the correlation coefficient for the best linear fit of the real concentrations to the model outputs was assumed as the criterion of model error. Correlation coefficients do not depend on measurement units and ranges of variables so it is an all-purpose type of prediction error and it may be used for the quality comparison of quite different models

The analysis was carried out using the programme Statistical Neural Networks. The analysed set of data was divided into three different subsets: the training subset (50% of cases), the verification subset (25% of cases) and the test subset (25% of cases). For ANN training two algorithms were used: Levenberg-Marquardt in the group of non-linear models, and Pseudo-Inverse in the case of linear models.

3 RESULTS AND DISCUSSION

3.1 *Regression models*

Generally, environmental regression models are not extremely precise. This results from the model concept. It is impossible in this type of models to recognize and exploit all important causes of a given modelled effect. This is why the precision difficulties in environmental modelling occur. In many applications, models are simplified on purpose in order to expand their usage. For example, the number of input variables may be restricted to a few very important ones, without the loss of the model accuracy (Hoffman 2003).

Air pollution prediction by autonomous models, restricted to the variables measured at air monitoring stations, is a concept that may be implemented at every air monitoring systems. In this concept some important information about the quantity, the time and the source of pollution emission is not introduced directly into models. The models are not fed with essential meteorological information of boundary layer structure, diffusion conditions and rainfalls, either. But the knowledge of the mentioned data might be indirectly and piecemeal introduced into the models by explanatory variables. For example, concentrations of primary air pollutants are correlated because they are usually derived from the same sources, i.e. fuel combustion processes. Even the O_3 concentration is correlated with concentrations of other pollutants, although ozone is a secondary pollutant.

Considering the above facts, it seems obvious that regression modelling can be useful but it cannot ensure a very high precision. Taking into account this likely weakness in prediction, the following classification of models is suggested (Hoffman 2004):

- highly precise models, if $r > 0.90$
- precise models, if $0.90 > r > 0.80$
- low precise models, if $0.80 > r > 0.50$
- useless models, if $r < 0.50$

In the first part of the study the accuracies of obtained regression models were compared. The correlation coefficient values are presented in Table 1, separately for both monitoring stations in Zabrze and in Kędzierzyn-Koźle, and separately for different pollutants. The errors of linear models were compared with the errors of their non-linear equivalents.

The obtained results show that non-linear models are always more precise than their linear equivalents. In the group of non-linear models there are four highly precise ones ($r > 0.90$), six precise ones ($0.90 > r > 0.80$), and only two low precise ones ($0.80 > r > 0.50$). Among linear models there are six precise ones, and six low precise ones, but there are no highly precise ones. The exchange of linear models into non-linear equivalents causes the graduation to the higher class of prediction precision in most cases (8 cases from 12).

Table 1. The comparison of correlation coefficient values calculated for regression linear and non-linear models (Zabrze94/97, KK94/99).

	Correlation coefficient, r			
	Zabrze94/97		KK94/99	
Predicted variable	Linear models	Non-linear models	Linear models	Non-linear models
O_3	0.86	0.92	0.84	0.89
NO	0.81	0.91	0.55	0.74
NO_2	0.78	0.88	0.71	0.77
SO_2	0.81	0.88	0.70	0.81
CO	0.89	0.93	0.79	0.84
PM	0.83	0.92	0.73	0.80

Table 2. The estimation of time series modelling accuracy for hourly concentrations, 24 steps, various lookaheads (Zabrze94/97).

	Correlation coefficient, r					
Lookahead	O_3	NO	NO_2	SO_2	CO	PM
1	0.96	0.86	0.91	0.95	0.94	0.92
2	0.91	0.71	0.82	0.91	0.87	0.85
3	0.87	0.62	0.76	0.88	0.81	0.79
4	0.83	0.55	0.71	0.85	0.77	0.74
5	0.80	0.51	0.68	0.84	0.74	0.70
6	0.78	0.49	0.65	0.82	0.71	0.67
8	0.76	0.46	0.62	0.79	0.68	0.63
12	0.74	0.42	0.58	0.75	0.63	0.58
24	0.75	0.37	0.54	0.70	0.57	0.53
72	0.67	0.29	0.38	0.57	0.44	0.38
240	0.62	0.25	0.32	0.52	0.39	0.31

The prediction abilities of models created for the data sets derived from different monitoring stations are distinct. In the case of the station situated in the vicinity of strong, local air pollution emission sources (Zabrze) prediction abilities are higher. For all pollutants, their concentration modelling gives better results at the Zabrze station than at Kędzierzyn-Koźle.

The ability to model ozone concentration is particularly encouraging, only slightly better at the station in Zabrze. Probably, the prediction of ozone concentrations does not depend strongly on locality. All the models of ozone concentrations belong to the highly precise and precise classes.

3.2 *Time series models*

The results of prognosis of hourly concentrations for different pollutants are presented in Table 2 (data set

Table 3. The estimation of time series modelling accuracy for hourly concentrations, 24 steps, various lookahead (KK94/99).

	Correlation coefficient, r					
Lookahead	O_3	NO	NO_2	SO_2	CO	PM
1	0.97	0.76	0.83	0.93	0.95	0.86
2	0.92	0.65	0.74	0.86	0.90	0.80
3	0.89	0.60	0.69	0.82	0.85	0.76
4	0.85	0.57	0.65	0.79	0.81	0.73
5	0.82	0.54	0.63	0.77	0.78	0.70
6	0.80	0.52	0.61	0.75	0.75	0.68
8	0.78	0.49	0.58	0.72	0.71	0.64
12	0.76	0.45	0.55	0.69	0.66	0.59
24	0.77	0.39	0.49	0.63	0.59	0.50
72	0.70	0.25	0.37	0.51	0.44	0.33
240	0.66	0.23	0.28	0.43	0.36	0.27

from Zabrze) as well as in Table 3 (data set from Kędzierzyn-Koźle). Modelling outcomes comprise the values of correlation coefficients. The lookahead was changed in the range from 1 to 240 hours. In earlier work the author showed that, for the hourly concentrations, the modelling accuracy apparently increased when the number of steps, rose to 24 hours (Hoffman 2004). Above this threshold the increase of precision is insignificant. A similar number of steps was assumed in this paper. All the compared time series models were generated with the same number of steps (equal to 24 hours).

The results obtained allow the estimation of how the accuracy changes depend on lookahead, thus they enable the determination of the horizon of reasonable prediction. Generally, decreased precision with the increase of lookahead is observed. For all pollutants, except ozone, the accuracy decreases quickly. Time-series modelling in the data set from Zabrze gives better results than the modelling in the data set from Kędzierzyn-Koźle. Prediction of NO, NO_2 and PM concentrations are worse in comparison with other pollutants. The best results were obtained in the case of ozone concentration. Ozone concentrations may often be predicted over even several days. For example, the value of correlation coefficient for the model that predicts ozone concentration in ten days (240 hours) is equal 0.66 (Table 3).

Using the model quality classification introduced for regression models, it appears that highly precise models ($r > 0.90$) may also be obtained by the time series analysis methods. In the data set from Zabrze, for lookahead equal to 1, highly precise models were generated for all pollutants except NO. And only for NO, non-linear regression modelling gives better results than time-series modelling. One-step-ahead time-series modelling is more advantageous than regression modelling for O_3, NO_2, SO_2, CO, PM. However, the precision of time-series prognosis quickly decreases with the every successive step and when the lookahead rises the time-series method becomes less advantageous than the regression method.

4 CONCLUSIONS

The analysis was carried out in order to compare and verify the methods of air pollutants concentration modelling. Some regression and forecasting methods based on artificial neural networks were compared. It can be stated that:

1. The application of autonomous regression models ensures relatively high precision prediction of air pollution. Promising results may be achieved by means of artificial neural networks, especially those of a non-linear type. A simple network structure of perceptrons ensures sufficient accuracy.
2. Precision of prediction depends on monitoring station location. Better results of regression modelling are obtained when there are strong sources of air pollution emission in the vicinity of the station.
3. Time-series models are precise, as well. Especially for short lookaheads, their accuracy may be comparable with the precision of regression models.
4. A decrease in the precision of time-series models with the increase of lookahead is observed. The loss of accuracy depends on the pollutant. The furthest reasonable prognosis may be made for ozone concentration. For other pollutants the horizon of reasonable prediction shortens in the order: SO_2, CO, PM_{10}, NO_2, NO.
5. Ozone is a secondary pollutant whose concentration changes quite regularly in the diurnal cycle. Therefore ozone concentration may by successfully modeled by means of time-series analysis. Modelling of levels of primary pollutants gives worse results because their diurnal cycles are less regular.

ACKNOWLEDGEMENTS

This work was carried out within the research project number 1 T09D 037 30, funded by the research budget of the Polish Government for the years 2006–2008.

REFERENCES

Abdul-Wahab, S.A., Al-Alawi, S.M. 2002. Assessment and prediction of tropospheric ozone concentration levels using artificial neural networks. *Environmental Modelling & Software* 17: 219–228.

Bennis, S., Berrada, F., Kang, N. 1997. Improving single-variable and multivariable techniques for estimating missing hydrological data. *Journal of Hydrology* 191: 87–105.

Gardner, M.W., Dorling, S.R. 1998. Artificial neural networks (the multilayer perceptron) – a review of applications in the atmospheric sciences. *Atmospheric Environment* 32: 2627–2636.

Hadjiiski, L., Geladi, P., Hopke, P. 1999. A comparison of modeling nonlinear systems with artificial neural networks and partial least squares. *Chemometrics and Intelligent Laboratory Systems* 49: 91–103.

Hauck, H., Kromp-Kolb, H., Petz, E., 1999. Requirements for the completeness of ambient air quality data sets with respect to derived parameters. *Atmospheric Environment* 33: 2059–2066.

Hoffman, S. 2003. Regression modelling of ground level ozone concentration. Environmental Engineering Studies. In Pawłowski L., Dudzińska M.R., Pawłowski A. (eds.), *Polish Research on the way to EU*. Kluwer Academic/Plenum Publishers, New York, 53–60.

Hoffman, S., Jasiński, R. 2004. Studies on NO_x concentration modeling in the air monitoring systems. In Dudzińska M.R. & Pawłowska M. (eds.), *Pathways of pollutants and mitigation strategies of their impact on the ecosystems*. Komitet 3Inżynierii Środowiska PAN, Lublin, 185–192.

Hoffman, S. 2006. Short-time forecasting of atmospheric NO_x concentration by neural networks. *Environmental Engineering Science* 23(4): 603–609.

Kolehmainen, M., Martikainen, H., Ruuskanen, J. 2001. Neural networks and periodic components used in air quality forecasting. *Atmospheric Environment* 35: 815–825.

Morabito, F.C., Versaci, M. 2003. Fuzzy neural identification and forecasting techniques to process experimental urban air pollution data. *Neural Networks* 16: 493–506.

Niska, H., Hiltunen, T., Karppinen, A., Ruuskanen, J., Kolehmainen, M. 2004. Evolving the neural network model for forecasting air pollution time series. *Artificial Intelligence* 17: 159–167.

Nosal, M., Legge, A.H., Krupa, S.V. 2000. Application of a stochastic. Weibull probability generator for replacing missing data on ambient concentrations of gaseous pollutants. *Environmental Pollution* 108: 439–446.

Tang, W.Y., Kassim, A.H.M., Abubakar, S.H. 1996. Comparaitive studies of various missing data treatment methods – Malaysian experience. *Atmospheric Research* 42: 247–262.

Turias, I.J., Gonzalez, F.J., Luz Martin, M., Galindo, P.L. 2006. A competitive neural network approach for meteorological situation clustering. *Atmospheric Environment* 40: 532–541.

Wang, D., Lu, W.Z. 2006. Interval estimation of urban ozone level and selection of influential factors by employing automatic relevance determination model. *Chemosphere* 62: 1600–1611.

Environmental Engineering – Pawłowski, Dudzińska & Pawłowski (eds)
© 2007 Taylor & Francis Group, London, ISBN13 978-0-415-40818-9

VOCs removal from exhaust gases using an electron beam accelerator

Janusz Licki
Institute of Atomic Energy, Otwock-Świerk, Poland

Andrzej G. Chmielewski, Anna Ostapczuk & Zbigniew Zimek
Institute of Nuclear Chemistry and Technology, Warsaw, Poland

ABSTRACT: Volatile organic compounds (VOCs) released into the atmosphere from various industrial processes cause serious and large scale environmental contamination. Several kinds of VOCs are very harmful to human health because of their carcinogenic effects and also produce toxic substances by photochemical oxidation in the atmosphere. Electron beam treatment is a promising technology for removing volatile organic compounds from vast quantities of off-gases containing low VOC concentrations. The laboratory plant for irradiation of gas mixtures containing VOC in a flow system by a high-energy electron beam was built in the Institute of Nuclear Chemistry and Technology in Warsaw. A study of removal efficiencies of selected VOCs and the influence of different parameters on the obtained efficiencies will be described.

Keywords: Volatile organic compounds, exhaust gases, electron beam, dose.

1 INTRODUCTION

Volatile organic compounds are released into the atmosphere from various industrial processes. Their main emission sources are: the mobile combustion sector, organic solvent application, petroleum industry, combustion of fossil fuels (industrial and non-industrial), waste treatment and disposal. Several kinds of VOCs are very harmful to human health and the environment and, released into the atmosphere, cause serious and large-scale environmental contamination. When exhaust gases, containing VOCs mixed with nitrogen oxides, are released into the atmosphere and irradiated by the ultraviolet (UV) component of sunlight, a complex chain of reactions converts them into products known as photochemical pollutants (ozone, aldehydes, hydrogen peroxide, peroxyacetyl nitrate, organic and inorganic acids and fine particles) (Ciccioli 1995). Among these, ozone is considered to be the most important because of the high concentrations produced and the wide range of effects it may have on human health, plant growth, material corrosion and climate change. This ozone is formed in the lower layer of the troposphere. The hydrocarbon reactivity is expressed in terms of photochemical ozone creation potential (POCP) units calculated with respect to ethylene as a reference compound (POCP = 100) (Derwent et al. 1996). The alkenes and aromatic hydrocarbons represent classes of VOCs with high potential for photochemical ozone production. VOCs released to the atmosphere affect large areas during long – distance redistribution through atmospheric pathways. The emission of VOCs should be controlled. The reduction of VOCs emission is a topic of growing interest in international conventions as well as in EU Directives.

Conventional treatment processes to control VOCs such as thermal oxidation, absorption, adsorption or catalytic oxidation require very high flow and emission making them expensive. These methods, which need a high reaction temperature, are not energetically economical when the concentration of VOC is low. If multiple pollutants are present in a gas stream at varying concentration then multiple control technologies may be needed, adding to the cost and complexity of the process. Thus, it is highly desirable to have a control technology capable of removing multiple gas-phase pollutants, thereby reducing cleanup costs and process complexity. The electron beam treatment is a promising technology for such a purpose.

2 MATERIALS AND METHODS

The electron beam irradiation of industrial exhaust gases (flue gases or other) leads to the formation of ions, radicals, atoms and excited species by radiolysis of the matrix gases (N_2, O_2, CO_2, H_2O). These species further react with O_2 and H_2O thereby forming OH^*, N^*, O^* and other radicals. The radicals oxidize

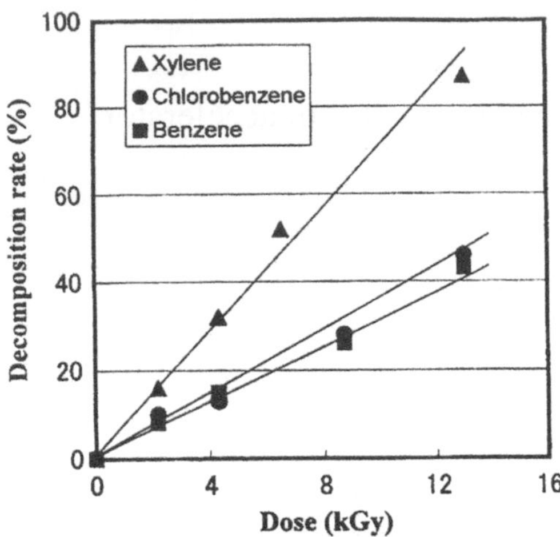

Figure 1. Decomposition of aromatics in batch system: xylene (93 ppm), chlorobenzene (102 ppm), benzene (87 ppm), H_2O (300 ppm) (Hirota et al. 2002).

the trace components such as SO_2, NO_x or VOCs in fast reaction chains. The oxidation products nucleate rapidly, if their vapour pressures are sufficiently low (e.g. H_2SO_4), to form an aerosol. The reactions occurring in the processes therefore resemble the chemistry proposed for the formation of photochemical smog. Studies of the electron beam treatment of VOCs have been pursued in laboratory plants in Japan (Hashimoto et al. 2000), Germany (Paur et al. 1991) and South Korea (Kim 2002). The purpose of this paper is to report on the construction of a Polish laboratory plant and the start of our investigations.

Many factors affect VOC removal by electron beam treatment. The removal efficiency depends on:

- The nature of the organic compounds involved. Fig. 1 (Hirota et al. 2002) presents the efficiencies obtained for three compounds. The study illustrates that the removal efficiencies of VOCs increase with the irradiation doses,
- The initial concentration of the VOCs. The removal efficiencies of aromatic VOCs increase as their initial concentration decrease (Fig. 2 (Paur 1997)),
- The humidity of the exhaust gases. Addition of water vapor increases the VOC removal efficiencies (Fig. 3 (Kim 2002)).

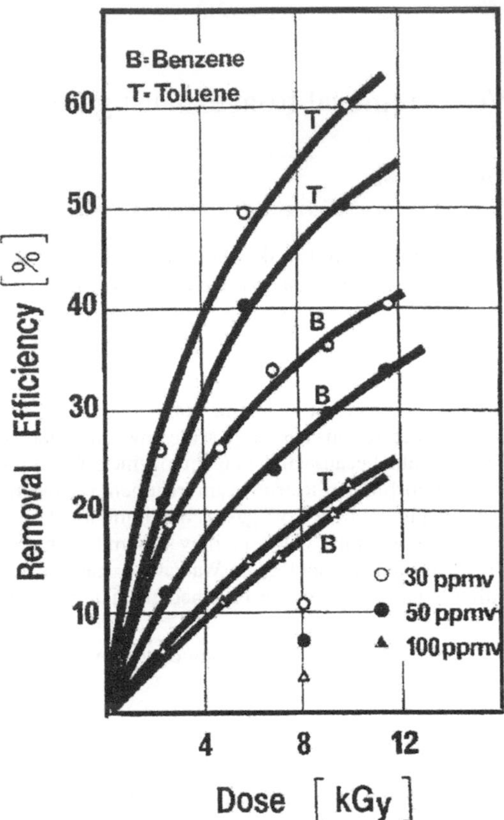

Figure 2. Decomposition of benzene and toluene in ambient air by treatment with EB (Paur 1997).

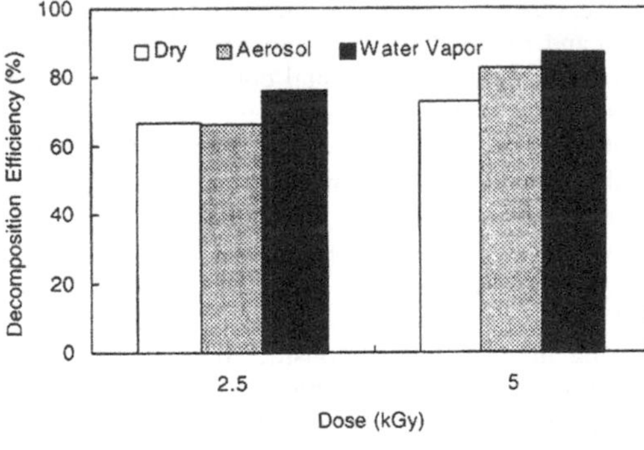

Figure 3. Effect of aerosol and water vapor on the decomposition of toluene at a flow system (Kim 2002).

3 RESULTS AND DISCUSSION

The laboratory plant for electron beam irradiation of gaseous mixtures containing selected VOCs was built in the Institute of Nuclear Chemistry and Technology in Warsaw (Fig. 4).

In the plant, gaseous VOC mixtures flow through the process vessel where after a full study of their decomposition including analysis of the aerosol formed is carried out. For this study benzene was selected as a typical VOC used in various industrial processes.

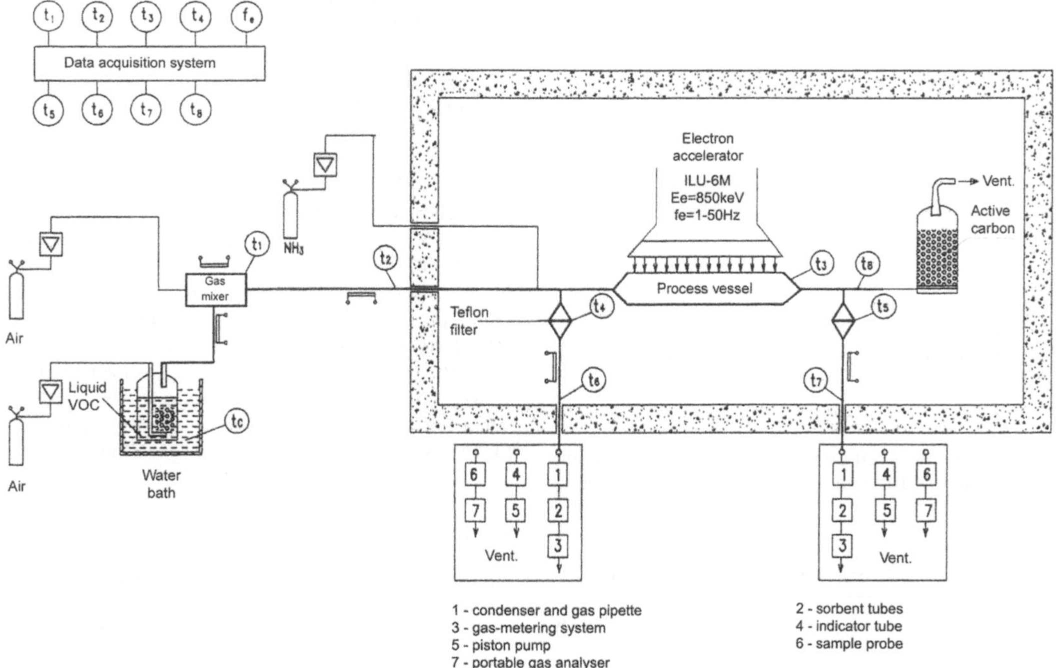

Figure 4. Flow diagram of electron beam irradiation of benzene/air mixture.

Gaseous benzene was generated by bubbling a carrier gas (synthetic air), through liquid benzene. The temperature of the liquid benzene was adjusted to 25°C in a water bath to keep its vapor pressure constant. The bubbled gas was diluted with the synthetic air from the second cylinder, as a dilution gas, into a gas mixer and introduced into the process vessel (reactor).

The concentration of benzene in the sample gas was adjusted by changing the flow rates of bubbling gas and dilution gas as well as the temperature of liquid benzene. The concentration of 100 ppm benzene was obtained at the following set of parameters:

flow of carrier gas – 1 l/h
flow of dilution gas – 800 l/h
temperature of liquid benzene – 22°C

The tubes through which the gas mixture was transported to the reactor and the process vessel were kept at temperature above the boiling point of benzene to avoid condensation of its vapor. A pulsed resonant electron accelerator ILU-6M with maximum beam power 20 kW was used for electron beam treatment. The gas mixture was irradiated with 0.8 MeV vertical electron beam in a stainless steel flow type process vessel. The vessel has a cuboidal shape with a total volume of 3 liters. The gas mixture in the reactor was exposed to the electron beam through a 50 μm thick titanium window. The dose deposited in the gas mixture was adjusted by changing the repetition rate of electron pulses up to 50 Hz. The average dose in the reactor was calculated from the dose distribution determined using a CTA film dosimeter (FTR-125, Fuji, Japan).

The gas mixture composition was determined at the inlet and outlet of the process vessel. These data were used for determination of the removal efficiency of benzene. Two independent heated sampling lines were installed. At the beginning of both lines heated (120°C) pre-weighed teflon filters (0.5 μm pore size) were installed for the collection of aerosols. After filtration, sample gas was transported to one of three different analytical setups. The main setup consisting of: a condenser, gas-pipette and gas adsorption tubes (two tubes of XAD-2 resin and one tube of activated carbon) was designed for gas analysis on a GC/MS system. The second system consisting of specific indicator tube and piston pump was designed for the approximate determination of the benzene concentration in the sample gas. The analyses of gaseous inorganic components of sample gas, like CO, NO, NO_2, CO_2 and O_2, were performed by flue gas analyzer (Lancom Series II, Land Combustion Ltd.).

The removal efficiency of benzene depends of the applied dose and its initial concentration. Fig. 5 shows experimental results.

Removal efficiencies of benzene increase as its initial concentration decreases and the irradiation doses

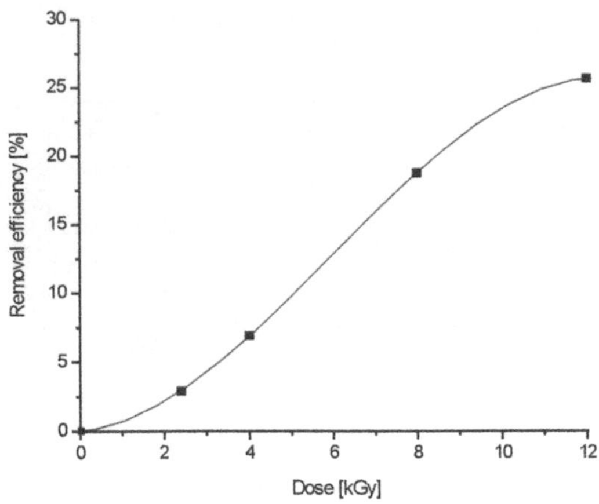

Figure 5. Removal efficiency of benzene. Initial benzene concentration of 105 ppm. Gas temperature in the process vessel was 100°C.

Table 1. Identified products of benzene decomposition.

Product	Identification technique
CO	Gas analyzer Land
Phenol	GC-FID
p-Benzoquinone	GC/MS
Oxepin	GC/MS
Benzaldehyde	GC/MS
Nitrobenzene	GC/MS and GC-FID
Biphenyl	GC/MS

increase. After electron beam irradiation, organic and inorganic gaseous products were formed. Table 1 presents the identified products. CO, CO_2, phenol, biphenyl and aerosols particles were found as products and were determined qualitatively. The experimental result showed that 30% of decomposed benzene was converted into aerosols.

The addition of ammonia to the gaseous mixture before its irradiation increases the removal efficiencies of benzene. Addition of a stoichiometric amount of NH_3 results in about 15% increase of benzene decomposition. These are the initial results and the complete studies with analysis of by-products formed in this case will be performed.

4 CONCLUSIONS

A laboratory plant for electron beam treatment of VOCs in a flow system was built in the INCT. Different analytical methods and monitoring systems were selected for analysis of sample gases before and after irradiation. Initial tests with electron-beam decomposition of vaporized benzene in air were carried out. Major products of the benzene decomposition were identified using a GC/MS system. Addition of ammonia to the gaseous mixture containing benzene before its irradiation increases its removal efficiency. Addition of a stoichiometric amount of NH_3 results in ca. 15% increase in benzene decomposition.

REFERENCES

Ciccioli, P. 1995. VOCs and air pollution. In Bloemen, H.J.T. Burn, J. (eds), *Chemistry and analysis of volatile organic compounds in the environment:* 92–174. London: Blackie Academic and Professional.

Derwent, R.G. Jenkin, M.E. Saunders, S.N. 1996. Photochemical ozone creation potentials for a large number of reactive hydrocarbons under European conditions. *Atmos. Environ.* 30: 181–199.

Hashimoto, S., Hakoda, T., Hirota, K. & Arai, H. 2000. Low energy electron beam treatment of VOCs. *Radiat. Phys. Chem.* 57: 485–488.

Hirota, K., Hakoda, T., Arai, H. & Hashimoto, S. 2002. Electron-beam decomposition of vaporized VOCs in air. *Radiat. Phys. Chem.* 65: 415–421.

Kim, Jo-Chun. 2002. Factors affecting aromatic VOC removal by electron beam treatment. *Radiat. Phys. Chem.* 65: 429–435.

Paur, H.R., Matzing, H. & Woletz, K. 1991. Removal of volatile organic compounds from industrial off-gas by irradiation induced aerosol formation. *J. Aerosol Sci.* 22: 509–512.

Paur, H.R. 1997. Decomposition of volatile organic compounds and polycyclic aromatic hydrocarbons in industrial off – gas by electron beams: A review. *Radiation technology for conservation of the environment, Proceedings of IAEA symposium held in Zakopane, Poland, 8–12 September* 1997: 67–86.

Environmental Engineering – Pawłowski, Dudzińska & Pawłowski (eds)
© 2007 Taylor & Francis Group, London, ISBN13 978-0-415-40818-9

Effect of alternative fuels co-incineration on NO_X emission from cement production

Jerzy Walaszek
Cement Plant "Ożarów" S.A., Ożarów, Poland

Marzenna R. Dudzinska
Institute of Environmental Protection Engineering, Lublin University of Technology, Nadbystrzycka, Lublin, Poland

ABSTRACT: Cement production is an energy-consuming process, therefore considerable effort to replace conventional fuels with alternative ones has been undertaken. Co-incineration of waste fuels is one such attempt. Substitution of fuels by organic wastes is profitable for the cement industry; however waste incineration is connected with high emissions of acid gases and organic substances. The influence of an alternative fuel (shredded used-tires) as an additive on the NO_X emission from a working cement plant installation in Ożarów, Poland, has been investigated. A decrease in NO_X emissions was found under technological conditions.

Keywords: Cement production, waste co-incineration, NO_X emission, emission reduction

1 INTRODUCTION

Cement belongs to the most valuable materials produced by man but at the same time it is one of the most material-consuming products. Fuels and raw materials traditionally used in the ordinary Portland cement manufacturing process are usually fossil fuels and minerals being mined and processed in locations close to the cement plant. Cement manufacturing requires huge quantities of energy to maintain high temperatures in cement kilns. Cement kilns are usually fired with non-renewable solid fuels. Exhaust gases contain many compounds harmful to the environment, such as nitrogen and sulphur oxides, polychlorinated dibenzo-p-dioxins and furans, as well as heavy metals (Dudzińska, Kozak, 2000). Cement production is responsible for 5% of the total annual carbon dioxide (CO_2) emission worldwide, but only 40% of this emission is formed as a product of solid fuel combustion; the remaining 60% comes directly from the chemical decarbonisation of the raw lime stone (raw material).

Forced by law requirements and economic incentives, the cement industry continually seeks new solutions enabling the reduced consumption of natural raw materials as well as minimising emissions (Walaszek, 2004). In 2002, the worldwide cement industry committed itself both to reduce carbon dioxide emissions and to support activities undertaken to stop climate change. Such a target can be achieved in two ways: the first of them consists in reducing fuel consumption and the second one – in reducing the clinker-to-cement ratio (Walaszek, 2003).

In recent years, fossil fuels and some raw materials used in the cement industry have been substituted by waste materials coming from agricultural and industrial processes. Used shredded tyres, solvents and sludge can be utilised as alternative energy sources instead of the solid fuels used for kiln firing. Substitution of fossil fuels by alternative ones is considered as an "environmentally friendly" method.

Public anxiety – apart from that connected with the possibility of leaching toxic substances from the final product – is related to those materials that can give higher emission levels than in the case of conventional fuel combustion. Considerable research has been carried out to check the possibility of the formation of dioxin and other hazardous substances as well as to the examination of combustion of wastes in the regular incinerators. Therefore, co-incineration of alternative fuels in cement production processes might result in increased emissions of other inorganic gases. Nitric oxides, partly responsible for acid rains, belong to such compounds.

1.1 *NO_X formation during clinker production*

Coal combustion is the source of heat energy delivered to the kiln system in the majority of cement plants in Poland and other countries. Despite alkaline conditions inside the cement kiln, nitric oxides, generally named "NO_X", as well as sulphur dioxide SO_2 are formed during the process, because of the natural occurrence of nitrogen and sulphur in fossil fuels. Up to 90% of nitric oxides formed during combustion of fuel in the kiln is nitrogen oxide (NO) and the remaining 5–10% is nitrogen dioxide (NO_2) (Nielsen and Jepsen, 1990).

Depending on the mechanism of NO_X formation during the clinker burning process we can distinguish three types of nitric oxides (Walaszek, 2002):

- thermal NO_X;
- instantaneous NO_X;
- fuel NO_X.

Thermal NO_X is formed when the temperature exceeds 1,200°C, and the reaction rate increase with the temperature increase. Crucial factors determining thermal NO_X formation are: the amount of oxygen delivered to the combustion process, combustion temperature, and the residence time in the high-temperature oxidising zone of the kiln. These reactions occur especially in the burner zone of the cement kiln.

Instantaneous NO_X is formed as a result of reactions between atmospheric nitrogen and free radicals derived from fuel. During the clinker burning process, instantaneous nitrogen oxides are formed in small quantities, thus their impact on the general process of NO_X formation can be neglected (their share in the total NO_X could be neglected).

Fuel NO_X is formed from the nitrogen compounds contained in fuel. The amount of these oxides does not depend on temperature to the same extent as thermal NO_X. Fuel NO_X has the largest share in the total formation of NO_X in the existing system of calciners installed in the clinker burning installation.

The reduction of the NO_X concentration in the emissions might be achieved by: improvement of the clinkering ability of the raw material feed, use of multiple-channel burners, combustion of fuel at the kiln inlet, injection of ammonia or carbamide into the cyclone preheater, use of special calciners, and automatic control of kiln operation (Kurdowski, 2004).

Multiple-channel burners that keep fuel in the flame axis reduce NO_X-emission by up to 20%. Keeping fuel in the flame axis also prevents it from entering the outer zone containing a high oxygen levels. At the same time multiple-channel burners do not allow secondary air to reach the flame core. Consequently, the predominant reaction taking place in the core is incomplete combustion, the main product of which is CO that in turn induces decomposition of the NO_X being formed (Enders, 1992). The multiple-channel burners also allow high flame stability when the concentration of primary air is low (ca. 4%–6%). A landmark in the technology of NO_X-emission reduction was the application of calciners with a reducing zone which decreases the concentration of NO_X in exhaust gas. Some well-known companies like – e.g. Kobe Steel and Nihon Cement Co. (Fujisawa and Matsuda, 1990) and FL Smidth (Tomsen and Jensen, 1998) have installed such calciners.

The content of NO_X in gas leaving the calciner depends on the following factors:

- temperature in the calciner,
- fuel grade (nitrogen content),
- concentration of NO_X brought into the calciner,
- quantity of oxygen consumed in combustion process.

The reduction in NO_X-emission achieved as a result of re-combustion in the calciner is mainly an effect of chain reactions involving gases formed in coal combustion and a result of catalytic effect of raw feed powder as well.

2 MATERIALS AND METHODS

The influence of co-incineration of coal and used shredded tyres as an alternative fuel on NO_X emission from the clinker production process has been examined at an operational kiln in the Ożarów Cement Plant in Poland. The kiln has a two-stream, four-stage cyclone preheater, calciner and a grate cooler of Colax type. The length of the kiln tube is 99 m with diameter 5.75 m. The maximal rotational speed is 3.5 rpm and the kiln capacity is 8000 tpd. The calciner lay-out is presented in Figure 1.

Calciner (A) is fired with coal dust, the dimensions of the reduction zone (E) are: $\sim 5.0 \times 3.5$ m. All the fuel is fed to the calciner's (E) reduction zone through a coal-fired burner (C); the raw meal is divided (point B) into two streams, about 70% of the raw meal is delivered directly into the calciner and the remaining 30% comes into the riser duct. Oxygen is fed to the reduction zone together with hot kiln gas (about 3–4% O_2), so the existing conditions are really very favourable for reduction of NO_X concentration (Walaszek and Kosciolek, 2000).

During normal operation the feed is delivered to the kiln system at its highest point using an elevator of the cyclone preheater. Raw meal flows through the preheater in the direction opposite to the flow direction of exhaust gas leaving the kiln and heating up the meal. After it leaves the third stage of the preheater, raw meal is divided into two streams and one stream of the meal comes into the riser duct running from the kiln to the calciner and the other one is fed to the upper part of the calciner. Fuel is delivered to the calciner's lower part. In this way it is possible to avoid formation of build-up in the riser duct and – at the same time – to create a reducing atmosphere in the lower part of the

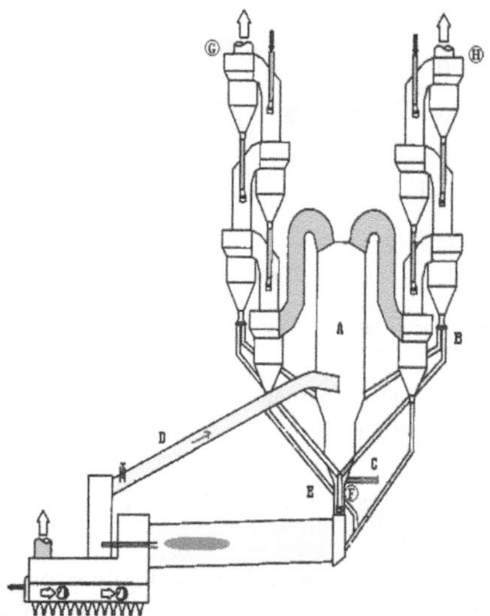

Figure 1. Calciner in Ożarów cement plant. A – calciner, B – distribution of raw meal being fed to the calciner, C – feeding with coal, D – combustion air coming from grate cooler (used in coal combustion process), E – calciner's reduction zone, F – NO_X measurement at the kiln inlet, G, H – NO_X measurement downstream A- and B-stream cyclones, I, J – NO_X measurement, stacks nos. 1 and 2.

calciner, which in consequence maintains low NO_X concentration in the exhaust gas. Coal is added to the lower part of the calciner and combustion air – (aka "tertiary air") – is supplied from the grate cooler. Raw meal comes into the kiln as a calcined material, its calcination ratio is 90–95%.

Due to the relatively low content of NO_2 in the combustion gas leaving the stack only NO is monitored. Monitoring results are shown in ppm (parts per million) of NO and can be related to the dry or wet form. Although only the NO concentration is measured, practically all cement plants have adopted a protocol according to which measurement results shown in ppm are multiplied by a factor of 2.05 and then presented in mg NO_2/Nm^3 (in wet or dry form).

2.1 *Test run*

In 2005 in the Ożarow Cement Plant some tests were carried out in order to examine the real influence of alternative fuels on emission levels. Emission to the atmosphere was measured at points I and J and NO_X concentrations in the installation were measured at points F, G and H (see Fig. 1.). The results of these measurements were compared with those obtained when the installation was fired with conventional fuels. Used tyres shredded into irregular pieces of ca. 70 mm thickness were used as alternative fuel. Alternative

Table 1. Alternative fuel composition.

Condition	Total moisture %	Ash %	Net calorific value kJ/kg	Total sulphur %
Ready-to-use	0.87	20.71	26.568	1.51

Table 2. Fuel consumption balance.

Basic run (Reference run)				
Conventional fuel		Alternative fuel	Portion of alternative fuel %	
The main burner t/h	Calciner t/h	Calciner t/h		
1.1	16.9	–	0	
Test run				
Conventional fuel		Alternative fuel fuel	Portion of alternative fuel %	
The main burner t/h	Calciner t/h	Calciner t/h	By weight	energy
11.0	13.7	3.7	13.0	14.2

fuel, fed using an autonomous proportioning system, was supplied to the calciner close to the existing conventional fuel burners.

The technical specification of the alternative fuel is presented in Table 1, while in Table 2 the fuel consumption balance is presented.

The percentage of alternative fuel by weight was calculated according to the following formula:

$$\textit{Percentage of alternative fuel} = C_z^{*}100/(C_k + C_z)$$

where:
C_k – quantity of alternative fuel;
C_z – quantity of conventional fuel.

The percentage of alternative fuel used as a substitute of process energy was calculated according to the formula:

$$\textit{Percentage of alternative fuel} = Q_z^{*}100/(Q_z + Q_k)$$

where:
Q_z – heat from combustion of alternative fuel;
Q_k – heat from combustion of conventional fuel.

The portion of alternative fuel used during the tests was 13% of fuel blend (by weight) and 14.2% as a substitute of energy essential for clinker burning process.

Table 3. NO_X emission.

	I K2 mg/Nm³	J K2 mg/Nm³	Average mg/Nm³	Emission standards mg/Nm³
Test	460	380	420	800
Background	585	455	520	

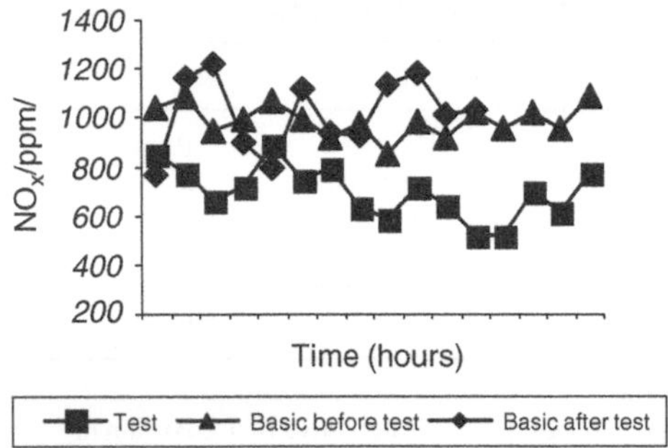

Figure 2. NO_X concentration at the kiln inlet.

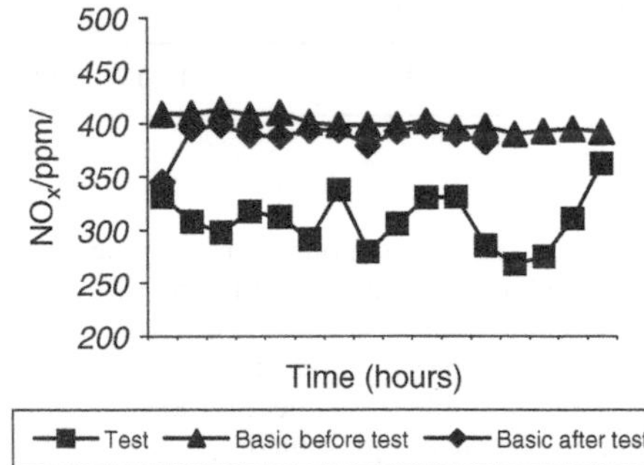

Figure 3. NO_X concentration downstream the kiln installation – A stream.

3 RESULTS AND DISCUSSION

NO_X concentrations were measured by a regular continuous monitoring system, available in the Ożarów Cement Plant, due to regulatory demands. All results were standardized to an oxygen concentration of 10%. Comparison of results of NO_X emission measurements taken at points I and J with emission standards is presented in Table 3. Concentrations of NO_X in the inlet (point F) and outlet (points G and H) of installation vs. time are presented in Figures 2–4.

According to Table 3 and Figures 2–4, an average decrease in NO_X concentration in the kiln system was observed at the following measuring points:

- F (kiln inlet): from 2,000 to 1,450 mg/Nm^3;
- G (downstream the installation): from 810 to 670 mg/Nm^3;

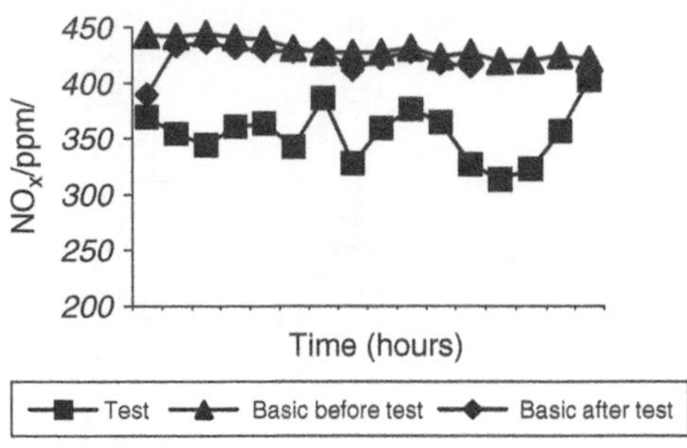

Figure 4. NO_X concentration downstream the kiln installation – B stream.

- H (downstream the installation): from 870 to 750 mg/Nm^3.

which resulted also in a decrease in emission to the atmosphere from 520 to 420 mg/Nm^3 on average. Such an emission level is definitely lower then the one imposed by emission standard and amounting to 800 mg/Nm^3 for processes of co-firing cement kilns with combustible wastes.

4 CONCLUSIONS

Due to both savings on raw materials that are energy sources and the potential of using the ever-growing mass of wastes, co-incineration of waste fuels and alternative fuels is becoming increasingly popular and will be applied more and more widely, as it allows savings in raw materials and energy, as well as to the utilisation of the growing amount of waste in the same processes. In a country such as Poland, where waste incineration is not popular and investment in new incinerators meets strong opposition, this method of waste removal seems particularly interesting.

Utilisation of wastes in cement kilns is profitable, not only from the economic point of view, but also brings about a reduction in the emission of some environmentally harmful gases. As shown earlier (*Dudzińska and Kozak, 2000*), this also applies to some waste materials enabling reduction in dioxin emissions. The results of tests carried out in the Ożarów Cement Plant have confirmed that combustion of alternative fuels can also bring about a decrease in NO_X emission. This finding is of great importance, because – as opposed to sulphur dioxide – decreasing the emission of nitric oxides being formed during combustion of conventional fuels requires large expenditure and is very difficult from the technological point of view. Combustion tests have revealed that independently of the nature of reducing properties of a calciner, further positive effects can also be achieved by means of co-firing with alternative fuels. The results obtained require – in order to confirm them – further tests using various alternative fuels.

REFERENCES

Dudzińska, M. R & Kozak, Z. 2000. PCDD/Fs Emission From Waste Utilisation During Clinker Production. In: *Thermal Solid Waste Utilisation in Regular and Industrial Facilities, Environmental Science Research*, vol. 58, Plenum-Kluwer, N. York, 41–50.

Enders, G. 1992, Improved Technologies for the Rational Use of Energy in the Cement Industry, Berlin, 327.

Fujisawa, T. & Matsuda, Y. New Suspension Preheater for Cement Burning – Dual Combustion and Denitration (DD) Process, Kolbeco Technical Bulletin, 1011.

Kurdowski, W. 2004, Methods for Reducing Emission of NO_XComing from Cement Kilns, *Cement – Lime – Concrete*, 148–151.

Nielsen, P. & Jepsen, O. 1990. Formation of SO_x and NO_X in Various Systems Using Thermal Processes, *World Cement* 12/1990: 528–537.

Tomsen, K. & Jensen, L.S. 1998. Fuller ILC – Low NO_X Calciner Commissioning and Operation at Lone Star St. Cruz in California, Zement Kalk Gypsum International, 10/1998: 542–550.

Walaszek, J. & Kościołek, R. 2000, The Biggest Kiln of Output of 8,000 tpd in Ożarów Cement Plant., *Cement – Lime – Concrete* 3/2000: 92–97.

Walaszek, J. 2002, Potential for Limitation of Nitric Oxides Emission in a Calciner-equipped Kiln, in the: *The 1st Environmental Engineering Congress – Reference materials*, Monographs of Polish Committee of Environmental Engineering of Polish Academy of Science, vol. 11: 213–230.

Walaszek, J. 2003, Use of Alternative Fuels in Cement Manufacturing Process, *Raw Materials and Construction Equipment,* BMP 4/2003: 17–19.

Walaszek, J. 2004, Some Aspects of the Need for Climate Protection, *Raw materials and construction equipment* BMP 1/2004:12–15.

Environmental Engineering – Pawłowski, Dudzińska & Pawłowski (eds)
© 2007 Taylor & Francis Group, London, ISBN13 978-0-415-40818-9

Activity of granular and monolithic catalysts during oxidation of selected organic chlorine compounds

Marek Kaźmierczak, Zbigniew Gorzka, Andrzej Żarczyński, Tadeusz Paryjczak & Marcin Zaborowski
Institute of General and Ecological Chemistry, Technical University of Lodz, Lodz

ABSTRACT: 1,1,2,2-tetrachloroethane (TChE) and propylene chlorohydrin (CHP) were oxidized in the temperature range from 300 to 600°C. The activities of three catalysts in the reactions: a granular one containing platinum on TiO_2-SiO_2 (Pel.Pt/Ti) and two monolithic catalysts, were compared. One of the monolithic catalysts contained platinum and rhodium on γ-Al_2O_3 (Mon.Pt-Rh) as the active components. Platinum on TiO_2 (Mon.Pt/Ti) was an active component in the case of the second monolithic catalyst. The combustion gases contained formaldehyde, chlorine, 2,3,7,8-TCDF and other 2,3,7,8-substituted PCDD/Fs congeners in trace concentrations. It was found that the oxidation of the substrates requires a temperature of ca 500°C (TChE, Pel.Pt/Ti), 550°C (TChE, Mon.Pt-Rh) and 400°C (CHP, Pel.Pt/Ti and Mon.Pt-Rh). TChE and CHP oxidation using Mon.Pt/Ti should be carried out at 650 and 550°C, respectively. The results of the investigation prove the high activity and resistance to deactivation of the applied catalysts.

Keywords: Thermocatalytic oxidation, 1,1,2,2-tetrachloroethane, propylene chlorohydrin, dioxins.

1 INTRODUCTION

Dioxin abatement methods are very important in thermal treatment of wastes containing chlorinated organic compounds (Dudzińska and Kozak 2001, Babushok and Tsang 2003, Everaert and Baeyens 2003, Tuppurainen et al. 2003, Quaß et al. 2004, Finocchio et al. 2006). However, the process needs to be optimized in order to reduce the concentration of these contaminants. The application of catalysts can be useful for decreasing dioxin emission as well as inducing a significant reduction of temperature during the treatment of organochlorine compounds (Kułażyński et al. 2002, Musialik-Piotrowska and Syczewska 2002). Moreover, the application of catalysts results in a decrease of apparatus and running costs in comparison with non-catalytic combustion which is generally carried out at temperatures >1100°C (European Parliament and the Council Directive 2000). Combustion gases clean-up from toxic chloro-organic compounds can be carried out using catalysts with granular or monolithic constitution (Agarval et al. 1992, Cybulski and Moulijn 1994, Lester 1999, Jeong et al. 2003).

The aim of the investigation was the selection of platinum catalysts suitable for an efficient destroying and environmentally safe process and applicable to both liquid and gaseous chloroorganic wastes in the temperature range from 300–600°C.

2 MATERIALS AND METHODS

1,1,2,2-tetrachloroethane (TChE) or propylene chlorohydrin (CHP) were selected as model compounds for the study. These compounds have significantly various chemical features and are important components of industrial chloro-organic wastes. The inlet mixture consisted of TChE (900 mg/m^3) or CHP (360 mg/m^3) and was dosed with water vapour (32,5 g/h) and air (215 l/h) to a tubular contact reactor (21 mm diameter) made of quartz and electrically heated. The activity of three catalysts at contact time 0.36 s was investigated:

- granular (Pel.Pt/Ti) containing 0.12% Pt (in relation to catalyst mass) on a carrier of titanium and silicon dioxides, produced by the Institute of Chemistry and Technology of Petroleum and Coal, Wrocław University of Technology,
- monolithic (Mon.Pt-Rh) with an intermediate layer of γ-Al_2O_3 containing 0.09% Pt and 0.04% Rh (in relation to catalyst mass) on a cordierite carrier (400 cells per square inch), produced by JMJ

Manufacturer of Catalysts, Puchalski an Krawczyk Company in Kalisz,

– monolithic (Mon.Pt/Ti) with an intermediate layer of TiO_2 containing 0.12% Pt (0.0028% Pt in relation to catalyst mass) on a cordierite carrier (200 cells per square inch), produced by "Blachownia" Institute of Heavy Organic Synthesis in Kędzierzyn Koźle.

After cooling down, the products of reaction and unreacted substances were determined using chromatographic and colorimetric methods, and automatic analysers. Concentrations of chlorine, formaldehyde, carbon monoxide and PCDD/Fs were analysed in the gaseous reaction products. Concentrations of chloride ions, formaldehyde and total organic carbon (TOC) were determined in liquid products. A Hewlett-Packard HP 6890 gas chromatograph with flame-ionization detector, a Knauer liquid chromatograph with spectro-photometric detector, a Madur GA-20 automatic gas analysers and a Shimadzu TOC-5050A automatic analyser were used in the analyses. Analyses of combustion gas samples for the content of PCDD/Fs were carried out in the Pulp and Paper Research Institute of Łódź, according to the obligatory standards (European Standard EN-1948).

3 RESULTS AND DISCUSSION

Results of the experiments are presented in Figs. 1–6 and Tables 1–3.

The effect of temperature and type of catalyst on TChE conversion is presented in Fig. 1 and Table 1. The conversion of TChE higher than 99% proceeds at 450°C in the presence of Pel.Pt/Ti catalyst, 550°C for Mon.Pt-Rh catalyst and at 650°C in the presence of Mon.Pt/Ti catalyst (Żarczyński et al. 2005).

The CHP conversion (Fig. 2, Table 2) was >99% at 375°C using the Pel.Pt/Ti catalyst, at 400°C in the presence of Mon.Pt-Rh catalyst (Kaźmierczak et al. 2005) and at 550°C in the presence of Mon.Pt/Ti catalyst. CHP is oxidised easier than TChE.

The oxidation of chloro-organic compounds always results in the formation of hydrogen chloride. Hydrogen chloride is separated from gaseous reaction products during their cooling and collected as dilute hydrochloric acid in the condensate (Tables 1 and 2). Under the conditions of oxidation process, the concentration of hydrochloric acid was about 7.0 g/l during TChE oxidation and 1.0 g/l during CHP oxidation (Kaźmierczak et al. 2005, Żarczyński et al. 2005).

The variation of formaldehyde concentration will be discussed separately. Its content in the condensate and combustion gases is presented in Figs. 3 and 4, respectively. For the TChE oxidation process, its concentration in the condensate from oxidations carried out in the presence of granular catalyst was 4.35 mg/l

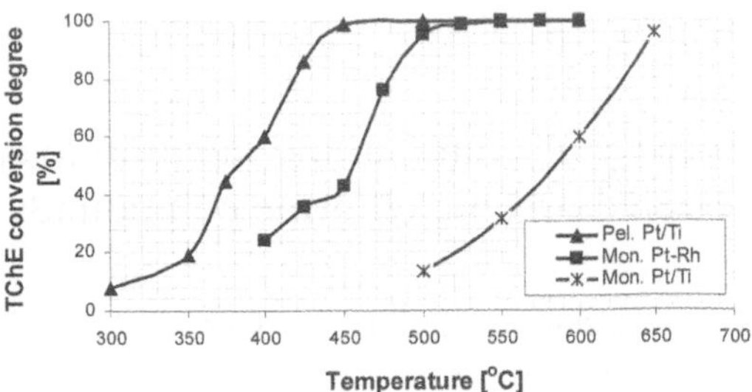

Figure 1. Dependence of TChE conversion degree on temperature and catalyst type.

Table 1. Concentrations of intermediate and final products in the condensate and combustion gases during TChE oxidation using catalysts.

	Catalyst; temperature		
Reaction products	Pel. Pt/Ti 450°C	Mon. Pt-Rh 550°C	Mon. Pt/Ti 550°C
Formaldehyde in a condensate, mg/l	0.89	0	0.83
Formaldehyde in combustion gases, mg/m^3	0	0	0
HCl in a condensate, mg/l	5400	7011	1686
Chlorine, mg/m^3	7.2	2.39	0.36
Carbon monoxide, mg/m^3	0	0	0
TChE conversion degree, %	99.1	99.1	31.8

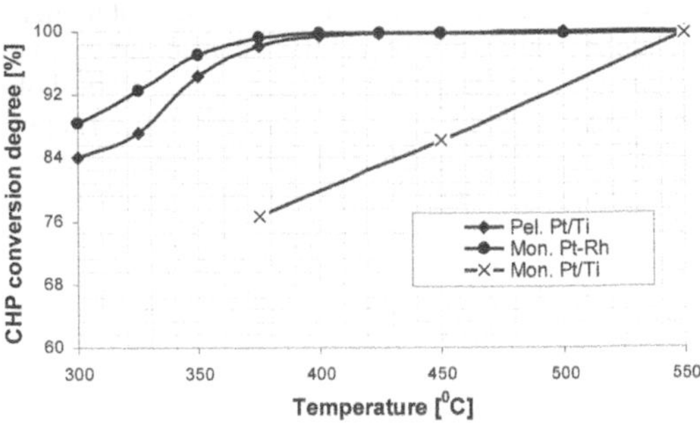

Figure 2. Dependence of CHP conversion on temperature and catalyst type (calculated as the loss of total organic carbon (X_{OWO})).

at 400°C and next decreased to 0.89 mg/l at 450°C and reach 0 mg/l in 500°C (Fig. 3, Table 1). In the reaction products of oxidations with Mon.Pt-Rh catalyst, formaldehyde did not occur (Żarczyński et al. 2005).

As easily seen from Fig. 3 and Table 2, for the process of CHP oxidation, the formaldehyde concentration in the condensate was 105.4 mg/l in 325°C in the presence of Pel.Pt/Ti catalyst and

Table 2. Concentrations of intermediate and final products in the condensate and combustion gases during CHP oxidation using catalysts.

Reaction products	Catalyst; temperature	
	Pel. Pt/Ti; 375°C	Mon. Pt-Rh; 375°C
Formaldehyde in a condensate, mg/l	89.93	27.9
Formaldehyde in combustion gases, mg/m^3	1.5	0.60
HCl in a condensate, mg/l	1139	849
Chlorine, mg/m^3	0.15	0.16
Carbon monoxide, mg/m^3	–	1–4
CHP in combustion gases, mg/m^3	0	0
CHP conversion degree (X_{owo}), %	99.2	98.7

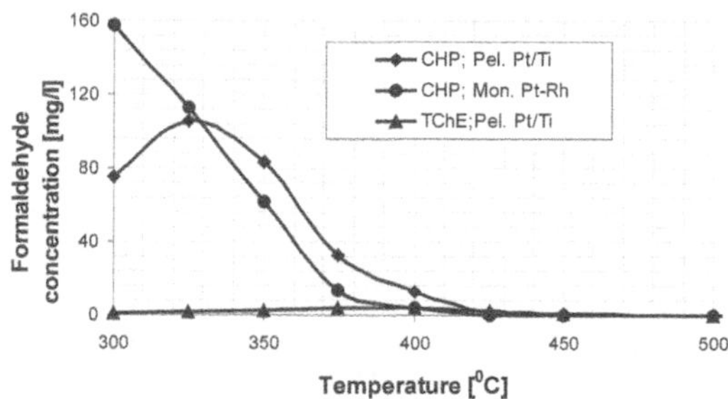

Figure 3. Dependence of formaldehyde concentration in condensate on temperature of CHP and TChE oxidation and catalyst type.

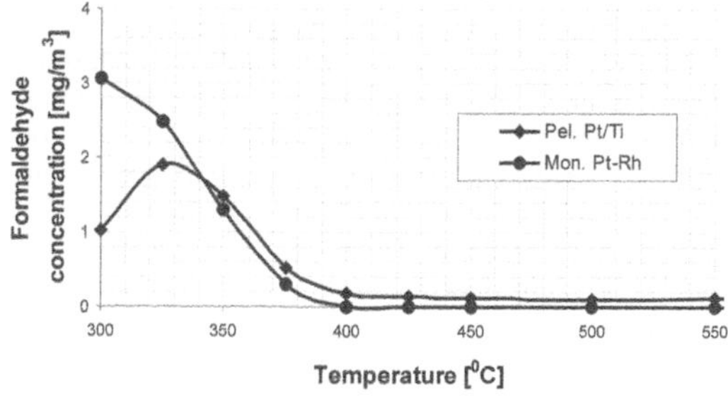

Figure 4. Dependence of formaldehyde concentration in combustion gases on temperature of CHP oxidation and catalyst type.

157.5 mg/l in 300°C in the presence of Mon.Pt-Rh catalyst (Kaźmierczak et al. 2005).

The combustion gases from the TChE oxidation did not contain formaldehyde if a catalysts was applied (Table 1). In the gaseous products of the CHP oxidation (Fig. 4, Table 2), the formaldehyde concentration was 1.89 mg/m^3 at 325°C in the presence of Pel.Pt/Ti

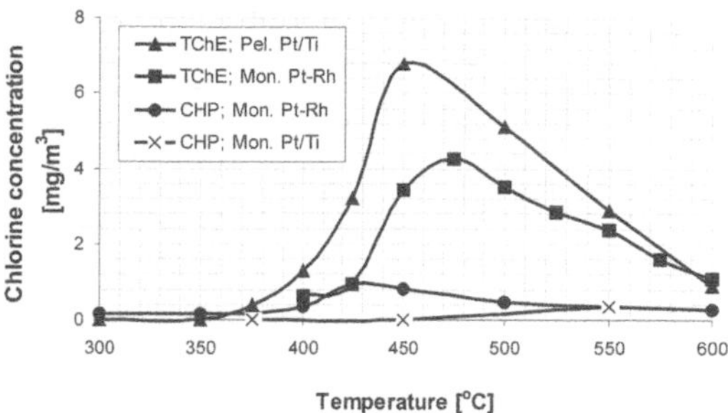

Figure 5. Dependence of chlorine concentration in combustion gases on temperature of CHP and TChE oxidation and catalyst type.

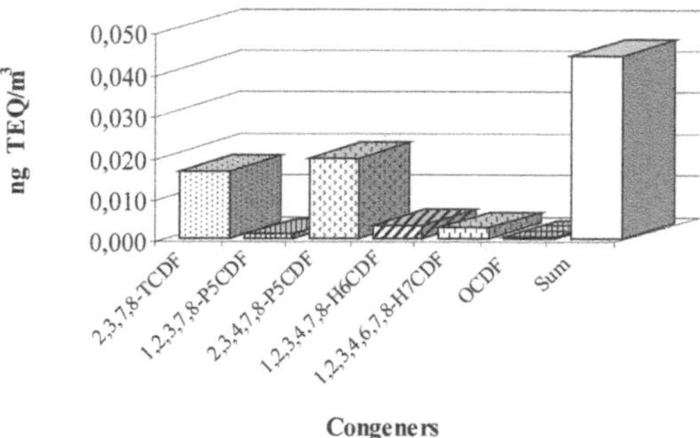

Figure 6. The content of dioxins in combustion gases from the CHP oxidation process using Mon.Pt-Rh catalyst at 375°C.

catalyst and 3.05 mg/m^3 in 300°C in the presence of Mon. Pt-Rh catalyst.

Changes in the concentration of formaldehyde presented in Figs. 3 and 4 are typical for intermediate reaction products.

Figure 5 presents the dependence of chlorine concentration in the combustion gases on the reaction temperature during TChE and CHP oxidation (also Tables 1 and 2). In the combustion gases, chlorine appeared although water vapour was present in the all tested temperature ranges. During the TChE oxidation, the chlorine concentration increased from 0 mg/m^3 at 300°C to the maximum value of 7.2 mg/m^3 at 450°C (Pel.Pt/Ti) and from 0.62 mg/m^3 at 400°C to 4.3 mg/m^3 at 475°C (Mon.Pt-Rh). Further temperature increase resulted in a decrease of chlorine concentration. During the CHP oxidation, the chlorine concentration increased from 0.15 mg/m^3 at 300°C to 0.93 mg/m^3 at 425°C (Pel.Pt/Ti) and decreased to 0.27 mg/m^3 at 600°C. In the CHP oxidation using Mon. Pt/Ti, the presence of chlorine (0.34 mg/m^3) was found only at 550°C.

The combustion gases were also analysed to determine their PCDD/Fs contents. Analytical results are presented in Fig. 6 and Table 3. The combustion

Table 3. Results of PCDD/Fs determination in combustion gases sampled during CHP and TChE oxidation using catalysts (Kaźmierczak et al. 2004, Żarczyński et al. 2005).

Oxidized compound; catalyst; temperature	2,3,7,8-PCDDs congeners Name	2,3,7,8-PCDFs congeners Name	PCDD/Fs total concentration ($ngTEQ/m^3$)
TChE; Pel. Pt/Ti; 450°C	1,2,3,4,7,8-H_6CDD 1,2,3,6,7,8-H_6CDD 1,2,3,7,8,9-H_6CDD 1,2,3,4,6,7,8-H_7CDD	1,2,3,7,8-P_5CDF 2,3,4,7,8-P_5CDF 1,2,3,4,7,8-H_6CDF 1,2,3,6,7,8-H_6CDF 1,2,3,7,8,9-H_6CDF 1,2,3,4,6,7,8-H_7CDF 1,2,3,4,7,8,9-H_7CDF	0.088
TChE; Mon. Pt-Rh; 550°C	Under detection limit 0,002 $ngTEQ/m^3$	2,3,7,8-TCDF	0.0035
TChE; Mon. Pt/Ti; 550°C	Under detection limit 0,002 $ngTEQ/m^3$	1,2,3,4,6,7,8-H_7CDF	0.0031
CHP; Pel. Pt/Ti; 375°C	1,2,3,7,8-P_5CDD 1,2,3,7,8,9-H_6CDD 1,2,3,4,6,7,8-H_7CDD	2,3,7,8-TCDF 1,2,3,4,7,8-H_6CDF 1,2,3,6,7,8-H_6CDF 1,2,3,7,8,9-H_6CDF 2,3,4,6,7,8-H_6CDF 1,2,3,4,6,7,8-H_7CDF 1,2,3,4,7,8,9-H_7CDF OCDF	0.034

gases from the TChE oxidation using catalysts (granular and monolithics) contained several PCDD/Fs congeners determined according to the European Standard EN-1948. Congeners from groups: P_5CDFs, H_6CDDs and H_7CDFs, were especially numerous. In some experiments 2,3,7,8-TCDF was observed but 2,3,7,8-TCDD was not found.

The limits of detection (LD) for particular PCDD/Fs groups were as follows:

TCDDs, TCDFs – above 0.01 ng/sample; P_5CDFs – above 0.01 ng/sample; H_6CDDs, H_6CDFs – above 0.01 ng/sample; H_7CDDs, H_7CDFs – above 0.02 ng/sample; OCDD, OCDF – above 0.02 ng/sample.

Dioxin concentrations in the combustion gases sample obtained during the TChE oxidation were: 0.088 $ngTEQ/m^3$ at 450°C (Pel. Pt/Ti), 0.0035 $ngTEQ/m^3$ (Mon. Pt-Rh) at 550°C and 0.0031 $ngTEQ/m^3$ (Mon. Pt/Ti) in 550°C. In the CHP oxidations using granular and monolithic catalysts, the dioxin concentrations at 375°C were 0.034 $ngTEQ/m^3$(Pel.Pt/Ti) and 0.044 $ngTEQ/m^3$ (Mon. Pt-Rh). These values were lower than the permissible value (0.1 $ngTEQ/m^3$).

4 CONCLUSIONS

The results of TChE and CHP oxidation prove the high activity and resistance to deactivation of the applied catalysts, however Mon.Pt/Ti catalyst appeared to be less active than the others. This catalyst contained only 0.0028% Pt and a lower number of cells per surface area unit than the Mon.Pt-Rh catalyst.

Comparison of dependences of substrate conversions on the temperature shows that slightly lower temperatures can be employed with the Pel.Pt/Ti catalyst than with the Mon.Pt-Rh one.

Oxidation of CHP was easier with all catalysts, however, at temperatures $<450°C$ high concentrations of formaldehyde were found in the reaction products. The products of TChE oxidation did not contain formaldehyde when the Mon.Pt-Rh catalyst was applied. Trace amounts were found if Pel.Pt/Ti and Mon.Pt/Ti catalysts were used.

The gaseous products of the TChE oxidation contained chlorine in concentrations higher than the products of the CHP oxidation. This substrate's oxidation did not result in 2,3,7,8-TCDD congener formation. However, combustion gases contained trace amounts of 2,3,7,8-TCDF and other PCDD/Fs. The toxicity equivalents of the combustion gas samples were lower than the admissible value of 0.1 $ngTEQ/m^3$.

The results of the investigation show the possibility of a decrease in the temperature of organic chlorine compounds treatment by several hundred °C in comparison with the non-catalytic high combustion temperature (1100–1500°C). The catalytic oxidation of the substrates requires temperatures of

500°C (TChE, Pel.Pt/Ti), 550°C (TChE, Mon.Pt-Rh) and 400°C (CHP, Pel.Pt/Ti and Mon.Pt-Rh). If the Mon.Pt/Ti catalyst is employed, the TChE oxidation can be carried out at 650°C and the CHP oxidation should proceed in 550°C or higher.

ACKNOWLEDGEMENTS

This work was supported by the Polish State Committee for Scientific Research under Grant No. 7 T09B 083 27.

REFERENCES

Agarval S. K., Spivey J. J., Butt J. B., 1992. Catalyst deactivation during deep oxidation of chlorocarbons. Appl. Catal. A: General, 82: 259–275.

Babushok V. I., Tsang W., 2003. Gas-phase mechanism for dioxin formation. Chemosphere, 51: 1023–1029.

Cybulski A., Moulijn J. A., 1994. Monoliths in heterogenous catalysis. Catal. Rev.-Sci. Eng. 36: 179–270.

Dudzińska M. R., Kozak Z., 2001. Polichlorowane dibenzo-p-dioksyny i dibenzofurany – właściwości i oddziaływanie na środowisko. Monografia 6, Komitet Inżynierii Środowiska PAN/Politechnika Lubelska, Lublin, Poland.

European Standard EN-1948, ICS 13.040.40: Stationary source emissions. Determination of the mass concentration of PCDD/Fs.

European Parliament and the Council Directive 2000/76/EC of 4 December 2000 on the incineration of waste OJ L 332, 28.12.2000, p. 91.

Everaert K., Baeyens J., 2003. The formation and emission of dioxins in large scale thermal processes. Chemosphere, 46: 439–448.

Finocchio E., Busca G., Notaro M., 2006. A review of catalytic processes for the destruction of PCDD and PCDF from waste gases. Appl. Catal. B: Environmental, 62: 12–20.

Jeong K.-E., Kim D.-C., Ihm S.-K., 2003. The nature of low temperature deactivation of $CoCr_2O_4$ and $CrOx/\gamma$-AL_2O_3 catalysts for the oxidative decomposition of trichloroethylene. Catal. Today 87: 29–34.

Kaźmierczak M., Gorzka Z., Żarczyński A., Paryjczak T., Zaborowski M., 2005. Aktywność zawierających platynę katalizatorów ziarnistego oraz monolitycznego w reakcjach utleniania tetrachloroetanu i chlorohydryny propylenowej, II Kongres Inżynierii Środowiska, Monografie Komitetu Inżynierii Środowiska PAN, 32: 1137–1145.

Kułażyński M., Van Omen J. G., Trawczyński J., Walendziewski J., 2002. Catalytic combustion of trichloroethylene over TiO_2-SiO_2 supported catalysts. Appl. Catal. B: Environmental, 36: 239–247.

Lester G. R., 1999. Catalytic destruction of hazardous halogenated organic chemicals. Catal. Today, 53: 407–418.

Musialik-Piotrowska A., Syczewska K., 2002. Catalytic oxidation of trichloroethylene in two-component mixtures with selected volatile organic components. Catal. Today 73: 332–342.

Quaß U., Fermann M., Bröker G., 2004. The European Dioxin Air Emission Inventory Project—Final Results. Chemosphere 54: 1319–1327.

Tuppurainen K., Asikainen A., Ruokojarvi P., Ruuskanen J., 2003. Perspectives on the formation of polychlorinated dibenzo-p-dioxins and dibenzofuranes during municipal solid waste (MSW) incineration and other combustion processes. Acc. Chem. Res. 36 (9): 652–658.

Żarczyński A., Gorzka Z., Paryjczak T., Kaźmierczak M., Szczepaniak B., 2005. Dioxins in the process of 1,1,2,2-tetrachloroethane oxidation with the application of monolithic catalysts. Pol. J. Chem. Technol. 7 (2): 100–104.

Żarczyński A., Kaźmierczak M., Gorzka Z., Zaborowski M., 2005. Dioksyny w procesie utleniania 1,1,2, 2-tetrachloroetanu w obecnożci wybranych katalizatorów. Chem. Inż. Ekol. 12 (S1): 113–122.

Environmental Engineering – Pawłowski, Dudzińska & Pawłowski (eds)
© 2007 Taylor & Francis Group, London, ISBN13 978-0-415-40818-9

The methodology of computing source-receptor air pollutant matrices for Integrated Assessment Models

Katarzyna Juda-Rezler
Warsaw University of Technology, Faculty of Environmental Engineering, Institute of Environmental Engineering Systems, ul. Nowowiejska, Warsaw, Poland

ABSTRACT: Polish sulfur dioxide emission to the air is still substantial and, according to the present predictions, will be the highest among EU countries in 2020. This may result in direct and indirect threat to human health and the environment. The current paper discusses the original Integrated Assessment Model for Air Pollution Problems named ROSE (Risk Of airborne Sulphur species on the Environment) developed and implemented specifically for Poland. An optimisation routine of the ROSE model can be used to identify emission control strategies that achieve environmental targets for ambient SO_2 and SO_4^{2-} concentrations as well as for sulfur deposition. For optimization calculations the source-receptor air pollutant matrices (SRMs) are necessary. However, when computing SRMs with the Eulerian grid air pollution model problems connected with non-linear effects appear. Analyses of such effects, the errors created by them as well as the method of calculations applied are presented.

1 INTRODUCTION

In the European Union, the assessment of environmental damage is currently performed according to the scheme D-P-S-I-R, which stands for Driving forces – Pressure – State – Impact – Responses. The environmentalist pressure connected with the emission of air pollutants to the atmosphere is still a serious issue in Poland, especially in relation to particulate matter, carbon dioxide and sulfur dioxide. Both the total emission of these pollutants in the country and the unit emissions are high, especially in comparison with other countries of the European Union. As for the driving forces, a significant role is played by the public power generation sector. According to the estimations by the Co-operative European Programme for Monitoring and Evaluation of the Long-range Transmission of Air Pollutants in Europe (EMEP – created within the framework of the Convention on Long-range Transboundary Air Pollution (LRTAP) of 1979), in the year 2020 the emission of SO_2 in Poland will be the highest among the EU countries and will account for 23% of the total emission of the EU (UNECE/EMEP 2004). It is necessary therefore to create and perfect tools for ambient air and ecosystems quality assessment and management. To this end Integrated Assessment Models for Air Pollution Problems (IAMs) are used. Such models provide decision support and facilitate the choice of optimal (within accepted objective functions) abatement strategies.

Among Integrated Assessment Models used in the world – the RAINS model developed at IIASA (International Institute of Applied Systems Analysis) is employed most frequently. In Europe, the RAINS-Europe model has been used in several policy contexts to identify cost-effective allocations of emission reductions to meet environmental policy targets. Optimisation results were used to guide international environmental negotiations on the Second Sulphur Protocol (1994) of the LRTAP Convention (Tuinstra et al. 1999), the Gothenburg Protocol to Abate Acidification, Eutrophication and Ground-level Ozone in 1999 (Amann et al. 1999) and the European Union's NEC (National Emission Ceilings) Directive (Amann & Lutz 2000, Wettestad 2002). Recently, an extension of the RAINS-Asia Integrated Assessment Model for acidification in Asia with an optimisation routine that can be used to identify cost-effective emission control strategies that achieve environmental targets for ambient SO_2 concentrations and sulfur deposition at least costs, has been presented (Carmichael et al. 2002, Cofala et al. 2004).

Based on the highly successful integrated assessment techniques used for transboundary air pollution in Europe and Asia, the Urban Scale Integrated Assessment Model (USIAM) has been developed to assist in urban air quality management. The USIAM model is able to objectively recommend specific emission reduction strategies from the perspective of maximising air quality standard compliance while minimising population exposure and cost of implementation. It has been implemented to examine annual average PM10 concentrations in major urban developments in general, and in the London metropolitan area in particular (Mediavilla-Sahagun & ApSimon 2003, Mediavilla-Sahagun & ApSimon 2006).

In this paper a methodology for the comprehensive assessment of the threat of sulfur species to the environment and human health in Poland and for creating optimum, with regard to environmental criteria, strategies for minimising this threat on the basis of the D-P-S-I-R scheme is presented. The original Integrated Assessment Model named ROSE (*Risk Of airborne Sulphur species on the Environment*) was developed and implemented specifically for Poland. ROSE (in Polish publications named BURZA) is comprised of a suite of models: an Eulerian grid air pollution model; statistical models for assessing environment sensitivity to the sulfur species and an optimisation model with modern evolutionary computation techniques. Environmental sensitivity to the sulfur species is assessed by calculating and mapping critical thresholds (above which damage may occur), i.e. the critical levels for SO_2 and SO_4^{2-} and so called "conditional critical loads" for acidifying sulfur. The exceedances of critical values (Ex) and current-to-critical value ratios (CCR) are employed for assessing environmental risk. The areas at risk from sulfur species are subsequently identified.

In the optimisation problem, the objective function is to minimise sulfur emission subject to the constraints specified for deposition and concentration.

The modelling area assumed for the ROSE model, which measures $900\,\text{km} \times 750\,\text{km}$ comprises the whole territory of Poland, and the applied computational grid (which is tied to the computational grid of the EMEP model) has a spacing $dx = dy = 30\,\text{km}$. For this area, an assessment of the current threat to the Polish natural environment caused by sulfur species has been performed (Juda-Rezler 2004a). The next task for the ROSE system was to generate and analyse scenarios for minimizing the threat. The goal was to minimize the existing risk by optimization of the electric energy production distribution. A multi-criterion objective function was formulated in order to describe the level of environmental risk. Results of the optimization calculus have been presented and discussed in Juda-Rezler (2004b). It was proven that environmental threat can be substantially minimized, while keeping total energy production at a constant level.

This paper concerns the methodology for carrying out calculations of the extent of atmospheric pollution caused by sulfur compounds for a particular scenario for minimising threats.

2 MATERIALS AND METHODS

2.1 *Assumptions about the optimisation process*

The extent of pollution on Polish territory is influenced by Polish sources of emission and sources from the neighbouring countries – first of all from Germany and the Czech Republic. Sources of emission from countries located to the east of Polish borders (above all from the Ukraine) as well as from other important European emitters – especially Great Britain, have a much smaller influence on the pollution in the country. A database on the SO_2 emission and parameters of the emission sources has been prepared for the year 1999 (Juda-Rezler 2004c).

Preparing scenarios for minimising the threat is obviously connected with the possibility to lower the pressure on the environment, and thus to reduce present emissions. Such a process will be termed controlling and sources of emission subject to analysis by the system ROSE will be called controllable sources. It is undisputable that the integrated system developed for the territory of Poland cannot assume the reduction of emission from the neighbouring countries. Therefore it will be assumed that pollution caused by the emission sources from beyond the territory of the country, ("uncontrollable sources"), will make up the so-called pollution background in the ROSE model.

When it comes to Polish sources of emission, theoretically it is possible to control (manage) all sources, for which data on SO_2 emission has been collected. The sources are as follows: emissions of the public power generation sector – electric power stations, thermal-electric power stations and thermal power stations, industrial emissions and municipal emissions. The power generation industry makes up the largest share in the emission of SO_2 in Poland – especially Large Combustion Plants (LCP), the total emission of which amounted to 774 Gg in 1999, which was 45% of total emission of the country, which for that year amounted to 1719 Gg. This comparison already suggests choosing LCP for emission control.

Another argument is the fact that out of all collected data on the emission of SO_2, the data concerning the power generation industry are the most reliable. Furthermore, performing optimisation calculations needs collecting additional, usually estimated data on sources of emission; at present it is possible to prepare such data practically only for power generation sources. Bearing in mind the above conditions, all power generation plants were initially classified as controllable sources. In the following part of this paper, the chimney of a power station, and not the whole station will be treated as the source of emission.

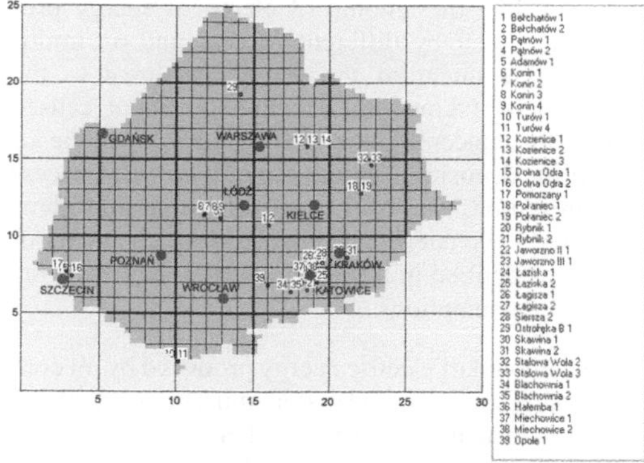

Figure 1. Area of modelling with the assumed computational grid (tied to the computational grid of the EMEP model) and locations of the controllable sources.

Accordingly, 103 power generation sources operated in 1999 – 39 chimneys of electric power stations, 46 chimneys of thermal-electric power stations and 19 chimneys of thermal power stations. In the next part of the paper, only chimneys of electric power stations were counted as controllable sources. Such a choice was caused, apart from the reasons connected with the extent of their total emission, by the role which particular plants play in the life of the country. So the operation of the electric power plant – as well as the emission of pollution by the plants, is closely related to the production of electric energy, and therefore can be quite precisely planned and determined by numbers. The remaining power plants also play the role of thermal plants, and the demand for heating changes according to meteorological conditions is quite hard to forecast.

Finally, the first accepted assumption for creating scenarios was a division of all sources of emission from the database into:

1 Controllable sources – i.e. those sources the emission of which it is possible to manage by using the ROSE model. Among these sources are all electric power stations in Poland, more specifically the chimneys of these electric power stations. There were 39 controllable sources; their location and names are presented in Figure 1.
2 Uncontrollable sources – i.e. those sources the emission of which remains constant, and the pollution that is caused forms the background pollution in the system. Sources of this type include all remaining emitters from the database, i.e. all foreign sources (Czech and German), Polish thermal-electric power stations and thermal stations, industrial sources and municipal sources.

Next, the analysis of assumptions which it is necessary (or possible) to make in order to control particular sources in order to lower the degree of threat to the environment in Poland was carried out. In optimisation models (e.g. Lowles et al. 1998) developed in Poland in recent years and dealing with SO_2 only, it was assumed that the only way of reducing the environmental threat was to lower pollutant emission by using available techniques of: carbon enrichment, building fluidal furnaces or installing devices for desulfurisation of the exhaust gases of a particular type. For the purpose of optimisation, classical methods of linear programming were applied. The target was, basically, not to exceed the limit of allowed annual average concentrations of SO_2 in the territory of the whole country. The optimisation calculus was supposed also to answer which of the possible technical solutions would be the cheapest. This procedure is now not up-to-date as Poland is obliged to meet the EU Directives with regard to environmental protection.

Directive 96/61/EC (IPPC – concerning Integrated Pollution Prevention and Control) assumes the use of the so called the Best Available Techniques (BATs). Directive 2001/80/EC (LPC – concerning Large Combustion Plants) determines the SO_2 emission limit values (ELVs) as follows: 400 mg/m_u^3, for the existing sources of emission, and 200 mg/m_u^3 which is a target limit for new sources (with a rated thermal input exceeding 300 MW_{th}).

In other integrated models used around the world – the RAINS model (Schöpp et al. 1999), or the ASAM model (ApSimon et al. 1994, ApSimon et al. 2002) – exceeding the critical load of acidity (Hettelingh et al. 2001) is considered as the criterion for the assessment of the degree of threat to the environment as regards sulfur and nitrogen compounds. For the optimisation, as in Polish models, linear programming techniques are used. The calculations performed by means of the RAINS model for the purpose of the Second Sulphur Protocol of the LRTAP Convention (Tuinstra et al. 1999), were aimed at finding cost-effective allocations of emission reductions in each particular country to meet European environmental policy targets, connected with acidification. The calculations of sulfur depositions (loads) for the year 1990, were performed using a Lagrangian model of long-range transport of air pollutants in Europe, developed by EMEP (Barret & Sandnes 1996). The protocol obliges all states to reduce SO_2 emissions by the year 2010 by such a percentage (with respect to the base emission year, 1980) that the 1990' exceedances of the sulfur critical loads will be lowered by 60% for each square of the EMEP model computational grid (150×150 km) covering Europe.

The means of minimising threat which have been used so far are becoming insufficient for the purpose of full realisation of EU directives, particularly their long-term objectives of not exceeding critical levels and loads and of effective protection of all people against recognised health risks from air pollution, which are to be achieved by the year 2020 (EU's NEC Directive, 2001/81/EC).

2.2 *Examples of scenarios*

The situation outlined above was an inspiration for the author to search for other solutions leading to lowering the degree of threat which was being caused by sulfur compounds to the environment in Poland. The first step was a statement – that the degree of threat cannot be determined by means of only one criterion. Therefore a multi-criteria assessment of the degree of threat to the environment was proposed, which took into consideration the threat caused by the concentrations of SO_2 and SO_4^{2-} in the air and sulfur depositions on the ground. Then the three following phenomena were analysed:

1 The same emission of SO_2 from particular sources causes a different degree of pollution to the environment. This results from the fact that depending on the location of the source and stack's parameters, taking into account meteorological conditions and the topography of the terrain, a particular emission causes different values of concentrations and depositions for each source and their different spatial distributions. In order to be able to control sources, it is necessary to perform calculations of the so-called the source-receptor matrices (SRMs).
2 The same degree of pollution to the environment causes a different degree of threat to the environment. This results from the fact that values and the spatial distribution of the threat to the environment by a certain pollutant depend on both the values of spatial distribution of concentrations/depositions of the certain pollutant and on the values and spatial distribution of allowed (critical) values for this pollutant. Therefore these sources – the emission of which causes the largest threat to the sensitive environment – should first be brought under control.
3 Each controllable source has a different unit emission Qj, i.e. emission resulting from producing a unit of energy, expressed in kg (SO_2)/MWh. At present, the unit emission from particular Polish electric power plants differs from plant to plant by as much as two orders of magnitude. It is worth mentioning that even after meeting the EU Directives, differences in unit emissions will be always present (although not as large as at present), because the type and characteristics of the fuel used play a decisive role. In connection with that it is obvious that the system of modelling should maximally use the energy production of these sources the unit emission of which is the smallest.

Taking the above conditions into account, the author has proposed a new methodology of minimising the threat to the environment. The methodology is based on controlling the amount of electrical energy produced in a year by different sources and on multi-criteria assessment of the degree of threat to the environment, taking into account the threat caused both by the concentrations of SO_2 and SO_4^{2-} in the air and sulfur deposition on the ground. For the process of optimisation, state-of-the-art evolutional algorithms were used (Michalewicz & Fogel 2004). The methodology was applied in the ROSE model, accepting the following assumptions:

1 The amount of electric energy produced by all controllable sources of the system must be constant and is assumed in advance. In other words, lowering the amount of the energy produced by one source must be compensated by the production of energy in another source. Because for each source the amount of produced energy (depending on the unit emission) marks a particular emission of SO_2, the proposed system is at the same time a system of controlling the emissions.
2 The amount of energy produced by particular sources cannot extend the maximum values for these sources, i.e. such values as would be reached if the source operated at full nominal power for a maximum time of use of the plant in the year. For a plant operating on brown coal (Bełchatów, Pątnów, Adamów, Konin, Turów) – this time amounts to 7250 h/year; for plants operating on black coal – 6500 h/year.
3 In the electric power network of the country electric power stations exist which must provide for energetic security of the country in emergency cases and the minimum power of which is precisely determined. These power plants are called "must runs".

Taking the above into account, for the experiments carried out with the use of the ROSE model, a scenario has been proposed in which we are looking for an optimum solution without considering the limitations resulting from the present activity of the electric power station. This means that we do not specify in advance any limitations with regard to the possibility of reducing energy production/ emission from particular sources, as long as the total assumed production of electric energy of all plants remains preserved.

We need to know which electric power plants must reduce the emission and by how much in order to achieve the assumed target of minimising threat caused by the sulfur compounds, and at the same time, meet the electric power demand at the assumed level. Only such an assumption will allow for the assessment of the role which particular electric power plants play in causing a threat to the environment and what real reductions in their production are needed to reach the optimum solution.

The realisation of the proposed scenario is possible, as long as there exist installed power surpluses in relation to present demands. This situation has been present in Poland for a few years now (about 60% of total installed power used). In the year 1999 the total production of electric power (by industrial electric power stations) amounted to 92.4 TWh, which was 58.93% of maximum production (when operating at nominal power over maximum time of use of the plant in the year).

This value has been assumed as a basis for further calculations – the assumed total demand for the production of electric power in industrial power stations. This means that all experiments controlling the operation of the plant carried out in order to minimise threat assume that the calculated total demand for power of the country must be maintained.

All simulation calculations (of pollution dispersion) and optimisation calculations in the ROSE model were carried out on data collected in the emission database for the year 1999. Moreover, for each controllable source, apart from the following basic data: parameters of the chimney (geographic coordinates, geometric height, emission of heat) and total annual emission of SO_2, the following estimations were made:

1 maximum hourly emission, (Q_h), which takes place when operating at full power,
2 maximum time of operation of the plant in the year at full (nominal) load (L_{ha}).

The above data are estimates and, as such, can contain some minor errors. These ought not to significantly affect the correctness and applicability of general conclusions derived from the presented research.

2.3 *Methodology of computing the state of pollution for a chosen scenario*

For uncontrollable sources calculations were performed using a numerical two-dimensional Eulerian grid model of the well-mixed layer: POLSOX-II (Juda-Rezler 2004c). In the final result, matrices containing the values of the pollution background: average annual concentrations and annual deposition of the modelled compounds for the computational grid of the ROSE model, were obtained.

Controlling the sources makes it necessary to perform calculations of source-receptor matrices $[A_{n,k}]$, in which are written the values $A_{n,k}$, representing concentration or deposition which a unit emission of n-th source (n = 1,2....,39) causes in the k-th square of the computational grid (k = 1,2....,750). In order to obtain such matrices, it is necessary to perform calculations concerning the spreading of the pollutions from each controllable source separately. Such calculations need special requirements for the air pollution model. The calculation methodology used is presented in the next chapter.

The obtained transfer matrices multiplied by the real (for the basis year) or sought (for the optimum solution) emission of a given source, give real/sought values of concentrations or depositions which this source caused/will cause in a given area (square of the computational grid). In the optimization process the value of emission is changed, and then – without the necessity to launch the model POLSOX-II again and using the computed transfer matrices, modified values of concentrations and depositions are obtained in a simple way.

For each of the controllable sources n and for each square of the computational grid k, the following calculations were performed:

$$Q_{n,\max} = \frac{Q_{n,h} \cdot L_{n,ha}}{1000} \tag{1}$$

$$Q_{n,akt} = \frac{Ster_n}{100} \cdot Q_{n,\max} \tag{2}$$

$$C_{El,k}(SO_2) = \sum_n \frac{CJ_{n,k}(SO_2) \cdot Q_{n,akt}}{1000} \tag{3}$$

$$C_{El,k}(SO_4) = \sum_n \frac{CJ_{n,k}(SO_4) \cdot Q_{n,akt}}{1000} \tag{4}$$

$$D_{El,k}(S) = \sum_n \frac{DJ_{n,k}(S) \cdot Q_{n,akt}}{1000} \tag{5}$$

$$C_k(SO_2) = C_{El,k}(SO_2) + C_{bac,k}(SO_2) \tag{6}$$

$$C_k(SO_4) = C_{El,k}(SO_4) + C_{bac,k}(SO_4) \tag{7}$$

$$D_k(S) = D_{El,k}(S) + D_{bac,k}(S) \tag{8}$$

where: $Q_{n,max}$ = maximum, for the source n, emission of SO_2 during a year [Mg]; $Q_{n,h}$ = maximum hourly, for the source n, emission of SO_2 [kg/h]; $L_{n,ha}$ = maximum number of hours of operation for the source n in a year, at nominal load [h]; $Q_{n,akt}$ = current emission of the source n [Mg]; $Ster_n$ = current steering for the source n (percentage of source use) [%]; $CJ_{n,k}(SO_2)$ = concentration of SO_2, which is caused by unit emission (1000 Mg) of the n-th controllable source in the k-th square [$\mu g/m^3$]; $CJ_{n,k}(SO_4)$ = concentration of SO_4^{2-}, which is caused by unit emission (1000 Mg) of the n-th controllable source in the k-th square [$\mu g/m^3$]; $DJ_{n,k}(S)$ = total deposition of S, which is caused by unit emission (1000 Mg) of the n-th controllable source in the k-th square [$g(S)/m^2$]; $C_{El,k}(SO_2)$ = concentration of SO_2, which is caused by the emission of all controllable sources in the k-th square [$\mu g/m^3$]; $C_{El,k}(SO_4)$ = concentration of SO_4^{2-}, which is caused by the emission of all controllable sources in the k-th square [$\mu g/m^3$]; $D_{El,k}(S)$ = total deposition of S, which is caused by the emission of all controllable sources in the k-th square [$g(S)/m^2$]; $C_{bac,k}(SO_2)$ = concentration of SO_2, which is caused by the emission of all uncontrollable sources in the k-th square [$\mu g/m^3$];

$C_{bac,k}(SO_4)$ = concentration of SO_4^{2-}, which is caused by the emission of all uncontrollable sources in the k-th square [$\mu g/m^3$]; $D_{bac,k}(S)$ = total deposition of S, which is caused by the emission of all uncontrollable sources in the k-th square [$g(S)/m^2$]; $C_k(SO_2)$ = concentration of SO_2, which is caused by the emission of all sources of emission in the k-th square [$\mu g/m^3$]; $C_k(SO_4)$ = concentration of SO_4^{2-}, which is caused by the emission of all sources of emission in the k-th square [$\mu g/m^3$]; $D_k(S)$ = total deposition of S, which is caused by the emission of all sources of emission in the k-th square [$g(S)/m^2$].

The index "current" refers here to real or sought values. By summing the obtained matrices of background concentrations/depositions with the matrices of concentrations/depositions for controllable sources obtained for real data, the state of pollution by sulfur compounds in the territory of Poland in the year 1999 has been determined. In the optimisation experiments carried out by summing the matrices of background concentrations/depositions with matrices of concentrations/depositions for controllable sources obtained for controlling proposed by the evolutional algorithm, the searched (after the optimisation) state of pollution in the territory of Poland was obtained.

2.4 *Methodology for computing source-receptor air pollutant matrices (SRMs)*

SRMs provide the important connection between emission from a given source and the state of pollution (concentration in the air, deposition) in the given receptor grid. The computation of the transfer matrix of the relation source – grid receptor is usually performed with the use of the Lagrangian air pollution models. There are two reasons for this: firstly, in the Lagrangian models the reference system is "assigned" to a moving particle of the air so in the model there is no advection part. This avoids errors connected with numerical diffusion. Secondly, the structure of the model enables computing SRMs for many sources simultaneously. Computing the SRMs with the use of the Eulerian models brings problems, among which the most important are:

1 Non-linearity brought by solving numerically the advection part of the model equations.
2 Non-linearity connected with chemical transformations considered in the model.
3 Time-consumption of the calculations connected with the necessity to perform calculations for each of the sources separately.

Taking the above into account, the Lagrangian model was considered to be advantageous for computing SRMs. This solution however was rejected. Firstly, in this paper the state of air pollution in Poland and threat to the environment resulting from it, was assessed based on the results of calculations of the Eulerian grid model (POLSOX-II), which were verified positively with the results of measurements (Juda-Rezler 2004c). Secondly, for the purpose of the optimisation mode of the ROSE model, the state of pollution is determined by summing the matrices of the background concentrations/depositions with the matrices of concentrations/depositions for controllable sources (summary matrices, Eqs. 6–8). Performing partial calculations by means of various models would result in lack of mass-conservation in the ROSE framework and errors difficult to estimate.

Therefore, SRMs for controllable sources were computed with the use of the model POLSOX-II; however in such a way as to reduce non-linear effects. In the case of this model, the non-linearity connected with chemical transformations does not exist, and non-linearity connected with numerical solving the advection part of advection-diffusion equation is connected with numerical errors. All numerical methods elaborated so far to solve the advection part of the differential equations in the air pollution models, contained – to a different degree – numerical errors connected with numerical diffusion or with generating unwanted oscillations, especially in the vicinity of point sources (Peters et al. 1995, Wind et al. 2002). In the model POLSOX-II the Galerkin finite element method, with a numerical filter for eliminating negative values in the results was used. The Galerkin finite element techniques found their broadest application in air pollution models because in the numerical tests carried out for the advection equation, they proved to be exact, consistent and stable. An additional advantage is their calculation speed and simplicity (Chock 1991). However, in the case of large gradients of concentration, numerical errors may occur, which are of great importance when the computing of transfer matrices is performed. These errors result from the fact that the sum of concentrations (depositions) computed for each of the N sources of emission individually is different to the concentration (deposition) computed for all N sources of emission.

Calculations performed simultaneously for all sources of emission contained in the database, have the lowest numerical error. Because these results were also positively verified by the measurement data (Juda-Rezler 2004c), they have been assumed as reference for further calculations. For the summary matrix obtained from Equation 6 for SO_2, with the use of SRMs computed in a "traditional" way, the mean bias (MB) in particular points of the computational grid ranged from −18% to +9% in comparison to the reference matrix. The mean normalised bias (MNB) amounted to 0.72%.

In order to reduce numerical errors, the so-called reverse method (Bartnicki 2000), used for determining the transfer matrices in the EMEP Eulerian model

(Wind et al. 2002), was applied in the ROSE model to compute SRMs for particular controllable sources. Transfer matrices for each of the controllable sources n, and for each of the compounds analysed in the paper are calculated as follows:

$$\left[A_{n,k}\right] = \left[A_{all,k}\right] - \left[A_{all-n,k}\right] \tag{9}$$

where $[A_{n,k}]$ denotes one of three SRMs (for concentrations of SO_2, SO_4^{2-} or deposition of S) for the source n; $[A_{all,k}]$ is one of three reference matrices (for concentrations of SO_2, SO_4^{2-} or deposition of S); $[A_{all-n,k}]$ refers to one of three matrices obtained from calculations performed for all sources from the database excluding the source n (for concentrations of SO_2, SO_4^{2-} or deposition of S).

In order to obtain unit SRMs $[AJ_{n,k}]$, matrices $[A_{n,k}]$ must be divided by the real (for the year 1999) emission of SO_2:

$$\left[AJ_{n,k}\right] = \frac{\left[A_{n,k}\right]}{Q_{n,99} \cdot 0.001} \tag{10}$$

where: $Q_{n,99}$ = emission of the source n in 1999 [Mg]; $[AJ_{n,k}]$ = Unit SRMs: $[CJ_{n,k}(SO_2)]$ = for concentrations of SO_2 which are caused by unit emission (1000 Mg) of the n-th controllable source in the computation grid ([$\mu g/m^3$]; $[CJ_{n,k}(SO_4]$ = for concentrations of SO_4^{2-} which are caused by unit emission (1000 Mg) of the n-th controllable source in the computation grid ([$\mu g/m^3$]; $[DJ_{n,k}(S)]$ = for total deposition of S which is caused by unit emission (1000 Mg) of the n-th controllable source in the computation grid [$g(S)/m^2$].

Because of the complexity of the calculation process, POLSOX-II is a demanding application with regard to processor speed and RAM memory. A simulation of the spreading of pollution performed for one year, which takes into account all sources of emission contained in the database, lasts about 70 minutes on a 2.2 GHz PC. The presented methodology for computing transfer matrices is therefore work- and time-consuming.

3 RESULTS

The use of the reverse method for computing SRMs for particular controllable sources, significantly reduced the occurrence of numerical errors. Summary matrices for particular variables, obtained by the use of SRMs computed in the way presented above, have been compared with reference matrices. For the summary matrix of SO_2 concentrations, the MB at particular points of the computational grid amounted to $-4\% \div +7\%$ in relation to the reference matrix, and the MNB amounted to 0.24%. For the summary matrix of SO_4^{2-}

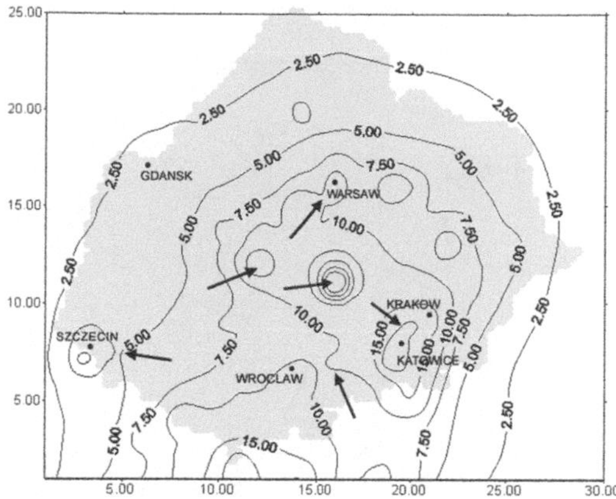

Figure 2. Reference results of calculations for the territory of Poland (in the assumed computational grid tied to the computational grid of the EMEP model). The variable – average annual concentration of SO_2 in 1999 [μg/m3]. Model POLSOX-II.

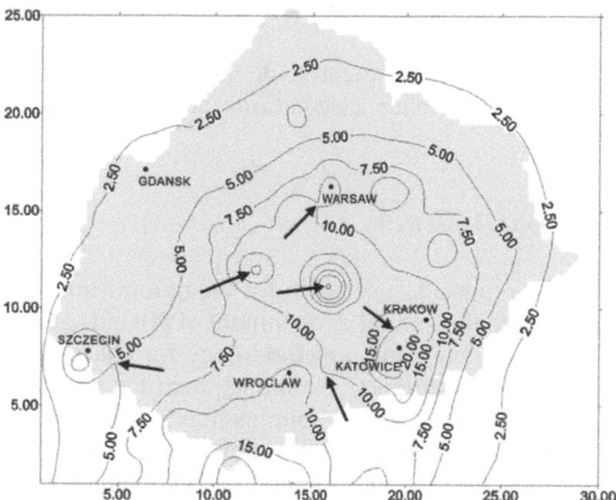

Figure 3. Results of calculations for summary matrix – traditional method. The variable – average annual concentration of SO2 in 1999 [μg/m3]. Model POLSOX-II.

concentrations, the MB at particular points of the computational grid amounted to $-2\% \div +2\%$ in relation to the reference matrix, and the MNB amounted to -0.15%. For the summary matrix of S deposition – the MB at particular points of the computational grid amounted to $-4\% \div 7\%$ in relation to the reference matrix, and the MNB amounted to 0.20%.

The effectiveness of the used methodology is confirmed by the analysis of the obtained distributions of concentrations and depositions.

The figures above present average annual concentrations of SO_2 for the territory of Poland obtained for the reference matrix (Figure 2), for the summary matrix obtained from Equation 6, by using SRMs computed in a "traditional" way (Figure 3) and for the

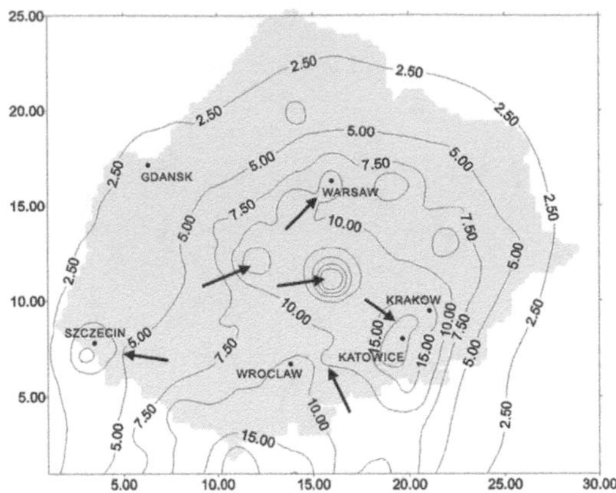

Figure 4. Results of calculations for summary matrix – reverse method. The variable – average annual concentration of SO2 in 1999 [μg/m3]. Model POLSOX-II.

summary matrix obtained from Equation 6, by using the SRMs computed with the use of the reverse method (Figure 4).

Areas, where the greatest deviations occurred in relation to reference calculations are marked with arrows.

4 CONCLUSIONS

The paper presents a scenario for the minimisation of the threat to the natural environment of Poland caused by SO_2 emission, and a methodology for calculating the state of pollution to the atmosphere for this scenario. Controlling energy sources makes it necessary to perform calculations of the source-receptor matrix for each of the modelled variables. Calculating SRMs by the use of Eulerian grid models is connected with numerical problems, resulting largely from nonlinearity brought mainly by numerically solving the advection part of the model equations. In order to avoid numerical errors, the so-called reverse method has been applied to compute the transfer matrix for particular controllable sources. This significantly reduced the occurrence of numerical errors. Summary matrices for particular variables, obtained by the use of SRMs computed as presented above, have been compared with reference matrices. The obtained results are satisfactory and provide for high accuracy of calculations. The methodology used enables computing SRMs by using Eulerian models. Models of this type are vital as they allow for incorporating all processes taking place in the atmosphere and are better tools than Lagrangian models for air quality assessment on the regional scale. The possibility of using the model POLSOX-II also for optimisation calculations is fundamental for the operation of the ROSE integrated assessment model. Performing partial calculations by means of different air pollution models would bring unknown and probably considerable error to the summary matrices and lack of mass-conservation in relation to reference matrices used for the assessment of threat.

The example scenarios presented in the previous paper (Juda-Rezler 2004b) evidently indicate significant capacity of emission control strategies that are targeted at minimizing the harmful effects of pollution, compared with traditional approaches that determine emission controls exclusively in relation to historical emission levels. For three analysed scenarios with multi-criterion objectives, the average SO_2 and SO_4^{2-} concentration were reduced by approximately 10%. Simultaneously, excess sulfur deposition for Poland was eliminated in 15% of grids and reduced by at least 20% in successive 60% grids. Total sulfur deposition was reduced by 10%. At the same time emission from LCP was reduced by approximately 25%, whilst the country's electricity output remained constant.

The optimisation mode of the ROSE framework appears as a powerful tool that can simultaneously address air and ecosystems pollution issues. Clearly, in such optimised strategies the distribution of electric energy production/emission reductions across the various sources is essentially determined by the type and stringency of the environmental targets. In the ROSE integrated assessment model, strategies for combined targets can be developed. At the present moment, the proposed methodology is optimisation with respect to environmental criteria solely and does not take into account techno-economic criteria connected with the production of electrical energy, nor the criteria connected with the transfer and distribution of this energy. Finally, any analysis of air pollution control strategies in Poland should be discussed among a wider range of specialists and decision makers.

Nevertheless, the presented methodology and the ROSE optimisation framework indicate the way toward looking for the best solutions for protecting human health and vegetation against air pollution problems. Certainly the framework should include other pollutants and environmental effects besides sulfur. To this end, the ROSE integrated assessment model should be extended to other essential pollutants and environmental effects, first of all to health impacts of both coarse and fine particles.

REFERENCES

Amann, M. & Lutz, M. 2000. The revision of the air quality legislation in the European Union related to ground-level ozone. *Journal of Hazardous Materials* 78 (1–3): 41–62.

Amann, M., Cofala, J., Heyes, C., Klimont, Z. & Schöpp, W. 1999. The RAINS model: a tool for assessing regional emission control strategies in Europe. *Pollution Atmospherique* 1999: 41–63.

ApSimon, H.M., Warren, R.F. & Wilson, J.J.N. 1994. The abatement strategies assessment model – ASAM: applications to abatement of sulphur dioxide emissions across Europe. *Atmospheric Environment* 28: 649–663.

ApSimon, H.M., Warren, R.F. & Kayin, S. 2002. Addressing uncertainty in environmental modelling: a case study of integrated assessment of strategies to combat long-range transboundary air pollution. *Atmospheric Environment* 36 (35): 5417–5426.

Barret, K. & Sandnes, H., 1996. Transboundary acidifying air pollution calculated transport and exchange across Europe, 1985–1995. In: K. Barrett & E. Berge (eds), *Transboundary Air Pollution in Europe*. MSC-W Status Report 1996, Meteorological Synthesizing Centre-West, Norwegian Meteorological Institute, Oslo, Norway.

Bartnicki, J. 2000. Non-Linear effects in the source receptor matrices computed with the EMEP Eulerian Acid Deposition Model. EMEP/MSC-W Note 4/00, The Norwegian Meteorological Institute, Oslo, Norway.

Carmichael, G.R., Streets, D., Calori, G., Amann, M., Jacobson, M., Hansen, J. & Ueda, H. 2002. Changing trends in sulphur emissions in Asia: implications for acid deposition, air pollution, and climate. *Environmental Science and Technology* 36 (22): 4707–4713.

Chock, D.P. 1991. A comparison of numerical methods for solving the advection equation – III. *Atmospheric Environment* 25A: 853–871.

Cofala, J., Amann, M., Gyarfas, F., Schoepp, W., Boudri, J.C., Hordijk, L., Kroeze, C. & Junfeng, L. 2004. Cost-effective control of SO2 emissions in Asia. *Journal of Environmental Management* 72 (3): 149–161.

Hettelingh, J.-P., Posch, M. & de Smet, P.A.M 2001. Multi-effect critical loads used in multi-pollutant reduction agreement in Europe. *Water, Air, and Soil Pollution* 130 (1–4): 1133–1138.

Juda-Rezler, K. 2004a. Risk assessment of airborne sulphur species in Poland. In: C. Borrego & S. Incecik (eds), *Air Pollution Modelling and its Application XVI*: 19–27. New York: Kluwer Academic/Plenum Publishers.

Juda-Rezler, K. 2004b. Application of an Integrated Assessment Modelling System in Minimization of the Environmental Risk in Poland. *Polish Journal of Environmental Studies* 13 (Supp. III): 276–278.

Juda-Rezler, K. 2004c. Modelling of the air pollution by sulphur species in Poland. *Environmental Protection Engineering* 30 (3): 53–71.

Lowles, I., ApSimon, H.M., Juda-Rezler, K., Abert, K., Brechler, J., Holpuch, J. & Grossinho, A. 1998. Integrated assessment models – tools for developing emission abatement strategies for the Black Triangle region. *Journal of Hazardous Material* 61 (1–3): 229–237.

Mediavilla-Sahagun, A. & ApSimon, H.M. 2003. Urban scale integrated assessment of options to reduce PM10 in London towards attainment of air quality objectives, *Atmospheric Environment* 37 (33): 4651–4665.

Mediavilla-Sahagun, A. & ApSimon, H.M. 2006. Urban scale integrated assessment for London: Which emission reduction strategies are more effective in attaining prescribed PM10 air quality standards by 2005? *Environmental Modelling & Software* 21 (4): 501–513.

Michalewicz, Z. & Fogel, D. 2004. Deriving evolutionary algorithms. In: *How to Solve It: Modern Heuristics*: 161–183. Berlin: Springer-Verlag.

Peters, L.K., Berkowitz, C.M. & Carmichael, G.R. 1995. The current state and future directions of Eulerian models in simulating the tropospheric chemistry and transport of chemical species: A review. *Atmospheric Environment* 29 (2): 189–222.

Schöpp, W., Amann, M., Cofala, J., Heyes, Ch. & Klimont, Z. 1999. Integrated assessment of European air pollution emission control strategies. *Environmental Modelling & Software* 14 (1): 1–9.

Tuinstra, W., Hordijk, L. & Amann, M. 1999. Using computer models in international negotiations: the case of acidification in Europe. *Environment* 41 (9): 32–42.

Wettestad, J. 2002. Clearing the air. Europe tackles transboundary pollution. *Environment* 44 (2): 32–40.

UNECE/EMEP, 2004, UNECE/EMEP activity data and emission database, WebDab 2004, http://webdab.emep.int.

Wind, P., Tarrasón, L., Berge, E., Slørdal, L.H., Solberg, S. & Walker, S-E. 2002. Development of a modelling system able to link hemispheric-regional and local air pollution, First Draft. Joint EMEP-W & NILU Note 5/2002, The Norwegian Meteorological Institute, Oslo, Norway.

Environmental Engineering – Pawłowski, Dudzińska & Pawłowski (eds)
 ISBN13 978-0-415-40818-9

Seasonal changes of PAHs levels in air dust particles taken in Lublin, Poland

Aneta Duda & Jacek Czerwiński
Lublin University of Technology, Lublin

ABSTRACT: Polycyclic aromatic hydrocarbons (PAHs) are present in the air either adsorbed on particulate matter or as vapor (minor part). Their oxidised derivatives and metabolites are known to be effective carcinogenic and mutagenic agents. In this study dust samples from Lublin have been collected for 24 months and analyzed for the PAHs content. For analysis of the seasonal changes in atmospheric PAHs content SEC clean-up, and final analysis with GC-MS was applied.

Keywords: Polycyclic Aromatic Hydrocarbons (PAHs), fine dust, SEC clean-up, GC-MS analysis.

1 INTRODUCTION

Polycyclic aromatic hydrocarbons (PAHs) are a ubiquitous class of hydrophobic and semivolatile organic compounds consisting of two or more fused benzene rings in linear, angular, or cluster arrangements. They are emitted, as combustion products, in anthropogenic and also in natural processes. They are present in the air, either adsorbed on particulate matter or as a vapor (minor part), and their role as atmospheric pollutants has been well established by several studies (Pereira 2000). PAHs, after chemical or metabolic processes, are effective carcinogenic/mutagenic agents, and as such they have become a major field of investigation. (Kot 2004, Wenzl 2006).

The importance of PAH in urban sites is mainly derived from their quantities produced by vehicle emissions, especially heavy duty diesel powered vehicles and the combustion of fuels for heating purposes.

Atmospheric particulate matter containing PAH is, in general, collected over Teflon, glass or quartz fiber filters, followed by an extraction based on Soxhlet, sonication, microwave, supercritical fluid or pressurized fluid (Cecinato 1999, Smith 2006, Wise 2004).

While Soxhlet extraction is time consuming and requires large amounts of toxic and expensive solvents, microwave and supercritical fluid extraction (SFE) require relatively expensive apparatus.

Sample preparation for the determination of persistent organic pollutants (also PAHs) from environmental matrices is difficult (de Pereira 2001, Wise 2004). The main reason is the complicated nature of these matrices, which contain large fractions of structures abundant in lipids and other compounds with lipophilic character. These lipids are comparable to many lipophilic organic pollutants (including PAHs) in terms of their physicochemical properties, e.g. solubility, molecular size. The co-extraction and co-elution of other lipophilic compounds and pollutants can lead to considerable interference during cleanup with overlapping between GC–MS analyte peaks and matrix peaks and consequently the faulty interpretation of analysis values (Smith 2006).

Recently, the extraction efficiency for PAHs and other organic pollutants from real plant matrices was improved by up to two orders of magnitude compared to conventional procedures such as Soxhlet or ultrasonic extraction by optimizing the pressurized extraction technique ASE (Accelerated Solvent Extraction). The high energy addition introduced by ASE temperature (up to 200°C) and pressure (up to 20 MPa) changes the solvent properties and leads to the specific separation of analytes and matrix along with the release of portions of bound pollutant residues. Lately, the application to POPs and PAHs and, in part, also an improvement in extraction efficiency from soils and sediments using ASE has also been reported. The higher extraction efficiency using the ASE technique also requires in some cases an improvement in the quality of the cleanup procedure.

If the conventional cleanup-procedure is employed to determine pollutants in complicated plant material, the quality of the GC–MS chromatograms is often inadequate for trouble-free analysis. To purify the extracts from chlorophyll, plant lipids and other substances more polar than PAHs, various procedures have previously been applied. Different chromatographic columns switched in sequence with aluminium oxide, Florisil and silica gel, and with activated and deactivated Florisil have been used. Silica inert solid-phase extraction (SPE) columns with multiple cleaning by solvents have been used. In addition, Florisil SPE columns have been employed and a series of complicated analytical cleanup steps was performed (Kot 2004, Wenzl 2006, Wise 2004).

Size-exclusion chromatography (SEC) was regarded as a suitable alternative method for improving previous cleanup procedures. It had not been used previously for the cleanup of PAHs or any other persistent pollutants from conifer needles, deciduous leaves, and mosses. In particular, the new SEC procedure needed to be tested on samples for which integrateble chromatograms were not obtained after either conventional Florisil cleanup or other methods. By choosing a suitable packing material with a very small pore size, a new, one-step, cleanup procedure for PAHs from the matrices described above is proposed.

2 MATERIALS AND METHODS

2.1 *Materials*

Dust samples were collected during the period of 24 months (from January 2004 till December 2005) on quartz fiber filters in the Botanic Garden of Lublin in the vicinity of the most important route Warsaw – board of Poland and the residential district Slawinek.

Solvents: hexane and dichloromethane "for residue analysis" (JT Baker – Germany) were used for the extraction and dilutions. A standard mixture of sixteen PAHs (10 μg/ml in acetonitrile: Promochem, Germany) according to US-EPA, was used for the preparation of GC-MS calibration solutions.

The following substances were used for the preparation of SEC calibration mixtures: elemental sulfur-puriss (POCh – Gliwice, Poland); perylene – GC standard (IChO – Warszawa, Poland), methoxychlor – GC standard (Riedel-deHaen – Germany), bis(ethylhexyl) phthalate – GC standard (Riedel-deHaen – Germany), corn oil – pharmacopeal grade. Helium (99.9996) was applied for GC-MS and for degassing of solvents. Argon (99.999) was used for gentle evaporation of extracts.

2.2 *Methods*

Extractions of dust samples adsorbed on filters were carried out in Soxleth apparatus with 200 ml of dichloromethane (for residue analysis JT Baker – Germany) for 8 hours. Then the extracts were

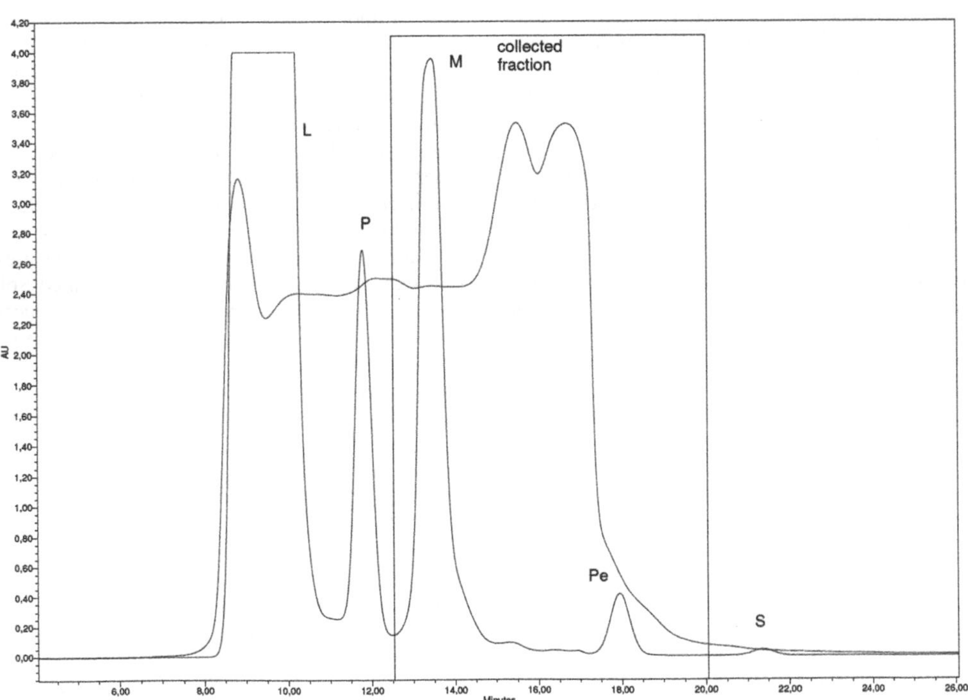

Figure 1. SEC-chromatograms of calibration standard mixture (thick line) and real sample (thin line). L – corn oil as lipid fraction marker; P – ethylhexylphthalate – phthalate fraction marker; M – methoxychlor – pesticide fraction marker; Pe – perylene – PAHs fraction marker; S – sulfur – low molecular compounds marker.

evaporated to 3 ml and passed through Millex-FG filter units (PTFE membrane, 0.22 μm pore size, 25 mm i.d. – Millipore, Bedford). Two ml of filtrated extract was injected to the SEC system.

Clean-up of extracts was performed with high resolution size exclusion chromatography (SEC) on Breeze 1525 system (Waters USA). The SEC system consists of following parts:

- Binary gradient pump Waters 1525
- Injection port Rheodyne with 2000 μl sample loop
- Envirogel GPC cleanup columns (coupled 19 × 150 mm & 19 × 300 mm) – Waters
- UV-V detector Waters M 2487 working @ $\lambda = 254$ nm
- Fraction collector Waters FC III
- Data acquisition – Breeze 3.30SPA.

The SEC-system was calibrated on a mixture containing the following compounds (all in dichloromethane):

- Corn oil – marker of lipids and waxes fraction
- bis(ethylhexyl) phthalate – marker of phthalates fraction
- methoxychlor – marker of pesticide fraction
- perylene – marker of PAH fraction
- elemental sulfur – low molecules fraction marker.

Mobile phase flow rate (dichloromethane) 5 mL/min. Collected fraction: 12.2–20 min.

Table 1. List of PAHs and the ions monitored (according to Soniassy 1994).

PAH Compounds	Abbreviation	Quantitative ions	Qualitative ions
Naphthalene	Naph	128	127,129
Acenaphthylene	Ace	152	151
Acenaphthene	Acy	154	153
Fluorene	Fl	166	165
Phenanthrene	Ph	178	176
Anthracene	A	178	176
Fluoranthene	Flu	202	200
Pyrene	Py	202	200
Benzo(a) anthracene	BaA	228	226
Chrysene	Chr	252	250, 253
Benzo(b) fluoranthene	BbF	252	250, 253
Benzo(k) fluoranthene	BkF	252	250, 253
Benzo(a)pyrene	BaP	252	250, 253
Indeno(1,2,3-cd) pyrene	IPy	276	277, 274
Benzo(ghi) perylene	BPe	276	277, 274
Dibenzo(a,h) anthracene	DBA	278	276, 274

Figure 1 presents SEC chromatograms of the calibration standard mixture (thick line) and a real sample (thin line).

Final determination were done with GCQ GC-MS (Finnigan – USA), configured as follows:

- AS-200 autosampler,
- split/splitless injector operated at 275°C
- DB-5ms capillary column 30 m × 0.25 mm, df 0.25 μm (J&W – USA)
- Temperature programming:
  60°C (1 min hold) 15°C/min to 160°C than 5°C/min to 320°C (10 min hold)
- Transfer line temperature 275°C
- Carrier gas He 5N (Lindegas Polska) @ 40 cm/s.

The system worked at *full-scan* mode (m/z 50–350 U) for determination of analytes Windows and In selected ion monitoring mode for quantitative analysis. Monitored ions for quantitative and qualitative analysis are given in a Table 1.

3 RESULTS

Chromatographic analysis of the liquid PAH standard and extracts of the dust samples resulted in the clean separation of peaks for all sixteen compounds. The total time of GC-MS analysis is around 50 min. Initially, two types of extractant were evaluated (dichloromethane and hexane). The PAH concentrations of CRM material found, averaged for each set of three extractions, are compared with certified values. Results of analysis of real samples are given in Table 2.

4 DISCUSSION

In Table 3 are presented the predominant PAHs and their most important sources.

The concentration levels found suggest that PAHs in atmospheric dust, largely stem from two main origins. In the heating season, heavy PAHs, suggestive of coal combustion, predominate. During the warm seasons, emission from diesels is the main source of PAHs in the atmosphere. Comparison with other places polluted by PAHs shows similar results (Chang 2006, Park 2002, Ravindra 2006) but not for unpolluted areas. Absence of low molecular weight PAHs (Nap, Ace, Acy) is probably due to their photoreactivity (Kot-Wasik 2004, Pereira 2000) and the possibility of oxidation reactions of these compounds.

The season of the year determines not only the quality and quantity but also the origin of PAHs present in the dust. This is exemplified in the samples collected in summer and winter. The atmospheric dust concentration is much higher in the latter season. Variations of concentration of PAHs between samples collected in winter and summer are not particularly large because,

Table 2. Seasonal variation of PAHs.

Sample	M_D [mg]	C_D [μg/m^3]	Concentration of PAHs [μg/g] Nap	Ace	Acy	Fl	Ph	A	Flu	Py	BaA	Chr	BbF	BkF	BaP	IPy	BPe
01 04	4.8	73	na	na	na	nd	nd	nd	nd	nd	40.6	64.4	50.8	nd	54.3	148.8	86.0
03 04	5.8	4	na	na	na	nd	nd	nd	nd	nd	30.4	50.9	25.0	nd	18.9	3.1	15.2
04 04	162.1	58	12.5	12.8	4.5	10.8	22.1	nd	45.2	39.9	95.2	55.6	35.1	41.1	106.8	25.4	19.2
05 04	82.6	30	nd	nd	10.0	nd	78.2	nd	64.2	57.8	133.5	40.9	256.7	67.9	98.9	61.2	33.6
08 04	141.0	49	na	na	na	nd	31.2	nd	36.2	37.7	17.6	28.5	150.0	28.0	52.4	36.7	18.8
09 04	108.5	39	na	na	na	nd	5.1	21.2	25.6	30.2	78.6	43.1	145.2	23.4	45.2	15.2	18.2
10 04	38.1	14	na	na	na	nd	79.1	12.2	26.0	120.1	194.0	246.1	53.0	10.1	247.2	149.7	69.3
11 04	148	53	nd	nd	nd	nd	63.3	26.4	227	221	25.8	65.8	90.8	28.9	36.4	54.1	54.2
12 04	181.2	63	nd	nd	nd	12.2	53.4	23.1	175	112	33.4	24.7	51.2	34.2	41.2	51.3	31.2
01 05	166.1	59.5	na	na	na	nd	nd	nd	nd	nd	51.6	41.2	37.8	nd	24.3	76.8	91.0
02 05	214.8	76.9	na	na	na	nd	nd	nd	nd	nd	40.6	64.8	51.8	nd	24.3	48.8	86.0
03 05	297.4	159.8	na	na	na	nd	nd	124.2	143.1	29.3	54.1	33.2	22.1	nd	18.9	31.4	26.2
04 05	198.0	124.0	23.1	15.2	21.5	13.1	54.4	1.6	39.2	47.6	79.1	61.7	33.6	34.9	102.2	26.1	32.2
05 05	124.6	69.4	17.1	22.2	nd	33.1	26.4	54.9	nd	76.2	55.4	69.1	69.7	54.1	19.0	25.1	nd
06 06	129.9	21.0	nd	11.2	8.3	7.1	16.1	32.1	11.0	64.1	33.2	46.1	54.8	31.1	nd	7.4	nd
08 05	119.4	19.4	56.8	17.4	10	nd	78.2	nd	45.9	21.3	125.4	43.9	63.1	49.1	78.1	46.2	nd
09 05	163.1	55.2	43	61.1	19.3	13.2	22.4	3.2	78.6	63.1	145.2	23.4	nd	15.2	18.2	46.0	19.3
10 05	243.0	67.7	na	na	na	nd	73.5	11.6	22.0	119.3	187	220.4	nd	21.3	231.0	131	75.2
11 05	295.1	87.0	nd	nd	nd	nd	52.2	24.4	208	124	26.8	47.8	100.8	27.9	32.9	63.2	47.1
12 05	490.3	283.7	nd	nd	nd	11.0	42.7	23.1	124	134	33.4	24.7	146.2	37.2	29.2	71.7	113.9
12 05a	439.1	264.2	nd	nd	nd	16.0	54.1	43.2	99.3	171.2	43.1	44.2	137.1	24.2	19.7	55.3	112.0

M_D – mass of collected dust.
C_D – concentration of collected dust in the air.
nd – not detected.
na – not analysed.

Table 3. Indicatory PAHs for various major sources (Chang 2006, Dobbins 2006, Launhardt 1998, Park 2002).

Category	Predominant PAH species
Stationary sources	
Steel industry	BaP, BaA, Pe, BeP, COR,
Cement	Acy, Acn, A
Power plant	CYC, DBA
Incineration	Py, Ph, Flu, IPy, Chr
Mobile sources	
Diesel vehicle emission	AcPy, FL, Flu, Ph, Pyr, Chr, BeP
Gasoline vehicle emission	Flu, Chr, IPy, BPe, CYC, COR
Combustion sources	
Incomplete combustion and pyrolysis of fuels	Fl, Pyr, Ph, BghiP, IPy
Combustion sources	Nap, Acy, Acn, Flu, Ph, Ant
Industrial oil burning	Pyr, Chr, Flu, BaP, CYC
Wood burning	BaP, Flu
Coal combustion	Ph, Flu, Py

for the concentration of PAHs in dust samples, diesels (in summer) and domestic heating (in winter) are the most responsible factors. However, the impact of PAHs in air dust samples is much higher. During the two-year observation period we noted the highest contribution of emitted PAHs during the frosty winter of 2005.

REFERENCES

Cecinato A., Marino F., Di Filippo P., Lepore L. & Possazini M. 1999. Distribution of n-alkanes, polynuclear aromatic hydrocarbons between the fine and coarse fractions of inhalable atmospheric particulates. *J. Chromatogr. A* 846: 255–264.

Chang K-F., Fang G-C., Chen J-C. & Wu Y-S. 2006. Atmospheric polycyclic aromatic hydrocarbons (PAHs) in Asia: A review from 1999 to 2004. *Environ. Pollut.* 142: 388–396.

Dobbins R.A., Fletcher R.A., Benner B.A. & Hoeft S. 2006. Polycyclic aromatic hydrocarbons in flames, in diesel fuels, and in diesel emissions. *Combustion & Flame*. 144 (4): 773–781.

Dodo G.H. & Knight M.M. 1999. Application of polydivinylbenzene liquid chromatography columns to remove lipid material from fish tissue extracts for the analysis of semivolatile organics. *J. Chromatogr. A* 859: 235–240.

Kot-Wasik A. 2004. Studies on fluorene stability in different liquid media. *Anal. Chim. Acta* 505: 289–299.

Kot-Wasik A. & Dąbrowska D. 2004a. Photodegradation and biodegradation study of benzo(a)pyrene in different liquid media *J. Photochem. Photobiol. A: Chemistry*. 168: 109–115.

Launhardt t., Strehler A., Dumler-Grald R., Thoma H. & Vierle O. 1998. PCDD/F and PAH-emission from house heating systems. *Chemosphere* 37: 2013–2020.

Olivella M.A., Ribalta T.G., de Febrer A.R., Mollet J.M. & de las Heras F.X.C. 2006. Distribution of polycyclic aromatic hydrocarbons in riverine waters after Mediterranean forest fires. *Sci. Total Environ.* 355 (1–3): 156–166.

Park S.S., Kim Y.J. & Kang C.H. 2002. Atmospheric polycyclic aromatic hydrocarbons in Seoul, Korea. *Atmospheric Environment* 36: 2917–2924.

Pereira Netto A.D., Moreira J.C., Dias A.E.X.O., Arbilla G., Ferreira L.F.V., Oliveira A. & Barek J. 2000. Evaluation of human contamination with polycyclic aromatic hydrocarbons (PAHs) and their nitrated derivatives (NHPAs): a review of methodology. *Quím. Nova*. 23 (6): 765–773.

de Pereira P.A., de Andrade J.B. & Miguel A.H. 2001. Determination of 16 priority polycyclic aromatic hydrocarbons in particulate matter by HRGC-MS after extraction by sonication. *Anal. Sci.* 17: 1229–1231.

Ravindra K., Bencs L., Wauters E., de Hoog J. & Deutsch F. 2006. Seasonal and site-specific variation in vapor and aerosol phase PAHs over Flanders (Belgium) and their relation with anthropogenic activities. *Atmos. Environ.* 40 (4): 771–785.

Smith K.E.C., Northcott G.L. & Jones K.C. 2006. Influence of the extraction methodology on the analysis of polycyclic aromatic hydrocarbons in pasture vegetation. *J. Chromatogr. A* 1116: 20–30.

Soniassy R., Sandra P. & Schlett C. 1994. *Water Analysis*, Hewlett Packard Company, Germany.

Wenzl T., Simon R., Anklam E. & Kleiner J. 2006. Analytical methods for polycyclic aromatic hydrocarbons (PAHs) in food and the environment needed for new food legislation in the European Union. *TRAC* 25: 716–725.

Wise S.A., Poster D.L., Schantz M.M., Kucklick J.R. & Sander L.C. 2004. Two new marine sediment standard reference materials (SRMs) for the determination of organic contaminants *Anal. Bioanal. Chem.* 378: 1251–1264.

Environmental Engineering – Pawłowski, Dudzińska & Pawłowski (eds)
© 2007 Taylor & Francis Group, London, ISBN13 978-0-415-40818-9

Aerosol particle concentration and the thermal conditions in a lecture room

Bernard Połednik, Andrzej Raczkowski & Sławomira Dumała
Lublin University of Technology, Lublin, Poland

ABSTRACT: This paper deals with the concentration of coarse aerosol particles and thermal conditions in a lecture room. The influence of indoor air temperature and humidity on the content of coarse aerosol particles of selected sizes is examined. The indoor air temperature, relative humidity and concentration of aerosol particles were measured continuously in an empty lecture room and in the same room occupied by students. The presented results revealed that the aerosol particle number concentration depends on indoor thermal conditions and the presence of students in the lecture room.

Keywords: Classroom aerosols, particle number concentration, indoor thermal conditions, specific enthalpy.

1 INTRODUCTION

Until now it has not been thoroughly explained how the content of aerosol particles in indoor air alters with fluctuations of indoor air temperature and humidity (Park et al. 2002, Kramer et al. 1999, Gorbunov et al. 1999). The mechanism of these changes is all the more unclear when the air temperature and humidity fluctuations result from people's presence. In the latter case, the influx of outdoor aerosols to the room is altered. The concentration of aerosol particles also changes as a result of breathing and human activities (Blondeau et al. 2005, Nazaroff 2004). People and their activities generate significant amounts of indoor aerosols, which potentially have an important influence on short-term expositions (Jankowska et al. 2004, Franck et al. 2001).

This work presents the results of coarse particle number concentration measurements in an empty lecture room and during classes under conditions of naturally changing indoor air temperature and humidity.

2 MATERIALS AND METHODS

The measurements of concentration of selected sizes of coarse aerosol particles, temperature and relative humidity of indoor air were performed in a lecture room located on the second floor of the Faculty of Environmental Engineering building of the Lublin University of Technology in Lublin, Poland. Inside the 86 m^3 cuboid room were located a lector's desk, tables and seats for 24 students. Gravitational ventilation of the room with two double pane windows was conducted by two ventilation openings (0.15×0.20 m). The particle number concentration measurements were performed using a four-channel laser counter ROYCO 243A with particle size thresholds 0.3 μm, 0.5 μm, 5 μm and 10 μm. The sample and delay time were set for 30 seconds and 15 minutes, respectively.

Continuous measurements were carried out during twelve weeks of the spring semester with different, naturally changed thermal parameters of indoor air in the alternately empty and occupied lecture room. The number of students present remained at 24 during the course of the whole experiment.

3 RESULTS AND DISCUSSION

Changes of concentration of aerosol particles in the lecture room with and without students present are shown in figure 1. The diagrams illustrate the integrated concentration changes caused by fluctuations of outdoor and indoor aerosols, as well as fluctuations caused by the presence and mobile activity of students in the room (Asmi et al. 1999). The highest amplitude changes occurred with the smallest particles.

According to the Kohler's theory the influence of indoor air temperature and humidity on the results of aerosol measurements has to be taken into consideration (Mikhailov et al. 2004). Relations between

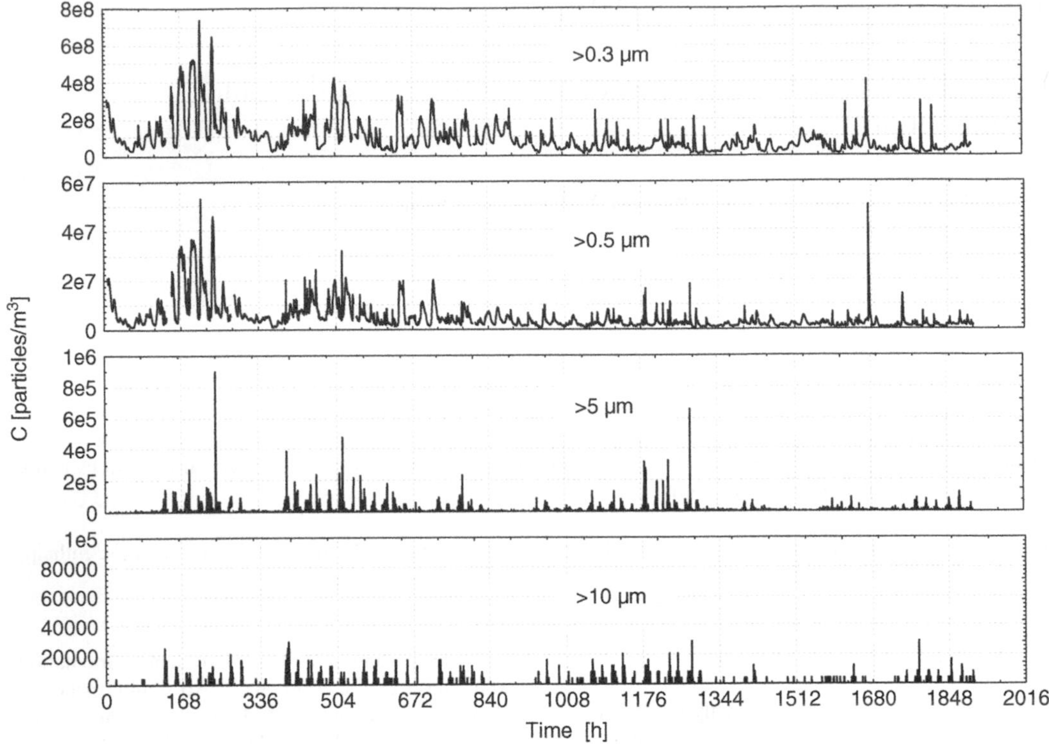

Figure 1. Changes of aerosol particle number concentration C in the lecture room air.

particle number concentration and air temperature and humidity in the unoccupied lecture room and in the presence of students are shown in the surface charts a and b in figure 2.

It can be seen that the course of concentration changes for small aerosol particles (>0.3 µm and >0.5 µm) has its minimum around 30–40% relative humidity. The concentration of these particles increases for higher and lower humidity with increasing and decreasing air temperature, respectively. To clarify these results more detailed research has to be done. Nevertheless, it could have substantial consequences when the number concentration of small particles is determined in changing indoor thermal conditions.

The content of bigger aerosol particles (>5 µm and >10 µm) is mainly influenced by the concentration of these particles in outdoor air and by the movement of students within the room (leaving and entering, presenting, etc) which can be considered as a substantial source of bulky aerosol particles in the lecture room.

The influence of indoor air thermal parameters on the concentration of measured aerosol particles in the lecture room is also shown in figure 3. Distributions of air specific enthalpy, for given aerosol particle number concentrations, in the lecture room without and with the presence of students are illustrated in diagrams a and b, respectively. The mean values of specific enthalpy h_m and the standard errors of the means *SE* are also inserted in the diagrams. The specific enthalpy was determined for the values of air temperature (rounded to nearest integer) and with the assumption that the air temperature was represented by the wet bulb temperature. This assumption, to some extent, allowed us to take into account the influence of air humidity on the results of aerosol measurements.

From the diagrams it can be seen that for all selected sizes of aerosol particles the mean values of air specific enthalpy in the lecture room without and with presence of students were about 67 kJ/kg and 71 kJ/kg and the standard errors were about 4.8 kJ/kg and 5.2 kJ/kg, respectively. This result may indicate that the quantification of aerosol particles in a given lecture room could be performed using the results of air temperature and air humidity measurements alone. Generally, it could be expected that for every initially measured

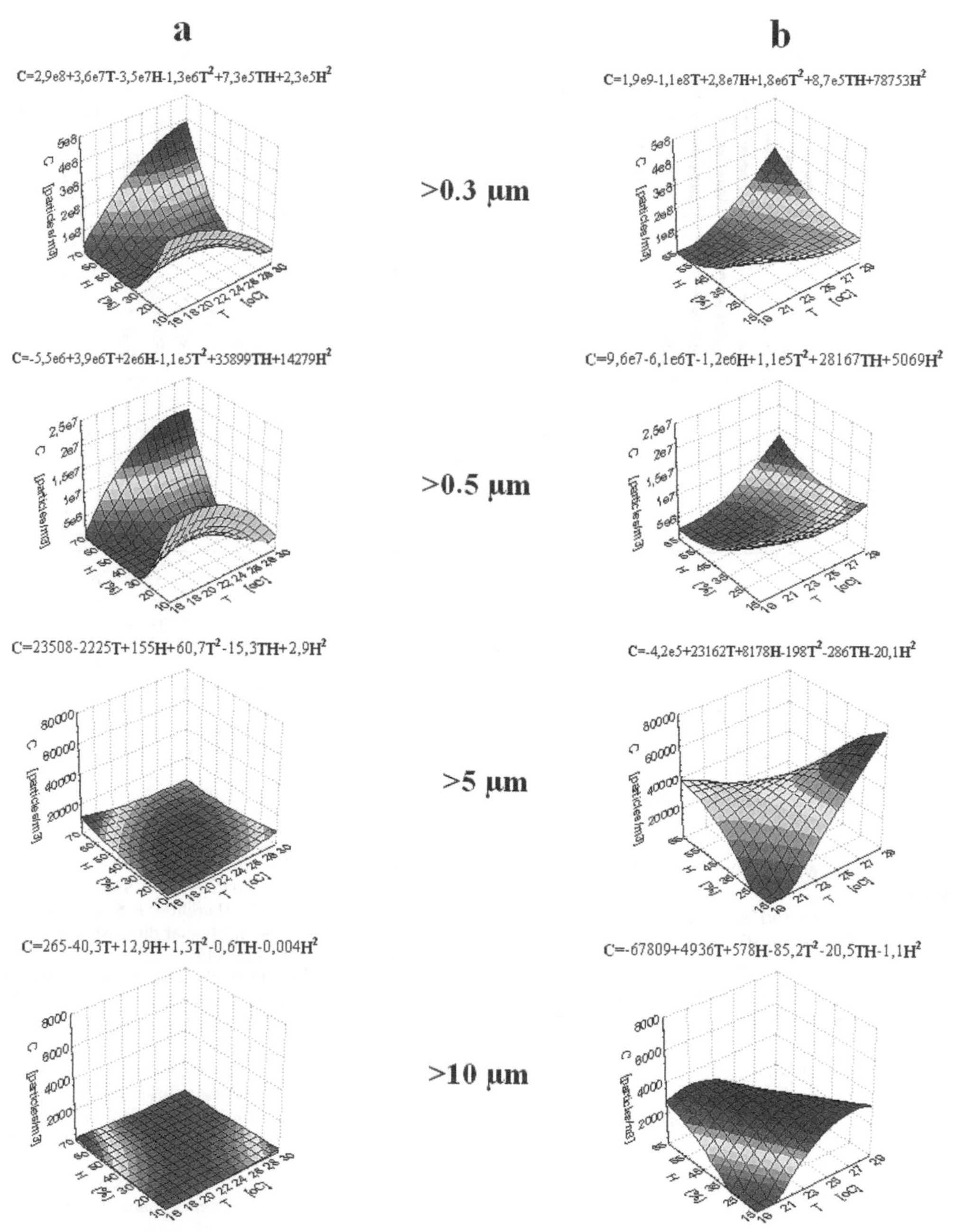

a
b
C=2,9e8+3,6e7T-3,5e7H-1,3e6T²+7,3e5TH+2,3e5H²
C=1,9e9-1,1e8T+2,8e7H+1,8e6T²+8,7e5TH+78753H²
>0.3 µm
C=-5,5e6+3,9e6T+2e6H-1,1e5T²+35899TH+14279H²
C=9,6e7-6,1e6T-1,2e6H+1,1e5T²+28167TH+5069H²
>0.5 µm
C=23508-2225T+155H+60,7T²-15,3TH+2,9H²
C=-4,2e5+23162T+8178H-198T²-286TH-20,1H²
>5 µm
C=265-40,3T+12,9H+1,3T²-0,6TH-0,004H²
C=-67809+4936T+578H-85,2T²-20,5TH-1,1H²
>10 µm
C [particles/m3]
H [%]
T [oC]

Figure 2. Relations between the indoor air temperature T, relative humidity H and aerosol particle number concentration C in the lecture room, a – without students, b – with students.

indoor environment the aerosol particle number concentration could, with predictable error, be determined simply on the basis of the thermal parameters of indoor air.

4 CONCLUSION

1. The content of aerosol particles in the lecture room air depends on the concentration of these particles

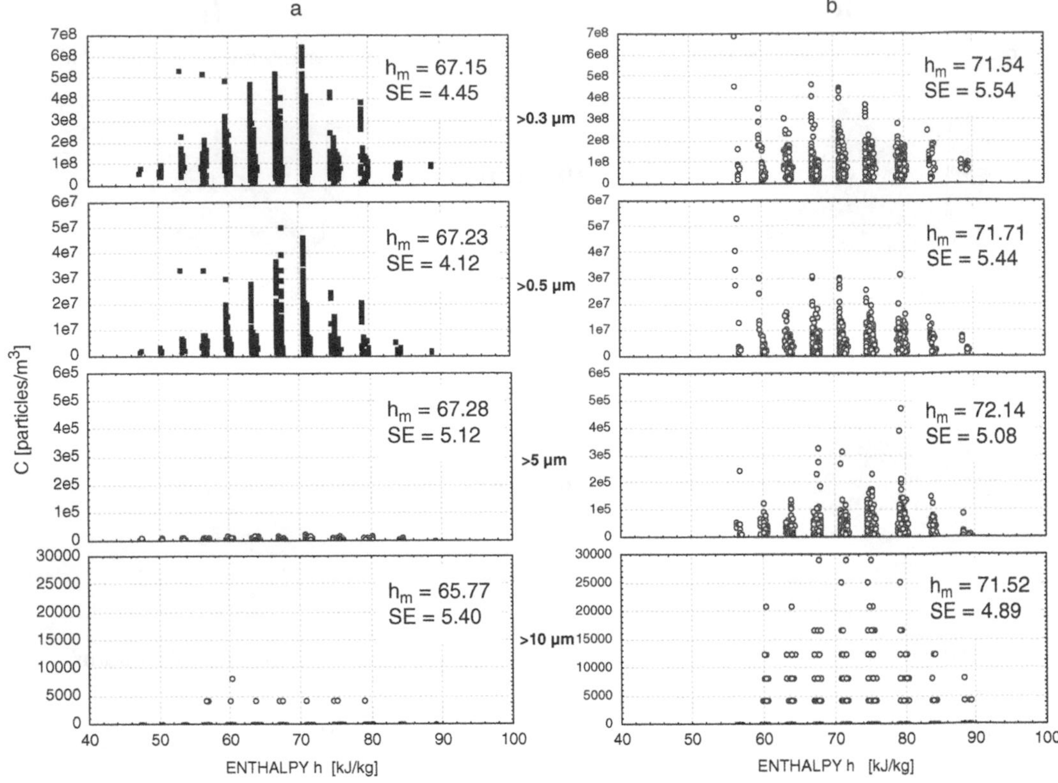

Figure 3. Distributions of specific enthalpy h in the lecture room, a – without students, b – with students.

in outdoor air and the movement of students about the room.

2. The temperature and humidity of the air in the lecture room has a significant influence on the small aerosol particle number concentration.
3. The mean values of indoor air specific enthalpy distributions are the same for all determined sizes of aerosol particles and depend on the occupancy of the lecture room.
4. The thermal conditions of the lecture room could provide a basis for prediction of the indoor air aerosol particle content.

REFERENCES

Asmi, A., Kulmala, M., Piriola, L. 1999. Modelling of indoor air aerosol size distributions with a sectional model. *Journal of Aerosol Science, Abstracts of the European Aerosol Conference*. 30: S719–S720.

Blondeau, P., Iordache, V., Poupard, O., Genin, D., Allard, F. 2005. Relationship between outdoor and indoor air quality in eight French schools. *Indoor Air*. 15: 2–12.

Franck, U., Manjarrez Colorado, M.W., Herbarth, O., Wehner, B., Wiedensohler, A. 2001. Indoor and outdoor fine particulates. *Journal of Aerosol Science, Abstracts of the European Aerosol Conference*. 1: S1077–S1078.

Gorbunov, B.,Clarke, A.G., Hamilton, R.S. 1999. Coagulation of soot particles and fractal dimension. *Journal of Aerosol Science*. 30: 445–446.

Jankowska, E., Kondej, D., Jankowski, T. 2003. Particle size distribution and aerosol concentration at office workplaces. *Journal of Aerosol Science, Abstracts of the European Aerosol Conference*. 1: 391–392.

Kramer, I., Pöschl, U., Niessner, R. 1999. Changes in the microstructure of aerosol particles after humidification. *Journal of Aerosol Science*. 30: 307–308.

Mikhailov, E., Vlasenko, S., Niessner, R., Pöschl, U. 2004. Interaction of aerosol particles composed of protein and salts with water vapor: hygroscopic growth and microstructural rearrangement. *Atmos. Chem. Phys.* 4: 323–350.

Nazaroff, W.W. 2004. Indoor particle dynamics. *Indoor Air*. 14: 175–183.

Park, S.H., Lee, K.W., Shimada, M., Okuyama, K. 2002. Change in particle size distribution of aerosol undergoing condensational growth: alternative analytical solution for the low Knudsen number regime. *Journal of Aerosol Science*. 33: 1297–1307.

Miscellaneous

Environmental Engineering – Pawłowski, Dudzińska & Pawłowski (eds)

Nitrogen retention in the Solina reservoir

Piotr Koszelnik & Janusz A. Tomaszek
Rzeszów University of Technology, Department of Chemistry and Environmental Engineering, Rzeszów, Poland

ABSTRACT: The paper presents the results of nitrogen mass balance for the Solina reservoir ecosystem. The aim of the research was to evaluate the total nitrogen retention (N_{ret}) in the reservoir. In the analysis the data from the balances conducted in the years 2001 to 2003 were juxtaposed with the earlier data obtained in the years 1999 to 2001. Since the largest amounts of nitrogen were retained in the reservoir from April to November each year, and previous studies suggested that N-retention was the lowest during winter, changes in nitrogen retention were analyzed in detail in the period from April to November. The maximum values of the nitrogen retention were always observed from mid-June to the end of August. A relationship was also noted between N_{ret} and mineral nitrogen concentration in superficial waters, and the total nitrogen load of the reservoir. The identified relationships suggest that nitrogen management on the catchment area plays a meaningful role in eutrophication of the Solina reservoir.

Keywords: Reservoir, eutrophication, nitrogen, mass-balance.

1 INTRODUCTION

Human activity has frequently exerted influence on the natural cycle of many elements in aquatic ecosystems. This is also true about nitrogen, which is one of the nutrients accelerating eutrophication (Vitousek et al. 1997, Seitzinger et al. 2002, Gonzales-Sagrario et. al. 2005). In the case of nitrogen, human interference leads to an undesirable increase in the nitrogen loads carried from rivers and streams to standing waters. Various studies conducted most intensively on the North-Atlantic watershed show that 30% to 50% of the anthropogenic nitrogen is exported to seas and oceans, where it accelerates eutrophication and causes nearly irreversible worsening of water quality (Howarth et al. 1996, Arheimer and Brendt 1998, Boyer et al. 2002). This means that at least half of the human-produced nitrogen yields to retention in lakes and reservoirs. (Billen et al. 1991, Seitzinger et al. 2002, Garnier et al. 1999). This elimination of some of the nitrogen load is positive from the point of view of the sea ecosystem, but it contributes to the decrease of inland water quality (Billen et al. 1991, Garnier et al. 1999, Tomaszek & Koszelnik 2003).

The term nitrogen retention (N_{ret}) is understood as the difference between the loads that flow into a water-body and run off it. N_{ret} is a function of many parameters; temperature, physical characteristics of the water-body (e.g. reservoir, lake) such as residence time, specific runoff, hydraulic load, as well as the general chemistry of water, availability of oxygen and especially the intensity of biochemical processes (Jensen et al. 1992, Windolf et al. 1996, Koszelnik & Tomaszek, 2003, Tomaszek & Koszelnik 2003).

The process is dependent on a large number of factors, but it is generally considered that nitrogen retention follows on from three main aquatic processes, i.e. sedimentation, uptake associated with the growth of vegetation, and denitrification. Reduction of the nitrate to N_2 plays a meaningful role in nitrogen retention in shallow and eutrophic lakes and reservoirs. A large trophy favours uptake of mineral nitrogen for the growth of water-organisms, while in deep lakes and reservoirs sedimentation of allochtonic matter may prevail (Jenssen et al. 1992, Callieri, 1993, Jansson et al. 1994, Mengis et al. 1997, Tomaszek & Czerwieniec, 2000, Seitzinger et al. 2002, Venohr et al. 2005, Wall et al. 2003 and 2005).

The goal of the present research was to determine the amount of nitrogen, which yielded to retention in the Solina reservoir during the years 1999–2003, and to perform an analysis and interpretation of parameters, which influenced the diversification of the retention.

2 MATERIALS AND METHODS

The Solina reservoir, Poland's biggest man-made lake, is situated on the upper San River in the Bieszczady

Mountains of southeastern Poland. Filled in 1968, it initially met the needs of the power industry, before also becoming a source of water supply and a center of recreation. Morphometric parameters of the reservoir are shown in Figure 1. The mean depth (22 m) and high value of the reservoir's volume to length of coastline ratio (157) ensure its immunity to degradation. The catchment area of the reservoir is characterized by a large slope, which in combination with its geological structure (bed-rock) determines the reservoir's low capacity of precipitation retention.

The greater part of the catchment area is covered by forest, and to a lesser extent by meadows and pastures. Croplands account for only a small fraction of the area. The drainage basin has a low population, with built-up areas concentrated mainly in the valleys near the mouths of rivers and streams. Industrial plants are entirely absent, while most of the resorts present have a recreational character, with leisure centers and bungalows. The tributaries of the reservoir differ one from another in their length (from 6.5 km to 92.4 km), areas of percolation (from 16.7 km^2 to 604 km^2) and mean flows (from 0.06 m^3s^{-1} to 10.4 m^3s^{-1}).

The nitrogen mass balance was conducted during the years 1999–2003. From March 1999 to September 2001 nitrogen retention was studied all year round. No work was undertaken in the winter months of 2001–2003. Samples were collected at eight points situated in estuarine sections of seven tributaries and at the dam-outflow as well as at the surface of the Solina reservoir (Fig. 1). Concentrations of nitrate, nitrite, ammonium and total nitrogen were determined in every sample using standard spectrophotometric methods (Tomaszek & Koszelnik, 2003). The flow rates at the minor inlets were calculated on the basis of the water-level gauge. Major tributaries and discharge were monitored by staff of the Water-Power Plant Solina-Myczkowce. Daily loads were expressed in mg N per sqm of the reservoir's surface area ($mgN\,m^{-2}d^{-1}$).

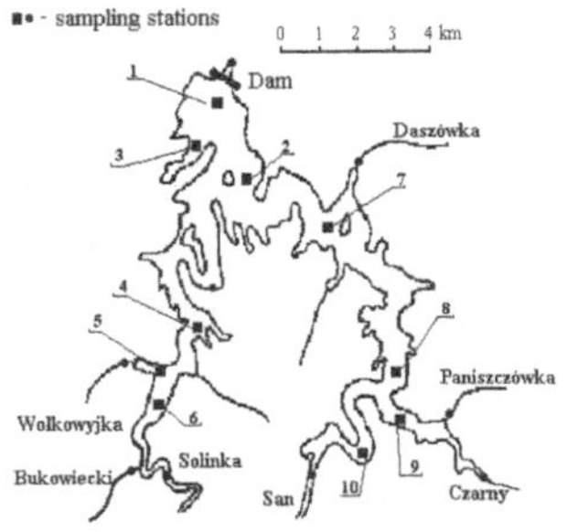

Parameters of the Solina reservoir:
Max. volume 502 x 10^6 m^3
Max. depth 60 m
Mean depth 22 m
Hydraulic retention time 215 days
Surface area 22 km^2
Length 26 km

Figure 1. Location of sampling sites and some of the physical features of the Solina reservoir.

3 RESULTS AND DISCUSSION

Figure 2 shows the changes in nitrogen retention in the Solina reservoir over the period of investigation. Outlines of the experiments along with the statistical data are presented in Table 1.

Large seasonal variations in nitrogen retention expressed as both residues in the reservoir (Fig. 2. N_{ret}, $mgN\,m^{-2}d^{-1}$) and in relation to loading (Fig. 3. $N_{ret\%}$, %) were recorded in all years. The highest amplitude was noted in the period between March and December 1999. when N_{ret} ranged between $-23.4\,mgN\,m^{-2}d^{-1}$ (September) and $411.1\,mgN\,m^{-2}d^{-1}$ (July). This constituted over 70% reduction in nitrogen loading for the maximum value of N_{ret}. The following year was characterised by a much smaller variation than the previous year (N_{ret} 39.2–285.0 $mgN\,m^{-2}d^{-1}$, $N_{ret\%}$ 16.5–67.3% of load), although the maximum was also observed in the summer. This effect most likely resulted from the climatic factor. The unusually warm spring of 2000 triggered off biological phenomena responsible for the uptake of mineral nitrogen for growth. Abnormally high values of N-retention were observed in winter, when a slow-down in the process of nitrogen accumulation in water-bodies usually takes place (Windolf et al. 1996). In 2001 maximum values of the nitrogen retention were noted in July as well; 250.0 $mgN\,m^{-2}d^{-1}$ although this time the reduction of the load was not so large (47.1%). Moreover, the standard deviation of only 7.1% suggests that the value of this parameter remained stable throughout the study. Minimum values were observed in October: 24.4%.

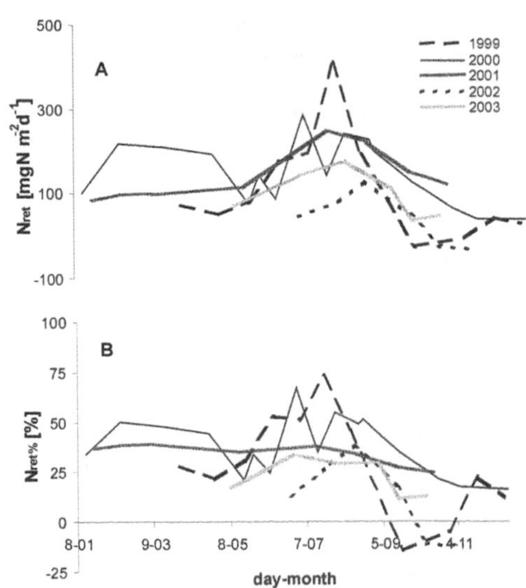

Figure 2. Seasonal variations in nitrogen retention expressed as a load (A) and in relation to loading (B).

In the following years the study was conducted with a lower frequency. The research concentrated on periods when the consumption of nitrogen should be the highest. In the year 2000, which was a very dry year, minimal hydraulic flow in tributaries was observed. This was due to the steady outflow indispensable for the proper functioning of the hydro-power plant. Those factors probably influenced the values of the N_{ret}, which were the lowest observed in the course of study.

The highest value, only 128.4 $mgN\,m^{-2}d^{-1}$ noted in August was merely 34% of the Solina reservoir's loading with nitrogen at that time. At the turn of the summer and in the autumn, when river flow was at its lowest, more nitrogen left the reservoir than flowed into it. A similar situation, caused by a similar set of circumstances, was observed in 1999. The mountainous watershed of the Solina Reservoir, which does not favour water retention, causes nitrogen loading to be contingent upon water supply which is directly proportional to the intensity of precipitation. Intensive atmospheric deposition leads to superficial supplying of water by nitrogen contained in both mineral and organic matter. In the last analysed season of 2003 N_{ret} ranged from 37.2 $mgN\,m^{-2}d^{-1}$ by the end of September to 175.9 $mgN\,m^{-2}d^{-1}$ in July. This confirmed the rule that the maximum nitrogen retention takes place during the summer.

Calculated annual mean values of both N_{ret} and $N_{ret\%}$ indicate that the most nitrogen was accumulated in the reservoir during the year 2001 and somewhat less in 2000; 152.7 $mgN\,m^{-2}d^{-1}$ and 35.5% as well as 149.7 $mgN\,m^{-2}d^{-1}$ and 29.3% respectively. Next in line are the years 2003 (98.2 $mgN\,m^{-2}d^{-1}$ and 22.4%) and 1999 (93.5 $mgN\,m^{-2}d^{-1}$ and 26.4%). The smallest values of both parameters were noted in 2002; N_{ret} 41.0 $mgN\,m^{-2}d^{-1}$ while $N_{ret\%}$ 12.4%. The climatic factor may be responsible for this quantitative and time diversity clearly observed during the years of study. An attempt to establish a relationship between nitrogen retention and hydraulic load of the reservoir did not produce the expected effect ($R = 0.32$), nevertheless good N_{ret} dependence on the inlet nitrogen load was found ($R = 0.81$, Fig. 3). In such cases the preliminary conclusions, which were drawn using considerably fewer data, were confirmed (Koszelnik & Tomaszek, 2003). On the basis of that one may say that the reduction of nitrogen loads from both point and non-point sources will be conducive to decreasing of the nitrogen retention in the Solina reservoir.

A relationship between N_{ret} and concentrations of nitrogen in superficial waters was also observed. A low level of nitrogen forms, and small variations in the concentrations of nitrogen forms were noted. The total nitrogen strength (N_{tot}) did not exceed the maximum value of 4.0 $mgN\,dm^{-3}$ (May 2003. sampling station no. 9). The sum of the nitrate, and nitrite and ammonium nitrogen (N_{miner}) as available nitrogen, averaged 60%, and ranged from 0.08 $mgN\,dm^{-3}$ (October 2002. sampling station no. 8) to 2.9 $mgN\,dm^{-3}$ (August 2003. sampling station no. 8).

Annual concentrations of both N_{tot} and N_{miner} were the highest in 2002 and the least in 1999 (table 1). A low correlation between mean concentration of total

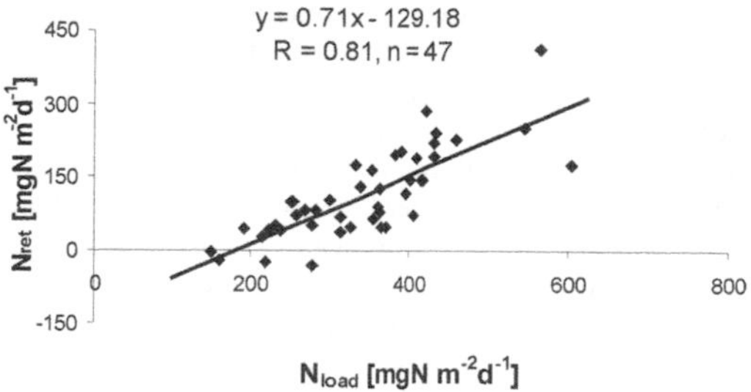

Figure 3. Relationship between nitrogen retention and load of nitrogen (for years 1999–2003).

Table 1. Nitrogen retention in the Solina reservoir during years 1999–2003. Also included are the mean total nitrogen and mean mineral nitrogen concentrations in superficial water.

Year	Months	No. of measur.	$[mgN\,d^{-1}m^{-2}]$ [%]	Mean	Min.	Max.	Std. dev.	N_{tot} $[mgN\,dm^{-3}]$	N_{miner}
1999	III–XII	11	N_{ret}	93.5	−23.4	411.1	125.2	1.92	1.31
			$N_{ret\%}$	26.4	−14.6	72.9	25.8		
2000	I–XII	16	N_{ret}	149.7	37.9	285.0	75.2	2.09	1.22
			$N_{ret\%}$	29.3	16.5	67.3	15.1		
2001	I–X	8	N_{ret}	152.7	97.9	250.0	57.0	2.03	1.24
			$N_{ret\%}$	35.5	24.4	47.1	7.1		
2002	VI–XI	6	N_{ret}	41.0	−33.0	128.4	62.0	2.45	1.43
			$N_{ret\%}$	12.4	−12.0	38.4	20.7		
2003	V–X	6	N_{ret}	98.2	37.2	175.9	55.4	2.44	1.39
			$N_{ret\%}$	22.4	12.0	34.0	9.5		

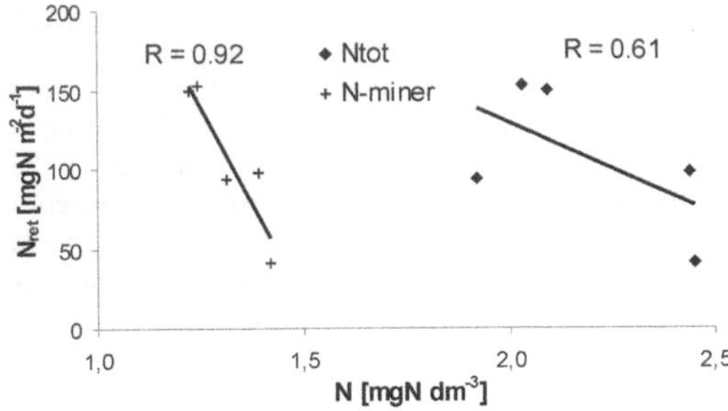

Figure 4. Relationship between N_{ret} and mean concentrations of nitrogen in superficial water.

nitrogen and mean nitrogen retention in individual years was noted (R = −0.61, Fig. 3). A higher correlation (R = −0.92. Fig. 4) was determined between N_{ret} and concentration of available nitrogen. In both cases the values of the Pearson correlation coefficient (R) testify to the inversely proportional dependence. On these grounds the conclusion was drawn that during the periods when nitrogen retention was low, its biological mechanisms, denitrification and assimilation might not have been very efficient. Consumption of ammonia and nitrate by water-organisms was so negligible that it moved the reaction equilibrium in the direction of substrate, therefore the concentrations of the mineral nitrogen in superficial waters of the Solina reservoir become somewhat higher. It is typical that very low retention was observed in the years when the reservoir's water level was at its lowest. This was connected with the low water supply and the fact that considerable parts of littoral zone, where the probability of nitrogen consumption is the greatest, were exposed.

4 CONCLUSIONS

Nitrogen retention in the Solina reservoir had seasonal character and its maximum values were observed during early summer. However, both the value of nitrogen loading held in the reservoir, and nitrogen retention changed within a relatively wide range.

Minimum values observed in the autumn are contingent on the hydrology of the tributaries. Annually 12–29% of the nitrogen run-off from the catchment was stopped in the reservoir during the duration of the project.

There is a correlation between the amount of nitrogen retention and nitrogen loading in the studied ecosystem. Therefore, a reduction in nitrogen production in the catchment area will lead to reduction in the reservoir's nitrogen retention.

Observation of the seasonal relationships between concentrations of nitrogen species in superficial waters and the value of N_{ret} shows that the process might positively influence the decreasing of concentrations. This is probably due to the periodical exposure of the high parts of the reservoir, where the mechanisms of nitrogen retention, especially denitrification, are able to use up considerable amount of mineral nitrogen.

ACKNOWLEDGMENTS

The research gained financial support from Poland's State Committee for Scientific Research.

REFERENCES

Arheimer, B. & Brendt, M. 1998. Modelling nitrogen transport and retention in catchments. *Ambio* 27(6): 378–386.

Billen, G., Lancelot, C. & Meybeck, M. 1999. N, P and Si Retention along the Aquatic Continuum from Land to Ocean, In: R.F.C. Mantoura, J.-M. Martin & R. Wollast (ed.), *Ocean Margin Processes in Global Change:* 19–44. New York: John Wiley & Sons Ltd.

Boyer, E.W., Goodale, C.L., Jaworski, N.A. & Howarth, N.W. 2002. Anthropogenic nitrogen sources and relationship to riverine nitrogen export in the northern USA. *Biogeochemistry:* 57/58: 137–169.

Callieri, C. 1997. Sedimentation and aggregate dynamics in Lake Maggiore, a large, deep lake in Northern Italy. *Mem. Ist. Ital. Idrobiol.* 56: 37–50.

Garnier, J., Leporcq, B., Sanchez, N. & Philippon, X. 1999. Biogeochemical mass balances (C, N, P, Si) in three large reservoirs of the Seine Basin (France). *Biogeochemistry* 47(2): 119–146.

Gonzales-Sagrario, M.A., Jeppesen, E., Goma, J., Søndergaard, M., Jensen, J.P., Lauridsen, T. & Landkildehus, F. 2005. Does high nitrogen loading prevent clear-water conditions in shallow lakes at moderately high phosphorus concentrations? *Freshwater Biology* 50: 27–41.

Howarth, R., Billen, G. & Swaney, D. 1996. Regional nitrogen budgets and riverine N and P fluxes for the drainages to the north Atlantic Ocean: natural and human influences. *Biogeochemistry* 35(1): 75–139.

Jansson, M., Andersson, R., Berggren, H. & Leonardson, L. 1994. Wetlands and lakes as nitrogen traps. *Ambio* 23(6): 320–325.

Jensen, J.P., Jeppesen, E., Kristensen, P., Christensen, P.B. & Søndergaard, M. 1992. Nitrogen Loss and denitrification as studied in relation to reductions in nitrogen loading in a shallow hypertrophic Lake Sobygard, Denmark. *Int. Revue ges. Hydrobiol.* 77(1): 29–42.

Koszelnik, P. & Tomaszek, J.A. 2003. The influence of reservoir morphometry on nitrogen retention, In: C.A Brebbia (ed.) *River Basin Management II, Progress in Water Resources* 7: 261–269. Southampton: Witt Press.

Mengis, M., Gächter, R., Wehrli, B. & Bernasconi, S. 1997. Nitrogen elimination in two deep eutrophic lakes. *Limnol. Oceanogr.* 42(7): 1530–1543.

Seitzinger, S.P., Styles, R.V., Boyer, E.W., Alexander, R.B., Billen, G., Howarth, R.W., Mayer, B. & van Bremer, N. 2002. Nitrogen retention in rivers: model development and application to watersheds in northern USA. *Biogeochemistry* 57/58: 199–237.

Tomaszek, J.A. & Czerwieniec, E. 2000. *In situ* chamber denitrification measurements in reservoir sediments: an example from southeast Poland. *Ecol. Engineering* 16: 61–71.

Tomaszek, J.A. & Koszelnik, P. 2003. A simple model of nitrogen retention in reservoirs. *Hydrobiologia* 504(1/3): 51–58.

Vitousek, P.M., Aber, J.D., Howarth, R.W., Likens, G.E., Matson, P.A., Schindler, D.W., Schlesinger, W.H. & Tilman D.G. 1997. Human alteration of the global nitrogen cycle: sources and consequences. *Ecol. Appl.* 7: 737–750.

Venohr, M., Donohue, I., Fogelberg, S., Arheimer, B., Irvine, K. & Behrend, H. 2005. Nitrogen retention in a river systems and effects of river morphology and lakes. *Wat. Sci. Tech.* 51(3–4): 19–29.

Wall, L.G., Tank, J.L., Royer, T.V., Kemp, M.J. & David, M.B. 2003. Can a reservoir act as a nitrogen sink? The role of denitrification in NO_3-retention within an agriculturally-influenced reservoir. *Bulletin of the North American Benthological Society* 20: 155.

Wall, L.G., Tank, J.L., Royer, T.V. & Bernot, M.J. 2005. Spatial and temporal variability in sediment denitrification within an agriculturally influenced reservoir. *Biogeochemistry* 76: 85–111.

Windolf, J., Jeppesen, E., Jensen, J.P. & Kristensen, P. 1996. Modelling of seasonal variation in nitrogen retention and in-lake concentration. A four-year mass balance study in 16 shallow Danish lakes. *Biogeochemistry* 33: 25–44.

Environmental Engineering – Pawłowski, Dudzińska & Pawłowski (eds)
 ISBN13 978-0-415-40818-9

Nickel and cobalt in bottom sediments of rivers in the upper Narew river basin

Elżbieta Skorbilowicz
Institute of Civil Engineering, Technical University of Bialystok, ul. Wiejska, Białystok

ABSTRACT: The aim of study was to evaluate the anthropogenic influence on the occurrence in nickel and cobalt in the bottom sediments of larger rivers in the basin of the upper Narew river. Samples of bottom sediments were taken once in 2003 at the edge opposite the main stream at sites where suspended material tends to be sedimented. Grains of less than 0.2 mm diameter were selected from bottom sediments using polyethylene sieves. After sample dissolution in nitric acid applying a microwave digester, the total contents of nickel and cobalt were determined in selected fractions by means of the AAS technique. Man's economic and living activities are the sources of nickel and cobalt deposited in bottom sediments of larger rivers. The bottom sediments of water-courses flowing through farming and drainage areas are richer in nickel and cobalt ions than water-courses in the river basins with large areas of forest used in farming.

Keywords: Heavy metals, rivers, sediments bottom.

1 INTRODUCTION

Deterioration of water quality as a result of anthropopressure is often connected with the enrichment of natural ecosystems with heavy metals (Chen et al. 2000, Tam and Wong 2000, Baptista Neto et al. 2000, Xiangdong et al. 2000, Tsail et al. 2003, Bojakowska 2003, Wiśniowska-Kielian & Niemiec 2004). Fine-grained, slime–silt sediments capture and deposit significant amounts of those elements in water. In the sediments deposited in the bottom metals are accumulated (Kwapuliński et al. 1993) The chemical analysis of sediments is a valuable source of information of the previous and current state of water environment quality (Rybosz-Maslowska et al. 2000, Ciszewski 2003).

Nickel easily enters fairly permanent chelate compounds and complex cations and anions which are not very mobile in bottom sediments at pH > 5.0. It also easily undergoes bioaccumulation especially in phytoplankton, which results in its quick introduction to the food chain, posing a threat to human and animal health. Cobalt, like nickel in the natural water environment, is not very stable in soluble form because it is firmly bound by the clay and hydroxide (Fe and Mn) fraction of bottom sediments as well as by phytoplankton (Kabata-Pendias 1993).

The aim of the investigation was to determine the total content of nickel and cobalt in bottom sediments of the rivers in the upper Narew river basin. It also aspired to show the relation between the type of river basin and the quantity of the investigated elements in bottom sediments.

2 MATERIALS AND METHODS

The upper Narew river basin is covered in Quaternary Pleistocene formations which consist of glacial formations of various facies and periglacial and fluvial formations. These are glacial water-glacial clays, silt, sands and gravel of central and northern Poland glaciation and peat, clays and sands of marginal lake and fluvial accumulation. River valleys are filled by the youngest Holocene formations: fen soils, peats, clays, gyttja and dune sands. The upper Narew river basin is agricultural and industrial in character. The food-processing industry predominates. This industry is mainly located near urban areas, especially near Białystok. The area of the tributary river basins range between 20 and 1856 km^2. The investigations involved the Narew river in ten test sections from Bondary to Tykocin, its minor and major tributaries, including the Supraśl with its tributaries. For each tributary two up to seven test sections were selected. Considering the type of river basins of the Narew's tributaries (vegetation, land use, point sources of pollution) the analysed rivers were divided into three groups: rivers in farming drainage basins, the ones covered mostly in forests and arable land. In the first group (drainage basins where arable land predominates) there are the following tributaries of the Narew: the Małynka, Olszanka,

Ruda, Łoknica, Krzywczanka, Strabelka, Czarna, Mieńka, Awissa, Turośnianka, Czaplinianka, Sokołda and Łanga. The following rivers have drainage basins where forests predominate (the second group): the Narewka, Płoska, Bakinówka, Świnobródka, Słoja, Starzynka, Poczepówka, Świdziałówka, Czarna – the tributary of the Supraśl, Czarna Rzeczka, Jurczycha, Krzemianka. The following rivers, which have point sources of pollution, are in the third group: the Biała – the tributary of the Supraśl, Dolistówka, Bażantarka, Rudnia, Orlanka, Biała – the tributary of the Orlanka, Nereśl, Targonka, Jaskranka and Horodnianka. The bottom sediments of the Narew and the Supraśl were analysed separately.

Samples of bottom sediments were collected for tests once in 2003, in the river bank zone, facing the current at locations where suspended material is deposited. The grain fraction <0.2 mm was isolated from sediments with a polyethylene sieve. After the samples were mineralized in nitric acid with a microwave mineraliser MARS 5 the total amount of nickel and cobalt was determined using ASA. The reaction was determined with a potentiometric method.

3 DISCUSSION OF THE RESULTS

The distribution of nickel and cobalt content in water sediments in Poland resembles its variability of these elements in soil. The concentration of nickel in sediments in the unpolluted areas remains below $10\,mg \cdot kg^{-1}$ (Lis Pasieczna 1995, Kabat-Pendias Pendias 1999). Our tests on the sediments of the upper Narew river basin showed that the concentration of nickel ranges from 1.2 to $21.4\,mg \cdot kg^{-1}$ (Table 1). Lis and Pasieczna (1995) quote the value of the median for nickel at 6 mg kg for surface water sediments in Poland. Using this content as the criterion it was found that in seventy two samples (54%) of sediments the concentration of nickel was lower than 6 mg kg, including twenty six results in the range $1.2–4\,mg \cdot kg^{-1}$. Concentrations above $6\,mg \cdot kg^{-1}$ were observed in sixty one samples (46%), including 1/3 samples (20) with a nickel content $>10\,mg \cdot kg^{-1}$. Considering the results in the light of natural geochemical content of nickel quoted by Bojakowska and Sokołowska (1998) 61% of sediment samples were labelled as contaminated. Nevertheless, it should be noted that even the highest content of nickel is still lower than $68\,mg \cdot kg^{-1}$. Turekian and Wedephol (1961) quote this value as the geochemical background.

The concentration of cobalt in the bottom sediments of rivers of the upper Narew drainage basin was within the range of $1.1–20.5\,mg \cdot kg^{-1}$ (Table 1). The geochemical background according to Turekian and Wedephol (1961) is $19\,mg \cdot kg^{-1}$. Using this criterion it was found that as many as 131 sediment samples (98%) can be labelled as uncontaminated. Lis and Pasieczna (1995) calculated that the median for cobalt for sediments of surface water in Poland is $3\,mg \cdot kg^{-1}$. Analysing the results it was found that in 124 samples (93%) the content of cobalt was higher than 3 mg kg, including 34 samples with cobalt concentrations higher than $10\,mg \cdot kg^{-1}$. The concentration $<3\,mg \cdot kg^{-1}$ was observed in only nine samples (7%). Work on the sediments in the Supraśl and the upper Narew also showed that there was more cobalt in the sediments of the Supraśl and more nickel in the sediments of the Narew. The median for cobalt was 7.5 mg kg in the Supraśl and for nickel $2.9\,mg \cdot kg^{-1}$. In the Supraśl the largest quantities of the investigated elements was found at the Fasty control point, past the Biała estuary (Co $10.1\,mg \cdot kg^{-1}$, Ni $5.8\,mg \cdot kg^{-1}$). The Biała, a tributary of the Supraśl, no doubt, affected the increased content of the investigated elements. It should be noted the majority of the Biała drainage basin is composed of urban areas, which is why its waters are permanently threatened with sewage dumping. Among others it includes dumping treated sewage, waters of the Dolistówka, the Bażantarka and other minor tributaries as well as pollution from agriculture and industry. The median for cobalt in the sediments of the Narew was $5.0\,mg \cdot kg^{-1}$ while for nickel $4.4\,mg \cdot kg^{-1}$. The largest content of cobalt ($10.9\,mg \cdot kg^{-1}$) in the Narew was found at Bondary, while the largest content of nickel ($7.6\,mg \cdot kg^{-1}$) was found at Złotoria control point, past the Supraśl estuary which carries a large amount of impurities into the Narew.

Considering the content of nickel and cobalt in sediments in rivers with different types of drainage basin use, it was found that the highest concentration of the investigated elements involved sediments in rivers whose drainage basins were mostly affected by sewers and farming. The median of nickel in this group of rivers was $6.7\,mg \cdot kg^{-1}$ and cobalt $10.4\,mg \cdot kg^{-1}$. The largest quantity of nickel was found in the Horodnianka ($21.4\,mg \cdot kg^{-1}$) flowing through an area in the vicinity of the Hryniewicze municipal dumping site. The most polluted river, considering the content of cobalt ($20.5\,mg \cdot kg^{-1}$), was the Bażantarka, a small water-course flowing through the city of Białystok. Nickel and cobalt enter the water along with waste, domestic and industrial sewage, territorial run-off and the source of metals in the run-off are dry and wet atmospheric deposits. Sewers are a source of practically all types of micro pollution identified in surface water. Many sections of rivers in with the Podlasie Province are polluted as a result of dumping untreated or inadequately treated domestic sewage and sewage from the food-processing industry. The conducted investigations into sediments in the upper Narew river basin showed that in the second group of

Table 1. Results of tests into grain fraction (<0.2 mm) in sediments of some rivers in the upper Narew river basin.

Rivers	Statistical parameters	Location	Content, mg · kg^{-1} s.m.		Reaction in H_2O
			Ni	Co	
The Narew N = 10		1. Bondary	5.1	10.9	7.2
		2. Narew	3.1	6.5	6.7
		3. Ploski	2.2	5.2	7.1
		4. Doktorce	4.4	2.7	7.2
		5. Uchowo	3.3	5.1	7.3
		6. Bokiny	3.5	3.1	7.1
		7. Rzędziany	5.2	4.9	6.9
		8. Złotoria	7.6	5.6	7.1
		9. Siekierki	4.3	3.9	7.2
		10. Tykocin	4.4	3.8	7.5
	Arithmetic mean		4.3	5.2	
	Deviation		1.5	2.3	0.2
	Min–max		2.2–7.6	2.7–10.9	6.7–7.5
	The median		4.4	5.0	
The Supraśl N = 7		1. Mościska	1.2	8.3	7.1
		2. Michałowo	1.9	7.5	7.5
		3. Zarzeczany	2.3	4.8	7.6
		4. Gródek	3.1	5.2	7.2
		5. Supraśl	3.7	5.1	8.1
		6. Wasilków	2.9	8.6	7.9
		7. Fasty	5.8	10.1	7.4
	Arithmetic mean		3.0	7.1	
	Deviation		1.5	2.1	0.4
	Min–max		1.2–5.8	4.8–10.1	7.1–8.1
	The median		2.9	7.5	
River basins where arable land predominates N = 47 Group I	Arithmetic mean		6.6	5.9	
	Deviation		2.7	2.6	0.4
	Min–max		3.4–16.2	1.7–13.9	5.9–7.9
	The median		5.8	5.8	
River basins where forests predominate N = 30 Group II	Arithmetic mean		6.9	7.5	
	Deviation		3.0	2.7	0.7
	Min–max		2.9–13.6	2.1–13.1	5.4–8.1
	The median		5.8	7.6	
Sewage and farming river basins N = 39 Group III	Arithmetic mean		7.8	10.4	
	Deviation		4.3	4.2	0.5
	Min–max		1.9–21.4	1.1–20.5	5.3–8.8
	The median		6.7	10.4	

rivers with drainage basins with majority of forests the median of nickel was 5.8 mg · kg^{-1} while the median of cobalt was 7.6 mg · kg^{-1}. The highest concentration of nickel 13.6 mg · kg^{-1} was found in sediments of the Swinobródka and cobalt (13.1 mgCo · kg^{-1}) in the Narewka. These are rivers flowing through the area covered almost entirely in forests. Rivers with large areas of forest have far more humus substances from forest soils (Gorniak & Zieliński 1998), and metals, in most cases, migrate into water from the land environment as mineral-organic complexes.

The results show that the lowest values of nickel and cobalt were found in river basins where arable land predominates, the median of nickel and cobalt was 5.8 mg · kg^{-1}. The reaction of the investigated sediments (Table 1) was neutral and slightly alkaline which is typical of fluvial sediments.

The investigation conducted into the sediments of the upper Narew river basin showed that the content of nickel and cobalt depends on the geochemical structure of the area and land use. The differences in the content of nickel and cobalt in the sediments of the investigated rivers resulted above all from the type of the river basin that surrounds it. It seems that the sediments of these water-courses are genetically closely related with the material of the drainage basins of more directly analysed rivers. Exceeding the geochemical background for nickel and cobalt in bottom sediments of the upper Narew river basin in most cases indicated point sources of pollution, mainly sewage dumping from areas not having any water-treatment facilities.

4 CONCLUSIONS

The current work shows that there was more cobalt in the sediments of the Supraśl and more nickel in the bottom sediments of the upper Narew.

The bottom sediments of water-courses flowing through the farming and drainage areas are richer in nickel and cobalt ions than the water-courses in the river basins with large areas of forest used in farming.

REFERENCES

Bojakowska, I. & Sokołowska, G. 1998. *Geochemiczne klasy czystości osadów wodnych*, Przeg. Geolog., 46, 1, 49–54.

Bojakowska, I. 2003. *Charakterystyka geochemiczna osadów Narwi i jej dopływów*, w: Mat. Konf.: Zagospodarowanie zlewni Bugu i Narwi w ramach zrównoważonego rozwoju, Warszawa – Popowo 23–24 maja, 137–146.

Baptista Neto, J.A., Smith, B.J., McAllister, J.J. 2000. *Heavy metal concentrations in surface sediments in a nearshore environment, Jurujuba Sound, Southeast Brazil*, Environmental Pollution, 109, 1–9.

Ciszewski, D. 2003. *Heavy metal in vertical profiles of the middle Odra River overbank sediments: evidence for pollution changes*. Water, Air, Soil Pollut., 143: 81–98.

Chen, J.S., Wang, F.Y., Li, X.D. & Song, J.J. 2000. *Geographical variations of trace elements in sediments of the major rivers in eastern China*, Environmental Geology, Springer-Verlag, 39 (12) November, 1334–1340.

Górniak, A. & Zieliński, P. 1998. *Wpływ lesistości zlewni na zawartość wód rzecznych województwa Biłostocego*, Przeg. Naukowy, SGGW, Warszawa, 16, 231–241.

Kabata-Pendias, A. 1993. *Biogeochemia chromu, niklu i glinu*, Chrom, nikiel i glin w środowisku – problemy ekologiczne i metodyczne, Zesz. Nauk PAN Kom. Człowiek i Środowisko, 9–14.

Kabata-Pendias, A. & Pendias, H. 1999. *Biogeochemia pierwiastków śladowych*, PWN, Warszawa, 364.

Lis, J. & Pasieczna, A. 1995. *Atlas geochemiczny Polski w skali 1: 2 500 000*, Państw. Inst. Geol., Warszawa, 72.

Kwapuliński, J., Wiechuła, D., Cebula, J., Bagier, M., Loska, K. & Szylman, E. 1993. *Specjacja niklu w osadzie dennym*, Chrom, nikiel i glin – problemy ekologiczne i metodyczne, Zesz. Nauk PAN Kom. Człowiek i Środowisko, 167–169.

Rybosz-Masłowska, S., Moraczewska-Majkut, K. & Krajewska, J. 2000. *Metale ciężkie w wodzie i osadach dennych zbiornika w Kozłowej Górze na górnym Śląsku*, Archiwum Ochrony Środowiska, 26: 127–140.

Tam, N.F.Y. & Wong, Y.S. 2000. *Spatial variation of heavy metals in surface sediments of Hong Kong mangrove swamps*, Environmental Pollution, 110, 195–205.

Tsail, J., Yu, K.C. & Ho, S.T. 2003. *Correlation of iron/iron oxides and trace heavy metals in sediments of five rivers, in southern Taiwan*, Diffuse Pollution Conference, Dublin, 14–25.

Xiangdong Li, Zhenguo Shen, Onyx, W.H. & Wal, Yok-sheung Li, 2000. *Chemical partitioning of heavy metal contaminants in sediments of the Pearl River Estuary*, Chemical Speciation and Bioavailability, 12(1), 17–25.

Wiśniowska-Kielian, B. & Niemiec, M. 2004. *Zawartość metali w osadach dennych rzeki Dunajec*, w: III Międzynarodowa Konferencja Naukowa "Toksyczne substancje w środowisku" Kraków, 7-8.09, 63.

Environmental Engineering – Pawłowski, Dudzińska & Pawłowski (eds)
© 2007 Taylor & Francis Group, London, ISBN13 978-0-415-40818-9

Fractionation of mercury in sediments of the Warta River (Poland)

Leonard Boszke
Department of Environmental Protection, Collegium Polonicum, Adam Mickiewicz University, Kościuszki Słubice, Poland

Artur Kowalski & Jerzy Siepak
Department Water and Soil Analysis, Adam Mickiewicz University, Drzymały Poznań, Poland

ABSTRACT: The paper reports results on the fractionation of mercury in relatively uncontaminated sediments of the Warta River, where the average mercury concentration in bulk sediments was $136 \pm 74\,\mathrm{ng\,g^{-1}}$ dry mass (range 57–340). The method of sequential extraction applied permits separation and determination of the following fractions of mercury: organomercury species, mobile fraction, acid soluble mercury species, mercury bound to humic acids and mercury bound to sulphides. Among the fractions separated, the largest were: mercury bound to sulphides ($58 \pm 17\%$, range 20–81%), humic matter ($23 \pm 9\%$, range 4–36) and organomercury compounds ($17 \pm 18\%$, range 0.03–65). The smallest fractions were water-soluble mercury ($2.1 \pm 0.9\%$, range 1.1–3.8) and acid soluble mercury ($0.4 \pm 0.1\%$, range 0.2–0.7).

Keywords: Mercury, Fractionation, Sequential extraction, Warta river, Poland.

1 INTRODUCTION

Recognition of mercury transformations and migration in the natural environment and development of adequate methods for its remediation require not only determination of the total mercury content but also the fractions of its different species (Boszke et al. 2002, 2003, Głosińska et al. 2001, Jackson, 1998a, Templenton et al. 2000, Thöming et al. 2000). Direct measurements of the concentration of particular mercury species in environmental samples are usually difficult or even impossible. Such methods as extended X-ray absorption fine structure spectroscopy (EXAFS) and wave dispersive X-ray microprobe spectroscopy (WD/XMP) permit direct identification of mercury species present in a given sample, but they are characterised by relatively high limits of detection which makes them only really useful in the speciation analysis of highly polluted samples having high total mercury concentrations (Gustin et al. 2002, Hesterberg et al. 2001, Kim et al. 2000, Sladek et al. 2002). Another approach used in the speciation analysis is based on mercury evaporation in a gradient of temperatures – SPTD (Solid Phase Thermo Desorption). In this method, different mercury species are converted to volatile forms at certain temperatures. Results obtained by mercury pyrolysis provide information about thermally related bonding strength and but do not allow any estimation of mercury mobility in the liquid phase (Biester & Scholz, 1997; Biester et al. 2002, Gustin et al. 2002, Sladek et al. 2003).

Another method is to subject solid-state matrices such as soil and sediment to sequential extraction by solutions of increasingly strong complexing ability (Calmano, 1983, Hall et al. 1996, Jackson, 1988, 1991, 1998a, b, Lechler et al. 1997, Tessier et al. 1979). This approach is based on the assumption that a given solution causes extraction of a certain operationally defined fraction of mercury species. The fraction includes a number of different but similar chemical species, as none of the solutions is selective. Although the extraction methods are based on chemical interactions in complex matrices and therefore yield results that are strongly influenced by matrix effect, they have the advantage of permitting determination of mercury species at relatively low concentrations (Biester & Scholz, 1997, Biester et al. 2002, Gustin et al. 2002, Sladek et al. 2002).

Metal species are traditionally classed in two broad categories based on the strength of the bonds with which they are attached to the matrix. The first group includes species so weakly bound to the solid phase that they can easily be replaced by others under natural conditions (e.g. metal cations can be replaced by other cations, including protons). This fraction is regarded in the conventional sequential extraction as the water-soluble, exchangeable, and carbonate bound fraction. In the classical sequential extractions, the metal species bound to oxides and oxyhydroxides, and those bound to organic matter are also called "bioavailable". For example, oxyhydroxides can be dissolved during reduction and organic matter can be decomposed, releasing the bound metals. The term "bioavailablity" is often regarded as synonymous with potential toxicity to organisms or potential ease of uptake by organisms. The second group comprises the metal species that are strongly bound to aluminosilicates and can be released only when the solid phase is destroyed. In the classical sequential extractions, the metal species bound to aluminosilicates are called "biounavailable" This term alludes to the fact that, because of their chemical properties in natural conditions, these species do not release free metal ions (Calmano, 1983, Jackson, 1988, 1991, 1998a, b, Templenton et al. 2000, Tessier et al. 1979).

According to the list of "Hard and Soft Acids and Bases" (Pearson, 1963), mercury being a "soft" metal, forms ligands with "soft" ligands such as sulphide and thiol groups. These complexes are characterised by stability constants a few orders of magnitude greater than those of the complexes of other metals. Consequently, mercury shows great affinity to organic matter rich in sulphur-containing ligands (thiol groups). Its complexes with sulphur-containing groups are so strong that they should be classified under the "biounavailable" fraction as conventionally defined (Barnett & Turner, 2001, Bloom et al. 2003, Lechler et al. 1997, Templeton et al. 2000). Mercury complexes formed with oxygen-containing compounds, such as those occurring on the surfaces of clayey minerals, oxides and oxyhydroxides are so weak that they should be classified as representing the relatively "bioavailable" fraction. Although HgS is strongly bound with the sediments it can be partly dissolved as a result of transformations involving bacteria or under the effect of oxidising conditions appearing as a consequence of bioturbation (Gobeil & Cossa, 1993, Hall et al. 1996, Gagnon et al. 1996). Moreover, the "bioavailable" fraction should include the water-soluble mercury species under the conditions of natural pH and salinity, e.g. when dissolved, Hg(II) is largely in the form of $HgCl_2$. This group of mercury species is described as mobile or relatively mobile and is thought to consist of ionic or electrically neutral for molecular weight species capable of being released from the solid phase into the ambient water. The term "biounavailable" fraction should refer to the compounds in which mercury is permanently bound under natural conditions, e.g. HgS. As follows from the above, the mercury species adsorbed on clay minerals or metal hydroxides or in inclusions in crystalline aluminosilicates are much less important than the analogous species of the other heavy metals (Barnett & Turner, 2001, Bloom et al. 2003, Gomez Ariza et al. 2000, Hall et al. 1996, Lechler et al. 1997, Templeton et al. 2000).

2 MATERIALS AND METHODS

2.1 *Sample collection and preparation for analysis*

Samples of sediments were collected from different sites of the Warta river in 2004 (Fig. 1). The samples were collected at the straight sections of the river course, at a distance of about 3 meters from the river bank and the depth of water at the collecting site varied from 50 to 100 cm. Sediment samples weighing ca. 2 kg were collected from the shore using a manual stainless steel scoop (0–20 cm) from each site. The samples were placed in a plastic vessel and transported to the lab where they were dried for a few weeks at room temperature in dust-free room to constant mass. The sample was then gently crushed in an agate mortar so as not to damage the structure of grains. Of these, portions of 250 g were collected and sieved through a copper sieve of mesh size 0.150 mm.

2.2 *Reagents and apparatus*

The analytical agents used for mercury determination were of highest available purity, made by Merck (Darmstadt, Germany), all dilutions were made with deionised water from Milli-Q (Milli-Q system, Millipore France), deaerated by bubbling with high purity

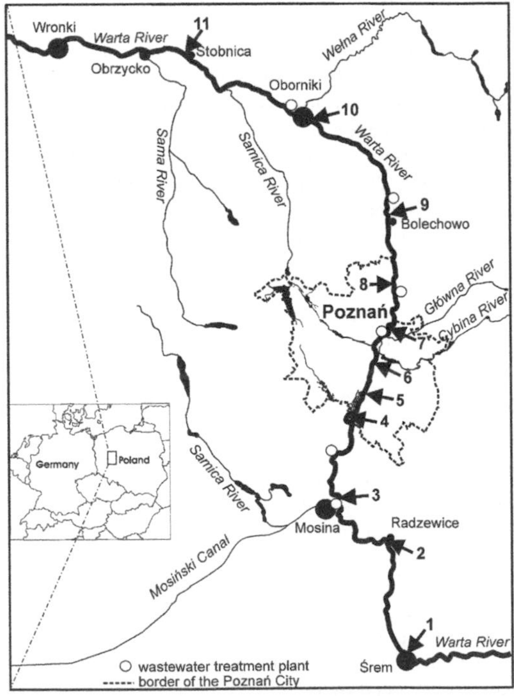

Figure 1. Sampling sites.

argon (99.99%) for about 12 hours. Laboratory vessels used were made of boron-silica glass of the highest quality. Before using in analyses the vessels were washed with a detergent, rinsed with tap water and rinsed three times with deionised water. Then the vessels were flooded with deionised water for about two hours and in an aqueous solution of ultrapure nitric acid (1:1 v/v) for 48 hours. After 48 hours the vessels were rinsed three times with deionised water, and left filled with deionised water for 24 hours. After drying, the vessels were kept in double, tight foil bags.

2.3 *Sequential extraction*

After reviewing the literature covering the determination of various species of mercury in solid samples, the following sequential extraction method was established (based partially on the work of Wallschläger et al. (1998a, b), Lechler et al. (1997), Bloom & Katon (2000), Renneberg & Dudas (2001)):

2.3.1 *Organomercury fraction (F1)*

A portion of 5.0 g of the sample was placed in a teflon centrifuge tube (50 mL). 30 mL of chloroform was added and the whole was mounted on a rotating shaker for three hours. Then, the tube was centrifuged for 15 minutes at 3000 rpm. The extract was filtered at the filtrating unit through cellulose acetate filters (diameter: 0.45 μm). The extraction was repeated with the next portion of chloroform (30 mL). The chloroform phase was poured into a 250 mL separator. The sediment left on the filter and in the flask were placed in a quartz vessel and dried in air in preparation for the next stage of sequential extraction (Fig. 2). Aqueous sodium thiosulphate solution (0.01 M; 10 mL) was placed in the separator to extract organomercury compounds from the chloroform. After shaking for three minutes, the separator was left to stand until separation of the organic and inorganic layers was complete (ca. 20 min.). From the sodium thiosulphate layer a portion of 5 mL of the solution was collected, placed in a measuring flask (50 mL) and treated with 20 μL 65% HNO_3. Then, 7.5 mL 33% HCl and 5 mL of a 1:1 solution of 0.0033 M $KBrO_3$/0.2,M KBr were added to the flask to oxidize all mercury species to Hg(II). Next, 30 μL 12% $NH_2OH \times HCl$ was added to the flask to remove free bromine. The measuring flask was filled with deionised water to the mark.

2.3.2 *Water-soluble fraction (F2)*

The dried sample obtained in stage 1 (F1) was placed in a teflon centrifuge tube for of 50 mL adding 30 mL of deionised water. Then the sample was shaken in a rotating shaker for 3 hours and centrifuged for 15 minutes at 3000 rpm. The extract was filtered off by a filter of cellulose acetate of 0.45 μm in diameter. A portion of 10 mL of the solution was collected and placed in a measuring flask of 50 mL in capacity, to which also 7.5 mL 33% HCl, 1 mL of the 1:1 solution of 0.033 M $KBrO_3$/0.2 M KBr and 30 μL of 12% $NH_2OH \times HCl$ were added. Next, the measuring flask was filled with deionised water. The sample was left in the test tubes for centrifuge for further analysis (Fig. 2).

2.3.3 *Acid-soluble fraction (F3)*

A portion of 25 mL 0.5 M HCl was added to the flasks with the samples obtained at stage 2 (F2). The, the flask was shaken in a rotating shaker for 1 hour and centrifuged for 15 minutes at 3000 rpm. The extract was filtered off by a cellulose acetate filter of 0.45 μm in diameter. 10 mL of the solution was placed in a measuring flask of 50 mL, to which 7.5 mL 33% HCl, 1 mL of 1:1 solution of 0.033 M $KBrO_3$/0.2 M KBr and 30 μL 12% $NH_2OH \times HCl$ were added. Next, the measuring flask was filled with deionised water. The sample was left in the test tube for centrifuge for further stages of the sequential analysis (Fig. 2).

2.3.4 *Humic fraction (F4)*

A portion of 30 mL of 0.2 M NaOH was added to the teflon test tubes for centrifuge containing the sediment sample obtained at stage 3 (F3). The test tube was shaken in a rotating shaker for 1 hour and centrifuged for 15 minutes at 3000 rpm. The extract was filtered off by cellulose acetate filter of 0.45 μm in diameter. 5 mL of the solution was collected and placed in the measuring flask of 50 mL in capacity, adding 7.5 mL 33% HCl, 1 mL of 1:1 solution of 0.033 M $KBrO_3$/0.2 M KBr and 30 μL 12% $NH_2OH \times HCl$. Next, the measuring flask was filled with deionised water. The precipitate left on the filter and in the flask was placed in a quartz vessel, dried in a lyophilising

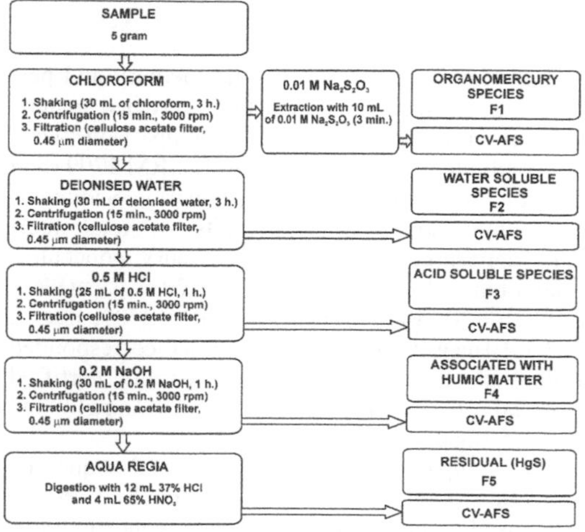

Figure 2. Scheme of sequential extraction.

cabinet, in order to prepare it for further steps of the sequential analysis (Fig. 2).

2.3.5 *Sulphide fraction (F5)*

About 1 g of the sample obtained at stage 4 (F4) was placed in a round-bottom flask of 250 mL in capacity, wetted with 0.5 mL of ultrapure water, adding 12 mL 37% HCl and 4 mL 65% HNO_3 and left at room temperature for 16 hours. Next, the round-bottom flask was connected to a partial condenser and a water cooler. The contents of the flask were slowly heated on a heater in temperature of boiling, until the conditions of vapour condensation kept for 2 hours to maintain the condensation zone in a cooler below 1/3 of its height. After cooling, the solution was filtered through a tissue into a 100 mL measuring flask and filled with deionised water to the mark. The same procedure was applied to determine the content of total mercury in bulk sediment.

2.4 *Mercury determination*

The mercury content of the solution was determined by cold-vapour atomic fluorescence spectroscopy (CV-AFS) using a Millennium Merlin mercury analyser (PS Analytical, England) after reduction of mercury(II) by $SnCl_2$. Calibration was performed with a mercury standard $HgNO_3$ (Merck, Germany) containing $980 \pm 020\,ng\,L^{-1}$ of mercury. The limit of detection of the method depends on the purity of the reagents used and reaches $0.1\,ng\,Hg\,L^{-1}$. The optimum range of mercury determination by this method is from 0.0001–100 $\mu g\,Hg\,L^{-1}$. When the concentration of mercury in the sample analysed was beyond the linear range of the method the analysis was repeated for the solution after appropriate dilution.

2.5 *Reference samples*

Along with determination of total mercury and performance of sequential extraction, certified materials were routinely analysed for comparison. They included: SRM 2711 (Montana Soil), SRM 2709 (San Joaquin Soil) and LGC 6137 (Estuarine Sediment) for comparison with the results of total mercury concentration. The total mercury concentrations obtained in our study $6060 \pm 70\,ng\,g^{-1}$ (SRM 2711, n = 3), $1440 \pm 30\,ng\,g^{-1}$ (SRM 2709, n = 4), $373 \pm 10\,ng\,g^{-1}$ (LGC 6137, n = 4) corresponded well with the values obtained for the certified materials: $6250 \pm 190\,ng\,g^{-1}$, $1400 \pm 80\,ng\,g^{-1}$ and $340 \pm 50\,ng\,g^{-1}$ respectively.

For comparison with the results of sequential extraction, only one certified sample (LGC 6137) was used. The sum of mercury concentrations obtained from particular fractions was $373\,ng\,g^{-1}$ dry mass, while the corresponding sum obtained for the certified material was $340\,ng\,g^{-1}$ dry mass. The method of sequential extraction is thus characterised by recovery of about 110% (range108–112%) and good reproducibility (Table 1).

Table 1. Statistical data on the fractionation of mercury in certified sediment LGC 6137.

LGC 6137 (n = 6)	Fractions (ng g^{-1})					Σ Fractions (ng g^{-1})
	F1	F2	F3	F4	F5	
Average	8.28	0.95	0.63	17.0	346	373
SD	0.59	0.02	0.03	0.9	5	5
Mediana	8.18	0.95	0.63	17.1	345	
RDS (%)	7	3	4	5	1	
Minimum	7.54	0.92	0.59	15.8	342	
Maximum	9.16	0.99	0.67	18.2	354	

3 RESULTS

The concentrations of total mercury and the contribution of particular mercury fractions in the total concentration are shown in Table 2.

The mean concentration of total mercury was $130 \pm 71\,ng\,g^{-1}$ dry mass (range 51–307) calculated as the sum of mercury concentrations in individual fractions and $136 \pm 74\,ng\,g^{-1}$ dry mass (range 47–310) in bulk sediments. The highest concentration of mercury was found in the Warta river sediment sample collected from the outlet of the municipal waste purification plant of Poznań. The fractions making the greatest contributions to the total mercury concentration were: mercury bound to sulphides 58 ± 17% (range 20–81%), humic matter 23 ± 9% (range 4–36) and compounds extracted by chloroform 17 ± 18% (<65). The fractions making the lowest contribution to the total mercury content were the water-soluble mercury 2.1 ± 0.9% (range 1.1–3.8) and acid soluble mercury species 0.4 ± 0.1% (range 0.2–0.7). The corresponding values determined in the certified reference standard LGC 6137 were similar, see Table 2. The absolute concentrations of mercury in particular fractions are $17 \pm 18\,ng\,g^{-1}$ (range 0.01–64) fraction F1; $3 \pm 2\,ng\,g^{-1}$ (range 1.3–10) fraction F2; $0.6 \pm 0.5\,ng\,g^{-1}$ (range 0.2–2.0) fraction F3; $31 \pm 20\,ng\,g^{-1}$ (range 4–36) fraction F4 and $80 \pm 59\,ng\,g^{-1}$ (range 20–232) fraction F5.

4 DISCUSSION

4.1 *Distribution of mercury in bulk sediments*

The concentrations of total mercury in the Warta river sediment samples taken at different sites show wide variation (Table 2). It seems that the differences in the spatial distribution of mercury follows from the

Table 2. The contribution of mercury from particular fractions in the total content of mercury in the samples of the river Warta sediments.

Sample number	Geographical coordinates		Fractions (%) F1	F2	F3	F4	F5	Σ Fractions (ng g^{-1})	Bulk sediment (ng g^{-1})	Water* (ng L^{-1})
1	N 52° 05′ 34.3″	E 17° 01′ 02.9″	65	1.9	0.4	13	20	100	103	17
2	N 52° 13′ 01.1″	E 16° 58′ 36.2″	25	2.2	0.4	29	44	136	138	13
3	N 52° 15′ 20.6″	E 16° 53′ 17.7″	11	1.1	0.4	31	56	154	165	27
4	N 52° 22′ 25.3″	E 16° 53′ 58.9″	32	2.3	0.5	21	44	58	60	26
5	N 52° 21′ 56.3″	E 16° 55′ 36.8″	9.3	2.4	0.4	36	52	85	80	20
6	N 52° 23′ 58.0″	E 16° 56′ 31.4″	8.7	3.8	0.5	27	60	51	47	22
7	N 52° 25′ 52.1″	E 16° 57′ 59.3″	7.1	1.5	0.7	18	73	121	121	32
8	N 52° 28′ 29.6″	E 16° 58′ 17.6″	0.0	3.2	0.6	21	76	307	310	36
9	N 52° 32′ 26.0″	E 16° 57′ 31.4″	13	2.0	0.4	4.1	81	105	105	15
10	N 52° 38′ 35.1″	E 16° 49′ 09.0″	6.7	1.1	0.4	31	61	181	203	13
11	N 52° 42′ 48.4″	E 16° 33′ 55.5″	6.4	1.3	0.2	24	68	137	162	12
Certified estuarine sediment – LGC 6137			2.2	0.3	0.2	4.6	93	373	340	

*From unpublished data.

local inflow of this element with industrial and municipal wastes from the cities on the river and in its catchment area. The city of Poznań seems to be an important source of mercury as the samples collected in the city are characterised by the highest concentrations of total mercury (310 ng g^{-1}). In the samples collected at other sites the total mercury concentration did not exceed 200 ng g^{-1}, which is assumed to be the upper limit of geochemical background (Kabata Pendias & Pendias, 1999). The values obtained in this study are comparable to those established in the sediments of the Rhine and Neckar rivers (Germany) and the Scheldt river (Belgium) equal to 400 ± 300, 300 ± 200 and 460 ng g^{-1} (vide Papina et al. 2000, Pilz & Yahya, 2000, Prange et al. 2002). Higher concentrations of mercury have been reported from other European rivers such as the Odra river (Poland) – 120–2999 ng g^{-1} (Boszke et al. 2004a, b), Elbe river (Germany) – 3700 ± 2800 ng g^{-1} (Prange et al. 2002) and the Yare river (UK) – 100–8130 ng g^{-1} (Birkett et al. 2002). In the Elbe river (Germany) the total mercury has been detected in concentrations up to 12,000 ng g^{-1} (Hintelmann & Wilken, 1995). In the bottom sediments from the lower Rhine river, the content of mercury in 1970 was over 10,000 ng g^{-1} and then it decreased to below 1000 ng g^{-1} in 1990 (Salomons & Förstner, 1984). Much higher concentrations of mercury have been detected in the sediments of the Nura river in Kazakhstan – 150,000–240,000 ng g^{-1} (Heaven et al. 2000) or different waterways in New Jersey (USA) – up to 29,600 ng g^{-1} (Bonnevie et al. 1993).

4.2 *Mercury in organomercury compounds*

Organomercury compounds are those in which mercury is bonded directly to the carbon atom e.g. $CH_3Hg(I)$ and $C_2H_5Hg(I)$. The extracting agents for these compounds were toluene (Miller et al. 1995), chloroform (Eguschi & Tomiyasu, 2002, Tomiyasu et al., 1996, 2000) and dichloromethane (Renneberg & Dudas, 2001). However, with the use of these solvents not only the organomercury compounds but also part of the mercury bound to organic matter is extracted (Eguschi & Tomiyasu, 2002). This explains the relatively large contributions of the mercury from this fraction in the total mercury concentrations, which varied from almost zero to 65% (Table 2). However, in the absolute values the mean concentration of mercury in this fraction is 17 ± 18 ng g^{-1} (range 0.08–64). An interesting observation is that the lowest concentrations of organomercury compounds and their lowest contribution in the total mercury content (Table 2) were found in the sample near the outflow of the city waste purification plant (site 8, Fig. 1). This would suggest that in the process of waste purification these compounds are effectively eliminated. Higher contributions of organomercury compounds were found in the samples of sediments collected both above and below the city of Poznań, and the maximum contribution of 65% was determined in the sample collected at site 1 (Fig. 1). The same sample revealed the highest concentration of organomercury compounds (64 ng g^{-1}), which would indicate that natural processes related to mercury methylation occur in the sediments of the Warta river.

In general, the contribution of organomercury compounds to the total content of mercury is low and greater in the bottom sediments polluted with mercury. For example, in the samples of bottom sediments from the Minamata Bay (Japan), the contribution of organomercury compounds varied from 1% to 4%, while in the Kagoshima Bay (Japan) it was from 7% to 37% (Egushi & Tomiyasu, 2002, Sakamoto et al. 1995). A greater contribution of ~30% of organomercury compounds in the total extractable organic matter was found in the sample of soil polluted with hydrocarbons (Renneberg & Dudas, 2001). In the samples of soil and sediments from the regions polluted with mercury e.g. near cinnabar mines or near chlor-alkali plants, the contribution of organomercury compounds to the total mercury concentration was very low (Bloom & Katon, 2000, Bloom et al. 2003, Martin-Doimeadios et al. 2000, Miller et al. 1995, Renneberg & Dudas, 2001), but their concentrations were comparable with those in the Warta river sediment. For example, comparable concentrations of organomercury compounds of 0.9–26 ng g^{-1} were determined in the marine bottom sediments from the Yatsushiro Sea in Japan (Tomiyasu et al. 2000), in the soil samples from the area strongly polluted with mercury near the cinnabar mine and processing plant 1–28 ng g^{-1} (Bloom & Katon, 2000, Bloom et al. 2003, Miller et al. 1995) or near a chlor-alkali producing plant 9 ng g^{-1} (Bloom & Katon, 2000, Bloom et al. 2003). Much higher concentrations of organomercury compounds were found near a gold mine 1000 ng g^{-1} (Miller et al. 1995).

4.3 *Water-soluble mercury*

The fraction known as the water-soluble mercury includes the mercury species present in pore water. Determination of mercury species present in this fraction involves filtering off pore water or addition of deionised water, shaking, filtering of water and determination of total mercury in the water extract (Biester & Scholz, 1997, Ching & Hongxiao, 1985, Renneberg & Dudas, 2001, Wallschläger et al. 1998a, b, Wasay et al. 1998). In the water phase mercury is not usually present in the form of water soluble ionic species but rather as species bound to organic matter or suspended mineral particles (Biester et al. 2002, Renneberg & Dudas, 2001, Wallschläger et al. 1996, 1998a). In our study the contribution of the water soluble mercury species was very small, equal on average to $2.1 \pm 0.9\%$ (range 1.1–3.8), which corresponds to the concentration of 3 ± 2 ng g^{-1} (range 1.3–10). The highest concentrations of water-soluble mercury species, similar to F1, were obtained in the sediment sample near the outlet of the city waste purification plant in Poznań (Table 2).

In general, the contribution of water-soluble mercury species is very small and often the concentrations determined are below the limit of detection or limit of determination of the method applied (Kot et al. 2002, Kot & Matyuskina, 2002). Contributions of the same order were determined in the river Ji Yun sediments in China (0.05–1.22%) (Ching & Hongxio, 1985) and in marine sediments from the Gdańsk Bay in Poland (<4%) (Bełdowski & Pempkowiak, 2003). The latter corresponds to mercury concentrations in this fraction of 150–250 ng g^{-1} (Ching & Hongxio, 1985) and 20 ng g^{-1} (Bełdowski & Pempkowiak, 2003), respectively. Contributions of the same order were also found in the soil polluted with mercury near a chlor-alkali plant, a timber conservation plant and cinnabar and gold mines: 0.13–7.6%, 0.03%, 0.001–0.19% and 0.032–5.3%, respectively (Biester & Scholz, 1997, Bloom et al. 2003, Bloom & Katon, 2000). However, in the strongly polluted river bottom sediments the concentrations of the water-soluble mercury species are much higher (Biester & Scholz, 1997, Bloom et al. 2003, Bloom & Katon, 2000, Miller et al. 1995, Wasay et al. 1998).

4.4 *Acid-soluble mercury*

Adsorption on clay minerals and bonding in their structure as well as bonding by iron and manganese hydroxides is of less importance for mercury species analysis than for other metals. This is explained by the fact that organic matter and sulphides are more effective in binding mercury. However, some inorganic compounds such as iron and manganese hydroxides can also play a role in the mobility of mercury. On the one hand, iron and manganese hydroxides provide excellent surfaces for the adsorption of organic matter to which mercury is bonded, while on the other hand, they can bind mercury themselves. Formation and dissolution of iron and manganese hydroxides is strongly controlled by the redox conditions and the content of oxygen in water and water reservoir sediments. In anaerobic conditions iron and manganese hydroxides dissolve, which releases heavy metals, including mercury (Gagnon et al., 1996, Guentzel et al. 1996, Hall et al. 1996, Luoma & Davis, 1983, Matilainen, 1995). The fraction of mercury bonded to iron and manganese hydroxides is determined after extraction with a solution of 0.1 M $NH_2OH \times HCl$ (pH = 2) (Wasay et al. 1998) or with a solution of 0.175 M $(NH_4)_2C_2O_4$ and 0.1 M CH_3COONH_4 (pH = 3.2) (Tamm's reagent) (Inacio et al. 1998). Some authors have used a solution of 0.5 N HCl (Lechler et al. 1997) or 0.3 M HCl (Ching & Hongxiao, 1985) to extract strongly bound Hg species including those bonded to iron and manganese hydroxides and carbonates. A solution of hydrochloric acid is not specific, so the fraction of mercury species soluble in it can include the species originally bound to organic matter and released to the solution in the process of simple protonation of specific sites in the organic matter (Lechler et al. 1997) and the mercury species strongly adsorbed on the surface of minerals (Lechler et al. 1997). According to another approach, the fraction of mercury species leached in the conditions present in the alimentary tract is determined (Barnett & Turner, 2001; Bloom et al., 2003). This fraction is called "bioavailable" and the extraction of this fraction was performed in conditions simulating those in the alimentary tract of mammals, that is, by treatment with a solution of HCl (pH = 2.5) (Barnett & Turner, 2001), mixture of HCl and CH_3COOH (pH = 2) (Bloom & Katon, 2000) or a mixture of 0.4 M glycine and 0.7 M HCl (Bloom et al. 2003).

In our study the contribution of the acid-soluble mercury fraction (F3) in the total mercury content was very small and not very variable – the mean contribution was $0.4 \pm 0.1\%$ (range 0.2–0.7), which corresponds to the mean concentration of 0.6 ± 0.5 ng g^{-1} (range 0.2–2.0) (Table 2). In general, the contributions of the mercury species bound to iron and manganese hydroxides in the total mercury concentration are low. In the sediments of the river Ji Yun (China) this contribution was from 0.34% to 1.55% (Ching & Hongxio, 1985). According to other authors the mean contribution of mercury in this fraction in the river Ji Yun and Zijan sediments (China) was 3.4%, and the maximum contribution of this fraction (13.1%) was found in the vicinity of waste outflow, the concentration of this mercury fraction in this sample was 262,000 ng g^{-1} (Peng & Wang, 1985). In the soil samples, the contribution of the acid-soluble mercury fraction was also small and varied from 0.3% to 2.3% in the samples collected near the chlor-alkali plant (Inacio et al. 1998) or 5.3% (Barnett & Turner, 2001). In the soil sample from the vicinity of a gold mine the contribution of mercury species extracted with a solution of HCl/CH_3COOOH, pH = 2 (stomach acid) was 1.2% (7750 ng g^{-1}) on the surface and 36% (15,280 ng g^{-1}) in a deeper layer (Bloom & Katon, 2000).

4.5 *Mercury bound to humic matter*

Organic matter is one of the most important components of bottom sediments and soils and it is to a significant degree responsible for binding metals. It has been indicated that organic matter can bind up to 95% of divalent mercury species (Meili, 1997). Not only Hg(II) but also methylmercury (strongly bound by sulphur in organic matter) (Hesterberg et al. 2001) have a strong affinity to organic matter. Binding of mercury in organic matter is mainly due to the reduced sulphur species in such functional groups as thiol group (R-SH), disulphide group (R-SS-R) or disulphane group (R-SSH) (Xia et al. 1999). Besides sulphur, oxygen and nitrogen also make a contribution to binding mercury in organic matter but their significance is lower (Hesterberg et al. 2001). A significant part of organic matter is the humus matter whose contribution in the total organic matter is ~25% in bottom sediments, 20% in marine water, 60% in the river water and 70% in marsh areas over the river catchment area (Weber, 1993). In order to determine the mercury species bound to humic and fulvic acids and soluble organic components, different authors have proposed sample treatment with the solutions of $Na_4P_2O_7$ (Henderson et al. 1998), KOH (Ching & Hongxiao, 1985) or NaOH (Biester et al. 2002, Lechler et al. 1997). In general, when the content of organic matter in the sample is high many authors recommend the use of additional extraction stages to determine the mercury species bound to different fractions of organic matter (Bełdowski & Pempkowiak, 2003, Ching & Hongxiao, 1985, Wallschläger et al. 1998a, b). In the our samples the contribution of mercury species bound to humus matter is (mean) $23 \pm 9\%$ (range 4–36), which corresponds to the mean concentration of 31 ± 20 ng g^{-1} (range 4–36) (Table 2). For comparison the contribution of mercury species bound to humic acids in the bottom sediments of the rivers Ji Yun and Zijan (China) ranged from 2.7% to 77.4% (Peng & Wang, 1985). The lowest contribution of the mercury species bound to organic matter was found in a sample from the region of waste release from a plant producing chlorine alkaline compounds. It amounted to 54,000 ng g^{-1}, while the concentration of mercury in the sample with the greatest contribution of the organic matter bound mercury was 4400 ng g^{-1} (Peng & Wang, 1985). Other authors (Ching & Hongxiao, 1985) reported the contribution of humic acid bound mercury in the bottom sediments from the same river Ji Yun as varying from 6.52% to 23.7% (800–46,650 ng g^{-1}). In the samples of marine bottom sediments the contribution of fulvic acid bound mercury was small and did not exceed a few percent (the maximum of 9%), while the contribution of humic acid bound mercury was much greater and in general amounted to about 20% (maximum of 54%) (Bełdowski & Pempkowiak, 2003). In general, greater contributions of mercury bound to organic matter are found in bottom sediments from not very polluted rivers and smaller in those from highly polluted ones. For example, in sediments uncontaminated with mercury sampled from the river Carson (USA) the contribution of mercury bound to humic matter was <40% but in the region highly polluted with mercury it decreased to a few percent (Lechler et al. 1997).

4.6 *Mercury bound to sulphides*

The fraction of mercury bound to sulphides is commonly referred to in the literature as residual as mercury rarely occurs with aluminosilicates (Lechler et al. 1997, Martin-Doimeadios et al. 2000). The residual fraction is determined in the last stage of the sequential extraction, which involves mineralisation of the post-extraction residue in a solution of aqua regia (Biester & Scholz, 1997, Bloom et al. 2003, Bloom & Katon, 2000, Ching & Hongxiao, 1985, Wasay et al. 1998). In the samples studied by us the contribution of mercury bound to sulphides was on average $58 \pm 17\%$ (range 20–81), which corresponds to the mean mercury concentration of 80 ± 59 ng g^{-1} (range 20–232) (Table 2).

Table 3. Correlation matrix.

	F1	F2	F3	F4	F5	F1 + F4	F2 + F3	F1 + F2 + F3 + F4	Total*	Water**
F1	1.00									
F2	−0.27	1.00								
F3	−0.28	**0.94**	1.00							
F4	−0.29	0.57	**0.63**	1.00						
F5	−0.49	**0.85**	**0.90**	**0.73**	1.00					
F1 + F4	0.55	0.28	0.32	**0.64**	0.24	1.00				
F2 + F3	−0.28	**1.00**	**0.95**	0.59	**0.87**	0.29	1.00			
F1 + F2 + F3 + F4	0.49	0.39	0.43	**0.68**	0.33	**0.99**	0.40	1.00		
Total*	−0.25	**0.84**	**0.89**	**0.84**	**0.95**	0.53	**0.86**	**0.61**	1.00	
Water**	−0.36	0.53	**0.64**	0.21	0.44	−0.10	0.56	0.03	0.36	1.00

N = 11; $p < 0.05$; * Total mercury concentration in bulk sediment; ** Total mercury concentration in bulk water.

In general, the contribution of mercury bound to sulphides is greater in the sediments and soil samples characterised by reduction conditions (Bełdowski & Pempkowiak, 2003, Lechler et al. 1997). For instance, the contribution of mercury bound to sulphides in the bottom sediments of the Gdańsk Bay (Poland) was close to 40%, with a maximum value of 96% in the bottom sediments characterised by reduction conditions (Bełdowski & Pempkowiak, 2003). In the soil profile, the contribution of residual mercury increased with increasing depth of the profile, from 8.1% in the surface layer (0–30 cm) to 69% in the layer at 50–80 cm (Lechler et al. 1997). A large contribution of mercury bound to sulphides of 71.4% (mean) (range 38.4–96.1) was found in the bottom sediments from the Kogushima Bay (Japan), where mercury sulphide is formed, among other routes, from H_2S and mercury chloride liberated from the fumarole gasses originating from the sea floor (Sakamoto et al., 1995). In the mercury polluted sediments from the river Ji Yun (China) the contribution of mercury bound to sulphides varied from 5.8% to 17.1% (800–76,520 ng g^{-1}) (Ching & Hongxiao, 1985). These values can be contrasted with the contribution of mercury bound to sulphides in mercury-polluted soil being of <99.1% in the vicinity of the plant producing chlorine alkaline compounds, <93.1% in the vicinity of the gold mine and only from 1.9% to 20.1% in the vicinity of the HgS mine (Bloom & Katon, 2000).

4.7 *Correlations*

The correlation coefficients of mercury concentration in different fractions and total mercury in sediment and in water of the Warta river are given in Table 3. Statistically significant positive correlations ($p < 0.05$) have been found between the total mercury concentration in sediment and mercury concentration in fractions F2, F3, F4 and F5. A statistically significant positive correlation has also been found between the total mercury concentration in bulk water and mercury concentration in fraction F3. Other correlation coefficients are given in Table 3.

5 CONCLUSIONS

Determination of mobility of trace metals including mercury, on the basis of the sequential extraction procedures arouses much controversy. Specificity and repeatability of the procedures depends on the chemical properties of a given element and chemical composition of the sample. Therefore, the quality of experimental results depends on such factors as the type and concentration of the reagent, extraction time, temperature or the ratio of the extraction solution volume to the mass of the sample and the processes of readsorption. Moreover, there are no natural reference materials to permit determination of the method's accuracy. Nevertheless, despite evident drawbacks related to incomplete selectivity of the extraction solutions, the methods based on extraction are still the main tool for speciation analysis. They are able to provide information on fractionation in certain analytical conditions.

As follows from our results, in the bottom sediments from the river Warta in the unpolluted sections, mercury occurs mainly in the form of mercury sulphide, hardly soluble and hardly bioavailable, and in forms bound to humic matter. We have also shown that the extremely toxic organic mercury species present in the river Warta bottom sediments are not of anthropogenic origin but appear as a result of natural methylation processes. Although the concentration of organomercury species extracted by chloroform in the samples studied is low, these species can pose a real threat to man and other living organisms because of their extreme toxicity and capability of bioaccumulation and biomagnification in the trophic chains.

ACKNOWLEDGMENT

This work was supported by the Polish Committee for Scientific Research (KBN) under grant No 3 T09D 079 26.

REFERENCES

Barnett, M.O. & Turner, R.R. 2001. Bioaccessibility of mercury in soils. *Soil & Sediment Contamination* 10: 301–316.

Bełdowski, J. & Pempkowiak, J. 2003. Horizontal and vertical variabilities of mercury concentration and speciation in sediments of Gdańsk Basin, Southern Baltic Sea. *Chemosphere* 52: 645–654.

Biester, H. & Scholz, C. 1997. Determination of mercury phase in contaminated soils. Mercury pyrolysis versus sequential extractions. *Environmental Science & Technology* 31: 233–239.

Biester, H., Muller, G. & Schöler, H.F. 2002. Binding and mobility of mercury in soils contaminated by emissions from chlor-alkali plants. *Science of the Total Environment* 284: 191–203.

Birkett, J.W., Noreng, J.M.K. & Lester, J.N. 2002. Spatial distribution of mercury in the sediments and riparian environment of the River Yare, Norfolk, UK. *Environmental Pollution* 116: 65–74.

Bloom, N.S. & Katon, J. 2000. Application of selective extractions to the determination of mercury speciation in mine tailings and adjacent soils. *Proceedings of Assessing and Managing Mercury from Historic and Current Mining Activities Conference*, San Francisco, CA November 28–30.

Bloom, N.S., Preus, E., Katon, J. & Hilter, M. 2003. Selective extractions to assess the biogeochemically relevant fractionation of inorganic mercury in sediment and soils. *Analytical Chimica Acta* 479: 233–248.

Bonnevie, N.L., Wenning, R.J., Huntley, S.L. & Bedbury, H. 1993. Distribution of inorganic compounds from three waterways in northern New Jersey. *Bulletin of Environmental Contamination & Toxicology* 51: 672–680.

Boszke, L., Głosińska, G. & Siepak, J. 2002. Some aspects of speciation of mercury in a water environment. *Polish Journal of Environmental Studies* 11: 285–298.

Boszke, L., Kowalski, A., Głosińska, G., Szarek, R. & Siepak, J. 2003. Selected factors affecting the speciation of mercury in the bottom sediments: an overview. *Polish Journal of Environmental Studies* 12: 5–13.

Boszke, L., Kowalski, A. & Siepak, J. 2004a. Grain size partitioning of mercury in sediments of the middle Odra river (Germany/Poland). *Water Air & Soil Pollution* 159: 125–138.

Boszke, L., Sobczyński, T., Głosińska, G., Kowalski, A. & Siepak, J. 2004b. Distribution of mercury and other heavy metals in the bottom sediments of the middle Odra river (Germany/Poland). *Polish Journal of Environmental Studies* 13: 495–502.

Calmano, W. 1983. Chemical extraction of heavy metals in polluted river sediments in central Europe. *Science of the Total Environment* 28: 77–90.

Ching-I. L. & Hongxiao, T. 1985. Chemical studies of aquatic pollution by heavy metals in China. In K.J. Irgolic & A.E. Martel (eds), *Environmental Inorganic Chemistry*: 359–371. Deerfield Beach, Florida: VCH Publishers.

Eguchi, T. & Tomiyasu, T. 2002. The speciation of mercury in sediments from Kagoshima Bay and Minamata Bay, Southern Kyusyu, Japan, by fractional extraction/cold-vapor AAS. *Bunseki Kagaku* 51: 859–864 (in Japanese with English abstract).

Gagnon, C., Pelletier, E., Mucci, A. & Fitzgerald, W.F. 1996. Diagenetic behavior of methylmercury in organic-rich coastal sediments. *Limnology & Oceanography* 41: 428–434.

Głosińska, G., Boszke, L. & Siepak, J. 2001. Evaluation of definition speciation and speciation analysis in Polish publications. *Chemia & Inżynieria Ekologiczna* 8: 1109–1119 (in Polish with English abstract).

Gobeil, C. & Cossa, D. 1993. Mercury in sediments and sediment pore water in the Laurentian Trough. *Canadian Journal of Fishery & Aquatic Science* 50: 1794–1800.

Gomez-Ariza, J.L., Giraldez, I., Sanchez-Rodaz, D. & Morales, E. 2000. Selectivity assessment of a sequential extraction procedure for metal mobility characterization using model phases. *Talanta* 52: 545–554.

Guentzel, J.L., Powell, R.T., Landing, W.M. & Mason, R.P. 1996. Mercury associated with colloidal material in an estuarine and an open ocean environment. *Marine Chemistry* 55: 177–188.

Gustin, M.S., Biester, H. & Kim, C.S. 2002. Investigation of the light-enhanced emission of mercury from naturally enriched substrates. *Atmospheric Environment* 36: 3241–3254.

Hall, G.E.M., Vaive, J.E., Beer, R. & Hoashi, M. 1996. Selective leaches revisited, with emphasis on the amorphous Fe oxyhydroxide phase extraction. *Journal of Geochemical Exploration* 56: 59–78.

Heaven, S., Ilyushenko, M.A., Tanton, T.W., Ullrich, S.M. & Yanin, E.P. 2000. Mercury in the Nura and its floodplain, Central Kazakhstan: I. River sediments and water. *Science of the Total Environment* 260: 35–44.

Hesterberg, D., Chou, J.W., Hutchison, K.J. & Sayers, D.E. 2001. Bonding of Hg(II) to reduced organic sulfur in humic acid as affected by S/Hg ratio. *Environmental Science & Technology* 35: 2741–2745.

Henderson, P.J., Mcmartin, I., Hall, G.E., Percival, J.B. & Walker, D.A. 1998. The chemical and physical characteristics of heavy metals in humus and till in the vicinity of the base metal smelter at Flin Flon, Manitoba, Canada. *Environmental Geology* 34: 39–58.

Hintelmann, H. & Wilken, R.-D. 1995. Levels of total mercury and methylmercury compounds in sediments of the polluted Elbe River: influence of seasonally and spatially varying environmental factors. *Science of the Total Environment* 166: 1–10.

Inacio, M.M., Pereira, V. & Pinto, M.S. 1998. Mercury contamination in sandy soils surrounding an industrial emission source (Estarreja, Portugal). *Geoderma* 85: 325–339.

Jackson, T.A. 1988: The mercury problem in recently formed reservoirs of northern Manitoba (Canada): effects of impoundment and other factors on the production of methyl mercury by microorganisms in sediments. *Canadian Journal of Fishery & Aquatic Science* 45: 97–121.

Jackson, T.A. 1991. Biological and environmental control of mercury accumulation by fish in lakes and reservoirs of northern Manitoba, Canada. *Canadian Journal of Fishery & Aquatic Science* 48: 2449–2470.

Jackson, T.A. 1998a. Mercury in aquatic ecosystems. In: W.J. Langston & M.J. Bebianno (eds), *Metal Metabolism in Aquatic Environments*: 77–158. London: Chapman & Hall.

Jackson, T.A. 1998b. The biogeochemical and ecological significance of interactions between colloidal minerals and trace elements. In: A. Parker & J.E. Rae (eds.), *Environmental Interactions of Clays*: 93–205. Berlin: Springer.

Kabata-Pendias, A. & Pendias, H. 1999. *Biogeochemistry of trace elements*. Warszawa: Państwowe Wydawnictwo Naukowe (in Polish).

Kim, C.S., Brown, G.E. & Rytuba, J.J. 2000. Characterization and speciation of mercury-bearing mine waste using X-ray absorption spectroscopy. *Science of the Total Environment* 126: 157–168.

Kot, F.S. & Matyushkina, L.A. 2002. Distribution of mercury in chemical fractions of contaminated urban soils of Middle Amur, Russia. *Journal of Environmental Monitoring* 4: 803–808.

Kot, F.S., Matyushkina, L.A., Nikorych, V.N. & Polivichenko, V.G. 2002. Mercury in chemical and thermal fractions of soils of Eastern Ukrainian Polissya. *Soil Science* 3: 94–101.

Lechler, P.J., Miller, J.R., Hsu, L.C. & Desilets, M.O. 1997. Mercury mobility at the Carson River superfund site, west-central Nevada, USA-interpretation of mercury speciation data in mill tailing, soils, and sediments. *Journal of Geochemical Exploration* 58: 259–267.

Luoma, S.N. & Davis J.A. 1983. Requirements for trace metal partitioning in oxidized estuarine sediments. *Marine Chemistry* 12: 159–181.

Martin-Doimedios, R.C.R., Wasserman, J.C., Bermejo, L.F.G., Amouroux, D., Nevado, J.J.B. & Donard, O.F.X. 2000. Chemical availability of mercury in stream sediments from the Almaden area, Spain. *Journal of Environmental Monitoring* 2: 360–366.

Matilainen, T. 1995. Involvement of bacteria in methylmercury formation in anaerobic lake waters. *Water Air & Soil Pollution* 80: 757–764.

Meili, M. 1997. Mercury in Lakes and Rivers. In: A. Sigel & H. Sigel (eds), *Mercury and Its Effects on Environment and Biology. Metal Ions in Biological Systems*: 21–51. New York: Marcel Dekker, Inc.

Miller, E.L., Dobb, D.E. and Heithmar, E.M. 1995. Speciation of mercury in soils by sequential extraction. *Presented at the USEPA Metal Speciation and Contamination of Surface Water Workshop*, Jekyll Island, GA.

Papina, T.S., Temerev, S.V. & Eyrikih, A.N. 2000. Heavy metals transport and distribution over the abiotic components of the river aquatic ecosystems (West Siberia, Russia). In J.O. Nriagu (ed), 11th *Annual International Conference on Heavy Metals in the Environment*, University of Michigan, School of Public Health, Ann Arbor, MI (CD-ROM).

Pearson, R.G. 1963. Introduction to hard and soft acids and bases. *Journal of American Chemical Society* 85: 3533–3539.

Peng, A. & Wang, Z. 1985. Mercury in river sediment. In: K.J. Irgolic & A.E. Martel (eds), *Environmental Inorganic Chemistry*: 393–400. Deerfield Beach, Florida: VCH Publishers.

Pilz, B. & Yahy, A.A. 2000. Development of the heavy-metal pollution in the sediments of the lower Neckar river during the past 25 years. In J.O. Nriagu (ed), 11th *Annual International Conference on Heavy Metals in the Environment*, University of Michigan, School of Public Health, Ann Arbor, MI (CD-ROM).

Prange, A., Furrer, R., Einax, J.W., Lochovsky, P., Kofalk, S. & Reincke H. 2002. Die Elbe und ihre Nebenflüsse: Belastung, Trends, Bewertung, Perspektiven. ATV-DVWK Deutsche Vereinigung für Wasserwirtschaft, Abwasser und Abfall e.V., Hennef., 168 pp. (in German).

Renneberg, A.J. & Dudas, M.J. 2001. Transformations of elemental mercury to inorganic and organic forms in mercury and hydrocarbon co-contaminated soils. *Chemosphere* 45: 1103–1109.

Sakamoto, H., Tomiyasu, T. & Yonehara, N. 1995. The content and chemical forms of mercury in sediments from Kagoshima Bay, in comparison with Minamata Bay and Yatsushiro Sea, Southwestern Japan *Geochemistry Journal* 29: 97–105.

Salomons, W. & Förstner, U. 1984. *Metals in Hydrocycle*. Berlin, Heidelberg: Springer Verlag.

Sladek, C. & Gustin, M.S. 2003. Evaluation of sequential and selective extraction methods for determination of mercury speciation and mobility in mine waste. *Applied Geochemistry* 18: 567–576.

Sladek, C., Gustin, M.S., Kim, C.S. & Biester H. 2002. Application of three methods for determining mercury speciation in mine waste. *Geochemistry Exploration Environmental Analysis* 2: 369–376.

Templeton, D.M., Ariese, F., Cornelis, R., Danielsson, L.-G., Muntau, H., Van Leeuwen, H.P. & Łobiński, R. 2000. Guidelines for terms related to chemical speciation and fractionation of elements. Definition, structural aspects, and methodological approaches (IUPAC Recommendations 2000). *Pure & Applied Chemistry* 72: 1453–1470.

Tessier, A., Campbell, P. & Bisson, M. 1979. Sequential extraction procedure for the speciation of particulate trace metals. *Analytical Chemistry* 51: 844–851.

Thöming, J., Stichnothe, H., Mangold, S. & Calmano, W. 2000. Hydrometalurgical approaches to soil remediation – process optimization applying heavy metal speciation. *Land Contamination & Reclamation* 8: 19–31.

Tomiyasu, T., Nagano, A., Sakamoto, H. & Yonehara, N. 1996. Differential determination of organic and inorganic mercury in sediment, soil and aquatic organisms by cold-vapor atomic absorption spectrometry. *Analytical Science* 12: 477–481.

Tomiyasu, T., Nagano, A., Yonehara, N., Sakamoto, H., Rifardi, Oki, K. & Akagi, H. 2000. Mercury contamination in the Yatsushiro Sea, south-western Japan: spatial variations of mercury in sediment. *Science of the Total Environment* 257: 121–132.

Wallschläger, D., Desai, M.V.M. & Wilken, R.-D. 1996. The role of humic substances in the aqueous mobilization of mercury from contaminated floodplain soils. *Water Air & Soil Pollution* 90: 507–520.

Wallschläger, D., Desai, M.V.M., Spengler, M., Windmöler, C.C. & Wilken, R.-D. 1998a. How humic substances dominate mercury geochemistry in contaminated floodplain

soils and sediments. *Journal of Environmental Quality* 27: 1044–1054.
Wallschläger, D., Desai, M.V.M., Spengler, M., Windmöler, C.C. & Wilken R.-D. 1998b. Mercury speciation in floodplain soils and sediments along a contaminated river transect. *Journal of Environmental Quality* 27: 1034–1043.
Wasay, S.A., Barrington, S. & Tokunaga, S. 1998. Retention form of heavy metals in three polluted soils. *Journal of Soil Contamination* 7: 103–119.
Weber, J.H. 1993. Review of possible paths for abiotic methylation of mercury(II) in the aquatic environment *Chemosphere* 51: 2063–2077.
Xia, K., Skyllberg, U.L., Bleam, W.F., Bloom, R.P., Nater, E.A. & Helmke, P.A. 1999. X-ray absorption spectroscopic evidence for the complexation of Hg(II) by reduced sulfur in soil humic substances. *Environmental Science & Technology* 33: 257–261.

Environmental Engineering – Pawłowski, Dudzińska & Pawłowski (eds)
© 2007 Taylor & Francis Group, London, ISBN13 978-0-415-40818-9

The effect of pressure on the binding of heavy metals by humic acid

Edward Mączka & Marek Kosmulski
Department of Electrochemistry, Lublin University of Technology, Lublin, Poland

ABSTRACT: High pressures (>10^7 Pa) depress the binding of metal cations by humic acid. The dilatometric effect of binding of Cu and Cd by humic acid in the presence of 0.5 mol dm^{-3} NaCl is 14-32 and 9-15 cm^3/mol, respectively, and it increases when the pH increases. Thus, at a pressure of 10^8 Pa (a 10 km water column), the binding constants are lower by a factor of 2–3 than at atmospheric pressure. This result is independent of the model of binding of metals by humic acid.

Keywords: Humic acid, heavy metals, adsorption, sea water.

1 INTRODUCTION

Krauskopf (1956) emphasized that the experimentally observed concentrations of heavy metals in sea water are lower by many orders of magnitude than the concentrations of heavy metals in saturated solutions of their hydroxides, basic carbonates, and of the other sparingly soluble salts that might be formed in sea water. He attributed the spontaneous removal of heavy metals from sea water to their binding by organic matter and to adsorption. Certain organic compounds show a high affinity to heavy metals, and this results in competition between the surface sites and organic matter for metal cations as well as in formation of surface complexes involving both organic ligands and heavy metal cations (Christl & Kretschmar 2001). Due to their high affinity towards heavy metals and to their abundance in the nature, humic acids are often used as representative organic compounds in laboratory studies of the binding of heavy metals by natural organic matter (Bianchini & Bowles 2002, Oste et al. 2002, Pempkowiak & Kosowska 1998, Pempkowiak et al. 1998, Zhou et al. 2005). A simple model involving a single solution reaction cannot quantitatively reproduce the binding of metal ions by humic acid over a wide range of concentrations and pH. More complex models involving multiple reactions (corresponding to different functional groups responsible for the binding of heavy metals by humic acid), electric-double-layer-interactions, etc., were more successful. The values of the binding constants characterizing the affinities of particular metal cations to humic acid are model-dependent.

The binding constant can be determined when the concentrations of particular species (bound and free metal ions) is known. Various experimental techniques have been used to determine the concentration of free heavy metals, not bound by organic matter (Abate & Masini 2001, Rozan & Benoit 1999, Trojanowicz et al. 1998, Voets et al. 2004), and the measurement by means of ion selective electrodes (ISE) is a direct, and relatively simple and frequently used method (Lu et al. 2000, Peijnenburg & Jager 2003, Marco 1996, Gismera et al. 2003). In the present paper, laboratory studies of binding of heavy metals by purified humic acid are reported.

The equilibrium constants of the reactions between heavy metals and organic compounds determined at atmospheric pressure, are not necessarily applicable for the conditions in deep oceans. The effect of pressure on the chemical equilibria in condensed phases becomes significant at pressures higher than about 10^7 Pa. For lower pressures, i.e. in the range 0–10^7 Pa, the equilibrium constants determined at atmospheric pressure are applicable, and the effect of pressure is negligible (Kosmulski & Mączka 2003, 2004). Deep oceans are an example of a system where the effect of pressure matters. Namely, the average depth of the ocean (ca. 3800 m) corresponds to a pressure of $3.8 \cdot 10^7$ Pa. Direct measurements of the binding of metal cations at pressures in the range 10^7–10^8 Pa are very difficult. Fortunately, the effect of pressure can be determined indirectly, from the following well-known thermodynamic relationship:

$$d \ln K/dp = -V^0/RT$$

where $V^0 = \Sigma V_{\text{products}} - \Sigma V_{\text{substrates}}$ is the standard volume of the reaction, and V_{products} and $V_{\text{substrates}}$ are standard molar volumes of the reactants, and the sum takes into account the stoichiometric coefficients of the reaction of interest.

In our previous papers (Kosmulski & Mączka 2003, 2004) we demonstrated that the high pressure in deep oceans substantially affects the equilibria of precipitation of heavy metal hydroxides and of binding of heavy metals by inorganic adsorbents. In these processes the volume of the system increases, thus the high pressure induces dissolution of metal hydroxides and desorption of heavy metal cations from inorganic adsorbents. Recently (Mączka & Kosmulski 2005), we presented a preliminary study of binding of heavy metal cations by unpurified humic acid, but the reproducibility of the results was limited. In the present study we used purified humic acid. Its preparation is described in detail in the experimental section. The equilibrium of this reaction is indirectly affected by the chemical speciation of metal ions in solution. Unfortunately, dilatometric data for the relevant solution reactions of metal ions, i.e. formation of inorganic complexes and hydrolysis are not available. Thus, we are only able to calculate the V^0 of the overall process rather than of actual reactions occurring in the system of interest.

2 EXPERIMENTAL

The sodium salt of humic acid used in the further experiments was prepared from Alfa Aesar humic acid (catalog number 41747). A portion of humic acid was shaken with acidified 0.5 mol dm^{-3} NaCl solution (pH 3) for 12 h and then centrifuged, and the supernatant was rejected. This procedure was repeated 20 times. Then the sediment was washed with a small amount of MilliQ water and lyophilized. Polypropylene beakers were used in the washing procedure. The term "sodium salt of humic acid" will be used in the further text to emphasize the difference between the purified product used in the present study and unpurified humic acid used in our previous study.

The dilatometric measurements were carried out at atmospheric pressure in a dilatometer of our own design (Mączka 2004), consisting of two chambers, 300 cm^3 each. Chamber 1 was filled with a solution of copper or cadmium nitrate (2.5–$5 \cdot 10^{-3}$ mol dm^{-3}) in 0.5 mol dm^{-3} NaCl. Chamber 2 was filled with a dispersion of 1 g of sodium salt of humic acid in 0.5 mol dm^{-3} NaCl. The water used was MilliQ quality, and the other reagents were reagent grade. The pH range 3–8 was studied. The final pH of the dispersion was adjusted by addition of 1 mol dm^{-3} NaOH solution to chamber 2. The solution and dispersion were degassed by means of an ultrasonic bath, and then thermostated at 25 ± 0.1°C by means of a Haake DC 30 thermostat. After at least 14 h of thermal equilibration the valves between the chambers were opened and the dispersion was stirred magnetically. The changes in the volume of the system were determined from the height of the liquid in a capillary on the top of the dilatometer. Once the volume of the system reached a constant value, the dilatometer was disassembled, and the pH of the dispersion as well as the equilibrium concentration of free heavy metal cations (not bound in organic complexes) in the solution were measured. The concentration of free heavy metal cations was measured by means of an ISE ion-meter (Denver Instruments Company, USA) with ECu-01 and ECd-01 ISE against a RL-100 reference electrode (all electrodes from Hydromet, Gliwice, Poland). The electrode was calibrated against a series of standards containing various concentrations of the metal of interest in 0.5 mol dm^{-3} NaCl, directly before the measurement. The operations involving the Cu-selective electrode were carried out in a dark box, because of the photosensitivity of the electrode (Antonelli 2001).

3 RESULTS AND DISCUSSION

The concentration range of heavy metal cations in the present study is higher by many orders of magnitude than typical concentrations encountered in the natural environment. This high concentration was necessary to induce a measurable dilatometric effect. It was shown (Kosmulski & Mączka 2003, 2004) that the experimentally determined values of ΔV of the precipitation of sparingly soluble metal hydroxides are only marginally different from the standard values which correspond to the infinite dilution. Very likely the values of ΔV of the complexation of heavy metal cations by humic acid determined in the present study are also approximately equal to the standard values.

The kinetics of the reaction between the sodium salt of humic acid and heavy metal cations are presented in Figures 1 (Cu) and 2 (Cd). In both figures the volume of the dispersion asymptotically increases, and a constant value is reached within a few hours. The error bars represent the resolution of the dilatometer, which is 0.5 mm^3. In a previous study of binding of Cu(II) and Ni(II) by unpurified humic acid (Mączka & Kosmulski 2005) the volume of the dispersion increased above the equilibrium volume in the initial phase of the reaction, and then it asymptotically decreased. Various possible mechanisms of this behavior were discussed.

Comparison of the results presented in Figures 1 and 2 on the one hand and those presented in the previous paper on the other suggests that in the case of unpurified humic acid, the binding of heavy metals is faster than the protonation and deprotonation of humic acid (which is governed by the rate of rearrangement of macromolecules, thus it is much slower than the reaction itself). In the present study the humic acid is chiefly in form of sodium salt thus the liberation of protons to the solution accompanying binding of heavy metal cations exercises a negligible dilatometric effect. Thus it is the ionic strength which substantially affects the kinetics of binding of the heavy metals by humic acid.

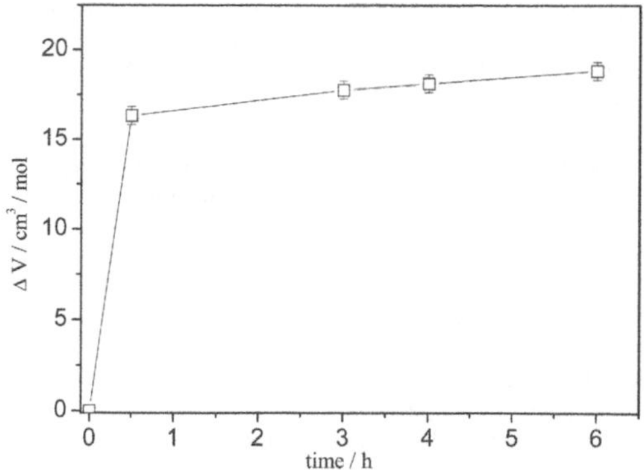

Figure 1. Kinetics of reaction in the system $Cu(NO_3)_2$ in 0.5 mol dm^{-3} NaCl and 1 g of sodium salt of humic acid in 0.5 mol dm^{-3} NaCl. Final pH 5.01.

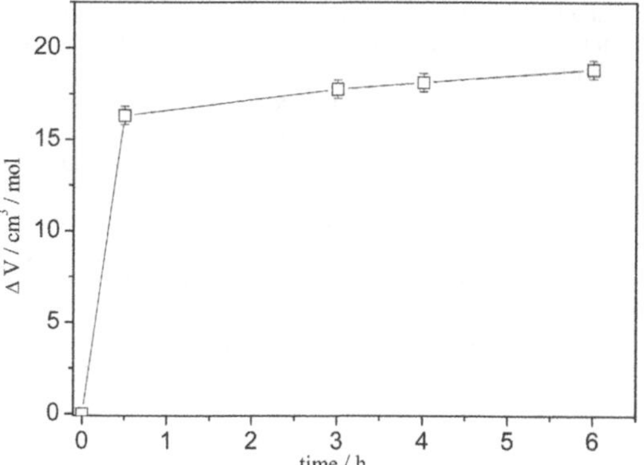

Figure 2. Kinetics of reaction in the system $Cd(NO_3)_2$ in 0.5 mol dm^{-3} NaCl and 1 g of sodium salt of humic acid in 0.5 mol dm^{-3} NaCl. Final pH 7.54.

Figure 3 shows the values of ΔV, i.e., the volume of reaction of binding of Cu(II) ions with sodium salt of humic acid as a function of the pH. The various symbols correspond to different equilibrium concentrations of Cu(II). The ΔV increases when the pH increases.

The values of ΔV found in the present study (14-32 cm^3/mol for Cu and 9–15 cm^3/mol for Cd, for different pH) are lower than corresponding values reported in our previous paper. Probably the difference is due to the fact, that in the present study an exchange between heavy metal cations and sodium was observed. The sodium cations liberated into the solution undergo hydration and thus they induce a negative dilatometric effect, which partially neutralizes the positive dilatometric effect due to the binding of heavy metal cations by humic acid. The increase in ΔV when pH increases is observed for both Cu (Fig. 3) and Cd, and is probably due to formation of

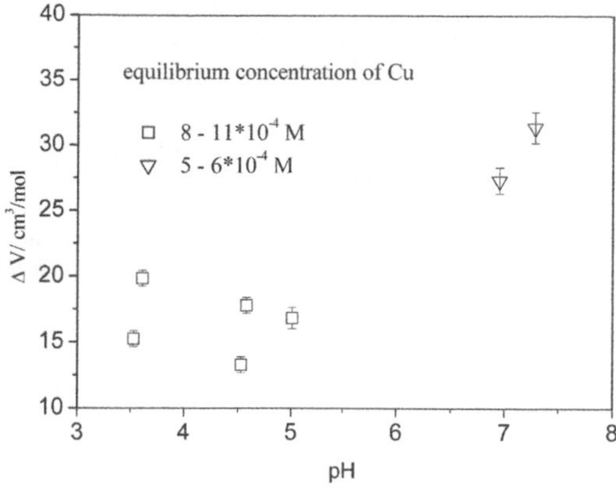

Figure 3. The effect of pH on the ΔV of reaction of Cu(II) with the sodium salt of humic acid in 0.5 mol dm^{-3} NaCl.

polynuclear complexes or to precipitation of heavy metal hydroxides at high pH. It should be emphasized that the values of ΔV in the previous study might be somewhat underestimated, because it was assumed that the excess of humic acid was sufficient to quantitatively bind the heavy metals, whereas, in the present study, the actual amount of bound heavy metals was determined experimentally.

The present results suggest that in the deepest ocean trenches, the binding constants of heavy metals by humic acid will be lower by a factor of 2–3 than the binding constants at atmospheric pressure. The presence of NaCl depresses the dilatometric effect with respect to that determined at low ionic strength. The relative decrease in the binding constants calculated from the dilatometric effect is independent of the assumed model of binding of metals by humic acid provided that only a single binding constant is used.

4 CONCLUSION

It was confirmed that the dilatometric effect of binding of heavy metals by humic acid is positive, of about 20 cm^3/mol. Therefore, the high pressures prevailing in deep oceans will induce liberation of heavy metal cations from their humic complexes. The magnitude of the dilatometric effect is correlated with the affinity of the heavy metals towards humic acid. Thus, copper, which shows a higher affinity then most other metals, induces a relatively strong dilatometric effect.

REFERENCES

Abate G. & Masini, J.C. 2001. Acid-basic and complexation properties of a sedimentary humic acid. A study on the Barra Bonita reservoir of Tietê river, São Paulo State, Brazil. *J. Braz. Chem. Soc.* 12: 109–116.

Antonelli, M.L., Calace, N., Centioli, D., Petronio, B.M. & Pietroletti, M. 2001. Complexing capacity of different molecular weight fractions of sedimentary humic substances, *Analytical Letters* 34(6): 989–1002.

Bianchini, A. & Bowles, K.C. 2002. Metal sulfides in oxygenated aquatic systems: implications for the biotic ligand model. *Comparative Biochemistry and Physiology Part C* 133: 51–64.

Christl, I. & Kretzschmar, R. 2001. Interaction of copper and fulvic acid at the hematite-water interface. *Geochimica et Cosmochimica Acta* 65: 3435–3442.

Gismera, M.J., Procopio, J.R., Sevilla, M.T. & Hernández L. 2003. Copper(II) ion-selective electrodes based on dithiosalicylic and thiosalicylic acids. *Electroanalysis* 15(2): 126–132.

Lu, X., Chen, Z., Hall, S.B. & Yang, X. 2000. Evaluation and characteristics of Pb(II) ion-selective electrode based on aquatic humic substances. *Analytica Chimica Acta* 418: 205–212.

Kosmulski, M. & Mączka, E. 2003. The effect of pressure on the sorption/precipitation of metal cations, and its possible role in spontaneous removal of heavy metal cations from sea water. *Colloids and Surfaces A* 223: 195–199.

Kosmulski, M. & Mączka, E. 2004. Dilatometric study of the adsorption of heavy metal cations on goethite. *Langmuir* 20: 2320–2323.

Krauskopf, K.B. 1956. "Factors controlling the concentrations of thirteen rare metals in sea-water". Geochim. Cosmochim. Acta 9: 1–32.

Marco, R.D. 1996. Surface studies of the jalpaite-based copper(II) ion-selective electrode membrane in seawater. *Marine Chemistry* 55: 389–398.

Mączka, E. 2004, Model studies of sorption in self-purification of ocean water, PhD thesis, Lublin University of Technology, Lublin (in Polish).

Mączka, E. & Kosmulski, M. 2005. The effect of pressure on the binding of heavy metals by organic matter. *Zeszyty Naukowe Politechniki Koszalińskiej* 22: 523–528 (in Polish).

Oste, L.A., Temminghoff, E.J.M., Lexmond, T.M. & van Riemsdijk, W.H. 2002. Measuring and modeling zinc and cadmium binding by humic acid. *Anal. Chem.* 74: 856–862.

Peijnenburg, W.J.G.M. & Jager, T. 2003. Monitoring approaches to assess bioaccessibility and bioavailability of metals: Matrix issues. *Ecotoxicology and Environmental Safety* 56: 63–77.

Pempkowiak, J. & Kosowska, A. 1998. Accumulation of cadmium by green algae *Chlorella vulgaris* in the presence of marine humic substances. *Environment International* 24: 583–588.

Pempkowiak, J., Piotrowska-Szpryt, M. & Kożuch, J. 1998. Rates of diagenetic changes of humic substances in Baltic surface sediments. *Environment International* 24: 589–594.

Rozan, T.F. & Benoit, G. 1999. Intercomparison of DPASV and ISE for the measurement of Cu complexation characteristics of NCM in freshwater. *Environ. Sci. Technol.* 33: 1766–1770.

Trojanowicz, M., Alexander, P.W. & Hibbert D.B. 1998. Flow-injection potentiometric determination of free cadmium ions with a cadmium ion-selective electrode. *Analytica Chimica Acta* 370: 267–278.

Voets, J., Bervoets, L. & Blust, R. 2004. Cadmium bioavailability and accumulation in the presence of humic acid to the zebra mussel, *Dreissena polymorpha, Environ. Sci. Technol.* 38: 1003–1008.

Zhou, P., Yan, H. & Gu, B. 2005. Competitive complexation of metal ions with humic substances. *Chemosphere* 58: 1327–1337.

Environmental Engineering – Pawłowski, Dudzińska & Pawłowski (eds)
© 2007 Taylor & Francis Group, London, ISBN13 978-0-415-40818-9

Partitioning of selected heavy metals between compartments of constructed wetlands for domestic sewage treatment

Hanna Obarska-Pempkowiak & Ewa Wojciechowska
Gdansk University of Technology, Narutowicza, Gdansk, Poland

ABSTRACT: The distribution of heavy metals in two constructed wetland systems was analysed. The spatial distribution of metals in the bottom sediments, metal contents in roots and above-ground plant tissues and seasonal changes of metal contents in different plant tissues were investigated. One of the systems was inhabited with *Phragmites australis* and the other one with three emerged macrophyte species: *Phragmites australis, Typha latifolia* and *Glyceria maxima.* It was found that the plants concentrated heavy metals in their roots. *P.australis* restricted transportation of metals to above-ground tissues. Leaves were the next (after roots) plant part regarding metals content in *P.australis* and *T.latifolia* but not in *G.maxima*. The course of seasonal changes of metal contents differed among three analysed plant species. Accumulation of metals in *G.maxima* increased during the vegetation season, in case of *P.australis*, the maximum of metal contents was observed in summer, while the contents of most metals in *T.latifolia* decreased during the vegetation period.

Keywords: Constructed wetland, sewage treatment, heavy metals.

1 INTRODUCTION

Constructed wetlands (CWs) have become a popular technology for the removal of nutrients from domestic sewage. Recently however, they have also been applied for heavy metals removal from domestic and industrial (mine) sewage (Gergsberg et al., 1986; Gries and Garbe, 1989; Dunbabin and Bowmer, 1992; Hawkins et al., 1997; Crites et al., 1997; Vymazal, 2003; Karathanasis and Johnson, 2003) as well as from landfill leachate (Peverly et al., 1995). Hydrophyte plants inhabiting the wetland systems can take up metals from soil and sediments. According to Ozimek (1988), the content of heavy metals in hydrophyte plants can be several times higher than in the surrounding environment. In many cases plants have no biological barriers protecting them against taking up excessive amounts of heavy metals. This most frequent mechanism of heavy metals uptake is called passive sorption. Usually there is a simple correlation between the content of a number of metals in the environment and in the plants. Cadmium, lead, molybdenum and nickel are accumulated passively in the plants. Some metals, for instance copper and zinc, are taken up selectively, since the plants utilize them in physiological processes. However, there is also a correlation between the concentration of these elements in the environment and in the plants (Ozimek, 1988).

Emerged macrophytes usually accumulate the highest amounts of trace metals in roots (Kufel and Kufel, 1985; Gries and Garbe, 1989; Peverly et al., 1995, Weis et al., 2004). This is due to the fact that the roots are stuck in the bottom sediments. The migration of heavy metals in plants is affected by plant metabolism and depends on the form and mobility of the metal ion. Most plants restrict the movement of metal ions into their photosynthetic tissues (above-ground parts) by the means of storage in cell walls and vacuoles or binding by metal-binding proteins (Weis et al., 2004). The metals necessary for plant growth are uptaken by roots and transported to other parts of the plant. The content of heavy metals in aquatic plants changes seasonally and is correlated to the lability of the metal form. Contents of mobile metals, for instance zinc and manganese, decrease during the vegetation season. According to Kufel and Kufel (1985) the contents of less mobile metals: cadmium, lead and cobalt remain unchanged, while contents of copper, chromium and nickel increase during the vegetation season.

In this article, the distribution of Cd, Zn, Cu, Pb, Fe and Mn in a surface flow wetland in Bielkowo near Gdańsk was analysed. The distribution of Pb, Cu and Cd in a vegetated submerged system in Przywidz near Gdańsk was also investigated. The spatial distribution of heavy metals in the wetland systems, contents of metals in various parts of plants and seasonal changes were analysed. The objective of the study was to find out whether the distribution of these elements in various parts of plants is similar or if distribution differs among metals.

2 STUDY SYSTEM

2.1 *Constructed wetland system in Bielkowo*

The wetland systems in Bielkowo (10 km from Gdansk) was constructed for protection of a drinking water intake for the city of Gdansk (at the Lake Goszyn) against surface and point sources of pollution. The facility is located in the bed of a stream tributary to the Radunia River, just before the river discharges to the Goszyn Lake.

The wetland system was designed as a reservoir surrounded with ground slopes, consolidated with fascine and turf. The total area of the reservoir is 6200 m^2; the volume is equal to 5000 m^3. Inside the reservoir a set of filtration dykes was constructed (Fig. 1).

The system consists of two sub-units:

- wet unit (pond) filled with water all the time (retention time 24 h and water flow 32 l/s)
- dry unit designed for storm water (maximal flow 640 l/s and retention time 0.5 h)

In the wet unit dykes of medium size sand (with hydraulic conductivity $k = 40 \div 86.4\,m/d$) are situated. The system, especially dykes, are inhabited by *Phragmites australis, Typha latifolia* and *Glyceria maxima*. On the bottom of the dykes there are plastic drainage pipes in order to collect infiltration water and direct it to the stream bed beneath the pond.

According to the design project, the dry section should be flooded with water only after heavy rainfalls, in order to retain in the system the first and probably the most contaminated wave of storm water. However, during the first two years of operation, due to mass algae blooming and lack of roots cultivating the ground, the surface of the dams was covered with a thick mat of biomass. This resulted in changing conductivity of the dams and flooding the "dry part" with water. During visits to the facility it was observed that the "dry part" was covered with water all the time and that no retention volume for stormwater run-off existed.

2.2 *Constructed wetland system in Przywidz*

The CW system in Przywidz (60 km south from Gdansk) was constructed as a pilot-scale system for treating domestic sewage. Before the inflow to CW system the sewage was pre-treated in an Imhoff tank and trickling biofilter. The inflow to the system was 22.5 m^3/d (150 Person Equivalent). The wetland system consists of two parts: a horizontal-flow vegetated submerged bed (VSB) (150 m^2) and a cascade filter

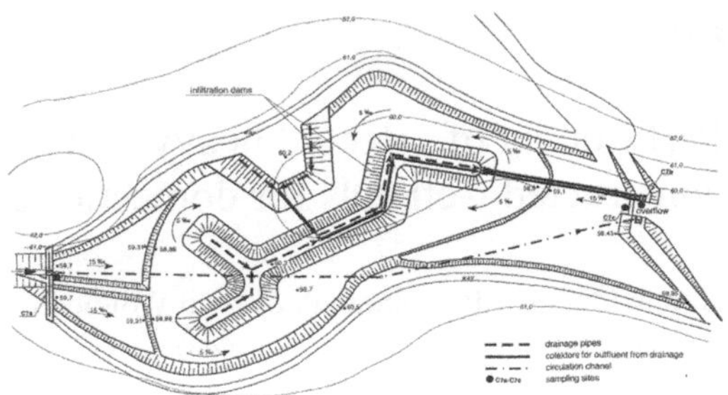

Figure 1. A scheme of the wetland system in Bielkowo.

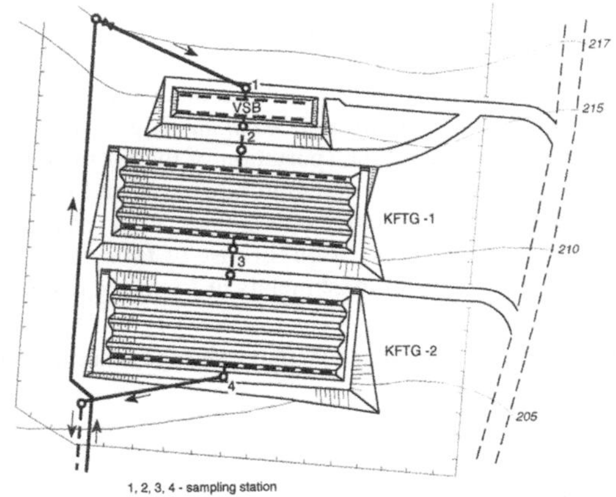

Figure 2. A scheme of the constructed wetland in Przywidz.

(KFTG) planted with *Phragmites australis* (720 m^2) (Fig. 2). The cascade filter consists of lower and upper cascades (KFTG-1 and KFTG-2, respectively). Each cascade is composed of three internal dams separated by four ditches. The length of each dam is 30 m. The bottom of each subsequent dam is placed 25 cm below the bottom of the preceding one.

Gravitational flow of sewage is secured by the filter situated on the slope of a hill. The sewage from the VSB filter inflow to the first ditch of the upper cascade, where aeration takes place. While the ditch is filling, the sewage percolates through the first dam to the ditch beneath and then it percolates through the second dam to the lower ditch and so on.

3 MATERIAL AND METHODS

The contents of heavy metals: Cd, Pb, Zn, Cu, Fe and Mn were measured in bottom sediments and in plant tissues. The measurements were performed three times during the vegetation season: at the beginning (May), in the middle (July) and at the end (October). The plant material, including stems, leaves and underground parts were collected from the whole system area in

Table 1. Comparison of certified and measured contents.

Metal	Recommended	Measured	Recovery (%)	RSD (%)
Cd	0.23 ± 0.04	0.19 ± 0.03	82	16
Cu	3.30 ± 0.25	3.50 ± 0.42	106	12
Pb	1.62*	1.80 ± 0.22	111	12
Zn	38.5 ± 1.9	37.6 ± 2.4	102	6
Mn	57.6 ± 2.1	55.1 ± 4.6	96	8
Fe	58.4 ± 3.9	57.1 ± 3.2	98	6

*information value.

Bielkowo and from VSB filter and the final ditches of KFTG1 and KFTG2 cascades in Przywidz. Samples of bottom sediments were collected in the same periods as samples of plant material (Obarska-Pempkowiak and Ozimek, 2003; Obarska-Pempkowiak, 2003).

Determination of the total content of heavy metals in the bottom sediments was performed according to the following procedure: 1.000 g subsamples of homogenized materials were digested with hydrofluoric acid ($2 \times 5\,cm^3$) followed by perchloric acid and nitric acid mixture 1:1 ($2 \times 5\,cm^3$). Excess acids were evaporated. The dry residue was dissolved in $0.1\,mol/dm^3$ HNO_3. The measurements were carried out in a model Video 11E spectrometer (flame atomic absorption spectrophotometry). Details of the procedure have been described elsewhere (Pempkowiak, 1991). The content of heavy metals in plant material was determined according to the same procedure – the only difference was that 0.5000 g subsamples were utilised, and hydrofluoric acid was not used. Suitable blanks were analysed in parallel with actual samples. Quality control was carried out by analysing certified materials. The recovery was in the range from 93+–3% (Fe), and 104+– 4% (Cd). The content of heavy metals in plant material was determined according to the same procedure – the only difference was that the hydrofluoric acid was not used.

Quality control was considered to be important. Certified material was analysed to access the accuracy and precision of determinations.

In the Table 1 a tested values of cadmium, lead, zinc, copper, iron and manganese are given in the second column. Recovery, third column, indicates satisfactory accuracy. Relative Standard Deviation, given in the fourth column, indicates that precision was no problem.

4 RESULTS

4.1 *Spatial distribution of heavy metals in CW systems*

In Tables 2 and 3 contents of four heavy metals (Cd, Cu, Pb and Zn) in the bottom sediments are presented.

Table 2. Average concentrations of heavy metals in surface sediments at the inflow and at the outflow of the constructed wetland in Bielkowo, mg/kg.

Metal	Inflow	Outflow
Cd	2.6	7.1
Cu	18.5	56.2
Pb	31.3	62.9
Zn	34.8	196

Table 3. Average concentrations of heavy metals in surface sediments from subsequent segments of CW system in Przywidz, mg/kg.

Metal	KFTG1 – ditch 1	KFTG1 – ditch 4	KFTG 2 – ditch 4
Cd	2.5	1	4
Cu	58	43	38
Pb	45	35	31
Zn	98	74	62

Contents of heavy metals were higher at the outflow than at the inflow, indicating that sedimentation and accumulation of organic matter took place mainly near the outflow of the pond. In the CW in Przywidz (Table 3) heavy metal contents in sediments were highest near the inlet to the system. Vymazal (2003) also reported the highest content of metals in the bottom sediments near the inlet to VSB systems. In Przywidz and in the systems analysed by Vymazal (2003) sedimentation took place right after the inlet. In CW Bielkowo, due to the different construction of the wetland system, sedimentation occurs near the outflow of the pond. These observations indicate that most of the metals are present in the suspended fractions and are removed from the wastewater during the sedimentation process.

4.2 *Distribution of heavy metals in plants. Seasonal changes*

Seasonal changes of Cu, Zn, Cd, Pb, Fe and Mn in different parts of *Glyceria maxima*, *Typha latifolia and Phragmites australis* in the CW system Bielkowo are presented in Tables 4, 5 and 6.

In general, the contents of heavy metals in the ground parts of plants (roots, rhizomes) were higher than in above-ground parts at both analysed CW systems. In CW Bielkowo higher contents of heavy metals in plant tissues were found at the outflow, which is consistent with higher accumulation of the metals in the bottom sediments.

In the CW Bielkowo the analysed plants (*Glyceria, Typha*) accumulated the highest amounts of heavy metals in rhizomes. There was one exception: in July and

Table 4. Seasonal changes of Cu, Zn, Cd, Fe and Mn in different parts of *Glyceria maxima*, mg/kg d.m.

		Inflow			Outflow		
Metal	Part of plant	May	July	October	May	July	October
Cu	leaves	**0.2***	**1.2**	**0.6**	**1.1**	**1.8**	**2.1**
		0.11–0.32**	0.53–2.31	0.45–3.38	1.02–1.84	1.54–2.57	0.71–3.76
	stems	**0.9**	**1.3**	**1.6**	**1.2**	**1.4**	**2.1**
		0.33–1.84	0.38–2.74	0.58–3.08	0.45–1.61	0.72–2.45	1.65–316
	roots	**2.1**	**3.6**	**3.5**	**3.8**	**4.3**	**4.1**
		1.00–3.11	2.87–4.05	1.90–4.23	1.74–6.22	0.95–5.54	2.10–7.06
Zn	leaves	**6.4**	**7.7**	**8.1**	**8.1**	**12.3**	**13.2**
		2.51–10.8	3.01–16.1	5.25–12.3	3.88–13.5	5.08–17.3	5.17–16.9
	stems	**10.4**	**15.6**	**22.8**	**15.6**	**10**	**28.4**
		4.45–14.8	9.16–24.1	13.8–35.1	10.5–20.8	5.36–43.1	18.3–60.1
	roots	**17.4**	**24.3**	**19.8**		**25.2**	**24.3**
		6.61–18.5	9.33–41.0	16.6–38.8		9.33–41.0	22.3–46.3
Cd	leaves	**0.10**	**0.13**	**0.14**	**0.14**	**0.14**	**0.19**
		0.08–0.12	0.06–0.35	0.05–0.32	0.09–0.20	0.09–0.15	0.07–0.27
	stems	**0.15**	**0.15**	**0.19**	**0.20**	**0.22**	**0.28**
		0.10–0.16	0.12–0.23	0.13–0.23	0.11–0.39	0.09–0.40	0.20–0.45
	roots	**0.19**	**0.22**	**0.20**	**0.23**	**0.11**	**0.22**
		0.08–0.28	0.11–0.51	0.09–0.27	0.12–0.26	0.08–0.33	0.10–0.33
Fe	leaves	**30.2**	**35.7**	**61.4**	**49.3**	**72.4**	**107.6**
		10.4–56.7	13.9–70.0	33.4–107	25.8–105	30.9–338	87.3–138
	stems	**8.1**	**12.5**	**31.3**	**16.6**	**31.7**	**59.6**
		1.10–16.0	1.34–30.0	16.07–64.7	8.22–35.3	10.6–58.6	22.6–89.1
	roots	**62.1**		**70.3**	**113**	**135**	**138**
		49.9–85.2		57.9–90.3	98.4–145	101–168	97.5–180
Mn	leaves	**49.1**	**57.3**	**84.5**	**113**	**241**	**252**
		10.3–102	11.7–210	70.8–100	51.5–160	73.0–506	71.1–371
	stems	**33.9**	**51.4**	**78.2**	**71.2**	**85.1**	**94.6**
		12.3–65.4	17.9–143	58.2–129	22.4–130	30.6–213	73.0–196
	roots	**285**		**315**	**680**	**545**	**730**
		158–550		12.6–466	134–1358	42.9–2192	49.5–2520

*average value.
**range.

October the content of Cd in *Glyceria* plants located at the outflow was higher in above-ground parts, especially in stems, than in roots. Also the content of Cd in roots decreased in July in comparison to May, indicating that this element was transported to the stems. In many cases the contents of metals in rhizomes did not change with time, suggesting that maximal content was reached. The contents of metals in *P. australis* was not measured.

The contents of analysed metal ions in above-ground parts of *Phragmites* and *Typha* was higher in leaves than in stems (except Cd and Zn in Typha). *G.maxima* had higher contents of Cu, Zn and Cd in stems than in leaves, while Fe and Mn were accumulated in higher amounts in the leaves.

In the CW in Bielkowo the course of seasonal changes of heavy metals contents was different for all analysed plant species. In case of *Glyceria* the content of metals in different parts of plants increased with time – it was the lowest in spring (May) and the highest in autumn (October), showing progressive accumulation of the metals during vegetation season. Zinc was an exception – the contents in roots and stems was lower in July than in May and again increased in October (in the leaves it increased during the season). In the case of *T.latifolia* (cattails) the highest contents of most metal ions (Cu, Zn, Fe, Mn) were found at the beginning of the vegetation season (in May). The contents of Cd and Pb were lowest in spring, reached a maximum in summer (July) and decreased towards the end of vegetation season (October). For *P.australis*, maximal contents of Cu, Zn, Cd and Pb were measured in July, while the Fe and Mn contents reached a maximum at the end of the vegetation season (October).

In the CW Przywidz *P.australis* accumulated the highest amounts of heavy metals (Cu, Pb and Cd) in rhizomes and roots. The average contents of metals in rhizomes was several times higher than in other tissues. Among the analysed metals Cu was most mobile and higher amounts were transported to the above-ground parts. The content of Cd was maximal at the

Table 5. Seasonal changes of Cu, Zn, Cd, Pb, Fe and Mn in different parts of *Typha latifolia*, mg/kg d.m.

		Inflow			Outflow		
Metal	Part of plant	May	July	October	May	July	October
Cu	leaves	**1.3*** 0.72–1.97**	**1.6** 0.86–2.49	**1.8** 0.84–2.23	**2.0** 0.85–2.83	**1.5** 1.28–2.51	**1.4** 0.39–2.16
	stems	**1.7** 0.60–4.09	**0.9** 0.61–1.26	**0.6** 0.28–1.46	**2.2** 1.33–3.01	**1.3** 0.72–2.45	**0.7** 0.35–1.74
	roots	**4.5** 3.85–6.54	**4.3** 3.46–6.89		**5.3** 4.96–7.33	**5.7** 4.84–7.18	
Zn	leaves	**13.2** 10.0–15.8	**8.1** 3.08–9.78	**5.3** 3.44–8.07	**16.1** 10.3–20.7	**10.9** 6.74–13.4	**11.4** 8.41–17.9
	stems	**20.0** 15.8–24.9	**4.5** 1.10–9.40	**3.9** 2.75–5.26	**24.9** 11.3–51.0	**15.6** 8.66–24.8	**6.4** 4.85–9.13
	roots	**40.1** 29.3–51.6			**65.8** 54.78–92.67		
Cd	leaves	**0.14** 0.10–0.15	**0.19** 0.17–0.24	**0.17** 0.10–0.28	**0.14** 0.10–0.16	**0.21** 0.13–0.30	**0.19** 0.15–0.25
	stems	**0.19** 0.15–0.22	**0.23** 0.09–0.027	**0.21** 0.16–0.32	**0.2** 0.12–0.30	**0.28** 0.11–0.49	**0.24** 0.23–0.35
	roots	**0.26** 0.18–0.37	**0.28** 0.11–0.51	**0.28** 0.19–0.37	**0.36** 0.11–0.46	**0.33** 0.20–0.39	**0.32** 0.27–0.57
Pb	leaves	b.d.l.***	**0.04** 0.01–0.08	b.d.l.	b.d.l.	**0.04** 0.01–0.10	b.d.l.
	stems	b.d.l.	**0.08** 0.03–0.12	b.d.l.	b.d.l.	**0.10** 0.10–0.13	b.d.l.
	roots	b.d.l.	**0.13** 0.10–0.18	b.d.l.	b.d.l.	**0.18** 0.16–0.20	b.d.l.
Fe	leaves	**125** 65.0–201	**51.2** 9.77–76.7	**34.6** 24.1–43.2	**160** 87.8–248	**65.1** 48.7–80.2	**33.7** 25.4–40.2
	stems	**96** 80.5–119	**49** 43.4–62.4	**27** 21.2–34.2	**65.1** 49.8–89.7	**32.8** 29.5–40.3	**16.5** 13.6–23.5
	roots	**260** 185–412			**415** 288–564		
Mn	leaves	**450** 307–594	**495** 324–609	**545** 371–713	**515** 436–605	**540** 457–653	**620** 340–987
	stems	**360** 263–428	**270** 33.5–379	**105** 68.1–157	**430** 200–650	**320** 158–357	**140** 65.8–320
	roots	**460** 313–706			**560** 498–703		

*average value.
**range.
***b.d.l. – below detection level.

beginning of vegetation season (similar situation as in case of *G.maxima* at the outflow from the CW system in Bielkowo). The contents of Pb and Cu reached a maximum in summer. In autumn (the end of vegetation season) the contents of all metals decreased. The contents of heavy metals in rhizomes were significantly higher in the VSB filter than in KFTG2, while the contents in stems and leaves were quite similar.

5 DISCUSSION

According to Baker and Walker (1989) plants employ three strategies regarding metal translocation to aboveground tissues: excluder strategy, indicator strategy and accumulator strategy. With the excluder strategy shoot concentrations are kept low until a critical soil concentration is reached. With the indicator strategy passive uptake and transport occurs and contents in plant tissues are proportional to soil metal concentration. Plants applying accumulator strategy are highly specialized metal hyperaccumulators and actively uptake metals.

In the current study we found that *Phragmites australis* (CW Przywidz) accumulated the largest amounts of metals in roots and rhizomes. This is in agreement with many literature reports (Kufel and Kufel, 1985; Gries and Garbe, 1989; Peverly et al., 1995; Weis et al., 2004). Hence, Weis et al. conclude that the plant employs the excluder strategy. The results

Table 6. Seasonal changes of Cu, Zn, Cd, Pb, Fe and Mn in above-ground parts of *Phragmites australis*, mg/kg d.m.

Metal	Part of plant	Outflow May	July	October
CU	leaves	**1.1*** 0.43–1.63**	**1.5** 0.97–2.03	**1.2** 0.49–2.24
	stems	**0.45** 0.20–0.80	**1.1** 0.70–1.41	**0.51** 0.25–0.90
Zn	leaves	**14.4** 9.50–20.1	**18.6** 13.9–25.6	**14.3** 10.6–18.3
	stems	**10.5** 3.10–17.9	**17.6** 11.6–23.2	**6.9** 3.75–7.30
Cd	leaves	**0.17** 0.09–0.25	**0.24** 0.09–0.38	**0.18** 0.10–0.25
	stems	**0.10** 0.05–0.16	**0.17** 0.09–0.26	**0.15** 0.08–0.20
Pb	leaves	**0.08** 0.03–0.13	**0.12** 0.09–0.16	**0.11** 0.08–0.13
	stems	**0.05** 0.01–0.10	**0.08** 0.04–0.10	**0.06** 0.05–0.07
Fe	leaves	**55.2** 35.2–77.2	**78.4** 40.6–164	**98.7** 63.5–138
	stems	**17.3** 5.00–26.2	**12.9** 5.81–26.2	**24.6** 15.2–41.1
Mn	leaves	**133** 33.0–240	**158** 79.9–293	**220** 42.2–371
	stems	**39.6** 18.2–73.8	**46.8** 29.9–72.8	**46.2** 18.5–129

*average value.
**range.

of our study from CW Przywidz show that though the contents of heavy metals in the bottom sediments were significantly higher near the inlet (VSB filter) and decreased along the treatment path, the metal contents in stems and leaves of *P.australis* collected from different segments of the system were comparable. On the other hand, rhizomes concentrations were significantly higher in the VSB filter than in KFTG2. It seems that *P.australis* collected from different segments of CW Przywidz transported more or less the same amounts of heavy metals to above-ground parts and that the content of metals in above-ground parts was not affected by the concentration in the environment (which was significantly higher in VSB filter). However, the contents of metals in above-ground parts were not constant during vegetation season – both at the CW Bielkowo and CW Przywidz a summer maximum of metal contents was observed.

Also the other analysed plant species: *G.maxima* and *T.latifolia* concentrated most metals in roots and rhizomes, though this observation does not regard all metals: Cd was transported to above-ground parts by both plant species and Zn only by *G.maxima*. According to Weis et al. (2004) the content of metals in plants is affected by the presence of other metals in the environment – for instance zinc distribution is affected by Cu.

Leaves are reported to be the next plant tissue (after roots) accumulating metals, followed by stems (Weis et al., 2004). Generally, the results of this study are in agreement with this since *T.latifolia* and *P.australis* had higher metal contents in leaves than in stems. *Glyceria maxima*, however had higher contents of Cd, Cu and Zn in stems than in leaves. Vymazal (2003), who analysed contents of heavy metals in various parts of *Phalaris arundinacea* in a wetland system in Nucice (the Czech Republic), reported that the contents of heavy metals in roots were significantly higher than in stems or leaves but no statistically significant difference between heavy metals contents in stems and leaves was observed.

6 CONCLUSIONS

1. Surface flow CWs inhabited with several macrophytes species are capable of removing heavy metals.
2. The highest contents of heavy metals were found in the ground parts of plants. In case of *G. maxima,* roots and rhizomes seem to contain the maximal amounts of heavy metals.
3. *T.latifolia* and *P.australis* contained higher amounts of metals in leaves than in stems, which was the opposite of *G.maxima*.
4. Accumulation of heavy metals in *G.maxima* increased during the vegetation season and reached its maximum in autumn.
5. The highest heavy metal contents in the season were observed in May. The contents of Cu, Zn, Fe and Mn in *T.latifolia* decreased during the vegetation season, while Cd and Pb contents were at their maximum in summer, when vegetation was most intensive.
6. The highest contents of heavy metals in *P.australis*, both in CWs Bielkowo and Przywidz, were observed in the middle of the vegetation period (July).
7. *P.australis* restricted transportation of heavy metals to above-ground tissues.

REFERENCES

Baker A.J., Walker P.L. 1989. Ecophysiology of metal uptake by tolerant plants. In: Shaw A.J. (Ed.), Heavy Metal Tolerance in Plants: Evolutionary Aspects. CRC Press, Boca Raton, FL: 156–177.

Crites R.W., Dombeck G.D., Watson R.C., Williams C.R. 1997. Removal of metals and ammonia in constructed wetlands. Wat. Environ. Res., 69 (2): 132–135.

Dunbabin J., Bowmer K.H. 1992. Potential use of constructed wetlands for treatment of industrial wastewater containing metals. The Science of the Total Environment, 3: 151–168.

Gersberg R.M., Es B.V., Lyon S.R. 1986. Role of aquatic plants in wastewater treatment by artificial wetlands. Wat. Res. 20(3): 363–368.

Gries C., Garbe D. 1989. Biomass and nitrogen, phosphorus and heavy metal content of *Phragmites australis* during the third growing season in a root zone wastewater treatment. Archiv f. Hydrobiologie. Stuttgart 1989, 117 (1): 97–105.

Hawkins W.B., Rodgers J.H. Jr., Gillespie W.B., Dunn A.W., Dorn P.B., Cano M.L. 1997. Design and construction of wetlands for aqueous transfers and transformations of selected metals. Ecotoxicology and Environmental Safety 36: 238–248.

Karathanansis A.D., Johnson C.M. 2003. Metal Removal by three aquatic plants in an acid mine drainage wetland. Mine Water and the Environment (2003) 22: 22–30.

Kufel I., Kufel . 1985. Heavy metals and eral nutrient budget in *Phragmites australis* and *Typha angustifolia*. Symp. Biol. Hungar. 1985, Vol. 29: 61–66.

Obarska-Pempkowiak H. 2003. Removal and retention of selected heavy metals in components of a hybrid wetland system. In: Mander U. and Jenssen P. (Eds): Constructed wetlands for wastewater treatment in cold climates. WIT Press Southampton, Boston, 2003: 300–309.

Obarska-Pempkowiak H., Ozimek T. 2003.Comparison of Usefulness of Three Emergent Macrophytes for Surface Water Protection against Pollution and Eutrophication: Case study Bielkowo, Poland. In: Vymazal J. (Ed.) Wetlands – nutrients, metals and mass cycling. Backhuys Publishers, Leiden, The Netherlands: 215–226.

Ozimek T. 1988. The role of emerged macrophytes in heavy metals cycling in aquatic ecosystems. Wiadomosci Ekologiczne 35/(1): 30–44 (in Polish).

Pempkowiak J. 1991. Enrichment factors of heavy metals in the Baltic surface sediments of dated with 210Pb and 137Cs, Environ. Int., 17, 421–428

Peverly J.H., Surface J.M., Wang T. 1995. Growth and trace metals absorption by *Phragmites australis* in wetlands constructed for landfill leachate treatment. Ecological Engineering 5(1995): 21–35.

Vymazal J. 2003. Distribution of iron, cadmium, nickel and lead in a constructed wetland receiving municipal sewage. In: Vymazal J. (Ed.): Wetlands – nutrients , metals and mass cycling. Backhuys Publishers, Leiden, 2003: 341–363.

Weis J.S., Glover T., Weis P. 2004. Interactions of metals affect their distribution in tissues of *Phragmites australis*. Environmental Pollution 131: 409–415.

Environmental Engineering – Pawłowski, Dudzińska & Pawłowski (eds)
© 2007 Taylor & Francis Group, London, ISBN13 978-0-415-40818-9

The influence of local sources of pollution on Cr, Ni, Cu and Co in sediments of some river basins of the upper Narew

Elżbieta Skorbilowicz
Institute of Civil Engineering, Technical University of Bialystok, ul. Wiejska, Bialystok

ABSTRACT: The following paper aims at determining the degree to which the sewage from a water treatment plant affects the content of chromium, nickel, copper and cobalt (total and soluble) in bottom sediments. Twelve water-courses were selected that receive treated sewage from water treatment plants which recycle sewage from towns and residential districts. Fractions applied in geochemical charting of <0.2 mm grain size were selected for analyses. In sediments, the total content of heavy metals was assessed after conducting mineralisation in nitric acid in a closed microwave system MARS 5. At the same time, in samples of bottom deposits, soluble metals were cold extracted with 10% HNO_3. The concentration of metals was determined using the ASA method. The extent of industrialisation, amount of sewage transported from water treatment plants and the amount of industrial waste are decisive factors which make water treatment plants the point source of bottom sediments pollution.

Keywords: Heavy metals, rivers, sediments bottom.

1 INTRODUCTION

The East of Poland is an area renowned for its natural wealth, however, human activity in the area brings about changes in the environment. As far as water ecosystems are concerned it is apparent in changing physico-chemical indicators of water quality, sediments and species and numbers of animals and plants. The major threat affecting the waters of reservoirs is contamination by heavy metals (Tam and Wong 2000, Baptista Neto et al. 2000, Chen et al. 2000, Xiangdong et al. 2000, Tsail et al. 2003) states that surface water is one of the most degraded elements of the environment in Poland and the main source of pollution is domestic and industrial sewage. Sewage is the source of practically all micro-impurities identified in surface water. Numerous sections of rivers in the Podlasie are polluted as a result of dumping untreated or inadequately treated domestic sewage or sewage from the food-processing industry. Metals which enter rivers may be subject to sorption in sediments because of the significant content of humus substances as well as mineral clay particles. In addition, their nearly neutral reaction also encourages their immobilisation. In general, along with the duration of a reservoir's existence grows the total content of heavy metals in bottom sediments. The knowledge of the content of heavy metals in the water and sediments while conducting detailed research into rivers may lead to a more comprehensive chemical analysis of water environment.

The following paper aims at determining the degree to which the sewage from water treatment plant affects the content of chromium, nickel, copper and cobalt (total and soluble) in the bottom sediments of selected river basins upper Narew.

2 MATERIALS AND METHODS

The research was conducted during summer 2004 in Podlasie. Twelve water-courses were selected that receive treated sewage from water treatment plants which recycle sewage from towns and residential districts. These are in most cases municipal water treatment plants which receive domestic sewage as well as sewage from the food-processing industry, mainly the dairy industry. Water treatment plants in Białystok, Bielsk Podlaski and Łapy partly also receive industrial waste. In order to simplify the interpretation of the results in selected rivers three sampling points were established: the 1st – control point – the source the other two points were situated on the left and the right river bank, past the sewage dump at a distance of 1 up to 10 m. The sample material was collected by the river bank, where suspended material is deposited. The representative sediment sample for each point was prepared by mixing several prime samples collected at different points in the river bed next to the banks within the distance range defined above, from the layer which was 0.05 m thick in amounts over 0.5 kg.

Preparing bottom sediment samples to determine metals involved drying until samples were air-dry and sifting with a polyethylene sieve with 0.2 mm mesh diameter. Fractions applied in geochemical charting of <0.2 mm grain size were selected for analyses. In sediments the total content of Cr, Ni, Cu and Co was assessed after conducting mineralisation in nitric acid in a closed microwave system MARS 5. At the same time, in samples of bottom deposits, soluble metals were cold extracted with 10% HNO_3. The concentration of metals was determined with the ASA method. The reaction of deposits in water was conducted with a potentiometer method.

3 RESULTS AND DISCUSSION

The concentration of total chromium in the investigated sediments was in the range of 2.1 to 57.1 $mg \cdot kg^{-1}$ (Table 1). Analysing the results in the light of natural geochemical content of chromium quoted by Bojakowska and Sokołowska (1998) at 5 $mg \cdot kg^{-1}$, it should be noticed that the quantity was not exceeded at five sites (12%). However, Lis and Pasieczna (1995) claim that the natural concentration of chromium in unpolluted river-bed sediments amounts to 10 $mg \cdot kg^{-1}$. In reference to the latter value the content of total chromium was exceeded at twenty three sites (60%). The largest amount of this element was found in sediments located past the treated water dump in rivers: the Awissa (57.1 $mg \cdot kg^{-1}$) the Horodnianka (47.3 $mg \cdot kg^{-1}$), the Biała, a tributary of the Supraśl (39.1 $mg \cdot kg^{-1}$) and the Sokołda (33.2 $mg \cdot kg^{-1}$). The lowest content of general chromium was in the range of 2.1 up to 12.9 $mg \cdot kg^{-1}$ and was found at investigation sites situated in the Orlanka, the Targonka and the Narewka river near the town of Białowieża. The content of soluble forms of chromium was between 0.5 up to 11.3 $mg \cdot kg^{-1}$ and only four sediment samples indicated concentration above 10 $mg \cdot kg^{-1}$. This relatively low content of the soluble form may indicate chromium of little mobility, that is also indicated by the percentage of the soluble form of chromium in the total amount which was in the range of 10 to 25%. Chromium is one of the least mobile elements in the natural environment.

The total concentration of copper in the investigated sediments was between 2.9 and 44.3 $mg \cdot kg^{-1}$ (Table 1). The content of copper below that of the geochemical background was found at almost every control site (source) (Bojakowska and Sokołowska 1998). After analysing the amount of copper in samples from particular rivers it can be stated that the sediments in the Rudnia river (source 3.7 $mg \cdot kg^{-1}$, 44.3 $mg \cdot kg^{-1}$) are the most polluted, the sediments of the Biała (a tributary of the Orlanka), the Biała (a tributary of the Supraśl), the Awissa and Horodnianka are less polluted. This is owing to the influence of a water treatment plant on polluting sediments with copper. In the other investigated rivers the arithmetic mean of the left and right bank past the sewage dump was within the 7.9 to 17.2 $mg \cdot kg^{-1}$ range. The sediments contained a small amount of soluble copper. The exception is the test material from the Rudnia river at Zabłudów (16.7 $mg \cdot kg^{-1}$). The soluble forms in most investigated sediments accounted for 15–30% of total copper. This may indicate binding of a major portion of the copper by an organic substance present in the bottom sediments. Only a small portion of the copper occurs in sediments in active, thus easily soluble forms. It is firmly bound by mineral and organic components in the sediments. However, it is an organic substance which performs the major role in binding.

The content of nickel in bottom sediments in thirty nine selected research sites was in the range of 3.1 to 41.9 $mg \cdot kg^{-1}$ (Table 1). The natural content of this element in sediments, according to Kabata-Pendias and Pendias (1999) and Lis and Pasieczna (1995) is 10 $mg \cdot kg^{-1}$. In the light of these results one could state that 33% of sediments in most control sites (source) are unpolluted, while 67% of the remaining sediments are polluted with this element. The amount of nickel in the investigated bottom sediments is directly affected by water treatment plants. The largest amount of nickel was found in sediments of the Horodnianka river at the town of Choroszcz (left bank 35.8 $mgNi \cdot kg^{-1}$, right bank 48.1 $mgNi \cdot kg^{-1}$). The arithmetic mean for nickel of the left and right bank past the sewage dump was approximately 20 $mgNi \cdot kg^{-1}$ in sediments of the Biała (Białystok), the Awissa, the Sokołda, the Narewka and the Narew. The content of soluble nickel was fairly diversified and ranged from 0.3 to 16.6 $mg \cdot kg^{-1}$. In two samples of sediment increased accumulation of soluble nickel was found: in the Horodnianka (right bank 16.6 $mg \cdot kg^{-1}$) and the Sokołda (right bank 12.3 $mg \cdot kg^{-1}$). Soluble forms in most investigated sediments accounted for 10–30% of total nickel. This metal easily creates chelate compounds and complex cations and anions.

The content of cobalt does not exceed 10 $mg \cdot kg^{-1}$ in unpolluted sediments (Bojakowska and Sokołowska 1998, Lis and Pasieczna 1995). The conducted tests showed fluctuations of the general content of this element – from 1.9 $mg \cdot kg^{-1}$ to 24.4 $mg \cdot kg^{-1}$ (Table 1). At twenty two sites the value of the geochemical background was exceeded by the amount of cobalt. The largest content of total cobalt in the collected sediments was found in the Biała river in Białystok (left bank 25.1 $mg \cdot kg^{-1}$, right bank 23.2 $mg \cdot kg^{-1}$), in the Narewka river (Narewka), the Awissa river and the Czarna river. The lowest content of total cobalt, ranging from 2.8 to 10.5 $mg \cdot kg^{-1}$

Table 1. The content of heavy metals in bottom sediments.

River	Town, amount of sewage carried away	Sampling sites*	Reaction	Content, $mg \cdot kg^{-1}$ s.m.							
				Cr		Ni		Cu		Co	
				Total	Soluble	Total	Soluble	Total	Soluble	Total	Soluble
Biała	Białystok	a	7.5	8.1	1.4	5.9	0.87	7.1	1.4	10.1	0.9
	5580 m^3/d	b	7.4	39.1	11.3	22.5	2.9	23.8	4.5	24.4	1.9
Biała	Bielsk Podlaski	a	6.9	7.1	1.1	8.1	1.1	8.6	1.3	8.1	0.9
	5000 m^3/d	b	7.2	24.9	7.0	17.5	2.0	22.1	4.1	10.5	1.3
Awissa	Łapy	a	7.9	11.6	2.2	5.1	1.9	4.1	1.3	4.1	1.1
	4423 m^3/d	b	7.3	57.1	9.1	24.2	6.1	22.7	5.9	18.5	3.3
Sokołda	Sokółka	a	7.4	8.1	2.1	5.1	1.4	2.9	1.5	5.3	1.4
	2057 m^3/d	b	6.8	33.2	10.4	25.1	10.7	17.2	2.8	12.2	1.9
Targonka	Mońki	a	7.8	2.1	0.4	7.1	1.9	3.3	0.9	1.9	0.9
	1650 m^3/d	b	7.6	8.3	1.7	10.6	3.4	9.9	3.7	2.8	1.4
Horodnianka	Choroszcz	a	6.9	11.6	2.2	10.1	2.8	8.1	1.2	3.3	0.8
	680 m^3/d	b	7.4	47.3	7.0	41.9	13.3	19.6	4.0	9.7	1.1
Czarna	Czarna Białostocka	a	6.8	7.1	2.1	8.2	2.9	2.9	0.8	8.2	1.2
	544 m^3/d	b	7.2	29.7	7.2	16.8	4.1	12.8	3.3	18.5	2.2
Narewka	Białowieża	e	6.6	9.9	1.8	11.1	1.5	3.2	0.4	6.3	0.6
	350 m^3/d	b	7.3	12.9	4.0	20.9	1.6	8.8	0.9	16.9	0.9
Supraśl	Gródek	a	7.1	19.1	2.2	3.1	0.5	3.2	0.6	7.1	0.7
	180 m^3/d	b	8.0	24.9	3.9	8.0	1.5	11.4	1.3	14.2	1.1
Rudnia	Zabłudów	a	7.1	9.6	2.9	5.9	2.4	3.7	1.1	2.1	0.4
	100 m^3/d	b	7.0	15.7	4.0	11.9	4.0	44.3	16.7	12.1	2.4
Narewka	Narewka	e	6.6	9.9	1.8	11.1	1.5	3.2	0.4	6.3	0.6
	52 m^3/d	b	6.9	16.2	3.8	21.5	3.6	12.8	3.3	20.2	3.6
Orlanka	Orla	a	6.9	2.9	0.7	4.1	1.2	3.1	0.8	9.1	1.2
	50 m^3/d	b	7.2	4.1	0.8	11.0	1.0	7.9	1.0	10.0	2.0
Narew	Narew	d	7.2	8.9	0.5	12.1	0.3	10.6	2.1	9.4	0.3
	40 m^3/d	b	7.8	22.0	2.0	20.8	0.7	17.2	2.6	12.0	0.5

*Key: a – river source, b – arithmetic mean from the right and left river bank, d – Babia Góra (asettlement), e – before Białowieża, total – total amount of an element, soluble – soluble amount of an element.

was found at the research sites situated in the Targonka, the Horodnianka, the Orlanka and the Biała (Bielsk Podlaski). The content of soluble forms of cobalt was relatively low and was within the range of 0.3 to 3.6 mg · kg and only three samples of sediment showed concentrations above 3 $mgCo \cdot kg^{-1}$. The largest amount of soluble cobalt was found in the sediments in the Narewka (Narewka) and the Awissa.

The sediment reaction in all investigated rivers was similar. It was neutral and slightly basic, thus typical of rivers.

The investigated water treatment plants are situated in some districts of Podlasie which is a region of unspoilt natural beauty. In recent years a number of water treatment plants were built or modernised for smaller towns, others will soon be constructed. However, in the existing treatment plants sewage is not treated adequately, the water dumped from the water treatment plants is a significant source of pollution. Our tests prove that water treatment plants affect the process of heavy metal accumulation in bottom sediments. The conducted statistical analysis showed correlations between the total amount of copper and cobalt in bottom sediment and the amount of sewage carried away from the water treatment plant with a correlation coefficient of 0.64 and 0.58. Anthropogenic sources of pollution of rivers in Podlasie is mainly sewage carried away from the major urban areas and factories as indicated by the research in Białystok (the Biała, a tributary of the Supraśl), Bielsk Podlaski (the Biała, a tributary of the Orlanka) and Łapy (the Awissa river). Major food-processing plants (dairy produce, fruit and vegetables processing, meat, mineral water production industry), plants producing construction materials, metal industry and power plants are all situated in these areas. In addition, in Białystok there is also textile industry (wool, cotton, carpet production), electronic industry, alcohol production industry, brewery, apart from the above there is a major sugar factory in Łapy. Geochemical research into sediments conducted in the remaining rivers show that most sediments contain amounts of metals similar to that of the geochemical background, which results from the geological structure of the area. In sediments from the Horodnianka a high content of nickel, chromium and copper was found while the bottom sediments from the Rudnia revealed an increased content of copper.

4 CONCLUSIONS

The extent of industrialisation, the amount of sewage removed by water treatment plants and the amount of industrial waste are decisive factors which make water treatment plants the point source of bottom sediments pollution.

The geochemical background value for bottom sediments was exceeded for all the investigated metals in the case of the Biała river, a tributary of the Supraśl.

REFERENCES

Baptista Neto, J. A., Smith, B. J. & McAllister, J. J. 2000. Heavy metal concentrations in surface sediments in a nearshore environment, Jurujuba Sound, Southeast Brazil, Environmental Pollution, 109, 1–9.

Bojakowska, I. & Sokołowska, G. 1998. Geochemiczne klasy czystości osadów wodnych, Przeg. Geolog., 46, 1, 49–54.

Chen, J. S., Wang, F. Y., Li, X. D. & Song, J. J. 2000. Geographical variations of trace elements in sediments of the major rivers in eastern China, Environmental Geology, Springer-Verlag, 39(12), November, 1334–1340.

Kabata-Pendias, A. & Pendias, H. 1999. Biogeochemia pierwiastków śladowych, PWN, Warszawa, 364.

Lis, J. & Pasieczna, A. 1995, Atlas geochemiczny Polski w skali 1: 2 500 000, Państw. Inst. Geol., Warszawa, 72.

Tam, N. F. Y. & Wong, Y. S. 2000. Spatial variation of heavy metals in surface sediments of Hong Kong mangrove swamps, Environmental Pollution, 110, 195–205.

Tsail, J., Yu, K. C. & Ho, S. T. 2003. Correlation of iron/iron oxides and trace heavy metals in sediments of five rivers, in southern Taiwan, Diffuse Pollution Conference, Dublin, 14–25.

Xiangdong Li, Zhenguo Shen, Onyx W. H. Wal & Yok-sheung Li. 2000. Chemical partitioning of heavy metal contaminants in sediments of the Pearl River Estuary, Chemical Speciation and Bioavailability, 12(1), 17–25.

Environmental Engineering – Pawłowski, Dudzińska & Pawłowski (eds)
ISBN13 978-0-415-40818-9

Selected methods to mitigate environment threats along Poland's Baltic Sea coast

Lidia Kruk-Dowgiałło, Radosław Opioła & Andrzej Osowiecki
Department of Ecology, Maritime Institute in Gdańsk, Dlugi Targ Gdańsk, Poland

Janusz Pempkowiak
Marine Biogeochemistry Dept., Institute of Oceanology, Powstanców Warszawy, Sopot, Poland

ABSTRACT: For many years the coastal zone of the Polish Baltic has been receiving sewage and pollution from the land based sources. Since the mid-1970s till the mid-1980s eutrophication processes were observed in the Gulf of Gdańsk. They were particularly intense in the Puck, Vistula and Szczecin lagoons. Eutrophication considerably affected the qualitative and quantitative structure of macrophytes and macrozoobenthos, diminished the population of commercial fish stock, caused massive algal and cyanobacterial summer blooms, impoverished sanitary conditions of the water, sediments and fish stocks etc. Two methods of marine environment mitigation, i.e. restoration of the biological resources by means of artificial reefs and elongation of underwater treated sewage collector pipes are discussed in the paper. In the authors' opinion additional measures, apart from nutrients and pollution load reduction, are necessary in order to improve the state of the environment of the Polish coastal zone.

Keywords: Threats mitigation, artificial reefs, underwater collectors, eutrophication, coastal zone.

1 INTRODUCTION

Environmental changes caused by several-decade long discharges of nutrients and toxic substances to the Baltic sea have considerably hampered fishery, as well as the tourist and recreational attractiveness of its coastal zone. This initiated subsequent deterioration of local communities' living standards and created social problems for local administrative bodies.

In this paper the authors discuss the present state, potential threats as well as activities aimed at restoration of the previous natural and economic values of the coastal zone. Two projects planned in the Gulf of Gdańsk, i.e. establishing artificial reefs as well as lengthening the underwater sewage collector from the sewage treatment plant have been selected for consideration.

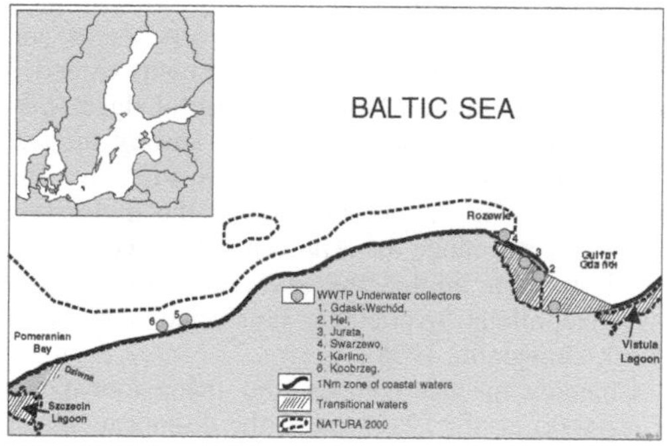

Figure 1. Map of the Polish Coastal zone of the Baltic Sea.

2 THE PRESENT STATE OF THE SOUTHERN BALTIC COASTAL ZONE

Human activity, both land and sea-based, constantly influences the coastal zone environment. Discharges of suspended matter, nutrients and toxic substances lasting for decades have led to permanent changes in the Baltic Sea coastal zone environment. The most drastic effects of deterioration can be observed in shallow water bodies with limited exchange of water with the open sea, such as the Vistula Lagoon, the Szczecin Lagoon, the Puck Lagoon, and some parts of the Gulf of Gdańsk, all included in the transitional waters category (Krzymiński et al., 2004), (Fig. 1, Table 1). The most spectacular effect of discharges from the land based sources is eutrophication. The process accelerated in 1960s and 1970s and decelerated in 1980s due to reduction of nutrient discharges. However, concentrations of nitrogen and phosphorus compounds

Table 1. General characteristics and evaluation of the environmental status of the Poland's Southern Baltic coastal areas.

Sea region	Area [km²]	Depth mean [m]	Depth max. [m]	Water volume [km³]	Watercourses Number	Watercourses Catchment [km²]	Trophic status
Vistula Lagoon (whole)	838	2.7	5.1	2.31	12	23 870	hypertrophy: Hot-Spot
Gulf of Gdańsk proper	4940	59	111	291	26	323 200	high
Puck Bay outer part	256	20.5	54	5.26	11	215	high
Puck Lagoon	102	3.13	9.4	0.32	4	694	high
Szczecin Lagoon (whole)	687	3.8	8.5	2.58	9	129 000	very high
Open coast: Rozewie to Dziwna mouth	~482	5	10	~2.4	8	17 359	moderate

in the isohaline layer have remained at such a high level that they are still responsible for excessive primary production of phytoplankton (Environment . . ., 2002). The main source of nutrients for the Gulf of Gdańsk is the Vistula River, the second largest river in the Baltic catchment area, and the Vistula Lagoon fed with nutrients by the Pregola River. The Vistula Lagoon is included in the Baltic Sea Hot Spots list. The open Baltic waters along the mid-coast, from the cap Rozewie to the Dziwna mouth are influenced by small local rivers. Nutrient levels in the Szczecin Lagoon and the Pomeranian Bay are closely related to loads discharged by the Odra River.

The first observed signs of eutrophication were development of brown algae mats (*Pilayella littoralis*) and disappearance of both *Fucus vesiculosus* and *Furcellaria lumbricalis*. The algal blooms extended both in time and area. Moreover, harmful algae: *Aphanizomenon flos-aquae, Nodularia spumigena* and *Dinophysis norvegica* increased considerably. The decrease of the photic zone led to the diminution of the depth range of Angiospermae occurrence, which in turn limited the associated communities of phytophilous invertebrate fauna and ichtiofauna (Kruk-Dowgiałło, 1998).

Changes in macrophytes and macrozoobenthos (Ciszewski et al., 1992) led to the disappearance of many commercial fish species, which were replaced with "fish weeds" for example *Gasterosteus aculeatus* and *Pungitius pungitius* mass-occurring in the 1980s and 1990s, *Neogobius melanostomus* dominating in the 1990s. Landings have decreased several times, and sanitary conditions of the beaches and coastal swimming areas have worsened (Sobol and Szumilas, 1992).

Despite a pronounced decrease in nutrients and loads of toxic substances discharged to the coastal zone in the last decade, hardly any improvement has been detected, apart from better sanitary conditions in the swimming areas (Kruk-Dowgiałło & Dubrawski, 1998a). High winter concentrations of nitrates resulting in the increase of the N:P ratio in comparison to 1970s accompanied by phosphates increase (attributed to anoxic conditions), indicate secondary eutrophication. The coastal zone can be also regarded as seriously contaminated with heavy metals and pesticides (Pempkowiak et al., 2000a; Pempkowiak et al., 2000b).

3 TYPES OF THREATS TO THE BALTIC COASTAL ZONE

For the purpose of this paper threats were clustered into three categories: i) caused by forces of Nature (climate changes, currents, waves and sea level rise); ii) originating from commercial and domestic activities (contaminants discharges via rivers and collectors, wet and dry atmospheric deposition, ship cargo transportation and unloading, dredging and silting works, mineral and oil extraction, renewable energy production and munitions dumping); iii) purposeful actions (military, terrorist activities, ballast and bilge water discharges, disobeying environmental regulations).

Most of these factors have influenced the coastal zone environment for a long period of time causing extensive changes and damage to the environment. Some problems originate from inadequate monitoring and limited knowledge of the effects brought about by specific threats. For example loads of contaminants discharged with river run-off and both legal and illegal collectors have been monitored inadequately. The State Monitoring Programme along the entire coast under Poland's jurisdiction comprises just twelve rivers (since 1988). From among 20 streams and rivers entering the Gulf of Gdańsk, only two are regularly monitored. The volume and quality of treated sewage are measured after the treatment process. Whenever the treatment facilities are situated at a distance from the coast this approach can be regarded as inadequate. Moreover, in the vicinity of collectors discharging treated sewage to the sea no obligatory environment impact assessments have been performed.

Atmospheric deposition of contaminants is monitored qualitatively and quantitatively on a small scale.

Table 2. The status of implemented and planned activities aimed at environmental threats mitigation: 1-conceptual, 2-decision-making, 3-implemented.

Mitigation activities	Activity description	Mitigation status		
		1	2	3
Sewage treatment plants	Biological reactors, chemical methods, i.e. reduction of phosphorus with iron sulphate;	X	X	X
Undersea collectors with diffusers	IV grade of sewage treatment: dispersion of treated sewage at a large distance from the shore and in possibly maximum volume of receptive medium;	X	X	X
Special areas	Protected by nature protecting acts – NATURA 2000, HELCOM BSPA, particularly sensitive – PSSA;	X	X	X
Ballast waters	Monitoring of invasive species and their utilization	X		
Artificial reefs[1]	Cleaning up environment and enhancement of bio resources renewal	X	~	
Restoration activities[2]	Improvement of the status of degraded sea bottom areas	X		
Re-introduction[2]	Improvement of the status of degraded undersea meadows	X		
Regulated channel	Increase in salinity and an water exchange between Bay of Gdańsk and Vistula Lagoon	X		

[1]Chojnacki (1995), Chojnacki and Ceronik (1992); [2]Kruk-Dowgiałło and Dubrawski (1998 b).

The impact of harbours on the marine coastal environment is not assessed on a regular basis, apart from occasional monitoring of larger harbours. No programme for such activities has been developed.

Similarly, no monitoring of sea transport has been implemented as yet, although the HELCOM Commission recommends continuous survey of sea accidents and ship collisions in the Baltic. Neither have ballast water discharges in the coastal zone been monitored. In the case of sediments extraction for artificial silting just the total mass of bottom sediments is recorded, without assessing loads of persistent organic pollutants, and heavy metals remobilized to sea water. Most often the assessment of the silting impact on the environment is based on the literature data, without any experimental verification of the assessment. The same applies to crude oil extraction. Fortunately, oil exploitation constitutes a smaller threat to the environment than hydrocarbons discharged with river run off, and treated sewage. Strangely enough, monitoring of the environment contamination caused by oil rigs, regarding both the scope and frequency of sampling, is the responsibility of operators (concession holders).

Construction of windmill farms in the Polish economic zone of the Baltic Sea has been temporarily ceased at the stage of the feasibility study. The studies have been carried out by investors who applied for concessions on construction of five small windmill farms covering altogether 50–60 km^2.

The present state of munitions and war gases (WG) dumped in the sea after the WWII is not monitored in Poland. The Helsinki Commission merely recommends preparation of instructions in case WG are fished out from the sea and make the public aware of their presence in the Baltic (Andrulewicz and Wielgat, 1998). Fishery can also be regarded as a threat to the marine environment since an undersize catch dumped to the sea can cause local contamination. However, the problem has not yet been recognized and no information on the impact is available. Polish legal regulations concerning protection of the environment still remain inadequate to the present day challenges. One of the major obstacles hampering the law enforcement is separation of responsibilities between land and marine administrative bodies. Also the legislation which is being developed separately for marine and land parts of the coast.

4 WAYS OF THREATS MITIGATION

Nowadays in Poland mainly technical activities (sewage treatment and collectors extending off shore) as well as administrative and legislative changes (protected areas) are carried out. Activities marked with grey (Table 2) are at the conceptual stage.

Of the eight activities aimed at diminishing the threats to the Gulf of Gdańsk marine environment which are ongoing or planned for implementation in the near future, two were selected for further consideration in this paper.

4.1 *Construction of artificial reefs*

The Code of Conduct for Responsible Fisheries, a document adopted by both fishermen and non-profit

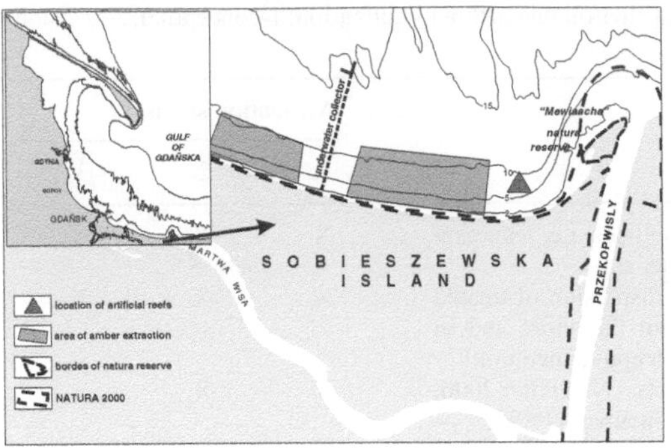

Figure 2. Location of artificial reefs in the Vistula River mouth area.

Figure 3. Elements of artificial reef system to be applied in the Vistula mouth, Gulf of Gdańsk (www.reefball.org).

organizations, recognizes artificial reefs as one of means of increasing fish stocks, thus enabling an increase in landings (Art. 8.11 of the Code). A number of European countries participate actively in the European Artificial Reef Research Network – EARRN established in 1995. The outcome is a set of publications characterizing the reefs functioning (Jensen and Collins, 2000; Sayer and Wilding 2002; Jensen 2002).

In areas where the reefs were established, soon abundant communities of fauna and flora developed. Apart from environment self-purification, this enhanced development of invertebrates and – as a consequence – created better feeding and living grounds for fish. The first artificial reefs along the Poland's coast were installed in 1980s in the estuary of the Odra River. This was followed by reef installations in the Pomeranian Bay (Chojnacki, 1995; Chojnacki & Ceronik, 1992), and in the Puck Bay (Kruk-Dowgiałło & Dubrawski, 1998b). The reefs made of concrete and/or fishing net was overgrown intensely by green algae (Enteromorpha), blue mussels (*Mytilus edulis trossulus*) and barnacle (*Balanus improvisus*).

Altogether three complexes of artificial reefs are planned to be constructed in the Polish coastal zone in order to enhance protection, and fish stock renewal. The program is sponsored by the Ministry of Fisheries and Rural Areas Development. The complex planned for the Gulf of Gdańsk consists of 50 modules (Figs. 2, 3). It will be a part of the world network of artificial reefs (REEF BALL) established by Reef Ball Development Group Ltd (Bradenton, USA; www.reefbal.org). The program aims at enhancing marine biota by establishing 1,000,000 Reef Balls throughout the word.

4.2 *Lengthening of treated sewage collectors*

Discharging treated sewage far from the shore is crucial for the coastal waters ecological status, especially sanitary conditions. Till now six far extending collectors were constructed along Poland's coast; two at the middle coast (Grzybów and Karlino), four in the Gulf of Gdańsk (Władysławowo, Hel, Jurata, Sobieszewo), (Fig. 1).

The underwater collector located in Karlino stretches 800 m into the sea. It has discharged treated sewage from the wood processing industry since 1972. The other middle coast underwater collector from municipal WWTP in Grzybowo extends 2200 m into the sea, the farthest segment of it acting as a 200 m long diffuser. Treated municipal sewage has been discharged there since 1980.

The municipal WWTP in Swarzewo used to discharge treated sewage to the Puck Lagoon (western part of the Gulf of Gdańsk). This considerably deteriorated the environment there. In 1998 a new collector extending 220 m into the open sea and distributing treated sewage via vertical diffuser was constructed in Władysławowo. A similar solution is planned for the Jurata, presently discharging treated sewage through a 1620 m collector to the Puck Bay.

However, the most spectacular achievement in sewage management in the region was the construction of the underwater collector discharging treated sewage from the Wschód WTTP in Gdańsk in 2002. Treated sewage is now discharged to the Gulf of Gdańsk through a 2500 m long pipe collector dispersing the sewage by two 213 m long diffusers at the depth of 12 m (Fig. 4). At present the Wschód WWTP is no longer on the list of environmental Baltic "Hot Spots".

One of the drawbacks of underwater collectors is their vulnerability due to the impact of natural forces such as water masses and sediment dynamics. Both waves and currents can unseal the system.

One of major problems is bio-fouling, which limits the effectiveness of diffusers. In case of the Wschód WWTP epiphytic biota was noticed at the outlet of diffusers after several months of operation (Fig. 5). On the other hand, underwater collectors create excellent conditions for the development of epiphytic biota by providing a hard substratum and constant supply of nutrients.

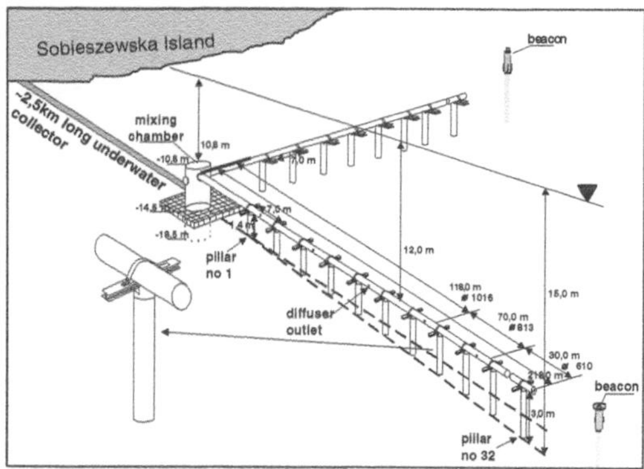

Figure 4. Diagram of the underwater collector and diffuser system of the Gdańsk-Wschód WWTP.

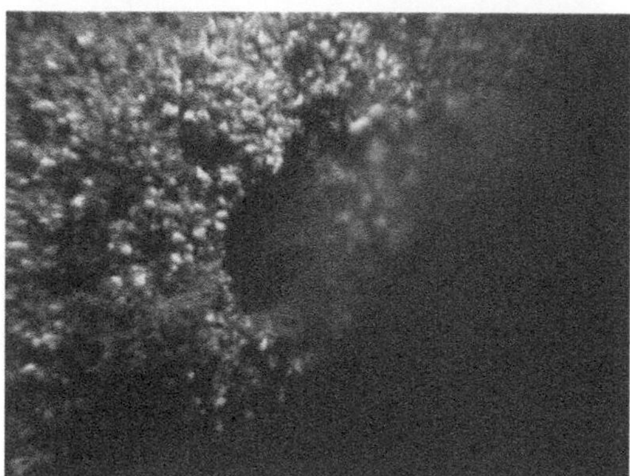

Figure 5. The state of the WWTP Gdańsk-Wschód diffuser outlet after a year of operation in the Gulf of Gdańsk (R. Opioła).

5 CONCLUSION

In Poland's Baltic Sea coastal zone the decrease in the loads of contaminants discharged is regarded as insufficient to improve the state of the marine environment. Further biotechnological activities are required to enhance the process of environment restoration.

Poland's accession to the EU created a new situation and impulses regarding the protection of coastal waters. Inventories of threats are obligatory and are to be elaborated according to WFD standards. At present, due to the lack of data, Polish coastal waters have been classified as "potentially threatened".

REFERENCES

Andrulewicz E., Wielgat M., 1998, Przewidywane losy amunicji chemicznej i bojowych środków trujących (BST) zatopionych w Morzu Bałtyckim, w: Broń chemiczna zatopiona w Morzu Bałtyckim, Materiały z sympozjum naukowego Akademii Marynarki Wojennej, Gdynia 22 kwietnia 1997, 97–113.

Chojnacki J.C., 1995, Effects of epibenthic succession on artificial reefs in the Pomeranian Bay, 14th Balt. Mar. Biol. Symposium, 12–18.09.1995, Parnu. p. 8.

Chojnacki J.C., Ceronik E., 1992, Artificial reefs of Pomeranian Bay (Southern Baltic) as biofiltration sites – preliminary results of pilot experiment, Mat.: Intern. Expert Conf. THE FUTURE OF THE BALTIC – ECOLOGY AND ECONOMICS Rostock, March, p. 22.

Ciszewski P., Ciszewska I., Kruk-Dowgiałło L., Rybicka D., Wiktor J., Wolska-Pyś M., Żmudziński L., Trokowicz D., 1992, Trends of long-term alternations of the Puck Bay ecosystem, [in]: Studia i Materiały Oceanograficzne Nr 60, National Scientific Committee on Oceanic Research PAS, PL ISDN 0208-421X, 33–84.

Environment of the Baltic Sea area 1994–1998, 2002, Baltic Sea Environment Proceedings No. 82B, Helsinki Commission.

EU WFD, 2000. http://www.eucc-d.de/infos/WaterFramework Directive.pdf

Jensen A.C., Collins K.J., Lockwood A.P.M. (eds.), 2000, Artificial reefs in European seas, Dordrecht, The Netherlands: Kluwer Academic Publishers, ISBN 0-7923-5845-7, p. 508.

Jensen A., 2002, Artificial reefs of Europe: perspective and future, ICES Journal of Science, 59, 3–13.

Kruk-Dowgiałło L., 1998, Phytobenthos as indicator of the state of environment of the Gulf of Gdańsk. Oceanological Studies No 4, Vol. XXVII, 105–123.

Kruk-Dowgiałło L., Dubrawski R., 1998a, The State of environment of the Gulf of Gdańsk coastal zone in autumn 1994 and summer 1995. Oceanolog. Stud. No 4, Vol. XXVII. PAN, IO UG: 137–158.

Kruk-Dowgiałło L., Dubrawski R., 1998b, A system of protection and restoration of the Gulf of Gdańsk. Bulletin of the Maritime Institute in Gdańsk. Vol. XXV, No 1, 45–69.

Krzymiński W., Kruk-Dowgiałło L., Zawadzka-Kahlau E., Dubrawski R., Kamińska M., Łysiak-Pastuszak E., 2004, Typology of Polish marine waters. (eds) Schernewski G i Wielgat M. Baltic Sea Typology. Coastline Reports 4, 39–48.

Pempkowiak J., Chiffoleau J-F., Staniszewski A., 2000a, Vertical and horizontal distribution of selected heavy metals in the Southern Baltic off Poland. Est. Coast. Shelf Sci., 51, 115–125.

Pempkowiak J., Tronczyński T., Pazdro K., 2000b, Spatial and temporal gradients of triazines in the Baltic Sea off Poland. Mar. Poll. Bull., 40, 1082–1093.

Sayer M., D., J., Wilding T., A., 2002, Planning, licensing and stakeholder consultation in an artificial reef development: the Loch Linnhe reef, a case study. ICES Journal of Science, 59, 178–185.

Sobol Z. i Szumilas T., 1992, An assessment of the effects of pollution In the Polish coastal area of the Baltic Sea 1984–1989, PAS, National Committee of Oceanic Research 61, 149–165.

The Code of Conduct for Responsible Fisheries, 1995, Food and Agriculture Organization of the United Nations, Rome.

Environmental Engineering – Pawłowski, Dudzińska & Pawłowski (eds)
© 2007 Taylor & Francis Group, London, ISBN13 978-0-415-40818-9

The influence of cell immobilization on petroleum hydrocarbons biodegradation

Katarzyna Piekarska
Wrocław University of Technology, Institute of Environmental Protection Engineering, Wybrzeże Wyspiańskiego Wrocław, Poland

ABSTRACT: Five strains of microorganisms are able to degrade diesel oil hydrocarbons isolated from water contaminated with petroleum compounds. These strains belonged to the following genera: *Pseudomonas sp.* M1, *Bacillus sp.* D1, *Rhodotorula sp.* P16, *Bacillus sp.* P17 and *Acinetobacter sp.* A2. The highest efficiency of biodegradation, about 94%, was achieved in case of a mixed culture containing two strains: *Bacillus sp.* D1 and *Pseudomonas sp.* M1. The mixed culture of these bacteria were immobilized in different polysaccharide carriers: agarose, carrageenan and sodium alginate. It was found that the immobilization of microorganism cells in 2% alginic acid sodium of high viscosity 14000 cps had a favourable effect on the biodegradation of diesel oil components, increasing the removal efficiency by 18.2% after 24 hours and by 24.3% after 72 hours of incubation in comparison to the removal efficiency by free cells.

Keywords: Diesel oil hydrocarbons, polysaccharides carriers, selected strains of microorganisms.

1 INTRODUCTION

Pollution of the environment with petroleum products is particularly dangerous because hydrocarbons are not only toxic but they can also be mutagenic and carcinogenic. They inhibit natural decomposition processes taking place in surface and ground waters, which results in the deterioration of water quality and loss of its value for drinking, household and recreation purposes (Alexander 2000). Therefore, this problem has become a subject of extensive research (Płaza et al. 2005). Among the processes of degradation of petroleum products, biological methods are receiving great attention (Leahy & Colwell 1990, Roling et al. 2003, Kaplan & Kitts 2004). Although petroleum and its products are used by microorganisms as food substrates, biodegradation is very slow in natural conditions. In the case of heavy pollution it may take many years, therefore it should be accelerated by man (Weber 1994, Vinas 2002, Abalos et al. 2004).

Immobilized cell technology has been widely applied in a variety of research and industrial applications (Cassidy & Trevors 1996, Willaert et al. 1996, Wang & Chao 2006). Immobilization of cells offers a number of advantages over free-cells. On the basis of numerous tests it was observed that deposited microorganisms can demonstrate a far greater metabolic activity and thus higher efficiency of the biodegradation processes than the aggregate of free-cells suspended in the environment. Furthermore, the immobilization of microorganisms allows the size of the equipment to be reduced for biological purification of sewage as well as increasing the concentration of microorganisms in the bioreactor (Kuncova et al. 2002, Kermanshahi et al. 2005, Tian et al. 2006). A variety of matrices have been used for cell immobilization, such as natural polymeric gels (agar, carrageenan, calcium alginate) and synthetic polymers (polyacrylamide, polyurethane, polyvinyl). Entrapment in natural polymeric gels has become a preferred technique for cell immobilization due to the toxicity problems associated with synthetic polymeric materials (Kutney et al. 1985, O'Reilly & Crawford 1989, Nicolov & Karamanev 1990, Reardon et al. 2002, Quek et al. 2006, Kim et al. 2006).

The subject of the present research was an attempt to immobilize the selected strains of bacteria in various polysaccharide carriers as well as their application in degradation of diesel oil components.

The scope of research covered the following: isolation and selection of active microorganisms able to degrade diesel oil components; identification of isolated microorganisms; evaluation of the intensity of the growth and the ability to degrade hydrocarbons by the isolated strains of microorganisms; immobilization of selected, active microorganisms with the use of polysaccharide carriers; evaluation of efficiency of biodegradation of diesel oil hydrocarbons by immobilized microorganisms.

2 MATERIAL AND METHODS

Diesel oil used in the experiment was purchased at a filling station in Wrocław. Chromatographic analyses showed that it was a mixture of 85 different hydrocarbons containing 10–24 carbon atoms per molecule. The prevailing group was the hydrocarbons with 16 carbon atoms. HPLC analysis revealed the presence of approximately 20% of aromatic compounds with signal spectra typical of monoaromatic, alkyl aromatic, naphtene alkyl aromatic, and di- and tricyclic aromatic compounds. The density of the oil was 0.804 g/cm^3.

The microorganisms degrading diesel oil hydrocarbons were isolated from water contaminated with petroleum compounds from the vicinity of filling stations and the Kerosene Products Centre in Wrocław.

The following microbiological substrates were used to isolate individual strains of microorganisms: agarized and liquid Siskina-Trocenko mineral medium (Siskina & Trocenko 1974) which was used to isolate prototrophic microorganisms which are able to use oil as their only source of carbon and energy, nutrient agar used to isolate as many saprophytic bacteria which are able to use oil in the presence of other organic substrates as possible, Sabourauda substrate used to isolate yeast strains (Gerhardt 1995). Diesel oil in the form of a film spread on the surface of the substrate was added to all isolation substrates.

In order to achieve highly active strains with a pronounced ability to utilize diesel oil the cultures were isolated after earlier growth of enrichment cultures (Vinas et al. 2002). Our method of growing enrichment cultures consisted of inoculation with a suspension of microorganisms taken from a previous culture of fresh mineral medium having had diesel oil hydrocarbons as its only source of carbon and energy.

The strains of microorganisms isolated in this way were subjected to further morphological, physiological and biochemical research in order to determine their taxonomic designation using Bergey's key (1975) (Gerhardt 1995).

The ability to degrade diesel oil by microorganisms was tested during a 31-day-long growth of strains on the liquid Siskinej-Trocenko mineral medium with diesel oil hydrocarbons contents of 1%, 5%, 10%, 25% and 50% vol. of the medium i.e. 0.804; 4.02; 8.04; 20.1 and 40.2 g of oil per dm^3. The tests were conducted in 250- cm^3 and 1000- cm^3 Erlenmeyer flasks. The total volume of the culture was 100 cm^3 in the case of individual strains, whereas in the case of the culture containing a mixed culture of microorganisms it was 500 cm^3. 2 cm^3 of individual strains of microorganisms suspended in the mineral medium of absorbance of 0.2 ($\lambda = 650$ nm) were placed in the flasks. The optical density of the tested cultures was measured on a SHIMADZU UV-VIS 1202 spectrophotometer. The process of chemooxidation was controlled in the flasks containing the same components but without microorganisms. All tests were performed at room temperature (22°C) on a rotary shaker (JW.ELECTRONIC WL-972; 90 rpm). After 31 days the loss of diesel oil components was detected in all cultures with the hexane extract method (Standard methods 1995). During the laboratory tests the total number of bacteria (Gerhardt 1995), the consumption of oxygen in the cultures were also measured with the use of WTW Oxi Top mercury-free measurement system (Instructions for WTW Oxi Top) as well as the emulsification activity of biomass. The emulgation activity of the test strains was measured in order to determine the level of biosurfactants produced by them (Rosenberg et al. 1979, Zhang et al. 2005). The measurement of the emulgation activity was conducted spectrophotometrically at $\lambda = 540$ nm. The reaction mixture was composed of 6 cm^3 of post-culture fluid, 2 cm^3 20 mM buffer Tris-HCl with pH 7.2, 0.5 cm^3 10 mM of solution of $MgSO_4$ and 0.2 cm^3 of diesel oil. The mixture before measurement was shaken for 1 hour on WL-972 JW.ELECTRONIC rotary shaker at 190 rpm.

The immobilization of the selected microorganisms was conducted in the following polysaccharide carriers purchased from Sigma: agarose (type VI), carrageenan (type I), alginate of viscosity of 3500 cps and alginate of viscosity of 14000 cps. The carriers were prepared as described in Chibata et al. (1986), Chibata et al. (1987) and Bucke (1987). The cells used in the immobilization process were prepared as follows: the cultures of the selected microorganisms were grown on the liquid Siskinej-Trocenko medium containing 100 cm^3 with 10% vol. diesel oil as the only source of carbon and energy. After a 6-day-long incubation the cultures were spun (19890x g, 15 min, 0°C, Sigma 2k15). The suspensions of cells were suspended in 1 cm^3 of a mineral medium and used in immobilization. The biosorbents were dissolved in the applicable amount of distilled water, autoclaved (15 min at 121°C), subsequently cooled to about 45°C and 1 cm^3 of bacterial suspension earlier prepared was suspended in it. Next, the uniform mixture was pipetted – drop by drop – to obtain a solution of gel and fixer, which was 0.02 M solution of calcium chloride in the case of sodium alginate and a solution containing 20 g of KCl and 0.15 g of $CaCl_2$ in 1 dm^3 of water in the case of carrageenan biosorbent and a mixture of alginate and carrageenan biosorbents. The grains of diameter of about 3 mm with the included biomass were hardened by keeping them in those solutions in the fridge for 24 hours in order to increase the mechanical strength of the carrier. After that time the grains were separated from the solution, rinsed with distilled water and moved to the nutrient medium. The agarized biosorbent was prepared by dissolving 6 g of agarose in 94 g of water (at <100°C to avoid disintegration of agarose by melting). Next, after cooling to about 45°C and after combining the gel with the bacterial biomass the gel was poured into Petri dishes. After cooling it was cut into cubes 3 mm × 3 mm × 3 mm. The cubes with included biomass were then moved to the nutrient medium.

The cultures of the immobilized bacteria were grown in sterile 1000 cm^3- Erlenmajer flasks on a JW.ELECTRONIC WL-972 rotary shaker machine (90 rpm). The total capacity of the culture was 500 cm^3. Biosorbents with the mounted biomass and 10% vol. diesel oil were placed in the flasks. At the same time cultures of free cells were grown in the same conditions and in the same concentration. After 24 and 72 hours the loss of diesel oil in the cultures was measured with the hexane extract method; the emulsification activity was also measured.

Next, in the same way, the cultures of the immobilized bacteria strains were prepared in 2% alginic acid sodium of viscosity of 14000 cps with 1%, 5%, 10%, 25% and 50% vol. diesel oil. Every day for 7 days the loss of diesel oil was measured in those cultures with the use of the hexane extract method, as well as emulsification activity and intensity of respiration processes with the use of WTW Oxi Top mercury-free measurement system.

3 RESULTS AND DISCUSSION

Five strains of microorganisms well adapted for degradation of hydrocarbon contaminants were isolated from the contaminated water. The strains include the following genera: *Pseudomonas sp.* M1, *Bacillus sp.* D1, *Rhodotorula sp.* P16, *Bacillus sp.* P17 and *Acinetobacter sp.* A2. Use of the hydrocarbon substrate by those strains was good (Tab. 1). A significant decrease of the concentration of diesel oil dependent on the concentration of hydrocarbon substances added to them after 31 days of the experiment was observed in all cultures. In cases in which the concentration of substrate in the culture was 1% vol., the loss of hydrocarbons was 15.7%–72.3%, when the concentration of diesel oil was 5% vol. it was 21%–67.3%, and when the

Table 1. Percentage loss of diesel oil after 31 days in the free culture of microorganisms on liquid mineral medium.

	Concentration of diesel oil in the culture		
Strain	1%	5%	10%
Rhodotorula sp. P16	43.2 ± 2.8	43.7 ± 1.8	19.2 ± 2.7
Bacillus sp. P17	15.7 ± 1.1	21.0 ± 2.3	2.2 ± 0.5
Acinetobacter sp. A2	63.0 ± 1.9	67.3 ± 1.5	25.3 ± 1.7
Pseudomonas sp. M1	51.2 ± 2.3	60.0 ± 2.7	23.1 ± 2.4
Bacillus sp. D1	72.3 ± 2.0	66.0 ± 1.4	40.1 ± 1.8

concentration of diesel oil in the culture was 10% vol. it was 2.2%–40.1%. The lowest ability to degrade hydrocarbon substances included in diesel oil was shown by the strain *Bacillus sp.* P17; depending on the initial concentration of the substrate: it was 15.7%, 21% and 2.2%. On the other hand, the largest loss of hydrocarbons was observed in the cultures of strains *Bacillus sp.* D1 and *Acinetobacter sp.* A2. It was 72.3%, 66%, 40.1% and 63%, 67.3% and 25.3% respectively. At the same time the biggest percentage loss of diesel oil was observed in the cultures of strains with 1% vol. concentration of the medium; the smallest one with the concentration of 10% vol. of substrate.

In cases in which the strains were isolated using the enrichment cultures method, after two months only two of the tested strains, namely *Bacillus sp.* D1 and *Pseudomonas sp.* M1 were present. This mixed culture of bacteria grew very well not only in low concentrations of hydrocarbon substances but also in higher concentrations, demonstrating a high respiration and emulgation activity as well as a high degree of biodegradation of the components of diesel oil (Tab. 2, Figs 1–3). After 31 days in the culture with two strains *Bacillus sp.* D1 and *Pseudomonas sp.* M1. the loss of hydrocarbons was between 22.3% (for 50% vol. of initial concentration of oil in the culture) and 93.8% (for 1% vol. of initial concentration of oil in the culture) (Tab. 2).

This mixed culture of bacteria was used also in the next stage of the research in which the ability to degrade hydrocarbons of diesel oil in free bacteria cells was compared with that of the immobilized bacteria in various polysaccharide carriers.

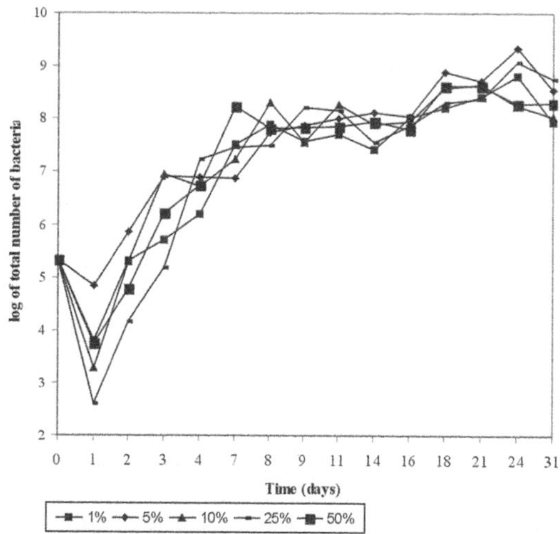

Figure 1. Growth of mixed culture of free cells of Pseudomonas *sp.* M1 and Bacillus *sp.* D1 on mineral medium with diesel oil hydrocarbons in the amount: 1%, 5%, 10%, 25% and 50%.

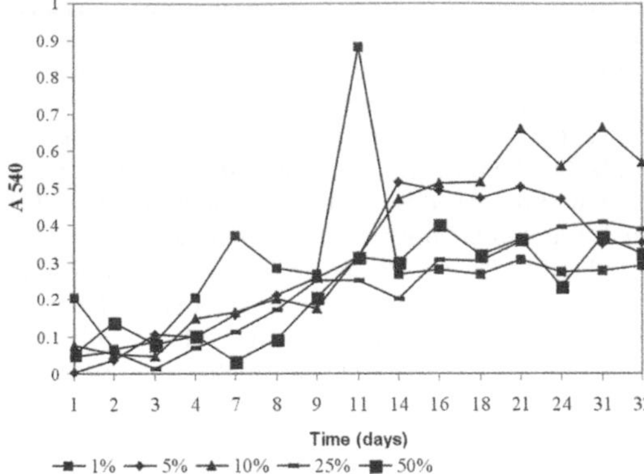

Figure 2. Emulsification activity of mixed biomass of free cells of Pseudomonas *sp*. M1 and Bacillus *sp*. D1 on mineral medium with diesel oil hydrocarbons in the amount: 1%, 5%, 10%, 25% and 50%.

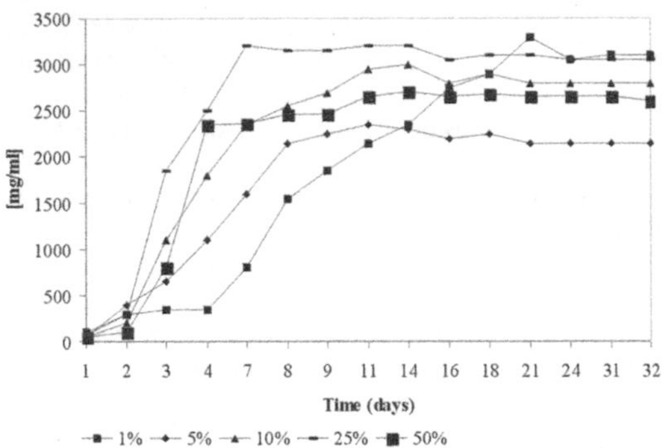

Figure 3. Concentration of oxygen in the mixed cultures of free cells of Pseudomonas *sp.* M1 and Bacillus *sp.* D1 growing on the mineral medium with diesel oil hydrocarbons in the amount: 1%, 5%, 10%, 25% and 50%.

In order to select the best carrier the immobilization was conducted in the following polysaccharide carriers: agarose, carrageenan, alginate of viscosity of 3500 cps, alginate of viscosity of 14000 cps, a mixture of alginate as well as carrageenan and alginate with polyvinyl alcohol. The results of the tests are presented in Table 3.

In most cases the immobilization of the cells of selected polysaccharide carriers increased the efficiency of diesel oil loss, with the exception of alginate of viscosity of 3500 cps prepared in the concentration of 3% and 4%, alginate of viscosity of 14000 cps in the concentration of 4% and carrageenan in the concentration of 4%. Although the increase of concentration of the polysaccharide carrier increases its mechanical strength, the possibility of unrestricted diffusion of extracellular enzymes, substrates and products of the reaction falls at the same time thus decreasing the biochemical activity of the tested microorganisms. On the other hand, the low concentration of polysaccharide reduced the integrity of the gel and caused its grains to dissolve and release the bacteria to the nutrient medium. This happened after 72 hours in the case of 1% alginate of viscosity of 3500 cps, 1% alginate of viscosity of 14000 cps an 1% carrageenan. 6% agarose, gel which developed from the combining of 1.5 g of alginate (3500 cps) and 0.5 g of carrageenan as well as 2 g of alginate (3500 cps) and 2 g of polyvinyl alcohol were also destroyed. However, adding 5 g polyvinyl alcohol to 2 g of alginate of viscosity of 3500 cps caused the nutrient medium to foam and become turbid.

Table 2. Percentage loss of diesel oil after 31 days in the free mixed culture of *Pseudomonas sp.* M1 and *Bacillus sp.* D1on liquid mineral medium.

Concentration of diesel oil in the culture				
1%	5%	10%	25%	50%
93.8 ± 3.2	72.1 ± 4.2	78.0 ± 3.7	47.5 ± 2.7	22.3 ± 1.5

The 2% alginate of viscosity of 14000 cps proved to be the best carrier. The culture containing 10% of diesel oil and the immobilized cells in this carrier indicated 42.4% loss of oil after 24 hours and 58.4% after 72 hours. After the same period the emulsification activity of the immobilized cells was 0.987 and 1.758 (A_{540}). For comparison the values for free cells were respectively: loss of oil – 24.2% and 34.1%, emulsification activity – 0.265 and 0.498 (A_{540}).

With the use of this carrier and the mixed culture of strains *Bacillus sp.* D1 and *Pseudomonas sp.* M1 immobilized cultures were prepared in which every day for 7 days the loss of diesel oil was measured with the use of the hexane extract method, as well as emulsification activity and intensiveness of respiration processes. Table 4 shows the results of the tests on diesel oil loss in culture tests. In all cases a greater efficiency of the decomposition of diesel oil by the immobilized cells was observed regardless of the given concentration of hydrocarbons.

The tests were conducted with the use of the Student's test on two levels of significance: 0.01 and 0.05; they demonstrated the statistical significance of the differences between percentage values of the removal of hydrocarbons by immobilized and free cells. Those results were confirmed by observations of the emulgation activity and by observation of the intensity of respiration processes of tested cultures. Both the level of production of surfactants (Fig. 4) and the respiration activity (Fig. 5) of the immobilized cells was greater than that of free cells in all tested concentrations of

Table 3. Loss of diesel oil and emulsification activity in the mixed culture of free and immobilized cells.

Type of carrier	Time [hours]	Loss of diesel oil [%]		Emulsification activity [A_{540}]	
		Free	Immobilized	Free	Immobilized
agarose 6%	24	28.3	32.4	0.372	0.410
	72	38.5	*39.1	0.619	*0.567
alginate 1%	24	31.2	34.8	0.398	0.498
3500 cps	72	39.4	*42.7	0.521	*0.956
alginate 2%	24	27.1	38.2	0.302	0.621
3500 cps	72	39.2	47.5	0.578	0.990
alginate 3%	24	25.4	19.4	0.298	0.120
3500 cps	72	37.8	21.3	0.502	0.213
alginate 4%	24	27.4	5.2	0.345	0.087
3500 cps	72	35.2	6.1	0.478	0.090
alginate 1%	24	29.6	31.5	0.384	0.398
14000 cps	72	39.8	*43.7	0.593	*0.989
alginate 2%	24	24.2	42.4	0.265	0.987
14000 cps	72	34.1	58.4	0.498	1.758
alginate 3%	24	25.1	34.1	0.276	0.456
14000 cps	72	37.6	41.2	0.567	0.998
alginate 4%	24	28.7	6.1	0.384	0.098
14000 cps	72	39.8	7.8	0.592	0.109
carrageenan	24	21.3	29.4	0.198	0.234
1%	72	32.1	*39.2	0.304	*0.578
carrageenan	24	28.9	37.2	0.398	0.504
2%	72	39.6	48.4	0.556	1.134
carrageenan	24	31.2	34.1	0.287	0.345
3%	72	39.2	42.7	0.567	1.087
carrageenan	24	26.1	7.8	0.289	0.102
4%	72	34.5	9.2	0.432	0.123
alginate (1.5 g)	24	28.5	39.1	0.354	0.987
3500 cps and carrageenan (0.5 g)	72	38.6	*42.6	0.543	*0.987
alginate (0.75 g)	24	23.1	37.4	0.201	0.965
14000 cps and carrageenan (0.25 g)	72	34.5	48.2	0.387	1.234
alginate (1.5 g)	24	28.7	35.6	0.321	0.567
3500 cps and polyvinyl alcohol (0.5 g)	72	39.8	48.1	0.587	1.123
alginate (2 g)	24	28.9	**31.4	0.376	**0.398
3500 cps and polyvinyl alcohol (5 g)	72	38.2	**42.3	0.598	**0.976
alginate (1 g)	24	28.2	39.4	0.345	0.401
14000 cps and polyvinyl alcohol (5 g)	72	39.0	45.5	0.599	1.123
alginate (1 g)	24	23.1	36.1	0.287	0.456
14000 cps and polyvinyl alcohol (10 g)	72	38.5	49.2	0.602	1.328

*carrier was destroyed.

**carrier in the culture medium caused its foaming efficiency of extraction process 85.4% non-biological loss of diesel oil due to chemooxidation 4.5%.

Table 4. Loss of diesel oil in the mixed culture of free and immobilized cells in 2% alginate of viscosity 14000 cps.

Day of culture	Loss of diesel oil [%] Concentration of diesel oil [%]				
	1	5	10	25	50
1*	29.8 ± 1.9	22.6 ± 2.6	21.7 ± 1.5	12.4 ± 2.8	7.5 ± 2.1
1**	34.6 ± 3.4	36.8 ± 2.9	45.2 ± 4.2	19.4 ± 2.8	15.5 ± 1.6
2*	48.5 ± 1.5	36.1 ± 4.5	26.7 ± 1.8	20.7 ± 4.5	8.0 ± 3.2
2**	52.8 ± 3.1	40.2 ± 2.3	49.1 ± 3.2	29.8 ± 2.1	17.0 ± 1.5
3*	63.1 ± 2.4	42.7 ± 3.4	33.8 ± 4.1	24.7 ± 2.8	9.5 ± 2.7
3**	72.1 ± 1.9	69.4 ± 1.8	57.3 ± 2.6	31.2 ± 1.9	18.0 ± 2.1
4*	71.5 ± 5.8	48.5 ± 3.8	47.8 ± 2.2	30.4 ± 3.7	12.5 ± 2.7
4**	85.4 ± 4.1	72.4 ± 2.7	64.2 ± 2.4	39.8 ± 3.2	20.8 ± 1.8
5*	79.5 ± 4.3	61.5 ± 3.4	55.3 ± 4.1	35.5 ± 2.1	15.0 ± 1.6
5**	92.3 ± 3.2	88.5 ± 2.8	70.9 ± 3.7	44.5 ± 3.2	25.0 ± 2.4
6*	80.1 ± 4.3	68.4 ± 1.7	58.6 ± 2.1	36.1 ± 5.1	17.0 ± 3.1
6**	94.5 ± 2.4	90.2 ± 4.1	72.3 ± 4.2	47.0 ± 2.3	28.4 ± 1.7
7*	85.1 ± 5.1	72.1 ± 3.9	60.2 ± 4.3	37.2 ± 3.9	18.5 ± 3.1
7**	96.1 ± 3.2	91.4 ± 2.7	77.4 ± 3.1	54.2 ± 2.3	32.1 ± 2.4

*Free cells.
**Immobilized cells.

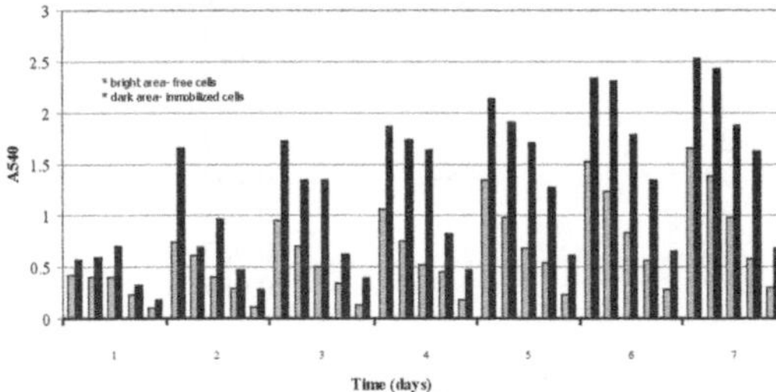

Figure 4. Emulsification activity in the culture of microorganisms free and immobilized in the 2% alginate of viscosity of 14000 cps in the present of different amounts of hydrocarbons (1%, 5%, 10%, 25% and 50%).

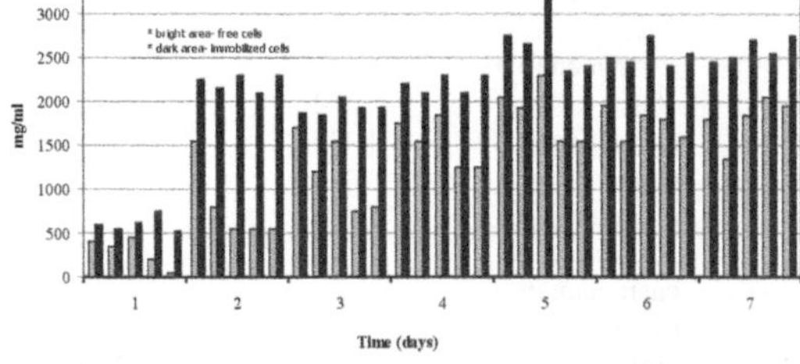

Figure 5. Concentration of oxygen in the culture of microorganisms free and immobilized in the 2% alginate viscosity (14000 cps) in the present of different amounts of hydrocarbons (1%, 5%, 10%, 25% and 50%).

diesel oil. In time the value of the tested parameters also increased and reached their highest values at the end of the work.

4 CONCLUSION

1. Strains of microorganisms were isolated from water contaminated with petroleum compounds enzymatically prepared for the use of various kinds of petroleum hydrocarbons as a source of carbon and energy. Those strains belonged to the following genera: *Pseudomonas sp.* M1, *Bacillus sp.* D1, *Rhodotorula sp.* P16, *Bacillus sp.* P17 and *Acinetobacter sp.* A2.
2. The highest efficiency of biodegradation of hydrocarbons was achieved when a mixture of strains *Bacillus sp.* D1 and *Pseudomonas sp.* M1 was used, and not individual active strains.
3. The tests performed on the cultures containing a mixed culture of bacteria *Bacillus sp.* D1 and *Pseudomonas sp.* M1 immobilized in various polysaccharide carriers, as well as diesel oil as the only source of carbon and energy, enabled the best gel for immobilization and biodegradation of hydrocarbons to be selected. The carrier was the 2% alginate (viscosity 14000 cps).
4. The immobilized cells in the selected gel improved the efficiency of degradation of hydrocarbons for a wide range of concentrations of diesel oil added to the culture (1%, 5%, 10%, 25% and 50%).
5. The tests of respiration activity and the tests regarding the ability to produce surfactants by free and immobilized cells of microorganisms confirmed a

higher activity for diesel oil hydrocarbon biodegradation in the case of the immobilized cells.

REFERENCES

Abalos, A.M.; Viñas, J.; Sabaté, M.; Manresa A. & Solanas A.M. 2004. Enhanced biodegradation of Casablanca crude oil by a microbial consortium in presence of a rhamnolipid produced by *Pseudomonas aeruginosa* AT10. *Biodegradation* 15:249–260.

Alexander, M. 2000. Aging, bioavailability, and overestimation of risk from environmental pollutants. *Environ. Sci. Technol.* 34: 4259–4265.

Bergey's, Manual of Determinative Bacteriology. 1975. The Williams and Wilkins Company, Baltimore.

Bucke, C. 1987. Cell immobilization in calcium alginate. *Methods Enzymol.* 135:175–189.

Cassidy, M.B.; Lee, H. & Trevors, J.T. 1996. Environmental applications of immobilized microbial cells: a review. *J. Ind. Microbiol.* 16, 79–101.

Chibata, I.; Tosa, T.; Sata, T. & Takata I. 1987. Immobilization of cells in carrageenan. *Methods Enzymol.* 135: 189–198.

Gerhardt, P.; Murray, R.G.; Costilow, R.N.; Nester, E.W.; Wood, W.A.; Krieg, N.R. & Phillips, G.B., editors. 1995. *Manual of methods for general and molecular bacteriology.* Washington, D.C.: American Society for Microbiology.

Kaplan, C.W. & C.L. Kitts. 2004. Bacterial succession in a petroleum land treatment unit. *Appl. Environ. Microbiol.* 70:1777–1786.

Kermanshahi, A.; Karamanev, D. & Margaritis, A. 2005. Biodegradation of petroleum hydrocarbons in an immobilized cell airlift bioreactor. *Water Res.* 39(15): 3704–14.

Kutney, J.P.; Choi, L.S.; Hewitt, G.M.; Salisbury, P.J. & Singh, M. 1985. Biotransformation of dehydroabietic acid with resting cell suspensions and calcium alginate-immobilized cells of Mortierella isabellina. *Appl Environ Microbiol.* 49(1): 96–100.

Kim, M.K.; Singleton, I.; Yin, C.R.; Quan, Z.X.; Lee, M. & Lee, S.T. 2006. Influence of phenol on the biodegradation of pyridine by freely suspended and immobilized Pseudomonas putida MK1. *Lett Appl Microbiol.* 42(5):495–500.

Leahy, J.G. & Colwell, R.R. 1990. Microbial degradation of hydrocarbons in the environment. *Microbiol Rev.* 54(3): 305–315.

Nicolov, L. & Karamanev, D. 1990. Change of microbial activity after immobilisation of microorganisms. DeBont JAM (Ed), *Physiology of Immobilised Cells*, Elsevier.

O'Reilly, K.T. & Crawford, R.L. 1989. Continuous limonin degradation by immobilized Rhodococcus fascians cells in K-carrageenan. *Appl Environ Microbiol.* 55(4): 866–870.

Płaza, G., Ulfig, K., Worsztynowicz, A., Malina, G., Krzemińska, B. & Brigmon, R.L. 2005. Respirometry for assessing the biodegradation of petroleum hydrocarbons. *Environ Technol.* 26(2):161–9.

Quek, E.; Ting, Y.P. & Tan, H.M. 2006. Rhodococcus sp. F92 immobilized on polyurethane foam shows ability to degrade various petroleum products. *Bioresour Technol.* 97(1): 32–8.

Reardon, K.F.; Mosteller, D.C.; Rogers, J.B.; DuTeau, N.M. & Kim, K.H. 2002. Biodegradation kinetics of aromatic hydrocarbon mixtures by pure and mixed bacterial cultures. *Environ Health Perspect.* 110 (Suppl 6):1005–1011.

Roling, W.F.; Head, I.M. & Larter, S.R. 2003. The microbiology of hydrocarbon degradation in subsurface petroleum reservoirs: perspectives and prospects. *Res Microbiol.* 154(5) :321–8. Review.

Rosenberg, E.; Zuckerberg, A.; Ovitz, C. & Gutnick, D.L. 1979. Emulsifier of *Arthrobacter sp.* RAG- 1: isolation and emulsifying properties. *Appl.Environ.Microbiol.* 37, 402–408.

Siskina, N.W. & Trocenko, J.A. 1974. Svoistva novogo stamma *Hyphomicrobium* ispolsuiscego odhouglerodnyie soiedimenia. *Microbiologia.* 5, 765–770.

Standard methods for the examination of water and waste water. 1995. 19th edition, American Public Health Association, Washington.

Wang, J.Y. & Chao, Y.P. 2006. Kinetics of p-cresol degradation by an immobilized Pseudomonas sp. *Appl Environ Microbiol.* 72(1): 927–931.

Weber, W.J. & Corseuil, H.X. 1994. Inoculation of contaminated subsurface soils with enriched indigenous microbes to enhance bioremediation rates. *Water Research* 28(6), 1407–1414.

Willaert, R.G.; Baron, G.V. & De Backer, L. 1996. *Immobilised living cell systems modelling and experimental methods.* Chichester, England: J.Wiley & Sons.

Viñas, M.; Grifoll, M.; Sabate, J. & Solanas, A.M. 2002. Biodegradation of a crude oil by three microbial consortia of different origins and metabolic capabilities. *J. Ind. Microbiol. Biotechnol.* 28:252–260.

Zhang, G.L.; Wu, Y.T.; Qian, X.P. & Meng, Q. 2005. Biodegradation of crude oil by Pseudomonas aeruginosa in the presence of rhamnolipids. *J Zhejiang Univ Sci B.* Aug; 6(8): 725–730. published online before print July 29, 2005.

Environmental Engineering – Pawłowski, Dudzińska & Pawłowski (eds)
© 2007 Taylor & Francis Group, London, ISBN13 978-0-415-40818-9

BTEX biodegradation by biosurfactant-producing bacteria

Grażyna A. Płaza*, Jacek Wypych & Krzysztof Ulfig
Institute for Ecology of Industrial Areas, Katowice, Poland

Christopher Berry & Robin L. Brigmon
Savannah River National Laboratory, Aiken, South Carolina, USA

ABSTRACT: Two biosurfactant-producing bacteria were isolated from an extremely polluted petroleum hydrocarbon contaminated site, and identified as *Ralstonia picketti* (BP-20) and *Alcaligenes piechaudii* (CZOR L-1B). Both bacteria were able to degrade BTEX hydrocarbons, but *Alcaligenes piechaudii* was found to the more efficient. The complete biodegradation of toluene and m + p-xylenes reached a maximum of 96% and 97%, respectively after 30 days of incubation. High depletion of m + p-xylenes was evident during the first stage of growth, and was almost 97%. The results showed the order of degradation to be: m + p-xylenes > toluene > o-xylene > ethylbenzene > benzene. Of particular interest is that hydrocarbon biodegradation by the bacteria was combined with their ability to produce biosurfactants and emulsification of hydrocarbons. The capacity of these natural microorganisms to produce biosurfactants and their ability to degrade hydrocarbons is promising for environmental restoration applications at hydrocarbon-contaminated sites.

Keywords: Monocyclic aromatic hydrocarbons, biodegradation, biosurfactants.

1 INTRODUCTION

Hydrocarbons, including petroleum and its products, grease, and halogenated compounds, are an important class of world-wide environmental pollutants. The presence of these hydrocarbons in the environment is of considerable public health and ecological concern due to their persistence, toxicity and ability to bioaccumulate through the food chain (Brigmon et al. 2002). BTEX compounds (benzene, ethylbenzene, toluene, and three isomers of xylene) are classified as environmental priority pollutants. They have some acute and long term toxic effects. In addition to its toxicity, benzene is know to be a carcinogen. Monoaromatic hydrocarbons are constituents of gasoline, diesel and jet fuels. They enter the water and soil environments due to accidental spills and leaking storage tanks.

Typically, contaminated sites are so variable that different technologies have to be developed to clean up the petroleum pollution. While chemical and physical methods have been historically used for petroleum cleanup, in recent years, one technology that is receiving increasing attention and success is bioremediation (Mulligan et al. 2001; Christofi & Ivshina, 2002; Singh & Cameotra 2004; Mulligan 2005). However, the limitations associated with bioremediation of petroleum hydrocarbons are their low water solubility and availability. This insolubility results in low bioavailability to soil microorganisms, which, in turn, can limit *in situ* biodegradation of these contaminants.

Surfactants and emulsifiers are compounds that can increase the solubility and dispersion of hydrophobic compounds. Increased availability of the hydrophobic contaminants to the hydrocarbon degrading microorganisms accelerates bioremediation. Biosurfactants (BS) are naturally produced by certain environmental microorganisms. A number of surfactants have been isolated from microbial cultures following growth of bacteria and fungi on a variety of aliphatic and aromatic hydrocarbons (Rosenberg & Ron 1999). These biosurfactants, which are generally extracellular, may be relatively simple glycolipids or complex high molecular weight substances. The advantages of these microbial products include their biodegradability, the diversity of structure and function for various applications, and the selectivity for specific hydrocarbons (Sukan & Kosaric 2000; Batista et al. 2006).

The aim of this study was to determine benzene, toluene, ethylbenzene, xylene isomers (BTEX) biodegradation in mixtures by two biosurfactant-producing bacteria isolated from a petroleum hydrocarbon contaminated site.

2 MATERIALS AND METHODS

2.1 *Investigation area*

More than a century of continuous use of a sulfuric acid-based oil refining method by the Czechowice-Dziedzice Oil Refinery (CZOR) in Poland has produced an estimated 120,000 tons of acidic, highly weathered, petroleum sludge. This waste has been deposited into three open waste lagoons, 3 meters deep and covering 3.8 hectares. In 1997, the smallest of the waste lagoons (0.3 hectare) was chosen for construction of aerobic biopile (Altman et al. 1997). The waste from the lagoon was removed, and 5000 tons of heavily petroleum contaminated soil (30 g/kg d.m. soil) was treated in the bioremediation process. Numerous COCs (contaminants of concern) were present at this site but the petroleum hydrocarbons were the main concern. The purpose was to evaluate novel technologies and applications for environmental restoration of soil heavily contaminated with petroleum waste by comparing bioremediation processes under active *vs.* passive aeration and the removal rates of both, easily biodegradable and recalcitrant petroleum hydrocarbons. The project focused on the application of cost-effective amendments for biostimulation, including additions of mineral NPK fertilizers and the surfactant, Rokafenol N8, to enhance hydrocarbon biodegradation. A simple and cost-effective *ex situ/on site* bioremediation technology was designed. Over the 20 month project, more than 81% (120 metrics tons) of petroleum hydrocarbons were biodegraded.

2.2 *Isolation of hydrocarbon-degrading bacteria*

10 g of mixed soil from the biopile were inoculated in 100 ml of the mineral medium (MM). The composition of the medium used was the following (g/l): NH_4NO_3 – 1; $MgSO_4 \cdot 7H_2O$ – 0.2; $CaCl_2 \cdot 2H_2O$ – 0.03; K_2HPO_4 – 1; KH_2PO_4 – 1. The medium was supplemented with 1 ml of the trace elements solution (Gerhardt 1981), and 1% (v/v) of crude oil as carbon and energy source. The incubation was performed at 30°C for 1 week. Development of bacterial colonies was obtained by a serial dilution-agar plating technique on standard methods agar (SMA, Biomerieux). Isolates were tested for their ability to grow on the solid mineral medium (MM) with different hydrocarbons. The following hydrocarbons: cyclohexane, hexadecane, xylene, benzene, toluene, heptane, decane, isooctane, hexane, mineral oil, pristane, and squalene were used as carbon and energy sources. All chemicals used were of analytical grade and purchased from Sigma-Aldrich Co. and Polish Chemical Reagents S.A., Gliwice. 200 μl of a selected hydrocarbon was placed on the filter paper, then the filter paper was overlaid by the mineral medium (MM) with 20 g of agar (DIFCO). 50 μl of bacterial suspension ($OD_{600\,nm} \sim 0.1$) were put on the agar surface. The incubation of Petri dishes was carried out at 30°C for two weeks.

2.3 *Screening of biosurfactant-producing bacteria*

The medium used for isolation and cultivation of biosurfactant-producing bacteria was as described by Abu-Ruwaida *et al.* (1991). The composition of the medium was the following (g/l): Na_2HPO_4 – 2.2; KH_2PO_4 – 1.4; $MgSO_4$ $7H_2O$ – 0.6; $(NH_4)_2SO_4$ – 3; yeast extract – 1; NaCl – 0.05; $FeSO_4 \cdot 7H_2O$ – 0.01, and 1 ml microelements solution (Gerhardt 1981). 500 μl of 24–hours bacteria culture (10^4–10^5 CFU/ml) as inoculum were added. 1% (v/v) of crude oil was used as the sole carbon source. Cultures of each bacterium were grown aerobically in 100 ml medium at 30°C for a period of seven days without shaking. The emulsification of hydrocarbons was determined by examining the presence of emulsion droplets in the cultures. After incubation the cultures were filtered by the filter paper (390, FILTRAK GmbH), and they were used to evaluate of biosurfactant production.

Hemolytic activity was carried out as described by Carrillo et al. (1996). Isolated strains were screened on blood agar plates containing 5% (v/v) blood and incubated at 30°C for 24–48 h. Hemolytic activity was detected as the presence of a clear zone around a colony.

Surface tension was measured with a ring – tensiometer (Krüss Digital-Tensiometer 10, Hamburg, Germany) at room temperature. Water (72 ± 0.2 mN/m) and medium (70 ± 0.3 mN/m) were used as controls.

All the experiments were carried out in triplicate.

2.4 *Identification of bacteria able to produce biosurfactant and degrade hydrocarbons*

Isolates, which grew well in the presence of hydrocarbons and produced biosurfactant were identified as follows. Isolates were grown on SMA plates, and a single colony of each isolate was resuspended and washed three times in 100 μl of the sterile water. The cell suspension was added to the PCR reaction mixture. PCR products were produced with whole-cell cultures described by Furlong et al. (2002). The PCR reaction was performed with puReTaq™ Ready-To-Go™ Polymerase Chain Reaction (PCR) beads (Amersham Biosciences) as previously described (Stefan & Atlas 1991). Beads are premixed.

2.5 *Growth of Ralstonia picketti SRS and Alcaligenes piechaudii SRS in liquid medium with benzene as carbon and energy source*

2 ml bacterial strains from overnight culture (10^4–10^5 CFU/ml) were transferred to 100 ml of sterile mineral medium (MM) with 1% (v/v) of benzene as the sole carbon and energy source. The cultures

were grown aerobically at 30°C in an orbital shaker at 150 rpm. The optical density of the cultures at 600 nm was continuously measured using a CECIL CE 2031 spectrophotometer during the incubation time. Growth curves were marked.

2.6 *BTEX biodegradation in aerobic conditions by Ralstonia picketti SRS and Alcaligenes piechaudii SRS*

2.6.1 *Incubation conditions*

Experiments were performed using 125 ml serum bottles containing 25 ml medium as described by Abu-Ruwaida et al. (1991). 500 μl of 24 h bacteria culture (10^4–10^5 CFU/ml) as inoculum were added. The vials were sealed with a Teflon-coated rubber and aluminium septum-cap. Headspaces of the bottles were flushed with oxygen. Finally, an initial BTEX mixture of 30 mg/l was added to the cultures using a glass microsyringe (Hamilton Co, # 701). The BTEX concentrations were chosen according to the literature with respect to their toxicity (Boyd et al. 1997). All hydrocarbons with 99.0% analytical standards were purchased from the Supelco Co. The incubation was carried out at 30°C by 30 days. Measurements of BTEX were done after 1, 2, 5, 10, 20 and 30 days of the incubation time. Sterile controls were prepared to evaluate hydrocarbon evaporation. The experiments were done in triplicate.

2.6.2 *Analytical procedure*

BTEX hydrocarbons concentrations were determined according to the method described by Wypych & Mańko (2002). HS-SPME-GC/MS was used. The analysis was carried out with a Star 3400 Cx gas chromatograph (equipped with a ^{63}Ni Electron Capture Detector); it was coupled to a mass spectrometer Saturn 3 and Autosampler 8200 Cx (Varian) with 10 ml autosampler vials. The chromatographic column with the phase DB624 and length 30 m × 0.32 mm ID (1.8 μm film thickness) was used. As the carrier gas helium was used: purity 99.999%, flowing capacity of 1.0 ml/min (in the temperature of 35°C). The gas chromatographic conditions were as follows: the oven temperature was held at 40°C for 10 minutes, than increasing 250°C by 10°C/min, and finally increasing by 5°C/min to 270°C. The total analysis time was 45 minutes. The injector temperature was 250°C. The MS operating conditions were the following: the mass range scanned was 30–250 amu at 1 sec/scan, temperature of ion trap was 170°C, multiplier voltage was 2700 V, ionization energy was 70 eV (electron impact mode EI). The transfer line temperature was 250°C. The temperature of the ECD detector (^{63}Ni) was 300°C.

Three types of SPME fiber were used (Supelco, Bellefonte) in an autosampler set of the following stationary phases: 100 μm polydimethylosiloxane,

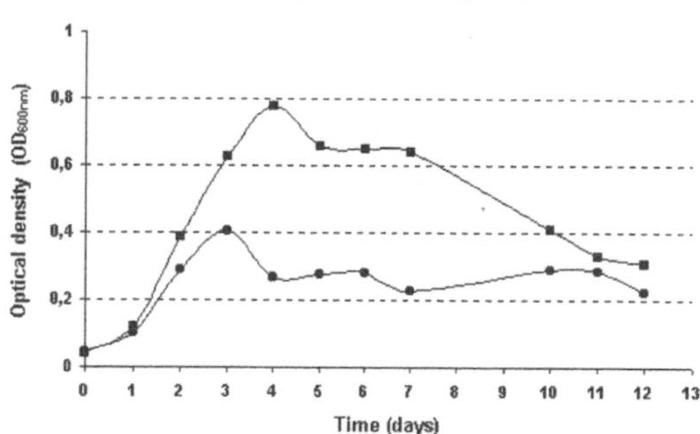

Figure 1. Growth of bacterial strains in liquid medium with benzene as carbon and energy source.

7 μm polydimethylosiloxane, 85 μm polyacrylate. The SPME fiber was conditioned prior to use in order to reduce bleeding by heating in a split/splitless injector with an open purger, in helium stream: fiber 100 μm PDMS in the temperature of 250°C (for 1 hour), fiber 7 μm PDMS at a temperature of 320°C (for 2 hours) and fiber 85 μm polyacrylate at a temperature of 300°C (for 2 h).

3 RESULTS AND DISCUSSION

Figure 1 presents the growth of bacteria strains in the liquid medium with benzene as energy and carbon source. Both bacteria grew very well in this medium, however *Alcaligenes piechaudii* growth was found to be better than that of *Ralstonia picketti*. Bacteria growth curves were typical for batch culture. The growth proceeded through a lag phase, which was observed between 1st and 2nd days of the incubation period, and went into a growth phase which was characterized by an exponential increase. The optimum of their growth was noted in third and fourth day of the incubation period. In the next days of the incubation and under experimental conditions the cultures were in balanced growth.

BTEX biodegradation rate is presented in Figure 2. Both bacteria were able to degrade BTEX hydrocarbons but *Alcaligenes piechaudii* was found to more efficient. Table 1 presents the removal of BTEX hydrocarbons by both bacteria, presented by % of removal after 30 days of the incubation. The complete biodegradation of toluene and m + p-xylenes for *Alcaligenes piechaudii* reached a maximum of 96% and 97%, respectively. Depletion of m + p-xylene was evident during the first stage of growth, and reached 97%. The results showed the order of degradation to be: m + p-xylene > toluene > o-xylene > ethylbenzene > benzene. No differences in hydrocarbon biodegradation rates between *Alcaligenes*

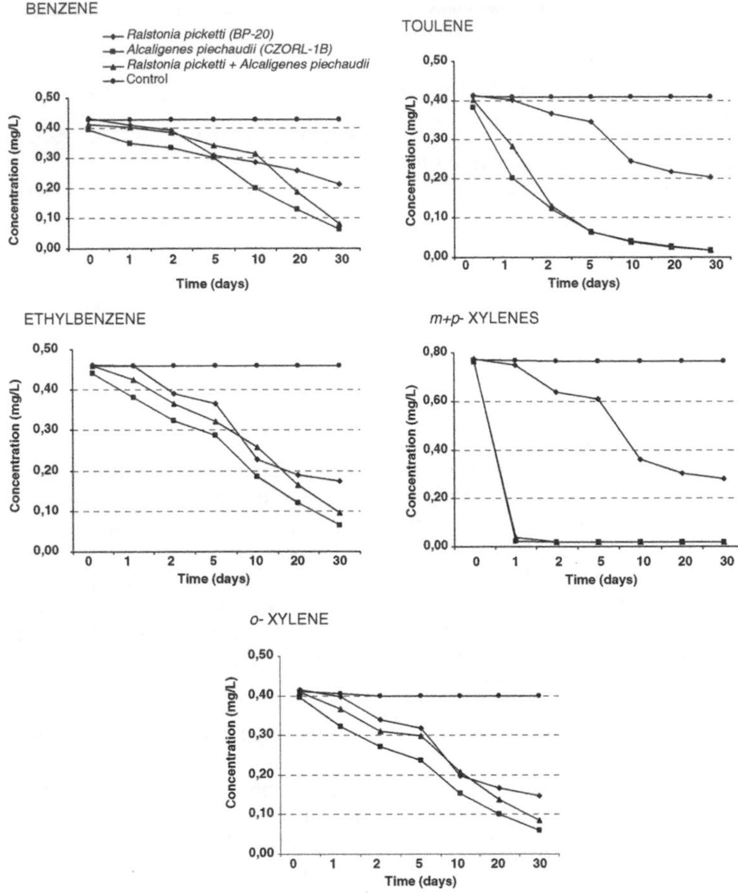

Figure 2. BTEX biodegradation by *Ralstonia picketti* and *Alcaligenes piechaudii* during the experiment course (mean values: SD ≤ 0.1 mg/l).

Table 1. BTEX biodegradation by isolated bacterial strains.

	Removal rate (%)*				
Bacteria	Benzene	Toluene	Ethylbenzene	m + p-xylenes	o-xylene
Ralstonia picketti	52	53	63	65	65
Alcaligenes piechaudii	84	96	85	97	85
Ralstonia picketti + *Alcaligenes piechaudii*	80	96	79	97	80

Standard deviation (SD) values were $\leq$2%; * % of biodegradation after 30 days of the incubation; The biodegradation rate was determined of BTEX degraded to the initial amounts.

piechaudii and mixed cultures *Ralstonia picketti* and *Alcaligenes piechaudii* were observed. Of particular interest is that the hydrocarbon biodegrading bacteria mechanism found here is combined with their ability to produce biosurfactants and increase BTEX bioavailability, probably by the changes of the surface active properties.

Ralstonia picketti SRS and *Alcaligenes piechaudii* SRS gave positive results to produce biosurfactants. They had hemolytic activity growing on blood agar.

Carrillo et al. (1996) and Youssef et al. (2005) found an association between hemolytic activity and surfactant production, and they recommended the use of blood agar lysis as a primary method to screen for bisurfactant activity.

The surface tensions were reduced to 61 mN/m and 55 mN/m by *Ralstonia picketti* and *Alcaligenes piechaudii,* respectively.

In the preliminary work the surface active properties, e.g. surface tension, emulsification, bacterial cell-surface hydrophobicity (BAH) and foamability of the culture filtrates of *Ralstonia picketti* SRS (BP 20) and *Alcaligenes piechaudii* SRS (CZOR L-1B) were evaluated (Płaza et al. 2005). The emulsification index (EI24) was almost 100% for all tested compounds (benzene, toluene, m + p-xylenes, petroleum oil, diesel oil) except diesel oil. *Ralstonia picketti* had a better foamability characteristic. Foam volume expressed as FV was 50 ml for *Ralstonia picketti* compared to *Alcaligenes piechaudii* where FV was 10 ml. The BAH (bacterial adhesion to hydrocarbons) measurements revealed higher adhesion of *Alcaligenes piechaudii* cells towards different hydrocarbons compared to *Ralstonia picketti* cells. The strains were found to have a surface hydrophobicity in the following order: aliphatic hydrocarbons, BTEX, and PAHs. The hydrophobicity of *Alcaligenes piechaudii* towards BTEX hydrocarbons (benzene, toluene, xylene) ranged from 53% to 68%, and was higher than the hydrophobicity of *Ralstonia picketti*. According to the literature the ability to adhere to bulk hydrocarbon is mostly a characteristic of hydrocarbon-degrading bacteria. The strains were found to be better emulsifiers than surface tension reducers. They produce water-soluble extracellular bioemulsifiers.

The properties, biosurfactant production and modulation of the cell surface hydrophobicity play an important role in efficient hydrocarbon assimilation/uptake, which mechanism is not fully understood. The increase in cell hydrophobicity contributes to the increase of the cell interaction with hydrocarbons enhancing the hydrocarbon availability and ultimately the degradation rate. The beneficial effect of biosurfactants on BTEX biodegradation has been observed. However, the relationship between the solubilization, bioavailability, and biodegradation is not yet known. Ron & Rosenberg (2001) paid attention to the fact that biosurfactants are produced by a wide variety of microorganisms, and they have very different chemical structures and surface properties. Different groups of biosurfactants have thus different natural roles in the growth and ability to biodegrade hydrocarbon of the producing microorganisms. Further studies on the relationship between adherence to hydrocarbons and growth of hydrocarbon-degrading bacteria are underway. Models for uptake of hydrocarbons consider the roles of dissolved molecules, contact of the cells with large oil droplets, or contact with fine oil droplets are presented by Neu (1996). Hommel (1990) concluded that microbial growth on hydrocarbons is accompanied by metabolic and structural alternations of the cell. The appearance of biosurfactant in the culture medium or attached to the cell boundaries is often regarded as a prerequisite for initial interactions of hydrocarbons with the microbial cells.

Three models of hydrocarbon transport to microbial cells are generally considered (Kosaric, 2000): (1) interaction of cells with more water-soluble hydrocarbons, (2) direct and predispensed complete reactions for performing PCR amplifications. With the exception of primers and template, the ambient temperature-stable beads provide all the necessary reagents to perform 25 μl polymerase chain reactions, e.g. stabilizers, BSA, dATP, dCTP, dGTP, dTTP, ~2.5 units of puReTaq DNA polymerase and reaction buffer. When a bead was reconstituted to a 25 μl final volume, the concentration of each dNTP was 200 mM in 10 mM Tris-HCl (pH 9.0), 50 mM KCl and 1.5 mM $MgCl_2$. A typical PCR contains $<1\ \mu g$ of template DNA and primers at a concentration of 0.2–1 μM. The samples were subjected to 30 cycles of 94°C for 1 min, 72°C for 2 min and 61°C for 1 min for denaturation, annealing and elongation steps, respectively. An initial denaturation step (95°C, 1 min) was used to ensure complete denaturation of the DNA. PCR amplification was performed in a Mastercycler® gradient machine (Eppendorf). The primers for the reactions were as follows: forward primer: 27f (5′-TTCCGGTTGATCCYGCCGGA-3′) and reverse primer: 1492 universal (5′- ACGGGCGGTGTGTRC-3′) (Furlong et al. 2002). Partial sequences of rRNA genes were obtained using an ABI 377 DNA Sequencer (Applied Biosystems, Foster City, CA) with an ABI PRISM BigDye terminator sequencing Lab kit at the Genome Analysis Facility in the Botany Department at the University of Georgia, Athens. Isolates were tentatively identified by similarity to sequences in the GenBank data base using the FastA algorithm (Pearson and Lipman, 1988) of the GCG software package (Genetics Computer Group, Wisconsin). (Kosaric, 2000): (1) interaction of cells with more water-soluble hydrocarbons, (2) direct contact of cells with large hydrocarbon drops – in this mechanism, microbial cells attach to the surface of hydrocarbon drops that are much larger than cells; the availability of substrate surface area for cell attachment is a limiting factor for microbial growth; biosurfactants/bioemulsifiers produced by hydrocarbon utilizing bacteria cause the dispersion of hydrocarbon droplets in the aqueous medium and thereby increase the surface area; addition of biosurfactants to the hydrocarbon medium stimulated growth of microorganisms, and (3) microbial cells interact with particles of solubilized, microemulsified hydrocarbons.

Depending on the specified organism, hydrocarbon uptake may take place through one or a combination of the presented mechanisms (Prabhu & Phale 2003).

4 CONCLUSIONS

The effects of biosurfactants on aliphatic and aromatic hydrocarbon biodegradation and their enhancement of bioremediation are well documented. However, the *mechanism* of enhanced biodegradation has not been established.

In this work BTEX biodegradation by two biosurfactant-producing bacteria was evaluated. This study has revealed that bacteria produced biosurfactants having the ability to degrade BTEX hydrocarbons can be used in different bioremediation applications. Probably, the uptake and transport of hydrocarbons, and consequently the speed and degree of their degradation were mostly improved by produced biosurfactants. However, there is no data on biosurfactant concentrations, or mode of action. In fact, we have no direct evidence that biosurfactants produced by *Ralstonia picketti* and *Alcaligenes piechaudii* were indeed involved in the degradation of the BTEX compounds. The properties, biosurfactant production and modulation of the cell surface hydrophobicity play an important role in efficient hydrocarbon assimilation/uptake, the mechanism of which is also not fully understood.

Multifunctional biosurfactants are known (Kosaric 2000). Biosurfactants are produced by a wide variety of microorganisms (bacteria, fungi), their chemical structures and surface properties are very different (Ron & Rosenberg 2001). Because their chemical structures and surface properties are so different, their role in mechanisms of assimilation/uptake, and in hydrocarbons biodegradation is difficult to generalize. The very diversity of biosurfactants makes it difficult to understand their role in these processes. In spite of this, the capacity of natural microorganisms to produce biosurfactants and to degrade hydrocarbons under a wide range of environmental conditions is promising for environmental restoration applications at hydrocarbon-contaminated sites. *Biosurfactants* have several potential advantages over *synthetic* surfactants for bioremediation applications. These advantages include biodegradability, low toxicity, temperature stability, solubility, pH range, high specificity for specific hydrocarbons, and the potential for *in situ* bioremediation production.

REFERENCES

Abu-Ruwaida, A.S., Banat, I.M., Haditirto, S., Salem, A., Kadri, M. 1991. Isolation of biosurfactant-producing bacteria. Product characterization and evaluation. *Acta Biotechnol.* 11, 315–324.

Altman, D.J., Hazen, T.C., Tien, A., Lombard, K.H., Worsztynowicz, A. 1997. *Czechowice Oil Refinery Bioremediation Demonstration Test Plan, WSRC-MS-97-21H* Westinghouse Savannah River Company, Aiken, S.C. DOE-NITS.

Batista, S.B., Mounteer, A.H., Amorim, F.R., Totola, M.R. 2006. Isolation and characterization of biosurfactant/bioemulsifier-production bacteria from petroleum contaminated sites. *Biores..Technol* 97, 868–875.

Boyd, E.M., Meharg, A.A., Wright, J., Killham, K. 1997. Assessment of toxicological interactions of benzene and its primary degradation products (catechol and phenol) using a *lux*-modified bacterial bioassay. *Environ.Toxicol.Chem.* 16, 8479–856.

Brigmon, R.L., Camper, D., Stutzenberger, F. 2002. Bioremediation of compounds hazardous to health and the environment—an overview. In: Singh, V. P. (Ed.), *Biotransformations: Bioremediation Technology for Health and Environmental Protection.* Elsevier Science Publishers, The Netherlands, pp. 1–28.

Carrillo, P.G., Mardaraz, C., Pitta-Alvarez, S.J., Giulietti, A.M. 1996. Isolation and selection of biosurfactant-producing bacteria. *World J. Microbiol. Biotechnol.* 12, 82–84.

Christofi, N., Ivshina, I.B. 2002. Microbial surfactants and their use in field studies of soil remediation. *J. App. Microbiol.* 93, 915–936.

Furlong, M.A., Singleton, D.R., Coleman, D.C., Whitman, W.B. 2002. Molecular and culture-based analyses of prokaryotic communities from an agricultural soil and the burrows and casts of the earthworm *Lumbricus rubellus*. *Appl. Environ. Microbiol.* 68, 1265–1279.

Gerhardt, P. 1981. *Manual of Methods for General Bacteriology*. American Society for Microbiology, Washington, DC 20006.

Hommel, R.K., 1990. Formation and physiological role of biosurfactants produced by hydrocarbon-utilizing microorganisms. *Biodegradation* 1, 107–119.

Kosaric, N. 2000. *Biosurfactants. Production. Properties. Applications*. Marcel Dekker, Inc.

Mulligan, C.N., Yong, R.N., Gibbs, B.F. 2001. Surfactant-enhanced remediation of contaminated soil: a review. *Eng. Geol.* 60, 371–380.

Mulligan, C.N. 2005. Environmental applications for biosurfactants. *Environ. Poll.* 133, 183–198.

Neu, T.R. 1996. Significance of bacterial surface-active compounds in interaction of bacteria with interfaces. *Microbiol. Rev.* 60, 151–166.

Pearson, W.R., Lipman, D.J. 1988. Improved tools for biological sequence comparison. *Proc. Natl Acad. Sci. USA* 85, 2444–2448.

Płaza, G., Ulfig, K., Brigmon, R.L. 2005. Surface active properties of bacterial strains isolated from petroleum-bioremediated soil. *Pol. J. Microbiol.* 54, 161–167.

Prabhu, Y., Phale, P.S. 2003. Biodegradation of phenanthrene by Pseudomonas sp. strains PP2: novel metabolic pathway role of biosurfactant and cell surface hydrophobicity in hydrocarbon assimilation. *Appl. Microbiol. Biotechnol.* 61, 343–351.

Ron, E.Z., Rosenberg, E. 2001. Natural roles of biosurfactants. *Environ. Microbiol.* 3, 229–236.

Rosenberg, E., Ron, E.Z. 1999. High- and low-molecular-mass microbial surfactants. *Appl. Microbiol. Biotechnol.* 52, 154–162.

Singh, P., Cameotra, S.S. 2004. Enhancement of metal bioremediation by use of microbial surfactants. *Bioch. Bioph. Res.Commun.* 319, 291–297.

Stefan, R.J., Atlas, S.R.M. 1991. Polymerase chain reaction: Applications in Environmental Microbiology. *Annu. Rev. Microbiol.* 45, 137.

Sukan, F.V., Kosaric, N. 2000. Biosurfactants. In: *Encyclopedia of Microbiology* vol.1, 2nd ed., Academic Press, pp. 618–635.

Youssef, N.H., Duncan, K.E., Nagle, D.P., Savager, K.N., Knapp, R.M., McInerney, M.J. 2004. Comparison of methods to detect biosurfactant production by diverse microorganisms. *J. Microbiol. Methods* 56, 339–346.

Wypych, J., Mańko, T. 2002. Determination of volatile organic compounds (VOCs) in water and soil using solid phase microextraction. *Chem. Anal.* 47, 507–512.

Environmental Engineering – Pawłowski, Dudzińska & Pawłowski (eds)
© 2007 Taylor & Francis Group, London, ISBN13 978-0-415-40818-9

The role of bacteria in the process of weathering of carbon wastes

Teresa Wlodarczyk
Institute of Agrophysics, Polish Academy of Sciences, Doswiadczalna Lublin, Poland

Zygmunt Strzyszcz
Institute of Basic Environment Engineering, Polish Academy of Sciences, M. Sklodowskiej-Curie Zabrze, Poland

ABSTRACT: Carbon wastes were inoculated with dominant bacteria isolated from investigated materials. SiO_2 transferred to solution and bacterial ability to produce organic acids was determined. The most intensively decomposed was montmorillonite, by *Arthrobacter* and *Micrococcus* genus. The intensity of decomposition of minerals depended on bacterial strains. The addition of black coal to rocks in monoculture cultivation had a positive impact on the amount of bacteria and the quality of decomposition. Much higher ability for decomposition of carbon rocks mixture was observed in bacterial culture with *Arthrobacter* and *Pseudomonas* than with *Micrococcus*. Chromatographic analysis indicated that bacteria strains had an ability to produce some volatile and non-volatile organic acids and differed in consideration of their number and quality. Weathering process of aluminosilicate is closely related to the kind of organic acids released by native bacteria, mainly volatile acids. The kind of cultivation medium affected the kind of realized acids within *Arthrobacter* genus.

Keywords: Carbon rocks, black coal, organic acids, bacterial strains.

1 INTRODUCTION

Mine wastes have been generated for several centuries and mining activity has accelerated significantly during the 20th century. Mine wastes constitute a potential source of contamination to the environment, as heavy metals and acids are released in large amounts. One kind of mine wastes are carbon wastes. The recultivation of carbon wastes is one of the most important environmental problems in Upper Silesia. According to the Main Statistical Department (Statistical Annual 2002), from 1990 to 2001 more than 110 Mg of industrial wastes were produced. Coal-mining was responsible for more than 80% of the total waste. Coal mining wastes are located in heaps and are the subject of biological reclamation by sodding, tree planting and afforesting. In the Silesia region there are 2,700 hectares of mine heaps waiting for reclamation. The stratigraphic disposition of coal mining wastes is different with regard to petrography, mineralogy and chemical composition. The Carboniferous rocks consist of sandstone, mudstone and clay stone with predomination of mudstones. These rocks are mainly composed of kaolinite, illite and montmorillonite (Strzyszcz 2003).

Microorganisms play an essential role in the weathering of different materials (Zagury et al. 2004; Severmann et al. 2006) A great variety of microorganisms has been found in mine wastes, and microbiological processes are usually responsible for the environmental hazard created by mine wastes (Ledin & Pedersen 1999). Biological decomposition of carbon wastes is very important in the initial stage of soil formation and depends, among other things, on the structure of the crystalline net of minerals, their chemical composition, and on the environmental conditions. Microorganisms are crucial in this process of soil formation (Aristovskaja 1973; Berthelin et al. 2006; Gierasimov 1973; Illmer 2006; Machill et al. 1997). The presence of active bacteria intensifies the weathering of the minerals (Garcia et al. 2005). The vital bacterial activity may be the cause of direct or indirect decomposition of mineral crystalline net and of the transition of elements into soluble state, available for living organisms (Aristovskaja et al. 1969; Aristovskaja 1980; Karavajko et al. 1980; Petsch et al. 2003; Rogers & Bennett 2004). A more universal and effective way is the direct activity of micro-organisms which, during the metabolic process, release active chemical compounds reacting strongly with the environment. These are mineral and organic acids and bases, as well as substances forming chelate compounds (Aristovskaja 1980; Rogers & Bennett 2004). In the initial stage of carbon wastes weathering, the main source of carbon for microorganisms is coal which comprises as much as ca. 30% in one of the largest heaps in Silesia.

The main purpose of this study was to estimate the growth dynamics of chosen native bacterial strains incubated on carbon rocks which contain mainly kaolinite, illite and montmorillonite and, additionally, after addition of black coal, to define the bacterial capacity for carbon rocks decomposition.

2 MATERIALS AND METHODS

The experiments were carried out on natural carbon wastes in the initial stage of storage (weathering), originated from the "Smolnica" heap – "fresh heap", and black coal taken from the "Szczygłowice" coalmine and disintegrated non-sterile and sterile carbon rocks: kaolinite, illite, montmorillonite and sandstone, and their mixtures simulating the average content of carbon rocks in the heap, taken directly from the coalmines "Szczygłowice", "Bogdanka", "Milowice" and "Brzeszcze", respectively. The participation fractions of carbon rocks in the mixture was as follows: kaolinite –50%, illite –35%, montmorillonite –10% and sandstone –5%. Additionally, 30% black coal was added.

The chemical composition of the investigated carbon rocks was measured according to the methods used for aluminosilicates (Jacob 1952), the mineralogical composition was measured using X-ray radiography analysis (Geigerflex apparatus, made by Rigaku). In the water extract prepared according to Arinuszkina (1970), the following were measured: SiO_2 and P_2O_5 colorimetrically, Ca_2^+ and Mg_2^+ complexometrically, K^+ and Na^+ photometrically.

The dominant bacteria were isolated from individual carbon rocks and their mixture, and from the "fresh heap". For bacterial strains isolation and incubation we used our own modification of mineral medium for siliceous bacteria, with the following composition: saccharose – 5 g, $(NH_4)_2SO_4$ – 1 g, Na_2HPO_4 – 2 g, $MgSO_4 \cdot 7H_2O$ – 0.5 g, $FeCl_3$ – 0.0005 g, $CaCO_3$ – 0.1 g, carbon rock – 1 g, agar – 20 g, in which potassium was replaced by a particular carbon rock depending on the variants of the experiment (Aleksandrov 1953).

Sterile powder of particular carbon rocks and their mixture with and without the addition of black coal were inoculated with bacterial cells characteristic for: illite – *Arthrobacter sp.* (16 K), and for "fresh heap" – *Arthrobacter sp.* (IVS),– *Pseudomonas stutzeri* (IS),– *Micrococcus sp.* (IIS, VS and VIS). The incubation lasted for 15 days at 28°C. During the incubation time, every five days the number of bacteria and the content of SiO_2 transferred to the solution on the last day of incubation were determined. In the 1st series of investigations on bacterial ability to produce organic acids, the mineral medium with the addition of 10 g glucose consisted of the following: $(NH_4)_2SO_4$ – 1 g, K_2HPO_4 – 0.1 g, $MgSO_4$– 0.3 g, $CaCl_2$ – 0.1 g, NaCl – 0.1 g, distilled water – 1000 ml (Aristovskaja & Kutuzowa 1968), and in the 2nd series the following medium (based on our own recipe) was used: $(NH_4)_2SO_4$ – 1 g, glucose – 5 g and kaolinite – 1 g, and the rest of the mineral nutrients were replaced with carbon rock. The sterile media were incubated with an approximate amount of cells. The cultivation was carried out for five days at 28°C. In the liquid culture four volatile acids: acetic, propionic, butyric and valeric, as well as 6 non-volatile acids: lactic, fumaric, succinic, malic, α-ketoglutaric and citric were determined using gas chromatography (Pye Unicam – model 104). The chromatograph was fitted with a flame ionization detector (FID) running at 250°C. The organic acids were separated on two columns – the first (2.1 m long) packed with a Chrom Q (100–120 mesh) maintained at 75–198°C for non-volatile organic acids and the second (1.5 m long) packed with Chromosorb 101 (100-120 mesh) maintained at 200°C for volatile organic acids. The carrier gas was argon flowing at a rate of 40 ml min^{-1}.

3 RESULTS AND DISCUSSION

The mineralogical composition of the investigated carbon rocks is presented in Table 1. The mudstone samples chosen for the experiments contained from 5 to 85% of kaolinite and 15 to 45% of illite. Only quartz and kaolinite were present in all of the investigated mudstones.

The investigated carbon rocks differed the most in their silica content and the content of the remaining components was comparable (Tab. 2).

The presence of silicon (SiO_2) in the solution was taken as indicating destruction of the mineral framework (Bigham et al. 2001). Among others, Ponomarieva (1980) assumed the silica present in carbon rocks solution as an indicator of mineral decomposition.

The study of the weathering of carbon rocks by native siliceous bacteria showed that montmorillonite was decomposed most intensively by both *Arthrobacter sp.* (16 K-dominant on illite; 229 mg SiO_2 in 1 l)

Table 1. Mineralogical composition of investigated carbon rocks [%].

Kind of mineral	Sandstone	Mudstone		
		Kaolinite	Illite	Montmorillonite
Quartz	95	5	20	10
Tridymite	5	–	–	–
Kaolinite	–	85	25	5
Siderite	–	5	–	–
Calcite	–	–	–	–
Illite	–	–	45	15
Smectite	–	–	10	70

and *Micrococcus sp.* (VS- dominant on fresh heap; 277 mg SiO_2 in 1 l). The intensive decomposition was accompanied by the highest growth dynamics of bacterial cells on montmorillonite during fifteen days of incubation (Tab. 3).

Zviagincev (1973) found that bacteria from *Micrococcus sp.* adsorbed on the minerals very strongly, especially on montmorillonite, which resulted in higher decomposition of the minerals. In our experiment less decomposition was observed in the bacterial culture with kaolinite in the case of *Arthrobacter sp.* (16 K) and with illite in the case of *Micrococcus sp.* (VS). *Arthrobacter sp.* (16 K) did not show any ability for sandstone decomposition while *Micrococcus sp.* (VS) – for kaolinite (Tab. 3). Generally, a higher ability for decomposition of carbon rocks was shown by *Micrococcus sp.* (VS) than by *Arthrobacter sp.* (16 K). Henderson & Duff (1963) claimed that it was difficult to find a correlation between the intensity of decomposition of minerals and their structure. In each analysed case the final results depended not only on the stability of the crystalline net but also the kind of decomposing factor as well as on the environmental conditions.

The addition of black coal for individual rocks in monoculture cultivation had a positive impact on the amount of bacteria and on the decomposition of the examined material. The number of bacterial cells increased many times in all investigated bacterial cultures and days of incubation. This effect was accompanied by higher weathering of carbon rocks in five out of the eight cases investigated (Tab. 3). The black coal addition changed the ability of the examined bacterial strains to decompose carbon rocks. This phenomenon was best seen in the case of sandstone where *Arthrobacter sp.* (16 K) decomposed it very intensively, while *Micrococcus sp.* (VS) lost its ability to weather minerals (Tab. 3).

Table 2. Some chemical and physicochemical characteristics of investigated carbon rocks [%].

Kind of determination	Sandstone	Mudstone		
		Kaolinite	Illite	Montmorillonite
SiO_2	96.93	44.14	65.64	60.05
P_2O_5	0.04	0.10	0.05	0.06
CaO	0.06	0.77	0.17	0.28
MgO	0.24	0.79	2.26	3.10
Na_2O	0.08	1.83	1.08	1.62
K_2O	trace	1.07	3.70	1.45
pH	8.61	6.40	8.95	9.7

Table 3. The decomposition of carbon rocks as affected by bacterial strains *Arthrobacter sp.* (16 K) and *Micrococcus sp.* (VS) and their growth dynamics expressed in percent of bacterial cells during 15 days of incubation compared to zero day taken as 100 percent.

Bacterial strain	Kind of carbon rocks and addition	% of bacterial cells compare to inoculation on the days of cultivation			SiO_2 mg L^{-1}
		5	10	15	
16 K	Sandstone	3	4	18	0
	Sandstone + black coal	4370	86	750	315.0
	Kaolinite	1	21	9	131.0
	Kaolinite + black coal	7010	490	62	26.0
	Illite	2520	53	120	71.0
	Illite + black coal	12160	10820	220	375.0
	Montmorillonite	13670	1510	226	229.0
	Montmorillonite + black coal	43900	14970	1900	210.0
VS	Sandstone	0.04	0.02	0.03	120.0
	Sandstone + black coal	230	3	270	0
	Kaolinite	37	39	27	0
	Kaolinite + black coal	46	42	58	255.0
	Illite	44	38	31	127.0
	Illite + black coal	830	690	440	315.0
	Montmorillonite	190	130	77	277.0
	Montmorillonite + black coal	160	200	91	645.0

Table 4. The decomposition of carbon rocks mixture as affected by bacterial strains *Arthrobacter sp.* (16 K and IVS), *Pseudomonas stutzeri* (IS) and *Micrococcus sp.* (IIS, VS and VIS) and their growth dynamics expressed in percent of bacterial cells during 15 days of incubation compared to zero day taken as 100 percent.

Name of bacterial strain	Kind of fertilisation	% of bacterial cells compare to inoculation on the days of cultivation			SiO_2 mg L^{-1}
		5	10	15	
Arthrobacter sp. (16 K)	control	25	1250	0	225.0
	black coal	180800	9800	350	1102.0
Arthrobacter sp. (IVS)	control	1490	1210	1480	150.0
	black coal	1070	2560	3050	600.0
Pseudomonas stutzeri (IS)	control	8000	4620	4540	165.0
	black coal	3615	11795	8640	817.0
Micrococcus sp. (IIS)	control	650	910	1070	37.0
	black coal	603	1300	11690	180.0
Micrococcus sp. (VS)	control	360	2830	2380	7.0
	black coal	4450	5520	4180	60.0
Micrococcus sp. (VIS)	control	3770	4010	4960	30.0
	black coal	3910	5990	8310	300.0

The decomposition of a carbon rock mixture, which simulated the natural heap, was carried out with the *Arthrobacter sp.* (IVS), *Pseudomonas stutzeri* (IS), *Micrococcus sp.* (IIS, VS and VIS) as representatives of "fresh heap" and, for comparison – *Arthrobacter sp.* (16 K).

The growth dynamics of the investigated bacteria was higher for the strains isolated from the "fresh heap", probably well adapted to growth on the carbon rocks mixture, than for *Arthrobacter sp.* (16 K) isolated from illite discarded from the coal-mine. Puente et al. (2006), who studied the potential rock weathering by bacteria, considered *Pseudomonas* one of the rock-weathering bacteria. The investigated bacteria dominant in the "fresh heap" showed increasing growth during the fifteen days of incubation while the number of *Arthrobacter sp.* (16 K) decreased to zero on the last day of incubation. This confirmed the good adaptation of strains isolated from the "fresh heap" to growth on the carbon rocks mixture (Tab. 3). The black coal addition to carbon rocks mixture effected an increase in the number of bacteria after the fifth day of incubation in all investigated variants. Three strains (IVS, IS and IIS) needed some time for adaptation after the addition of C substrate. During the first five days of bacteria cultivation a decrease was observed in their number compared with the control. The highest positive response to black coal addition was observed in the case of growth of *Arthrobacter sp.* (16 K) (Tab. 3).

A much higher ability for decomposition of carbon rocks mixture by the bacterial strains mentioned above was observed in bacterial culture with *Arthrobacter* and *Pseudomonas sp.* than with *Micrococcus sp.* (Tab. 3). According to Kalinowski et al. (2004) *Pseudomonas fluorescens* and *Pseudomonas stutzeri* can produce short-chain organic acids and element-specific ligands (siderophores) that are able to change pH and enhance chelation, which results in increased mobilisation of many trace elements.

The weathering of carbon rocks was apparently higher in their mixtures enriched with black coal than in the control variants. Very high response to black coal addition was noticed in the bacterial culture *Micrococcus sp.* (VS and VIS) where SiO_2 released into the solution was about ten times higher then in the controls. The highest ability to weather carbon rock mixture was that of *Arthrobacter sp.* (16 K) in the control as well as with black coal addition (Tab. 4).

Generally, the studies showed that all the strains had the ability to decompose bed rocks. In the opinion of Wagner & Schwartz (1967), full mineral decomposition is possible under suitable conditions. They found that from aluminosilicates sodium and potassium cations are released at the beginning, then calcium and magnesium, and at the end silicon and aluminium.

According to Rogers & Bennett (2004), silicate minerals contain nutrients necessary for microbial growth, but whether the microbial community benefits from their release during weathering is unclear.

In our investigations, the sterile mineral bed without added potassium was replaced by particular carbon rocks providing surrogate sources of K for bacterial growth.

As was shown, one of the most important factors determining carbon rock weathering is the species affiliation of the bacteria. The decomposition of the same mudstone was different depending on the bacterial strains. It seems that the differences in weathering of the same rocks by bacteria originated from

Table 5. The ability of some examined bacterial strains to form organic acids.

Name of bacterial strain	Volatile acids				Non-volatile acids					
	ac	pr	bu	va	la	fu	su	ma	α-k	ci
Arthrobacter sp. (16 K)	+	+	−	+	+	−	+	+	+	+
Arthrobacter sp. (16 K)/K[#]	+	+	−	+	+	−	+	−	+	−
Arthrobacter sp. (IVS)	+	−	−	−	+	−	+	+	+	+
Arthrobacter sp. (IVS)/K[#]	+	−	+	+	+	−	+	−	+	−
Pseudomonas stutzeri (IS)	+	−	−	−	+	+	+	−	+	−
Pseudomonas stutzeri (IS)/K[#]	+	−	−	−	+	−	+	−	−	−

ac. acetic, pr. propionic, bu. butyric, va. valeric , la. lactic, fu. fumaric, su. succinic, ma. malic, α-k. α-ketoglutaric, ci. citric
[#] incubation with kaolinite addition.
(+) – presence of organic acids;
(−) – absence of organic acids.

their microbial metabolism. The ability of acid formation and its release to the environment is wide spread among autotrophic as well as heterotrophic bacteria. To check bacterial ability to produce organic acids, chosen strains were cultivated on full mineral medium with glucose (cultivation I) and on a medium constituting the source of N and C where the rest of the mineral nutrients were replaced by carbon rock – kaolinite (cultivation II).

The chromatographic analysis of post-cultivation liquid indicated that the chosen bacteria strains did indeed produce some volatile and non-volatile organic acids but differed in their number and nature (Tab. 5). Considering the bacteria growing on the full mineral medium, *Arthrobacter sp.* (16 K) produced the highest number of organic acids, both volatile (3) and non-volatile (5). The rest of the investigated strains (slightly less active) produced only one volatile acid and five – *Arthrobacter sp.* (IVS) and four – *Pseudomonas stutzeri* (IS) non-volatile acids. It seems that the number of produced organic acids, especially volatile acids, played an important role in the decomposition of aluminosilicate. Berthelinand & Dommergues (1975) found that decomposition of granite sand depended on the quality of organic acids produced by microorganisms, especially on the amount of produced volatile acids. Our investigations showed that the degradation of bed rocks by bacterial strains depended not only on the amount of produced acids but also on their quality. Probably acetic acid among the volatile acids played the most important role in decomposition of carbon rocks because only this acid was produced by all of the investigated strains. In the case of non-volatile acids, it seems that three acids: lactic, succinic and α-ketoglutaric played a very important role in carbon rocks weathering (Tab. 5). All these acids were produced by the tested bacterial strains. It should be emphasized that the two *Arthrobacter* strains (16 K and IVS) produced exactly the same kind of non-volatile acids, but differences were observed in volatile acids production, which confirmed the conclusion that volatile acids played a more important role in the decomposition of aluminosilicate than non-volatile acids. Smyk (1970) investigatedmicrobial weathering of aluminosilicate and found that a part of the indigenous bacterial strains isolated, especially *Arthrobacter sp.* (which produced α-ketoglutaric acid), had the ability of decomposition of minerals, including montmorillonite. Rogers & Bennett (2004) stated that the addition of native microbial consortium to a microcosms containing silicates or glass with iron inclusions correlated to accelerated weathering and release of Si into the solution.

Generally, the different qualitative composition of metabolite (organic acids, volatile and non-volatile) produced by the bacteria may be, among other things, the cause of differences in the degree of decomposition of the same mineral by bacteria belonging to an identical strain. After cultivation II, the chromatographic analysis of post-cultivation liquid showed that the bacterial strains produced different numbers and kinds of organic acids. The bacteria belonging to *Arthrobacter* strains (16 K and IVS) produced three, but different, volatile acids and three non-volatile acids, but the same. It seems that modification of the medium resulted in a differentiation of only volatile acids within the genus of *Arthrobacter* but did not influence the production of non-volatile acids within the investigated *Arthrobacter sp.* (Tab. 5).

4 CONCLUSIONS

1) The biological degree of decomposition of carbon wastes depended on the kind of carbon rocks, bacterial activity, and on the nutrient availability of the rocks.
2) The dominant bacteria belonged to the genus of: *Arthrobacter, Pseudomonas, Micrococcus* and *Bacillus*.
3) All the chosen indigenous bacterial strains isolated from carbon rocks and "fresh heap" showed the ability to weather minerals.

4) Differences were observed in the activity of carbon rocks and their mixture decomposition by bacteria from *Arthrobacter sp.*, *Pseudomonas stutzeri* and *Micrococcus sp.*
5) Bacteria belonging to *Arthrobacter* and *Pseudomonas* appeared to be the most active in the decomposition process.
6) The addition of black coal to individual rocks in monoculture cultivation had a positive effect on the amount of bacteria and on the decomposition of examined materials.
7) The weathering of aluminosilicate is closely related to the kind of organic acids released by native bacteria, mainly volatile acids.
8) The kind of cultivation medium affected the differentiation of volatile acids within the *Arthrobacter* genus but did not influence the production of non-volatile acids within the investigated *Arthrobacter sp.*

REFERENCES

Aleksandrov, V.G. 1953. *Siliceous bacteria*. State Publisher Agricultural Literature. Moscow. (in Russian).

Arinuszkina, E.V. 1970. *Handbook of Chemical Soil Analysis*. MGU Publisher, Moscow University. Moscow. (in Russian).

Aristovskaja, T.V. 1973. *Some aspects of geochemical soil microorganisms activity as a part of biogeocenoses. Problem of Biogeocenology*. Science. Moscow. (in Russian).

Aristovskaja, T.V. 1980. *Microbiology of Soil Forming Process*. Science, Leningrad. (in Russian).

Aristovskaja, T.V., Daragan, A.J., Zykina, L.V. & Kutuzova, R.S. 1969. Mikrobiologiczeskije factory migracji niekotorych mineralnych elementov w poczvach. *Poczv*. 9: 95–104. (in Russian).

Aristovskaja, T.V. & Kutuzowa, R.S. 1968. Silica mobilization from sparingly soluble natural compounds by microbiological factors. *Soil Science* 12: 59–66. (in Russian).

Berthelin, J. & Dommergues, Y. 1975. Role de produits du metabolisme Microbien dans la solubilisation des mineraux d'une arene granitique. *Rev. Ecol. Biol. Sol.* IX(3): 397–406.

Berthelin, J., Ona-Nguema, G., Stemmler, S., Quantin, C., Abdelmoula, M. & Jorand, F. 2006. Bioreduction of ferric species and biogenesis of green rusts in soils. *Comptes Rendus Geosciences* 338: 447–455.

Bigham, J.M., Bhatti, T.M., Vuorinen, A. & Tuovinen, O.H. 2001. Dissolution and structural alterating of phlogopite mediate by proton attack and bacterial oxidation of ferrous iron. *Hydrometallurgy* 59(2–3): 301–309.

Garcia, C., Ballester, A., González, F. & Blázquez, M.L. 2005. Pyrite behaviour in a tailings pond. *Hydrometallurgy* 76(1–2): 25–36.

Gierasimov, I. P. 1973. Primary soil processes as a structure for soils genetic diagnostic. *Soil Science* 5: 102–113. (in Russian).

Henderson, M.E.K. & Duff, R.B. 1963. The release of metallic and silicate ions from minerals, rocks and soil by fungal activity. *J. Soil Sci.* 14(2): 236–246.

Illmer, P. 2006. A commercially available iron-chelating agent, Desferal, promotes Fe- and Al-mobilization in soils. *Soil Biology and Biochemistry* 38(6): 1491–1493.

Jacob, J. 1952. *Chemische Analyse der Gesteine und silikatischen Mineralen*. Birkhäuser Verlag, Basel.

Kalinowski, B.E., Oskarsson, A., Albinsson, Y., Arlinger, J., Ödegaard-Jensen, A., Andlid, T. & Pedersen K. 2004. Microbial leaching of uranium and other trace elements from shale mine tailings at Ranstad. *Geoderma* 122(2-4): 177–194.

Karavajko, G.I., Krucko, V.S., Melnikova, E.O., Avakjan, Z.A. & Ostruszko, J.I. 1980. Role of microorganisms in decomposition of spodumen. *Microb.* XLIX(3): 547–551. (in Russian).

Ledin, M. & Pedersen, K. 1999. The environmental impact of mine wastes – Roles of microorganisms and their significance in treatment of mine wastes. *Earth-Science Reviews* 41(1–2): 67–108.

Machill, S., Althaus, K., Krumbein, W.E. & Steger, W.E. 1997. Identification of organic compounds extracted from black weathered surfaces of Saxonean sandstone, correlation with atmospheric input and rock inhibiting microflora. *Organic Geochemistry* 27(1–2): 79–97.

Petsch, S.T., Edwards, K.J. & Eglinton, T.I. 2003. Abundance, distribution and δ13C analysis of microbial phospholipid-derived fatty acids in a black shale weathering profile. *Organic Geochemistry* 34(6): 731–743.

Ponomarieva, V.V. & Plotnikova, T.A. 1980. *Humus and Pedogenesis*. Science. Leningrad. (in Russian).

Puente, M.E., Rodriguez-Jaramillo, M.C., Li, C.Y. & Bashan, Y. 2006. Image analysis for quantification of bacterial rock weathering *Journal of Microbiological Methods*, 64(2): 275–286.

Rogers, J.R. & Bennett, P.C. 2004. Mineral stimulation of subsurface microorganisms: release of limiting nutrients from silicates. *Chemical Geology* 203: 91–108.

Severmann, S., Mills, R.A., Palmer, M.R., Telling, J.P., Cragg, B. & Parkes, R.J. 2006. The role of prokaryotes in subsurface weathering of hydrothermal sediments: A combined geochemical and microbiological investigation. *Geochimica et Cosmochimica Acta* 70(7): 1677–1694.

Smyk, B. 1970. Microbiological decomposition of silicate and aluminosilicate. *Postepy Mikrobiologii* IX(1): 121–136. (in Polish).

Statistical Annual (MSD). 2002. (in Polish).

Strzyszcz, Z. 2003. Geowissenschaftliche Aspekte der bodenlosen Rekultivierung von Sandgrubenrestlöcher und Berghalden in Südpolen – Ein Überblick. *Z. Geol. Wiss.* 31(H2): 129–142. Berlin.

Wagner, M. & Schwartz, W. 1967. Geomikrobiologische Untersuchungen. IX Verwertung von Gestein und Mineralpurven als Mineralsalzquelle für Bakterien. *Z. Allgem. Mikrobiol.* Bd 7(H2).

Zagury, G. J., Oudjehani, K. & Deschênes, L. 2004. Characterization and availability of cyanide in solid mine tailings from gold extraction plants. *Science of The Total Environment* 320(2-3): 211–224.

Zviagincev, D.G. 1973. *Relationship between microorganisms and hard surfaces*. Publisher of Minsk University. Minsk. (in Russia).

Environmental Engineering – Pawłowski, Dudzińska & Pawłowski (eds)
© 2007 Taylor & Francis Group, London, ISBN13 978-0-415-40818-9

Retentional abilities of soils in eroded environment

Marcin Widomski & Henryk Sobczuk
Lublin University of Technology, Faculty of Environmental Engineering, Nadbystrzycka, Lublin, Poland

ABSTRACT: This article presents the results of an analysis of the influence of water erosion on the water-retention characteristics (pF curves) of loess soil in a chosen catchment in Wyżyna Lubelska. The authors have chosen the "Euro – East" fruit farm in Olszanka near Krasnystaw (Poland) as a research object. The results of the field and laboratory studies of composition and pF curves of erosional soils supplemented with numerical calculations of pF curves in full range are presented. The Casagrande – Proszynski method of determination of soil composition, the sand box method and numerical calculations for obtaining pF curves were applied as the main research methods. The results of laboratory work and numerical calculations allowed the conclusions placed in the article. A noticeable and spatially variable influence of water erosion on soil composition and water-transport characteristics at different depths under ground level and at different location on the slope was observed.

Keywords: Water erosion, soil deterioration, pF curve, numerical simulation.

1 INTRODUCTION

Soil and land deterioration caused by water erosion is one of the main threats to the agricultural and woodland environments in Poland as well as in many similar (geologically and climatically) countries (Ananda & Heratrh 2003, Cerdan et al. 2002, Fullen 2003, Józefaciuk 1999). In the late nineties of the last century about 29% of the total area of Poland was endangered by the effects of water erosion. It can also be said that 21% of arable lands and 8% of woodlands are under threat in the whole country.

Wyżyna Lubelska, one of the main uplands in Poland, located in the SE part of the country, belongs to a group of regions described as the most threatened by the effects of water erosion by Józefaciuk (Józefaciuk 1995). Water erosion, at a medium or high rate, was noted on about 30% of its area. On the highest noted single surface – washes of soil suspension in the 80s and the 90s reached the level of $5870\,kg{\cdot}km^{-2}$ and $6480\,kg{\cdot}km^{-2}$ respectively (Turski et al. 1991, Dechnik, Filipek 1996).

There are several main factors increasing the effectiveness of the destructive influence of water erosion (Józefaciuk 1999, Valentin et al. 2005, Askoy & Kavvas 2005):

- climatic conditions which have an impact on the surface flow rate: precipitation, frequency of heavy rainfalls, thickness of snow cover;
- surface features determining the speed of surface flow and the volume of erosive surface-wash: hillside slopes, their length and exposure;
- soil characteristics: its erodibility, related to the mechanical and mineral composition of soils;
- use of soils: manner of conducting agricultural activity.

The water erosion phenomenon causes many dangerous and destructive changes in soils and landscapes. The most important are (Lado, Ben-Hur 2004, Van Oost et al. 2005, Linczar et al. 1991, Józefaciuk 1995):

- changes in the mechanical and mineral composition of soil; which reduce its water conductivity and infiltrational abilities, as well as the storage volume and water availability for plants;
- changes and transformations of ground surface – destruction of the top layers of soil by rill and slope erosion, creation of erosional gullies – the effect of gully erosion;
- changes in the water balance of erosional catchment and shortening of the water cycle;
- soil impoverishment resulting from increased removal of nutrients from soil layers.

Many studies, carried out in the 1990s and focused on the influence of water erosion on quantitative and qualitative composition of loess soils in upland regions of Poland proved that changes caused by erosion are quite dangerous. Linczar (Linczar et al. 1991) found that after a long presence of erosional activity on loess soil the variation of different granulometric fractions was significantly higher – especially the content of colloidal clay. This leads to increased compactness of the highest layers of loess soil. This phenomenon is directly connected to changes of water-transport characteristics of soils.

This article presents the results of studies focused on the influence of water erosion on the water-retention characteristics of loess soils in a chosen catchment in Wyżyna Lubelska. The authors present the results of the field and laboratory studies of composition and pF curves (in range of pF = 0–2,0) of erosional soils supplemented with numerical calculations of pF curves in range of pF = 2,0–4,3.

2 MATERIALS AND METHODS

This paper contains results of studies ran on a field object located in Olszanka near Kraśniczyn – the SE part of Wyżyna Lubelska. This object is nowadays used as a fruit-growing and cultivable farm with an area of about 200 hectares managed by the "Euro – East" firm. The farm is situated on the upland region Działy Grabowieckie reaching 300 m above sea-level, with 100 meter deep gullies. The minimum and maximum ground elevations are between 255 and 295 metres above sea-level. The soil cover on Olszanka object is mainly composed of eroded Quaternary covers represented by layers of loess and clay. The Cretaceous deposits located directly beneath the Quaternary cover are comprised of chalky clays and chalk rocks. The chalky sediments are not homogenous but they occur in layers of different thickness, hardness and colour.

Soils present in Olszanka can be described as leached brown soils of different structure resulting from intensive denudation of slopes. On lower and upper parts of wide valleys soils of II quality class can be found, III b and VI a classes were noted on steep slopes.

The research object is located in the Climatic Field of Lublin characterised by a continental climate, yearly precipitation of 600 mm, yearly medium temperature +7.4°C solar radiation intensity – 135.328 $W \cdot m^{-2}$ and a 218 day vegetation period (Józefaciuk et al. 1996).

The described research object in Olszanka is located in a zone highly endangered by the effects of water erosion. The authors checked results of archival studies concerning the problem of water erosion in the area of the present farm in Olszanka from the last century. Before WWII water erosion was noted (with different intensity) on about 65% of the whole area of today's farm. In the years 1946–1950, when the object was used as a cattle pasture, thanks to the development of sod, the endangered area was reduced to about 6%. Then, in its time as a State Agricultural Farm, as a result of the method of cultivation, the area of soils destroyed by water soil erosion was growing rapidly, reaching 96%. This was the main reason to transform the cultivatable farm into a fruit-growing farm in 1977 (Józefaciuk el at. 1976, Józefaciuk 1996). The mentioned transformation was linked to implementing soil protection against the effects of water erosion. Works started in the late 70's have been continued till present times – the result is an original, anti-erosion system to test the effectiveness of which was the main aim of the research programme No 1564/T09/2001/21.

The article presents the results of studies investigating changes caused by water erosion phenomenon in quantitative composition of eroded soils. The influence of water erosion on water – transport characteristics such as the pF curve was also observed. The main aim of the research was the assessment of the influence of erosion on the composition and transport characteristics for different layers of eroded soils located in different locations on a chosen slope of the field object in Olszanka. The studies covered field and laboratory measurements and numerical calculations of pF curves for chosen samples of eroded soils.

The methodology of the research included:

- collecting samples of eroded soils, from different levels – 0–5 cm (top, surface layer) and 35–40 cm below the ground level; located on one of the slopes on Olszanka object;
- analysis of composition of selected samples of soils – Casagrande – Proszynski method (BN-76/9180-06);
- measurements of pF curves in range pF = 0–2 for chosen samples of soils – sand box (Kowalik 1973);
- numerical calculations of pF curves for the described soils in the range pF = 0–4,3 using computer program written by H. Sobczuk (Sobczuk 1999);
- analysis of results and drawing conclusions.

The presented studies were run on one chosen slope located near one of antierosional installations placed in the field object in Olszanka. The chosen slope was also equipped with the system of terraces decreasing the speed of surface water flow and increasing its infiltration into lower levels. Each terrace was additionally equipped with a ditch filled with sand as shown on Fig. 1. This system was installed to increase the soil's infiltrational abilities by changing its water conductivity due to the enrichment of the soil profile with materials of higher conductivity. The longitudinal section of the chosen slope, with dimensions, anti-erosion system elements and marked measurement points is presented in Fig. 1. The points numbered

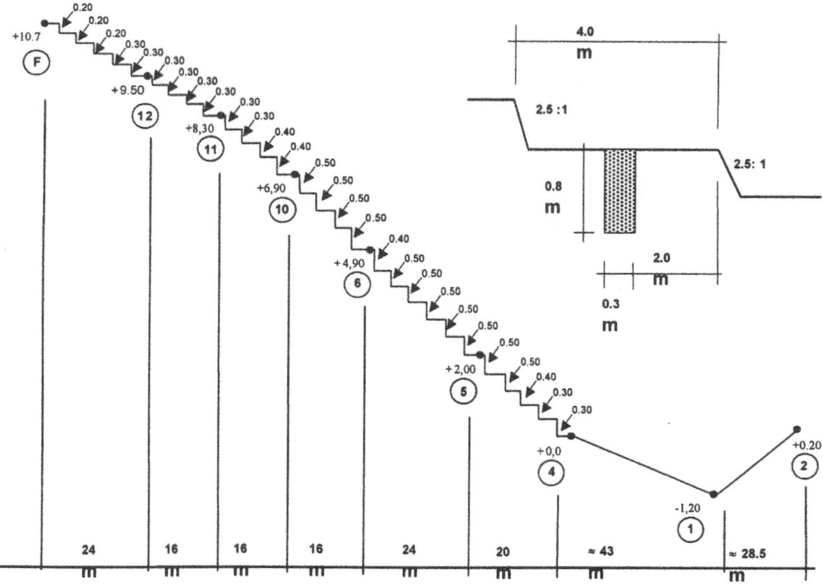

Figure 1. The scheme of the longitudinal section of the studied slope.

Table 1. Composition of eroded soils on the chosen slope.

Point	Depth (cm)	Content [%] of individual factions [mm]						Type of soil
		1–0,1	0,1–0,05	0,05–0,02	0,02–0,005	0,005–0,002	<0,002	
1	0–5	28	5	35	17	5	10	Light alluvial
	35–40	13	16	37	18	5	11	Strong alluvial
6	0–5	9	8	43	22	8	10	Strong alluvial
	35–40	14	9	37	20	8	12	Loess
12	0–5	12	9	41	19	9	10	Loess
	35–40	18	9	35	15	7	16	Strong alluvial

1, 6 and 12 were chosen, for the research, as most representative.

3 DISCUSSION

Table 1 presents the quantitative composition of soils from two different levels measured at points 1, 6 and 12 using the Casagrande – Proszynski method (BN-76/9180-06).

Table 1 shows differences in composition between two neighboring layers of eroded soils in three different locations on the chosen slope: the top, the middle of the slope and the slope base (Fig. 1).

The mentioned differences are clearly visible not only between the top surface (0–5 cm) and the lower level (35–40 cm) but also among the three locations – points No 1, 6 and 12; i.e. the highest content of fractions of 1–0,05 mm and the lowest of 0,05–0,002 mm were noted in both layers of point 1, located on the base of the slope. The composition of the top surface layers is interesting – the higher the sampling point the higher the content of clay – at lower locations clays were washed away by surface flows. In the middle of the slope the highest content of 0,05-0,005 mm faction was noted.

Figure 2 shows water characteristics of the chosen soil (points 1 & 12) obtained by numerical estimation of the results of laboratory measurements.

The laboratory work covered the range of 0–100 cm and the numerical estimation was carried out up to 200 cm with use of Mualem's method (Mualem 1976, Zaradny 1990) – coefficients of determination for each calculation were higher than $R^2 = 0.7$. Each graph

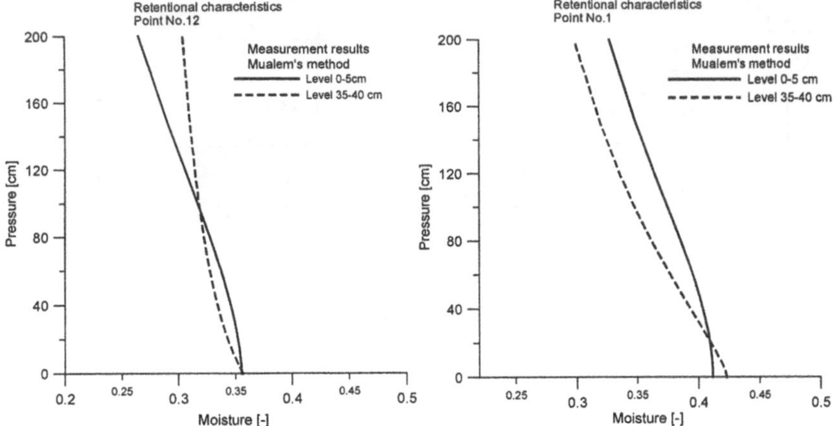

Figure 2. Retentional characteristics for points No. 1 and 12.

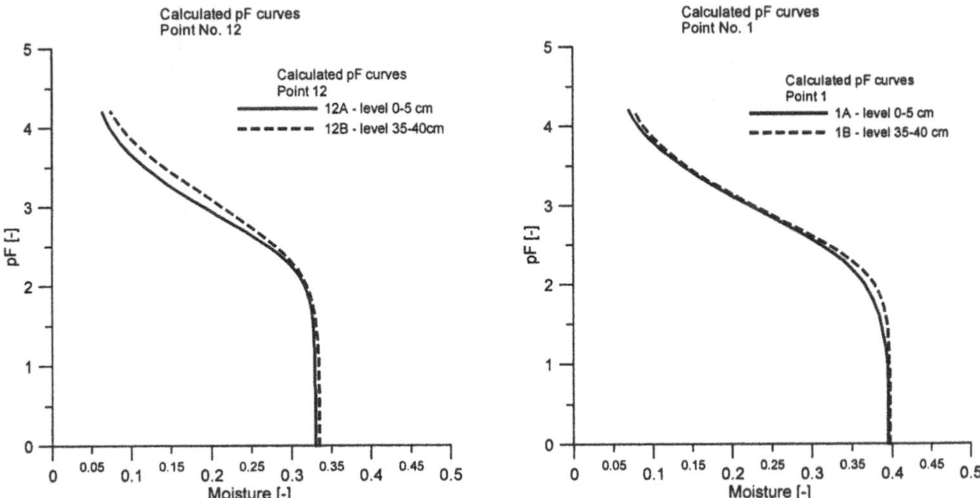

Figure 3. Calculated pF curves for researched points on the chosen soil profile.

presents two curves – each for a different layer – 0–5 cm and 35–40 cm below ground level measured at the same point (Fig. 1).

The difference in retentional curves between two chosen points is clearly visible. The water content in soils localized in point 12 is (at the same pressure) about 5% of moisture content lower than in point number 1. The inclination of the curve for 35–40 cm is also higher in point No. 1. These results show clear connection with the quantitative composition of soils presented in Tab. 1.

Differences between curves for each level in the same point can be also found:

- the soil porosity, according to the measurements, for each studied level, was different at point 1 and nearly the same for point 12;
- the water content at point No. 1 at 35–40 cm level diminishes with the increase of pressure – the difference in comparison with the surface level reaches 3.5% of moisture;
- water retention abilities for the 35–40 cm level at point 12 are higher than at surface level – 5% of difference in moisture content.

The authors would like to call the reader's attention to the obvious fact that measurement results are liable to methodological errors.

The pF curves for each layer (0–5 & 35–40 cm below. ground level) at both points (1 & 12) calculated by software by H. Sobczuk (Sobczuk 1999) are presented on Fig. 3.

The analysis of calculations allows some conclusions to be drawn. The retentional abilities, as was in the case of the measurements results, are higher at point No. 1 – the difference in water content reaches 5% up to pF = 3. For higher pressure (pF > 3) pF curves become nearly similar. The shape of curves at both discussed points is similar up to pF = 2,5. After this, a noticeably higher declination of the pF curves at point No. 1 was noted. This means that at point No. 1 the effective water retention (pF = 2,0–2,7 (Olszta, Zawadzki 1991)) reaches higher values than at point No. 12.

The differences between pF curves for both levels at the same point can also be noted. The retentional abilities are slightly higher at the 35–40 cm level at both points. Moreover, in the case of the pF curve at point No. 1 water content values at the same pressure are higher in the deeper layer – range pF = 1–2,5. The opposite situation was noted in point No. 12 – noticeable and clear growth of retentional abilities in the lower layer in comparison with the ground surface can be observed.

4 CONCLUSIONS

The analysis of results of work focused on the quantitative composition of eroded soil, laboratory measurement of retentional abilities and numerical calculations of pF curves which allow us to advance the following conclusions:

- water-erosion changes the quantitative composition of top surface layer soils, noticeably reducing its retentional abilities;
- noticeable differences in quantitative composition of soils from two different levels measured in three points located on different places on the eroded slope (No. 1,6 and 12) were observed;
- differences in measured retentional curves for points located on the top and the base of the slope were noted, leading to the lowering of the retentional abilities of soil localized on the top of the hill;
- the water content values measured at point No. 12 (the top part of the slope) are lower for the top eroded layer, which is directly connected to changes in the quantitative composition of soil caused by water-erosion phenomenon;
- calculated pF curves show that effective water retention is reduced in soils located on the top part of the eroded slope (about 0,06 [-] of moisture content);
- there is a need for future research covering more points located at different locations of the eroded slope and focusing on the influence of water-erosion on the water-transport characteristics of soils;
- long-term, periodical research should be undertaken to evaluate the dynamics of changes in soils caused by water-erosion.

REFERENCES

Aksoy, H., Kavvas, M.L. 2005. A review of hillslope and watershed scale erosion and sediment transport models. *Catena* 64 (2005) 247–271.

Ananda, J., Herath, G. 2003. Soil erosion in developing countries: a socio-economic appraisal *Journal of Environmental Management* 68 (2003) 343–353.

BN-76/9180-06 *Gleby i utwory mineralne. Pobieranie próbek i oznaczanie składu mechanicznego.*

Cerdan, O., Le Bissonnais, Y., Couturier, A., Bourennane, H., Souchere, V. 2002. Rill erosion on cultivated hillslopes during two extreme rainfall events in Normandy, France. *Soil & Tillage Research* 67 (2002) 99–108.

Dechnik, I., Filipek, M. 1996. Wpływ następczy procesów erozji wodnej na niektóre właściwości fizykochemiczne gleb. Ogólnopolskie Sympozjum Naukowe *Ochrona agrosytemów przed erozją.* 115–122, IUNG Puławy, Puławy.

Fullen, M.A. 2003. Soil erosion and conservation in northern Europe. *Progress in Physical Geography* 27.3 (2003) 331–358.

Józefaciuk, A., Józefaciuk, Cz. 1995. *Erozja agrosystemów.* Biblioteka Monitoringu Środowiska. Warszawa.

Józefaciuk, Cz. 1975. Rekultywacja i melioracje przeciwerozyjne gruntów obiektu Olszanka – Zakład Ogrodniczy PPGR Żułów. *Studium przedprojektowe, Projekt ST – TITR w Lublinie.* Puławy.

Józefaciuk, Cz., Józefaciuk, A. 1999. *Ochrona gruntów przed erozją.* Poradnik IUNG, Puławy.

Józefaciuk, Cz., Józefaciuk, A., Nowocień, E., Rubaj, J. 1996. Ocena sadowniczego zagospodarowania silnie urzeźbionych gruntów lessowych na przykładzie obiektu Olszanka. Ogólnopolskie Sympozjum Naukowe *Ochrona agrosytemów przed erozją.* 229 – 243. IUNG Puławy, Puławy.

Kowalik, P. 1973. *Zarys fizyki gruntów.* Wyd. Politech. Gdań., Gdańsk.

Lado, M., Ben-Hur, M. 2004. Soil mineralogy effects on seal formation, runoff and soil loss, *Applied Clay Science* 24 (2004) 209– 224.

Linczar, M., Chodak, T., Linczar, S.E. 1991. Wpływ procesów erozji na skład ilościowy i jakościowy fazy stałej niektórych gleb lessowych Dolnego Śląska, *Erozja i jej zapobieganie*, 21–30, AR Lublin

Mualem, Y. 1976. A new model for predicting the hydraulic conductivity of unsaturated porous media. *Water Resources Res.*, 12, 3: 513–522.

Olszta, W., Zawadzki, S. 1991. Właściwości retencyjne gleb, metody ich wyznaczania oraz sposoby wykorzystania w melioracji. *Materiały instruktażowe IMUZ* 94. Falenty.

Sobczuk, H. 1999. *Opis stanu fizycznego gleby jako ośrodka nieuporządkowanego na przykładzie krzywych retencji wody.* Monografia Acta Agrophisyca, Lublin.

Turski, R., Paluszek, J., Słowińska – Jurkiewicz, A. 1991. Wpływ rzeźby terenu na stopień zerodowania i właściwości fizyczne gleb lessowych. *Erozja gleb i jej zapobieganie*, s. 47–61, AR, Lublin.

Valentin, C., Poesen, J., Li, Y. 2005. Gully erosion: Impacts, factors and control. *Catena* 63 (2005) 132–153.
Van Oost, K., Van Muysen, W. , Govers, G., Deckers, J., Quine. T.A. 2005. From water to tillage erosion dominated land form evolution. *Geomorphology* 72 (2005) 193–203.
Zaradny, H. 1990. Matematyczne metody opisu i rozwiązań zagadnień przepływu wody w nienasyconych i nasyconych gruntach i glebach. Prace IBW PAN nr 23, Gdańsk.

Environmental Engineering – Pawłowski, Dudzińska & Pawłowski (eds)
© 2007 Taylor & Francis Group, London, ISBN13 978-0-415-40818-9

The influence of a chosen parameter on the hydraulic conductivity calculation

Małgorzata Iwanek, Dariusz Kowalski & Wenanty Olszta
Department of Environmental Protection Engineering, Lublin University of Technology, Lublin, Poland

ABSTRACT: The unsaturated hydraulic conductivity coefficient $k(h)$ is one of the basic parameters connected with the soil-water flow. There are several algorithms to predict this value and the van Genuchten-Mualem model is among them.

The objective of this paper is to estimate the parameter L value for the mentioned model and compare results of $k(h)$ calculations for the estimated L with those for the conventionally used value $L = 0.5$. Moreover, the authors determined the influence of the exponent L on the accuracy of the calculated parameter $k(h)$ in comparison with empirical values.

In all considered hydrogenic soil monoliths we found $L < 0$. We also noticed the existence of border suction head value corresponding with pF2, below which calculated $k(h)$ value for negative L was closer to empirical value than for $L = 0{,}5$. It is still necessary to improve the accuracy of fitting of calculated and measured $k(h)$ results for chosen soil profiles; here, the exponent L can be of great importance.

Keywords: Hydraulic conductivity coefficient, van Genuchten-Mualem model, exponent L.

1 INTRODUCTION

The unsaturated hydraulic conductivity coefficient $k(h)$ is one of the basic parameters connected with water flow in porous media. In recent years, in connection with the development of computer technology, it gained special importance as one of the basic input quantities in the most often applied models of water flow and pollution migrations in soils (Feddes et al., 1988, Poulsen et al., 1998, Crescimanno and Garofalo, 2005).

To predict the unsaturated hydraulic conductivity coefficient in the quickest and cheapest way, calculation methods in the suction head or moisture function are available. There are several algorithms of this kind, and the van Genuchten-Mualem model is among them. It is applied very often (e.g. Bruckler et al., 2002, Schmalz et al., 2003, Schaap and van Genuchten, 2005, Schwärzel et al., 2006). The main difficulty of this method lies in accurate estimation of water retention function parameters and correct determination of the dimensionless constant that accounts for pore tortuosity and pore connectivity, called in this article the exponent L (formula 2).

The authors' previous investigations showed that there is a large discrepancy between $k(h)$ values obtained from laboratory tests and those calculated using the van Genuchten model, in the full considered range of pressure head, for hydrogenic soil monoliths. In calculations conducted by the mentioned method, investigators (e.g. Kodešová et al., 1998, Poulsen et al., 2002, Schwartz and Evett, 2002, Tyner and Brown, 2004) usually assumed, as proposed by Mualem (1976), $L = 0{,}5$ in expression (2) as an optimal value. On the other hand, the published results of some investigations (Vereecken, 1995, Schaap and Leij, 2000) proved that this constant may have assumed different values, both positive and negative and, moreover, that it was related to the organic matter content (Wösten et al., 1995).

The objective of this paper is to estimate the exponent L values and to determine the its influence on the $k(h)$ values, calculated using the van Genuchten-Mualem model. These calculations were verified with the results of an empirical investigation, carried out for chosen hydrogenic soils, available thanks to the IMUZ – Lublin.

2 MATERIALS AND METHODS

Results of laboratory analysis, including water retention characteristics – the so-called pF-curves, saturated and unsaturated hydraulic conductivity coefficients for chosen peat soils monoliths, were taken as a basis for the investigations. Results of empirical determination and profiles descriptions were available thanks to the IMUZ – department in Lublin.

The following soil profiles were used in the tests: Krowie Bagno 15 (Mt I aa), Krowie Bagno 7 (Mt I ab), Krowie Bagno 476 (Mt I bb), Hanna-Holeszów 64/80(Mt I bc), Krowie Bagno 10 (Mt I aa), Sosnowica III(Mt I ba), Krowie Bagno 16 (Mt I bb) and Zienki Bukaciarnia 2 (Mt I b1).

Laboratory analysis were conducted using methods, that are valid in the IMUZ – pF curves were determined using the sand box method, saturated hydraulic conductivity coefficient (k_s) – using Wit's apparatus, unsaturated hydraulic conductivity coefficient $k(h)$ – by drying monolith method.

Comparison of k_s determination for hydrogenic soils, made by the IMUZ-Lublin, IMUZ-Bydgoszcz and University of Agriculture in Lublin, implies that using the Wit's apparatus to determine k_s for peat soils often causes overestimation of this parameter in comparison to field methods. In this connection, in the current paper the authors assumed $k(h)$ values received by drying monoliths method for the lowest measured suction head as k_s values. This method is recommended by some authors (e.g. Durner, 1984).

In numerical tests van Genuchten-Mualem model was taken to calculate the unsaturated hydraulic conductivity coefficient (van Genuchten, 1980):

$$k(h) = k_r \cdot k_s \tag{1}$$

$$k_r = \left[\frac{1}{1+(\alpha \cdot h)^n}\right]^{\left(1-\frac{1}{n}\right)\cdot L} \left\{1 - \left[\frac{(\alpha \cdot h)^n}{1+(\alpha \cdot h)^n}\right]^{\left(1-\frac{1}{n}\right)}\right\}^2 \tag{2}$$

where: $k(h)$ [mm/d], k_s [mm/d] – unsaturated and saturated hydraulic conductivity coefficient, respectively, k_r – dimensionless relative hydraulic conductivity coefficient, h [cm] – pressure (suction) head, α [cm^{-1}] ($\alpha > 0$) – parameter related to the inverse of the air entry pressure, L – dimensionless constant that accounts for pore tortuosity and pore connectivity, n ($n > 1$) – dimensionless measure of the pore-size distribution (van Genuchten and Nielsen, 1985, Schaap and Leij, 2000).

The water retention curve parameters α and n, were estimated using nonlinear regression analysis on the basis of the relationship:

$$\frac{\theta_s - \theta_r}{\theta - \theta_r} = \left[1 + (\alpha \cdot h)^n\right]^{\left(1-\frac{1}{n}\right)} \tag{3}$$

where: θ [cm$^3 \cdot$ cm^{-3}] – volumetric water content corresponding to the pressure head h [cm], θ_r and θ_s [cm$^3 \cdot$ cm^{-3}] – residual and saturated water content, respectively.

The exponent L value was estimated for four among eight hydrogenic soil monoliths (Krowie Bagno 10, Sosnowica III, Krowie Bagno 16 and Zienki Bukaciarnia 2) using the formula (2) in nonlinear regression analysis. Received values were used to predict $k(h)$ according relationship (2) for the rest of soil samples (Krowie Bagno 15, Krowie Bagno 7, Krowie Bagno 476, Hanna-Holeszów 64/80). The unsaturated hydraulic conductivity coefficient $k(h)$ was also calculated for these samples with $L = 0$ and $L = 0.5$. This latter value was proposed by Mualem (1976).

Statistical calculations were conducted using the STATISTICA program with the significance level 0.05.

The accuracy of calculated and empirical results fitting was tested using the average relative error calculated by the formula

$$s = \frac{1}{n_p} \sum_{i=1}^{n} \delta_i \tag{4}$$

where: n_p – number of measurements, δ_i – relative measurement error according to relationship:

$$\delta_i = \left|\frac{k_{pom_i} - k_{obl\,i}}{k_{pom_i}}\right| \cdot 100\% \tag{5}$$

where: k_{pom}, k_{obl} – measured and calculated value of unsaturated hydraulic conductivity coefficient, respectively.

3 RESULTS AND DISCUSSION

A comparison of measured and calculated results, including pF-curves and hydraulic conductivity is shown in Fig. 1.

The shape of water retention curves (Fig. 1A) shows the good fit of the approximate function to measured values of the retention capacity. A minor discrepancy occurred only for the Krowie Bagno 15 monolith, for volumetric water content near 80 cm^3/cm^3.

Comparison of measured and calculated unsaturated hydraulic conductivity values (Fig. 1B) indicated discrepancies in all cases, over the whole range of considered suction heads.

Results of θ_r, α and n estimation, gained on the basis of regression analysis, using the nonlinear least squares method, are shown in Table 1.

R-squared coefficient values were higher than 0.97 in all cases. Parameters α and n, appeared to be highly statistically significant, contrary to θ_r, the estimation of which exceeded the assumed significance level. The authors' previous analysis also indicates that the θ_r value does not affect $k(h)$ calculations results essentially.

The next part of the research was the estimation of exponent L value, based on the expression (2). The results are shown in Table 2.

In all considered cases, it is seen that $L < 0$ and the values also appeared statistical significant.

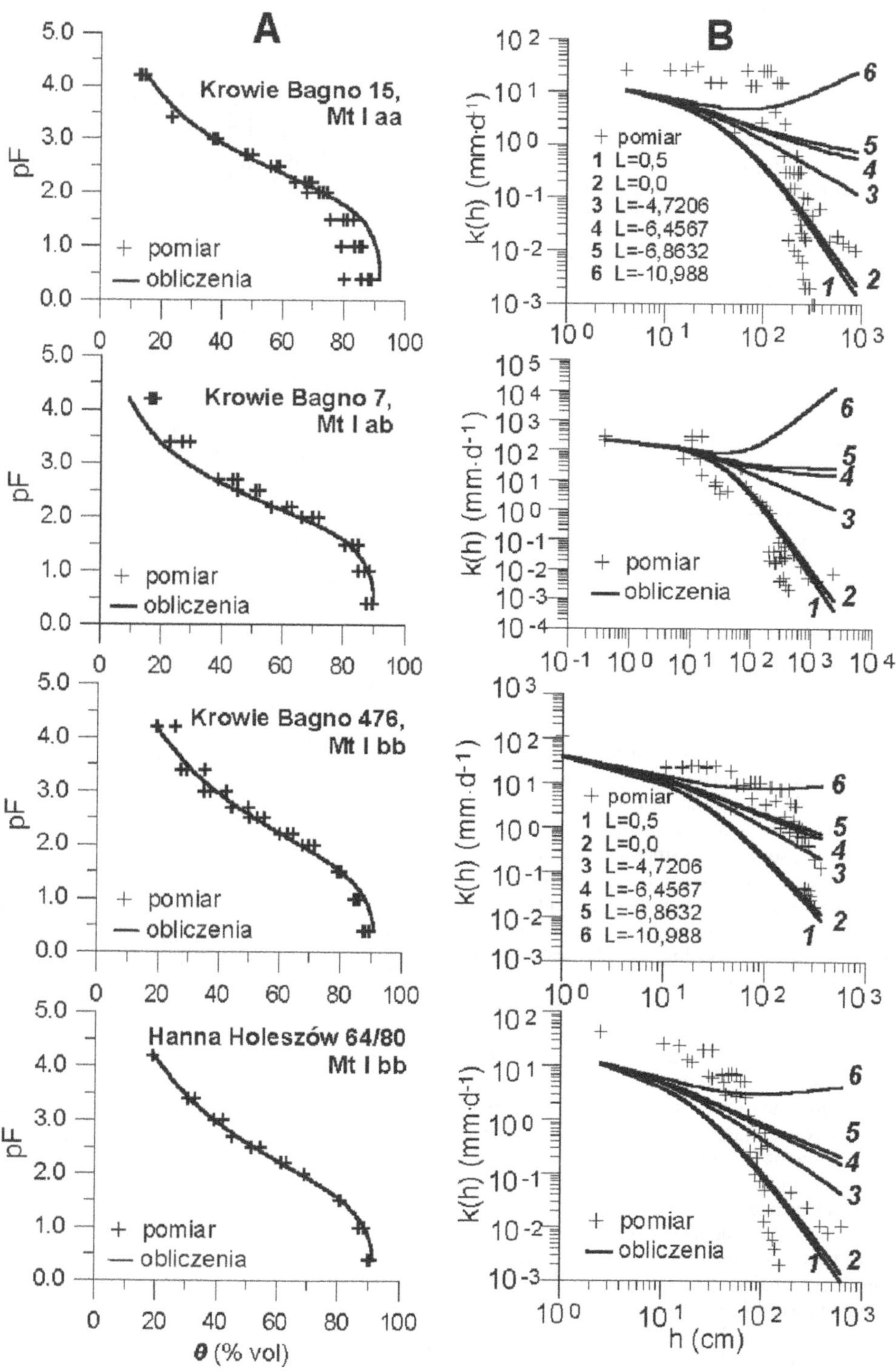

Figure 1. Water retention curves for tested soil profiles (A). Diagram of calculated and measured unsaturated hydraulic conductivity coefficient values in dependence on suction head h for tested soil profiles (B).

Table 1. Parameters of function (3) approximating water retention curve.

Soil sample	Estimated parameters		
	θ_r	α	n
Krowie Bagno 15 – Mt I aa	0.000	0.013	1.343
Krowie Bagno 7 – Mt I ab	0.086	0.016	1.413
Krowie Bagno – Mt I bb	0.006	0.029	1.252
Hanna Holeszów 64/80 – Mt I bc	0.000	0.027	1.256
Krowie Bagno 10 – Mt I aa	0.000	0.018	1.312
Sosnowica III – Mt I ba	0.000	0.017	1.300
Krowie Bagno 16 – Mt I bb	0.000	0.009	1.367
Zienki Bukaciarnia 2 – Mt I b1	0.000	0.021	1.221

Table 2. Exponent L values estimated on the base of formula (3).

Soil sample	Exponent L	Correlation coefficient R
Krowie Bagno 10 – Mt I aa	−6.45674	0.72813
Sosnowica III – Mt I ba	−4.72060	0.68473
Krowie Bagno 16 – Mt I bb	−6.86325	0.51975
Zienki Bukaciarnia 2 – Mt I b1	−10.98800	0.36390

There are negative L values among literature data (e.g. Schuh and Cline,1990, Yates et al., 1992). But if Mualem's assumption is correct, that expression $[(\theta-\theta_r)/(\theta_s-\theta_r)]^L$ (symbols as in formulae (2) and (3)) accounts for pore tortuosity and continuity, the expression should have values <1 (Schaap and Leij, 2000). The base of a power in this expression is between 0 and 1, so the above-mentioned assumption would be satisfied, if the power were positive. In that case, the results suggest treating the exponent L as an empirical parameter without physical interpretation.

Further research, according to the established methodology, were conducted to determine how the exponent L influences the results of $k(h)$ calculations. To this end, $k(h)$ values calculated according to formula (2) for L equals 0,5, 0, −4,72060, −6,45674, −6,86325, −10,98800, respectively, were compared with measured $k(h)$ values for Krowie Bagno 15, Krowie Bagno 7, Krowie Bagno 476 i Hanna Holeszów 64/80 samples. The error of calculation results in comparison with the empirical ones was determined according to relationship (5).

On the basis of the conducted calculations and analysis for all tested soil monoliths, it was claimed that exponent L essentially affects fitting conformity of model and empirical $k(h)$ values. Moreover, for each tested sample except the Krowie Bagno 7 sample, it the existence of a border suction head value was observed, below which the value of relative error δ_i of the unsaturated hydraulic conductivity coefficient for negative L was smaller than for $L=0,5$ (Fig. 2) and above which

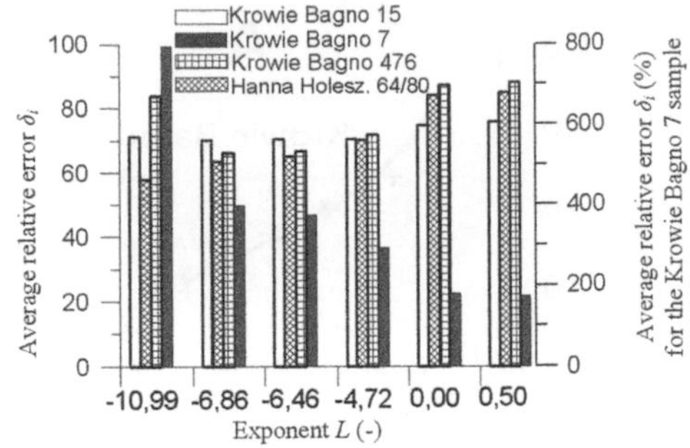

Figure 2. Average relative error δ_i value according to formula (2) for various values of L for suction head corresponding with pF < 2.

this error value was much larger (even by 2–3 orders of magnitude). These border values equal respectively: for Krowie Bagno 15 sample – $h=166$ cm, Krowie Bagno 476 – $h=102$ cm and Hanna Holeszów 64/80 – $h=94$ cm, and it can be assumed that they correspond with pF2 closely. This suggests that during $k(h)$ calculation, it would be sensible to separate and separately consider two cases: for pF ≤ 2 and pF > 2.

The unsaturated hydraulic conductivity coefficient for the Krowie Bagno 7 sample was determined for 40 suction head values between 0.4 and 2253 cm and for negative L a smaller δ_i was obtained than for $L=0.5$; only for 4 cases.

One of the considered monoliths did not show as many regularities as the others, so it is necessary to conduct further research using more soil samples to confirm or exclude the existence of the mentioned regularities. Moreover, the authors managed to achieve a decrease in the relative error δ_i for negative L, but its average value calculated according to formula (4) is still high (Fig. 2). In that case it is necessary to look for methods of increasing the relative error; this will be the purpose of further research.

4 CONCLUSIONS

To recapitulate – the van Genuchten-Mualem relationship appeared to approximate the pF curve of tested profiles well and two of three parameters estimated on its basis can be deemed as statistically significant. The exponent L determined on the basis of formula (2) also appeared statistically significant and possessed negative values for all tested soil monoliths.

Calculations and analysis conducted in the range of tested peat soils led to the formulation of the following conclusions:

- it is necessary to improve conformity of calculated and measured $k(h)$ results for chosen soil profiles, taking into consideration exponent L values,

- evaluation of the influence of the exponent L value on the agreement between model and empirical $k(h)$ results for the soils considered, proved it be significant,
- existence of a border suction head value corresponding with pF2, below which the calculated $k(h)$ value for negative L was closer to the empirical value than for $L = 0,5$, was affirmed for three of the four tested monoliths,
- though the present results reveal a certain regularity, they were obtained for only four soil samples, so further research on more diverse testing material, including peat soils, is necessary before more solid conclusions can be reached.

REFERENCES

Bruckler L., Bertuzzi P., Angulo-Jaramillo R., Ruy S., 2002. Testing an infiltration method for estimating soil hydraulic properties in the laboratory. *Soil Sci. Soc. Am. J.* 66, 384–395.

Crescimanno G., Garofalo P., 2005. Application and evaluation of the SWAP model for simulatong water and solute transport in a craking clay soil. *Soil Sci. Soc. Am. J.* 69, 1943–1954.

Durner W., 1994. Hydraulic conductivity estimation for soils with heterogenous pore structure. *Water Resour. Res.* 30, 211–223.

Feddes R.A., Kabat P., van Bakel P.J.T., Bronswijk J.J.B., Halbertsma J., 1988. Modelling soil water dynamics in unsaturated zone – state of the art. *J. Hydrol.* 100, 69–111.

Kodešová R., Gribb M.M., Šimůnek J., 1998. Estimating soil hydraulic properties from transient cone permeameter data. *Soil Sci.* 163, 436–453.

Mualem Y., 1976. A new model for predicting the hydraulic conductivity of unsaturated porous media. *Water Resour. Res.* 12, 513–522.

Poulsen T.G., Moldrup P., Jacobsen O.H., 1998. One-parameter models for unsaturated hydraulic conductivity. *Soil Sci.* 163, 425–435.

Poulsen T.G., Moldrup P., Iversen B.V., Jacobsen O.H., 2002. Three-region Campbell model for unsaturated hydraulic conductivity in undisturbed soils. *Soil Sci. Soc. Am. J.* 66, 744–752.

Schaap M.G., Leij F.J., 2000. Improved prediction of unsaturated hydraulic conductivity with the Mualem-van Genuchten model. *Soil Sci. Soc. Am. J.* 64, 843–851.

Schaap M.G., van Genuchten M.Th., 2005. A modified Mualem-van Genuchten formulation for improved description of the hydraulic conductivity near saturation. *Vadose Zone J.* 5, 27–34.

Schmalz B., Lenartz B., van Genuchten M.Th., 2003. Analysis of unsaturated water flow in a large sand tank. *Soil Sci.* 168, 3–14.

Schuh W.M., Cline R.L., 1990. Effect of soil properties on unsaturated hydraulic conductivity pore-interaction factors. *Soil Sci. Soc. Am. J.* 54, 1509–1519.

Schwartz R.C., Evett S.R., 2002. Estimating hydraulic properties of a fine-textured soil using a disc infiltrometer. *Soil Sci. Soc. Am. J.* 66, 1409–1423.

Schwärzel K., Šimůnek J., Stoffregen H., Wessolek, van Genuchten M.Th., 2006. Estimation of the unsaturated hydraulic conductivity of peat soils: Laboratory versus field data. *Vadose Zone J.* 5, 628–640.

Tyner J.S., Brown G.O., 2004. Improvements to estimating unsaturated soil properties from horizontal infiltration. *Soil Sci. Soc. Am. J.* 68, 1–6.

van Genuchten M.Th., 1980. A closed-form equation for predicting the hydraulic conductivity of unsaturated soils. *Soil Sci. Soc. Am. J.* 44, 892–898.

van Genuchten M.Th., Nielsen D.R., 1985. On describing and predicting the hydraulic properties of unsaturated soils. *Annales Geophysicae* 3(5), 615–628.

Vereecken H., 1995. Estimating the unsaturated hydraulic conductivity from theoretical models using simple soil properties. *Geoderma* 65, 81–92.

Wösten J.H.M., Finke P.A., Jansen M.J.W., 1995. Comparison of class and continuous pedotransfer functions to generate soil hydraulic characteristics. *Geoderma* 66, 227–237.

Yates S.R., van Genuchten M.Th., Warrick A.W., Leij F.J., 1992. Analysis of measured, predicted, and estimated hydraulic conductivity using the RETC computer program. *Soil Sci. Soc. Am. J.* 56, 347–354.

Environmental Engineering – Pawłowski, Dudzińska & Pawłowski (eds)
© 2007 Taylor & Francis Group, London, ISBN13 978-0-415-40818-9

Rationalization of district heating networks using artificial neural networks

Piotr Ziembicki
Institute of Environmental Engineering, University of Zielona Góra, Poland

Paweł Malinowski
Institute of Air-Conditioning and District Heating, Wrocław University of Technology, Poland

ABSTRACT: This article provides a short review of research trends utilizing artificial neural networks as aids to the design and operation of heating networks. In the first part of the article, the capabilities of artificial neural networks are described in general, and the kinds of problems that can be solved with their use are outlined, as well as the kinds of issues to which artificial neural networks may not be applied. Also, the main stages of neural model development have been defined and the most frequently used neural networks structures and training algorithms have been described. In the second part, results of the study are presented, aiming to demonstrate that the utilization of artificial neural networks allows the minimisation of points for heating network monitoring and, at the same time, to achieve a topographic distribution that allows an overview of heating system status to be obtained.

Keywords: Heating networks, monitoring, neural networks, neural classification.

1 INTRODUCTION

The obligations assumed by the European Union countries, including Poland, associated with the implementation of the Kyoto Protocol, which include, among other things, actions against climate change and international cooperation for sustainable development on a global level, require a reorientation of energy policy. The key to fulfilling the above-mentioned obligations is to develop new and renewable sources of energy as well as to improve the energy efficiency of the existing systems through optimisation of combined heat and power generation, rationalization of heating networks used for distribution of a heating medium, and heating substations.

The rationalization and improvement of heating networks efficiency is a very complex task due to the multitude of existing parameters which must be optimised as well as the necessity to use advanced modelling and simulation methods together with monitoring and telemetry systems of heating networks. This article presents research results on the use of artificial neural networks in district heating network optimisation.

The term optimisation can be defined as an activity aimed to achieve the best possible result in given conditions and at established evaluation criteria. One of the most important and indispensable stages of the optimisation process is the construction of a system model to serve as its image. An abstract mathematical model, which is most often used, can be considered as a set of variables and relations among them.

An analysis of a complex heating system operating in diverse and often unpredictable conditions involves the creation of complex mathematical models (Savola and Keppo, 2005). In many cases, however, the present level of knowledge does not make it possible to map all essential features of the system. Accurate computational methods are not available, either (Larsen et al., 2002). In these circumstances, new modelling techniques are continuously being sought, and there is a growing interest in artificial neural networks, which are a modelling technique capable of mapping complex non-linear functions. The properties of artificial neural networks make it possible to create mathematical models of the complex systems which operate in a manner that cannot be fully anticipated. In this context, and due to the huge amounts of data to be dealt with, the selection of neural simulation and classification as a tool for modelling and analysis of the heating system is justified (Mihalakakou et al., 2002; Kalogirou, 2001).

2 MATERIALS AND METHODS

As in their biological archetype, the processing of information in artificial neural networks is accomplished through the network of computing nodes (neurons) and their interconnections. The basic features which distinguish them from other data processing systems include ability to work effectively even when they are partly damaged, parallel and distributed processing (in hardware realization), ability to make generalizations, interpolation and prediction, as well as their low susceptibility to errors in data sets. However, the most important feature of neural networks is their ability to adapt (or ability to learn) making possible such a selection of parameters that will allow the adaptation of network operation to solving a specific problem. When designing a neural network, no operational algorithm is defined for its particular elements but only its architecture, original values of weights and learning method (Korbicz et al., 1994).

Most of the recently created and used networks have the layer structure in which an input layer, an output layer and hidden layers are distinguished in respect of the travelling direction of processed signals. In general, they can be divided into feed-forward, recurrent and cellular neural networks.

In the cellular neural networks, neurons are arranged in the shape of a square or rectangle so it may be stated that they have an internal geometrical structure. The location of each neuron may be described with two indices indicating the row and column number, as in a matrix. The interconnections between processing elements (artificial neurons) involve only neurons from the neighbourhood of defined radius. These connections are in most cases non-linear, described by a set of differential equations.

Typical examples of such a network include self-organizing maps of attributes called Kohonen networks, which feature, apart from their structure, the non-supervised training algorithm, in other words, based on learning patterns that contain input data only (without the corresponding output data). The training of Kohonen networks is based on the competition idea in which only one processing element is activated – the winner.

When taking the type of activation function as a division criterion, RBF networks (Radial Basis Functions) and Probabilistic Neural Networks (PNN) are distinguished, and Generalized Regression Neural Networks (GRNN) are distinguished in respect of the purpose criterion. PNN networks are designed to solve classification problems, and GRNN networks are used for solving regression problems.

The behaviour of a neuron network is determined to a great extent by its structure and weight vectors on the connections between particular neurons. It is not possible to determine those vectors a priori, so in use there are iterative multi-stage methods of neuron network "programming", which is called neural network training. In order to allow the application of these methods it is necessary to supplement the artificial neuron network model with a weight changing processor and error detector.

The essential problem, from the researchers' point of view, is the classification issue defined as the transformation of a point from the input space to the output space, which is called the classification space. The purpose of the classification network is to estimate the probability that a case belongs to a given class; in fact, during the training process, such a network learns to estimate the function of probability density represented by the data gathered. Modern methods of probability density function estimation are based on the nucleus approximation in which simple (nucleus) functions are located where any available case occurs, and then they are added together in order to obtain the estimator of total probability density function (Bishop, 1998).

Artificial neural networks are used very often to solve many categories of problems in science and engineering. In the district heating area there were many research programs aimed at implementation of artificial neural network as a tool for modelling district heating systems and processes of heat supply. Kalogirou presented the application of neural networks and other artificial intelligence methods and techniques to model and control combustion processes in different types of devices as well as in furnaces (Kalogirou, 2003). In another research paper (Mihalakakou et al., 2002) material was presented about the implementation of neural network based prediction to the problem of heat and cold supply for an apartment building in Athens. Similar research was described by Olofsson in (Olofsson et al., 1998). In-Ho (In-Ho and Kwang-Woo, 2004) presented an interesting utilization of artificial neural network capabilities for the prediction of room temperature drop time. A neural model was built with the use of data generated in the computer simulation carried out on the basis of construction and operation data of a variety of objects. The simulation of the model and its optimisation allowed the determination of optimal neural network structure and the best training algorithm, which would allow the prediction of room temperature drop time. Mahmoud dealt with a similar problem in his paper (Mahmoud and Ben-Nakhi, 2003), in which he utilized artificial neural networks for the optimisation of air-conditioning system switching time in public buildings. The construction and training of their neural model were carried out on the basis of simulation data of two buildings in Kuwait prepared with the use of computer programme and climatic data. Six different neural models were built and their simulation and optimisation made it possible to select the best neural network structure intended for the prediction of air-conditioning system switching time. In the paper (In-Ho et al., 2003) authors presented the application of neural networks to control the heating installation of an office building. In addition, artificial neural networks allows the solution of more general problems.

In the paper (Sozen et al., 2004) the implementation of neural networks together with meteorological database that helps to find the best economical location for solar systems, was presented. The issues associated with solar energy and the optimisation of solutions associated with their utilization were also addressed by Kalogirou. The purpose of the study presented in his paper (Kalogirou, 2004) was the utilization of artificial intelligence, and especially artificial neural networks and genetic algorithms for the optimisation of solar systems with the aim to maximize financial savings achieved as a result of their utilization. The neural model built on the basis of data obtained during the computer simulation of solar systems allowed him to achieve substantial financial savings from the tested solutions and considerably reduced the time required for analyses of the proposed solutions.

Also, Cetiner presented the utilization of SSNs for optimization of solar systems in his study (Cetiner et al., 2005), in which he built an artificial neural network intended for the analysis of different factors, such as weather conditions and solar radiation intensity, on the production of usable hot water. The neural model was trained with the use of experimental data obtained from the usable hot water preparation system which had been built earlier. Aydinalp presented in his paper (Aydinalp et al., 2004) the analysis of possibilities to implement and implementation of two neural models intended for modelling of energy consumption in multi-family residential buildings. The neural models built in the study were used for the analysis of energy consumption for heating and preparation of usable hot water. The results obtained from the neural simulation were more correct and adequate than those obtained with the use of classical (non-neural) modelling methods.

An overview of different applications of artificial neural networks associated with energy systems was presented by Kalogirou (Kalogirou, 2000). The author divided neural models according to their application area into modelling of: solar systems for steam production, usable hot water preparation systems using solar collectors, HVAC systems, conditions of solar radiation intensity and wind velocity, power systems, and also into issues associated in general with prediction and forecasting. The publication presents a number of implementations of artificial neural networks in the energy systems with the most interesting examples given here only, due to the limited space of this paper, which include: utilization of SSNs for heat demand analysis at the limited object data available, prediction of air infiltration flow into tested building, prediction of energy consumption for heating in passive solar buildings with respect to the structure of building partitions, short-term prediction of electric energy demand only on the basis of data obtained from the system operation.

3 RESULTS AND DISCUSSION

The principal goal of the research carried out by the authors was to check whether the use of artificial neural networks would allow performing the analysis of heating network monitoring points leading to the reduction of their number and, at the same time, achieving such topographic distribution that would allow obtaining the full picture of the heating system status. The positive answer to the above question would allow us to limit the necessary number of telemetry chambers for the existing heating network.

The execution of the research project required neural classification of heating system state to the specific, predefined category. Due to the specificity of problem for which the neural model was built, a number of tests were performed aimed at finding both optimal neural network and the best set of training data as well as the set of heating network state categories.

In the designed neural model, the issue of defining heating network state categories was particularly difficult due to the complexity of relationship between the network parameters. The relationship between flow, pressure, feeding temperature and cooling, as well as the impact of such elements as irregularities in domestic hot water consumption, seasonal technologies, and night or weekend temperature reductions, all make it impossible to define the precise rules of qualification to particular class. Meteorological data, such as data containing hourly distribution of external temperatures for each day of the tested period, allowed the solving of this problem through the determination of specific state sets (network state categories) on the basis of external temperature, type of day (R – working day, W – holiday) and time of the day (N – night time, D – daytime). After assuming such a criterion for heating network state image generation, it was possible to build the set of states being a compilation of the above conditions. Another motivating argument for such a selection of parameters was the practical inability to analyse the very large numbers of data available.

When solving classification problems it is necessary to define a variable containing the name and number of the category which network input parameters belong to. Usually, these are non-numerical variables, which considerably complicates operations performed on them. The most frequently found type of non-numerical data comprises nominal values which may be presented in a numerical way. They can be of two-state nature, which may be represented through the conversion into numerical state (values 0 and 1).

Operations on multi-state data are more difficult. They may be represented by a single numerical value (e.g. 0, 1, 2) but this can cause unnecessary prioritising of nominal values, which may falsify their true meaning. The one-of-N coding is a better solution, in which one neuron corresponds to only one of N possible values of the variable under examination. The specific state of a nominal variable is represented by the signal value of 1, appearing on the neuron corresponding to that state while the values on all the other neurons are 0. The number of numerical variables is then equal to the number of possible values of the nominal variable. The problem will occur in the case of nominal variables with a large number of possible states, because the number of numerical variables required for one-of-N coding is not acceptable due to the very large size of resulted network, which makes the training process very difficult.

The construction and testing of neural models of the heating network was started with the determination of two research directions. The first one focused on the model in which single neural networks classified all categories of heating network states. In the utilized neural networks, the one-of-N coding was applied and such a model was called an integrated neural model. The task of the other research direction was the creation of self-contained neural networks able to classify group categories of the heating network states. In the utilized neural networks, two-state (or maximum four-state) coding was applied and such a model was called a partial neural model.

The telemetry data obtained, and the computer simulation model built were used to define a series of training, validating and testing data sets. The designed neural models were based on both the training data sets built on the basis of computing results generated from the simulation model (neural models DSZ.1, DSZ.2, DSC.ZT01 ÷ DSC.ZT19), and on the basis of measuring data obtained from the monitoring of the tested heating system (neural models DTZ.1 ÷ DTZ.4, DTC.ZT01 ÷ DTC.ZT19). The letters in the names of particular neural models have the following meaning: DS – simulation data, DT – telemetry data, Z – integrated model, C – partial model, ZT – external temperature range, e.g. DTZ.1 – integrated neural model No. 1, based on data originating from heating system telemetry, DSC.ZT03 – partial neural model, based on simulation data, qualifying heating network states for external temperatures $-16°C < T_z \leq -14°C$ (part of temperature range definition is presented in Table 2).

In both types of neural models, the training, validating and testing data were analysed with different training methods and in the presence as well as in the absence of defective data. In both of them, a variety of neural network types were tested, such as MLP, RBF, linear networks and PNN, among the others. In total, 812 neural networks were tested, allowing the formulation of conclusion that both perceptron and linear-type neural networks were completely unsuitable for heating network state classification, and that the best classification quality in most neural model variants was shown by probabilistic neural networks (PNN).

ZM_SELKC	KLASA	01FZ01	01FZ03
U	02RN	794,64	91,21
U	02RN	797,28	90,81
W	02RN	795,78	91,52
T	02RN	797,27	91,46
U	01RN	797,27	91,46
U	01RN	797,27	91,46
U	01RD	797,71	91,59
W	01RD	797,71	92,92
T	01RD	798,17	92,82
U	02RD	799,37	92,42
U	03RD	799,46	91,28
U	04RD	799,62	87,23
U	05RD	787,08	88,23
U	06RD	769,48	92,38
U	06RD	769,48	92,38
U	06RD	769,48	92,38

Figure 1. Part of training data for integrated neural network.

In the neural models based on the computer simulation of a tested heating network, parameters originating from chambers: 01FZ03, 04FZ02, 13FZ03, 14FZ05 and 15FZ02 were verified as defective data. The sets of measurement data were also analysed in respect of defective data. As a result, the parameters from chambers: 01FZ03, 04FZ02, 04FZ03, 12FZ05, 13FZ03, 15FZ04, 04PZ02, 04PZ03, 04TZ02 were removed. The reasons for the removal of data originating from the above measuring points were: a considerable number of missing readouts (probably caused by instrument or communication failures), a large number of values considerably deviating from the parameter average value, and long periods of constant values of the recorded parameter (probably caused by failures of the recording devices). Due to periodically occurring missing readouts in different chambers (at the same hours) the selected whole days (or selected day hours) of readouts were also removed.

Figure 1 shows part of one of the training patterns (containing also validation and test data) for the integrated neural model. The file allows the training of the qualifying network for all heating network state categories.

The column "ZM_SELEKC" in the training data set includes a selection variable used for the separation of training, validation and testing data. Symbol of case assignment: "U" – network training, "W" – validation, "T" – testing. In the column "KLASA", categories of heating network state images are included, which are also a categorized output variable of neural networks, and the next columns contain parameters in particular chambers.

In the columns "Neural model features" of Table 1 and Table 2, characteristics of training data used for training particular neural networks are presented, providing the number of variables and patterns, and

Table 1. Part of the list of integrated neural model variants.

Simulation data		Telemetry data	
Variant	Neural model features	Variant	Neural model features
DSZ.1	Selected days of heating season 10.96 ÷ 05.97 *Include defective data* 56 variables 936 patterns	DTZ.1	Heating seasons: 01.96 ÷ 04.96 10.96 ÷ 04.97 10.97 ÷ 04.98 *Include defective data* 64 variables 12000 patterns
DSZ.2	Selected days of heating season 10.96 ÷ 05.97 *Exclude defective data* 51 variables 936 patterns	DTZ.2	Heating seasons: 01.96 ÷ 04.96 10.96 ÷ 04.97 10.97 ÷ 04.98 *Exclude defective data* 55 variables 11807 patterns
...	...	...	...

Table 2. Part of the list of partial neural model variants.

Simulation data		Telemetry data	
Variant	Neural model features	Variant	Neural model features
DSC.ZT01	Selected days of heating season 10.96 ÷ 05.97 *Exclude defective data* Classification only states of network for external temp.: $T_z \leq -18°C$ 51 variables 5 patterns	DTC. ZT01	Heating seasons: 01.96 ÷ 04.96 10.96 ÷ 04.97 10.97 ÷ 04.98 *Exclude defective data* Classification only states of network for external temp.: $T_z \leq -18°C$ 55 variables 5 patterns
DSC.ZT02	Selected days of heating season 10.96 ÷ 05.97 *Exclude defective data* Classification only states of network for external temp.: $-18°C < T_z \leq -16°C$ 51 variables 7 patterns	DTC. ZT02	Heating seasons: 01.96 ÷ 04.96 10.96 ÷ 04.97 10.97 ÷ 04.98 *Exclude defective data* Classification only states of network for external temp.: $-18°C < T_z \leq -16°C$ 55 variables 39 patterns
...	...	...	...

information on defective data presence. Due to the limited space, the complete list of neural model variants as well as of training data are not presented.

The analysis of the results obtained from neural simulation of the heating network allowed the selection of neural networks which were then utilized for further research efforts. An example structure of the neural network selected for further testing is presented in Figure 2. Validation error was the main acceptance criterion for network selection. In the case the values of the above parameter were similar, better qualification quality or less complicated structure (smaller number of layers and neurons in a layer) were decisive during the selection of networks. Neural networks tested in the partial neural model (except the rejected networks) showed similar values of the validation error and similar qualification quality. Due to the clear symptoms of network overlearning or insufficient number of learning patterns, 10 networks from the partial model based on simulation data and 2 networks from the partial model trained by telemetry data were rejected.

The current research allowed the selection of neural networks that achieved average classification quality above 0.94. This means that, in average, more than 94% of state images presented to the neural networks during the training process were classified correctly. The validation and testing processes resulted in respectively 65% and more than 62% of correctly classified network categories. The constructed models achieved such parameters for complete input data, i.e. measured (or computer simulated) parameters in all telemetry chambers functioning in the tested heating network.

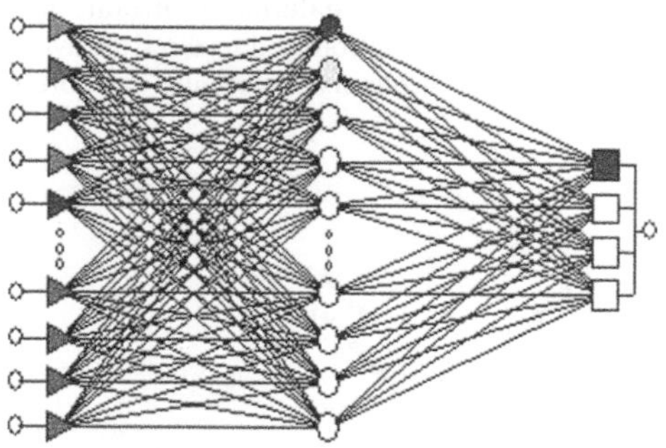

Figure 2. Probabilistic neuron network selected for further testing.

The determination of the minimum number and optimal topographic arrangement of telemetry chambers was carried out by eliminating particular input data from the neuron networks (which were, at the same time, the parameters measured or simulated in particular telemetry chambers of the heating network) and testing the classification quality. It was assumed that, with respect to the economic effect, only removing measurements of all parameters (flow, pressure and temperature) in a given telemetry chamber would make sense.

The criterion deciding on acceptance or rejection of the variant of number and topographic arrangement of telemetry chambers was the classification quality of state images of heating network by the neural network. It was assumed that an average decrease in classification quality by more than 15% meant that the telemetry chambers remaining in the heating system should continue to function.

The neural models were trained and tested using measurements of network medium parameters performed in all telemetry chambers. In order to analyse the possibility of removing the control and measurement instruments from the selected chambers, eight variants of heating network arrangements were prepared. In each of these variants, the number and topographic arrangement of telemetry chambers were adjusted. The economic effect can be achieved in the case of complete elimination of telemetry and control in a given network location. Therefore, it was assumed that the elimination of measurements in a chamber would mean no readouts of flow rate, pressure and temperature.

Figure 3 shows one of the diagrams of the tested heating network with the removed telemetry chambers presented in grey.

The neuron simulation of the prepared variants of number and arrangement of telemetry chambers was performed separately for all neural networks. Classification results were placed in the summary table (Table 3) in which, in order to perform an additional analysis of results obtained during the neural simulation, a parameter was introduced to show the average drop in classification quality of particular categories of

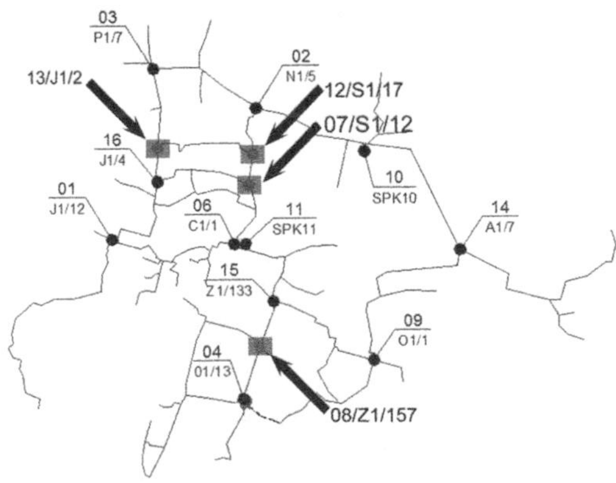

Figure 3. Variant IV: removed measurements in chambers 13, 07, 08, 12.

heating network state images in a given variant of the telemetry chamber number and topographic arrangement. The calculation of this parameter values was done by determination of average drop in qualification quality in a given variant for each of the heating network state image category.

The analysis of simulation results shows that, for each neural network, the classification quality of heating network state images decreases with an increase of number of removed telemetry chambers. The maximum average decrease in classification quality occurred in Variant VIII for neural network No. 20 (PNN 63:63-6000-73:1 – according to scheme: I:N-N-N:O, where: I – number of input variables, O – number of output variables, N – number of neurons in

Table 3. Average percentage drops in qualification quality of network state images.

Neural model	Average percentage decrease of classification quality							
	I	II	III	IV	V	VI	VII	VIII
Integrated neural model – based on the simulating data								
DSZ.1	7	15	24	32	38	43	45	48
DSZ.2	13	26	37	45	51	53	55	56
Integrated model – based on the telemetry data								
DTZ.1	17	27	49	56	60	64	66	67
DTZ.2	17	27	49	51	56	60	63	64
DTZ.3	6	25	40	48	56	61	65	66
DTZ.4	9	29	46	50	56	62	65	67
Partial model – based on the simulating data								
DSZ.ZT05	0	2	2	2	2	0	0	0
DSC.ZT06	2	2	2	10	10	12	13	13
DSC.ZT10	0	0	0	0	2	−2	−2	−2
DSC.ZT11	6	8	14	23	29	36	45	46
DSC.ZT12	1	1	1	5	6	5	15	25
DSC.ZT14	4	4	3	6	9	16	22	23
DSC.ZT15	3	5	10	12	19	20	32	37
DSC.ZT17	−5	−5	−5	4	9	17	14	13
DSZ.ZT19	2	2	8	8	12	13	15	17
Partial model – based on the telemetry data								
DTC.ZT03	4	6	9	7	11	12	27	29
DTZ.ZT04	1	3	3	9	8	5	17	22
DTC.ZT05	1	3	6	7	14	21	27	33
DTC.ZT06	0	0	0	1	12	14	19	29
DTC.ZT07	1	1	2	3	8	15	26	39
DTC.ZT08	0	1	6	18	21	28	37	48
DTC.ZT09	1	3	6	7	11	18	21	32
DTC.ZT10	1	5	16	20	25	36	43	48
DTC.ZT11	0	5	26	32	35	39	46	49
DTC.ZT12	2	8	23	32	35	40	43	45
DTC.ZT13	1	8	22	39	41	44	47	48
DTC.ZT14	1	4	30	40	46	52	57	60
DTC.ZT15	2	4	21	29	34	38	40	47
DTC.ZT16	0	1	12	17	23	32	40	46
DTC.ZT17	2	5	5	10	14	21	37	42
DTC.ZT18	1	7	11	13	20	29	33	39
DTC.ZT19	1	3	5	5	6	18	44	50

particular layers), in DTZ.1 neural model. The maximum average decrease in classification quality (for all neural networks and variants functioning in a given class of neural model) was recorded for the integrated neural model based on telemetry data, and this amounted to 48%. In the partial model based on the data obtained from computer simulation, this factor amounted to 9%.

The analysis of simulation results allowed the determination of variants of number and topographic arrangement of telemetry chambers in which the monitoring of parameters make it possible to map the state of tested heating network through the network or network group functioning in a given model.

4 CONCLUSIONS

The following conclusions may be drawn from the study:

(1) The study has demonstrated that probabilistic neural networks are well suited for the analysis of any heating network, including heating networks in the design phase.
(2) The constructed neural model for the classification of defined heating network state image categories (being in fact the sets of network heat media parameters) allows the quick determination of the whole heating network state on the basis of measurements in a few selected telemetry chambers.
(3) In the partial neural model constructed on the basis of the data obtained from the tested heating system, the acceptable variant is the one in which the parameters measured in three telemetry chambers No. 13, 07 and 08 were removed.
(4) In the partial neural model constructed on the basis of the data obtained from computer simulation of the heating system, the acceptable variant is the one in which the parameters measured in telemetry chambers No.13, 07, 08 and 12 were removed. The removal of these four telemetry chambers means the decrease by 25% of all telemetry chambers currently functioning in the tested heating system.
(5) The analysis of classification results in variants with a different number and arrangement of measuring chambers has demonstrated that it is necessary to construct a separate, dedicated neural model for each variant of number and topographic arrangement of telemetry chambers.

REFERENCES

Aydinalp M., Ugursal V.I., Fung A.S., Modelling of the space and domestic hot-water heating energy-consumption in the residential sector using neural networks, 2004, Applied Energy 79, p. 159–178

Bishop Ch., Neural networks for pattern recognition, 1998, Clanedron Press

Cetiner C., Halici F., Cacur H., Taymaz I., Generating hot water by solar energy and application of neural network, 2005, Applied Thermal Engineering 25, p. 1337–1348

In-Ho Y., Myoung-Souk Y., Kwang-Woo K., Application of artificial neural network to predict the optimal start time for heating system in building, 2003, Energy Conversion and Management 44, p. 2791–2809

In-Ho Y., Kwang-Woo K., Prediction of the time of room air temperature descending for heating systems in buildings, 2004, Building and Environment 39, p. 19–29

Kalogirou S.A., Application of artificial neural-networks for energy systems, 2000, Applied Energy 67, p. 17–35

Kalogirou S.A., Artificial neural networks in renewable energy systems applications: a review, 2001, Renewable and Sustainable Energy Reviews 5, p. 373–401

Kalogirou S.A., Artificial intelligence for the modeling and control of combustion processes: a review, 2003, Progress in Energy and Combustion Science 29, p. 515–566

Kalogirou S.A., Optimization of solar systems using artificial neural-networks and genetic algorithms, 2004, Applied Energy 77, p. 383–405

Korbicz J., Ouchowicz A., Uciński D., Artificial neural networks. Basics and implementations (in Polish: Sztuczne sieci neuronowe. Podstawy i zastosowania), 1994, PLJ

Larsen H.V., Palsson H., Bohm B., Ravn H.F., Aggregated dynamic simulation model of district heating networks, 2002, Energy Conversion and Management 43, p. 995–1019

Mahmoud M.A., Ben-Nakhi A.E., Architecture and performance of neural networks for efficient A/C control in buildings, 2003, Energy Conversion and Management 44, p. 3207–3226

Mihalakakou G., Santamouris M., Tsangrassoulis A., On the energy consumption in residential buildings, 2002, Energy and Buildings 34, p. 727–736

Olofsson T., Andersson S., Östin R.A, A method for predicting the annual building heating demand based on limited performance data, 1998, Energy and Buildings 28, p. 101–108

Savola T., Keppo I., Off-design simulation and mathematical modeling of small-scale CHP plants at part loads, 2005, Applied Thermal Engineering 25, p. 1219–1232

Sozen A., Arcaklio E., Ozalp M., Kanit E.G., Use of artificial neural networks for mapping of solar potential in Turkey, 2004, Applied Energy 77, p. 273–286

Environmental Engineering – Pawłowski, Dudzińska & Pawłowski (eds)
© 2007 Taylor & Francis Group, London, ISBN13 978-0-415-40818-9

Efficiency characteristics of flat-plate solar collectors

Alicja Siuta-Olcha
Lublin University of Technology, Faculty of Environmental Engineering, Lublin, Poland

ABSTRACT: The paper presents a simulation analysis of thermal steady states of a flat-plate solar collector. The computational results of thermal resistances and the temperature field distribution in the cross section of a collector have been obtained on the basis of simulation software. The method of testing was developed by means of the probability procedure. The theoretical characteristics of the collector efficiency are also presented. The influence of weather conditions, such as: ambient temperature, solar radiation intensity, wind velocity on the collector efficiency have been analysed. The dependence of the thermal efficiency on the reduced temperature difference has been shown. Moreover, the influence of thermal resistances on fluid temperature increase has been determined.

Keywords: Solar energy, solar collector, thermal performance, test methods.

1 INTRODUCTION

Thermal properties of flat-plate solar collectors depend mainly on the instantaneous efficiency value that is defined as the ratio of the useful gain over some specified time period to the incident solar energy over the same time period (Duffie & Beckman 1991). Construction and material parameters and operating conditions of a collector influence this thermal efficiency. Efficiency characteristics can be determined on the basis of experiments carried out either outdoors or indoors, in laboratory conditions with the use of a solar simulator. The principles for collector efficiency tests in steady states or in quasi–steady states are unified and collected in relevant standards (Fischer et al. 2004). The alternative test method is the dynamic research on collectors in transient states (Nayak & Amer 2000). The measuring stands for both tests can be the same. One should remember that the determination of efficiency curves on the basis of experiments does not guarantee fully repeatable important random operation parameters, such as: insolation, ambient temperature, wind velocity. Theoretical test methods, at a certain stage, can replace expensive and time consuming experiments. Also a collector performance has been described when entropy generation is minimum (Saha & Mahanta 2001). The thermal behavior of a copolymer solar collector has been modeled and optimised using a finite difference model (Cristofari et al. 2002, Cristofari et al. 2003).

The paper presents the description of efficiency characteristics of a flat-plate fluid solar collector determined on the basis of a probabilistic computational method.

2 MATERIALS AND METHODS

The subject of this research is the flat-plate solar collector with the parameters presented in Table 1. The computational procedure of collector thermal efficiency in steady states has been developed using the Exodus procedure (Siuta-Olcha 2004). The mathematical description of this method allows temperature values at characteristic points of the collector area, to be determined. The one–dimensional model of the collector has been established. The mathematical description of the model takes into account the complex thermal exchange combining convection, radiation and conduction.

The energy balance of the steady state operation specifies solar gain by an absorber plate, useful thermal energy received by the fluid flowing through collector channels and the overall heat loss to the surroundings. The collector efficiency is described by Equation 1:

$$\eta = \frac{\int \dot{Q} d\tau}{A \cdot \int G d\tau} \tag{1}$$

Equation 1 can be written for the steady-state conditions in the following form:

$$\eta = \frac{\dot{Q}}{A \cdot G} \tag{2}$$

The fundamental equation describing thermal efficiency of solar collectors versus its optical properties, construction and meteorological parameters is based on Hottel–Whillier–Bliss's equation and its modifications (Duffie & Beckman 1991, Henden et al. 2000, Eisenmann et al. 2004):

$$\eta = (\tau \cdot \alpha) - \frac{U(T_{p,m} - T_a)}{G} \quad (3)$$

$$\eta = F_R\left[(\tau \cdot \alpha) - \frac{U(T_{f,i} - T_a)}{G}\right] \quad (4)$$

$$\eta = F'\left[(\tau \cdot \alpha) - \frac{U(T_{f,m} - T_a)}{G}\right] \quad (5)$$

$$\eta = \frac{\dot{m} \cdot c_f(T_{f,o} - T_{f,i})}{A \cdot G} \quad (6)$$

The efficiency characteristics determined according to the above formulae give straight line images. The transmittance–absorptance product is not constant under operational conditions but depends rather on transmission properties, the number and the thickness of glass covers, absorber plate emittance and the angle of solar incidence on the collector surface.

Results of efficiency tests yield an equation in the form of second-order polynomial curve:

$$\eta = a - a_1 \cdot T^* - a_2 \cdot G \cdot (T^*)^2 \quad (7)$$

Table 1. Construction and material parameters of the tested collector.

Parameter	Description
Dimensions: length × width × depth, mm	2015 × 1010 × 90
Gross area of collector, m^2	2.04
Aperture area of collector, m^2	2.00
Absorber area of collector, m^2	1.73
Glass cover thickness, mm	4
Absorber plate thickness, mm	0.3
Insulation thickness, mm	50
Collector box thickness, mm	1
Tube diameter, mm	8
Collector fluid conduits diameter, mm	18
Front cover material	solar tempered glass
Tube and absorber material	copper
Covering layers on absorbers	black galvanic chromium on a nickel surface
Insulation material	mineral wool + Al foil
Collector box material	aluminium
Number of transparent covers	1
Glass transmittance	0.91
Absorber plate absorptance	>0.96
Absorber plate emittance	0.08 ÷ 0.10
Fluid flow, dm^3/h	60 ÷ 120
Max efficiency of a collector related to absorber surface	0.80

The efficiency curve is a parabola because the coefficient of overall loss to the ambient surroundings actually increases slightly together with the absorber plate temperature increase.

3 RESULTS AND DISCUSSION

Computational results have been obtained on the basis of the computer program SprPE.PAS, written in our laboratory. The analysis assumes the collector tilt angle of 45° and constant optical parameters. Efficiency values have been calculated by means of Equation 6. Computer simulations enable the determination of the influence of particular thermal resistances of the collector on the resulting increase in fluid temperature. Figures 1–2 present the dependence of fluid temperature increase on collector thermal resistances in relative values for two flow fluxes. The thermal resistance between the glass cover and the absorber plate is the most important influence on the fluid temperature rise. The temperature of the medium increases together with the increase of thermal resistance between (i) the absorber and the surroundings and (ii) between the front cover and the surroundings but less rapidly on account of the good thermal insulation of the collector.

The temperature of the medium flowing out of the collector decreases when the thermal resistance between the absorber and the medium increases. At higher fluid flows the influence of particular thermal resistances on medium temperature is smaller.

Figure 3 presents the efficiency curves of the investigated collector versus reduced temperature difference. The collector characteristic obtained by the static method is described by the curve equation of the

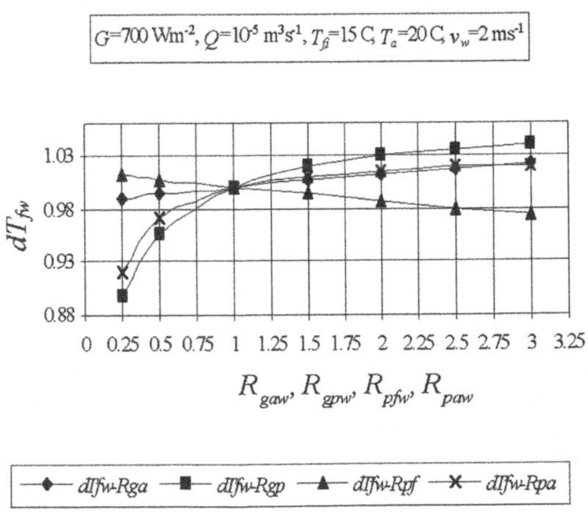

Figure 1. The influence of thermal resistances on the medium temperature increase in relative values.

second-order with the correlation coefficient of 0.998 (Pluta 2001):

$$\eta = 0.794 - 3.4 \cdot T* - 0.0158 \cdot G(T*)^2.$$

On the basis of dynamic tests (Pluta 2001), the thermal efficiency characteristic is given by the following formula:

$$\eta = 0.8097 - 3.36 \cdot T* - 0.0193 \cdot G(T*)^2.$$

The incident solar energy on the collector is related to the absorber surface in both sets of results. Theoretical computations give a gross efficiency curve (assuming, according to the collector efficiency definition Equation 2, that solar energy reaches the gross collector area), according to the approximate equation:

$$\eta = 0.74 - 4.5487 \cdot T* - 0.0016385 \cdot G(T*)^2.$$

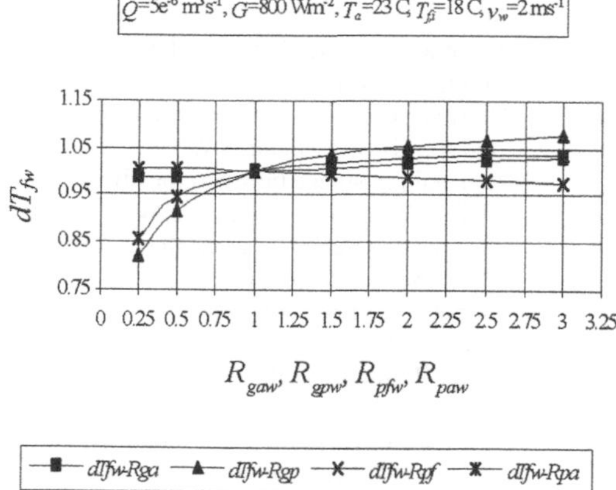

Figure 2. The influence of thermal resistances on the medium temperature increase in relative values.

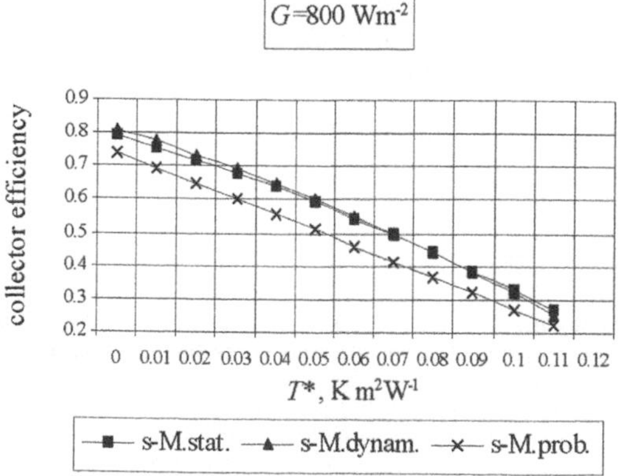

Figure 3. The efficiency curves of the investigated collector versus reduced temperature difference.

The most significant influence on the thermal performance is exerted by the solar radiation intensity. In the case when the constant value of mean surplus of fluid temperature over ambient temperature is maintained, the thermal efficiency of collectors increases together with insolation along the curves presented in Figure 4. The higher the temperature surplus, the lower the efficiency. Figure 4 presents also changes of useful energy flux gained by 1 m^2 of collector surface for the conditions considered. The presented dependencies are important in practice because they determine directly the insolation conditions and temperature levels at which the collector can perform. In the case of a linear increase in fluid temperature, an efficiency drop can be expected at higher solar irradiance on the collector plane.

The influence of ambient temperature is presented in Figure 5. The efficiency increases together with this temperature in the manner also determined by the irradiance level. The efficiency increase is accompanied by the increase of thermal resistance between the glass cover and the ambience and between the absorber plate and the ambience. This results in a lower overall heat loss coefficient.

The influence of wind velocity on the thermal efficiency of a collector is presented in Figure 6. The maximal efficiency occurs under wind-free conditions. This

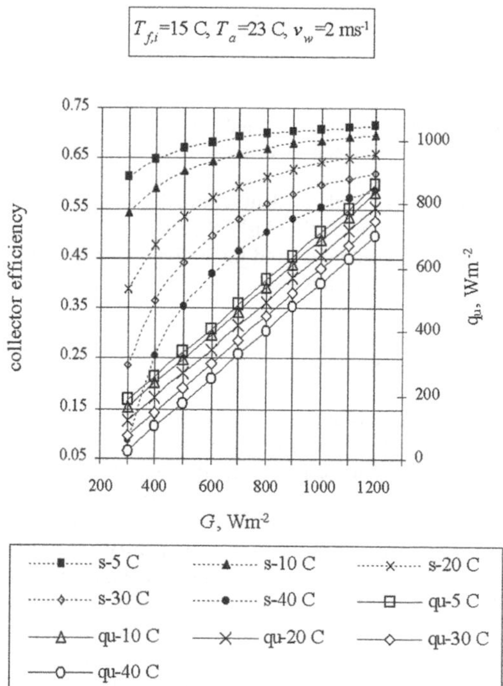

Figure 4. The influence of solar irradiance on thermal efficiency of the collector and on useful energy flux.

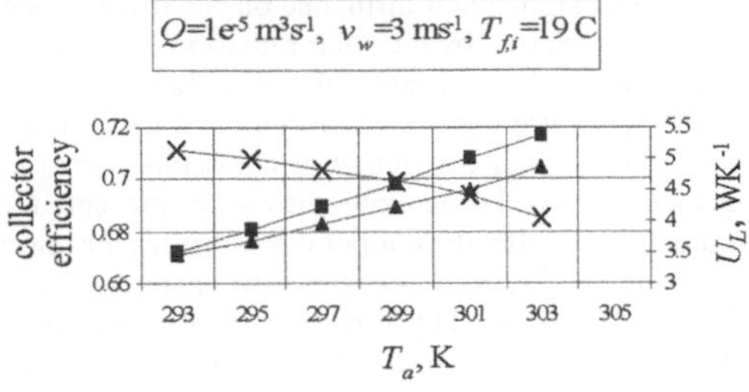

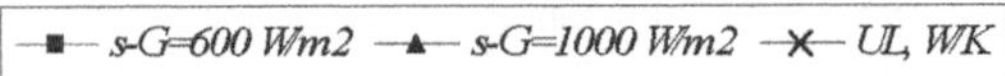

Figure 5. The influence of ambient temperature on thermal efficiency of the collector and on overall heat loss coefficient.

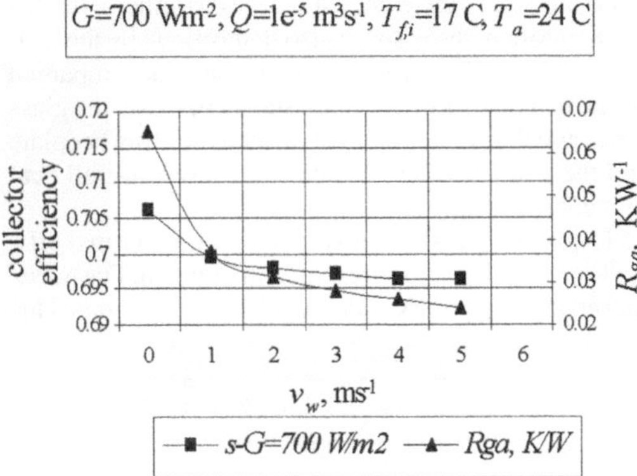

Figure 6. The influence of wind velocity on the thermal efficiency of the collector and on thermal resistance R_{ga}.

dependence is mostly evident within the wind velocity range of $0 \div 3\,\mathrm{ms}^{-1}$. Similarly, the thermal resistance between the front cover and the ambience drops with the wind velocity increase.

4 CONCLUSIONS

The proposed computing procedure makes it possible to perform the precise analysis of thermal states in steady conditions. The determination of efficiency characteristics by means of computations enables the determination of optimal material and construction parameters for collectors and their qualitative evaluation. The thermal characteristic curves obtained from laboratory tests and those from computations using the Exodus procedure at $T^* \in (0.02; 0.11)$ are practically identical. The thermal resistance between the absorber plate and the glass cover is important to the fluid temperature increase. The thermal resistance of the fluid is the most important influence on its heating up dynamics. The research considers mainly the dependence of the gained energy on meteorological parameters, such as solar irradiance, ambient temperature, wind velocity and direction.

NOMENCLATURE

- a, a_1, a_2 = algebraic constants, reference T^*;
- A_c = solar collector area, m^2;
- A_p = absorber plate area, m^2;
- c_p = specific heat of the fluid, $\mathrm{Jkg^{-1}K^{-1}}$;
- F' = collector efficiency factor;
- F_R = collector heat removal factor;
- G = global solar irradiance on collector plane, $\mathrm{Wm^{-2}}$;
- $\dot{m}$ = mass flow rate of heat transfer fluid, $\mathrm{kgs^{-1}}$;
- Q = medium volume flux through collectors, $\mathrm{m^3s^{-1}}$;
- $\dot{Q}u$ = useful energy gain, W;
- $\dot{q}u$ = energy flux density received by the fluid, $\mathrm{Wm^{-2}}$;
- R_{ga} = thermal resistance between the front cover and the surroundings, $\mathrm{KW^{-1}}$;
- R_{gp} = thermal resistance between the front cover and the absorber plate, $\mathrm{KW^{-1}}$;
- R_{pa} = thermal resistance between the absorber plate and the surroundings, $\mathrm{KW^{-1}}$;
- R_{pf} = thermal resistance between the absorber plate and the fluid, $\mathrm{KW^{-1}}$;
- T_a = ambient temperature, K;
- $T_{f,i}$ = inlet fluid temperature, K;
- $T_{f,m}$ = mean fluid temperature, K;
- $T_{f,o}$ = outlet fluid temperature, K;
- $T_{p,m}$ = mean absorber plate temperature, K;
- T^* = reduced temperature difference, $\mathrm{m^2KW^{-1}}$;
- U_L = overall heat loss coefficient of a collector to the surroundings related to 1 m^2 of collector area, $\mathrm{Wm^{-2} \cdot K^{-1}}$;
- U_c = overall loss coefficient of a collector, $\mathrm{WK^{-1}}$;
- v_w = wind velocity, $\mathrm{ms^{-1}}$;
- η = collector efficiency;
- τ = time, s;
- $(\tau \cdot \alpha)$ = effective transmittance-absorptance product.

REFERENCES

Cristofari, C. et al. 2002. Modelling and performance of a copolymer solar water heating collector. *Solar Energy* 72 (2): 99–112.

Cristofari, C. et al. 2003. Influence of the flow rate and the tank stratification degree on the performances of a solar flat-plate collector. *International Journal of Thermal Sciences* 42 (5): 455–469.

Duffie, J.A. & Beckman, W.A. 1991. *Solar Engineering of Thermal Processes*. 2nd ed. New York: John Wiley & Sons.

Eisenmann, W. et al. 2004. On the correlations between collector efficiency factor and material content of parallel flow flat-plate solar collectors. *Solar Energy* 76 (4): 381–387.

Fischer, S. et al. 2004. Collector test method under quasi-dynamic conditions according to the European Standard EN 12975-2. *Solar Energy* 76 (1–3): 117–123.

Henden, L. et al. 2002. Thermal performance of combined solar systems with different collector efficiencies. *Solar Energy* 72 (4): 299–305.

Nayak, J.K. & Amer, E.H. 2000. Experimental and theoretical evaluation of dynamic test procedures for solar flat-plate collectors. *Solar Energy* 69 (5): 377–401.

Pluta, Z. 2001. Performance characteristics of solar collectors from dynamic measurements (in Polish). *Proceedings of the Symposium Technical, Ecological and Economic Aspects of the Renewable Energy Engineering*: 203–210. Warsaw: SGGW.

Saha, S.K. & Mahanta, D.K. 2001. Thermodynamic optimization of solar flat-plate collector. *Renewable Energy* 23 (2): 181–193.

Siuta-Olcha, A. 2004. Qualitative evaluation of flat plate solar collectors by the probability method (in Polish). *Engineering and Protection of Environment* 7 (2): 209–225. Czestochowa.

Environmental Engineering – Pawłowski, Dudzińska & Pawłowski (eds)
 ISBN13 978-0-415-40818-9

Fuzzy linguistic model of solar radiation impact on building heat demand

Paweł Malinowski & Iwona Polarczyk
Institute of Air-Conditioning and District Heating, Wrocław University of Technology, Poland

Piotr Ziembicki
Institute of Environmental Engineering, University of Zielona Góra, Poland

ABSTRACT: The presented article provides a review of research work utilizing fuzzy logic to build a model of a horizontal solar radiation impact on single-family building heat demand. In its first part the methods and capabilities of fuzzy logic modelling have been described in general, and kinds of problems that were addressed to solve with its use have been outlined. In the next part of the article, results of the study have been presented, aimed to demonstrate methodology and to provide a structural description of the linguistic model of solar radiation impact to be used in a predictive fuzzy controller for building heating control. The fulfilment of study objectives has created a basis of the methodology that allows addressing the above issue for any fuzzy controller for control of thermal comfort in buildings.

Keywords: Fuzzy modelling, fuzzy logic control, thermal comfort regulation, solar radiation.

1 INTRODUCTION

In recent years, fuzzy logic has found significant applications in a number of disciplines and fuzzy control, in particular, has emerged as one of the most interesting areas in the application of fuzzy set theory. The known applications of fuzzy controllers with a solar radiation component include such projects as heat exchange, water heating, thermal comfort control (Dounis, Manolakis, 2001), solar plant control (Silva and others, 2003), energy optimisation models (Ari and others, 2005), industrial robot control, etc. This article provides a structural description of the linguistic model of a horizontal solar radiation (Bellocchi and others, 2002) impact to be used in a predictive fuzzy controller (Gomez and others, 2002) for building heating control (Calvino and others, 2004).

2 MATERIALS AND METHODS

Fuzzy system models are divided into two main categories of basically different capacity to represent various types of information: Linguistic Models (LM) and Takagi-Sugeno-Kanga Models (TSK). The linguistic model describes a system using a set of "IF-THEN" rules with vague predicates; a set of rules replaces a normal set of formulae used to characterize the system. The linguistic model is a knowledge based model. Its ability to map the reality depends on the existence of a rule base and reasoning mechanism. In this article, due to limits on text length, a method for rule base generation has been presented only, and the reasoning mechanism is not considered. At the current stage of theory development, the construction of a linguistic model based on the direct approach is more an art of intuition and experience than a strict theory. The linguistic description is created on the basis of a-priori knowledge about the system. The source for linguistic rule generation is an expert's knowledge about the system. As there are no general principles for this method of fuzzy modelling, the construction process of the linguistic model (LM) for solar radiation impact on the heat demand of a single-family house will be illustrated.

3 RESULTS AND DISCUSSION

The modelled object is assumed to be a one-storey, single-family house with a cellar. Two zones are selected within the object: zone 1 – ground floor (living level) and zone 2 (housekeeping level). The house is assumed to contain the following rooms: 3 bedrooms, living room, kitchen with dining room, bathroom, and toilet.

Additional information:

- house usable area 110.5 m^2
- glazing area amounts to 15% of usable area

- infiltration, Equivalent Leakage Area of 430 cm^2 is assumed
- house orientation North(entrance)-South(living room)

The simulation of particular house variant performance was carried out using the DOE-2 software. Data were available for a period of one year from 1 January to 31 December, and they were collected in 1-hour intervals. Parameters considered in the simulation were as follows:

- d1: date (month, day)
- d2: time (hour), measuring step of 1 h
- d3: Text – external temperature [°C]
- d4: I_{gh} – horizontal solar radiation [W/m^2]
- d5: I_{gs1} – solar radiation on surface of wall 1 [W/m^2]
- d6: I_{gs2} – solar radiation on surface of wall 2 [W/m^2]
- d7: I_{gs3} – solar radiation on surface of wall 3 [W/m^2]
- d8: I_{gs4} – solar radiation on surface of wall 4 [W/m^2]
- d9: V_{ent} – wind velocity [m/s]
- d10: T_{z1} – temperature in zone 1 [°C]
- d11: T_{z2} – temperature in zone 2 [°C]
- d12: Q_{chauf} – energy demand for heating [kW]

The value of horizontal solar radiation is a function of the solar position in respect of the measuring point, and this can be easily expressed as a function of time of day for the known geographical position of the measurement site. The proposed model construction is based on the following easily available and intuitive linguistic variables:

- DAY TYPE (division according to the value of horizontal solar radiation)
- DAY HOUR

The linguistic variables formulated above take the following values (fuzzy variables) further used in the modelling process:

- DAY TYPE
  - 1st variable CLEAR SKY
  - 2nd variable SUNNY
  - 3rd variable SO-SO
  - 4th variable CLOUDY
  - 5th variable HEAVY CLOUDS
- DAY HOUR
  - 1st variable and next as clock hours from 1.00 to 24.00

The membership function values of particular fuzzy variables for the given linguistic variables were assumed as shown in Figure 1.

In the case of the impact of horizontal solar radiation on house heating performance, the knowledge base contained the results of hundreds of computer simulations carried out with an aid of the DOE-2 software. Those simulations were carried out for the earlier presented model variant of a single-family house using various sets of real meteorological data. Then, in the course of further analysis, the change curves of house heat demand in a function of day hour were determined and the day classification was carried out with respect to their membership to specific values of the day type linguistic variable.

The "IF-THEN" fuzzy rules were formulated as a result of the analysis of knowledge base data collected during the computer simulations joining together the two input linguistic variables (DAY TYPE and DAY HOUR) and one output variable (house heat demand). The rules were formulated according the so-called "fuzzy graph" method (Kajl and others, 1995). The basis for this method is the graph shown in Figure 2, presenting the model house heat demand values as a function of day type and day hour.

This graph was created as a result of standard stochastic processing of computer simulation data collected in the knowledge base. The curves of model house heat demand were determined as the expected values of simulation results obtained with the use of various real meteorological data.

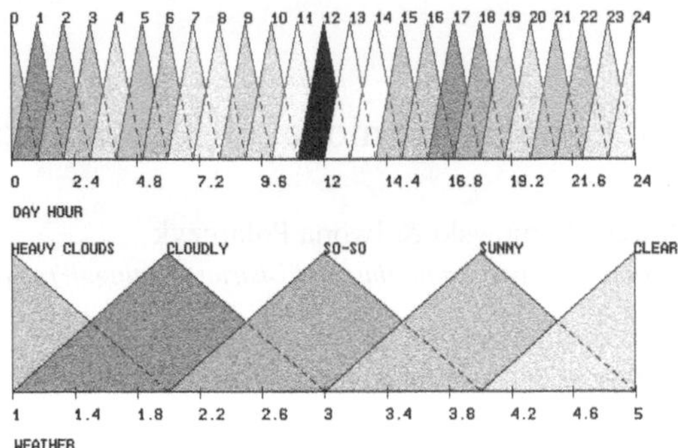

Figure 1. Graphs of membership function values for input linguistic variables (DAY HOUR, DAY TYPE).

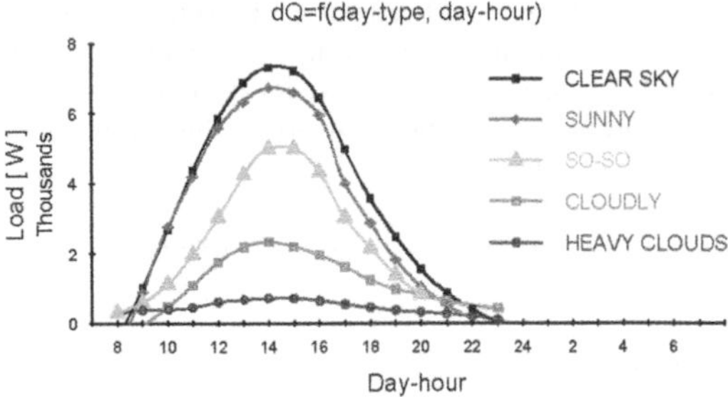

Figure 2. Change curves of house heat demand for particular DAY TYPES.

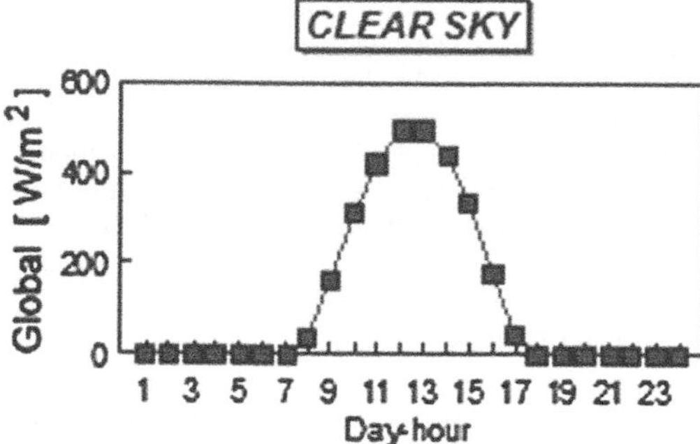

Figure 3. Solar radiation change curves. Day Type – Clear sky.

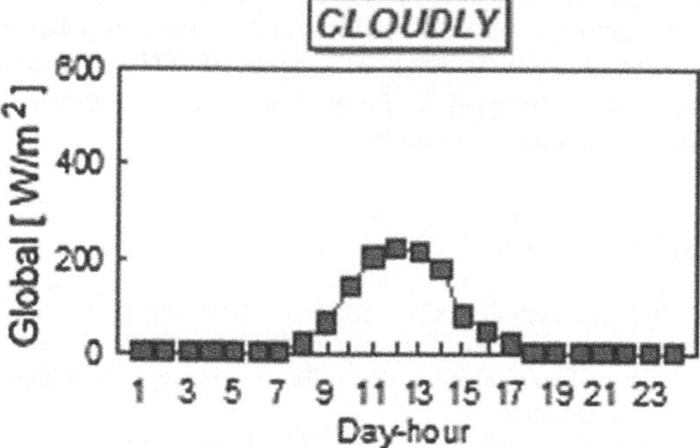

Figure 4. Solar radiation change curves. Day Type – Cloudy.

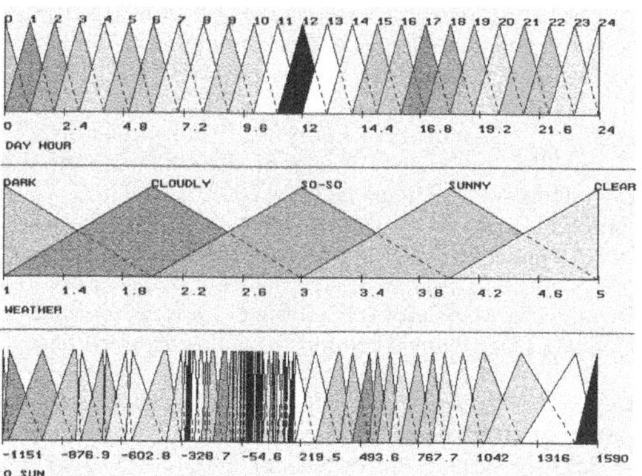

Figure 5. Fuzzy model graphical presentation of solar radiation impact on changes in single-family house heat demand.

Figures 3 and 4 shows change curves of horizontal solar radiation in a function of day hour for the selected, defined Day Types.

The whole fuzzy model of the modelled single-family house heat performance in January may be presented graphically in Figure 5.

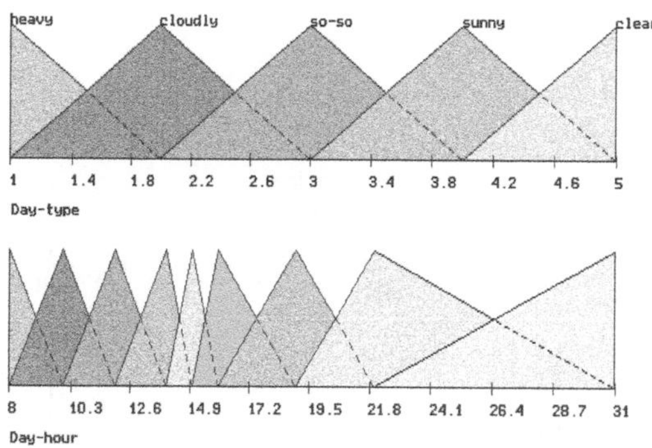

Figure 6. Graphs of membership functions for post-optimisation input linguistic variables.

In the above presented model (Figure 5) optimisation of the rule number for single-family house performance under the horizontal solar radiation impact was carried out. The main activities were concentrated on changing fuzzy variables of the DAY HOUR linguistic variable. A new linguistic variable, DAY TIME was introduced, and instead of twenty four one-hour fuzzy variables, nine sectional variables were introduced to define particular times of day.

- DAY TIME
  - 1st variable MORNING about 8.00
  - 2nd variable BEFORE NOON about 11.00
  - 3rd variable NOON about 12.00
  - 4th variable TWO O'CLOCK about 14.00
  - 5th variable THREE O'CLOCK about 15.00
  - 6th variable AFTERNOON about 16.00
  - 7th variable EVENING about 19.00
  - 8th variable TEN O'CLOCK about 22.00
  - 9th variable NIGHT about 24.00

Graphs of the new membership function values for post-optimisation fuzzy variables are presented in Figure 6.

The rule base for the so-defined fuzzy variables comprises forty five "IF-THEN" fuzzy rules, and its graphical interpretation looks is shown in Figure 7.

The stage of model verification is integrally connected with the phase of model construction, and the verification presented here is some kind of summary of the previous activities. The essential part of each verification is the problem of criteria that can make it possible to confirm the adequacy of proposed model. The basic criterion assumed in the examples presented is pragmatic compatibility, i.e. comparison of the model and modelled system output values at the same external conditions affecting the system and the model. In the case of fuzzy modelling of solar radiation impact on single-family house heat performance presented in this article, the verification problem was solved by

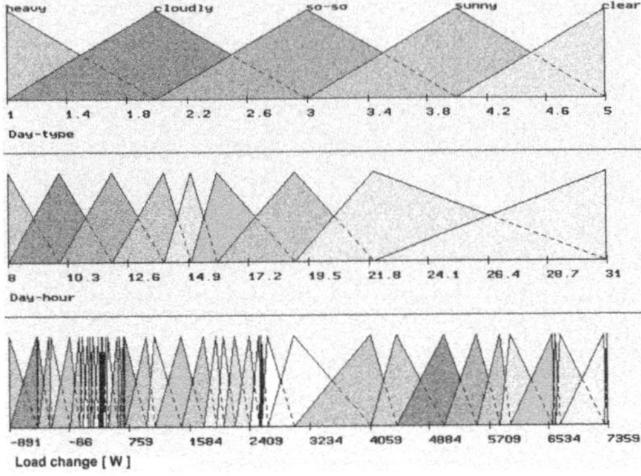

Figure 7. Fuzzy model graphical presentation of solar radiation impact on changes in single-family house heat demand after optimisation of rule number.

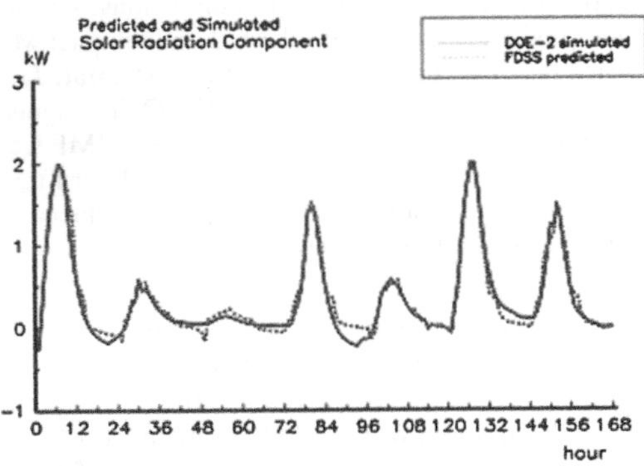

Figure 8. Comparison of curves of solar radiation impact on building heat demand change, predicted by fuzzy model and simulated by DOE-2 software.

comparing its results to the simulation results generated by the DOE-2 software, or to another model which was the basis for the construction of fuzzy model on this stage of research.

Standard deviations for the total building heat demand, between the values predicted by the fuzzy model with one hour advance and those obtained from the simulation, range from 194 W to 161 W. Variance factors are between 5.14% and 4.26% respectively.

In order to compare visually the compatibility of the constructed fuzzy model with the data obtained from the DOE-2 simulation software, the curve representing the values from the fuzzy model and the curve obtained from the simulation have been plotted together in Figure. The data presented cover the period of one week (7 days). The results are presented in longer time intervals as this allows the verification of structural correctness of the constructed fuzzy model. The period presented has been selected randomly from the collection of available meteorological data.

4 CONCLUSIONS

A method for constructing a microphone fuzzy linguistic model of horizontal solar radiation impact on building heat demand has been presented. The obtained linguistic model simulation data are compared with the simulation data from the advanced, complex DOE-2 software, and the results achieved are satisfactory. The proposed linguistic model has been constructed on the basis of simple "IF-THEN" rules and is easy to apply in the predictive fuzzy controller automatic control system.

REFERENCES

Calvino F., Gennusa M.L., Rizzo G., Scaccianoce G., The control of indoor thermal comfort conditions: introducing a fuzzy adaptive controller, Energy and Buildings 36, 2004, p. 97–102

Silva R.N., Rato L.M., Lemos J.M., Time scaling internal state predictive control of a solar plant, 2003, Control Engineering Practice 11–12, p. 1459–1467

Ari S., Cosden I.A., Khalifa H.E., Dannenhoffer J.F., Wilcoxen P., Isik C., Constrained fuzzy logic approximation for indoor comfort and energy optimisation, Fuzzy Information Processing Society NAFIPS 2005, 2005, p. 1222–1233

Gomez V., Casanovas A., Fuzzy logic and meteorological variables: a case study of solar irradiance, Fuzzy sets and systems vol. 126, issue 1(2002), 2002, p. 121–128

Bellocchi G., Acutis M., Fila G., Donatelli M., An indicator of solar radiation model performance based on a fuzzy expert system, 2002, Agronomy Journal 94, p. 1222–1233

Dounis A.I., Manolakis D.E., Design of a fuzzy system for living space thermal comfort regulation, Applied Energy 2001

Kajl S., Malinowski P., Balaziński M., Czogała E., Predicting Building Thermal Performance Using Fuzzy Decision Support System, Proceedings of FUZZ-IEEE/IFES 95, Yokohama 1995, Japan, March 1995

Kajl S., Malinowski P., Balaziński M., Czogała E., Fuzzy Logic and Neural Networks Approach to Thermal Descriptions of Buildings, Second European Congress on Intelligent Technologies and Soft Computing, EUFIT 95, Aachen, Germany, August 1995

Environmental Engineering – Pawłowski, Dudzińska & Pawłowski (eds)
© 2007 Taylor & Francis Group, London, ISBN13 978-0-415-40818-9

Environmental damage mapping in historical stonework

Rudolf Plagge
Dresden University of Technology, Institute of Building Climatology, Zellescher Weg Dresden, Germany

Henryk Sobczuk
Lublin University of Technology, Faculty of Environmental Engineering, Lublin, Nadbystrzycka, Poland

ABSTRACT: In the paper the measurement of the influence of environmental and climatic factors on historical stonework is described from the point of view of moisture and salt content distribution within the building structure. As the object for demonstrating the investigation method the Finca Marina Manresa (ca. 500 yrs) on Mallorca was chosen. To detect the moisture and salt content of the building walls at numerous locations along a masonry transect the TDR-technique was applied and used for mapping. Time domain reflectometry (TDR) is a non-destructive method to determine the volumetric water content within a building structure. In the paper, the different steps of a measurement, including sensor installation and the development of suitable calibration functions are presented. Spatial distribution analysis of measured parameters allows the mechanisms of salt and moisture damage in the masonry to be discerned, from which suggestions for effective restoration can follow.

Keywords: Environmental damage, historical building, moisture measurement.

1 INTRODUCTION

The Finca Marina Manresa on the island of Mallorca, Spain was built around 1500 as a one-storey bastion with moats as protection against pirates. The building was extended in 1900 by an upper storey and used as a luxury dwelling. In 2000 the building was completely renovated and modernised. A general view of the building is presented in Figure 1.

The influence of environmental and climatic factors on the historical stonework has caused damp and salt damage to the 120 cm thick north wall (Fig. 2). It was a recurring problem, necessitating an investigation of the exact causes.

Up to now, experts have seen the damage as the result of hygroscopic condensation, induced by salt particles carried by sea breezes which enter the building.

Using TDR provides an alternative to taking samples by drilling, thus helping to preserve the historical structure.

The damp and salt content were to be measured and mapped at many points on the north wall of the Finca Marina Manresa using TDR. Comparison of the results

Figure 1. North wall view of Finca Marina Manresa on Mallorca.

Figure 2. Salt damage to the North wall of Finca Marina Manresa.

lead to conclusions about the causes of the damp and salt damage and provided hints to possible solutions.

2 MATERIALS AND METHODS

Time Domain Reflectometry is a non-destructive dielectric measuring method which determines water content. It is based on the measurement of the travel time of an electromagnetic wave along a metal rod, the so-called wave-guide, which is inserted into the material to be tested. The probes used have a diameter of 2 mm and are placed 16 mm apart. The rods are 100 mm long and are connected at one end to a special circuit board which converts the wave resistance electronically. They are connected by an RG58 or semirigid coaxial cable to the measuring instrument. The circuit board is sealed in a 15 cm long plastic tube with epoxy resin, to protect the electronics from damp.

The time of the electromagnetic pulse travel along the probe rods indicates the apparent dielectric constant, ε_a, of the material. The more moist the material is, the slower the electromagnetic pulse and the bigger the dielectric constant. As the pulse moves in the sensor system, reflections occur at the entry and exit points of the probe, which can be made visible on the screen of a sampling oscilloscope. This pair of pulse reflections is read into the internal computer of the TDR measuring instrument and processed there. The computer automatically takes specific data, such as the geometry of the probes and their electric characteristics, into account. Further information about the technology used can be found in Plagge (2003).

In addition, the TDR probes can be used to measure the electrical conductivity, σ_a, of a material moistened with the pore water solution. The conductivity is established by direct electrical measurement of the resistance. This is based on the voltage division of two resistances, where one resistance, R_2, that of the TDR probe, is within the material to be measured and the other, R_1, is a series-connected control resistance of known resistivity. Analogous to the TDR technology, voltage pulses are also generated and sent through the cable – sensor system. However, in this case, it is not the reflection images but the voltage drop in the TDR probe which is registered and processed. This procedure takes into account the reduction of the reflections in the coaxial cable and the sensor probes. Measuring the voltage drop, U_1 and U_2, at the resistances gives information about the electric resistance of the sensor and is used to determine the conductivity. The unknown probe resistance of the TDR probe is calculated using equation 1:

$$R_2 = \frac{R_1 U_2}{U_1 - U_2} \qquad (1)$$

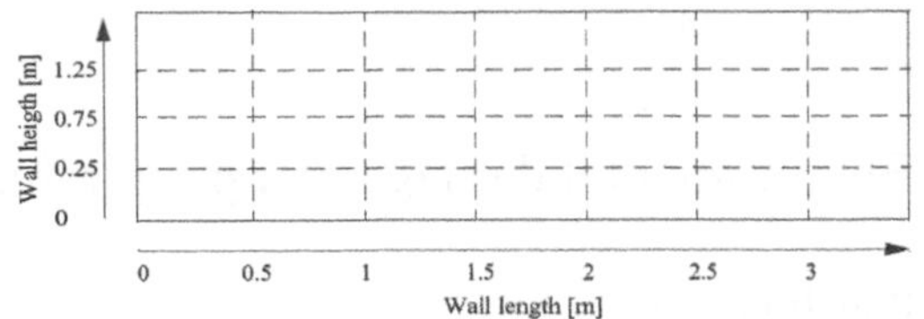

Figure 3. Diagram of the grid measurements used for the Finca walls.

The specific resistance measured corresponds with the conductivity of the material and dimension of the probe rods and their distance. Because the geometry of the probes affects the result, it must be taken into account. The details are shown in the paper by Plagge (2003). The concentration of salt can be deduced from a mixed dielectric component approach using a calibration function, Plagge et al. (1997) and Plagge (2003). In many cases, it is sufficient to know the total concentration of salts in the pore water Suchorab et al. (2005). Since the volume measured is identical, the method is particularly suited to investigating the temporal and spatial variability assessment of the measured values.

In order to assess the effects of damp and salt, the walls of three north facing rooms: L-left – room with the fireplace, M-medium – entrance hall, and R-right – dining room, were examined to provide the spatial distribution of moisture and salinity. It was necessary to divide the wall area into a grid with the distance suitable to reach the required accuracy. First, the wall was measured and defined grid points were marked, 50 cm apart: see the diagram in Fig. 3. The wall was 23.5 m long and 3.5 m high, resulting in 336 grid points to measure moisture and salinity.

In order to measure the damp and salt content of the walls, the TDR probe were inserted into the wall. Holes were drilled for the two 2 mm thin and 100 mm long probe rods in vicinity of the each grid point. Titanium nitrite- coated carbide drills of different lengths were used. Starting with the shortest drill, two parallel holes were drilled in the walls at the points marked, using a drilling gauge. Progressively longer drills were then used until the hole reached the desired depth.

As the walls were built of the porous local sandstone, the Marez, with plaster, the drilling was relatively easy. All the holes were drilled within 2 days, wearing out 17 carbide drills in the process. A calibrated TDR-2 probe was inserted in the holes and the computer controlled measurements were started. The visual control of the reflection images provided an additional check of the proper installation. In a few isolated cases, larger hollow spaces in the walls were tested, which are shown as a second reflection in the diagram. Since the measurements represent an average value along the whole probe, the maximal water

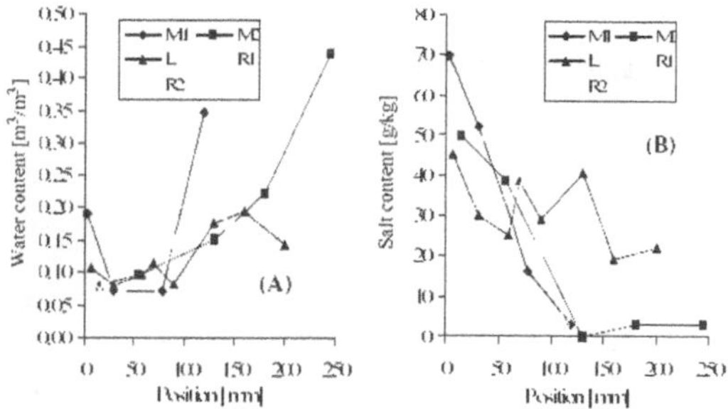

Figure 4. Results of the water content (A) and salt concentrations (B) in the core drillings at five sampling points in the areas L, M and R.

content shown in these cases was underestimated. In these doubtful cases, a further sample was taken close to the desired grid points.

The result of measurement of moisture content by TDR methods depends also on solid phase density and its dielectric properties. In order to avoid errors, a suitable calibration procedure was employed.

In the simplest case, capillary porous material can be described as a mixture of the three phases, liquid, solid and gaseous. Because water with $\varepsilon_{\text{water}} \sim 81$ has a much higher dielectric permittivity than mineral building materials, for example, $\varepsilon_{\text{brick}} \sim 3\text{–}6$ or air, $\varepsilon_{\text{air}} \sim 1$, the water content can be determined by a calibration function. The mixed dielectric permitivity calculated according to model proposed in Tinga et al. (1973) is given by:

$$\varepsilon_a = [\theta\varepsilon_w^\beta + (1-\phi)\varepsilon_s^\beta + (\varphi-\theta)\varepsilon_g^\beta]^{1/\beta} \qquad (2)$$

and the parameters for this are the dielectric constants of the liquid, solid and gaseous phases and are equivalent to the water content, θ, and the porosity, ϕ. From the equation (2), the volume of the three phases can be determined. In the equation (2): β represents a geometric factor which depends on the order factor of the 3-phase mixture and particle orientation in the electric field produced. The parameters are adjusted by reference measurements fit by the least squares method.

In the present work, five core drillings were taken, where the TDR measurements have already been made. These core samples served as a reference water content measurements, which were determined gravimetrically in the laboratory. The position of the individual drilling points are shown in Figs 5, 6 and 7. The results of the core measurements can be seen in Fig. 4. Determining the water content by equation 2 simply requires the measurement of porosity or density and the knowledge of the dielectric constant of the dry material. The property of the thin layer of plaster is ignored here, because it has only a negligible influence on the results.

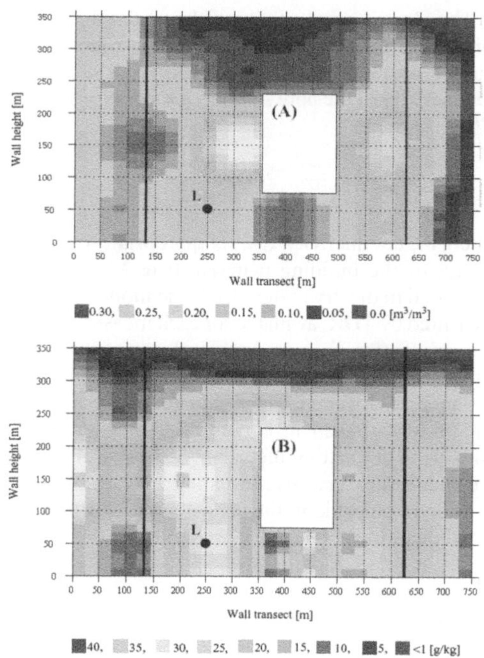

Figure 5. Water content (A) and salt content (B) in transect of the room with the fireplace, L.

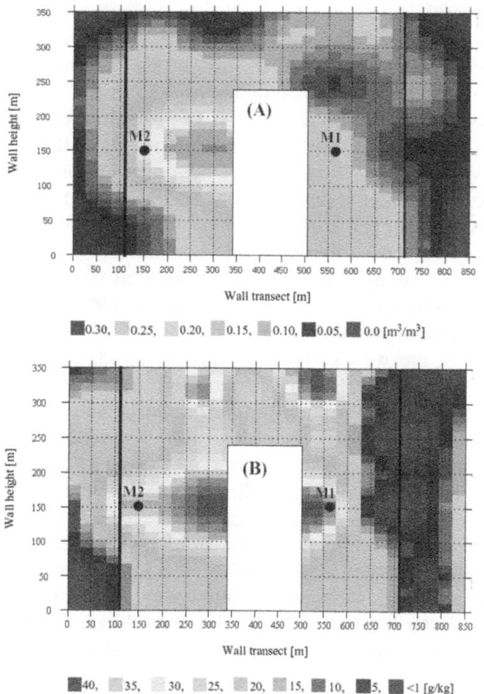

Figure 6. Water content (A) and salt content (B) in the transect of the entrance hall, M.

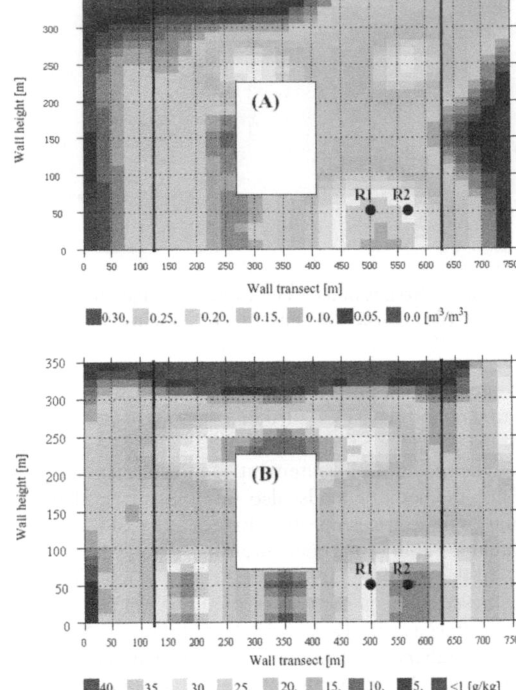

Figure 7. Water content (A) and salt content (B) in the transect of the dining room, R.

The electric conductivity of the material is a combination of the conductivity of the pore water and the surface conductivity of the matrix. To determine the conductivity of the pore solution, a model from Plagge (2003) is used.

$$\sqrt{\sigma_a} = \sqrt{\sigma_\theta}\theta T + \sqrt{\sigma_s} \tag{3}$$

where the conductivity, $\surd\sigma_S$, is appropriate to the solid matrix of the building material, here Marez, and is measured in the dry material. θ is the moisture content measured by TDR, available for each measuring point. The parameter T is an empirical correction factor, which is determined by the reference measurements of the core drillings. The segments of the cores are cut up and shaken in bottles of distilled water for 10 days. After setting the balance, the dissolved salts are chemically analysed, revealing a clear dominance of sodium chloride in the present case, whereas only small quantities of sulphate are found.

3 RESULTS

Figure 4 shows the results for the measurement of core drillings in areas L, M and R.

The core drillings showed generally rising water content with increasing depth (Fig. 4). The moisture content is reduced through evaporation at the wall surface. Near the surface, the water content varies between 0.07 and $0.17\,m^3/m^3$; inside the walls, the content rises to a maximum of $0.47\,m^3/m^3$. In the core sample M2, the Marez is almost saturated, apart from the pores. Additional measurements of the relative humidity with capacitive humidity sensors in the core drillings revealed nearly 100% humidity in all areas. If the moisture content of the walls was caused only by the sodium chloride content and the water vapour condensation, the water content would be at $<0.02\,m^3/m^3$ at 75% humidity and could not reach the level measured.

On the wall surface, the salt accumulates due to solute flow to the surface and water evaporation and reaches concentration around 40–50 g NaCl per kg material, which is equivalent to 4–5 mass-% of the Marez. In the gypsum plaster, up to 7 mass-% could be found. According to the leaflet of the WTA, the International Association for Science and Technology of Building Maintenance and Monument preservation, a maximum chloride level of 0.1 mass% is given, above which salt removal followed by the application of plaster is recommended. Thus the salt content in the areas near the surface is 40 to 70 times above allowed levels.

The test results of the TDR measurements for the north facing walls of the room with the fireplace, the entrance hall and the dining room are shown in Figs 5, 6 and 7. Since only the values for discrete positions in the transect are available, it was necessary to use a suitable procedure to interpolate and smooth the data for visual presentation. In the graphs presented here, a locally weighted regression algorithm is used, which delivers useful results for measurements that have non-Gaussian distribution or non-monotonous progression (Cleveland 1979).

The visual presentations cover the north-facing walls with the connecting interior walls up to the double doors, or the outer walls of the corner rooms to the window embrasures of the outer side walls. The beginning and end of the outer wall sections are each marked by thick vertical lines. The white areas show the windows or the terrace door.

According to the reasoning presented above, the locally high water content is not just the result of the hygroscopic condensation of water vapour onto NaCl. There must be another source of moisture, through which water enters the construction, dissolves NaCl, and transports it to the surface, where the salt crystallises and causes the visible damage. Possible sources of moisture include the leaky balcony, driving rain striking the outer walls, or the rainwater from the gargoyles which are directed at the side walls.

The random distribution of moisture content can only be adequately explained by the structure of the outer walls. In the past, thick walls were usually constructed with an inner and an outer shell, between which all the left-over mortar and rubble were thrown when the structure was complete. The hollow spaces which were discovered during the core drilling lead to the conclusion that the walls treated here were built according to this method.

This means that water which enters via the balcony does not just spread through capillaries, but also flows downwards by gravity until it reaches a thicker layer or window lintel. From there it spreads to the inside and causes damp and salt efflorescence through evaporation in places where the wall is less waterproof.

A further suggestion must be added to the assumption that the salt content of the outer walls was caused by the tiny salty drops of seawater carried by the wind over the centuries, namely the possibility that salty sand from the nearby beach and sea water were used for the mortar when the wall was built. Thus loose sand with seashells and lumps of limestone were found between the inner and outer shells of the walls. It is also possible that the Marez stone itself was quarried close to the coast and therefore was already contaminated with salt.

The extremely high salt content found in the walls is the result of a high degree of evaporation of rainwater that flows in from the balcony to the inner surface of the walls, leaving transported salt there. The water entry is limited to certain points, so it is necessary to view the rainwater control and the sealing of the balcony separately. The concept for sealing the balcony involved an inner drainage which led water to the outside. Between the side boundaries of angular edge bars of galvanised steel and the outer wall of the upper storey is a lower level of screed above an impervious layer. Both the impervious layer and the screed on top of it form a longitudinal gutter which should direct the rainwater to the two gargoyles on the end wall and from these to the eaves. Teak boards (35 mm thick), screwed onto laths bedded in the screed, provide a walkway.

However, the drainage concept given fails to fulfil elementary demands.

- The drainage capacity is too small, given the expected amount of rain. The gargoyles have a diameter of only 3 cm^2 and are thus too small by a factor of 20 and are easily blocked by falling leaves. In addition, the gargoyles direct the water onto the western and eastern outer walls.
- The laths are laid parallel to the outer walls, acting as a dam between the inner lath and the outer wall, as well as between the outer lath and the angular edge bars.
- The rubber-type seal is not glued to the angular edge bars, but only laid loosely against them. The seal is only as high as the teak boards and not glued to the walls. The edges of the sealing strips are not properly glued.
- The sills of the balcony doors are only 2 cm above the surface of the teak boards or above the angular edge bars.

Satisfactory drainage of the balcony is impossible because of the poor design and execution. Even in moderate rainfall, the water collects between the inner lath and the outer wall and the outer lath and the angular edge bars, reaching the upper surface of the laths and flowing between the angular edge bars and the non-glued seal into the lower structure. Heavy rainfall of 10 l/m^2h overloads the gargoyles and the balcony floods, causing the water to flow over the seal at the edge into the outer wall and thus into the ceilings and the outer wall of the ground floor. Subtropical rainfall of >50 l/m^2h overloads the system even more, because the water running down the outer walls flows as far as the door sill in the upstairs rooms.

Even near the coast, heavy rain contains little salt, but it can dissolve the wind-borne salts already present and transport them into the outer walls with the water. This effect is fairly small, so the faulty balcony seal can not be solely responsible for the salt content of the outer walls. However, the dissolution of the salt crystals in the outer walls, their transport into the inner surface and all the damage this causes, necessitating the repairs to the balcony, are the result of the inflow of water due to the faults in water guidance and the sealing of the balcony. The fact that humidity of more than 75%, which is often the case on Mallorca, dissolves some salts anyway, does not alter these facts.

4 CONCLUDING REMARKS

Old historical building often have relatively thick stone walls that undergo periodical wetting and drying. The water entering through the wall surface dries relatively quickly, but water that enters the wall structure from the top can penetrate deeply into the construction and thus remain there for a long time causing structural deterioration. The TDR moisture and salinity measurement has proven to be a good technique for moisture and salt distribution mapping. Knowing the moisture and salt distribution within the stonework, one can predict how the water enters the structure and suggest possible methods to prevent further damage. The measurement method described here can be applied for non-destructive diagnosis of the water distribution of historical thick stone constructions.

REFERENCES

Cleveland, W.S. 1979: Robust locally weighted regression and smoothing scatterplots. *Journal of the American Statistical Association*, 74, 829–836.

Plagge, R., P. Häupl & J. Grunewald 1997: TDR-SS-Methode, *ein Meßverfahren zur kombinierten Bestimmung der Materialfeuchte und der elektrischen Leitfähigkeit kapillar poröser Medien*. 9. Feuchtetag am 17./18. September 1997 in Weimar. 9, 161–173.

Plagge, R., 2003: Bestimmung von Materialfeuchte und Salzgehalt in kapillar porösen Materialien mit TDR. *Kolloquium mit Workshop Innovative Feuchtemessung in Forschung und Praxis,* Karlsruhe, 3.-4. Juli 2003, pp28.

Suchorab, Z., D. Barnat-Hunek, H. Sobczuk, Adaptation of reflectometric Techniques for moisture measurement of rock walls on the example of Janowiec castle., *in Mat. Of International Workshop City of Tomorrow and Cultural Heritage – Pomerania Outlook,* Dec. 8–9, 2005 Gdańsk.

Tinga, W.R., W.A.G. Voss & D.F. Blossey 1973: Generalized approach to multiphase dielectric mixture theorie. *J. Appl. Phys*. 44, 3897–3902.

Environmental Engineering – Pawłowski, Dudzińska & Pawłowski (eds)
© 2007 Taylor & Francis Group, London, ISBN13 978-0-415-40818-9

Study of the catalytic decomposition of PCDD/Fs on a V_2O_5-WO_3/Al_2O_3-TiO_2 catalyst

Grzegorz Wielgosiński
Technical University of Łódź, Wólczańska Łódź, Poland

ABSTRACT: Results of studies on the catalytic decomposition of PCDD/Fs in flue gases from incineration plants are discussed in the paper. The studies were carried out in a specially built pilot plant system whose most important element was a catalytic reactor with a monolithic vanadium-tungsten catalyst (V_2O_5 and WO_3) on TiO_2/γ-Al_2O_3 carrier. The experimental system was connected to an operating hazardous waste incineration plant (F.U.H. "EKO-TOP", Rzeszów) equipped with a rotary kiln of capacity ca. 3 500 Mg/year. The effect of temperature and catalyst load on the decomposition reaction was analyzed. Promising yields of PCDD/Fs decomposition were obtained – 97,5% efficiency TEQ (WHO) reduction at temperatures >260°C, at a mean catalyst load of about 3 500 1/h. The content of PCDD/Fs in flue gas was analyzed by gas chromatography, according to the standard EN-1948.

Keywords: PCDD/Fs destruction, vanadium-tungsten catalyst, dechlorination and oxidation.

1 INTRODUCTION

Incineration of such heterogeneous materials as wastes, whether municipal, industrial, medical or sewage sludge, is a source of emission of many chemical substances to the atmosphere, and the presence of substances containing both organic and inorganic chlorine compounds in the mass of waste to be incinerated, is a source of pollutants, namely polychlorinated dibenzo-p-dioxins (PCDDs) and polychlorinated dibenzofurans (PCDFs), due to which incineration plants have got a bad press. A survey of the literature (Wielgosiński 2001) on dioxin formation allows us to assume that dioxins can be formed in practically any thermal process, i.e. in particular those occurring at high temperatures if, in the incineration medium, there is organic matter and chlorine.

Methods for reducing dioxin emissions from technological processes can be roughly divided into two groups – primary and secondary.

The primary methods include directly intervening with a technological process and creating such conditions that the amount of produced dioxins be as small as possible (Wielgosiński 2003). Among the primary methods, the most important one is to avoid the presence of chlorine in thermal processes. Dioxin formation is not an equilibrium process and even a trace amount of chlorine and organic matter in the elevated temperature zone (200–700°C) must cause the formation of dioxins. Unfortunately, the presence of chlorine is often unavoidable. Temperature is also important – dioxins are not so stable as they are generally thought to be and they decompose at ca. 700°C. Unfortunately, an optimum range of temperature for *de novo* synthesis is around 300–350°C, so during cooling of gases from 700°C down to ca. 200°C they can be synthesised. The amount of dioxins formed as a result of *de novo* synthesis is inversely proportional to the rate of gas cooling (heat extraction), so, as presented above, very efficient heat recovery systems (heat exchangers) should be constructed and gases should be cooled by so-called quenching, i.e. cold water injection. Additionally, the *de novo* synthesis is supported by the presence of carbon monoxide, soot and dust containing metals (e.g. copper, aluminium or zinc). Hence, the primary methods for emission reduction include flue gas after-burning in order to minimise the presence of carbon oxide and soot, and hot gas dedusting, or selection of such flow rates through the heat recovery systems that dust contained in gases can not be deposited on them. Zones formed in the heat recovery system where dust and soot are deposited and cause a significant increase of the amount of dioxins produced in the *de novo* synthesis. In incineration processes, of primary importance is a correct incineration technique that would ensure the most complete incineration (Ishikawa 1997).

Among the secondary methods to reduce emission of dioxins, the most important are the following three methods:

- adsorption on activated carbon (on a solid bed or a jet method),
- catalytic decomposition of dioxins on a vanadium catalyst,
- filtration-catalytic method "REMEDIA®",
- absorption-adsorption method "ADIOX®".

In Poland, the only method to decrease dioxin concentration in flue gases that is used at present (exclusively in installations constructed at the end of 1990s) is to remove dioxins by adsorption. This technique of dioxin removal enables a significant decrease of the concentration of dioxins in flue gases, but as a result, a spent adsorbent saturated with dioxins remains, and its storage or utilisation is a problem that has not found a solution until now. A consequence is a transfer of the dioxins to another place in the environment, and not their destruction.

The catalytic method has no such disadvantage. During the operation of systems for catalytic denitrification of flue gases (SCR) it was observed that dioxins occurring in the flue gases were also decomposed effectively.

Lately, a new method for the removal of dioxins from flue gases has been developed by GORE Associates and called "REMEDIA®" (Pranghofer 2001). The method combines adsorption, dust removal and catalytic decomposition. The efficiency of dioxin removal exceeds 95%. According to the authors of this technology, it is slightly cheaper than the method with sorbent injection and fabric filter (about 60% cheaper than the catalytic systems). The small number of industrial applications makes it impossible to verify this information for the time being, although recently it has been reported that the method appeared to be very efficient in one of Belgium's municipal waste incineration plants – IVRO Roeselare with capacity ca. 47 000 Mg/year (Bonte 2002).

The absorption-adsorption method ADIOX® has been developed recently by the Swedish company Götaverken Miljö. It uses the so-called "memory effect" (Kreisz 1997) observed in waste incineration plants which consists of absorption of dioxins in plastic (polypropylene) elements of flue gas treatment system and their further desorption from packing elements to the stream of flue gases where their concentration decreases significantly as a result, for instance, of applying the primary methods of emission reduction. The ADIOX® method consists of adding activated carbon to polypropylene used in the construction of elements for wet flue gas treatment (mainly packing). Dioxin molecules that were absorbed in polypropylene are adsorbed additionally on the carbon surface and eliminated efficiently from the stream of flue gases in a permanent way. The method was used successfully in several municipal waste incineration plants (Kloding, Thisted, Denmark and Umeå, Sweden) (Andersson 2002). However, there has been no information on how the spent packing elements of absorption columns that contained significant amounts of dioxins were treated.

Studies by Hagenmeier (Hagenmeier 1989) showed that the vanadium-tungsten (V_2O_5-WO_3) catalyst on a titanium oxide carrier (TiO_2) decomposed PCDD/Fs with the release of CO_2, H_2O and HCl. This catalyst is generally used in selective catalytic reduction (SCR) of nitric oxides (with the addition of ammonia). Later studies (Hagenmeier 1990, Fahlenkamp 1991, Ide 1996) showed that, on a properly prepared V_2O_5-WO_3/TiO_2 catalyst in the presence of ammonia, it was possible to reduce the emissions of both nitric oxides and PCDD/Fs.

2 MATERIALS AND METHODS

Studies on the catalytic decomposition of 1,2-dichlo-robenzene were carried out using a specially constructed pilot-plant apparatus, whose the most important element was a catalytic reactor with a monolithic catalyst in the form of $150 \times 150 \times 100$ mm cubes. The specific area of the catalyst was around $1050\,m^2/m^3$. The catalyst was made from cordierite onto which an active layer of the following composition was deposited:

Al_2O_3	–	64% wt.
TiO_2	–	26% wt.
V_2O_5	–	6.6% wt.
WO_3	–	3.4% wt.

The reactor made it possible to carry out the process in the temperature range 100–400°C, at variable catalyst load and gas flow rate through the reactor. It had four separate gas flow lines which allowed one catalyst cube to be placed horizontally, and up to four cubes vertically one over the other.

A schematic diagram of the experimental set-up is shown in Fig. 1.

Investigation of the catalytic decomposition of dioxins is very expensive and difficult. It is practically infeasible in laboratory conditions. The reason is the problem with producing a stream of gases containing a stable amount of polychlorinated dibenzo-p-dioxins and polychlorinated dibenzofurans in laboratory conditions. Another problem is the necessity of taking gas samples of volume of the order of several (2–8) m^3 to enable precise determination of dioxin concentrations. Hence, in this study the author decided to carry out research in a pilot-plant installation connected to a real operating hazardous waste incineration plant, where in the stream of untreated flue gas the concentration of dioxins is relatively high. The plant selected was a hazardous waste thermal incineration plant (F.U.H.

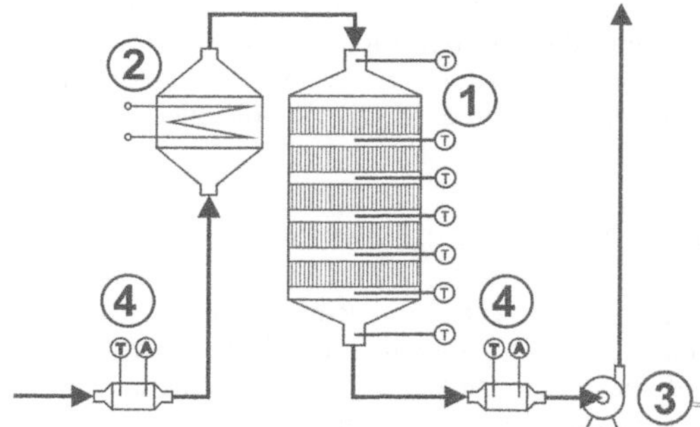

Figure 1. Schematic diagram of the experimental set-up. (1) – catalytic reactor, (2) – electric heater, (3) – fan, (4) – parameter control point (temperature, pressure, flow and concentration).

"EKO-TOP", Rzeszów), equipped with a rotary kiln of capacity ca. 3 500 Mg/year.

3 RESULTS AND DISCUSSION

Before starting the main measurements, preliminary tests were made using 1,2-dichlorobenzene as a dioxin "simulator". It was found that the rate of catalytic dioxin decomposition was not affected by mass transfer processes in the gas phase (Wielgosiński 2004). Hence, it was assumed that this conclusion would be true also in the case of PCDD/Fs.

The effect of temperature and catalyst load on the efficiency of PCDD/Fs decomposition on the applied vanadium-tungsten catalyst was investigated. With flue gas stream (ca. 4 200 m^3/h) 200 m^3/h flue gases were taken up at the maximum and directed through a heater to the catalytic reactor. The concentration of 17 PCDD/Fs at the reactor inlet and outlet was analysed. The content of PCDD/Fs in the gas phase was determined according to the EN-1948 method. The PCDD/Fs concentration at the inlet ranged from 150 to 450 ng/m^3, depending on sample.

The effect of process temperature on the efficiency of PCDD/Fs decomposition on the vanadium-tungsten catalyst is illustrated in Fig. 2.

Figure 3 shows dependence of PCDD/Fs decomposition efficiency on the catalyst load given as the quotient of gas volume flow rate and the catalyst volume (1/h).

We managed to obtain a very high efficiency of PCDD/Fs decomposition reaching as high as 97.5% TEQ (WHO) reduction in the best sample, at the temperature ca. 260°C and catalyst load around 3 500 1/h. The results (90–95%) were not much lower for the catalyst load equal to 5 900 1/h.

Figure 4 shows a profile of PCDD/Fs congeners at the inlet and outlet from the reactor, and Figure 5

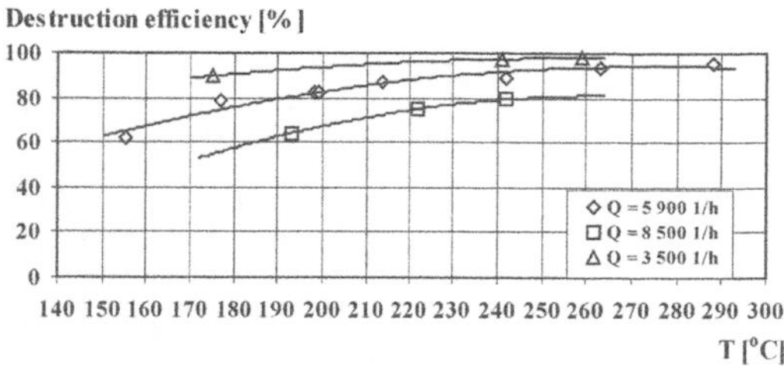

Figure 2. Dependence of decomposition degree of all 17 PCDD/Fs congeners on process temperature as a function of catalyst load.

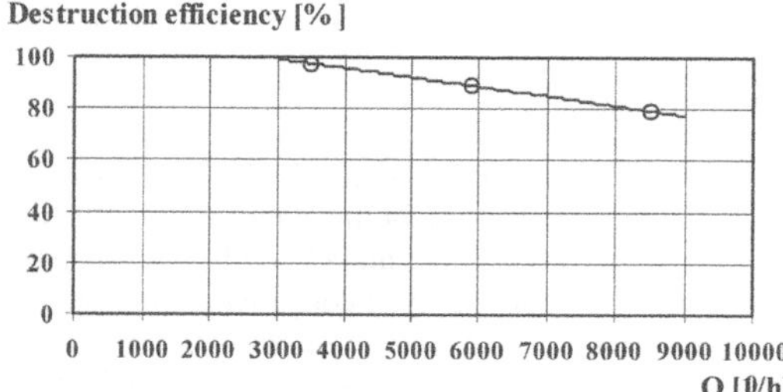

Figure 3. Dependence of decomposition degree of all 17 PCDD/Fs congeners on catalyst load at the temperature 242°C.

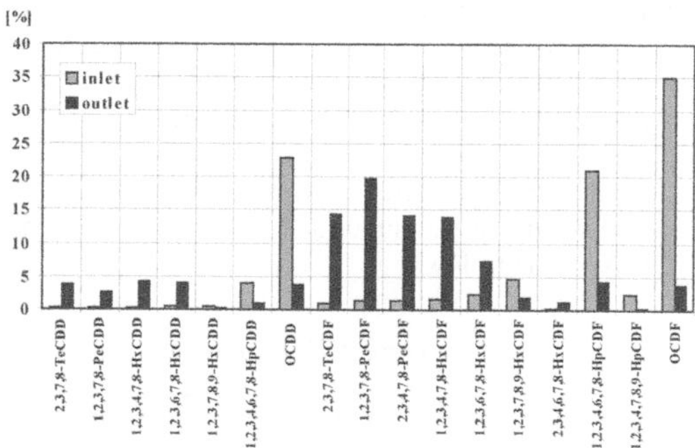

Figure 4. Profiles of congeners at the inlet and outlet of the catalytic reactor in the best sample.

illustrates the degree of decomposition for subsequent PCDD/Fs congeners in the best sample.

When analysing results for individual congeners, we concluded that the highest efficiency is obtained for those PCDD/Fs congeners which contain the highest number of chlorine atoms in the molecule. This is not in agreement with the observations made by Weber (Weber 1999), who found that most easily decomposed (dechlorinated and oxidised) were those PCDD/Fs congeners which had the smallest number

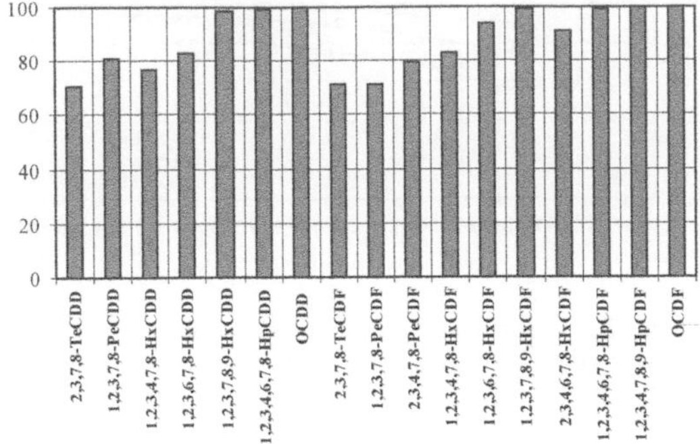

Figure 5. Decomposition degree of subsequent PCDD/Fs congeners on the catalyst in the bests sample.

of chlorine atoms in the molecule. There is a relatively simple explanation for this phenomenon. Weber carried out his work under model conditions, decomposing a specified PCDD/Fs congener. In our study, the investigation covers real flue gases from a hazardous waste incineration plant with all PCDD/Fs congeners. If, after Weber, the less chlorinated congeners are most easily decomposed and PCDD/Fs undergo stepwise dechlorination, we will observe a loss of the most highly chlorinated dioxins. However, in less chlorinated species, two opposite processes will control the relative concentrations – on one hand, loss of a given congener caused by decomposition and, on the other, an increase in the amount of a given congener due to destruction of more highly chlorinated molecules.

Apparently, the general reaction scheme of catalytic decomposition of polychlorinated dibenzo-p-dioxins (PCDDs) is as follows:

$$OctaCDD \xrightarrow[catalyst]{} HeptaCDD + HCl$$

$$HeptaCDD \xrightarrow[catalyst]{} HexaCDD + HCl$$

$$HexaCDD \xrightarrow[catalyst]{} PentaCDD + HCl$$

$$PentaCDD \xrightarrow[catalyst]{} TetraCDD + HCl$$

$$TetraCDD \xrightarrow[catalyst]{} TriCDD + HCl$$

$$TriCDD \xrightarrow[catalyst]{} DiCDD + HCl$$

$$DiCDD \xrightarrow[catalyst]{} MonoCDD + HCl$$

$$MonoCDD \xrightarrow[catalyst]{} DD + HCl$$

$$DD \xrightarrow[catalyst, O_2]{} CO_2 + H_2O$$

So, it may be assumed that an analogous decomposition reaction sequence holds for the polychlorinated dibenzofurans (PCDFs). However, it cannot be excluded that processes of oxidation occur parallel to the dechlorination processes, although it is more probable that the oxidation reaction is slower than dechlorination and so, first we observe dechlorination, and then oxidation of PCDD/Fs molecules.

4 CONCLUDING REMARKS

Studies on the catalytic decomposition of polychlorinated dibenzo-p-dioxins and polychlorinated dibenzofurans in the stream of flue gases generated by an operational hazardous waste incineration plant were carried out. The dependence of the obtained degree of PCDD/Fs decomposition on reaction temperature and catalyst load was determined. A very high efficiency of PCDD/Fs decomposition, reaching even 97.5% in the best case, was obtained at ca. 260°C and a catalyst load ca. 3 500 1/h. Similar though slightly lower results (90–95%) were attained for the catalyst load 5 900 1/h.

REFERENCES

Andersson S., Kreisz S., Hunsinger H., 2002, PCDD/F removal from flue gases in wet scrubbers – a novel technique, *Organohalogen Compounds*, 58, 127–160.

Bonte J.L., Fritsky K.J., Plinke M.A., Wilken M., 2002, Catalytic destruction of PCDD/F in a fabric filter: experience at a municipal waste incinerator in Belgium, *Waste Management*, 22, 421–426.

Fahlenkamp H., Mittlebach G., Hagenmeier H., Brunner H., Tichaczek K.-H., 1991, Katalytische Oxidation. Eine Technik zur Verminderung der PCDD/PCDF-Emission aus Müllverbrennungsanlagen auf kleiner 0,1 ng TE/m^3, *VGB Kraftwerkstechnik*, 7, 71.

Hagenmeier H., 1989, Katalytische Oxidation halogenierter Kohlenwasserstoffe unter besonderer Berücksichtigung des Dioxinproblems, *VDI Berichte*, 730, 239.

Hagenmeier H., Mittelbach G., 1990, Versuche zum Katalytischen NO_x und Dioxin Abbau im Abgas einer Hausmüllverbrennungsanlage, *VGB Kraftwerkstechnik*, 6, 70.

Ide Y., Kashiwabara K., Okada S., Mori T., Hara M., 1996, Catalytic decomposition of dioxin from MSW incinerator flue gas, *Chemosphere*, 32, 189–198.

Ishikawa R., Buekens A., Huang H., Watanbe K., 1997, Influence of combustion conditions on dioxin in industrial scale fluidized bed incinerator. Experimental study and statistical modeling, *Chemosphere*, 35, 465–477.

Kreisz S., Hunsinger H., Vogg H., 1997, Technical plastics as PCDD.F absorbers, *Chemosphere*, 1997, 34, 1045–1052.

Pranghofer G., 2001, Destruction of polychlorinated dibenzo-p-dioxins and dibenzofurans in fabric filters: recent experiences with a catalytic filter system. 3rd International Symposium on incineration and flue gas treatment technologies "INCINERATION'2001", Bruxelles 2001.

Weber R., Sakurai T., Hagenmeier H., 1999, Low temperature decomposition of PCDD/PCDF, chlorobenzenes and PAHs by TiO_2-based V_2O_5-WO_3 catalyst, *Applied Catalysis B*, 20, 249–256.

Wielgosiński G., 2001, Dioxin formation in waste incineration processes. V International Conference "Dioxins in industry and environment", Cracow University of Technology, Kraków 2001.

Wielgosiński G., 2003, Emissions from waste incineration plants – primary methods of emission reduction, *Energy Policy*, 6, 131–140.

Wielgosiński G., Grochowalski A., Machej T., Pająk T., Ćwikaąlski W., 2004, Catalytic destruction of PCDD/Fs. Effect of temperature and catalyst loading on efficiency of 1,2-dichlorobenzene destruction on V_2O_5-WO_3/Al_2O_3-TiO_2 catalyst, *Organohalogen Compounds*, 66, 1082–1086.

Environmental Engineering – Pawłowski, Dudzińska & Pawłowski (eds)
© 2007 Taylor & Francis Group, London, ISBN13 978-0-415-40818-9

Gas holdup and bubble velocity determination in an air-lift column

Daniel Zając & Roman Ulbrich
Opole University of Technology, ul. S. Mikolajczyka, Opole, Poland

ABSTRACT: The subject of this work is an experimental hydrodynamic analysis of two-phase gas-liquid flow carried in an air-lift reactor with internal circulation. Analysis was made for six constructional variants of the apparatus which differed in the depth of immersion and width of the riser. The goal of this work was reached using a digital image processing and analysis method. Air-water flow was registered by high speed CMOS video camera. Gray level value of the images was the basic parameter which allowed for hydrodynamic investigations. The fluctuations of gray level value in interrogation areas together with its statistical analysis allowed identification of the recorded flow structures. Gas holdup was determined using a bed expansion method with the assistance of edge detection procedures due to position and shape of free surface of the two phase bed. The velocity of the gas phase was determined by a digital particle image velocimetry technique.

Keywords: Bubble column, air-lift reactor, gas holdup, DPIV.

1 INTRODUCTION

Bubble columns are mass transfer and reaction devices in which gas is brought into contact and reacts with a liquid phase. Sometimes the gas also reacts with dissolved or suspended components in the liquid phase. Due to their advantages (low energy consumption, simple construction, lack of any mechanically operated parts, high heat and mass transfer coefficients) and various modifications, such devices are commonly used in the chemical industry, biotechnology and environmental engineering (waste treatment). Airlift reactors are special type of bubble columns which offer directional fluid circulation. The circulation is stabilized by inserted loop (external or internal). Air lift reactors permit the processing of large amounts of gas and provide a homogenous flow regime. The high rate of circulation means shorter mixing times and hence the lack of any significant concentration gradient (Deckwer 1995). This is a very important feature for biotechnological processes, in which it is important to provide constant composition of the material surrounding the biomass.

In the literature papers can be found which concern identification of flow regimes with the use of differential pressure signals (Gourich et al. 2006), measurements of the axial distribution of gas holdup using γ rays (Haibo et al. 2005) or electrical tomography and gas disengagement technique (Fransolet et al. 2005). Among many experimental results there are also papers concerned with the modeling of bubble column reactors. Haut & Cartage (2005) have used a mathematical model for the estimation of the mass transfer rate under a heterogeneous regime and Wiemann & Mewes (2005) have calculated flow fields in two and three phase bubble columns.

The aim of this work concerns an essential problem which can arise when analyzing plant and design data. Namely, it is connected with the existence of complex relationships between hydrodynamic factors and geometric size, gas distribution, operating conditions and physical properties of the phases. The reactor construction (size and gas distributor) together with process parameters (physical properties of the phases, solubility) and adjustable operating conditions (flow rate, pressure, temperature, type of the flow), have a decisive influence on non-adjustable operating conditions such as phase holdups, interfacial area, phase velocities etc., which are measurable indicators of a reactor's performance. The aim of this work includes an experimental analysis of the hydrodynamics of two-phase gas-liquid flow operated in an air-lift reactor with external circulation (flow structures identification, gas holdup determination and gas phase velocity determination).

Analysis was carried out for six different constructional variants of a rectangular apparatus, which differed in depth's immersion B and width A of the riser (Fig. 1). In other words, devices with a different ratio of aeration to recirculation zones were analysed. Because in the recirculation zone a two-phase mixture has a lower fraction of gas bubbles (countercurrent flow causes degassing), it is crucial to maximize the flow velocity in this region. Improper selection of the

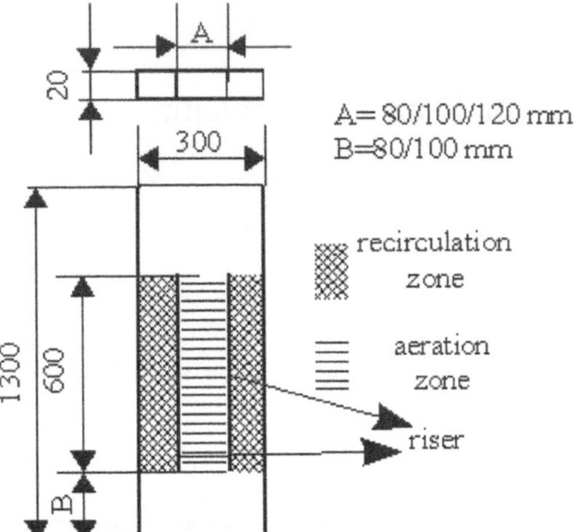

Figure 1. Physical model of analyzed air-lift reactor.

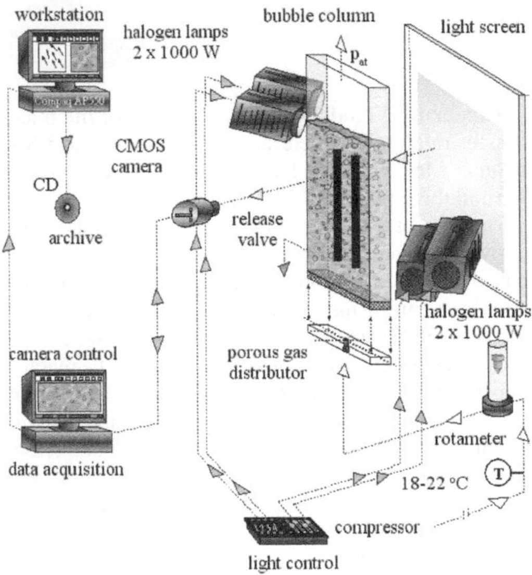

Figure 2. Experimental setup.

riser's geometry can slow down the phase velocities in this area and hence can, for example, lead to anoxia of the microorganisms suspended in the liquid phase.

2 MATERIALS AND RESULTS

A sketch of the experimental set-up is shown in Figure 2. The two-phase air-water flow was operated in a rectangular column made of plexiglass. The column's dimensions are 1300 × 300 × 20 mm (Fig. 1). Gas distribution was carried out by means of porous material of pore diameter 0.1 mm. The gas rate u_g based on the empty reactor cross-sectional area was within the 1.0–12.5 cm/s range.

The column was illuminated by four halogen lamps (1000 W power each) connected to the control box. Two-phase flow was recorded by a high speed CMOS HCC-1000 camera (VDS Vosskühler) with the sampling frequency $f_s = 100$ Hz. For each set gas rate, 10 s process' duration was recorded as a sequence of 1000 frames at resolution 1024 × 512 pixels and gray level scale.

The gray level of the recorded frames was the basic parameter for further analysis.

In so-called interrogation areas, with the size of 20 × 100 pixels, mapped on the each image from the particular sequences (coordinates were set in the middle of the aeration zone) two features were extracted:

- the mean gray level value B_0,
- boundary line length L_0.

The mean gray level value represents the mean superficial gas fraction in the interrogation area of $(l, t),(r, b)$ coordinates and is defined as

$$B_0 = \frac{1}{(r-l)(b-t)} \sum_{m=l}^{r} \sum_{n=t}^{b} G_{m,n} \tag{1}$$

where $G_{m,n}$ = gray level value of the pixel of (m, n) coordinates. The boundary line length represents the local dispersion of the gas bubbles (high dispersion causes the increase of this parameter) and is defined as

$$B_0 = \frac{1}{(r-l)(b-t-2)} \sum_{m=l-1}^{r-1} \sum_{n=t-1}^{b-1} edge(G_{m,n}) \tag{2}$$

where

$$edge(G_{m,n}) = \begin{cases} 1 \;\; when \bigvee_{k\in\{-1,0,1\}} \bigvee_{l\in\{-1,0,1\}} G_{m,n} \neq G_{m+k,n+l} \\ 0 \;\; when \bigwedge_{k\in\{-1,0,1\}} \bigwedge_{l\in\{-1,0,1\}} G_{m,n} = G_{m+k,n+l} \end{cases}$$

The results of the calculation in such interrogation areas are time series of the above-mentioned features/parameters. The stochastic nature of the phenomena of two-phase flows allows statistic analysis of measurements' results to provide valuable information about the process. Hence, from the obtained time series basic statistic parameters and functions were calculated.

The gas holdup ε_g was measured with the use of a modified bed expansion method. The modification is that the height of the dispersed two-phase bed is replaced by its area, while the latter is calculated from recorded images using edge detection procedures and numeric integrating (Fig. 3).

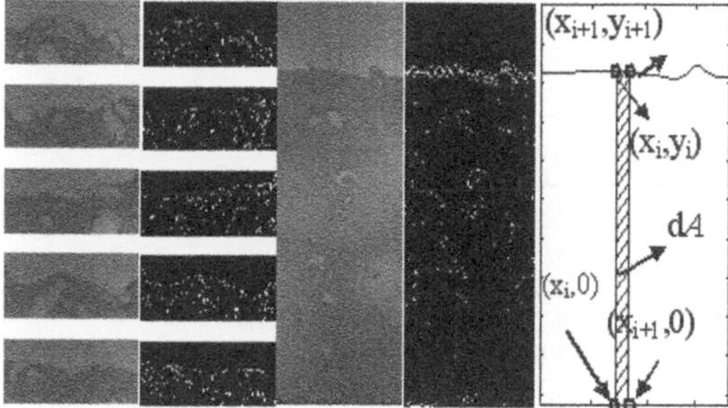

Figure 3. Gas holdup calculation method. From left: free surface recorded and it's edge detection result, an example of recorded image and it's edge detection result, numeric trapezoid integration.

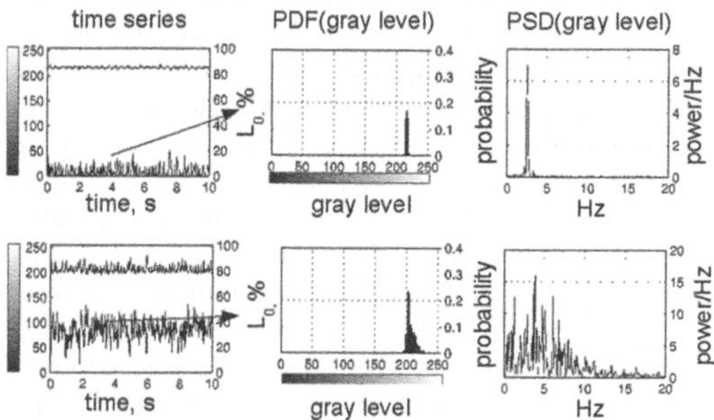

Figure 4. Time series and their statistic functions of homogenous flow regime, up (superficial gas flow rate 0.06 m/s) and heterogeneous (b) flow regimes (superficial gas flow rate 0.12 m/s), down.

The last part of the research was the optimization of geometric size of the riser due to gas velocity maximization in recirculation zones. As mentioned, the devices with six different riser sizes were analyzed (Fig. 1). Gas velocity in recirculation zones was measured using the Particle Image Velocimetry technique with opflow algorithm (Quénot et al. 1996). The results of these measurements are velocity fields of the gas phase in the bubble column.

3 RESULTS AND DISCUSSION

As mentioned previously, the obtained time series of mean gray level value and boundary line length fraction were extracted from each frame in particular sequences of two phase flow recordings (Fig. 4).

On the basis of this, it was possible to distinguish homogenous flow from heterogeneous flow. Additionally, having the time course of gray level change allows two main statistical functions to be accessed; PDF (probability density function) and PSD (power spectrum density). These provide better identification of

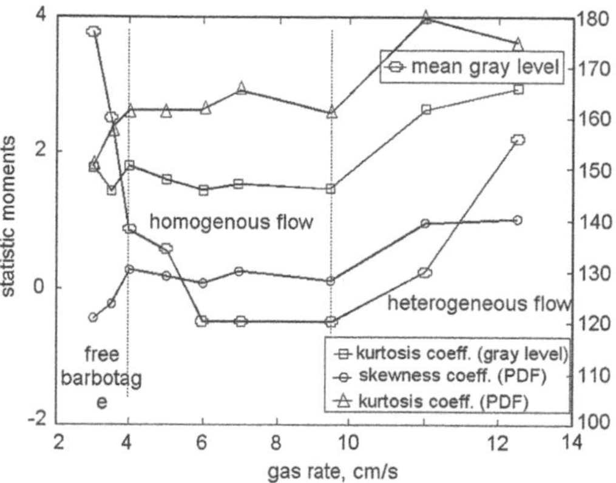

Figure 5. Statistic moments as a function of gas rate u_g for different flow regimes.

the flow patterns. PSD functions can provide also the information about the generation frequency of bubbles.

The statistic functions and time series mentioned above provide however only a qualitative description. In order to fully identify and quantitatively describe the recorded flow regimes, skewness and kurtosis coefficients were also calculated.

The results are shown in Figure 5. It is now possible to strictly define particular flow regimes ranges (including free barbotage phenomena). It is clear that the homogenous flow regime is situated between 2.0 and 7.5 cm/s of the air flow rate.

During the research it has been stated that the widest range of homogenous flow appearance occurs in the largest aeration area ($A = 120$ mm). In all six measured variants of the column, in the recirculation zones the flow structure was of bubble form. Only starting from $u_g = 10$ cm/s gas rate were large bubbles and small plugs noticed. The results of the statistical analysis of the recorded time series are presented in Table 1.

The next part of the research included the measurements of gas holdup, which depended very strongly on the flow regime. In bubble columns, the primary flow regime is bubbly flow. Therefore, fluctuations in gas holdup are not so dominant as in other devices in which two-phase flow occurs. Despite this, during the investigations it was found that the cross-sectional area of aeration zone had a visible influence on gas holdup fluctuations (Fig. 6).

Comparing the apparatus with the smallest aeration area ($A = 80$ mm) to the apparatus with the highest aeration area ($A = 120$ mm) the difference reaches 5%. The measured range of gas holdup was 0.05 to 0.40. Generally, highest aeration area causes the increase gas holdup in all reactors (Fig. 6). The second geometric parameter – the depth of immersion B had only minimal influence on the gas holdup

Table 1. Statistical parameters of recorded flow structures.

Gas flow rate u_g, cm/s	statistical parameters STD	Range	Skewness	Kurtosis	Flow pattern
1.0	1.45	11	0.08	2.86	free barbotage
1.5	2.04	17	1.05	5.78	homogenous
2.0	2.74	19	1.28	5.45	homogenous
3.0	3.69	22	1.18	4.25	homogenous
4.0	4.94	29	1.21	4.08	homogenous
5.0	6.00	34	1.14	3.79	homogenous
7.5	6.41	37	1.13	3.89	homogenous
10.0	7.61	39	0.81	3.03	heterogeneous
12.5	8.31	43	0.68	2.63	heterogeneous

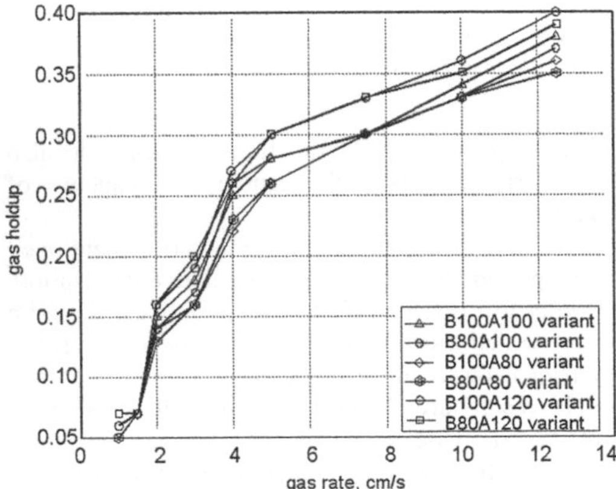

Figure 6. Gas holdup ε_g as a function of gas rate u_g for different sizes of the riser.

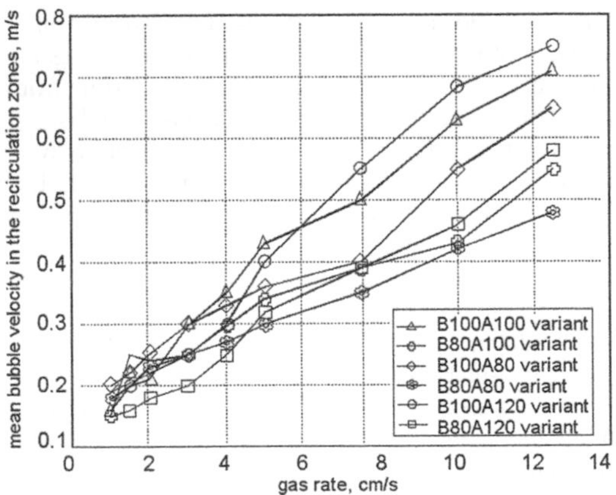

Figure 7. Mean velocity of the gas phase in recirculation zones as a function of gas rate for different riser's size.

values (1% maximal differences between particular variants).

The mean gas velocity in the recirculation zone was calculated from the fifteen instantaneous velocity fields for each recoded sequence. The first conclusion which comes from Figure 7 results is that depth's immersion B has a decisive influence on the bubble velocity in zones mentioned. It is clear that there are obvious differences between $B = 80$ and $B = 100$ mm variants. The second conclusion is that the size of the riser has also a clear significance during the establishment of gas velocity. In other words, in the apparatus with the largest aeration cross-sectional area (smallest recirculation zones) the highest bubbles' velocity was measured. The lowest velocities were observed in the variant with $A = 80$ mm. In all measured geometric variants, the gas velocity increases proportionally with the gas flow rate rise. However the highest proportionality factor is visible in the courses which reflect velocities in variants with $B = 100$ mm.

The highest gas mean velocities in recirculation zones were measured in the $B = 100/A = 120$ mm geometric variant (0.76 m/s). Whereas the lowest velocities can be observed in $B = 80/A = 80$ mm.

The change of gas velocities caused by geometric modifications of the riser had also an influence for gas holdup determination. A general rule is that different velocities cause an entrainment of different number of the bubbles.

4 CONCLUSIONS

- statistical analysis of gray level time series led to identification of flow structures,
- the DIP method permitted an analysis of the complex hydrodynamics of the two-phase flow, operating in a bubble column,
- a noticeable influence of the immersion depth B of the riser on the gas velocity in recirculation zones was observed, the highest velocities were measured for $B = 100$ mm over the bottom,
- the riser's width A had also the influence on the gas velocity, the highest values were measured in the apparatus with the biggest aeration zone ($A = 120$ mm),

– the highest gas velocity in the recirculation zone was observed in the $B = 100/A = 120$ mm variant of the riser's geometry (0.75 m/s), the lowest velocity in $B = 80/A = 80$ mm version (0.59 m/s),
– the highest values of gas holdup were measured for the largest aeration zone ($A = 120$ mm).

REFERENCES

Deckwer, W-D. 1995. *Bubble column reactors.* Chichester: Wiley & Sons.

Gourich, B., Vial, C., Essadki, A.H., Allam, F., Soulami, M.B., Ziyad, M. 2006. Identification of flow regimes and transition points in a bubble column through analysis of differential pressure signal—Influence of the coalescence behavior of the liquid phase. *Chemical Engineering and Processing* vol. 45: 214–223

Jin, H., Yang, S., Guo, Z., He G., Tong Z. 2005. Analysis of gas holdup in bubble columns with non-Newtonian fluid using electrical resistance tomography and dynamic gas disengagement technique. *Chemical Engineering Journal* vol. 115. 45–50

Fransolet, E., Crine, M., Marchot, P., Toye, D. 2005. Analysis of gas holdup in bubble columns with non-Newtonian fluid using electrical resistance tomography and dynamic gas disengagement technique. *Chemical Engineering Science* vol. 60: 6118–6123

Hauta, B., Cartage, T. 2005. Mathematical modeling of gas–liquid mass transfer rate in bubble columns operated in the heterogeneous regime. *Chemical Engineering Science* vol. 60: 5937–5944

Wiemann, D., Mewes, D. 2005. Calculation of flow fields in two and three-phase bubble columns considering mass transfer. *Chemical Engineering Science* vol. 60: 6085–6083

Quénot, G.M., Pakleza, J., Kowalewski, T.A. 1996. Particle image velocimetry with optical flow. *Experiments in Fluids* vol. 25: 177–189

Environmental Engineering – Pawłowski, Dudzińska & Pawłowski (eds)
© 2007 Taylor & Francis Group, London, ISBN13 978-0-415-40818-9

Application of lichens for the determination of precipitation pH by the exposure method

Andrzej Kłos, Małgorzata Rajfur & Maria Wacławek
Chair of Biotechnology and Molecular Biology, Opole University, ul. Kard. B. Kominka Opole, Poland

Witold Wacławek
Chair of Chemical Physics, Institute of Chemistry, Opole University, ul. Oleska Opole, Poland

ABSTRACT: The work presents preliminary results of research related to a method of determination of atmospheric precipitation pH using the lichen cationactive layer composition exposed together with calcium sulphate. The studies were conducted in the city centre of Opole. *Hypogymnia physodes* lichens were used in the studies. Types and different phases of precipitation were considered. The results were compared with pH values measured using a pH-meter.

Keywords: Lichens, ion exchange, lichenoindication, pH of precipitation.

1 INTRODUCTION

Lichens have been known as bioindicators of environment pollution for over a hundred years. One of the first and most frequently referred to examples is the studies by William Nylander (1822–1899), a Finnish biologist, demonstrating the dependences between the variety and number of lichens, and the level of atmospheric pollution (Vitikainen 2001). Nowadays, bioindicators, including lichens, are recognised as pillars of modern environmental monitoring (Hauck 2005). However, one of the unsolved problems of biomonitoring is the lack of a uniform research procedure.

As demonstrated by many publications (Wolterbeek 2003, Seaward 2005), attempts to standardise research procedures have been made, but they are not fully clear-cut and have raised many disputes. An important research subject seems to be determination of individual processes occurring between the environment and the bioindicator, and the analysis of influence of abiotic factors on dynamic equilibria in bioindicator structure (homeostasis). They relate to, *inter alia*: climate influence: wind direction and intensity (Freitas et al. 2001), precipitation intensity (Reis et al. 2003) and the impact of exposure time (Reis et al. 1999) as well as interrelations between the concentration of cations accumulated in lichens and present in their surrounding. The studies focus on, *inter alia*: the effect of precipitation acidity on macroelements accumulated in lichens: K, Mg and Ca (Hyvärinen & Crittenden 1996); the effect of precipitation acidity on sorption of heavy metals e.g.: nickel (Tarhanen et al. 1999) and uranium (Haas et al. 1998); and the effect of magnesium and calcium on manganese ion sorption in lichen structure (Hauck et al. 2002).

A frequent research topic is the qualitative and semi-quantitative determination of environment pollution based on the analysis of diversity and number of lichens present in a given area. Such studies make use of the methods based on different lichen scales, e.g. (Hawksworth & Rose 1970), modified for geographical areas, depending on species of lichens that are found there. They are often referred to air pollution with sulphur dioxide that is the main cause of precipitation acidity.

A more accurate method of forming lichen maps is determination of the so-called IAP (Index of Atmospheric Purity): $IAP = \Sigma F_i$, where F_i means frequency of occurrence (maximum 10), calculated as the number of squares in a network (a 30×50 cm square, divided into ten 15×10 cm areas) where every species is found. Application of the method requires comparison of lichens growing on similar substrate. High importance is ascribed to the species and age of a given tree (Conti & Cecchetti 2001).

This article presents results of laboratory studies and initial results of field studies related to the quantitative method using lichens to determine precipitation acidity. The method has been developed on the basis of results from previously work related to mutual equilibria of hydrogen ions and metal cations: naturally found in lichen environment: Na^+, K^+, Mg^{2+} and Ca^{2+} (Kłos et al. 2004, Kłos et al. 2005a, Kłos et al. 2006). The studies were conducted in the city centre of Opole.

1.1 *Physicochemical basis of the method*

Previous studies demonstrated that one of the cation sorption mechanisms in lichens is ion exchange (Kłos et al. 2005a). This mechanism is also suggested by other authors (e.g.: Schwartzman et al. 1991, Gadd 1993, Brown & Brumelis 1996, Figueira et al. 2003). It was shown that ion exchange occurred between the environment (water solution) and the lichen cationactive layer. The cationactive layer was defined as a part of lichen structure (super-cellular structure) that reversibly binds cations present in the lichen environment as a result of ion exchange. The determined sorption capacity of *Hypogymnia physodes* lichens is $c^* = 0.145$ m/g d.m. (Kłos et al. 2005a). This concentration, expressed in mol/g of dry mass of lichens, is defined by the following equation (1).

$$c^* = 10^3 \times c \times z \qquad (1)$$

where: c – concentration, expressed in m/g of d.m., z – ion valence.

Similarly, we define ion concentrations in a solution; however, here c^* unit is M.

Concentrations expressed in this way and other values calculated on their basis allow comparison of mutual relations between cations of different valence.

The process secondary or simultaneous towards the ion exchange process, is the permanent embedding of cations in to the lichen structure. For example, radiocaesium studies in the *Cetraria islandica* lichen structure demonstrated the presence of ^{137}Cs in proteins, polypeptides and saccharides extracted from lichens (Nedić et al. 1999). In lipids, it was found that the location of Cu and Pd that indicated the mechanism of metal cation complexation in lichens (Takani et al. 2002).

The method of determination of precipitation pH was based on results of studies of equilibria and the ion exchange kinetics of the following cations naturally found in the lichen environment: H^+, Na^+, K^+, Mg^{2+} and Ca^{2+}. Mutual relations of cation pairs can be described by the expression for the equilibrium constant of the heterophase double displacement reaction (dependence 2).

$$y^* = (K^*_{A/B} \times x^*) \times [1 + (K^*_{A/B} - 1) \times x^*]^{-1} \qquad (2)$$

where: K^* – is the equilibrium constant, y^*, x^* – are mole fractions, respectively, in lichens in the solution, A, B – are cations.

We can also determine the reaction rate constant using a mathematical description of second-order reaction, for $c^*_A \neq c^*_B$, (dependence 3).

$$\log (x_A/y_B) = 2{,}303^{-1} \times k^* \times t \times (c^*_{A,0} - c^*_{B,0}) \qquad (3)$$

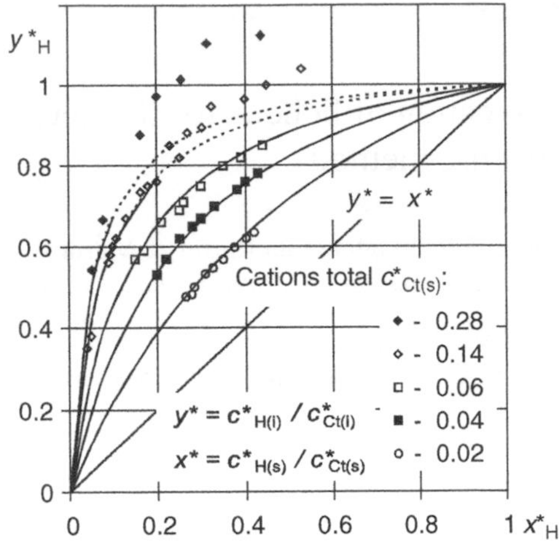

Figure 1. Equilibriums of Na/H ion exchange.

where: k^* – is the reaction rate constant, t – is time, $c^*_{A,0}$, $c^*_{B,0}$ – are initial A and B ion concentrations, respectively, in the solution and in the lichen cationactive layer.

Figure 1 illustrates graphical interpretation of changes of the equilibrium constant $K^*_{Na/H}$ between hydrogen and sodium ions in the solution and in the lichen cationactive layer depending on the total cation concentration in the solution (Kłos et al., in press).

The figure shows interesting discrepancies between the assumed course of the function and the experimental points in the region, where hydrogen ion concentration c^*_H in the state of equilibrium is higher from 0.126 M (pH < 3.9). The discrepancies probably result from destruction of the lichen thallus.

The equilibria, determined according to the dependence (2), show that the concentration of mobile hydrogen ions bound in the lichen cationactive layer $c^*_{H(i)}$ depends on the concentration of the ions $c^*_{H(s)}$ and on the concentration and type of cation total $c^*_{Ct(s)}$ in precipitation with which the lichens are in contact.

Studies of the kinetics of the process (dependence 3) prove that the lichen – solution equilibrium is achieved after approximately 30 minutes, i.e. in natural conditions.

2 MATERIALS AND METHODS

For the purpose of laboratory studies, *Hypogymnia physodes* lichens (approximately 2.5 g d.m.), collected and separated from mechanical impurities, were exposed (ca. 1 hour) to synthetic precipitation of known pH and salinity. Depending on the method of salinity determination, approximately 0.2 g of calcium fluoride or sulphide was added to the lichens.

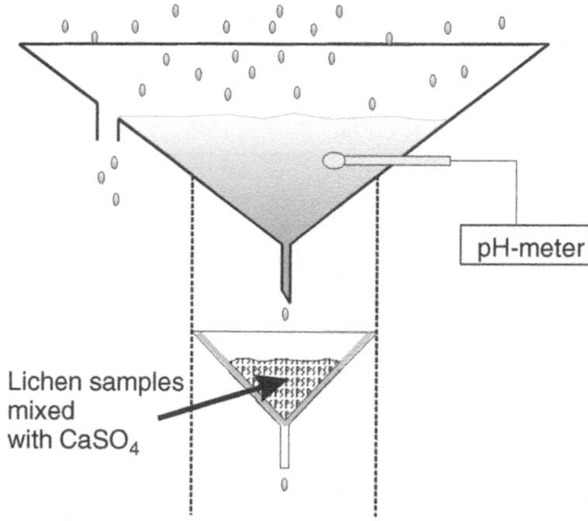

Figure 2. Analytical system.

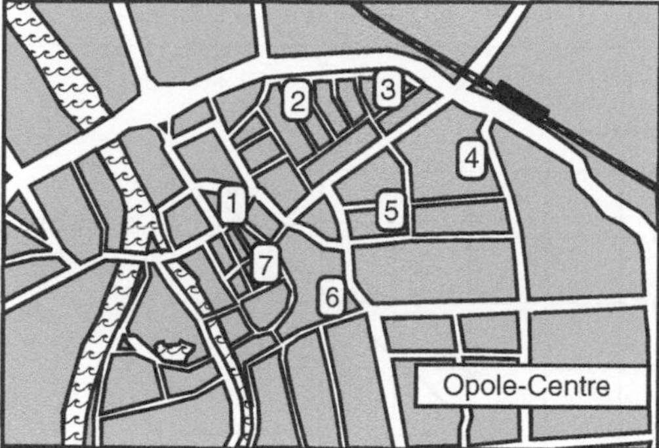

Figure 3. Locations of lichen exposure in the city centre of Opole.

The analytical set-up (Fig. 2) was constructed in such a way that the intensity of lichen sprinkling amounted approximately to 1 ml/min. Because of the presence of an overflow system, practical use of the set provides constant intensity of sprinkling regardless of rain intensity. Next, dried samples of lichens (1 g d.m.), in order to determine the number of hydrogen ions bound in the lichen cationactive layer, were dipped in a solution of copper sulphate ($c^* = 2.0$ M; 200 ml). During the ion exchange process, changes of conductivity and pH of the solution were measured.

In field studies, identically prepared samples placed in analytical sets, were exposed to precipitation. To average salinity, approximately 0.2 g of calcium sulphate was added to lichens.

The analytical set-up was arranged in such a way that the intensity of lichen sprinkling was ca. 0.6 ml/min (in laboratory studies, the sprinkling intensity reached approximately 1.0 ml/min) regardless of rain intensity.

Lichens were exposed for four hours in seven locations in the city (Fig. 3); in the case of locations 1–8 lichen samples were exposed successively for 30 minutes and then analysed, every 15 minutes; for comparative purposes, the potentiometric pH measurements of precipitation were conducted at location 1.

After exposure, drying at 30 °C to obtain dry matter, lichen samples (1 g d.m.) were immersed in copper sulphate (200 ml; $c^* = 2.0$ M). The initial (pH_0) and final (pH_1) pH of the solution was measured. According to the dependence (4), the hydrogen ion concentration increase was calculated allowing the determination of pH of precipitation in accordance with the linear dependence $\Delta c^*_{H(Cu)} = f(pH_{precipit.})$ shown in Figure 5 (exposure to $CaSO_4$).

$$\Delta c^*_{H(Cu)} = 10^3 \times (10^{-pH,1} - 10^{-pH,0}) \quad (4)$$

The number of hydrogen ions desorbed in 200 ml of the solution ($n^*_{H(Cu)} = 0.2 \times \Delta c^*_{H(Cu)}$) corresponds to the hydrogen ion concentration bound in 1 g of d.m. of lichens $c^*_{H(j)}$ [m/g d.m.] used for the experiments. The course of hydrogen ion desorption is consistent with the equation (5).

$$Cu^{2+} + 2HR \leftrightarrows 2H^+ + CuR_2 \quad (5)$$

where: R – anion in the cation active layer.

Equipment purchased from the ELMETRON company of Zabrze, Poland, was used in the studies: portable pH-meter CP215 and stationary pH-meter CP501, the measurement uncertainties were 0.05 and 0.02 pH units, respectively.

3 RESULTS AND DISCUSSION

The results of laboratory experiments in which artificial precipitation was used and the results of pH precipitation measurements made under field conditions are presented separately below.

3.1 *Laboratory studies*

For studies with the lichen exposure method, we used the method consisting in averaging of precipitation salinity by saturation of the natural precipitation with a slightly soluble salt (Kłos et al. 2005a). In laboratory studies, lichen samples were mixed with calcium fluoride or sulphate. This allowed us to obtain a relatively constant calcium ion concentration in the water in which the lichens had been in contact. For samples prepared in this way, the concentration of ions in the lichen cationactive layer depended on their concentration in precipitation, and the influence of natural salinity of precipitation could be omitted.. The graphic interpretation (Fig. 4) shows the difference in the course of the function $c^*_{H(i)} = f(pH_{precipit.})$ depending on the type and concentration of salts contained in the solution with which the lichens were in contact.

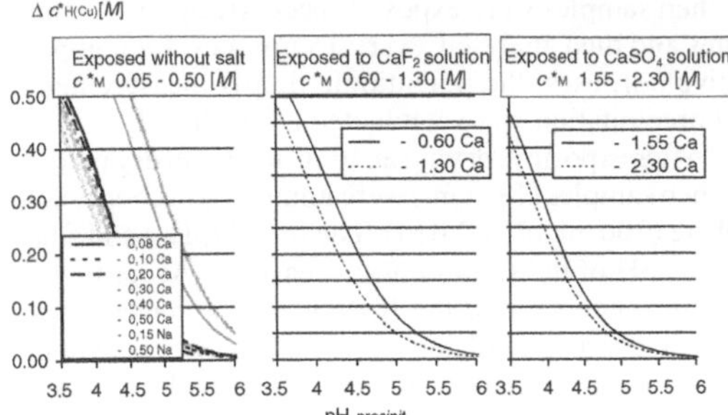

Figure 4. Course of the function $\Delta c^*_{H(Cu)} = f(pH_{precipit.})$ depending on the type and concentration of salts contained in the solution with which lichens are in contact.

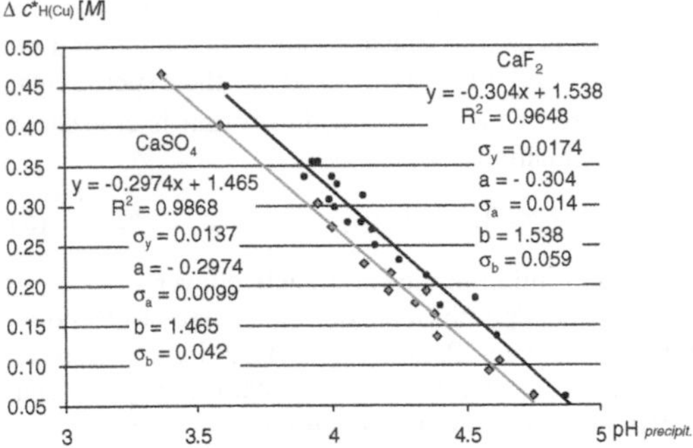

Figure 5. Linear dependences $\Delta c^*_{H(Cu)} = f(pH_{precipit.})$ possible to be used in determination of the precipitation pH.

The hydrogen ion concentration in the lichen cationactive layer was determined by desorption of these ions from lichens in the copper sulphate solution ($c^* = 2.0$ M) and determination of the increase of hydrogen ion concentration in the solution using a pH-meter (dependence 4).

As seen in the diagram (Fig. 4), salinity averaging shows a good approximation to the linear dependence $\Delta c^*_{H(Cu)} = a \times pH_{precipit.} + b$, regarding changes of pH of precipitation from 3.5 to approximately 5. Parameters of the straight lines are shown in Figure 5.

The mean standard deviation σ for a series of measurements controlled in the laboratory amounts to 0.054 and 0.050 of the determined pH value, respectively, for lichens exposed with CaF_2 and $CaSO_4$.

The maximum uncertainty of a single measurement, resulting from the uncertainty of pH measurements in the solution of copper sulphate (±0.04 of pH), referred to the threshold salinity value (Table 1), in the range of linear dependences $\Delta c^*_{H(Cu)} = a \times (pH_{precipit.}) + b$, amounts to: ±0.18 of pH value in the case of application of CaF_2 and ±0.10 of pH value in the case of application of $CaSO_4$.

Table 1. Measured and determined pH values by presented method.

t [min]	pH measured	pH_0	pH_1	$\Delta c^*_{H(Cu)}$ [M]	determined pH
15	4.15	–	–	–	–
30	3.93	5.01	3.51	0.299	3.92
45	4.01	–	–	–	–
60	4.04	5.04	3.56	0.266	4.03
75	4.01	–	–	–	–
90	4.09	5.02	3.64	0.219	4.19
105	4.11	–	–	–	–
120	4.09	5.01	3.61	0.245	4.10
135	4.13	–	–	–	–
150	4.11	5.01	3.59	0.247	4.09
165	4.27	–	–	–	–
180	4.44	5.02	3.86	0.128	4.49
195	4.23	–	–	–	–
210	4.36	5.01	3.72	0.181	4.32
225	4.18	–	–	–	–
240	4.42	5.03	3.84	0.135	4.47

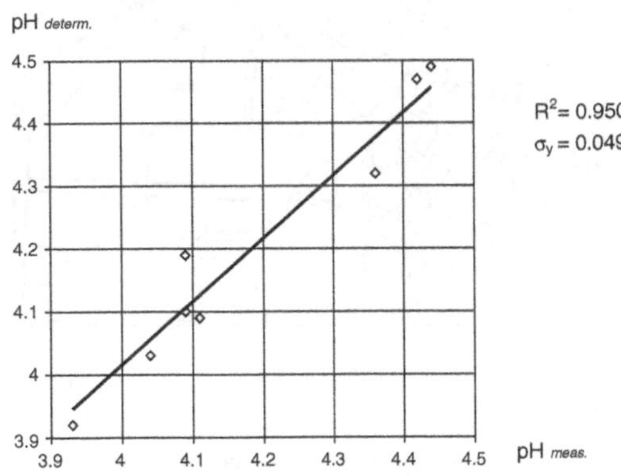

Figure 6. Comparison of measured and determined pH values of precipitation.

3.2 *Field studies*

Table 1 presents the data related to the pH values of precipitation measured and determined on the basis of the analysis of hydrogen ion concentration bound in the lichen cationactive layer. The linear dependence: $\Delta c^*_{H(Cu)} = -0.2974 \times pH_{precipit.} + 1.465$, determined in the laboratory and using Figure 5, was used for the calculations.

Figure 6 compares pH values of precipitation with the values measured using a pH-meter.

The mean standard deviation σ for a series of control measurements (0.049) corresponds to the mean standard deviation determined in laboratory conditions (0.050).

Table 2 presents calculated pH values of precipitation after four hours of exposure, determined on the

Table 2. The pH values of precipitation determined on the basis of results of the studies regarding hydrogen ion concentration accumulated in lichens.

Measurement location	1	2	3	4	5	6	7	average.	σ_{pH}
pH of precipitation	4.35	4.65	4.44	4.31	4.28	4.40	4.47	4.41	0.12

basis of results of the studies regarding hydrogen ion concentration accumulated in lichens.

In the examined area (ca. 2 km^2) pH values determined on the basis of the lichen composition analysis were in the range 4.28–4.65. The average pH value of precipitation amounted to 4.41.

It can be assumed that hydrogen ion concentration in the examined precipitation had similar values, as demonstrated by the measurement, because of a small area under investigation.

Small differences in pH values of precipitation determined using the described method (Table 2) demonstrate that the maximum uncertainty of pH determination on the basis of a single measurement is ca. 0.2 of a pHunit, with mean standard deviation $\sigma = 0.12$.

Because of the diversity of abiotic factors, which can be eliminated in a laboratory, the studies will be continued in field conditions.

4 CONCLUSIONS

The methods for the determination of the pH of precipitation presented here were developed on the basis of studies of the equilibrium (Kłos et al. 2006) and kinetics of ion exchange (Kłos et al. 2005b) occurring between the cationactive layer and the water solution with which lichens are contacted.

As was demonstrated, the concentration of mobile hydrogen ions bound in the lichen cationactive layer $c^*_{H(i)}$ depends on the concentration of these ions $c^*_{H(s)}$ and on the total concentration and type of cations $c^*_{Ct(s)}$ in the precipitation with which lichens are in contact.

The method described consists of averaging the salinity of the precipitation with which the lichens are in contact by saturation the precipitation with calcium sulphate mixed with exposed lichens. It allows a relatively constant precipitation salinity (in the tested system, c^*_{Ct} was 1.55–2.30 M), to be obtained regardless of the natural salinity, which, in turn, enables pH of precipitation on the basis of the linear dependence $\Delta c^*_{H(Cu)} = f(\text{pH}_{precipit.})$ to be determined (Fig. 5). Determination of mobile hydrogen ion concentration in the lichen cationactive layer ($n^*_{H(Cu)} = 0.2 \times \Delta c^*_{H(Cu)}$) consists of desorption of the ions in a copper sulphate solution.

Initial studies indicated that the tested method can be used for determination of precipitation pH. The mean standard deviation σ of the values measured using a pH-meter, for a series of control measurements (Fig. 6), amounted to 0.049, with the correlation coefficient value of $R^2 = 0.950$.

By applying the salinity averaging method (lichen exposure method), the precipitation pH can be determined within the range of 3.5–4.8; however, long-term contact with precipitation of $\text{pH} < 3.9$ results in destruction of the thallus of the studied lichen. The method can be successfully used to monitor precipitation acidity in urban and industrial areas.

REFERENCES

Brown, D.H. & Brumelis, G.A. 1996. Biomonitoring method using the cellular distribution of metals in mosses. *Sci. Total Environ.* 187(2): 153–161.

Conti, M.E. & Cecchetti, G. 2001. Biological monitoring: lichens as bioindicators of air pollution assessment – a review. *Environ. Pollut.* 114: 471–492.

Gadd, G.M. 1993. Interaction of fungi with toxic metals. *New Phytol.* 124: 25–60.

Figueira, R. et al. 2003. Development and calibration of epiphytic lichens as saltfall biomonitors. *Proc. intern. workshop biomonit. atmos. pollut. BioMAP II, Vienna, 28 August–3 September 2000*, Vienna: IAEA.

Freitas, M.C. et al. 2001. Use of lichen transplants in atmospheric deposition studies. *J. Radioanal. Nucl. Chem.* 249: 307–315.

Haas, J.R. et al. 1998. Bioaccumulation of metals by lichens: Uptake of aqueous uranium by Peltigera membranacea as a function of time and pH. *Am. Mineral.* 83: 1494–1502.

Hauck, M. et al. 2002. Manganese uptake in the epiphytic lichens Hypogymnia physodes and Lecanora conizaeoides. *Environ. Exp. Bot.* 48: 107–117.

Hauck, M. 2005. Epiphytic lichen diversity on dead and dying conifers under different levels of atmospheric pollution. *Environ. Pollut.* 135: 111–119.

Hawksworth, D.L. & Rose, F. 1970. Qualitative scale for estimating sulphur dioxide air pollution in England and Wales using epiphytic lichens. *Nature.* 254: 145–148.

Hyvärinen, By M. & Crittenden, P. D. 1996. Cation ratios in Cladonia portentosa as indices of precipitation acidity in the British Isles. *New Phytol.* 132: 521–532.

Kłos, A. et al. 2004. Assessment of precipitation pH value based on the composition of cationactive layer of lichens. *Ann. Polish Chem. Soc.* 3(1): 1124–1127.

Kłos, A. et al. 2005a. Ion equilibrium in lichen surrounding. *Bioelectrochemistry.* 66: 95–103.

Kłos, A. et al. 2005b. Ion exchange kinetics in lichen environment. *Ecol. Chem. Eng.* 12: 1353–1365.

Kłos, A. et al. 2006. Method of assessment of precipitation pH based on analysis of the composition of cationactive layer of lichens. *Electrochim. Act.* 51(24): 5053–5061

Nedić, O. et al. 1999. Organic cesium carrier(s) in lichen. *Sci. Total Environ.* 227: 93–100.

Reis, M.A. et al. 1999. Lichens (*Parmelia sulcata*) time response model to environmental elemental availability. *Sci. Total Environ*. 232: 105–115.

Reis, M.A. et al. 2003. Surface-layer model of lichen uptake, modelling Na response. *Proc. intern. workshop biomonit. atmos. pollut. BioMAP II, Vienna, 28 August–3 September 2000*, Vienna: IAEA.

Schwartzman, D.W. et al. 1991. An ion-exchange model of lead- 210 and lead uptake in a foliose lichen; application to quantitative monitoring of airborne lead fallout. *Sci. Total Environ*. 100: 319–336.

Seaward, M.R.D., in press. Biomonitors of environmental pollution: an appraisal of their effectiveness. *Proc. XIV centr. europ. conf. ECOpole'05, Jamrozowa Polana, 20–22 October 2005*. Opole: TChIE.

Takani, M. et al. 2002. Spectroscopic and structural characterization of copper(II) and palladium(II) complexes of a lichen substance using acid and its derivatives. Possible forms of environmental metals retained in lichens. *J. Inorg. Biochem*. 91: 139–150.

Tarhanen, S. et al. 1999. Membrane permeability response of lichen Bryoria fuscescens to wet deposited heavy metals and acid rain. *Environ. Pollut*. 104: 121–129.

Wolterbeek, B. 2003. Biomonitoring of trace element air pollution: principles, possibilities and perspectives. *Proc. intern. workshop biomonit. atmos. pollut. BioMAP II, Vienna, 28 August–3 September 2000*, Vienna: IAEA.

Vitikainen, O. 2001. William Nylander (1822–1899) and Lichen Chemotaxonomy. *Bryologist* 104(2): 263–267.

Environmental Engineering – Pawłowski, Dudzińska & Pawłowski (eds)

Speciation of lead in industrially polluted soils – a comparison of two extraction methods

Barbara Gworek & Igor Kondzielski
Institute of Environmental Protection, Krucza, Warsaw, Poland

Andrzej Mocek
Department of Soil Sciences, Poznań Agricultural University, Mazowiecka, Poznań, Poland

ABSTRACT: The main aim of this study was to compare two different sequential fractionation procedures, in order to assess their applicability to Pb speciation in industrially polluted soils. The compared speciation procedures were McLaren & Crawford's and Tessier's. The soil profiles used in this study, representing two types of soil: deluvial humous soil and deluvial brown soil, were sampled in the vicinity of the "Legnica" Copper Smelter (Legnicko-Głogowski Copper Mining Region).

Comparison of the methods revealed better applicability of the Tessier procedure for the determination of lead in industrially polluted deluvial soils. This was demonstrated, especially in the case of oxidisable, reducible and residual fractions. This conclusion was further supported by the results of the correlation between the lead content in a given fraction and some basic properties of the examined soils.

Keywords: Lead, contaminated soils, sequential extraction, Tessier's procedure, McLaren & Crawford's procedure.

1 INTRODUCTION

The pollution of soils by heavy metals is nowadays considered to be one of the most important ecological problems. Because of their toxicity in elevated concentrations heavy metals may negatively influence the functioning of the ecosystems. They also pose a serious threat to the human health. It is therefore necessary to assess the environmental risk posed by them to living organisms. The most useful tool for this purpose is speciation analysis. This enables the identification and quantification of the forms in which the heavy metals occur in polluted matrices. Although the term is generally applied to the qualitative and quantitative analysis of metals in solid matrices – soils and sediments, there is an increasing number of publications presenting the results of speciation of elements in other environmental matrices, like water or living organisms and their tissues (Das et al. 1995, Rubio & Rauret 1996).

The speciation of heavy metals in soils as an analytical tool has been widely used since the early 1980's. Its career started in 1978, when Tessier et al. published the results of their work on the fractionation of heavy metals in sediments. Their procedure has become the first fully operational multielement fractionation procedure, and remains, despite some modifications, one of the most commonly used fractionation procedures. It should be mentioned that the idea of the fractionation of heavy metals in various solid matrices predates the Tessier procedure. Prior to the development of this fractionation method there had already existed several others. These were, however, designed for the speciation of one or more chosen elements. The main advantage of Tessier's procedure lays in its universality – it can be used for almost all heavy metals in almost all solid environmental matrices.

Some of the fractionation methods predating Tessier's method, like the McLaren & Crawford procedure, are still in use, and like the former, they have undergone several modifications in order to ameliorate them.

Tessier's procedure, despite its many drawbacks, is still one of the most commonly used sequential fractionation procedures and, together with the three-step BCR procedure developed later, is recommended on the EU level (Ure 1996, Gomez Ariza et al. 2000, Gleyzes et al. 2002).

Lead is one of the topmost elements on the list of heavy metals, posing a serious threat to the environment and human health. The reason for this is its high acute and chronic toxicity. In addition, although some researchers dispute this statement, it is so far considered as not having any biochemical function. For this reason there is a substantial need to monitor the content of lead in various environmental matrices as well as the forms in which it occurs. This is also a case of soils, where excessive lead, despite its relative immobility in this matrix, can influence its functioning and have an impact on the quality of ground and surface waters.

Several fractionation procedures, (one-step and multistep), were developed in order to monitor behaviour of heavy metals in soils. They have been the objects of intensive studies to find the best one for this purpose.

The main task of the present study was to establish the compatibility of the results obtained using two speciation procedures: McLaren & Crawford's and Tessier's, during the qualitative determination of lead forms in the profiles of two industrially polluted deluvial soils.

Although the McLaren & Crawford's procedure has not been in use for more than a decade now, having been replaced by other, modern and better procedures, like that of Tessier, there is much data on the speciation of heavy metals, including lead, obtained with its help. These results are often very early results of heavy metals speciation for the given area. Lately, in order to obtain better results the speciation procedures were changed, often for the Tessier's procedure, which remains one of the most commonly used speciation procedures. Therefore, in order to establish the eventual continuity of the studies despite the change of the speciation procedure, it is necessary to determine the eventual compatibility of the speciation methods, in this case McLaren & Crawford's and Tessier's, and the extent of this compatibility. This, in consequence, should show whether the results obtained using McLaren & Crawford's method may be considered as equal to those obtained by Tessier's procedure, what means in practice, for many areas, much longer periods of continuous monitoring of soil pollution by lead occurring in the given area. Otherwise the results obtained using McLaren & Crawford's speciation procedure must be regarded as of limited, only informative value.

Such was also the case of our studies on the dynamics of changes of lead forms in industrially polluted soils in the Legnicko-Głogowski Copper Mining Region. Changing the speciation procedure from McLaren & Crawford's to Tessier's we observed the change in obtained results. By performing the speciation of lead simultaneously in several soil samples from this area using the two procedures we hoped to check whether the observed differences would be so significant that it is impossible to compare the two sets of results, or if it is nevertheless possible to consider the continuity of the study preserved.

2 MATERIALS AND METHODS

The research was carried out on six soil profiles taken from the vicinity of "Legnica" Copper Smelter in the Legnicko-Głogowski Copper Mining Region. Two types of soils were examined: deluvial brown soil (Eutric Cambisol; two profiles) and deluvial humic soil (Glyeic Phaozem; four profiles) Sampling area characteristics and sampling area procedures are described in detail by Mocek et al. (1991) and Gworek & Mocek (2001).

Sampled profiles were divided into the genetic horizons, dried and sieved. From each of the horizons, samples were taken for further analysis.

Prior to the speciation analysis the following parameters were determined for each dried soil sample:

- granulometric composition using the Casagrande method modified by Prószyński et al. (Mocek et al. 1997), after the separation of the sample's skeletalparts,
- soil pH in H_2O and KCl using the potentiometric method (Mocek et al. 1997),
- organic carbon content using Tiurin's method (Mocek et al. 1997),
- CEC using Metson's method (Mocek et al. 1997),
- content of Fe- and Mn oxides using Mehra-Jackson's method (Mehra & Jackson 1960).

Lead speciation in dried soil samples was performed using McLaren & Crawford's and Tessier's methods. It followed the schemes presented in Table 1 (Gomez Ariza et al. 2000, Gleyzes et al. 2000, Ponizovsky & Mironenko 2001, McLaren & Crawford 1973, Tessier et al. 1979).

For McLaren & Crawford's procedure there was one deviation from the scheme presented in the Table 1. It consisted in the rinsing of the examined soil samples with water prior to their extraction with calcium chloride solution. The obtained water fraction was collected, combined with the $CaCl_2$ fraction and analysed as one.

The extracts were analysed for the lead content using Flame – and GF-AAS analysis.

Simultaneously the soil samples were examined for the total lead content after their digestion in 30% HF.

The obtained results were tabulated and presented in graphical form as a percentage of total lead content in each fraction for each genetic horizon of the examined soils.

Finally for each speciation procedure the correlation between some basic soil properties and the lead content in each fraction was determined. This was performed using the least squares method. The results of these calculations are also presented in tabular form below.

Table 1. Metal fractions and fractionation procedures in McLaren & Crawford's and Tessier's methods.

	Tessier's method		McLaren & Crawford's method	
Fraction °	Metal form – name of the fraction	Extracting agent	Metal form – name of the fraction	Extracting agent
I	Water soluble and exchangeable	1 M $MgCl_2$	Water soluble and exchangeable	0.05 M $CaCl_2$
II	Bound to carbonates (acidic)	1 M CH_3COONa + CH_3COOH	Specifically adsorbed and exchangeable (acidic)	2.5% CH_3COOH
III	Bound to Fe and Mn oxides (reducible)	0.04 M $NH_2OH{\cdot}HCl$ in 25% CH_3COOH	Specifically bound to organic matter (oxidisable)	0.1 M $K_4P_2O_7$
IV	Specifically bound to organic matter (oxidisable)	0.02 M $HNO_3 + H_2O_2$ and 3.2 M CH_3COONH_4 in 20% HNO_3	Occluded in oxides (reducible)	1.0 M $C_2H_2O_4$ + 0.175 M $(NH_4)_2C_2O_4$
V	Residual	concentrated HCl and HNO_3 (3:1)	Residual	30% HF + concentrated $HClO_4$

Table 2. Basic properties of soils used in the study.

Profile No.	Genetic horizon	Depth cm	C org. %	Colloidal clay	Fe-OX mg·kg^{-1} d. w.	Mn-OX mg·kg^{-1} d. w.	pH H_2O	pH KCl	T CEC cmol(+)/kg
Deluvial humous soil *Gleyic Phaeozems*									
1	Ad	0–3	2.21	9.0	1326.2	132.1	6.50	6.15	20.40
	A	4–15	2.36	10.0	1186.5	131.6	6.75	6.45	19.16
	ACgg	16–68	1.20	9.0	1238.6	166.4	6.60	6.20	15.35
	Cgg	69–100	0.51	11.0	1170.6	98.4	6.90	6.40	14.60
2	Ad	0–3	2.26	9.0	1381.3	145.2	6.80	6.40	21.50
	A	4–46	1.88	8.0	1460.6	105.2	6.85	6.40	18.32
	Cgg	47–98	0.62	9.0	1783.3	120.6	6.90	6.50	12.46
3	Ad	0–3	2.64	8.0	1366.4	90.3	6.65	6.20	24.12
	A	4–27	2.33	10.0	1429.4	91.5	6.80	6.40	21.76
	AC	28–64	0.82	10.0	1356.5	79.9	6.70	6.35	16.15
	Cgg	65–100	0.40	11.0	1146.4	278.6	6.80	6.35	12.56
4	Ad	0–3	2.48	11.0	1421.1	128.6	6.50	6.00	20.62
	AC	4–58	2.10	7.0	1444.3	126.1	6.70	6.20	18.44
	Cgg	59–130	0.74	10.0	1232.6	163.5	7.10	6.50	13.16
Deluvial brown soil Eutric*Cambisols*									
5	Ad	0–3	1.41	10.0	1092.2	94.9	6.70	6.10	13.27
	A	4–20	1.18	8.0	1110.4	114.0	6.90	6.40	10.86
	Bbr	21–48	0.42	18.0	937.6	86.8	6.95	6.50	10.30
	C	49–120	0.34	11.0	855.2	87.7	6.90	6.55	8.14
6	Ad	0–3	1.34	9.0	1134.3	98.2	6.50	6.00	10.63
	A	4–16	1.20	9.0	1215.4	92.4	6.60	6.15	10.24
	Bbr	17–45	0.38	21.0	1084.2	83.3	6.65	6.30	8.80
	C	46–100	0.27	11.0	967.5	81.5	6.80	6.40	7.35

3 RESULTS AND DISCUSSION

All the results are summarised in Tables 2–5. Table 2 presents the results of the analysis of examined soils properties. The results of the speciation analysis are presented in Tables 3 and 4, in Table 3 for McLaren & Crawford's speciation method and in Table 4 for Tessier's procedure. Table 5 contains the results of correlation calculations for both procedures. The results of the speciation analysis for both procedures and both soil types are additionally presented on Figures 1–4. They are displayed in the form of bar graphs showing the distribution of lead in each fraction for the given horizon.

Table 3. Lead fractions determined by McLaren and Crawford's method in deluvial soils (mg·kg^{-1} d. w.). Mean for n = 6 ± sd.

Profile No.	Genetic horizon	Depth cm	Fractions I	II	III	IV	V	Σ FI–FV	Total Pb
Deluvial humous soil *Gleyic Phaeozems*									
1	Ad	0–3	6.6 ± 0.9	38.5 ± 4.6	256.9 ± 21.3	37.2 ± 4.9	114.6 ± 12.4	**453.8**	**386.4 ± 49.9**
	A	4–15	8.1 ± 1.4	7.1 ± 1.8	23.4 ± 2.8	7.3 ± 1.3	45.5 ± 5.4	**91.4**	**109.2 ± 12.4**
	ACgg	16–68	8.6 ± 2.0	5.3 ± 1.7	10.8 ± 2.1	6.0 ± 1.4	16.1 ± 2.3	**46.8**	**48.3 ± 6.2**
	Cgg	69–100	7.0 ± 1.3	4.5 ± 0.9	5.0 ± 1.2	8.9 ± 2.0	9.5 ± 2.1	**34.9**	**32.7 ± 5.1**
2	Ad	0–3	8.2 ± 2.4	10.2 ± 2.6	68.9 ± 5.4	34.0 ± 5.2	130.1 ± 15.6	**251.4**	**270.4 ± 34.1**
	A	4–46	7.9 ± 1.9	12.4 ± 2.9	127.4 ± 10.6	33.1 ± 4.7	115.1 ± 9.4	**295.9**	**286.2 ± 33.7**
	Cgg	47–98	6.0 ± 1.6	2.8 ± 0.6	5.4 ± 1.7	12.4 ± 2.8	42.6 ± 5.1	**69.2**	**49.9 ± 5.3**
3	Ad	0–3	5.8 ± 1.4	16.1 ± 3.1	183.6 ± 17.4	36.4 ± 4.4	123.9 ± 14.2	**365.8**	**352.4 ± 44.3**
	A	4–27	5.8 ± 2.1	11.8 ± 2.6	173.8 ± 14.3	34.0 ± 3.9	105.3 ± 9.7	**330.7**	**293.3 ± 36.2**
	AC	28–64	7.1 ± 2.4	2.4 ± 0.6	15.0 ± 2.9	12.6 ± 2.8	25.7 ± 3.1	**62.8**	**56.2 ± 4.9**
	Cgg	65–100	6.9 ± 1.8	1.9 ± 0.2	5.3 ± 1.1	9.6 ± 1.3	12.4 ± 2.6	**36.1**	**40.3 ± 4.2**
4	Ad	0–3	7.1 ± 2.3	18.6 ± 3.6	178.1 ± 20.4	56.4 ± 6.2	172.3 ± 16.7	**432.5**	**396.1 ± 46.3**
	AC	4–58	6.4 ± 1.8	2.6 ± 0.4	8.8 ± 2.6	11.9 ± 2.9	36.3 ± 5.4	**66.0**	**60.2 ± 7.7**
	Cgg	59–130	5.4 ± 1.9	1.8 ± 0.5	2.5 ± 0.4	8.3 ± 2.0	7.5 ± 2.5	**25.5**	**28.3 ± 3.4**
Deluvial brown soil Eutric *Cambisols*									
5	Ad	0–3	6.0 ± 1.8	25.3 ± 2.9	104.5 ± 9.6	48.5 ± 5.3	65.4 ± 6.1	**249.7**	**214.7 ± 32.2**
	A	4–20	6.9 ± 2.1	14.2 ± 2.4	92.9 ± 8.2	40.4 ± 3.9	38.3 ± 4.3	**192.7**	**193.2 ± 21.7**
	Bbr	21–48	6.1 ± 1.4	2.4 ± 0.8	15.3 ± 3.0	11.5 ± 2.1	13.9 ± 2.4	**49.2**	**52.6 ± 5.8**
	C	49–120	4.2 ± 1.1	2.4 ± 0.4	7.0 ± 2.1	10.3 ± 2.0	13.2 ± 2.1	**37.1**	**40.9 ± 5.1**
6	Ad	0–3	4.0 ± 0.8	18.4 ± 3.1	146.1 ± 11.1	42.6 ± 5.6	58.4 ± 6.7	**269.5**	**260.4 ± 33.7**
	A	4–16	4.6 ± 1.1	11.7 ± 2.0	87.5 ± 9.4	38.1 ± 4.2	40.1 ± 5.2	**182.0**	**137.2 ± 14.4**
	Bbr	17–45	4.2 ± 1.1	2.6 ± 0.4	18.2 ± 3.2	9.6 ± 2.7	10.7 ± 2.6	**45.3**	**40.6 ± 5.8**
	C	46–100	2.6 ± 0.9	2.5 ± 0.4	6.3 ± 1.9	9.2 ± 2.0	8.8 ± 1.1	**29.4**	**20.4 ± 3.1**

As the results of the soil analyses presented in the Table 2 show, both types of soil used in the present study may be characterised as follows: they are neutral (pH = 6.0–6.5), containing less than 2% of organic carbon and about 10% of colloidal clay, with CEC ranging from ~7 to ~20 cmol (+)/kg.

All these soil properties will influence the results of the speciation of lead in the examined soils. A comparison of the results shows in both cases the lack of significant differences between the total content of lead determined in a single-step extraction and the sum of its content in all fractions. Thus it was assumed that the latter values can be taken for the further considerations. This fact confirms also the correctness of the performed speciation analysis.

As might be expected the total content of lead within the examined soil profiles decreased with depth. The explanation for this is the accumulation occurring in the topmost horizon. More interesting are the differences in the total content of lead in the same profiles and genetic horizons obtained using these two procedures. It is 5–40% higher in the case of McLaren & Crawford's procedure. A possible explanation of this fact is the difference in strength of the extractants used in the determination of the total content of the lead in analysed soil samples. The mixture of HF and $HClO_4$ of McLaren & Craword's procedure is much stronger than the *aqua regia* used in Tessier's procedure. On the other hand, the results of the total lead content obtained for the same genetic horizons of different profiles were much more uniform when Tessier's procedure was applied. This tendency is much better demonstrated in the case of deluvial humous soil, which has generally higher contents of organic carbon, colloidal clay, iron and manganese oxides, and subsequently higher CEC value. This, itself, indicates the dependency of obtained results on the soil parameters' and subsequently the appropriateness of the choice of fractionation procedure.

Examining the results of the speciation of lead using McLaren & Crawford's and Tessier's procedure in detail it can be generally stated that the contents of lead in each fraction were greater in the case of McLaren & Crawford's procedure, although the uniformity of the results was lower than for the Tessier procedure. The only exception was the reducible fraction i.e. that bound to the Fe and Mn oxides, which is determined in the third step in Tessier's procedure, while in case of McLaren & Crawford's procedure it is step four of the extraction. The reasons for this change of order as well as in the results will be explained later.

The content of lead in the fraction FI is generally higher for McLaren & Crawford's procedure. As this can be not explained by the difference in the ionic

Table 4. Lead fractions determined by Tessier's method in deluvial soils ($mg \cdot kg^{-1}$ d. w.). Mean for $n = 6 \pm sd$.

Profile No.	Genetic horizon	Depth cm	Fractions F I	F II	F III	F IV	F V	**Σ FI–FV**	**Total Pb**
Deluvial humous soil *Gleyic Phaeozems*									
1	Ad	0–3	8.5 ± 1.1	36.4 ± 3.9	200.7 ± 24.0	40.7 ± 5.4	50.2 ± 6.9	**336.5**	**386.4 ± 49.9**
	A	4–15	6.6 ± 0.9	29.7 ± 4.2	40.6 ± 4.8	8.2 ± 1.2	11.4 ± 2.1	**96.5**	**109.2 ± 12.4**
	ACgg	16–68	1.0 ± 0.1	2.8 ± 0.3	26.7 ± 3.4	2.3 ± 0.4	6.0 ± 0.9	**38.1**	**48.3 ± 6.2**
	Cgg	69–100	1.0 ± 0.1	2.9 ± 0.4	3.7 ± 0.4	2.2 ± 0.1	14.7 ± 2.0	**24.5**	**32.7 ± 5.1**
2	Ad	0–3	4.5 ± 0.6	13.2 ± 2.0	121.2 ± 14.0	75.2 ± 9.1	23.5 ± 2.7	**237.6**	**270.4 ± 34.1**
	A	4–46	4.2 ± 0.6	10.9 ± 2.1	130.7 ± 15.2	35.8 ± 4.6	57.1 ± 6.2	**238.7**	**286.2 ± 33.7**
	Cgg	47–98	1.0 ± 0.1	1.9 ± 0.3	8.5 ± 1.0	3.2 ± 0.4	24.0 ± 3.1	**38.6**	**49.9 ± 5.3**
3	Ad	0–3	3.3 ± 0.4	22.6 ± 4.0	120.3 ± 14.1	114.7 ± 12.1	78.3 ± 9.1	**339.2**	**352.4 ± 44.3**
	A	4–27	4.5 ± 0.4	17.2 ± 3.7	172.0 ± 20.1	35.9 ± 4.2	45.2 ± 5.1	**274.8**	**293.3 ± 36.2**
	AC	28–64	0.8 ± 0.1	6.4 ± 0.6	14.4 ± 2.1	2.4 ± 0.2	19.4 ± 2.7	**43.4**	**56.2 ± 4.9**
	Cgg	65–100	1.0 ± 0.1	3.8 ± 0.3	14.2 ± 2.7	1.0 ± 0.1	15.2 ± 2.1	**35.2**	**40.3 ± 4.2**
4	Ad	0–3	9.1 ± 1.6	38.9 ± 4.7	80.4 ± 8.1	90.7 ± 10.2	90.9 ± 10.6	**310.0**	**396.1 ± 46.3**
	AC	4–58	3.6 ± 0.4	7.8 ± 0.9	12.7 ± 1.6	8.4 ± 0.9	24.0 ± 3.1	**56.5**	**60.2 ± 7.7**
	Cgg	59–130	1.1 ± 0.1	3.2 ± 0.3	8.1 ± 1.0	1.6 ± 0.3	7.2 ± 1.1	**21.2**	**28.3 ± 3.4**
Deluvial brown soil Eutric *Cambisols*									
5	Ad	0–3	6.0 ± 0.9	10.4 ± 1.8	130.8 ± 14.4	29.2 ± 3.1	29.6 ± 3.6	**206.0**	**214.7 ± 32.2**
	A	4–20	6.4 ± 1.4	10.1 ± 1.9	109.8 ± 12.1	24.9 ± 3.8	24.0 ± 3.1	**175.2**	**193.2 ± 21.7**
	Bbr	21–48	1.8 ± 0.2	5.6 ± 0.9	12.4 ± 2.1	6.4 ± 2.1	19.3 ± 2.7	**45.5**	**52.6 ± 5.8**
	C	49–120	2.0 ± 0.2	5.5 ± 0.6	6.0 ± 0.8	4.5 ± 0.6	17.8 ± 2.1	**35.8**	**40.9 ± 5.1**
6	Ad	0–3	6.6 ± 1.1	19.2 ± 2.6	147.3 ± 16.2	34.2 ± 4.1	23.7 ± 3.6	**231.0**	**260.4 ± 33.7**
	A	4–16	7.2 ± 1.0	15.6 ± 2.1	20.9 ± 3.2	40.1 ± 6.0	20.4 ± 3.9	**104.2**	**137.2 ± 14.4**
	Bbr	17–45	4.2 ± 0.8	5.5 ± 0.9	13.5 ± 2.1	10.6 ± 2.5	7.2 ± 1.1	**41.0**	**40.6 ± 5.8**
	C	46–100	2.0 ± 0.3	4.1 ± 0.7	4.0 ± 0.7	6.0 ± 1.0	5.1 ± 0.9	**21.2**	**20.4 ± 3.1**

Table 5. Significant correlation between soil basic properties and lead fractions determined by McLaren & Crawford (M) and Tessier et al. (T) methods.

Soil property	Fractionation method	Fraction F I	F II	F III	F IV	F V	Total content
Colloidal clay content	M	−0.19	0.21	−0.12	−0.20	0.14	0.34(*)
	T	−0.20	0.48(*)	0.01	0.15	0.36(*)	
C_{org} content	M	0.07	0.21	0.69(**)	0.04	0.84(**)	0.48(*)
	T	0.24	0.01	0.03	0.51(*)	0.19	
T CEC	M	0.51(*)	0.13	−0.20	0.03	0.36(*)	0.18
	T	−0.07	−0.18	−0.17	−0.09	0.40(*)	
Fe-OX content	M	−0.19	0.12	0.01	−0.17	0.22	0.54(**)
	T	−0.27	−0.06	0.15	0.11	0.35(*)	

(*) $P = 0.05$; (**) $P = 0.01$;

strength of the extractants used (it is well known that magnesium chloride solution has higher ionic strength than the calcium chloride solution, thus the amount of lead extracted by the former extractant should be greater than for the later) this phenomenon is possibly due to the fact that the first step in case of McLaren & Crawford's procedure actually consists of two steps (extraction with water (a very brief stage), immediately followed by extraction with calcium chloride). This may also result in the equilibrium subsequently reflected in the other stages. That, in turn, might led to some unpredictability of the results in McLaren & Crawford's procedure. In some cases there was an increase of the concentration of lead in the fraction FI down the soil profile – a feature not generally observed in Tessier's extraction, where there was a general decrease of lead content with the depth.

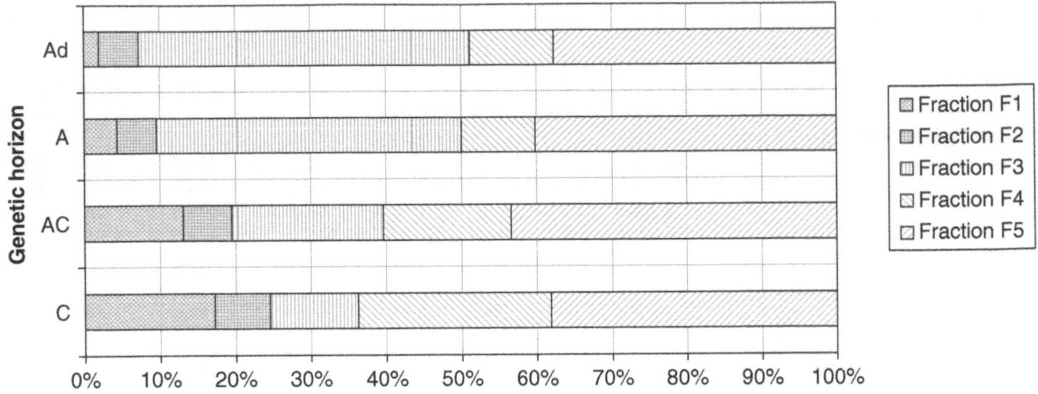

Figure 1. Percentage share of lead fraction determined by McLaren & Crawford's method in genetic horizons of Gleyic Phaozems.

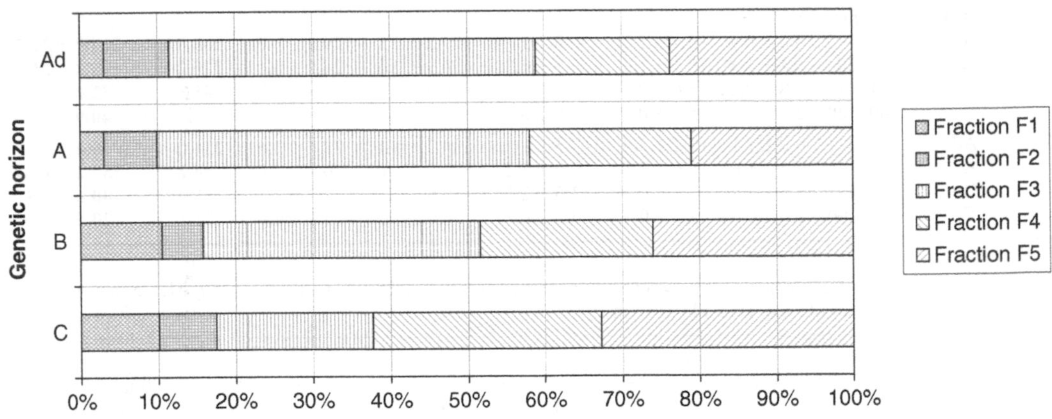

Figure 2. Percentage share of lead fraction determined by McLaren & Crawford's method in genetic horizons of Eutric Cambisols.

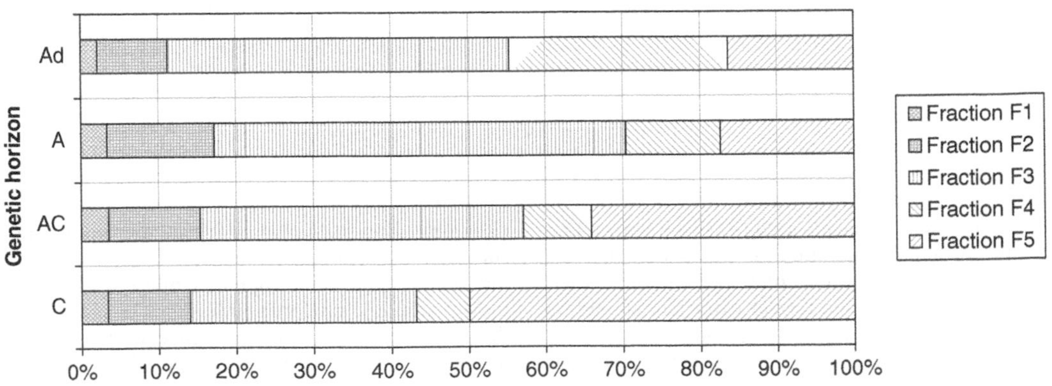

Figure 3. Percentage share of lead fraction determined by Tessier's method in genetic horizons of Gleyic Phaozems.

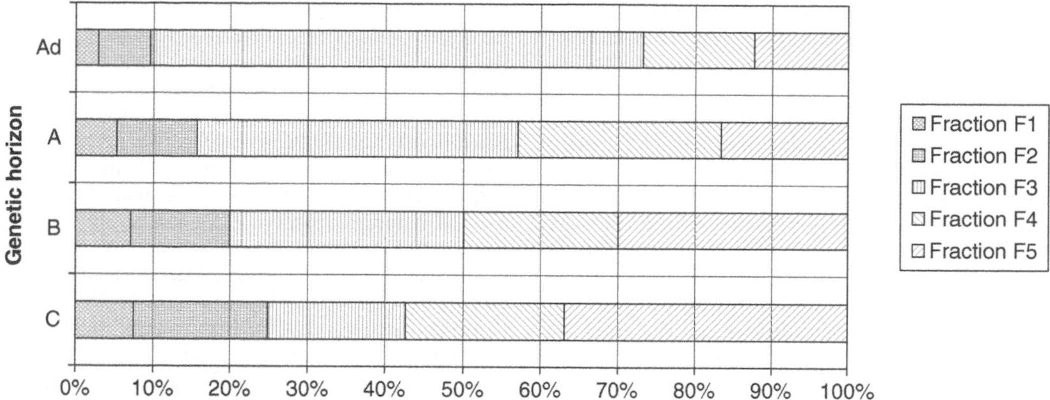

Figure 4. Percentage share of lead fraction determined by Tessier's method in genetic horizons of Eutric Cambisols.

Both extraction procedures show significant content of lead in the second, acidosoluble fraction. A possible explanation of this fact is that the extracting agents used here are able not only to desorb the metal ions linked to the carbonates, but also other lead minerals which might be present, like lead oxides, sulphates, other minerals or their mixtures. These mineral forms of lead are reported to be quite abundant on the areas where the pollution of soils by lead results from smelting activity. Therefore, they may attribute to the lead extracted in the second step in deeper genetic horizons. This claim may be further supported when some basic properties of the examined soils, especially pH and Fe-Mn oxides content, are considered.

Fraction II together with fraction I constitute 8–20% of the total pool of this element in examined soil samples. In both procedures these values are of comparable magnitude, although Tessier's extraction procedure gives slightly higher values compared to McLaren & Crawford's method, which may be again be attributed to the extractant's strength. The forms determined in these two steps being much easily mobilisable, in comparison to the forms determined in the remaining steps, and thus bioavailable, they can pose a significant risk to the organisms living in the examined area as well as to the surface and groundwater.

Quite high levels of lead were also determined in the oxidisable fractions, also called "specifically bound to organic matter" (FIII for McLaren & Crawford and FIV for Tessier). The content of lead in this fraction, although decreased with depth, was still significant even in the deeper layers. It can be explained by the contribution of sulphides, that were in this step oxidised simultaneously with the organic matter. Consequently the amount of lead extracted in this step was quite high down the examined horizons. This was despite the fact that, as the theory says and what prove the results given in the Table 2, the content of the organic matterdecreases with the depth. The contribution of the sulphides is quite easy explicable by taking into account the fact that in industrially polluted areas, especially those impacted by metal mining and smelting activities, the pollution of soils by sulphur is quite common.

In the case of Tessier's procedure the extracting agent used oxidises not only the organic substances, to which the metal ions are bound, but also the sulphides present in the sample. This seems not to be the case of McLaren & Crawford's procedure, because the pyrophosphate ions only promote the dispersion of the organic colloids, but in very basic solutions (pH = 10) amorphous oxides are also extracted. This may be also the case in this study and may partly explain the results obtained in the next – reducible fraction.

In case of the reducible fraction i.e. that bound to Fe and Mn oxides, the amount of lead extracted using Tessier's procedure was greater than in the case of McLaren & Crawford's. The explanation for this phenomenon is the partial dissolution of the amorphous oxides in the previous step of McLaren & Crawford's procedure. This does not occur in Tessier's procedure, because this step is prior to the extraction of lead from oxidisable fraction and thus preliminary steps do not impact Fe and Mn oxides. Moreover, resulting from the extracting agent's composition, there is no readsorption of lead on the oxides. This also proves that, in case of the examined soils, Tessier's procedure is a better choice because it better reflects this soil property of both types of examined soils.

Finally, some attention must be given to the residual fraction. As mentioned previously, the extractants used in this step in McLaren & Crawford's procedure are very strong, much stronger than those used in Tessier's procedure, and are able to dissolve all compounds present in soil, including silicates. Also in this fractionation procedure sulphides remained almost totally untouched in previous stages. Now they are practically dissolved, together with other practically insoluble lead forms. All this contributes to the higher content of lead in the residual fraction as well as to some uniformity of the results, not observed in case of Tessier's procedure, where the share of the residual fraction increases with depth. Another reason for the surprising uniformity of the lead content in the residual fraction despite the depth may be the strength of the extracting agent. This mixture is commonly used to dissolve very resistant materials (for example rocks), but many authors have reported that it gives quite unpredictable results in the case of soils, especially if the extraction time regimes are not strictly observed. Thus, it is generally not recommended for soil analyses. The comparison and evaluation of the obtained results, as presented above, clearly indicates that of two tested fractionation procedures the Tessier method gives a more accurate picture of the quantitative distribution of lead forms in examined soils. Hence, this method can be considered as more suitable for the quantification of lead forms in industrially polluted deluvial soils.

Further support for this statement is provided in Table 5, where the results of the assessment of correlation between lead content in each fraction and some basic properties of examined soils are presented. The results presented above clearly indicate that Tessier's procedure better reflects the relationship between examined soils properties and lead content in each fraction. It also gives more reliable results in terms of the theoretical bases of the speciation than it is in case of McLaren & Crawford's procedure. Thus, the former may be generally considered more appropriate for determination of lead forms in industrially polluted soils having similar characteristics to those examined in the present study. Consequently, it can be regarded as the better analytical tool for the assessment of the risk posed by lead in metal polluted soils.

4 CONCLUSIONS

Comparison of the results of lead fractionation in two industrially polluted deluvial soils using Tessier's and McLaren & Crawford's fractionation procedures indicate the generally greater applicability of the Tessier procedure. This statement is confirmed mainly by the results of the fractionation, their greater uniformity and better fitting to the theoretical bases of the fractionation scheme. This is demonstrated most clearly for the residual fraction, where, in the case of the Tessier procedure, the content of lead increases with depth from about 20 to about 40% for both types of soils while in McLaren & Crawford's procedure it remains at a similar level of 25–30% for Eutric Cambisols and ~40% for Glyeic Phaozems. This lack of agreement between the theoretical bases and obtained results can be stated in case of McLaren & Crawford's procedure also for the other extraction steps. Also the statistical evaluation of the results, i.e. the determination of the correlation between lead content in each of the fractions and some basic soil parameters, confirms this statement. In the case of Tessier's procedure the existence of medium to strong correlation (about 0.5 and higher) was observed between lead content in the acidosoluble fraction (FII) and colloidal clay content (0.48) as well as the content of this element in reducible fraction (FIV) and organic carbon content (0.51). Some correlation was also observed between lead content in residual fraction and colloidal clay content (0.36), CEC (0.40) and FeOX content (0.35) in the examined soils. In the case of McLaren & Crawford's procedure a medium correlation was found only between lead content in FI and CEC (0.51) and lead content in FIII and C_{org} content (0.69). The existence of the very strong correlation between lead content in residual fraction (FV) and the C_{org} content, seen for the McLaren & Crawford's procedure, further supports the thesis of the inadequacy of this speciation method for the quantification of lead forms in industrially polluted deluvial soils.

Therefore, the Tessier's fractionation procedure should be recommended as the more appropriate analytical tool in risk assessment for lead as a pollutant in soils having similar characteristics as those employed in this study.

REFERENCES

Das A.K., Chakraborty R., Cervera M.L., de la Guardia M. 1995. Metal speciation in soil matrices, *Talanta*, 42, 1007–1030.

Gleyzes C., Tellier S., Astruc M. 2002. Fractionation studies of trace elements in contaminated soils and sediments: a review of sequential extraction procedures, *Trends in Analytical Chemistry.* 21 (6 + 7), 451–467.

Gomez Ariza J.L., Giraldez I., Sanchez-Rodas D., Morales E. 2000. Comparison of the feasibility of three extraction procedures for trace metal partitioning in sediments from south-west Spain, The *Science of the Total Environment*, 246, 271–283.

Gworek B., Mocek A. 2003 Comparison of sequential extraction methods with reference to zinc fractions in contaminated soils, *Polish Journal of Environmental Studies*, 12(1), 41–48.

McLaren R.G., Crawford D.V. 1973. Studies on soil copper. 1. The fractionation of Cu in soils, *Journal of Soil Science*, 24, 443.

Mehra O.P., Jackson M. L. 1960. Iron oxide removal from soil clays by dithionite-citrate system buffered with sodium dicarbonate, *Clays and Clay Minerals*, 5, 317–327.

Mocek A., Drzymała S., Maszner P. 1997 *Genesis, analysis and classification of soils* Poznań, Poznań Agricultural University Press (in Polish).

Mocek A., Kociałkowski Z. W., Mocek I. 1991. The fractions of lead in deluvial soils from sanitary zone of "Legnica" lead smelter, *Roczniki Akademii Rolniczej w Poznaniu*, **CCXXVI**, 123–130 (in Polish).

Ponizovsky A., Mironenko E. 2001. Speciation and Sorption of Lead(II) in Soils In Iskandar I. K., Kirkham M. B., (eds). *Trace elements in soil: bioavailability, flux and transfer*, Boca Raton (Fla.): CRC Press LCC.

Rubio R., Rauret G. 1996. Validation of the methods for heavy metal speciation in soils and sediments, *Journal of Radioanalytical and Nuclear Chemistry,* 208 (2), 529–540.

Tessier A., Campbell P.G.C., Bisson M. 1979. Sequential extraction procedure for the speciation of particular trace metals, *Analytical Chemistry.*, 51, 844.

Ure A.M. (1996). Single extraction schemes for soil analysis and related applications, *The Science of the Total Environment*, 178, 3–10.

Environmental Engineering – Pawłowski, Dudzińska & Pawłowski (eds)
© 2007 Taylor & Francis Group, London, ISBN13 978-0-415-40818-9

Temperature effect on the oxidation of 1,3-dichloro-2-propanol with Pt and Rh monolith

Andrzej Żarczyński, Zbigniew Gorzka, Marcin Zaborowski & Marek Kaźmierczak
Institute of General and Ecological Chemistry, Technical University of Lodz, Lodz

Małgorzata Michniewicz
Pulp & Paper Research Institute, Lodz, Poland

ABSTRACT: Experiments were carried out on laboratory scale with continuous method using apparatus with tubular quartz reactor containing monolithic catalyst and inserted in an electric furnace. Aqueous solution of 1,3-dichloro-2-propanol (DCHP) at concentration of 2.5 g/l was applied in experiments. The monolithic catalyst with cordierite support containing platinum (0.09%) and rhodium (0.04%) deposited at intermediate layer with γ-Al_2O_3 was used in the experiments. Contact time was 0.36 s and the temperature was ranging from 300 to 600°C. The aim of the investigation was determination of optimum parameters of total DCHP oxidation at the catalyst.

Almost total decomposition of DCHP was achieved in the presence of the catalyst and the temperature of 375°C. The possibility of a decrease in the treatment temperature (from 1200–1500°C to at least 450°C) of waste chloroorganic derivatives of propane with the structure similar to the structure of chlorohydrins, was found due to the application of the catalyst.

Keywords: Thermocatalytic oxidation, 1,3-dichloro-2-propanol, dioxins emission, monolithic catalysts.

1 INTRODUCTION

Industrial processes, including the synthesis of propane derivatives, e.g. propylene oxide and epichlorohydrin, result in liquid chloroorganic wastes which contain among other compounds 1,2-dichloropropane, propylene chlorohydrin, 2,2′-dichlorodiisopropyl ether and 1,2,3-trichloropropane (Milchert et al. 1996). In Poland, these wastes are incinerated in the temperature range from 1200 to 1500°C and with contact times of 2–3 s. The high temperature and relatively long time of the reaction prevents the formation and emission of polychlorinated dibenzo-p-dioxins and dibenzo-furanes (PCDD/Fs) to the environment (Przondo & Rogala 1996, Tuppurainen et al. 2003). Determination of optimum parameters in the thermal utilisation of waste chlorine compounds is very important due to the toxic properties of PCDD/Fs (Fiedler 1996, Grochowalski & Wybraniec 1996, Dudzińska & Kozak 2001, Everaert & Baeyens 2003, Tuppurainen et al. 2003, Finocchio et al. 2006).

Thermocatalytic oxidation carried out in the significantly lower temperature range (300–600°C) can be an alternative of the above mentioned treatment of liquid and gaseous chloroorganic wastes. The oxidation of organic compounds with various chemical structures can be carried out using catalysts with granular structure (Spivey 1987, Spivey & Butt 1992, Petrosius et al. 1993, Windawi & Wyatt 1993, Park et al. 2004, Lichtenberg & Amiridis 2004, Kaźmierczak et al. 2003, Kaźmierczak et al. in press). The applicability of monolith catalysts has been widespread over recent years. These catalysts are commonly used in automotive vehicles and often in industry (Boos et al. 1992, Cybulski & Moulijn 1994, Nijhuis et al. 2001, Musialik-Piotrowska & Syczewska 2002, Koyer-Gołkowska et al. 2004, Kaźmierczak et al. in press) due to their low resistance to gas flow and lack of sensitivity to dust in comparison with granular catalysts.

The aim of the investigation was the determination of optimum parameters for the total oxidation of 1,3-dichloro-2-propanol (DCHP) using a catalyst. Previous work has proved the possibility of complete degradation of chlorohydrin propylene molecules at 300°C and oxidation of 95–99 high percentage of intermediate products in the temperature range from 300 to 450°C depending on the type of the granular catalyst used (Kaźmierczak et al. 2003, Kaźmierczak et al. in press).

Table 1. Characteristics of tested monolithic catalyst.

Catalyst parameter	Platinum-rhodium catalyst
Carrier	ceramic, cordierite
Intermediate layer	γ-Al_2O_3
Active components	0.09% Pt and 0.04% Rh in the catalyst mass
Shape of cells, side length [mm]	square, 1.0
Cell density [cell/cm^2]	62
Wall thickness [mm]	0.1–0.2
Size of a catalyst:	
diameter [mm],	21
length [mm]	76
Producer	JMJ Manufacturer of Catalysts, Puchalski and Krawczyk Company, Kalisz

2 MATERIALS AND METHODS

Experiments were carried out on the laboratory scale with a continuous method. The main part of the apparatus consisted of a tubular quartz reactor containing a monolithic catalyst. The reactor was placed inside an electrically heated furnace. The catalyst at cordierite carrier, containing platinum and rhodium spread over an intermediate layer composed of γ-Al_2O_3, was applied in the experiments. The catalyst specifications are presented in Table 1. The contact time was 0.36 s. An aqueous solution of DCHP (2.5 g/l) was the subject of the investigation. It was supplied to an evaporator with the flow rate of 32.5 g/h. Simultaneously, air was pumped to the evaporator with the flow rate of 215 dm^3/h. The oxidation of the reaction mixture was carried out in the temperature range from 300 to 600°C.

Cooled combustion gases and aqueous condensate were the reaction products. A concentration of chloride ions in the condensate was determined. Products arising from the complete oxidation of the substrate, i.e. formaldehyde, chlorine, carbon monoxide, total organic carbon (TOC) and PCDD/Fs, were also analysed.

Samples of cooled combustion gases were analysed in order to determine the content of PCDD/Fs formed during the reaction carried out at 375 and 450°C. Analyses were made according to current analytical standards (European Standard EN-1948) in the Environmental Protection Laboratory in Pulp & Paper Research Institute of Łódź.

3 RESULTS AND DISCUSSION

Results of the experiments are presented in Figures 1-7 and Table 2.

The dependence of DCHP conversion expressed as loss in total organic carbon (X_{TOC}) in liquid products

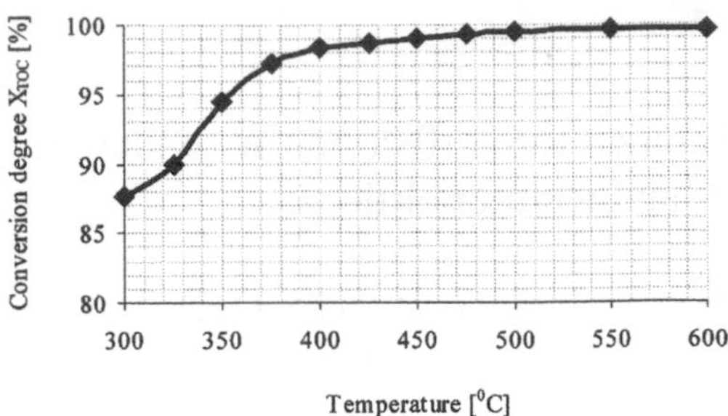

Figure 1. DCHP oxidation degree dependence on reaction temperature.

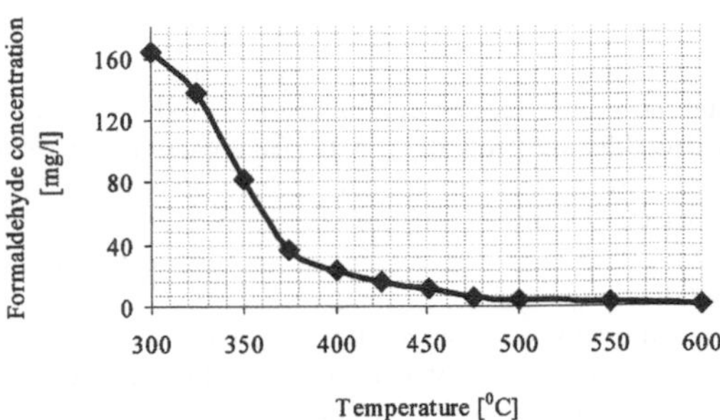

Figure 2. A dependence of formaldehyde concentration in a condensate on temperature of DCHP oxidation.

compared to the TOC value in the substrate solution (710 mgC/l), with a reaction temperature in the range from 300 to 600°C is presented in Figure 1. Oxidation of carbon in the substrate proceeded in 87.5% in the temperature of 300°C with application of the catalyst. In order to achieve a value of $X_{TOC} > 99\%$, it was necessary to apply a temperature of 500°C or higher. Carbon dioxide, water vapour and hydrochloric acid were the final products of the reaction. Intermediate products were as follows: formaldehyde, chlorine and carbon monoxide.

Figure 2 presents results of the analysis of formaldehyde in the condensate obtained in the catalysed process. The content of formaldehyde decreased with an increase in the temperature, from the average value of 163.7 mg/l (300°C) to 2.61 mg/l (600°C).

Combustion of chloroorganic compounds always causes formation of hydrogen chloride and often gaseous chlorine. Hydrogen chlorine separated from the combustion gases is obtained as dilute hydrochloric acid in the condensate. The effect of the temperature during the substrate oxidation on the concentration of chloride ions in the condensate is presented in Figure 3. The concentration of chloride ions in the tested temperature range was on the average 1500 mg/l.

The effect of temperature in DCHP oxidation on the concentration of chlorine in combustion gases is

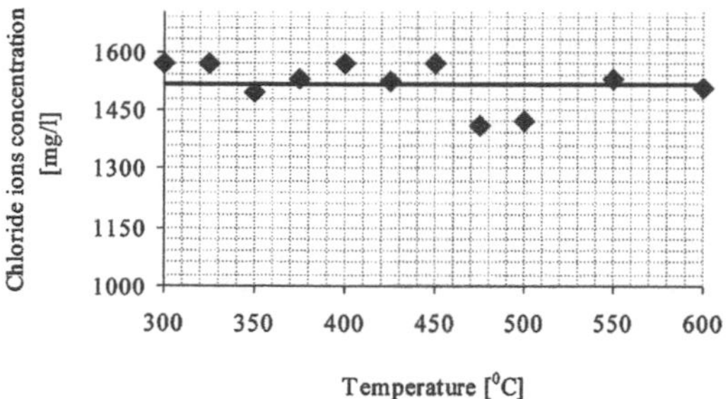

Figure 3. A dependence of chloride ions concentration in a condensate on temperature of DCHP oxidation.

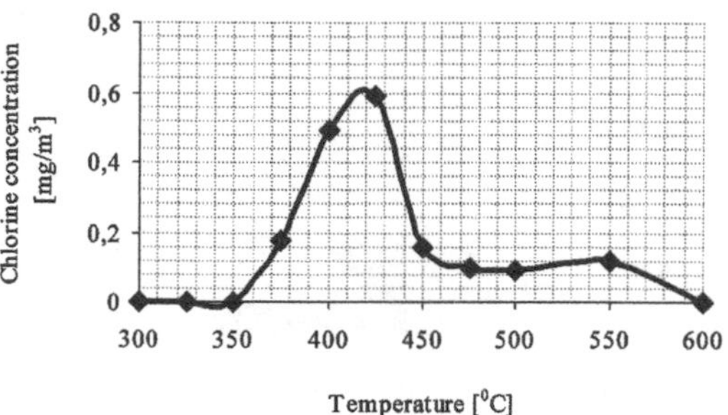

Figure 4. A dependence of chlorine concentration in combustion gases on temperature of DCHP oxidation.

presented in Figure 4. Chlorine was present in combustion gases in the temperature range from 350 to 600°C with the maximum concentration of 0.59 mg/m^3 in the temperature of 425°C. Chlorine is a reactive product of dechloration and oxidation of chloroorganic compounds. It can be also formed in the Deacon reaction as a result of hydrogen chloride oxidation:

$$2HCl + 0.5\ O_2 \leftrightarrow Cl_2 + H_2O + 57.18\ kJ$$

The value of the reaction equilibrium constant decreases with increasing temperature – the higher the temperature, the equilibrium constant is shifted more to the left side. The presence of water vapour in the reaction zone significantly reduces formation of chlorine. The concentration of chlorine in the reaction zone is negligible and does not cause deactivation of the tested catalyst.

Figure 5 presents the results of formaldehyde analysis in the combustion gases in the temperature range from 300 to 600°C using the catalyst. The formaldehyde concentration decreased from 2.88 mg/m^3 at 325°C to 0 mg/m^3 at 600°C.

Carbon monoxide is the main product of incomplete oxidation of organic compounds during high-temperature combustion as well under catalytic combustion. Its concentration decreased from 6 to 1.5 ppm

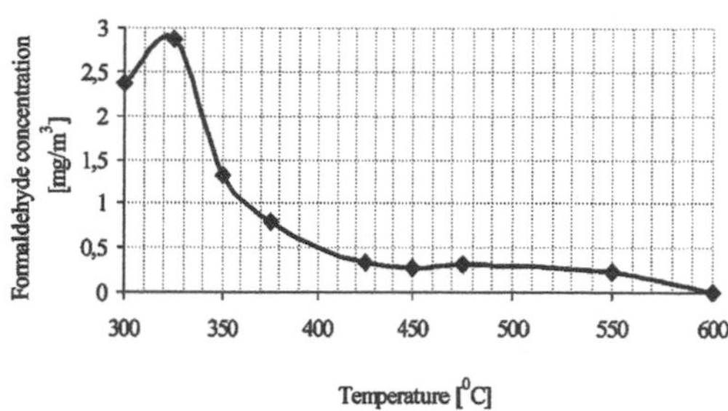

Figure 5. Dependence of formaldehyde concentration in combustion gases on temperature of DCHP oxidation.

in the combustion gases obtained during catalytic DCHP oxidation in the range from 300 to 375°C.

Table 2 presents concentrations of final and intermediate products in the condensate and combustion gases from DCHP oxidation at 375 and 450°C. Under these conditions the substrate was 96% (375°C) and 98% (450°C), oxidised. Concentrations of chloride ions in the condensate were about 1500 mg/l after the stabilisation of catalyst work. The value of TOC was about 26.48 mgC/l at 375°C and 15.45 mgC/l at 450°C. The concentration of formaldehyde in the condensate was on the average 37.74 mg/l at 375°C and 6.96 mg/l at 450°C. The concentration of chlorine in the combustion gases obtained in these two temperatures was 0.19 and 0.31 mg/m^3, respectively.

The presence of PCDD/Fs congeners was found in the combustion gases from the process of DCHP oxidation (Figs. 6 and 7). The combustion gases from the reaction carried out at temperatures of 375 and 450°C contained six and four PCDD/Fs congeners, respectively. These congeners had a chlorine atom connected with a carbon atom in 2,3,7,8 position.

The concentrations of 2,3,7,8-TCDF and 2,3,4,7,8-P_5CDF were especially high −0.0115 and 0.011 ng TEQ/m^3 (375°C), respectively. These congeners have significant toxicity equivalent factors (TEF_i), currently accepted as 0.1 and 0.5. It was found that the toxicity equivalent in a sample of combustion gases was 0.0289 ngTEQ/m^3 at 375°C and 0.01203 ngTEQ/m^3 at 450°C. These values did not exceed the standards for incineration plants in Poland and Europe. The standard value is 0.1 ngTEQ/m^3 (European Parliament and the Council Directive 2000). An increase in the temperature from 375 to 450°C caused more than a two-fold decrease in the toxicity equivalent in a sample of the combustion gases and a decrease in the number and types of detected PCDD/Fs congeners.

Accumulation of carbon deposit on the surface of the tested monolithic catalyst was not found, although experiments were carried out in the range of relatively low temperatures (300–600°C). This is connected with the presence of excess oxygen (from

Table 2. Average concentrations of intermediate and final products in the condensate and combustion gases and calculation of result in the process of DCHP oxidation with application of monolithic catalyst in the temperatures of 375 and 450°C.

Reaction products	Results of analyses and calculations	
	T = 375°C	T = 450°C
DCHP w combustion gases, mg/m³	0	0
TOC, mgC/l	26.48	15.45
Formaldehyde in a condensate, mg/l	37.74	6.96
Formaldehyde in combustion gases, mg/m³	0.91	0
Carbon monoxide, ppm	0–2	0
Chlorine, mg/m³	0.19	0.31
HCl in a condensate, mg/l	1400–1580	1400–1550
HCl in combustion gases, mg/m³	<5	<5
DCHP conversion, %	96.3	97.9

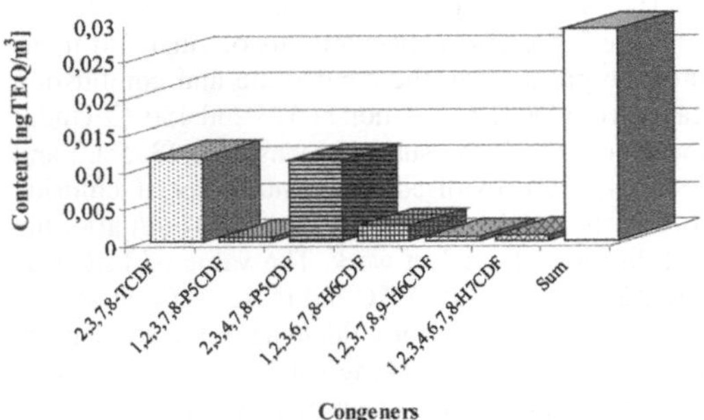

Figure 6. The content of dioxins in combustion gases from DCHP oxidation process with application of catalyst at 375°C.

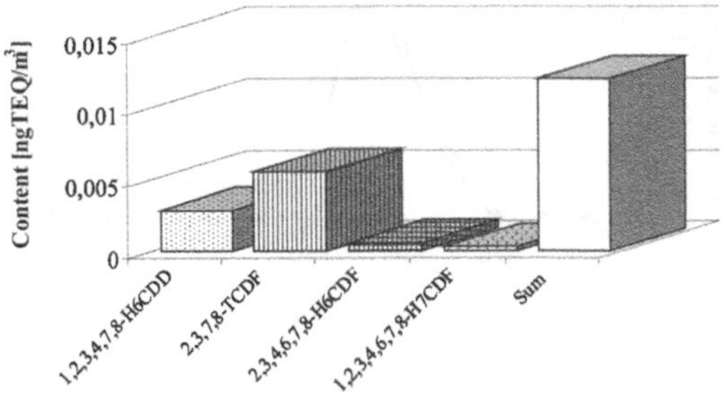

Figure 7. The content of dioxins in combustion gases from DCHP oxidation process with application of catalyst at 450°C.

air) and water vapour from evaporation of the oxidized solution. Catalyst activity did not decrease during one year of operation including also periodical oxidation of other chloroorganic compounds, e.g. 1,1,2,2-tetrachloroethane and propylene chlorohydrin (Żarczyński et al. 2005, Kaźmierczak et al. in press).

The results of the catalyst experiments show the possibility of threefold decrease in temperature during the treatment of waste chloroorganic derivatives of propane with the structure similar to the structure of chlorohydrins, in comparison with the high-temperature combustion (1200–1500°C).

Although the concentration of PCDD/Fs in the combustion gases was acceptable, the temperature of 375°C was too low because it did not ensure effective oxidation of intermediate products, e.g. formaldehyde. The experimental results prove that this temperature should be at least 450°C.

4 CONCLUSION

Almost total destruction of DCHP was achieved in the presence of monolithic catalyst in the temperature of 375°C. The final products of DCHP oxidation were as follows: carbon dioxide, water vapour and hydrochloric acid. Formaldehyde, chlorine, carbon monoxide and PCDD/Fs were the intermediate products. An increase in the temperature caused a decrease in the concentration of intermediate products, i.e. formaldehyde in the condensate, and formaldehyde and carbon monoxide in the combustion gases. The concentration of chloride ions in the tested temperature range was, on the average ca. 1500 mg/l. The catalyst was not deactivated in the experiments carried out during one year.

Increasing the temperature from 375 to 450°C significantly decreased the toxicity equivalent in the combustion gases and the number of detected PCDD/Fs congeners as well as a partial change in their nature.

The fundamental conclusion drawn from the experimental results concerns the possibility of decreasing the treatment temperature of waste chloroorganic derivatives of propane with the structure similar to the structure of chlorohydrins. This decrease, to at least 450°C, is due to the application of the monolithic platinum-rhodium catalyst in comparison to high-temperature combustion (1200–1500°C).

ACKNOWLEDGEMENTS

This work was supported by the Polish State Committee for Scientific Research under Grant No. 7 T09B 083 27.

REFERENCES

Boos R., Budin R., Harlt H., Stock M., Würst F., 1992. PCCD- and PCDF- destruction by a SCR-Unit in a municipal waste incinerator. *Chemosphere* 25(3): 375–382.

Cybulski A., Moulijn J.A., 1994. Monoliths in heterogenous catalysis. *Catal. Rev.-Sci. Eng.* 36: 179–270.

Dudzińska M.R., Kozak Z., 2001. *Polichlorowane dibenzo-p-dioksyny i dibenzofurany – właściwości i oddziaływanie na środowisko.* Monografia 6, Lublin: KIŚ PAN/Politechnika Lubelska.

European Parliament and the Council Directive 2000/76/EC of December 2000 on the incineration of waste. OJ No. 332, p. 91, 2000/12/28.

European Standard EN-1948, ICS 13.040.40. Stationary source emissions. Determination of the mass concentration of PCDD/Fs. Part 1. Sampling, Part 2. Extraction and clean-up, Part 3. Identification and Quantification.

Everaert K., Baeyens J., 2003. The formation and emission of dioxins in large scale thermal processes. *Chemosphere,* 46: 439–448.

Fiedler H., 1996. Sources of PCDD/PCDF and impact on the environment. *Chemosphere,* 32: 55–64.

Finocchio E., Busca G., Notaro M., 2006. A review of catalytic processes for the destruction of PCDD and PCDF from waste gases. *Appl. Catal. B: Environmental,* 62: 12–20.

Grochowalski A., Wybraniec S., 1996. Levels of PCDDs and PCDFs in flue gas and fly ash from coal combustion in power plant. *Chem. Anal. (Warszaw),* 41: 27–35.

Kaźmierczak M., Zaborowski M., Żarczyński A., Gorzka Z., Michniewicz M., 2003. Effect of temperature and triethanoloamine addition on emission of toxic compounds in the process of propylene chlorohydrin oxidation. *Ann. Pol. Chem. Soc.* 2: 788–793.

Kaźmierczak M., Gorzka Z., Żarczyński A., Paryjczak T., Zaborowski M., Activity of granular and monolithic catalysts during oxidation of selected organic chlorine compounds. *"Environmental Engineering Studies in Poland"* CRC Press (in press).

Koyer-Gołkowska A., Musialik-Piotrowska A., Rutkowski J. D., 2004. Oxidation of chlorinated hydrocarbons over Pt-Rh-based catalyst. Part 1. Chlorinated methanes. *Catal. Today,* 90: 133–138.

Lichtenberger J., Amiridis M.D., 2004. Catalytic oxidation of chlorinated benzenes over V_2O_5/TiO_2 catalysts. *J. Catal.,* 223: 296–308.

Milchert E., Rudnicki J., Stefańska J., 1996. Utylizacja odpadów z produkcji epichlorohydryny glicerynowej. *Przem. Chem.* 75: 212–214.

Musialik-Piotrowska A., Syczewska K., 2002. Catalytic oxidation of trichloroethylene in two-component mixtures with selected volatile organic components. *Catal. Today,* 73: 332–342.

Nijhuis T. A., Beers A. E., Vergunst T., Hoek I., Kapteijn F., Moulijn J. A., 2001. Preparation of monolithic catalysts. *Catal. Rev.,* 43(4): 345–380.

Park J.-N., Lee Ch.-W., Chang J.-S., Park Ch.-H., 2004. Catalytic oxidation of trichloroethylene over Pd-loaded sulfated zirconia. *Bull. Korean Chem. Soc.,* 25(9): 1355–1360.

Petrosius S.C., Drago R.S., Young V., Gruneeald G.C., 1993. Low – temperature decomposition of some halogenated hydrocarbons using metal oxide/porous carbon catalysts. *J. Am. Chem. Soc.,* 115: 6131–6137.

Przondo J., Rogala J., 1996. Przemysłowa instalacja spalania ciekłych odpadów chloroorganicznych w Z. Ch. "Rokita" S.A. *Przem. Chem.,* 75: 98–101.

Spivey J. J., 1987. Complete catalytic oxidation of volatile organics. *Ind. Eng. Chem. Res.,* 26: 2165–2180.

Spivey J.J., Butt J.B., 1992. Literature review. Deactivation of catalysts in the oxidation of volatile organics compounds. *Catal. Today,* 11: 465–500.

Tuppurainen K., Asikainen A., Ruokojarvi P., Ruuskanen J., 2003. Perspectives on the formation of polychlorinated dibenzo-p-dioxins and dibenzofuranes during municipal solid waste (MSW) incineration and other combustion processes. *Acc. Chem. Res.,* 36 (9): 652–658.

Windawi H., Wyatt M., 1993. Catalytic destruction of halogenated volatile organic compounds, *Platinum Metals Rev.,* 4: 186–193.

Żarczyński A., Gorzka Z., Paryjczak T., Kaźmierczak M., Szczepaniak B., 2005. Dioxins in the process of 1,1,2, 2-tetrachloroethane oxidation with the application of monolithic catalysts. *Pol. J. Chem. Technol.,* 7(2): 100–104.

Environmental Engineering – Pawłowski, Dudzińska & Pawłowski (eds)
© 2007 Taylor & Francis Group, London, ISBN13 978-0-415-40818-9

Adsorption of silver ions on chitosan hydrogel beads

Zofia Modrzejewska & Roman Zarzycki
Technical University of Process and Environmental Engineering, Department of Environmental Engineering Systems, Łodź, ul. Wólczańska

Stanisław Biniak
Nicolaus Copernicus University Department of Fundamentals of Chemistry, Toruń, ul. Gagarina

ABSTRACT: The aim of the work was to check the possibility of using the adsorption properties of hydrogel chitosan granules for the removal of silver ions. The adsorption from silver nitrate $AgNO_3$ and sulfate Ag_2SO_4 was investigated. The authors compared the adsorption ability of silver ions on the chitosan formed into hydrogel granules and on chitosan acetate – an initial solution from which beads were formed. The removal of silver ions by means of chitosan salt in the form of chitosan acetate is definitely lower than in the case of hydrogel chitosan beads made from the same salt. The hydrogel beads remove silver ions better; most probably water present in the structure facilitates diffusion of ions in a porous structure. The chemical character of adsorption is confirmed by IR spectral studies.

Keywords: Chitosan, adsorption, Ag (I).

1 INTRODUCTION

Silver finds numerous applications in many areas, starting from electronic and electric industry (due to its high electric conductivity and resistance to corrosion) and ending in the pharmaceutical industry and medicine (it has bactericidal properties). However, it is generally used in the production of silver halides for photographic purposes. Silver alloys are used in the production of jewellery and coins. The annual world demand for silver is about 25,000 tons, of this about a third is used in photochemical and photographic industry, only a half of it being recovered.

The main source of silver recovery is the treatment of wastewater from electroplating workshops and film processing. During processing, especially in fixation, ca. 80% of the silver contained in the photographic film transfers to the fixer solution – in one dm^3 of spent fixer, there are from 2 to 6 grams of pure silver.

Demand for silver grows from year to year, particularly in the electronics industry. Hence, silver recovery is connected with environmental protection and also has an economic aspect. The following methods are used to recover silver from liquid silver-containing substances arising from film processing: electrolysis, ion exchange, precipitation, adsorption and replacement by metals. A variety of ion exchange called ion floatation has also been developed. Silver can also be recovered using fermentation processes. For silver recovery from spent photographic materials (prints, films, etc.) and their wastes, instead of incineration, pyrolysis and other chemical and biochemical methods are applied.

The aim of research presented in this study was to verify the possibility of using the adsorption properties of chitosan in the removal of silver ions. Chitosan $(1 \rightarrow 4)$-2 amino-2-deoxy-β-D-glucane is a natural polymer, a product of chitin$(1 \rightarrow 4)$-2 acetamido-2-deoxy-β-D-glucane deacetylation. Its chemical structure is shown in Fig. 1.

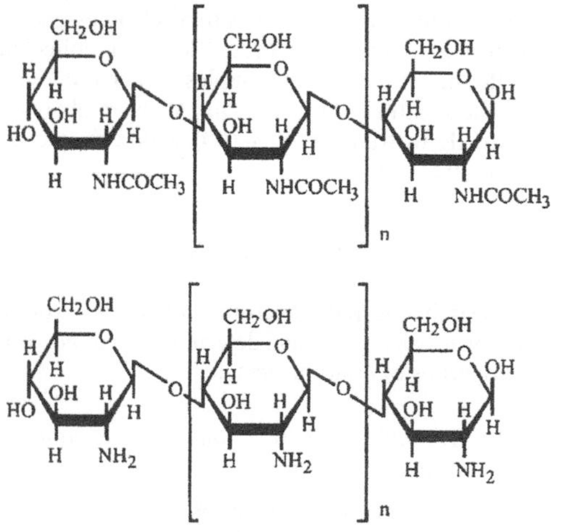

Figure 1. Chemical structure of chitin and chitosan.

Due to the presence of reactive amino and hydroxide groups it is used as an adsorbent. In the extensive literature on the adsorption of metal ions on chitosan, there are only a few publications (Atia, Asem 2005; Bailey 1999; Huang 2004; Ketrin, 2006; Sakamoto Hidefumi 2003; Songkroah, 2004; Varma, 2004; Zarzycki 2002; Yi Ying, 2003; Yoshizuka Kazuharu, 2000) that refer to silver, hence the problem seems to be very up-to-date. The adsorption of silver from silver nitrate $AgNO_3$ and sulphate Ag_2SO_4 was investigated. The adsorption ability of chitosan formed into hydrogel beads was compared with that of chitosan acetate – the precursor solution from which the beads were produced.

2 MATERIALS AND METHODS

2.1 *Characteristics of the adsorption bed*

As an adsorbent porous hydrogel chitosan beads of diameter 3×10^{-3} m were used. They were produced by the phase inversion method from a 4% chitosan solution of mean molecular weight $2 \cdot 10^5$ D and deacetylation degree 68%. 4% acetic acid was used as a chitosan solvent. Granules were formed in up to 10% sodium hydroxide and left in it for 24 h. Then, they were washed with distilled water until reaching neutral pH of the water in which the beads were left.

2.2 *Characteristics of the tested solution*

Silver adsorption was tested in water solutions of silver nitrate $AgNO_3$ and sulphate Ag_2SO_4 at silver concentrations of up to 10 mmole Ag/dm^3.

2.3 *Analytical methods*

The concentration of silver ions was determined using an EAg/S-01 sulphide-silver electrode and by the ICP method of emission mass spectrometry.

The information on the structural changes taking place during the adsorption of Ag^+ ions, was collected using IR spectroscopy with samples prepared on KBr discs. The spectra of chitosan samples were obtained using an IR Instrument (MB-100) with a frequency range of 400–4000 cm^{-1}.

3 RESULTS AND DISCUSSION

The research covered: determination of adsorption isotherms for silver ions and determination of adsorption properties of chitosan acetate.

3.1 *Determination of adsorption and desorption isotherms for Ag^+ ions*

Adsorption was carried out in an immobilised bed, in the system combined with a mixer (shaker, at amplitude oscillation of 8 mm), at temperature 293 K, at initial pH = 5. The adsorbent was chitosan beads (20 g), equivalent to about 1 g of pure chitosan. Tests were carried out on 0.25 dm^3 samples, i.e. at the ratio $(m/V) = 4$. Figures 2 and 3 shows the adsorption in time for silver nitrate and silver sulphate, respectively.

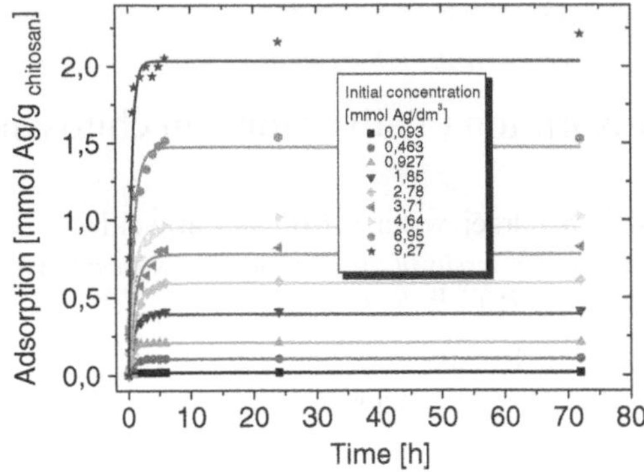

Figure 2. Silver adsorption in time – silver sulphate.

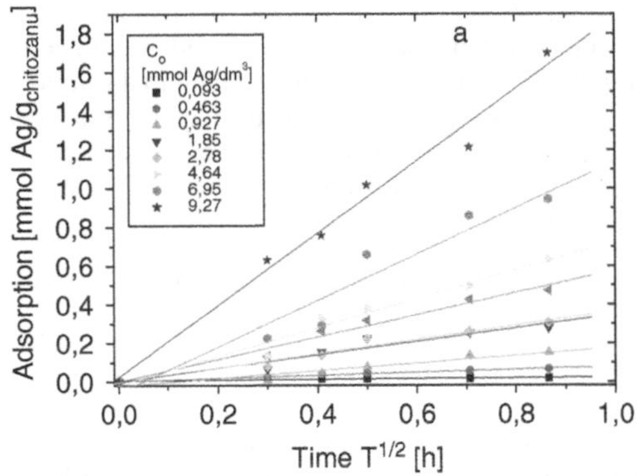

Figure 2a. Linearized adsorption process – silver sulphate.

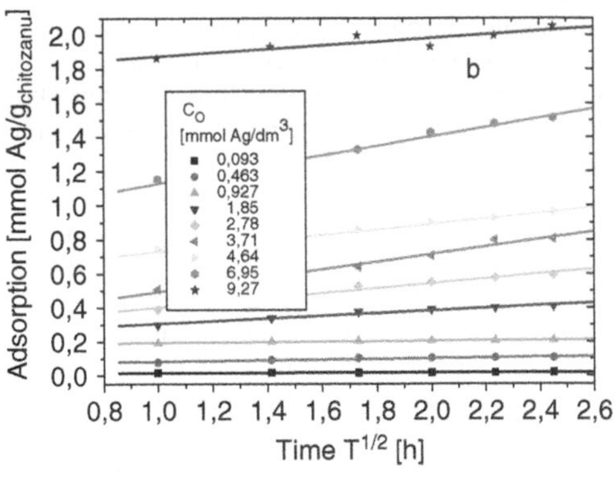

Figure 2b. Linearized adsorption process – silver sulphate.

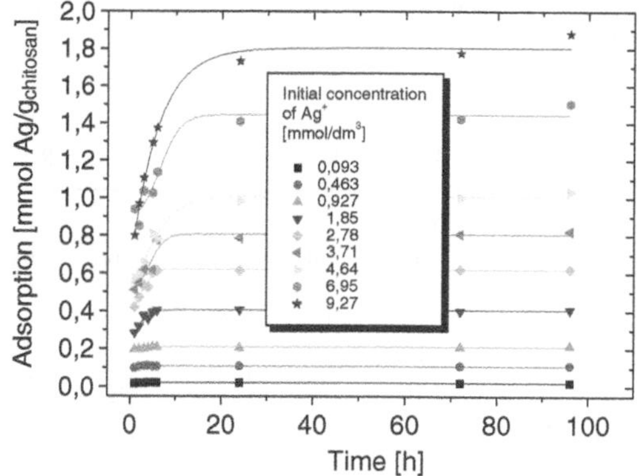

Figure 3. Silver adsorption in time – silver nitrate.

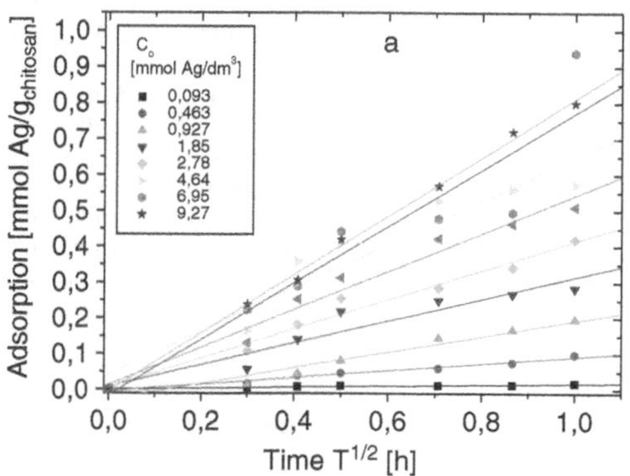

Figure 3a. Linearized adsorption process – silver nitrate.

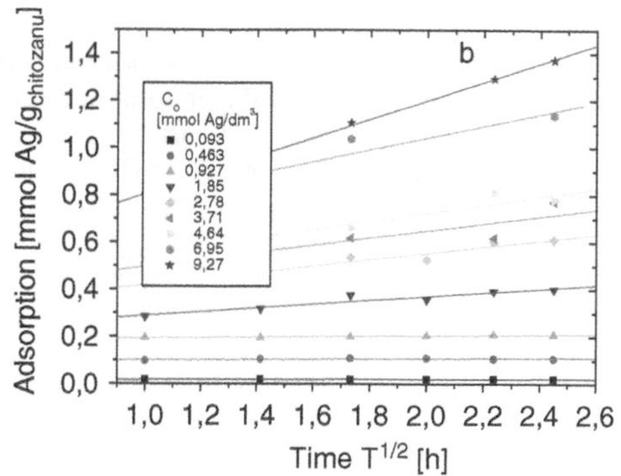

Figure 3b. Linearized adsorption process – silver nitrate.

The derived isotherms are shown in Figs. 4 and 5, respectively.

The process of adsorption was followed by desorption from solutions containing 20 g humid chitosan beads each, and 250 ml of silver solution at an appropriate concentration. 50 ml of the solution was

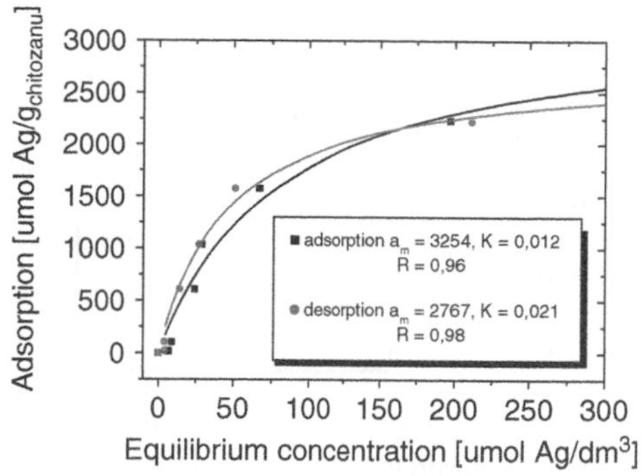

Figure 4a. Adsorption – desorption isotherm. Ag^{+} from sulphate – Langmuir equations.

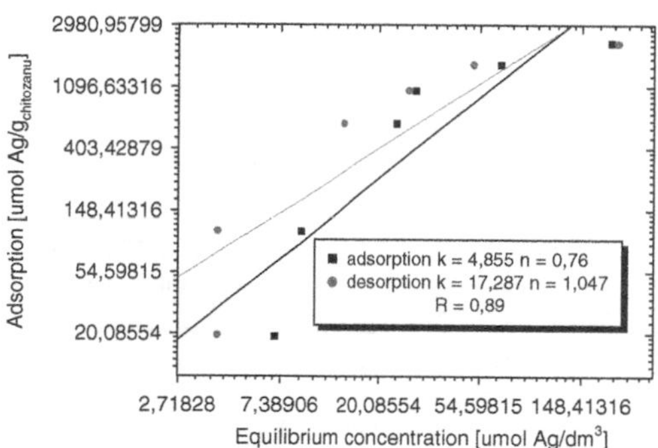

Figure 4b. Adsorption – desorption isotherm. Ag^{+} from sulphate – Freundlich equations.

replaced by 50 ml of distilled water at pH concordant with the initial pH of the solutions. The obtained desorption isotherms were compared with adsorption isotherms and presented in Figs. 4 and 5.

The rate of adsorption at the tested range of concentration was described by the exponential pseudo first-order equation:

$$\frac{d a_t}{d t} = k_1 \left(a_e - a_t\right) \qquad [1]$$

or pseudo second-order equation:

$$\frac{d a_t}{d t} = k_2 \left(a_e - a_t\right)^2,$$

which upon integration, assumes the form:

$$a_t = a_m \left(1 - e^{-kt}\right) \qquad [2]$$

$$a_t = a_e \frac{a_e k_2 t}{1 + a_e k_2 t}. \qquad [3]$$

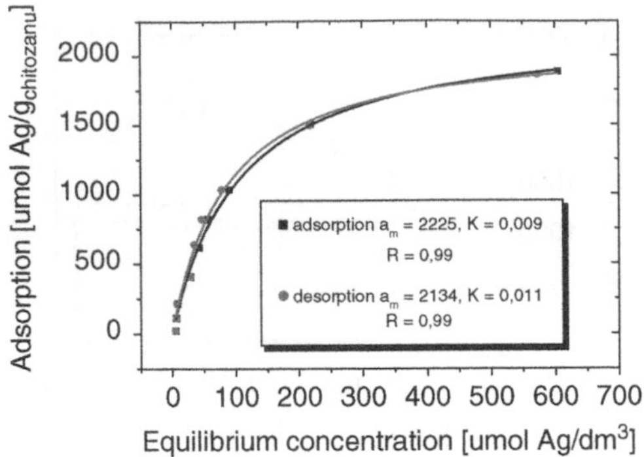

Figure 5a. Adsorption – desorption isotherm.
Ag^+ from nitrate – Langmuir equations.

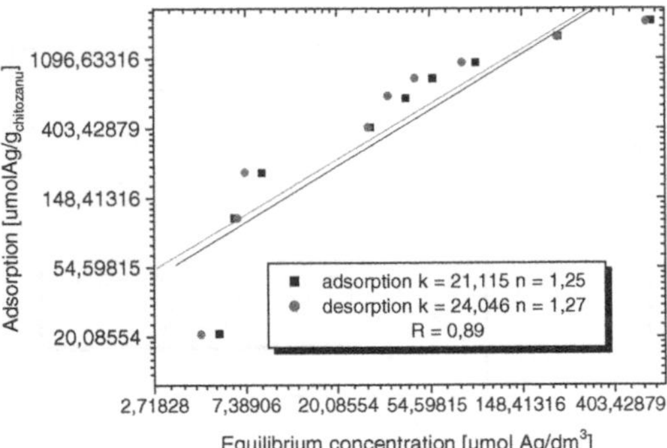

Figure 5b. Adsorption – desorption isotherm.
Ag^+ from sulphate – Freundlich equations.

Parameters of the equations are given in Table 1.

Since the models do not reflect fully the process rates and no conclusions can be drawn on the mechanisms, a model proposed by Weber and Morris was used. In this model the change of adsorption in time is the function:

$$a = \left(\frac{D \cdot t}{r^2}\right)^{\frac{1}{2}} \qquad [4]$$

where r is the bead radius and D is the effective diffusion coefficient. Linearized curves

$$a_t = f\left(t^{\frac{1}{2}}\right) \qquad [5]$$

are illustrated in Fig. 2a and b and Figs. 3a and b.

The process of adsorption can be divided into three stages:

- rapid adsorption during the first hour (Figs. 2a and 3 a),
- slow process lasting from the first to the sixth hour controlled probably by inner diffusion (Figs. 2b and 3 b),
- a very slow stage lasting up to 24 h when the equilibrium state is reached in the case of concentrations up to 5 mmole/dm^3, or even several days in the case of higher concentrations.

Table 1. Parameters of the equations describing the time dependence of the process.

	$a_t = a_m(1-e^{-kt})$			$a_t = a_e \frac{a_e k_2 t}{1+a_e k_2 t}$		
	q_e	k_1	R	q_e	k_2	R
Ag_2SO_4 [mmole/dm^3]						
0.093	0.018	10.7	0.92	0.018	11.90	0.94
0.463	0.105	1.09	0.84	0.11	14.7	0.85
0.927	0.207	1.2	0.76	0.22	7.79	0.75
1.85	0.392	1.3	0.73	0.42	5.14	0.78
2.78	0.59	0.8	0.88	0.65	1.69	0.88
3.71	0.76	0.9	0.81	0.85	1.54	0.85
4.64	0.95	1.04	0.8	1.04	1.41	0.83
6.95	1.471	1.05	0.82	1.6	0.95	0.84
9.27	2.033	1.68	0.81	2.18	1.19	0.85
$AgNO_3$[mmole/dm^3]						
0.093	0.021	1.7	0.91	0.217	217	0.92
0.463	0.109	2.25	0.91	0.11	78.5	0.87
0.927	0.205	2.9	0.81	0.208	54.9	0.68
1.85	0.393	1.06	0.80	0.41	5.26	0.91
2.78	0.59	0.94	0.77	0.63	2.89	0.91
3.71	0.76	0.74	0.63	0.81	1.5	0.86
4.64	0.95	0.49	0.69	1.01	0.81	0.9
6.95	1.31	1.7	0.49	1.41	0.72	0.75
9.27	1.74	0.36	0.86	1.84	0.3	0.97

In the first hour, Ag^+ ions are combined with the reactive groups present in the chitosan molecule (NH_2^-, OH^-) near the outer bead surface. They can block the pores on the outer surface so that the next Ag^+ ions have a longer diffusion path to the active groups inside the bead structure, hence the process slows down in consecutive hours. Assuming that the state of equilibrium is settled after 48 h, the isotherms were determined.

They are shown in Figs. 4a, 5a and 4b, 5b and described by the Langmuir [6]and Freundlich [7] equations

$$a^* = \frac{a_m \cdot K \cdot C}{1 + K \cdot C} \qquad [6]$$

$$a^* = k \cdot C^{\frac{1}{n}} \qquad [7]$$

It follows that hydrogel chitosan beads possess good absorption properties towards Ag^+ ions. Adsorption

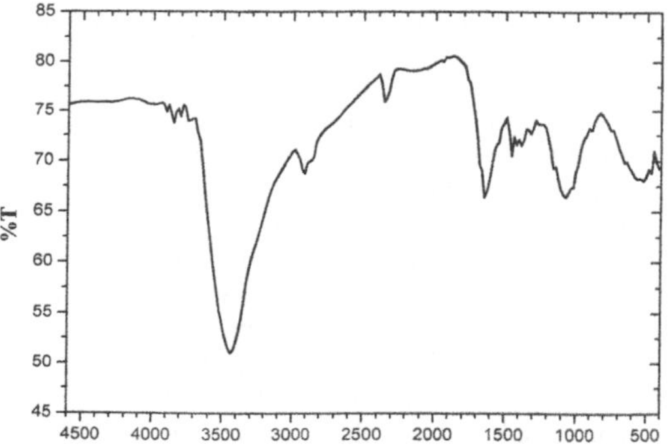

Figure 6. Chitosan – IR spectrum.

isotherms in the range of low equilibrium concentrations (to 10 $mmole_{Ag}/dm^3$) are well describe by the Langmuir isotherms. A slightly higher adsorption of Ag^+ ions was obtained using sulphate. The isotherms obtained in the adsorption process overlap the isotherms from desorption which is evidence of the chemical character of the process.

The equilibrium adsorption of Ag^+ from sulphates is about 3, and for nitrates 2.2 mmole $Ag/g_{chitosan}$. The constant K reflecting the interrelationship between the adsorbent and adsorptive is practically the same for sulphates and nitrates and is equal to 0.01. A higher constant was obtained in the process of desorption (for sulphate). This again confirms the chemical character of the process.

The chemical character of adsorption is also confirmed by IR spectral studies. In the transmittance IR spectra of the chitosan sample (Fig. 6) the band of stretching O-H vibration (3600–3100 cm^{-1}) was due to structural hydroxylic groups. The asymmetry of this bond at lower wave numbers indicates the presence of strong hydrogen bonds and N-H (amine) structures. The presence of bands characteristic of CH_2 structures (2900 and 2870 cm^{-1}) confirms the existence of some aliphatic species in the polymer structure (symmetric and asymmetric CH_2 stretching in the pyranose ring). Below 1700 cm^{-1} the overlapped bands characteristic of N-acetylated amine at 1660 cm^{-1} (C = O), protonated amino groups (NH_3^+) at 1630 cm^{-1} and amine group (N-H bending vibration) at 1560 cm^{-1} can be observed; these correspond to the primary and secondary amines connected with pyranose rings. Another broad (but smaller) band in the 1470–1300 cm^{-1} range consists of a series of overlapping bands ascribable to the deformation vibration of hydroxyl groups and in-plane vibration of C-H in various ring structures. The partially resolved peaks forming the band in the 1250–1000 cm^{-1} range can be assigned to ether-like oxygen bridges (C-O-C) symmetric stretching vibrations and O-H vibrations in different structural environments.

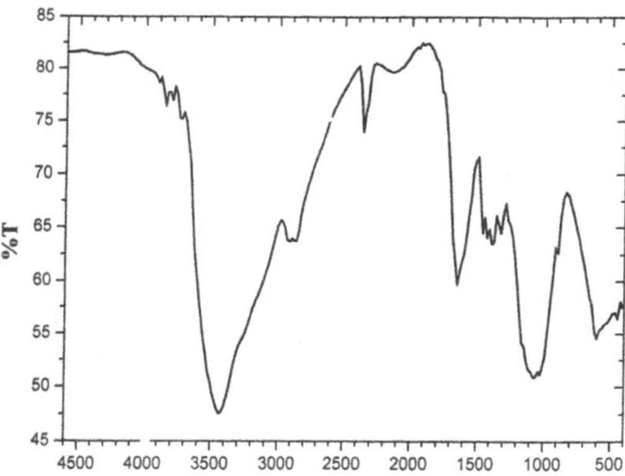

Figure 7. Chitosan-Ag – IR spectrum.

The spectral changes in the transmittance IR spectra of polymer tested after Ag^+ ions adsorption are shown in Fig 7. First, an increase in the intramolecular interactions and hydroxyl band intensity can be observed in 3600–3100 cm^{-1} wave number range. A change in the band range from 3000 cm^{-1} is also observed. During adsorption, the pH of the adsorptive increases, hence the amount of protonated amino groups in chitosan adsorbent increases, and the amino groups react with Ag^+ ions. This is probably the reason why the separation of peaks at this band in the spectrum is less visible after absorption. A change is also observed in the bands characteristic of CH_3COH groups (850, 840 cm^{-1}) and a saccharide structure, i.e. 890 and 1150–1040 cm^{-1}. The band related to the presence of oxygen bridges O-C-O at 1050 cm^{-1} is equalised with the band 1150 cm^{-1} and the bands 850, 840 cm^{-1} are combined. This may be evidence of stable β-glucosamine bonds between glucosamine and N-acetylamine molecules and of bonds containing CH_3COH groups.

The interpretation of IR spectra provides only a qualitative assessment. It can be presumed that -OH, NH and CH_3OH groups take part in chelating chitosan with Ag^+ ions. The possibility of redox reaction (Ag^+ ions reduction and oxidation of some hydroxyl moieties) should also be taken into consideration. The contribution of particular groups requires further XPS studies.

3.2 *Determination of adsorption properties of chitosan acetate*

To compare possibilities of the removal of silver ions by means of chitosan in different forms, Ag^+ adsorption on hydrogel beads and chitosan salt was studied in parallel. The chitosan salt selected was a 4% chitosan acetate, i.e. the solution from which the hydrogel chitosan beads were produced. The following procedure was assumed: to silver solutions of concentrations 20 and 100 mg Ag/dm^3 (0.18 and 9.27 $mmole/dm^3$),

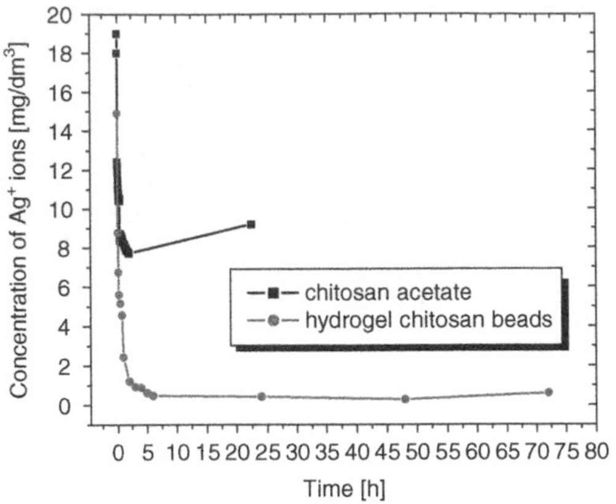

Figure 8. Adsorption of Ag^+ from different forms of chitosan; silver nitrate 20 mg/dm^3.

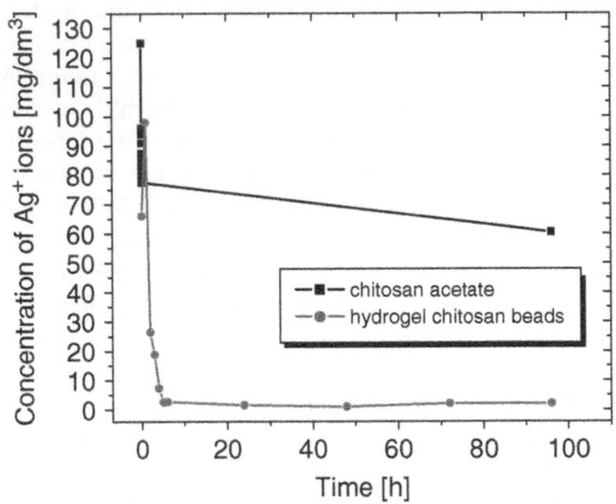

Figure 9. Adsorption of Ag^+ from different forms of chitosan; silver nitrate 100 mg/dm^3.

25 ml of chitosan acetate at pH = 4 were added. The amount of chitosan in the added salt was equal to the amount of chitosan present in the hydrogel. Results are shown in Figs. 8 and 9.

The investigations indicated that chitosan in the form of hydrogel had the better adsorption properties. A lower adsorption for chitosan salts is most probably due to the presence of protonated amino groups in the chitosan molecule. In the case of low Ag^+ concentrations in the solution, the complex of these ions with chitosan in the form of salt is not stable.

4 CONCLUSIONS

The following conclusions can be drawn from the research:

1. Hydrogel chitosan beads show higher adsorption than chitosan in the form of a salt – chitosan acetate.
2. Adsorption ability of hydrogel towards Ag^+ ions depends slightly on the presence of co-ions.
3. Most probably the process of adsorption in the first hour is dominated mainly by a chemical reaction, and then controlled by diffusion into the porous hydrogel structure.
4. The chemical character of adsorption is confirmed by IR spectral studies. It can be presumed that -OH, NH and CH_2OH groups take part in chelating chitosan with Ag^+ ions. The possibility of redox reaction (reduction of Ag^+ ions and oxidation of some hydroxyl moieties) can be considered.
5. The results show that studies on developing a method of recovery of Ag^+ ions stemming from electroplating or film processing should be continued.

REFERENCES

Atia, Asem A., Adsorption of silver(I) and gold(III) on resins derived from bisthiourea and application to retrieval of silver ions from processed photo films, 2005, *Hydrometallurgy*, 80, 1–2, 98–106.

Bailey S.E., Olin T.J., Bricka R.M., 1999, A review of potentially low-cost sorbents for heavy metals, *Wat. Res.*, 33, 2469–2479.

Huang, Haizhen, Yuan, Qiang, Yang, Xiurong, 2004, Preparation and characterization of metal–chitosan nanocomposites, *Colloids and Surfaces B: Biointerfaces*, 39, 1–2, 25, 31–37.

Ketrin K., Rosi, Takayanagi, Toshio, Oshima, Mitsuko, Motomizu, Shoji, Synthesis of a chitosan-based chelating resin and its application to the selective concentration and ultratrace determination of silver in environmental water samples, *Analytica Chimica Acta* Volume: 558, Issue: 1–2, February 3, 2006, pp. 246–253.

Sakamoto Hidefumi, Ishikawa Junichi, Koike Masaki, Doi Kunio, Wada Hiroko, 2003, Adsorption and concentration of silver ion with polymer-supported polythiazaalkane resins, *Reactive and Functional Polymers*, 55, 3, 299–310.

Songkroah, C., Nakbanpote, W., Thiravetyan, P., 2004, Recovery of silver-thiosulphate complexes with chitin, *Process Biochemistry*, 39, 11, 30, 1553–1559.

Varma, A.J., Deshpande, S.V., Kennedy, J.F., 2004, Metal complexation by chitosan and its derivatives: a review, *Carbohydrate Polymers*, 55, 1, 1, 77–93.

Zarzycki R., Sujka W., Dorabialska M., Modrzejewska Z., 2002, Desorption of Ag^+ ions on chitosan beads, CHISA Praga.

Yi, Ying, Wang, Yuting, Liu, Hui, 2003, Preparation of new crosslinked chitosan with crown ether and their adsorption silver ion for antibacterial activities, *Carbohydrate Polymers*, 53, 4, 1, 425–430.

Yoshizuka Kazuharu, Lou Zhengrong, Inoue Katsutoshi, 2000, Silver-complexed chitosan microparticles for pesticide removal, *Reactive and Functional Polymers*, 44, 1, 14, 47–54.

Environmental Engineering – Pawłowski, Dudzińska & Pawłowski (eds)
© 2007 Taylor & Francis Group, London, ISBN13 978-0-415-40818-9

Author index

Printed and bound by CPI Group (UK) Ltd, Croydon, CR0 4YY
01/11/2024
01782599-0002